BIOLOGY

Exploring the Diversity of Life

Third Canadian Edition

Volume Two

Peter J. Russell

Paul E. Hertz

Beverly McMillan

M. Brock Fenton
Western University

Heather Addy
University of Calgary

Denis Maxwell
Western University

Tom Haffie
Western University

Bill Milsom
University of British Columbia

NELSON
EDUCATION

Biology: Exploring the Diversity of Life, Third Canadian Edition, Volume 2

by Peter J. Russell, Paul E. Hertz, Beverly McMillan, M. Brock Fenton, Heather Addy, Denis Maxwell, Tom Haffie, and Bill Milsom

Vice President, Editorial Higher Education:
Anne Williams

Senior Publisher:
Paul Fam

Marketing Manager:
Leanne Newell

Developmental Editor:
Toni Chahley

Photo Researcher and Permissions Coordinator:
Kristiina Paul

Senior Production Project Manager:
Natalia Denesiuk Harris

Production Service:
Integra Software Services Pvt. Ltd.

Substantive Editor:
Rosemary Tanner

Copy Editor:
Julia Cochrane

Proofreader:
Integra Software Services Pvt. Ltd.

Indexer:
Integra Software Services Pvt. Ltd.

Design Director:
Ken Phipps

Managing Designer:
Franca Amore

Interior Design:
Dianna Little

Cover Design:
Courtney Hellam and Liz Harasymczuk

Cover Images:
(front) DAVID DOUBILET/National Geographic Creative; (back) kentoh/Shutterstock.com

Compositor:
Integra Software Services Pvt. Ltd.

*For, and because of, our generations
of students.*

About the Canadian Authors

M.B. (BROCK) FENTON received his Ph.D. from the University of Toronto in 1969. Since then, he has been a faculty member in biology at Carleton University, then at York University, and then at Western University. In addition to teaching parts of first-year biology, he has also taught vertebrate biology, animal biology, and conservation biology, as well as field courses in the biology and behaviour of bats. He has received awards for his teaching (Carleton University Faculty of Science Teaching Award; Ontario Confederation of University Faculty Associations Teaching Award; and a 3M Teaching Fellowship, Society for Teaching and Learning in Higher Education), in addition to recognition of his work on public awareness of science (Gordin Kaplan Award from the Canadian Federation of Biological Societies; Honourary Life Membership, Science North, Sudbury, Ontario; Canadian Council of University Biology Chairs Distinguished Canadian Biologist Award; The McNeil Medal for the Public Awareness of Science of the Royal Society of Canada; and the Sir Sanford Fleming Medal for public awareness of Science, the Royal Canadian Institute). He also received the C. Hart Merriam Award from the American Society of Mammalogists for excellence in scientific research. Bats and their biology, behaviour, evolution, and echolocation are the topics of his research, which has been funded by the Natural Sciences and Engineering Research Council of Canada (NSERC). In November 2014, Brock was inducted as a Fellow of the Royal Society of Canada.

HEATHER ADDY received her Ph.D. in plant–soil relationships from the University of Guelph in 1995. During this training and in a subsequent postdoctoral fellowship at the University of Alberta, she discovered a love of teaching. In 1998, she joined the Department of Biological Sciences at the University of Calgary in a faculty position that emphasizes teaching and teaching-related scholarship.

In addition to teaching introductory biology classes and courses in plant and fungal biology, she is currently the coordinator of the Teaching Skills Workshop Program for the Faculty of Science. Her pedagogical focus is on collaborative learning methods such as team-based learning and the incorporation of peer mentors and class representatives in undergraduate education. She received the Faculty of Science Award for Excellence in Teaching in 2005, a Students' Union Teaching Excellence Award in 2008 and 2014, and one of the inaugural University of Calgary Teaching Awards in 2014.

DENIS MAXWELL received his Ph.D. from the University of Western Ontario in 1995. His thesis, under the supervision of Norm Huner, focused on the role of the redox state of photosynthetic electron transport in photo acclimation in green algae. Following his doctorate, he was awarded an NSERC postdoctoral fellowship. He undertook postdoctoral training at the Department of Energy Plant Research Laboratory at Michigan State University, where he studied the function of the mitochondrial alternative oxidase. After taking up a faculty position at the University of New Brunswick in 2000, he moved in 2003 to the Department of Biology at Western University. He currently serves as Associate Chair—Undergraduate Education and teaches first-year biology.

TOM HAFFIE is a graduate of the University of Guelph and the University of Saskatchewan in the area of microbial genetics. Tom has devoted his 30-year career at Western University to teaching large biology classes in lecture, laboratory, and tutorial settings. He led the development of the innovative core laboratory course in the biology program; was an early adopter of computer animation in lectures; and, most recently, has coordinated the implementation of personal response technology across campus. He holds a University Students' Council Award for Excellence in Teaching, a UWO Pleva Award for Excellence in Teaching, a Fellowship in Teaching Innovation, a Province of Ontario Award for Leadership in Faculty Teaching (LIFT), and a Canadian national 3M Fellowship for Excellence in Teaching.

Bill Milsom

BILL MILSOM (Ph.D., University of British Columbia) is a professor in the Department of Zoology at the University of British Columbia, where he has taught a variety of courses, including first-year biology, for over 30 years. His research interests include the evolutionary origins of respiratory processes and the adaptive changes in these processes that allow animals to exploit diverse environments. He examines respiratory and cardiovascular adaptations in vertebrate animals in rest, sleep, exercise, altitude, dormancy, hibernation, diving, and so on. This research contributes to our understanding of the mechanistic basis of biodiversity and the physiological costs of habitat selection. His research has been funded by NSERC, and he has received several academic awards and distinctions, including the Fry Medal of the Canadian Society of Zoologists, the August Krogh Award of the American Physiological Society, the Bidder Lecture of the Society for Experimental Biology, and the Izaak Walton Killam Award for Excellence in Mentoring. He has served as the President of the Canadian Society of Zoologists and as President of the International Congress of Comparative Physiology and Biochemistry.

About the U.S. Authors

PETER J. RUSSELL received a B.Sc. in Biology from the University of Sussex, England, in 1968 and a Ph.D. in Genetics from Cornell University in 1972. He has been a member of the Biology faculty of Reed College since 1972 and is currently a professor of biology, emeritus. Peter taught a section of the introductory biology course, a genetics course, and a research literature course on molecular virology. In 1987 he received the Burlington Northern Faculty Achievement Award from Reed College in recognition of his excellence in teaching. Since 1986, he has been the author of a successful genetics textbook; current editions are *iGenetics: A Molecular Approach, iGenetics: A Mendelian Approach*, and *Essential iGenetics*. Peter's research was in the area of molecular genetics, with a specific interest in characterizing the role of host genes in the replication of the RNA genome of a pathogenic plant virus, and the expression of the genes of the virus; yeast was used as the model host. His research has been funded by agencies including the National Institutes of Health, the National Science Foundation, the American Cancer Society, the Department of Defense, the Medical Research Foundation of Oregon, and the Murdoch Foundation. He has published his research results in a variety of journals, including *Genetics, Journal of Bacteriology, Molecular and General Genetics, Nucleic Acids Research, Plasmid*, and *Molecular and Cellular Biology*. Peter has a long history of encouraging faculty research involving undergraduates, including cofounding the biology division of the Council on Undergraduate Research in 1985. He was Principal Investigator/Program Director of a National Science Foundation Award for the Integration of Research and Education (NSF–AIRE) to Reed College, 1998 to 2002.

PAUL E. HERTZ was born and raised in New York City. He received a B.S. in Biology from Stanford University in 1972, an A.M. in Biology from Harvard University in 1973, and a Ph.D. in Biology from Harvard University in 1977. While completing field research for the doctorate, he served on the biology faculty of the University of Puerto Rico at Rio Piedras. After spending two years as an Isaac Walton Killam Postdoctoral Fellow at Dalhousie University, Paul accepted a teaching position at Barnard College, where he has taught since 1979. He was named Ann Whitney Olin Professor of Biology in 2000, and he received The Barnard Award for Excellence in Teaching in 2007. In addition to serving on numerous college committees, Paul chaired Barnard's Biology Department for eight years and served as Acting Provost and Dean of the Faculty from 2011 to 2012. He is the founding Program Director of the Hughes Science Pipeline Project at Barnard, an undergraduate curriculum and research program that has been funded continuously by the Howard Hughes Medical Institute since 1992. The Pipeline Project includes the Intercollegiate Partnership, a program for local community college students that facilitates their transfer to four-year colleges and universities. He teaches one semester of the introductory sequence for biology majors and pre-professional students, lecture and laboratory courses in vertebrate zoology and ecology, and a year-long seminar that introduces first-year students to scientific research. Paul is an animal physiological ecologist with a specific research interest in the thermal biology of lizards. He has conducted fieldwork in the West Indies since the mid-1970s, most recently focusing on the lizards of Cuba. His work has been funded by the NSF, and he has published his research in such prestigious journals as *The American Naturalist, Ecology, Nature, Oecologia*, and *Proceedings of the Royal Society*. In 2010, he and his colleagues at three other universities received funding from NSF for a project designed to detect the effects of global climate warming on the biology of Anolis lizards in Puerto Rico.

BEVERLY MCMILLAN has been a science writer for more than 25 years. She holds undergraduate and graduate degrees from the University of California, Berkeley, and is coauthor of a college text in human biology, now in its tenth edition. She has also written or coauthored numerous trade books on scientific subjects and has worked extensively in educational and commercial publishing, including eight years in editorial management positions in the college divisions of Random House and McGraw-Hill.

Brief Table of Contents

Contents

Ryan M. Bolton/Shutterstock.com

Daniel Hebert/Shutterstock.com

Nancy Kennedy/Shutterstock.com

Noppharat05081977/iStock/Thinkstock

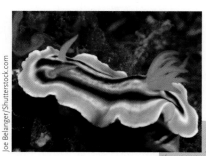

Preface

Welcome to an exploration of the diversity of life. The main goal of this text is to guide you on a journey of discovery about life's diversity across levels ranging from molecules to genes, cells to organs, and species to ecosystems. Along the way, we will explore many questions about the mechanisms underlying diversity as well as the consequences of diversity for our own species and for others.

At first glance, the riot of life that animates the biosphere overwhelms our minds. One way to begin to make sense of this diversity is to divide it into manageable sections on the basis of differences. Thus, in this book for instance, we highlight the divisions between prokaryotic and eukaryotic organisms, plants and animals, protostomes and deuterostomes. We also consider features found in all life forms to stress similarities as well as differences. We examine how different organisms solve the common problems of finding nutrients, energy, and mates on the third rock from our Sun. What basic evolutionary principles inform the relationships among life forms regardless of their different body plans, habitats, or life histories? Unlike many other first-year biology texts, this book has chapters integrating basic concepts such as the effects of genetic recombination, light, nutrition, and domestication across the breadth of life from microbes to mistletoe to moose. As you read this book, you will be referred frequently to other chapters for linked information that expands the ideas further.

Evolution provides a powerful conceptual lens for viewing and understanding the roots and history of diversity. We will demonstrate how knowledge of evolution helps us appreciate the changes we observe in organisms. Whether the focus is the conversion of free-living prokaryotic organisms into mitochondria and chloroplasts or the steps involved in the domestication of rice, selection for particular traits over time can explain the current condition.

We hope that Canadian students will find the subject of biology as it is presented here accessible and engaging because it is presented in familiar contexts. We have highlighted the work of Canadian scientists, used examples of Canadian species, referred to Canadian regulations and institutions, and highlighted discoveries made by Canadians.

Many biology textbooks use the first few chapters to review fundamentals of chemistry and biochemistry as well as information on the scientific method. Instead of focusing on this background information, we have used the first chapter, in particular, to immediately engage students by conveying the excitement that is modern biology. We have put important background information in the centre of the book as a distinct reference section entitled *The Chemical, Physical, and Environmental Foundations of Biology*. With their purple borders, these pages are distinct and easy to find and have become affectionately known as *The Purple Pages*. These pages enable information to be readily identifiable and accessible to students as they move through the textbook rather than being tied to a particular chapter.

In addition to presenting material about biology, this book also makes a point of highlighting particular people, important molecules, interesting contexts, and examples of life in extreme conditions. Science that appears in textbooks is the product of people who have made careful and systematic observations that led them to formulate hypotheses about these observations and, where appropriate, design and execute experiments to test these hypotheses. We illustrate this in most chapters with boxed stories about how particular people have used their ingenuity and creativity to expand our knowledge of biology. We have endeavoured to show not just the science itself but also the process behind the science.

Although biology is not simply chemistry, specific chemicals and their interactions can have dramatic effects on biological systems. From water to progesterone, amanitin, and DDT, each chapter features the activity of a relevant chemical.

To help frame the material with an engaging context, we begin each chapter with a section called "Why It Matters." In addition, several chapters include boxed accounts of organisms thriving "on the edge" at unusual temperatures, pressures, radiation dosages, salt concentrations, and so on. These brief articles explain how our understanding of "normal" can be increased through study of the "extreme."

Examining how biological systems work is another theme pervading this text and underlying the idea of diversity. We have intentionally tried to include examples that will tax your imagination, from sea slugs that steal chloroplasts for use as solar panels, to hummingbirds fuelling their hovering flight, to adaptive radiation of viruses. In each situation, we examine how biologists have explored and assessed the inner workings of organisms from gene regulation to the challenges of digesting cellulose.

Solving problems is another theme that runs throughout the book. Whether the topic is gene therapy to treat a disease in people, increasing crop production, or conserving endangered species, both the problem and the solution lie in biology. We will explore large problems facing planet Earth and the social implications that arise from them.

Science is by its nature a progressive enterprise in which answers to questions open new questions for consideration. Each chapter presents Questions for Discussion (also mentioned below) to emphasize that biologists still have a lot to learn—topics for you to tackle should you decide to pursue a career in research.

Study Breaks occur after most sections in the chapters. They contain questions written by students to identify some of the important features of the section. At the end of each chapter is a group of multiple-choice self-test questions, the answers to which can be found at the end of the book. Questions for Discussion at the end of each chapter challenge you to think more broadly about biology. You are encouraged to use these in discussions with other students and to explore potential answers by using the resources of the electronic or physical library.

To maximize the chances of producing a useful text that draws in students (and instructors), we sought the advice of colleagues who teach biology (members of the Editorial Advisory Board). We also asked students (members of the Student Advisory Boards) for their advice and comments. These groups read draft chapters, evaluated the effectiveness of important visuals in the textbook, and provided valuable feedback, but any mistakes are ours.

We hope that you are as captivated by the biological world as we are and are drawn from one chapter to another. But don't stop there—use electronic and other resources to broaden your search for understanding, and, most important, observe and enjoy the diversity of life around you.

M. Brock Fenton
Heather Addy
Denis Maxwell
Tom Haffie
Bill Milsom

London, Calgary, and Vancouver
January 2015

New to This Edition

This section highlights the changes we made to enhance the effectiveness of *Biology: Exploring the Diversity of Life,* Third Canadian Edition. Every chapter has been updated to ensure currency of information. We made organizational changes to more closely link related topics and reflect preferred teaching sequences. New features in the text have been developed to help students actively engage in their study of biology. Key chapters have been extensively revised to provide a full treatment of the subject matter that reflects new developments in these specialized fields. Enriched media offerings provide students with a broad spectrum of learning opportunities.

Organizational Changes

We divided "DNA Technologies and Genomics" into two chapters: "DNA Technologies" with a new emphasis on the role of synthetic biology and updated gene therapy, and "Genomes and Proteomes," which provides more focused coverage of genome evolution, personal genomics, and metagenomics.

In response to reviewers' comments, we moved Unit Seven: Ecology and Climate Change, and Unit Eight: Biology in Action from the end of the book to after Unit Six: Diversity of Life to enhance the flow of topics in the first half of the textbook.

The latter half of the book has been streamlined and reorganized to completely separate the discussion of the systems and processes of plants and animals. We divided "Plant and Animal Nutrition" into separate chapters dealing with plants and animals, respectively. "Plant Signals and Responses to the Environment" is a new chapter that includes more in-depth treatment of this subject matter. While students need to realize that plants and animals face the same challenges, they also need to consider that plants and animals respond to these challenges in different ways, reflecting some very fundamental differences. The material in these sections has been reorganized to fit the style in which the material is presented in most introductory biology courses. But it is our hope that as students read the chapters in Units 9 and 10, they will think about how differences in animal and plant growth, development, reproduction, and other processes relate to the fundamental difference between being motile—obtaining the carbon needed to build bodily structures from the food they eat (animals)—and being sessile—obtaining the carbon needed to build body structures from the air and sunlight they capture (plants).

Finally, we have a new Unit Eleven: Systems and Processes: Interacting with the Environment, which contains a new chapter, Chapter 52: Conservation and Evolutionary Physiology, which draws together themes from the book in a series of engaging case studies.

New Features

New Volume Introductions

Providing students with an overview of the content of each volume, our new volume introductions set the stage for the presentation of the material in the chapters that follow. These introductions enhance the flow of information across the textbook by setting the units and chapters within the broader context of the discipline as a whole.

New Chapter Dealing Exclusively with Genomes and Proteomes

In response to feedback from reviewers, and with the careful guidance of our Editorial Advisory Board, we have developed an entirely new chapter on Genomes and Proteomes. As sequencing costs continue to fall, the volume of available sequences continues to accumulate in databanks. What sequences are present in a given biological sample? What does a particular sequence mean? What can comparing sequences reveal about evolution? What can one's own sequence predict about one's future health? The new chapter addresses these questions, and more, in a Canadian context.

Extensively Revised and Rewritten Unit Five: Evolution and Classification

Given the importance and dynamism of this topic, Unit Five: Evolution and Classification has been extensively revised and rewritten. With the help and guidance of reviewers, subject matter experts, and our Editorial Advisory Board, we have rewritten this unit, reorganizing the coverage of topics, and expanding our coverage from four to five chapters. This new unit presents and carefully builds upon key concepts in evolution, providing students with a strong foundation in this essential area. Unique to this textbook, Chapter 21: Humans and Evolution forms a capstone for the unit, drawing together the key themes presented in the chapters that precede it.

Chapters in Unit 5: Evolution and Classification are as follows:

- Chapter 17: Evolution: The Development of the Theory
- Chapter 18: Microevolution: Changes within Populations
- Chapter 19: Species and Macroevolution
- Chapter 20: Understanding the History of Life on Earth
- Chapter 21: Humans and Evolution

New Chapter 52: Conservation and Evolutionary Physiology

Our ability to design conservation measures to protect species threatened by environmental change is based on an understanding of the physiological processes presented in the preceding chapters. This knowledge is required to predict the consequences of environmental change on a species' survival and of the measures that will or will not be beneficial as conservation policy. In this chapter, we discuss our understanding of how the physiological processes presented in individual chapters are integrated within organisms and then present a few case histories to illustrate how we can use this information to understand the evolution of physiological processes and to inform conservation policies. We have carefully selected only a few examples of the many that exist (such as ways to reduce mortality in salmon fisheries due to by-catch, explanations for high plant biodiversity on nutrient-poor soils, and the use of physiology to predict the behaviour of such invasive species as cheatgrass, and suggest strategies for the control of these organisms), allowing instructors to expand this treatment with other examples of their own choosing.

Improved Illustrations

A priority for the third Canadian edition was improved illustrations throughout the book. Members of our Student Advisory Board at the University of Alberta reviewed key illustrations throughout the book and provided helpful guidance on improving their clarity and effectiveness. Many figures have been revised and redrawn in response to this feedback.

Major revisions to selected chapters are listed below:

Chapter 4: Energy and Enzymes
- More accurate description of ATP hydrolysis
- More complete treatment of coupled reactions
- New section focused on the flow of energy through the biosphere
- More accurate concept of "energy spreading" in defining entropy

Chapter 7: Photosynthesis
- Added section looking at photosynthesis from a global perspective
- New "Why It Matters" focused on biofuel production
- Improved figures, including one illustrating leaf gas exchange

Chapter 8: Cell Cycles
- Refocused cell cycle control onto checkpoints
- New Concept Fix related to segregation of cytoplasmic elements
- New reference to stem cells in plants and animals as examples of asymmetric cell division (new Figure 8.19)

Chapter 14: Control of Gene Expression
- New emphasis on the role of epigenetic regulation

Chapter 15: DNA Technologies
- Updated DNA sequencing
- New emphasis on synthetic biology and updated gene therapy

Chapter 16: Genomes and Proteomes (NEW)
- Highlights human variation and personal genomics
- Discusses functional and comparative genomics, including genome evolution, metagenomics, and synthetic life
- Explains modern technologies such as next-generation sequencing and microarrays

Chapter 17: Evolution: The Development of the Theory (NEW)
- Explores the development of the theory of evolution from Aristotle's knowledge of the natural world, to Darwin, to the Modern Synthesis
- Introduces variation and selection, major components of Darwin's theory

Chapter 18: Microevolution: Changes within Populations (NEW)
- Expands the topics of variation and selection
- Introduces different types of variation, as well as a variety of selection agents such as mutation and nonrandom mating
- Discusses the role played by population biology in evolution along with the Hardy–Weinberg Principle

Chapter 19: Species and Macroevolution (NEW)
- Begins with a thorough discussion of species, species concepts, and speciation
- Explores reproductive isolating mechanisms and the variety of geographic speciation over time

Chapter 20: Understanding the History of Life on Earth (NEW)
- Outlines the history of life on Earth by exploring the geological time scale, the fossil record, and the lives of prehistoric organisms
- Ties the discussion of the history of life to phylogeny and phylogenetics as they apply to the evolution of birds

Chapter 21: Humans and Evolution (NEW)
- Draws together key themes in the previous chapters and sets them within the context of human evolution

Chapter 22: Bacteria and Archaea
- Updated "Why It Matters" captures recent research on the gut microbiome and its role in human health
- New "Life on the Edge" box deals with poikilohydric plants

Chapter 23: Viruses, Viroids, and Prions: Infectious Biological Particles
- Updated information on prions to reflect current research
- Modified colour scheme of figures to increase clarity

Chapter 24: Protists
- Updated to include the evolutionary tree of eukaryotes showing "supergoups"
- Revised classification of protist groups according to this new evolutionary tree

Chapter 25: Fungi
- Clarified placement of fungi that lack sexual stage in fungal classification

Chapter 26: Plants
- Revised section on evolutionary relationships among bryophytes to reflect current research findings

Chapter 30: Population Interactions and Community Ecology
- New "Why It Matters" dealing with morphological specializations.

Chapter 31: Ecosystems
- New table outlining the characteristics of the terrestrial ecozones of Canada

Chapter 32: Conservation of Biodiversity
- New section on white-nose syndrome and bats

Chapter 34: Organization of the Plant Body
- Revised "People behind Biology" to highlight new research with greater significance related to provision of clean drinking water

Chapter 35: Transport in Plants
- Revised and clarified material on water potential

Chapter 37: Plant Nutrition (NEW)
- Revised section detailing effects of nutrient deficiency on plant structure

Chapter 38: Plant Signals and Responses to the Environment
- Updated information on brassinosteroids and plant chemical defences

Chapter 48: Animal Nutrition
- New "People behind Biology" featuring Frederick Banting and George Best

Chapter 51: Defences against Disease
- New material on *E. coli*
- New section on parasite–host interactions and the successes and limitations of vaccines

Chapter 52: Conservation and Evolutionary Physiology (NEW)
- Integrates the different physiological processes described in the preceding chapters as they interact in plants and animals in nature
- Explores the evolution of physiological processes and the manner in which an understanding of this can inform conservation issues

THINK AND ENGAGE LIKE A SCIENTIST

MindTap™

Engage, Adapt, and Master!

Stay organized and efficient with **MindTap**—a single destination with all the course material and study aids you need to succeed. Built-in apps leverage social media and the latest learning technology to help you succeed. Our customized learning path is designed to help you engage with biological concepts, identify gaps in your knowledge, and master the material!

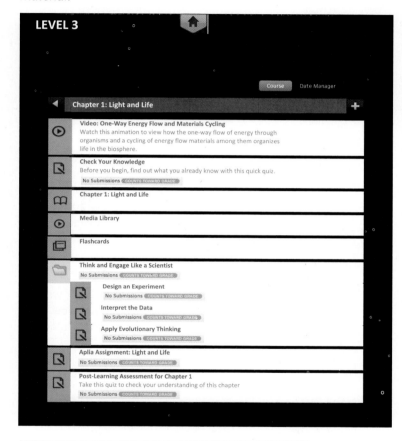

Engage!
The learning path for each chapter begins with an engaging video designed to pique your interest in the chapter contents.

Adapt!
Take the tutorial quiz to assess gaps in your knowledge. Strengthen your knowledge of concepts by reviewing the eBook, animations, and instructive videos. Reinforce your knowledge with glossary flashcards.

Master!
Think and engage like a scientist by taking short-answer quizzes:

- **Apply Evolutionary Thinking** questions ask you to interpret a relevant topic in relation to the principles of evolutionary biology.

- **Design an Experiment** challenges your understanding of the chapter and helps you deepen your understanding of the scientific method as you consider how to develop and test hypotheses about a situation that relates to a main chapter topic.

- **Interpret the Data** questions help you develop analytical and quantitative skills by asking you to interpret graphical or tabular results of experimental or observational research experiments for which the hypotheses and methods of analysis are presented.

Test your mastery of concepts with Aplia for Biology, an interactive online tool that complements the text and helps you learn and understand key concepts through focused assignments, an engaging variety of problem types, exceptional text/art integration, and immediate feedback.

And assess your knowledge of chapter concepts by taking the **Review Quiz**.

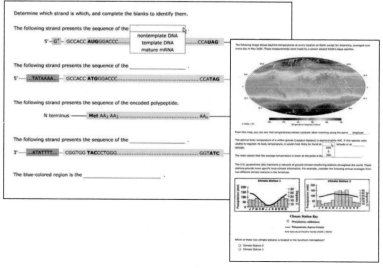

THE BIG PICTURE

Each chapter of *Biology: Exploring the Diversity of Life,* Third Canadian Edition, is carefully organized and presented in digestible chunks so you can stay focused on the most important concepts. Easy-to-use learning tools point out the topics covered in each chapter, show why they are important, and help you learn the material.

Study Plan The Study Plan provides a list of the major sections in the chapter. Each section breaks the material into a manageable amount of information, building on knowledge and understanding as you acquire it.

Why It Matters For each chapter we provide a brief contextual overview, outlining the main points that follow.

Study Breaks The Study Breaks encourage you to pause and think about the key concepts you have just encountered before moving to the next section. The Study Break questions are written by Canadian students for their peers across the country and are intended to identify some of the important features of the section.

STUDY BREAK

1. Define light.
2. What structural feature is common to all pigments?

MOLECULE BEHIND BIOLOGY 3.1

L1 Ligase Ribozyme

In the RNA world, RNA served as the molecule of information storage as well as structure and catalysis. To replicate RNA, individual nucleotide triphosphate monomers need to be joined, or ligated, together to form the RNA polymer. Today, this ligation reaction, carried out by a group of protein enzymes called polymerases, can result in RNA strands being many thousands of nucleotides in length. How this polymerization reaction would have been catalyzed in an RNA-only world stumped scientists for years.

Using what is called *in vitro* evolution and selection, scientists recently produced a range of synthetic ribozymes that do not currently exist in nature. One of these synthetic ribozymes is called the L1 ligase ribozyme, and it has been shown to catalyze the joining of two RNA monomers. This finding clearly suggests that, although not currently found in nature, a ribozyme capable of ligating nucleotides together may have existed on primitive Earth.

Molecule behind Biology "Molecule behind Biology" boxes give students a sense of the exciting impact of molecular research. From water to progesterone, amanitin, and DDT, each chapter features the activity of a relevant chemical.

PEOPLE BEHIND BIOLOGY 19.1

Barcode of Life Project

Today, most systematists working on living organisms include molecular characters as part of the data set when determining evolutionary relationships. Molecular data include nucleotide base sequences of DNA and RNA or the amino acid sequences of the proteins for which they code. Technological advances have automated many of the necessary laboratory techniques, and analytical software makes it easy to compare new data with information filed in data banks, for example, the Barcode of Life project.

In 2003, Paul Hebert, of the Biodiversity Institute of Ontario at the University of Guelph, proposed a new system of species identification using a sequence of approximately 600 DNA base pairs of the mitochondrial gene cytochrome oxidase I. He anticipated a database that would "provide a new master key for identifying species, one whose power will rise with increased taxon coverage and with faster, cheaper sequencing." He compared the system to the barcode system for retail products such as grocery items, calling it the Barcode of Life (**Figure 1**).

Molecular sequences have some practical advantages over organismal characters. First, they provide abundant data because every base in a nucleic acid can serve as a separate, independent character for analysis. Moreover, because many genes have been conserved by evolution, molecular sequences can be compared between distantly related organisms that share

no organismal characters. Molecular characters can also be used to study closely related species that have only minor morphological differences. Finally, many nucleic acids are not directly affected by the developmental or environmental factors that cause nongenetic morphological variations (see Chapter 18).

But there are drawbacks to using molecular characters in taxonomy. There are only four alternative character states (the four nucleotide bases) at each position in a DNA or RNA sequence (see Chapter 11). If two species have the same nucleotide base substitution at a given position in a DNA segment, their similarity may have evolved independently. As a

result, systematists often find it difficult to verify that molecular similarities were inherited from a common ancestor.

For organismal characters, biologists can establish that similarities are homologous by analyzing the characters' embryonic development or details of their function. But molecular characters have no embryonic development, and biologists still do not understand the functional significance of most molecular differences. Despite these disadvantages, molecular characters represent the genome directly, and researchers use them with great success in phylogenetic analyses and in embryology.

Figure 1

People behind Biology "People behind Biology" boxes in most chapters contain boxed stories about how particular people have used their ingenuity and creativity to expand our knowledge of biology. The purpose of these boxes is to recognize that advances in biology are accomplished by people.

Life on the Edge "Life on the Edge" boxes provide accounts of organisms thriving "on the edge" at unusual temperatures, pressures, radiation dosages, salt concentrations, and so on. These boxes explain how our understanding of "normal" can be increased through study of the "extreme."

The Chemical, Physical, and Environmental Foundations of Biology
While many textbooks use the first few chapters to introduce and/or review, we believe that the first chapters should convey the excitement and interest of biology itself. We therefore placed important background information about biology and chemistry in the reference section entitled *The Chemical, Physical, and Environmental Foundations of Biology*, in the centre of the book. With their purple borders, these pages are distinct and easy to find and have become affectionately known as **The Purple Pages**.

CONCEPT FIX **Concept Fix Icons** Concept Fixes draw on the extensive research literature dealing with misconceptions commonly held by biology students. Strategically placed throughout the text, these short segments help students identify—and correct—a wide range of misunderstandings. ⬡

Your study of biology focuses not only on *what* scientists now know about the living world but also on *how* they know it. Use these unique features to learn through example how scientists ask scientific questions and pose and test hypotheses.

Throughout the book, we identify recent discoveries made possible by the development of new techniques and new knowledge. In "People behind Biology" 48.1, page 1208, for example, learn how luck and good timing led to the isolation of insulin and a treatment for diabetes.

The Chemical, Physical, and Environmental Foundations of Biology Also known as *The Purple Pages*, these pages enable information to be readily identifiable and accessible to students as they move through the textbook, rather than tied to one particular chapter. *The Purple Pages* keep background information out of the main text, allowing you to focus on the bigger picture.

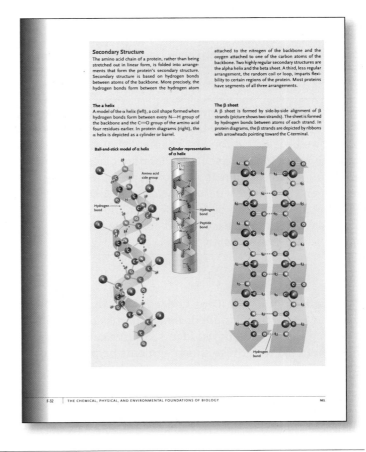

Research Figures Located throughout the book, research figures contain information about how biologists formulate and test specific hypotheses by gathering and interpreting data.

Research Method

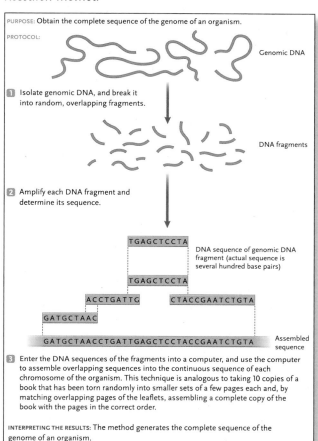

PURPOSE: Obtain the complete sequence of the genome of an organism.

PROTOCOL:

Genomic DNA

1 Isolate genomic DNA, and break it into random, overlapping fragments.

DNA fragments

2 Amplify each DNA fragment and determine its sequence.

TGAGCTCCTA

DNA sequence of genomic DNA fragment (actual sequence is several hundred base pairs)

TGAGCTCCTA

ACCTGATTG   CTACCGAATCTGTA

GATGCTAAC

GATGCTAACCTGATTGAGCTCCTACCGAATCTGTA

Assembled sequence

3 Enter the DNA sequences of the fragments into a computer, and use the computer to assemble overlapping sequences into the continuous sequence of each chromosome of the organism. This technique is analogous to taking 10 copies of a book that has been torn randomly into smaller sets of a few pages each and, by matching overlapping pages of the leaflets, assembling a complete copy of the book with the pages in the correct order.

INTERPRETING THE RESULTS: The method generates the complete sequence of the genome of an organism.

Research Method Figure 16.1 Whole-genome shotgun sequencing.

Experimental Research

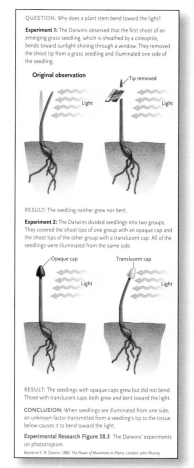

QUESTION: Why does a plant stem bend toward the light?

Experiment 1: The Darwins observed that the first shoot of an emerging grass seedling, which is sheathed by a coleoptile, bends toward sunlight shining through a window. They removed the shoot tip from a grass seedling and illuminated one side of the seedling.

Original observation   Tip removed

Light   Light

RESULT: The seedling neither grew nor bent.

Experiment 2: The Darwins divided seedlings into two groups. They covered the shoot tips of one group with an opaque cap and the shoot tips of the other group with a translucent cap. All of the seedlings were illuminated from the same side.

Opaque cap   Translucent cap

Light   Light

RESULT: The seedlings with opaque caps grew but did not bend. Those with translucent caps both grew *and* bent toward the light.

CONCLUSION: When seedlings are illuminated from one side, an unknown factor transmitted from a seedling's tip to the tissue below causes it to bend toward the light.

Experimental Research Figure 38.3 The Darwins' experiments on phototropism.

Based on C. R. Darwin. 1880. *The Power of Movement in Plants.* London: John Murray.

Observational Research

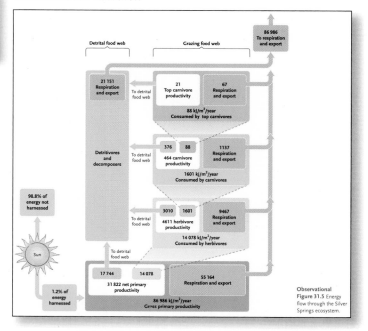

Observational Figure 31.5 Energy flow through the Silver Springs ecosystem.

VISUAL LEARNING

Spectacular illustrations—developed with great care—help you visualize biological processes, relationships, and structures.

Illustrations of complex biological processes are annotated with numbered step-by-step explanations that lead you through all the major points. Orientation diagrams are inset on figures and help you identify the specific biological process being depicted and where the process takes place.

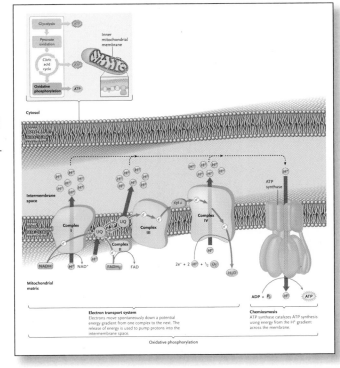

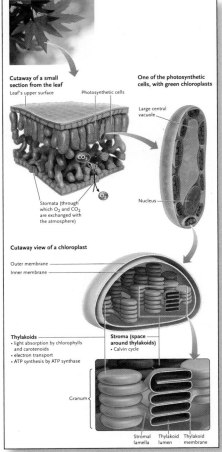

From Macro to Micro Multiple views help you visualize the levels of organization of biological structures and how systems function as a whole.

Electron micrographs are keyed to selected illustrations to help clarify biological structures.

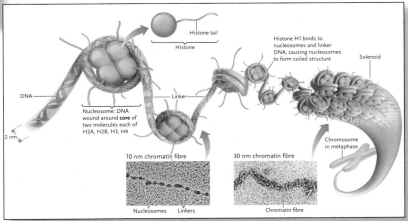

End-of-chapter material encourages you to review the content, assess your understanding, think analytically, and apply what you have learned to novel situations.

Review Key Concepts This brief review often references figures and tables in the chapter and summarizes important ideas developed in the chapter.

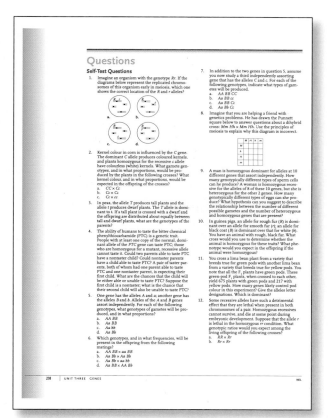

Self-Test Questions These end-of-chapter questions focus on factual content in the chapter while encouraging you to apply what you have learned. Answers to the Self-Test questions have been compiled and placed at the end of the book.

Questions for Discussion

1. The eyes of brown-eyed people are not alike but rather vary considerably in shade and pattern. What do you think causes these differences?

2. Explain how individuals of an organism that are phenotypically alike can produce different ratios of progeny phenotypes.

3. ABO blood type tests can be used to exclude paternity. Suppose a defendant who is the alleged father of a child takes a blood-type test and the results do not exclude him as the father. Do the results indicate that he is the father? What arguments could a lawyer make based on the test results to exclude the defendant from being the father? (Assume the tests were performed correctly.)

Questions for Discussion These questions enable you to participate in discussions on key questions to build your knowledge and learn from others.

Succeed in the course with these dynamic resources!

MindTap

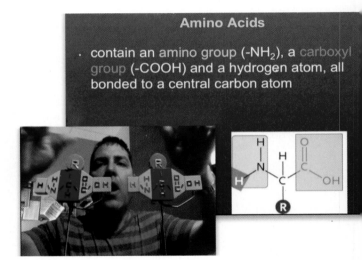

Amino Acids

- contain an amino group ($-NH_2$), a carboxyl group ($-COOH$) and a hydrogen atom, all bonded to a central carbon atom

MindTap™ **Stay organized and efficient with MindTap—a single destination with all** the course material and study aids you need to succeed. **Built-in apps leverage social media and the latest learning technology.**

For example:

- ReadSpeaker will read the text to you.
- Flashcards are prepopulated to provide you with a jump start for review—or you can create your own.
- You can highlight text and make notes in your MindTap Reader. Your notes will flow into Evernote, the electronic notebook app that you can access anywhere when it's time to study for the exam.

Self-quizzing allows you to assess your understanding.

Also available in the MindTap for Biology are engaging and informative videos that accompany the Purple Pages. From matter to polypeptides, Todd Nickle of Mount Royal University (pictured) will walk you through these foundational concepts, strengthening your understanding and helping you build a strong base of knowledge and understanding for biology.

Visit www.nelson.com/student to start using **MindTap.** Enter the Online Access Code from the card included with your textbook. If a code card is *not* provided, you can purchase instant access at NELSONbrain.com.

Aplia for Biology

aplia™ **Strengthen your understanding of biology with Aplia™!**

This innovative, easy-to-use, interactive technology gives you more practice, with detailed feedback to help you learn with every question!

Aplia's focused assignments and active learning opportunities (including randomized questions, exceptional text/art integration, and immediate feedback) get you involved with biology and help you think like a scientist. For more information, visit **www.aplia.com/biology.**

Interactive problems and figures help you visualize dynamic biological processes and integrate concepts, art, media, and homework practice.

Study Guide

The Study Guide (ISBN: 978-0-17-660144-7) from the second Canadian edition has been adapted for the third Canadian edition by Dora Cavallo-Medved of the University of Windsor, Roy V. Rea of the University of Northern British Columbia, and Julie Smit, of the University of Windsor. The Study Guide helps students integrate the concepts within the text and provides study strategies, interactive exercises, self-test questions, and more.

MindTap

MindTap **Engage, adapt, and master!**
Engage your students with the personalized teaching experience of **MindTap**. With relevant assignments that guide students to analyze, apply, and elevate thinking, **MindTap** allows instructors to measure skills and promote better outcomes with ease. Including interactive quizzing, this online tutorial and diagnostic tool identifies each student's unique needs with a pretest. The learning path then helps students focus on concepts they're having the most difficulty mastering. It refers to the accompanying MindTap Reader eBook and provides a variety of learning activities designed to appeal to diverse ways of learning. After completing the study plan, students take Aplia problem sets and then take a posttest to measure their understanding of the material. Instructors have the ability to customize the learning path, add their own content, and track and monitor student progress by using the instructor Gradebook and Progress app. The **MindTap** resources have been developed by Dora Cavallo-Medved of the University of Windor and Reehan Mirza of Nipissing University.

Aplia for Biology

aplia **Get your students engaged and motivated with Aplia for Biology.**
Help your students learn key concepts via Aplia's focused assignments and active learning opportunities that include randomized, automatically graded questions, exceptional text/art integration, and immediate feedback. Aplia has a full course management system that can be used independently or in conjunction with other course management systems such as Blackboard and WebCT. **Visit www.aplia.com/biology.**

The Aplia course for *Biology: Exploring the Diversity of Life,* Third Canadian Edition, was prepared by Anna Rissanen of Memorial University, and Todd Nickle and Alexandria Farmer of Mount Royal University.

Instructor's Resource DVD: The ultimate tool for customizing lectures and presentations.

Select instructor ancillaries are provided on the *Instructor's Resource DVD* (ISBN 978-0-17-660143-0) and the Instructor Companion site.

 The **Nelson Education Teaching Advantage (NETA)** program delivers research-based instructor resources that promote student engagement and higher-order thinking to enable the success of Canadian students and educators.

To ensure the high quality of these materials, all Nelson ancillaries have been professionally copy-edited.

Be sure to visit Nelson Education's **Inspired Instruction** website at **www.nelson.com/inspired/** to find out more about NETA. Don't miss the testimonials of instructors who have used NETA supplements and seen student engagement increase!

NETA Test Bank: This resource was written by Ivona Mladenovic of Simon Fraser University. It includes over 2500 multiple-choice questions written according to NETA guidelines for effective construction and development of higher-order questions. The Test Bank was copy-edited by a NETA-trained editor and reviewed by David DiBattista for adherence to NETA best practices. Also included are true/false, essay, short-answer, matching, and completion questions. Test Bank files are provided in Microsoft Word format for easy editing and in PDF for convenient printing, whatever your system.

cognero The **NETA Test Bank** is available in a new, cloud-based platform. **Nelson Testing Powered by Cognero®** is a secure online testing system that allows you to author, edit, and manage test bank content from any place you have Internet access. No special installations or downloads are needed, and the desktop-inspired interface, with its drop-down menus and familiar, intuitive tools, allows you to create and manage tests with ease. You can create multiple test versions in an instant and import or export content into other systems. Tests can be delivered from your learning management system, your classroom, or wherever you want. Nelson Testing Powered by Cognero can be accessed through www.nelson.com/instructor.

NETA PowerPoint: Microsoft PowerPoint® lecture slides for every chapter have been created by Jane Young of the University of Northern British Columbia. There is an average of 80 slides per chapter, many featuring key figures, tables, and photographs from *Biology: Exploring the Diversity of Life,* Third Canadian Edition. The PowerPoint slides also feature "build slides"—selected illustrations with labels from the book that have been reworked to allow optimal display in PowerPoint. NETA principles of clear design and engaging content have been incorporated throughout, making it simple for instructors to customize the deck for their courses.

Image Library: This resource consists of digital copies of figures, short tables, and photographs used in the book. Instructors may use these jpegs to customize the NETA PowerPoint slides or create their own PowerPoint presentations.

NETA Instructor's Manual: This resource was written by Tamara Kelly of York University and Tanya Noel of the University of Windsor. It is organized according to the textbook chapters and addresses key educational concerns, such as typical stumbling blocks students face and how to address them. Other features include tips on teaching using cases as well as suggestions on how to present material and use technology and other resources effectively, integrating the other supplements available to both students and instructors. This manual doesn't simply reinvent what's currently in the text; it helps the instructor make the material relevant and engaging to students.

Day One: Day One—ProfInClass is a PowerPoint presentation that you can customize to orient your students to the class and their text at the beginning of the course.

TurningPoint®: Another valuable resource for instructors is **TurningPoint® classroom response software** customized for *Biology: Exploring the Diversity of Life*, Third Canadian Edition, by Jane Young at the University of Northern British Columbia. Now you can author, deliver, show, access, and grade, all in PowerPoint, with no toggling back and forth between screens! JoinIn on TurningPoint is the only classroom response software tool that gives you true PowerPoint integration. With JoinIn, you are no longer tied to your computer. You can walk about your classroom as you lecture, showing slides and collecting and displaying responses with ease. There is simply no easier or more effective way to turn your lecture hall into a personal, fully interactive experience for your students. If you can use PowerPoint, you can use JoinIn on TurningPoint! (Contact your Nelson publishing representative for details.) These contain poll slides and pre- and posttest slides for each chapter in the text.

Prospering in Biology

Using This Book

The following are things you will need to know in order to use this text and prosper in biology.

Names

What's in a name? People are very attached to names—their own names, the names of other people, the names of flowers and food and cars and so on. It is not surprising that biologists would also be concerned about names. Take, for example, our use of scientific names. Scientific names are always italicized and Latinized.

Castor canadensis Kuhl is the scientific name of the Canadian beaver. *Castor* is the genus name; *canadensis* is the specific epithet. Together they make up the name of the species, which was first described by a person called Kuhl. "Beaver" by itself is not enough because there is a European beaver, *Castor fiber,* and an extinct giant beaver, *Castoides ohioensis*. Furthermore, common names can vary from place to place (*Myotis lucifugus* is sometimes known as the "little brown bat" or the "little brown myotis").

Biologists prefer scientific names because the name (Latinized) tells you about the organism. There are strict rules about the derivation and use of scientific names. Common names are not so restricted, so they are not precise. For example, in *Myotis lucifugus, Myotis* means "mouse-eared" and *lucifugus* means "flees the light"; hence, this species is a mouse-eared bat that flees the light.

Birds can be an exception. There are accepted "standard" common names for birds. The American Robin is *Turdus migratorius*. The common names for birds are usually capitalized because of the standardization. However, the common names of mammals are not capitalized, except for geographic names or patronyms (*geographic* = named after a country, e.g., Canadian beaver; *patronym* = named after someone, e.g., Ord's kangaroo rat).

Although a few plants that have very broad distributions may have accepted standard common names (e.g., white spruce, *Picea glauca*), most plants have many common names. Furthermore, the same common name is often used for more than one species. Several species in the genus *Taraxacum* are referred to as "dandelion." It is important to use the scientific names of plants to be sure that it is clear exactly which plant we mean. The scientific names of plants also tell us something about the plant. The scientific name for the weed quack grass, *Elymus repens,* tells us that this is a type of wild rye (*Elymus*) and that this particular species spreads or creeps (*repens* = creeping). Anyone who has tried to eliminate this plant from their garden or yard knows how it creeps! Unlike

for animals, plant-naming rules forbid the use of the same word for both genus and species names for a plant; thus, although *Bison bison* is an acceptable scientific name for buffalo, such a name would never be accepted for a plant.

In this book, we present the scientific names of organisms when we mention them. We follow standard abbreviations; for example, although the full name of an organism is used the first time it is mentioned (e.g., *Castor canadensis*), subsequent references to that same organism abbreviate the genus name and provide the full species name (e.g., *C. canadensis*).

In some areas of biology, the standard representation is of the genus, for example, *Chlamydomonas*. In other cases, names are so commonly used that only the abbreviation may be used (e.g., *E. coli* for *Escherichia coli*).

Units

The units of measure used by biologists are standardized (metric or SI) units, used throughout the world in science.

Definitions

The science of biology is replete with specialized terms (sometimes referred to as "jargon") used to communicate specific information. It follows that, as with scientific names, specialized terms increase the precision with which biologists communicate among themselves and with others. Be cautious about the use of terms because jargon can obscure precision. When we encounter a "slippery" term (such as *species* or *gene*), we explain why one definition for all situations is not feasible.

Time

In this book, we use CE (Common Era) to refer to the years since year 1 and BCE (Before the Common Era) to refer to years before that.

Geologists think of time over very long periods. A geological time scale (see *The Purple Pages,* pp. F-50–F-51) shows that the age of Earth could be measured in years, but it's challenging to think of billions of years expressed in days (or hours, etc.). With the advent of using the decay rates of radioisotopes to measure the age of rocks, geologists adopted 1950 as the baseline, the "Present," and the past is referred to as BP ("Before Present"). A notation of 30 000 years BP (^{14}C) indicates 30 000 years before 1950 using the ^{14}C method of dating.

Other dating systems are also used. Some archaeologists use PPNA (PrePottery Neolithic A, where A is the horizon or stratum). In deposits along the Euphrates River, 11 000 PPNA appears to be the same as 11 000 BP. In this book, we use BCE or BP as the time units, except when referring to events or species from more than 100 000 years ago. For those dates, we refer you to the geological time scale (see *The Purple Pages*, pp. F-50–F-51).

Sources

Where does the information presented in a text or in class come from? What is the difference between what you read in a textbook or an encyclopedia and the material you see in a newspaper or tabloid? When the topic relates to science, the information should be based on material that has been published in a scholarly journal. In this context, "scholarly" refers to the process of review. Scholars submit their manuscripts reporting their research findings to the editor (or editorial board) of a journal. The editor, in turn, sends the manuscript out for comment and review by recognized authorities in the field. The process is designed to ensure that what is published is as accurate and appropriate as possible. The review process sets the scholarly journal apart from the tabloid.

There are literally thousands of scholarly journals, which, together, publish millions of articles each year. Some journals are more influential than others, for example, *Science* and *Nature*. These two journals are published weekly and invariably contain new information of interest to biologists.

To collect information for this text, we have drawn on published works that have gone through the process of scholarly review. Specific references (citations) are provided, usually in the electronic resources designed to complement the book.

A citation is intended to make the information accessible. Although there are many different formats for citations, the important elements include (in some order) the name(s) of the author(s), the date of publication, the title, and the publisher. When the source is published in a scholarly journal, the journal name, its volume number, and the pages are also provided. With the citation information, you can visit a library and locate the original source. This is true for both electronic (virtual) and real libraries.

Students of biology benefit by making it a habit to look at the most recent issues of their favourite scholarly journals and use them to keep abreast of new developments.

M. Brock Fenton
Heather Addy
Denis Maxwell
Tom Haffie
Bill Milsom

London, Calgary, and Vancouver
January 2015

Acknowledgements

We thank the many people who have worked with us on the production of this text, particularly Paul Fam, Senior Publisher, whose foresight brought the idea to us and whose persistence saw the project through. Thanks go to those who reviewed the second Canadian edition text to provide us with feedback for the third Canadian edition, including

Declan Ali, University of Alberta
Eric Alcorn, Acadia University
Patricia Chow-Fraser, McMaster University
Kimberley Gilbride, Ryerson University
Roberto Quinlan, York University
Matthew Smith, Wilfrid Laurier University
Christopher Todd, University of Saskatchewan
Kenneth Wilson, University of Saskatchewan

We are also grateful to the members of the Editorial Advisory Board and the Student Advisory Boards for the third Canadian edition, who provided us with valuable feedback and alternative perspectives (special acknowledgements to these individuals are listed below). We also thank Richard Walker at the University of Calgary and Ken Davey at York University, who began this journey with us but were unable to continue. We thank Carl Lowenberger for contributing Chapter 51, Defences against Disease. We are especially grateful to Toni Chahley, Developmental Editor, who kept us moving through the chapters at an efficient pace, along with Shanthi Guruswamy and Alex Antidius, Project Managers, and Natalia Denesiuk Harris, Senior Production Project Manager. We thank Kristiina Paul, our photo researcher, for her hard work with the numerous photos in the book, and Julia Cochrane for her careful and thoughtful copy-editing. Finally, we thank Leanne Newell, Marketing Manager, for making us look good.

Brock Fenton would like to thank Allan Noon, who offered much advice about taking pictures; Laura Barclay, Jeremy McNeil, Tony Percival-Smith, C.S. (Rufus) Churcher, and David and Meg Cumming for the use of their images; and Karen Campbell for providing a critical read on Chapter 48, Putting Selection to Work.

Heather Addy would like to thank Ed Yeung for generously providing many images and assistance with revision of figures, and Cindy Graham, David Bird, Fengshan Ma, and William Huddleston for providing feedback and valuable suggestions for improving several chapters.

Tom Haffie would like to acknowledge the cheerful and insightful editorial work of Jennifer Waugh on Chapter 16.

The authors are all indebted to Johnston Miller, whose extensive background research anchored our Concept Fixes in the education literature.

We thank Rosemary Tanner, our substantive editor, for bringing her considerable skills to this edition. Her insight and expertise have brought a new level of precision and clarity to key chapters in the book. She is a consummate professional, and we are all very grateful for her patience and good humour.

It is never easy to be in the family of an academic scientist. We are especially grateful to our families for their sustained support over the course of our careers, particularly during those times when our attentions were fully captivated by bacteria, algae, fungi, parasites, snakes, geese, or bats. Saying "yes" to a textbook project means saying "no" to a variety of other pursuits. We appreciate the patience and understanding of those closest to us that enabled the temporary reallocation of considerable time from other endeavours and relationships.

Many of our colleagues have contributed to our development as teachers and scholars by acting as mentors, collaborators, and, on occasion, "worthy opponents." Like all teachers, we owe particular gratitude to our students. They have gathered with us around the discipline of biology, sharing their potent blend of enthusiasm and curiosity and leaving us energized and optimistic for the future.

Editorial and Student Advisory Boards

We were very fortunate to have the assistance of some extraordinary students and instructors of biology across Canada who provided us with feedback that helped shape this textbook into what you see before you. As such, we would like to say a very special thank-you to the following people:

Editorial Advisory Board

Peter Boag, Queen's University
Julie Clark, York University
Brett Couch, University of British Columbia
Robert Edwards, University of Calgary
Mark Fitzpatrick, University of Toronto
Jon Houseman, University of Ottawa
Andrew Laursen, Ryerson University
Todd Nickle, Mount Royal University
Cynthia Paszkowski, University of Alberta
Roy Rea, University of Northern British Columbia
Matthew Smith, Wilfrid Laurier University
Christopher Todd, University of Saskatchewan

Student Advisory Boards

University of Alberta

Mount Royal University

Paul Fam/Nelson Education Ltd.

Paul Fam/Nelson Education Ltd.

Depicted above are members of the University of Alberta Student Advisory Board: (left to right) Shargeel Hayat, Mark Kamprath, Esperance Madera, Maxwell Douglas, Kina Tiet, Dana Miller, Jessica Lamont. Not pictured: Punit Virk.

Depicted above are members of the Student Advisory Board at Mount Royal University: (left to right) Joshua Loza, Lorne Sobcyzk, Hala Al Sharbati, Nikolai Heise, Karishma Rahmat, Todd Nickle, Kathleen Malabug, Katherine Obrovac, Kristin Milloy, Michael Rudolf, Dorothy Hill, Maria Fernanda Ochoa.

EVOLUTION, ASPECTS OF DIVERSITY, ECOLOGY, AND APPLIED BIOLOGY

M.B. Fenton

Venus flytrap (*Dionaea muscipula*). A leaf specialized to trap insects shows the edges lined with bristles, black trigger hairs (trichomes) that spring the trap, and the red colour that attracts prey.

This volume consists of four units, progressing from mechanisms involved in diversification (Unit Five) to explorations of the diversity of life (Unit Six). The ecological and environmental contexts and patterns associated with diversification (Unit Seven) are considered along with two facets of applied biology (Unit Seven), specifically conservation of biodiversity and how humans have used selection to produce food. The impacts of evolution and diversification of the human evolutionary lineage establish our evolutionary roots and illustrate the fundamental impact our species has had on global diversity. On the positive side, humans have turned evolutionary processes to our advantage in domesticating other organisms to our advantage, but the negative impact of these outcomes often is alarmingly obvious.

You will see common themes in the chapters in this volume including fitness, natural selection, adaptive advantages, diversification, and evolution. The central role of the environment in evolutionary history and ongoing evolution also is apparent. The vital importance of interactions among organisms is another recurring theme.

Each chapter in this volume connects directly and indirectly to other chapters in the book, echoing the fundamental theme "exploring the diversity of life." The connections hinge on fundamental concepts and facts of life, such as those explored in Volume I. These relate to energy and metabolism, the structures of cells, and the way that cells operate. Genetics is a central pillar of biology, and advances in genetics have greatly increased our understanding of evolutionary history the relationships among organisms. Also essential are connections to the ways in which multicellular organisms operate (Volume III).

In each unit covered in this volume, it is clear how the science of biology has progressed, from observation and description, to generation of hypotheses about diversity, and to specific testable predictions arising from the hypotheses. Methodological advances and new techniques have allowed more rigorous testing of hypotheses and related predictions about how life has diversified and continues to diversify on Earth.

The solutions that plants have evolved to use animals to gain access to nitrogen illustrates the some of the principles underlying this volume. In plants, the repeated appearance of mechanisms for obtaining nitrogen by trapping and digesting animals reflects the selective advantage(s) of this way of life. The diversity of carnivorous plants and the appearance of this way of life in different evolutionary lineages of plants have always intrigued many students of biology, including Charles Darwin. In 1875, Darwin noted that there were at least 20 genera of carnivorous plants distributed in 10 families.

The diversity of animal traps in plants **(Figure 1)** is well known and there are both terrestrial and aquatic carnivorous plants. Sticky traps also appear in species in the families Droseraceae and Lentbulariaceae. Bucket traps, in which the insect (or other prey) falls into a pool of water contained in a modified leaf and drowns, have appeared independently in at least four families of plants. Active traps, where part of the plant moves to catch the animal, occur in Droseraceae and Lentbulari-aceae, two families with both terrestrial and aquatic species. A recent (2012) addition to the list of carnivorous plants comes from low-nutrient, sandy soils in Brazil. Three species in the genus *Philcoxia* (family Plantaginaceae) use underground adhesive leaves to trap and digest nematode worms. In 2013, a bromeliad-like plant (*Paepalanthus bromelioides*; family Eriocaulaceae) emerged as a species on its way to becoming carnivorous. Analysis of isotopes of Nitrogen in this plant revealed 15N that was derived from termites. These plants grow on termite nests, so they had not directly captured the termites.

Isotopic analysis also revealed that some pitcher plants in the genus *Nepenthes* have specialized pitchers that allow them to collect urine and feces from mammals, specifically tree shrews and roosting bats. Isotopic analysis also revealed that many other pitcher plants acquired their nitrogen from symbionts such as rotifers, living in the pitcher water. Bacteria and rotifers consumed the animals that drowned in the pitchers and it was their excretory products that provided nitrogen to the plants.

The waterwheel plant, a relatively close relative of Venus flytraps, is widespread globally, occurring in Europe, Africa, Asia, and Australia. Venus fly-traps, however, naturally occur only in North and South Carolina in the United States. The long-term survival of both of these specialized species is a matter of concern for conservationists.

In short, studies of carnivorous plants provide rich examples of evolution, diversification, and outcomes of interactions among organisms. The many different plants with carnivorous habits provide further examples of parallel and convergent evolution that extend to details of digestion of prey and the operation of traps connect with material presented in Volume III.

a. b. c. d. e.

M.B. Fenton

Figure 1
A sampling of terrestrial insectivorous plants. Included are plants with adhesive leaves **(a, b),** and pitcher plants **(c, d and e).** Butterwort (*Pinguicula* spp. - a) has adhesive leaves that have caught insects. A leaf of a sundew (*Drosera* spp. - b) bear adhesive droplets at the ends of trichomes. The pitcher plants include c - *Darlingtonia* from California, d - *Sarracinea* from Ontario, and e - *Nepenthes* from the East Indies.

An Australian sheep blowfly (family: Calliphoridae, *Lucilia cuprina*) (6–9 mm long) (left). Maggots of Australian sheep blowflies (right).

© Julian Money-Kyrle/Alamy

Shuchi Arora, Carl Bachhuber, Chu Sing Lim, "Maggot Metabolites and their combinatory effects with antibiotic on Staphylococcus aureus. Annals of Clinical Microbiology and Antimicrobials 2011, 10:6, doi:10.1186/1476-0711-10-6.

17

Evolution: The Development of the Theory

WHY IT MATTERS

Australian sheep blowflies occur throughout Australia in habitats ranging from urban to rural settings, in semi-arid open lands as well as forests and woodlands. These flies have spread widely throughout the world, including to eastern Canada. They are considered pests because females deposit their eggs in open wounds on livestock such as sheep (called "fly strikes"). The eggs hatch into maggots, which eat flesh and damage the wool. A single female can produce hundreds of eggs. These flies complete their life cycle—egg, larva, pupa, and adult—in about 11 to 21 days.

Fly strikes by Australian sheep blowflies cost the wool industry over $A150 million a year ($A stands for Australian dollars). Tools for controlling the blowflies include bait traps and at least five kinds of chemical insecticides. Organophosphate insecticides are effective because they affect the nervous system and rapidly kill both adult flies and maggots.

Some Australian sheep blowflies have a mutation that makes them resistant to organophosphate insecticides. This resistance is based on a single mutation that appeared in a few individuals of this species. The DNA of resistant flies differs slightly from that of susceptible flies. When sprayed with the insecticide, resistant flies have enhanced fitness: they survive to reproduce, while the others around them die. The flies' short life cycle and high reproductive rate ensure the rapid spread of the resistant mutation in the fly population. Put another way, selection associated with the pesticide changes the population structure of sprayed flies, quickly eliminating nonresistant individuals.

This example shows how the combination of a genetic-based change (a small mutation) and strong selective pressure (lethal organophosphate insecticide) has altered the population of Australian sheep blowflies over time. This also shows how human interference can result in evolution by natural selection.

Evolution is the main unifying concept in biology, explaining how the diversity of life on Earth arose and how species change over time in response to changes in their abiotic and biotic environment. Our knowledge of how populations of organisms change over time has been enriched with data obtained using techniques of molecular biology, particularly those relating to genetics. However, we will begin with a history of the ideas and thinking that led to the work of Charles Darwin and Alfred Wallace. In 1858, these men independently published descriptions of variation within species (intraspecific variation) and proposed a process called natural selection to explain how species changed over time.

17.1 Evolution: What Is It?

Evolution means gradual change (from the Latin *evolere* = to unfold or unroll). **Biological evolution** refers to gradual change of populations or organisms over time, measuring time in generations rather than years. For example, while it appears that the sheep blowfly populations evolved quickly, it took many 11-day-long generations.

Many people associate evolution with dinosaurs and other fossils, implying that evolution happened in the past. However, organisms continue to evolve today. As in the past, new species are formed and others go extinct. In fact, ongoing evolution helps to explain the variation within and among the populations of a given species. This sometimes makes it difficult to determine when populations comprise distinct species (see Chapters 18 and 19).

Although biological evolution refers to gradual change, not all gradual change is biological evolution. Aging is a feature of life that involves gradual change. Family pictures record how you have changed since you were born. However, changes that occur over the lifetime of a single organism are not evolutionary. Gradual change is also a feature of the environment. Seasonal changes such as the fall of leaves, the growth of new sprouts, and animal reproductive behaviours are examples of gradual change. However, none of these changes is an example of evolution.

STUDY BREAK

1. What is evolution? What are selection pressures?
2. Why is the growth and development of an individual not evolution?

17.2 Pre-Darwin Knowledge of the Natural World

17.2a Early Views of Organisms

The Greek philosopher Aristotle (384–322 BCE) was a keen observer of nature and natural history. He examined the form and variety of organisms in their natural environments and believed that both inanimate objects and living organisms had fixed characteristics. Careful study of their differences and similarities enabled Aristotle to create a ladderlike classification of nature from the simplest to the most complex: minerals ranked below plants, plants below animals, animals below humans, and humans below the gods of the spiritual realm. Many of Aristotle's writings on living organisms were considered to be true for almost two millennia.

By the fourteenth century, Aristotle's classification was being merged with the biblical and other accounts of creation, at least in the Christian and Islamic worlds. At that time, Europeans thought that all of the different kinds of organisms had been specifically created by a god. The different kinds could never change or become extinct, and new kinds could never arise. Biological research was dominated by **natural theology**, which sought to name and catalogue all of God's creation. Careful study of each species could identify its position and purpose in the *Scala Naturae*, or Great Chain of Being, as Aristotle's ladder of life was called. This approach to nature and history was clear in the later work of Carolus Linnaeus (1707–1778), whose efforts were *ad majorem Dei gloriam* (for the greater glory of God) (Chapter 19).

By 1600, the English philosopher and statesman Sir Francis Bacon (1561–1626) established the importance of observation, experimentation, and finding evidence to support a proposal or theory (= inductive reasoning). Other scientists proposed theories to describe how physical events (mechanistic) worked, notably Nicolaus Copernicus (1473–1543), Galileo Galilei (1564–1642), René Descartes (1596–1650), and Sir Isaac Newton (1643–1727). The collective work of these scientists gave rise to three new disciplines—biogeography, comparative morphology, and geology—promoting a growing awareness of change.

Biogeography. As long as naturalists encountered organisms only from Europe and surrounding lands, the *Scala Naturae* was easily followed. But global explorations in the fifteenth through seventeenth centuries provided naturalists with thousands of unknown plants and animals from around the world. Although some were similar to European species, others were new and very strange.

Studies of the world distribution of plants and animals, now called **biogeography**, raised puzzling questions. Was there no limit to the number of species

African ostrich
(Struthio camelus)

Johan Swanepoel/Shutterstock.com

South American rhea
(Rhea americana)

Kenneth W. Fink/Science Source

Australian emu
(Dromaius novaehollandiae)

S.Cooper Digital/Shutterstock.com

Figure 17.1
These three species of large, flightless birds, with greatly reduced wings, appear very similar. In fact, they occupy similar habitats in geographically separated regions. Recent analysis of mitochondrial genomes reveals that all of the major lineages of large flightless birds (known as *ratites*) evolved from ancestors that had dispersed by flight.

created by God? Where did all these species fit in the *Scala Naturae*? If they had all been created in the Garden of Eden, why did some species have limited geographical distributions, whereas others were widespread? And why were some species found in Africa or Asia different from those found in Europe, whereas other species from far-flung locations were similar to each other **(Figure 17.1)**?

Comparative Morphology. When naturalists compared the morphology (anatomical structure) of organisms, they discovered interesting similarities and differences. For example, the front legs of pigs, the flippers of dolphins, and the wings of bats differ markedly in size, shape, and function **(Figure 17.2)**. But these appendages have similar locations in the mammals' bodies; all are constructed of bones, muscles, and skin; and all develop similarly in the animals' embryos. If these limbs were specially created for different means of locomotion, naturalists wondered, why didn't the

Creator use entirely different materials and structures for walking, swimming, and flying?

Natural theologians countered this argument by stating that the body plans were perfect, and there was no need to invent a new plan for every species. But a French scientist, George-Louis Leclerc (1707–1788), le Comte de Buffon, was still puzzled by the existence of body parts with no apparent function. For example, he noted that the feet of pigs and some other mammals have two toes that never touch the ground (pig digits 2 and 5 in Figure 17.2). If each species was anatomically perfect for its particular way of life, Buffon asked, why did useless structures exist?

Buffon proposed that some animals must have *changed* since their creation; he suggested that **vestigial structures**, these useless parts he observed, must have functioned in ancestral organisms. Buffon offered no explanation of how functional structures became vestigial, but he clearly recognized that some species were "conceived by Nature and produced by Time."

Geology. Georges Cuvier (1769–1832), a French zoologist, realized that the layers of fossils he found represented organisms that had lived at sequential times in the past. He suggested that abrupt changes between geologic layers marked dramatic shifts in ancient environments. Cuvier and his followers developed the theory of **catastrophism**, reasoning that each layer of fossils represented the remains of organisms that had died in a local catastrophe such as a flood. Somewhat different species then recolonized the area, and when another catastrophe struck, they became a different set of fossils in the next higher layer.

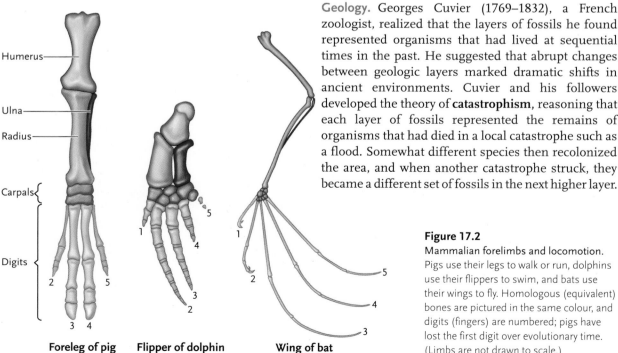

Foreleg of pig   Flipper of dolphin   Wing of bat

Figure 17.2
Mammalian forelimbs and locomotion.
Pigs use their legs to walk or run, dolphins use their flippers to swim, and bats use their wings to fly. Homologous (equivalent) bones are pictured in the same colour, and digits (fingers) are numbered; pigs have lost the first digit over evolutionary time. (Limbs are not drawn to scale.)

From Fins to Fingers

The early embryonic development of the limbs of fishes and tetrapods is similar. The limbs start as buds of mesoderm (see Chapter 42), which thicken by increased cell division. As the buds elongate, cartilage is deposited at localized centres, the precursors of later limb bones. In fishes, bones develop along a central axis from base to tip **(Figure 1a)**. In tetrapods, centres of cartilage formation generate the long bones of the limb and the five digits of the foot (or hand; Figure 1b).

Biologists used molecular techniques to assess the patterns of development and determine if the digits of tetrapods were modifications of the bones radiating from the central axis in fish. In tetrapods with paired fore- and hindlimbs, groups of **homeobox** genes control their development. A comparison of *HoxD* genes in zebrafish (*Danio rerio*) and previously available data from birds and mammals revealed the details of development. Using the DNA from a rodent *HoxD* gene as a probe, researchers searched for similar genes in fragmented zebrafish DNA. After cloning and sequencing, it was clear that the *HoxD-11*, *HoxD-12*, and *HoxD-13* genes in zebrafish are arranged in the same order as they are in rodents.

Paolo Sordino, Franks van der Hoeven, and Denis Duboule tested the activity of *HoxD* genes in developing zebrafish using a nucleic acid probe that could pair with mRNA products of the genes. The probe was linked to a blue dye molecule so that the cells in which a particular *HoxD* gene was active would appear blue in the light microscope. In zebrafish, the *HoxD* genes became active in cells along the posterior side of the central axis (Figure 1c). As fin development neared completion, the activity of *HoxD* genes dropped off.

Using the same approach in tetrapods, the researchers found that the *HoxD* genes were activated in two distinct phases (Figure 1d). In phase 1, gene activity was restricted to the posterior half of the limb, as it had been in zebrafish. This period of activity corresponded to the development of long limb bones. In phase 2, the *HoxD* genes became active in a band of cells perpendicular to the central axis. Here, cartilage centres formed the bones of the digits that developed in an anterior–posterior band. These patterns differed from those in zebrafish, suggesting morphological novelty in tetrapods. These changes in *HoxD* activity must have preceded the development of tetrapod limbs.

a. Fishes

b. Tetrapods

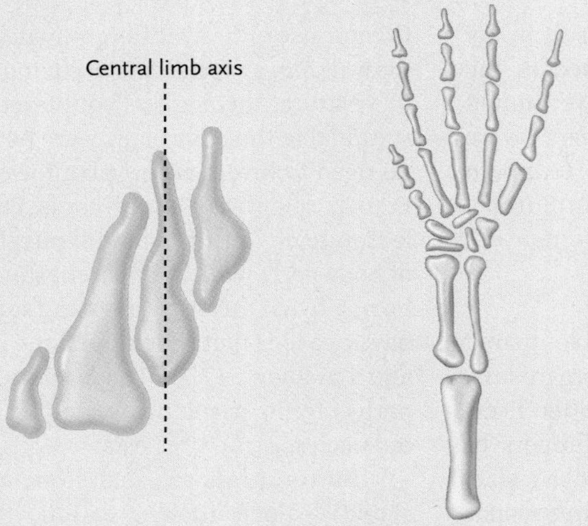

Central limb axis

Bones in the fin of a fish develop from centres of cartilage formation along a central axis (dashed line).

Bones in the limb and digits of a tetrapod also develop from centres of cartilage formation in the central axis.

c. Fishes

d. Tetrapods

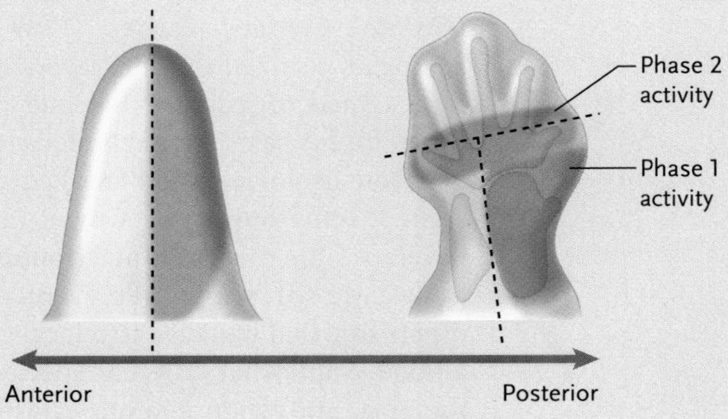

Phase 2 activity

Phase 1 activity

Anterior

Posterior

During development of the fin in fishes, *HoxD* genes become active in cells posterior to the central axis of the fin (shown in blue).

During development of the limb and digits in tetrapods, *HoxD* genes first become active in cells posterior to the central axis of the limb (blue). Later, these genes are active in a band of cells perpendicular to the central axis of the limb (green).

Figure 1
Fins versus fingers, Hox *genes in action.*

Today we know that huge catastrophes, such as the impact of an asteroid, can cause the extinction of tens of thousands of species. One example was the extinction event that marked the disappearance of dinosaurs and the end of the Cretaceous (66 mya [million years ago]).

17.2b Our View of Earth Changes

Bishop James Ussher (1581–1656), a theologian and scholar, calculated the age of Earth by counting the number of generations mentioned in the Bible. He came up with the year 4004 BCE as the year of Creation. Dr. John Lightfoot (1602–1675), the Vice-Chancellor of Cambridge University, continued with this research to come up with a more precise time. He concluded that Earth had been created on October 23, 4004 BCE, at 9:00 in the morning. The idea that Earth was about 6000 years old persisted for several centuries.

In 1795, the Scottish geologist James Hutton (1726–1797) argued that slow and continuous physical processes, acting over very long periods of time, produced Earth's major geologic features. The movement of water in a river slowly erodes the land and deposits sediments near the river's mouth. Given enough time, erosion creates deep canyons, and sedimentation results in thick topsoil on flood plains. Hutton's **gradualism**, the view that Earth changed *slowly* over its history, contrasted sharply with Cuvier's catastrophism.

The English geologist Charles Lyell (1797–1875) championed and extended Hutton's ideas in an influential series of books, *Principles of Geology: An Attempt to Explain the Former Changes of the Earth's Surface by Reference to Causes Now in Operation*. Lyell argued that the geologic processes that sculpted Earth's surface over long periods of time, such as volcanic eruptions, earthquakes, erosion, and the formation and movement of glaciers, are exactly the same as the processes we observe today. This concept, **uniformitarianism**, undermined any remaining notions of an unchanging Earth. Because geologic processes proceed very slowly, it must have taken millions of years, not just a few thousand, to mould the landscape into its current configuration.

STUDY BREAK

1. What did Buffon, Cuvier, and Lyell contribute to our knowledge of life?
2. How do the concepts of gradualism and uniformitarianism in geology undermine the belief that Earth is only about 6000 years old?

17.3 Biological Evolution

17.3a Lamarck

Jean Baptiste de Lamarck (1744–1829) proposed the first comprehensive theory of biological evolution based on specific mechanisms. He proposed that a metaphysical "perfecting principle" caused organisms to become better suited to their environments. In Lamarck's theory, simple organisms evolved into more complex ones, moving up the ladder of life. Microscopic organisms were replaced at the bottom by spontaneous generation (in which living organisms arise from nonliving material, such as dirt or dead organisms). Lamarck theorized that two mechanisms fostered evolutionary change. According to his principle of use and disuse, body parts grow in proportion to how much they are used, as anyone whose exercise regime includes lifting heavy weights well knows. Conversely, unused structures get weaker and shrink, like the muscles of an arm immobilized in a cast. According to his second principle—the inheritance of acquired characteristics—changes that an organism acquires during its lifetime are inherited by its offspring. Thus, Lamarck argued that long-legged wading birds, such as herons, are descended from short-legged ancestors that stretched their legs to stay dry while feeding in shallow water **(Figure 17.3)**. Their offspring inherited slightly longer legs, and after many generations, their legs became extremely long.

Today we know that Lamarck's proposed mechanisms do not cause evolutionary change. Although muscles do grow larger through continued use, structural changes acquired during an organism's lifetime are not inherited by the next generation. Despite the shortcomings of his theory, Lamarck made four important contributions to the development of an evolutionary worldview:

1. He proposed that all species change through time.
2. He recognized that changes are passed from one generation to the next.

Figure 17.3

A great blue heron (*Ardea herodias*). Like many other wading birds, herons have long, stiltlike legs. Lamarck hypothesized that as wading birds stretched their legs while feeding, successive generations of their offspring would have progressively longer legs.

3. He suggested that organisms change in response to their environments.
4. He hypothesized the existence of specific mechanisms that caused evolutionary change.

All four of these ideas became cornerstones of Darwin's later theory of evolution by natural selection. Perhaps Lamarck's most important contribution was that he fostered discussion. By the mid-nineteenth century, most educated Europeans were talking about evolutionary change, whether they believed in it or not.

17.3b Darwin

Documenting variation and describing how selection works were Charles Darwin's central contributions to our knowledge of evolution, putting it in context and explaining its significance. In 1831, Darwin set sail on HMS *Beagle* as a naturalist. The *Beagle* first sailed westward to map the coastline of South America and then on to circumnavigate the globe **(Figure 17.4).** The *Beagle*'s voyage lasted nearly five years. While on the voyage, Darwin read Charles Lyell's *Principles of Geology.* He began to see rock formations through Lyell's eyes and to apply gradualism and uniformitarianism to the living world.

Because of his seasickness, Darwin seized every chance to go ashore. He collected plants and animals in Brazilian rain forests and fossils in Patagonia. He hiked the grasslands of the pampas and climbed the Andes in Chile. He discovered fossils in Argentina that resembled organisms inhabiting the same region today. For example, despite an enormous size difference, living armadillos and fossilized glyptodonts had similar body armour, but they were unlike any other species known to science **(Figure 17.5).** If both species had been created at the same time and both were found in South America, why didn't glyptodonts still live alongside armadillos? Darwin later wondered whether armadillos might be living descendants of the now-extinct glyptodonts.

On the Galápagos Islands **(Figure 17.6),** Darwin found "strange and wonderful creatures," including giant tortoises, small finchlike birds, and lizards that dove into the sea to eat algae. Darwin noticed that the

Figure 17.5

Ancestors and descendants. Darwin hypothesized that even though an extinct glyptodont (top) probably weighed 300 to 400 times as much as a living nine-banded armadillo (*Dasypus novemcinctus*), their obvious resemblance suggested that they are related.

animals and plants on different islands varied slightly in form. Indeed, experienced sailors could easily identify a tortoise's island of origin by the shape of its shell. Moreover, many species resembled those on the distant South American mainland. Why did so many different species of organisms occupy one small island cluster? Why did these species resemble others from the nearest continent? Darwin later hypothesized that the plants and animals of the Galápagos were descended from South American ancestors and that the appearance of individuals making up populations changed after being isolated on a particular island.

As he visited the different islands, Darwin collected a diverse group of finches **(Figure 17.7).** He noticed great variability in the shapes of their bills, but he incorrectly assumed that birds on different islands belonged to the same species. Thus, he did not record the island where he captured each specimen. Luckily, the *Beagle*'s captain, Robert Fitzroy, had more thoroughly documented his own collection, allowing Darwin to study the relationships and geographical distributions of a dozen species. As Darwin reviewed the data, he began to focus on two aspects of a general problem. Why were the finches on a particular island slightly different from those on nearby islands, and how did all these different species arise?

17.3c Developing the Theory of Natural Selection

Having grown up in rural England, Darwin was well aware that "like begets like"; that is, offspring resemble

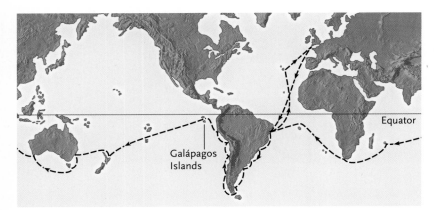

Figure 17.4

Darwin's voyage: map of the path followed by HMS *Beagle*, 1831–1836.

a. The Galápagos

b. Galápagos tortoise

Gabor Nedecky/Shutterstock.com

c. Marine iguana

Ryan M. Bolton/Shutterstock.com

Darwin

Wolf

Pinta

Marchena Genovesa

Santiago

Equator

Bartolomé

Seymour

Rábida Baltra

Fernandina Pinzón

Santa Cruz

Santa Fe

San Cristóbal

Tortuga

Isabela

Española

Floreana

d. Blue-footed booby

Chris Howey/Shutterstock.com

Figure 17.6

The Galápagos Islands. Between 3 and 5 mya, volcanic eruptions created the Galápagos Islands **(a)** about 1000 km west of Ecuador. The islands were named for the giant tortoises **(b)** found there (in Spanish, *galapa* means tortoise). This tortoise (*Geochelone elephantopus*) is native to Isla Santa Cruz. **(c)** Marine iguanas (*Amblyrhynchus cristatus*) dive into the Pacific Ocean to feed on algae. **(d)** A male blue-footed booby (*Sula nebouxii*) engages in **courtship display.**

their parents. Plant and animal breeders had applied this basic truth of inheritance for thousands of years. By selectively breeding individuals with the characteristics they wanted, breeders enhanced those traits in future generations.

For example, the English people in the 1800s loved their dogs. Earlier, most dogs were working dogs, bred for hunting, rat catching, sheep herding, and guarding. In the nineteenth century, however, people wanted more decorative dogs and lap dogs. For example, dog

a. Warbler finch
(Certhidea olivacea)

b. Common cactus-finch
(Geospiza scandens)

© Images & Stories/Alamy

c. Large ground-finch
(Geospiza magnirostris)

© Krystyna Szulecka/Alamy

d. Woodpecker finch
(Camarhynchus pallidus)

Stubblefield Photography/Shutterstock.com

Tierbild Okapia/Science Source

Figure 17.7

Bill shape and food habits. The 13 finch species that inhabit the Galápagos Islands are descended from a common ancestor, a seed-eating ground finch that migrated to the islands from South America. **(a)** The warbler finch uses its slender bill to probe for insects in vegetation. **(b)** The common cactus-finch has a medium-sized bill suitable for eating cactus flowers and fruit. **(c)** The large ground-finch uses its thick, strong bill to crush cactus seeds. **(d)** The woodpecker finch hammers at tree bark with its bill; then it uses cactus spines, held in its bill, to probe for wood-boring insects such as termites.

breeders crossed Clydesdale terriers and Skye terriers with each other and with other small terrier breeds. These crosses resulted in the tiny (2–3 kg) Yorkshire terrier, a very popular breed even today.

Although the mechanism of heredity was not yet understood, selective breeding was applied countless times to produce bigger beets, plumper pigs, and prize-winning pigeons. Darwin, who bred pigeons himself, called this process **artificial selection**, since humans were selecting the characteristics they wanted in the offspring by choosing parents with those traits. Darwin could see that selection could operate in nature, but he puzzled about how it worked. He reasoned that if a person could select different characteristics when breeding organisms, then nature could do so as well. This he called **natural selection**, and he defined it as the "principle by which each slight variation [of a trait], if useful, is preserved." In other words, natural selection is the process by which characteristics that better enable organisms to adapt to specific environmental pressures will tend to increase in succeeding generations in a population. Organisms with those characteristics are better able to survive and can reproduce in greater numbers than those without the characteristics.

During the 1840s and 1850s, Darwin led a reclusive life, accumulating evidence of evolutionary change and trying to identify the mechanisms that caused it. He read Thomas Malthus's famous *Essay on the Principles of Population*. Malthus, an English clergyman and economist, observed that England's population was growing much faster than the country's agricultural capacity. This situation meant that individuals competed for food and some would inevitably starve. Darwin applied Malthus's argument to organisms in nature. He observed that many species typically produce many more offspring than are needed to replace the parent generation, yet the world is not overrun by any one species, be it sunflowers, earthworms, tortoises, or bears. Darwin calculated that if its reproduction went unchecked, a single pair of elephants (the slowest-breeding animal known) would leave roughly 19 million descendants after only 750 years. Instead, some members of every population survive and reproduce, whereas many others die without reproducing.

Darwin made several major observations **(Figure 17.8):**

- Individuals within populations vary in size, form, colour, behaviour, and other characteristics.
- Many of these variations are passed on from parent to offspring.
- Some of the inherited variations enable some individuals to survive and reproduce better than others. A modern example is resistance to organophosphorus insecticides, where resistant flies leave many young, and nonresistant flies leave few, if any, descendants (see "Why It Matters"). In this way, favourable hereditary traits become more common in the next generation.
- If the next generation is subjected to the same process of selection, these favourable traits will become even more common.

Because this process is analogous to artificial selection, Darwin called this mechanism natural selection, which means that individuals with certain inherited traits leave more offspring than do individuals without those traits. As an evolutionary mechanism, natural selection favours **adaptive traits**, hereditary characteristics that make organisms more likely to survive and reproduce under a given set of environmental conditions. And by favouring individuals that are well adapted to the environments in which they live, natural selection can cause populations to change through time. For example, each species of Galápagos finch (Figure 17.7) has a distinctive bill. Variations in bill size and shape make some birds better adapted for crushing seeds and others for capturing insects. Imagine an island where large seeds were the only food available; individuals

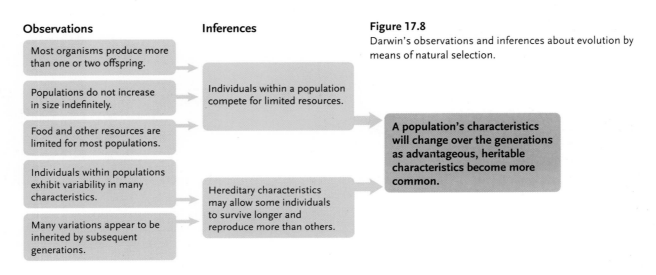

Observations

Most organisms produce more than one or two offspring.

Populations do not increase in size indefinitely.

Food and other resources are limited for most populations.

Individuals within populations exhibit variability in many characteristics.

Many variations appear to be inherited by subsequent generations.

Inferences

Individuals within a population compete for limited resources.

Hereditary characteristics may allow some individuals to survive longer and reproduce more than others.

A population's characteristics will change over the generations as advantageous, heritable characteristics become more common.

Figure 17.8
Darwin's observations and inferences about evolution by means of natural selection.

PEOPLE BEHIND BIOLOGY 17.1

Peter and Rosemary Grant

Peter and Rosemary Grant are evolutionary biologists who work at Princeton University. They have been responsible for ground-breaking work on cactus finches (*Geospiza conirostris*), a species of Darwin's finches from Genovesa Island in the Galapagos **(Figure 1)**. Of particular note are their long-term studies that have advanced our understanding of variation and how it can change over time.

The Grants analyzed a likely case of disruptive selection on the size and shape of the bills of cactus finches. During normal weather cycles, the finches eat ripe cactus fruits, seeds, and exposed insects. During drought years food is scarce and the birds also search for insects by stripping bark from the branches of bushes and trees. About 70% of cactus finches on Genovesa died during the long drought of 1977. Survivors exhibited high variability in their bills. Birds that stripped bark from branches to look for insects had particularly deep bills.

Those that opened cactus fruits to expose the fleshy interior had especially long bills. Thus, birds with extreme bill phenotypes appeared to feed efficiently on specific resources. This established disruptive selection on the size and shape of bills because the selective advantage of large bills did not apply under some environmental conditions. Intermediate bill morphologies may be favoured during nondrought years when insects and small seeds are abundant.

Geospiza conirostris

Birds with long bills open cactus fruits to feed on the fleshy pulp.

Birds with intermediate bills may be favoured during nondrought years when many types of food are available.

Birds with deep bills strip bark from trees to locate insects.

Figure 1

Disruptive selection as illustrated by cactus finches, in which these birds show variability in the sizes and shapes of their bills.

with a stout bill would be more likely to survive and reproduce than would birds with slender bills. These favoured individuals would pass the genes that produce stout bills to their descendants, while those with slender bills would not reproduce because of starvation. After many generations, the bills might resemble those of *Geospiza magnirostris* (see Figure 17.7c). Natural selection also changes nonmorphologic characteristics of populations; for example, insect populations that are exposed to insecticides develop resistance to these toxic chemicals over time (see "Why It Matters").

Darwin realized that natural selection could also account for striking differences between populations and, given enough time, for the production of new species. For example, suppose that small insects were the only food available to finches on a particular island. Birds with long, thin bills might be favoured by natural selection, and the population of finches might eventually possess a bill shaped like that of *Certhidea olivacea* (see Figure 17.7a). Considering that many characteristics affect survival and reproduction, natural selection would cause the populations to become more different over time, a process called **evolutionary divergence.**

17.3d Wallace

Alfred Russel Wallace was a contemporary of Darwin's who studied organisms in the Amazonian rain forest and in the East Indies. Like Darwin, Wallace was a keen observer of nature who kept careful notes and drawings of his observations. He travelled extensively in the Amazon rain forest and the Malay Archipelago (modern-day Indonesia), collecting specimens and describing the geology of the areas. Wallace's work provides one of the more interesting parts of the story of Darwin's development and publication of his theory.

On 18 June 1858, Darwin received a letter from Wallace in which he outlined his ideas about how populations of organisms change over time. Wallace's work and ideas mirrored Darwin's own research. To his credit, Darwin forwarded Wallace's manuscript to Charles Lyell, who had been encouraging Darwin to publish his theory. On 1 July 1858, papers by Darwin and Wallace were presented to the Linnaean Society of London. On 24 November 1859, Darwin's book, *On the Origin of Species by Means of Natural Selection,* was published. As evidenced by the conclusions of two different people, evolution via natural selection clearly was an idea whose time had come.

17.3e Impact of the Theory of Evolution by Natural Selection

It would be hard to overestimate the impact on western thought of Darwin's and Wallace's theory. In *The Origin*, Darwin proposed natural selection as the mechanism that drives evolutionary change. In fact, most of *The Origin* was an explanation of how natural selection acted on the variability within groups of organisms, preserving advantageous traits and eliminating disadvantageous ones.

Darwin argued that all the organisms that have ever lived arose through **descent with modification**, the evolutionary alteration and diversification of ancestral species. He envisioned this pattern of descent as a tree growing through time **(Figure 17.9)**. The base of the trunk represents the ancestor of all organisms. Branching points above it represent the evolutionary divergence of ancestors into their descendants. Each limb represents a body plan suitable for a particular way of life, smaller branches represent more narrowly defined groups of organisms, and the uppermost twigs represent living species. Biologists still apply this analogy today when studying phylogenies (see page 455, Chapter 20).

Four characteristics distinguish Darwin's theory from earlier explanations of biological diversity and adaptive traits:

1. Darwin provided purely physical, rather than spiritual, explanations for the origins of biological diversity.
2. Darwin recognized that evolutionary change occurs in groups of organisms rather than in individuals: some members of a group survive and reproduce more successfully than others.
3. Darwin described evolution as a multistage process: variations arise within groups, natural selection eliminates unsuccessful variations, and the next generation inherits successful variations.
4. Like Lamarck, Darwin understood that evolution occurs because some organisms function better than others *in a particular environment*.

Evolution was a popular topic in Victorian England, and Darwin's theory was both praised and ridiculed. Nevertheless, Darwin's painstaking logic and careful documentation convinced many readers that evolution really does take place. The major stumbling block for some readers was that Darwin had no clear idea of how a variant arose or how it was passed from one generation to the next.

Although Darwin had not speculated about the evolution of humans in *Origin of Species,* he did in another book, *The Descent of Man, and Selection in Relation to Sex,* published in 1871. Needless to say, certain influential Victorians were not amused by the suggestion that humans and apes shared a common ancestry.

STUDY BREAK

1. What observations that Darwin made on his round-the-world voyage influenced his later thoughts about evolution?
2. Describe an example of artificial selection not included in the text.
3. How did Darwin's understanding of artificial selection enable him to envision the process of natural selection?
4. What is natural selection?

17.4 Evolutionary Biology since Darwin

One of the remarkable features of Darwin's and Wallace's work is that they developed the foundational concept of natural selection without any understanding of a mechanism for how traits were inherited. This was because neither of them understood genetics, a scientific field that was also in its infancy. At about the same time as Darwin's book was published, Gregor Mendel published his work on inheritance in pea plants. However, this study was not well known in England until 1900. At that time, scientists thought that Darwin's and Mendel's theories conflicted. One problem was that Darwin had used complex characteristics, such as the structure of bird bills, to illustrate how natural selection worked. We now know that several genes often control such traits (see page 290, Chapter 13 and page 314, Chapter 14). By contrast, Mendel had studied simpler characteristics, such as the height of pea plants. A single gene often controls simple traits, which is one reason Mendel could interpret his experimental results so clearly. Biologists initially had a hard time applying Mendel's

Present

Time

Origin of life

Figure 17.9

The Tree of Life. Darwin envisioned the history of life as a tree. Branching points represent the origins of new lineages; branches that do not reach the top represent extinct groups.

straightforward experimental results to Darwin's complex examples. Today, however, it is accepted throughout the scientific community that phenotypic variation among organisms reflects genetic differences.

17.4a The Modern Synthesis

In the early twentieth century, Thomas Hunt Morgan of Columbia University determined that genes are carried on chromosomes. His experiments enabled geneticists and mathematicians to forge a critical link between Darwin's and Mendel's ideas (see pages 410 and 411, Chapter 18). The new discipline, **population genetics**, recognized the importance of genetic variation as the raw material of evolution. Population geneticists constructed mathematical models, which applied equally well to simple and complex traits, to predict how natural selection and other processes influence a population's genetics.

In the 1930s and 1940s, a unified theory of evolution, called the **modern synthesis**, integrated data from biogeography, comparative morphology, comparative embryology, genetics, paleontology, and taxonomy within an evolutionary framework. The authors of the modern synthesis focused on evolutionary change within populations. Although they considered natural selection the primary mechanism of evolution, they acknowledged the importance of other processes (such as genetic drift; see Chapter 18). Proponents of the modern synthesis also embraced Darwin's idea of gradual change and de-emphasized the significance of mutations that changed traits suddenly and dramatically.

The modern synthesis also tried to link the two levels of evolutionary change that Darwin had identified: microevolution and macroevolution. **Microevolution** describes the small-scale genetic changes that populations undergo, often in response to shifting environmental circumstances; a small evolutionary shift in the size of the bill of a Galapagos finch is an example of microevolution. **Macroevolution** describes larger-scale evolutionary changes observed in species. According to the modern synthesis, macroevolution results from the gradual accumulation of microevolutionary changes. Researchers have recently begun to unravel the genetic mechanisms that establish a relationship between these two levels of evolutionary change (see page 456, Chapter 20). Research since the discovery of the structure of DNA has led to a more thorough understanding of mutations. It also led to the science of molecular genetics.

Nowadays, biologists understand that biological evolution involves the combination of heritable changes in individuals plus selective pressures, the environmental pressures felt by organisms. Selective pressures operate on the phenotype of organisms, improving or reducing the success of individuals within a population that have a certain inherited trait. This trait can make these individuals more or less adapted to their environments and consequently more or less likely to produce viable offspring. As we have seen, these heritable traits can be expressed anywhere from the molecular level to the whole organism. Heritable changes, combined with selective pressures, can produce new types (species) of organisms from existing ancestors.

17.4b Further Evolutionary Research

Since the emergence of the modern synthesis, scientists have assembled a huge body of evidence from many biological disciplines that indicate that biological evolution is a fact of life on Earth.

Adaptation by Natural Selection. Biologists interpret the products of natural selection as evolutionary adaptations. For example, the wings of birds, which have been modified by evolutionary processes over millions of years, have an obvious function that helps these animals survive and reproduce. Throughout this book, you will encounter many examples of adaptive structures in plants and animals that have been modelled by natural selection. Sometimes, however, natural selection operates on a short time scale, as illustrated by the development of pesticide resistance in insects (see "Why It Matters").

The Fossil Record. Because evolution results from the modification of existing species, Darwin's theory proposes that all species that have ever lived are genetically related. The fossil record documents such continuity in morphological characteristics, providing clear evidence of ongoing change in **biological lineages**, evolutionary sequences of ancestral organisms and their descendants (see Chapter 20). For example, the evolution of modern birds can be traced from a dinosaur ancestor through fossils such as *Archaeopteryx lithographica* (**Figure 17.10, p. 402**). This species, discovered only two years after *The Origin* was published, resembled both dinosaurs and birds. Like the small carnivorous dinosaur *Dromaeosaurus*, *Archaeopteryx* walked on its hind legs and had teeth; long fingers with claws on its forelimbs; and a long, bony tail. Like modern birds, it had enlarged flight feathers on its forelimbs. Recently discovered fossils reveal that many bird ancestors were feathered.

Historical Biogeography. The study of the geographical distributions of plants and animals in relation to their evolutionary history is generally consistent with Darwin's theory of evolution. Species on oceanic islands often closely resemble species on the nearest mainland, suggesting that the island and mainland species have a common ancestry. Moreover, species on a continental land mass are clearly related to one another and are often distinct from those on other continents. For example, monkeys in South America have long, prehensile tails and broad noses, traits that

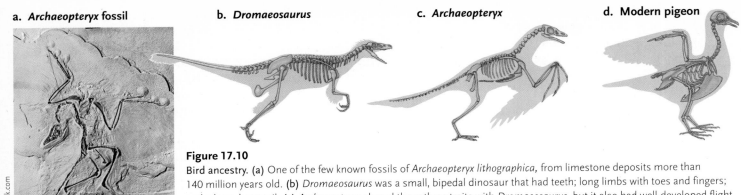

a. *Archaeopteryx* fossil   **b.** *Dromaeosaurus*   **c.** *Archaeopteryx*   **d.** Modern pigeon

Figure 17.10
Bird ancestry. **(a)** One of the few known fossils of *Archaeopteryx lithographica*, from limestone deposits more than 140 million years old. **(b)** *Dromaeosaurus* was a small, bipedal dinosaur that had teeth; long limbs with toes and fingers; and a long, bony tail. **(c)** *Archaeopteryx* shared these three traits with *Dromaeosaurus*, but it also had well-developed flight feathers in its forelimbs, a characteristic that it shares with modern birds. **(d)** Modern birds, such as the pigeon, have long limbs similar to those of *Dromaeosaurus* and *Archaeopteryx*, but their fingers and bony tails are greatly reduced, and a horny bill has replaced the teeth in the mouths of their ancestors.

they inherited from a shared South American ancestor. By contrast, monkeys in Africa and Asia evolved from a different common ancestor in the Old World, and their shorter tails and narrower noses distinguish them from their American cousins.

Comparative Morphology. Analyses of the structure of living and extinct organisms are based on the comparison of **homologous traits**, characteristics that are similar in two species because they inherited the genetic basis of the trait from their common ancestor. For example, the forelimbs of all four-legged vertebrates are homologous because they evolved from a common ancestor with a forelimb composed of the same component parts (see Figure 17.2, p. 393). Even though the shapes of the bones are different in pigs, dolphins, and bats, similarities in the three limbs are apparent. The differences in structural details arose over evolutionary time, allowing pigs to walk, dolphins to swim, and bats to fly. The arms of humans and the wings of birds are constructed of comparable elements, suggesting that they, too, share a common ancestor with the three species illustrated.

17.4c Molecular Techniques

Molecular techniques provide biologists with powerful tools for exploring all aspects of life—and evolutionary biology is no exception. From Darwin's time until the mid-twentieth century, biologists tried to discern the evolutionary history of animals by comparing their embryos and patterns of development **(Figure 17.11)**. The early embryos of related species are often strikingly similar, but morphological differences appear as the embryos grow and develop their adult forms (see Chapter 42). For example, the early embryos of most four-legged vertebrates (such as lizards, mammals, and birds) develop "limb buds" from which the legs or wings grow. Forelimbs and their supporting structures grow at the base of the neck, just in front of the ribcage, and hindlimbs grow right behind the ribcage (Figure 17.11a). Similarities in the limb buds and the positions of the limbs in

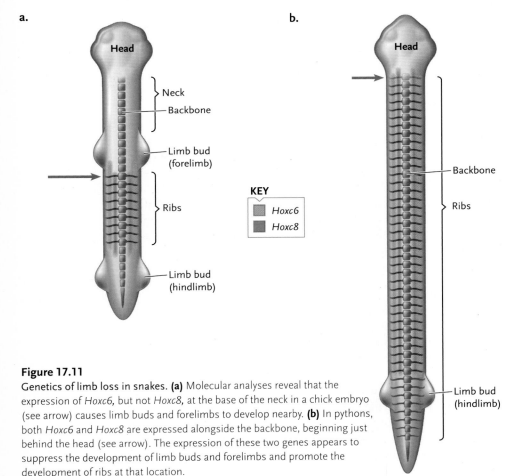

KEY

| | Hoxc6 |
| | Hoxc8 |

Figure 17.11
Genetics of limb loss in snakes. **(a)** Molecular analyses reveal that the expression of *Hoxc6*, but not *Hoxc8*, at the base of the neck in a chick embryo (see arrow) causes limb buds and forelimbs to develop nearby. **(b)** In pythons, both *Hoxc6* and *Hoxc8* are expressed alongside the backbone, beginning just behind the head (see arrow). The expression of these two genes appears to suppress the development of limb buds and forelimbs and promote the development of ribs at that location.

these animals provide evidence of their descent from a shared ancestor. Differences in their adult structures are caused by additional genetic instructions that have evolved over time.

However, most snakes show no traces of limbs or necks; their ribcages are positioned right behind their heads (Figure 17.11b). The fossil record shows us that snakes evolved from four-legged ancestors in stages: early snakes had small hindlimbs, and the most recently evolved snakes have no limbs at all. Only the most ancient living snakes, pythons and boas, have any traces of limbs—vestigial hindlimbs, which appear as a pair of tiny claw-like structures near the base of the tail. Observational studies of their embryos reveal that most living snakes never develop limb buds; by contrast, pythons and boas develop hind limb buds, which grow only slightly as the animal develops.

Two regulatory genes, *Hoxc6* and *Hoxc8*, determine whether forelimbs or ribs grow at a particular site along an animal's backbone. The *Hox* genes either activate or suppress other genes that direct the development of these structures. Forelimbs—but not ribs—grow just in front of the tissues where only *Hoxc6* is expressed. By contrast, ribs—but not forelimbs—grow where both *Hoxc6* and *Hoxc8* are expressed (Figure 17.11a).

Martin J. Cohn of the University of Reading and Cheryll Tickle of University College London reported that in pythons (primitive snakes), both *Hoxc6* and *Hoxc8* are expressed all along the backbone, beginning at the base of the skull; as a result, a python's ribcage develops right behind its head, and no limb buds or forelimbs develop (Figure 17.11b). Thus, snakes have no forelimbs or necks because a mutation causes the expression of *Hoxc8* to extend into a more forward region of the animal's body. All descendants of the ancestor with that original mutation now lack necks and forelimbs. Cohn and Tickle's research also suggests that the second stage in snake evolution, the reduction or complete absence of hindlimbs, is caused by other genetic variations that appeared some time after the altered expression pattern of *Hoxc8*. Thus, molecular research has identified the genetic changes that caused snakes to lose their forelimbs and necks before losing their hindlimbs.

STUDY BREAK

1. What types of data provide evidence that evolution has adapted organisms to their environments and promoted the diversification of species?
2. How have molecular techniques enhanced the study of evolutionary biology? Give an example not included in the textbook.

17.5 Variation and Selection

As Darwin observed, individuals exhibit variation in many characteristics. Natural selection acts at that individual level. The inherited characteristics of an individual may allow it to survive longer and, therefore, reproduce more than others without that characteristic (Figure 17.8). These individuals have enhanced fitness: they are better adapted to their environment. The particular traits that make these individuals more fit are considered adaptive: they help the individual survive. Not all traits are adaptive; some have no bearing on whether an individual survives or not. We will examine some examples that illustrate the relationships between variation, selection, and evolution.

17.5a Plant Poisons

Convergent evolution is the independent evolution of similar traits in unrelated species, such as the wings of insects, birds, and bats. The convergent evolution of solutions to threats to survival is a recurring theme among organisms. For example, most insects cannot ingest cardenolides (also known as cardiac glycosides), which are produced by plants such as milkweed (*Asclepias* spp.) and dogbane (*Apocyanum* spp.) to protect them from herbivorous insects. Cardenolides are toxic because they block an essential transmembrane carrier, the sodium–potassium pump, Na–K ATPase (see page 110, Chapter 5). Insects from several distinct evolutionary lineages **(Figure 17.12, p. 405)** have independently acquired the ability to ingest cardenolides, giving them access to more food resources. In monarch butterflies (*Danaus plexippus*), a simple mutation, producing a substitution of asparagine for histidine on the Na–K ATPase protein, confers protection from cardenolides. At least 17 other species of insects (representing 15 genera and 4 orders) have variations in the gene sequences that encode for the Na–K ATPase protein and also eat plants with cardenolides; these variations have evolved independently in these individual species. This situation demonstrates convergence across 300 million years of insect evolution and diversification. The convergence is one molecular solution permitting consumption of cardenolides.

Two advantages are associated with the ability to ingest cardiac glycosides. First, this ability provides herbivorous insects access to a greater variety and amount of food. Second, many insects that ingest cardiac glycosides and other toxic plant products use these molecules in their own defence. The monarch butterfly is an excellent example of protection conferred by ingested cardenolides.

17.5b Venom

Many species of animals produce venom that they use in defence (to deter or distract predators) or offence (to immobilize and digest prey). Venom that produces

The Woolly Mammoth's Closest Living Relative

Based on morphological evidence, paleobiologists suspected that woolly mammoths (*Mammuthus primigenius*) were more closely related to living Asian elephants (*Elephas maximus*) than to living African elephants (*Loxodonta africana*). When they recovered samples of mammoth DNA, researchers knew they could settle the matter.

In 2006, Hendrick Poinar of McMaster University and colleagues sequenced 13 million base pairs of mitochondrial and nuclear DNA extracted from the jawbone of a woolly mammoth (*Mammuthus primigenius*), a species that has been extinct for at least 4000 years. The mammoth, which died 27 000 years ago, had been preserved in a Siberian ice cave. When the researchers compared its DNA sequences to those from a living African elephant (*Loxodonta africana*), they discovered that more than 98% of the sequence was identical in the two species, confirming their close evolutionary relationship.

Svante Pääbo and his colleagues at the Max Planck Institute for Evolutionary Anthropology, Leipzig, Germany, and researchers at institutions in England, Germany, and the United States, used molecular techniques—specifically, PCR amplification, cloning, and sequencing of mitochondrial DNA (see Chapter 15)—to analyze the complete mitochondrial genome sequence of a woolly mammoth, and compared it to the sequences from two living elephant species.

Based on known mitochondrial DNA sequences from the African and Asian elephants, they designed and synthesized 46 pairs of PCR primers **(Figure 1)**. Assuming homology with the elephant sequences, these primer pairs were predicted to amplify the entire circular mitochondrial genome of the mammoth (yellow circle) as overlapping DNA fragments. All of the primer pairs were used in a single PCR, and the resulting amplified DNA fragments were cloned and sequenced. Based on their overlaps, the sequences could be arranged in a circle (blue and red), which showed that the researchers had succeeded in amplifying the entire mitochondrial genome.

The mammoth DNA sequences were more similar to those from the Asian elephant than those from the African elephant. Thus, Asian elephants are the woolly mammoth's closest living relatives. This conclusion also makes biogeographic sense because woolly mammoths and Asian elephants occupied the same land mass (Asia), whereas African elephants live on a different continent.

Source: J. Krause et al. 2006. Multiplex amplification of the mammoth mitochondrial genome and the evolution of the Elephantidae. *Nature* 439:724–727.

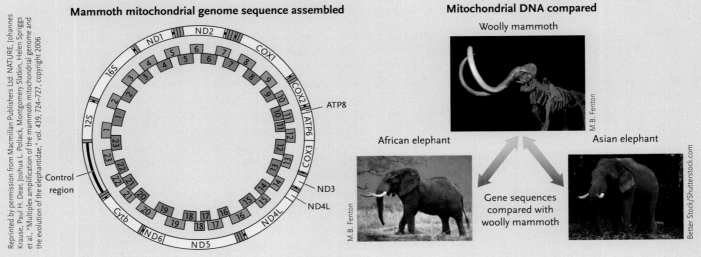

Figure 1

(a) Mammoth mitochondrial genome sequence assembled. *(b)* Mitochondrial DNA sequences compared.

intense, incapacitating pain is used in defence and delivered by stinging or biting. Bark scorpions **(Figure 17.13,** *Centuroides sculpturatus***)** produce defensive venom that deters attacks by mammals such as house mice, rats, and humans. The scorpion's venom works by activating the Nav1.7 voltage-gated Na$^+$ channels on pain receptors connected to the central nervous system. Grasshopper mice (*Onychomys torridus*) regularly catch and eat bark scorpions but show little response to their stings. In these predatory rodents, Nav1.8 acts as an analgesic by inhibiting the movement of Na$^+$ ions.

In the natural world, bark scorpions have many predators, not just grasshopper mice, and these mice eat many other species of arthropods. Defensive venom therefore provides an advantage to bark scorpions in many, but not all, defensive situations. But in this example, the resistance of grasshopper mice to the venom provides them access to otherwise inaccessible

M.B. Fenton

Jarrod Erbe/Shutterstock.com

Figure 17.12
A monarch butterfly and the milkweed that its caterpillars eat. Cardenolides are found in the white latex sap. In this picture a drop of sap sits at a break in the leaf.

food, increasing their fitness and, again, changing the incidence of resistance in the population.

17.5c Sexual Selection

With animals having separated sexes there will in most cases be a struggle between the males for possession of the females. The most vigorous individuals, or those which have most successfully struggled with their conditions of life, will generally leave most progeny. But success will often depend on having special weapons or means of defence, or on the charms of the males; and the slightest advantage will lead to victory.

So Darwin wrote in *On the Origin of Species*. In animals that are sexually dimorphic (males look different from females), heritable traits involved in competition between males or in attracting females can be subject

Matthew and Ashlee Rowe

Figure 17.13
Bark scorpion and grasshopper mouse.

to very strong selection. **Sexual selection**, which is a type of natural selection, involves variation in reproductive success that is influenced by sexually dimorphic characters. Sexual selection can involve phenotypic variations in morphology; in behaviour; and even in **pheromones**, chemicals used to attract mates.

Variation in the morphology of males can influence the competition for females. In guppies (*Poecilia reticulata*), males with clawlike structures on their intromittent organ (penis equivalent) transfer more sperm during copulation than males lacking the "claws." In water striders (*Rheumatobates rileyi*), some males have specialized structures on their antennae that allow them to grasp and subdue resistant females. The gene family *distal-less* (DLX) controls these specialized structures, connecting morphology and function and associated advantages to underlying genetics and inheritance.

In some animals, exaggeration of morphological features that distinguish males from females are proxies for male quality, and female choice contributes to sexual selection. In stalked-eyed flies **(Figure 17.14a, p. 406),** males with widely separated eyes are more attractive to females than males with shorter eye stalks. Male hammerheaded bats (Figure 17.14b) call to attract females to mating sites. The inflated skulls and voice boxes of males are associated with their ability to attract females. The selective pressure on these characteristics has led to their gradual enlargement or enhancement.

In other animals, sexual selection is driven by competition among males rather than female choice. Males of many species threaten or fight for dominance, such as the Siamese fighting fish (*Betta splendens*), octopus, and crabs. Using their long horns, male impala fight for access to females (Figure 17.14c). Larger, older males have larger horns and mate more often than younger, smaller males.

The act of mating, by itself, does not ensure reproduction. Furthermore, producing more offspring does not really "count" as success (fitness) until they have also reproduced. Also underlying sexual selection is the reality that females and males do not incur the same costs and benefits during reproduction (see also page 1012,

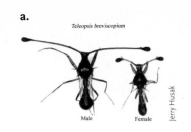
a.
Teleopsis breviscopium

Male   Female

Jerry Husak

b.
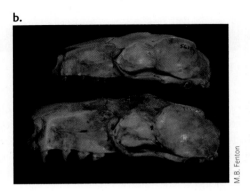
M.B. Fenton

c.

Courtesy of Naas Rautenbach

Figure 17.14
A sampling of sexual dimorphism in animals. **(a)** In stalked-eyed flies, the male's eye stalks are much longer than those of the female. **(b)** The skulls of male hammerheaded bats (*Hypsignathus monstrosus*) are much larger than female skulls (top). **(c)** Male impala (*Aepyceros melampus*) have horns but females do not.

Chapter 42). A comparison of mammals and birds illustrates the point. Birds lay eggs and both males and females may feed the young. However, female mammals alone bear the costs of pregnancy and lactation. If females mate with a strong, vigorous male, their offspring may be more adapted and better able to survive.

The tools provided by modern genetics allow biologists much more precision when it comes to measuring success in reproduction. For instance, people studying birds can now observe the behaviour of individual males and females, including copulations, and then use genetic techniques to determine which males sired which offspring. This situation shows direct connections between behaviour and reproductive output of males and of females.

STUDY BREAK

1. What is resistance? Why is it important? (See "Why It Matters.")
2. What is sexual selection? How does it relate to natural selection?
3. How do interactions between bark scorpions and grasshopper mice illustrate selection?
4. Why is heritability important in the definition of selection?

Putting Evolution in Perspective

Evolution has been the unifying theory in biology, drawing together virtually every aspect of the discipline. The scope of evolution has been introduced in this chapter, with a focus on variation, selection, selective advantages, artificial selection, and natural selection. Variation at the molecular and organismal level has been discussed in the context of selective advantages. We have also considered the contributions of Darwin and Wallace, who moved biological evolution from a vague concept to one with a clear explanation of how evolution could work. This chapter sets the stage for the balance of the unit on evolution.

Review

aplia™

To access course materials such as Aplia and other companion resources, please visit www.NELSONbrain.com.

17.1 Evolution: What Is It?

- Biological evolution refers to gradual change of populations or organisms over time; measuring time in generations rather than years.
- Organisms continue to evolve today.
- Not all gradual change is biological evolution.

17.2 Pre-Darwin Knowledge of the Natural World

- Well before Darwin published his theory of evolution by natural selection, changes in scientific thought and methods paved the way for its appearance. Fields such as geology contributed through documentation of the fossil record and the discovery that Earth was very, very old.
- Remember the contributions of people such as Aristotle, Bacon, Buffon, Cuvier, Ussher, Hutton, and Lyell.

17.3 Biological Evolution

- Lamarck's ideas became cornerstones of Darwin's later theory of evolution by natural selection.
- His voyage on HMS Beagle introduced Charles Darwin to many variations in the natural world.
- Darwin understood the process of artificial selection and reasoned that if a person could select different characteristics when breeding organisms, then nature could do so as well. He defined natural selection as the "principle by which each slight variation [of a trait], if useful, is preserved."
- Remember the contributions of people such as Lamarck, Malthus, Darwin, and Wallace.

17.4 Evolutionary Biology since Darwin

- Scientists working in population genetics developed theories of evolutionary change by integrating Darwin's ideas with Mendel's research on genetics.
- A unified theory of evolution, called the modern synthesis, integrates data from biogeography, comparative morphology, comparative embryology, genetics, paleontology, and taxonomy within an evolutionary framework.
- Phenotypic variation among organisms reflects genetic differences.
- Microevolution describes the small-scale genetic changes that populations undergo, often in response to shifting environmental circumstances.

- Macroevolution describes larger-scale evolutionary changes observed in species.
- According to the modern synthesis, macroevolution results from the gradual accumulation of microevolutionary changes.
- Studies of adaptation, the fossil record, historical biogeography, and comparative morphology provide compelling evidence of evolutionary change.
- Molecular techniques have extended the achievements of the modern synthesis, allowing precise analysis of the genetic basis of evolutionary change and the genetic relatedness of living and extinct organisms.

17.5 Variation and Selection

- Darwin observed that individuals exhibit variation in many characteristics.
- Natural selection acts at that individual level.
- The inherited characteristics of an individual may allow it to survive longer and, therefore, reproduce more than others without that characteristic. These individuals have enhanced fitness: they are better adapted to their environment.
- The particular traits that make these individuals more fit are considered adaptive: they help the individual survive.
- Not all traits are adaptive; some have no bearing on whether an individual survives or not.

Questions

Self-Test Questions

1. Which of the following statements about evolutionary studies is incorrect?
   a. Biologists study the products of evolution to understand processes causing it.
   b. Biologists design molecular experiments to examine evolutionary processes operating over short time periods.
   c. Biologists study variation in homologous structures among related organisms.
   d. Biologists examine why a huge variety of species may inhabit a small island.

2. Natural selection acts on
   a. species
   b. genera
   c. individuals
   d. subspecies

3. The belief that evolution is progressive or goal oriented is called:
   a. gradualism
   b. uniformitarianism
   c. taxonomy
   d. orthogenesis
   e. the modern synthesis

4. The wings of birds, the forelegs of pigs, and the flippers of whales are examples of
   a. vestigial structures
   b. homologous structures
   c. acquired characteristics
   d. artificial selection

5. Which of the following statements is NOT compatible with Darwin's theory?
   a. Evolution has altered and diversified ancestral species.
   b. Evolution occurs in individuals rather than in groups.
   c. Natural selection eliminates unsuccessful variations.
   d. Evolution occurs because some individuals function better than others in a particular environment.

6. Which of the following does NOT contribute to the study of evolution?
   a. population genetics
   b. inheritance of acquired characteristics
   c. the fossil record
   d. comparative morphology

7. Which of the following could be an example of microevolution?
   a. a slight change in a bird population's song arising from a small genetic change in the population
   b. the evolution of many species of finch from a common ancestor
   c. the sudden disappearance of an entire genus
   d. the direct evolutionary link between living primates and humans

8. Which of the following ideas proposed by Lamarck was NOT included in Darwin's theory?
   a. Organisms change in response to their environments.
   b. Changes that an organism acquires during its lifetime are passed to its offspring.
   c. All species change with time.
   d. Changes are passed from one generation to the next.

9. Medical advances now allow many people who suffer from genetic diseases to survive and reproduce. These advances
   a. refute Darwin's theory
   b. disprove descent with modification
   c. reduce the effects of natural selection
   d. eliminate adaptive traits

10. Which of the following ideas is NOT included in Darwin's theory?
    a. All organisms that have ever existed arose through evolutionary modifications of ancestral species.
    b. The great variety of species alive today resulted from the diversification of ancestral species.
    c. Natural selection drives some evolutionary change.
    d. Natural selection preserves advantageous traits.
    e. Natural selection eliminates adaptive traits.

Questions for Discussion

1. Would Charles Darwin have had the same inspiration about natural selection and evolution had he visited other islands? Think of the situation on Hawaii, on Easter Island, or on Tristan da Cunha. Why would the island matter?

2. Explain why the characteristics we see in living organisms adapt them to the environments in which their ancestors lived rather than to the environments in which they live today. Give examples of this situation.

3. Why was there debate about the age of Earth?

Humpback whales were seriously depleted in the mid-1900s.

18

Microevolution: Changes within Populations

WHY IT MATTERS

As a result of more than 250 years of commercial whaling, humpback whales (*Megaptera novaeangliae*) experienced a disastrous population decline—from about 125 000 to 5000 individuals. After an international agreement limited whaling in 1966, humpbacks have rebounded strongly; about 80 000 individuals now form three distinct populations in the North Atlantic, North Pacific, and Southern oceans. Yet, conservation biologists wondered if because the population decreased to only 5000 animals, the genetic variability in the whales may have been reduced. Such a loss could have adverse effects on the population's reproductive capacity, resistance to disease, and ability to survive unfavourable environmental changes. A situation in which a population regrows from a small number is called a bottleneck (see Section 18.3c).

In the early 1990s, a large group of researchers working in Hawaii, the continental United States, Canada, Australia, South Africa, Mexico, and the Dominican Republic measured genetic variability in surviving humpback populations. They studied mitochondrial DNA (mtDNA) because it is small, easily extracted, and easily analyzed. Almost all of the variability in mtDNA comes from chance mutations that occur at a steady rate, rather than from genetic recombination (see Section 9.3). Since mtDNA is haploid, the mutation rate is more or less constant. Except for variations produced by mutations that occurred after the population bottleneck—which can be estimated from the mutation rate and subtracted from the total variation detected—the amount of variability in mtDNA should reflect the amount present in the population during the bottleneck.

The researchers obtained skin samples from 90 humpback whales distributed among the populations. They extracted mtDNA from each sample and amplified a

463-base-pair segment of mtDNA that includes most of the variable nucleotide positions. They then determined the DNA sequences of the segments. (DNA sequencing is described in Section 16.1.)

The researchers were surprised to find that the mtDNA sequence variation was relatively high in most of their samples. However, a subpopulation of the North Pacific population near Hawaii showed no variability at all in the mtDNA segment examined. One possible hypothesis for this result is that the Hawaiian subpopulation originated recently, perhaps during the twentieth century. There is indirect support for that idea: whaling records from the nineteenth century list no humpback sightings around Hawaii, and the native Hawaiian people have no legends or words describing whales of the humpback type (baleen whales). Perhaps the Hawaiian subpopulation was started by a few whales with limited genetic variability, an example of the founder effect (see Section 18.3c).

With the exception of this Hawaiian subpopulation, humpback whales appear to have retained genetic variability comparable to that seen in other animals. Humpbacks have a potential life span of about 50 years. Thus, some individuals still alive at the time of the study had been born before the most intense period of commercial hunting in the mid-twentieth century; those individuals provided a reservoir of genetic variability from the old populations. These results suggest that the hunting ban came just in time to prevent a significant loss of genetic variability in humpback whales.

The evolution of the humpback whales is an example of **microevolution**, which is a change in frequencies of alleles or heritable phenotypic variants in a population over time. A **population** of organisms includes all the individuals of a single species that live together in the same place and time. Today, when scientists study microevolution, they analyze variations—the differences between individuals—in natural populations. They determine how and why these variations are inherited. Darwin recognized the importance of heritable variation

within populations; he also realized that natural selection can change the pattern of variation in a population from one generation to the next. Scientists have since learned that microevolutionary change results from several processes, not just natural selection, and that sometimes these processes counteract each other.

In this chapter, we first examine the extensive variation that exists within natural populations. We then take a detailed look at the most important processes that alter genetic variation within populations, causing microevolutionary change. Finally, we consider how microevolution can fine-tune the functioning of populations within their environments.

18.1 Variation in Natural Populations

In some populations, individuals vary dramatically in appearance, but the members of most populations look pretty much alike **(Figure 18.1)**. However, even those that look alike, such as the *Cerion* snails in Figure 18.1b, are not identical. With a scale and ruler, you could detect differences in their weight as well as in the length and diameter of their shells. With suitable techniques, you could also document variations in their individual biochemistry, physiology, internal anatomy, and behaviour. All of these are examples of **phenotypic variation**, differences in appearance or function among individuals of a population. If a difference is heritable, it is passed from generation to generation.

18.1a Phenotypic Variation

Darwin's theory recognized the importance of heritable phenotypic variation. Today, microevolutionary studies often begin by assessing phenotypic variation within populations. Most characters exhibit **quantitative variation**: individuals differ in small, incremental ways. If you measured the height of everyone in your biology class, for example, you would see that height varies almost continuously from your shortest to your tallest classmate. Humans also exhibit quantitative variation in the length of their toes, the number of hairs on their heads, and their weight, as discussed in Section 10.2.

We usually display data on quantitative variation in a bar graph or, if the sample is large enough, as a curve **(Figure 18.2)**. The width of the curve is proportional to the variability—the amount of variation—among individuals, and the *mean* describes the average value of the character. As you will see shortly, natural selection often changes the mean value of a character or its variability within populations.

a. European garden snails
(Cepaea nemoralis)

b. Bahaman land snails
(Cerion christophei)

George Bernard/Science Source

Carnegie Museum of Natural History

Figure 18.1
Phenotypic variation. **(a)** Shells of European garden snails from a population in Scotland vary considerably in appearance. **(b)** By contrast, shells of land snails from a population in the Bahamas look very similar.

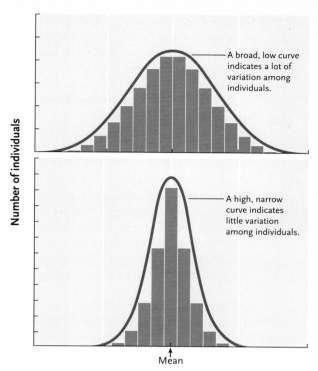

A broad, low curve indicates a lot of variation among individuals.

A high, narrow curve indicates little variation among individuals.

Mean

Measurement or value of trait

Figure 18.2

Quantitative variation. Many traits vary continuously among members of a population, and a bar graph of the data often approximates a bell-shaped curve. The mean defines the average value of the trait in the population, and the width of the curve is proportional to the variability among individuals.

Other characters, such as those Mendel studied (see Chapter 10), exhibit **qualitative variation:** they exist in two or more discrete states, and intermediate forms are often absent. Snow geese, for example, have *either* blue *or* white feathers **(Figure 18.3)**, no pale blue. The existence of discrete variants of a character is called a **polymorphism** (*poly* = many, *morphos* = form); we describe such traits as *polymorphic*. The *Cepaea nemoralis* snail shells in Figure 18.1a are polymorphic

© Morales/AgeFotostock

Figure 18.3

Qualitative variation. Individual snow geese (*Chen caerulescens*) are either blue or white. Although both colours are present in many populations, geese tend to associate with others of the same colour.

in background colour, number of stripes, and colour of stripes. Biochemical polymorphisms, like the human A, B, AB, and O blood types (described in Section 10.2), are also common.

We describe phenotypic polymorphisms quantitatively by calculating the percentage or *frequency* of each trait. For example, if you counted 123 blue snow geese and 369 white ones in a population of 492 geese, the frequency of the blue phenotype would be 123/492 or 0.25, and the frequency of the white phenotype would be 369/492 or 0.75.

Phenotypic variation within populations may be caused by genetic differences between individuals, by differences in the environmental factors that individuals experience, or by an interaction between an individual's genetics and the environment. As a result, genetic and phenotypic variations may not be perfectly correlated. Organisms with different genotypes often exhibit the same phenotype. For example, recall Gregor Mendel's experiments with pea plants (see Chapter 10). Plants with homozygous recessive alleles for flower colour had white flowers, while plants with homozygous dominant alleles or with heterozygous alleles had purple flowers. Plants with purple flowers have two different genotypes, even though they exhibit the same phenotype.

Conversely, organisms with the same genotype sometimes exhibit different phenotypes. For example, the acidity of soil influences flower colour in some plants **(Figure 18.4, p. 412)**. Knowing whether phenotypic variation is caused by genetic differences, environmental factors, or an interaction of the two is important because *only genetically based variation is subject to evolutionary change.* At the same time, *it is the phenotype of an individual organism, rather than its genotype, that is successful or not.* In other words, natural selection operates on the phenotype, not the genotype. And it operates on the whole phenotype, not just one gene at a time.

Knowing the causes of phenotypic variation also has important practical applications. Suppose, for example, that one field of wheat produced more grain than another. If a difference in the availability of nutrients or water caused the difference in yield, a farmer might choose to fertilize or irrigate the less productive field. But if the difference in productivity resulted from genetic differences between the plants in the two fields, a farmer might plant only the more productive genotype. Because environmental factors can influence the expression of genes, an organism's phenotype is frequently the product of an interaction between its genotype and its environment. In our hypothetical example, the farmer may maximize yield by fertilizing and irrigating the more productive genotype of wheat.

How can we determine whether phenotypic variation is caused by environmental factors or by genetic differences? We can test for an environmental cause experimentally by changing one environmental

©iStock.com/mcswin

©iStock.com/jdkapp12

Figure 18.4

Environmental effects on phenotype. Soil acidity affects the expression of the gene controlling flower colour in the common garden plant *Hydrangea macrophylla*. When grown in acidic soil, it produces deep blue flowers. In neutral or alkaline soil, its flowers are bright pink.

variable and measuring the effects on genetically similar subjects. You can try this yourself by growing some cuttings from a single ivy plant in shade and other cuttings from the same plant in full sun. Although the parent plant and all its cuttings have the same genotype, the cuttings grown in sun will produce smaller leaves and shorter stems than those grown in the shade.

Breeding experiments can demonstrate the genetic basis of phenotypic variation. For example, Mendel inferred the genetic basis of qualitative traits, such as flower colour in peas, by crossing plants with different phenotypes. Although simple crosses will not reveal the genetic basis of variations in quantitative traits, these characteristics will respond to artificial selection if the variation has some genetic basis. For example, researchers observed that individual house mice (*Mus domesticus*) differ in activity levels, as measured by how much they use an exercise wheel and how fast they run. John G. Swallow and his colleagues at the University of Wisconsin, Madison, used artificial selection to produce lines of mice that exhibit increased wheel-running behaviour, demonstrating that the observed differences in these two aspects of activity level have a genetic basis.

Breeding experiments are not always practical, however, particularly for organisms with long generation times. Ethical concerns also mean these techniques are unthinkable for humans. Instead, researchers sometimes study the inheritance of particular traits by analyzing genealogical pedigrees, as discussed in Section 11.2, but this approach often provides poor results for analyses of complex traits.

18.1b Genetic Variation

An **allele** is one member of a gene pair that occupies a single location (locus) on a chromosome (see Section 10.2c). A gene can have more than one possible allele, and occasionally several alleles. In diploid organisms, only two of these alleles are present in any gene pair, and haploid organisms have only one of each type of allele.

Genetic variation, the raw material moulded by microevolutionary processes, has two potential sources: the production of new alleles and the rearrangement of existing alleles. Most new alleles probably arise from small-scale mutations in DNA. The rearrangement of *existing* alleles into new combinations can result from larger-scale changes in chromosome structure or number, as well as from several forms of genetic recombination, including crossing-over between homologous chromosomes during meiosis, independent assortment of nonhomologous chromosomes during meiosis, and random fertilizations between genetically different sperm and eggs. This shuffling of alleles into new combinations can produce an extraordinary number of novel genotypes in the next generation. By one estimate, more than 10^{600} combinations of alleles are possible in human gametes, yet there are fewer than 10^{10} humans alive today. So unless you have an identical twin, it is extremely unlikely that another person with your genotype has ever lived or ever will.

18.1c Natural Selection and Phenotypic Variation

Above, you learned that natural selection operates on an organism's phenotype. Biologists measure the effects of natural selection on phenotypic variation by recording changes in the mean and variability of characters over time (see Figure 18.2). Three modes of natural selection have been identified: directional selection, stabilizing selection, and disruptive selection **(Figure 18.5).**

Directional Selection. Traits undergo **directional selection** when individuals near one end of the phenotypic spectrum have the highest relative fitness. Directional selection shifts a trait away from the existing mean and toward the favoured extreme (see Figure 18.5a). After selection, the trait's mean value is higher or lower than before, and variability in the trait may be reduced.

Directional selection is very common. For example, predatory fish promote directional selection for larger body size in guppies when they selectively feed on the smallest individuals in a guppy population. Most cases of artificial selection are directional, aimed at increasing or decreasing specific phenotypic traits. Humans routinely use directional selection to produce domestic animals and crops with desired characteristics, such as the small size of Chihuahuas and the intense heat of chili peppers.

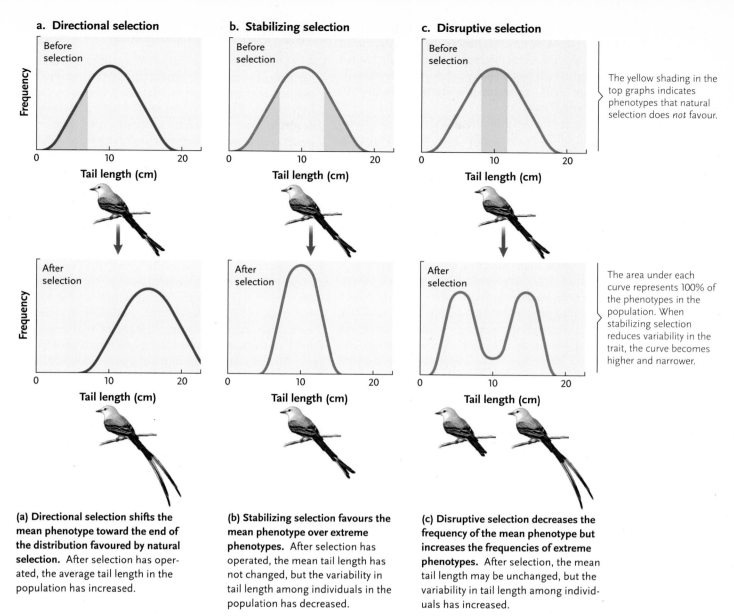

a. Directional selection

Before selection

Frequency

Tail length (cm)
0   10   20

b. Stabilizing selection

Before selection

Frequency

Tail length (cm)
0   10   20

c. Disruptive selection

Before selection

Frequency

Tail length (cm)
0   10   20

The yellow shading in the top graphs indicates phenotypes that natural selection does *not* favour.

After selection

Frequency

Tail length (cm)
0   10   20

After selection

Frequency

Tail length (cm)
0   10   20

After selection

Frequency

Tail length (cm)
0   10   20

The area under each curve represents 100% of the phenotypes in the population. When stabilizing selection reduces variability in the trait, the curve becomes higher and narrower.

(a) Directional selection shifts the mean phenotype toward the end of the distribution favoured by natural selection. After selection has operated, the average tail length in the population has increased.

(b) Stabilizing selection favours the mean phenotype over extreme phenotypes. After selection has operated, the mean tail length has not changed, but the variability in tail length among individuals in the population has decreased.

(c) Disruptive selection decreases the frequency of the mean phenotype but increases the frequencies of extreme phenotypes. After selection, the mean tail length may be unchanged, but the variability in tail length among individuals has increased.

Figure 18.5
Three modes of natural selection. A hypothetical example using tail length of birds as the quantitative trait subject to selection.

Stabilizing Selection. Traits undergo **stabilizing selection** when individuals expressing intermediate phenotypes have the highest relative fitness (Figure 18.5b). By eliminating phenotypic extremes, stabilizing selection reduces genetic and phenotypic variation and increases the frequency of intermediate phenotypes. Stabilizing selection is probably the most common mode of natural selection, affecting many familiar traits. For example, very small and very large human newborns are less likely to survive than those born at an intermediate weight.

Stabilizing selection can result from multiple selective forces acting on the same trait but in opposite directions. This pattern is seen in the gallmaking fly (*Eurosta solidaginis*), a small insect that feeds on the tall goldenrod plant (*Solidago altissima*). When a fly larva hatches from its egg, it bores into a goldenrod stem, and the plant responds by producing a spherical growth deformity called a gall. The larva feeds on plant tissues inside the gall. Galls vary dramatically in size; genetic experiments indicate that gall size is a heritable trait of the fly, although plant genotype also has an effect.

Fly larvae inside galls are subjected to two opposing patterns of directional selection. On one hand, a tiny wasp (*Eurytoma gigantea*) parasitizes gallmaking flies by laying eggs in fly larvae inside their galls. After hatching, the young wasps feed on the fly larvae, killing them in the process. However, adult wasps are so small that they cannot easily penetrate the thick walls of a large gall; they generally lay eggs in fly larvae occupying small galls. Thus, wasps establish directional selection that favours flies that produce large galls, which are consequently less likely to be parasitized. On the other hand, several bird species open galls to feed on mature fly larvae; these predators preferentially open large galls, fostering directional selection in favour of small galls.

In about one-third of the populations surveyed, wasps and birds attacked galls with equal frequency, and flies producing galls of intermediate size had the highest survival rate. The smallest and largest galls—as well as the genetic predisposition to make very small or very large galls—were eliminated from the population.

Disruptive Selection. Traits undergo **disruptive selection** when extreme phenotypes have higher relative fitness than intermediate phenotypes (see Figure 18.5c). Thus, alleles producing extreme phenotypes become more common, promoting polymorphism. Under natural conditions, disruptive selection is much less common than directional selection and stabilizing selection. "People behind Biology" in Chapter 17 describes a good example of disruptive selection in Galápagos finches.

STUDY BREAK

1. If a population of skunks includes some individuals with stripes and others with spots, would you describe the variation as quantitative or qualitative?
2. In the experiment on house mice described above, how did researchers demonstrate that variations in activity level had a genetic basis?
3. What factors contribute to phenotypic variation in a population?
4. Which mode of natural selection increases the representation of the average phenotype in a population?

18.2 Population Genetics

To predict how certain factors may influence genetic variation, population geneticists first describe the genetic structure of a population. They then create hypotheses, formalized in mathematical models, to describe how evolutionary processes may change the genetic structure under specified conditions. Finally, researchers test the predictions of these models to evaluate the ideas about evolution that are embodied within them.

18.2a Genetic Structure of Populations

Populations are made up of individuals of the same species, each with its own genotype. In diploid organisms with pairs of homologous chromosomes, an individual's genotype includes two alleles (either two copies of the same allele or two different alleles) at each gene locus. The sum of all alleles at all gene loci in all individuals is called the population's **gene pool**.

To describe the structure of a gene pool, scientists first identify the genotypes in a representative sample and calculate **genotype frequencies**, the percentages of individuals possessing each genotype. They can then calculate **allele frequencies**, the **relative abundances** of the different alleles. For a locus with two alleles, scientists use the symbol p to identify the frequency of one allele, and q to identify the frequency of the other allele.

The calculation of genotype and allele frequencies for the two alleles at the gene locus governing flower colour in snapdragons (*Antirrhinum* spp.) is straightforward **(Table 18.1)**. This locus is easy to study because it exhibits incomplete dominance (see Section 10.2). Individuals that are homozygous for the C^R allele ($C^R C^R$) have red flowers, those homozygous for the C^W allele ($C^W C^W$) have white flowers, and heterozygotes ($C^R C^W$) have pink flowers. Genotype

Table 18.1	Calculation of Genotype Frequencies and Allele Frequencies for the Snapdragon Flower Colour Locus

Because each diploid individual has two alleles at each gene locus, a sample of 1000 individuals has a total of 2000 alleles at the C locus.

Flower Colour Phenotype	Genotype	Number of Individuals	Genotype Frequencies[1]	Total Number of C^R Alleles[2]	Total Number of C^W Alleles[2]
Red	$C^R C^R$	450	450/1000 = 0.45	2 × 450 = 900	0 × 450 = 0
Pink	$C^R C^W$	500	500/1000 = 0.50	1 × 500 = 500	1 × 500 = 500
White	$C^W C^W$	50	50/1000 = 0.05	0 × 50 = 0	2 × 50 = 100
	Total	1000	0.45 + 0.50 + 0.05 = 1.0	1400	600

To calculate allele frequencies, use the total of 1400 + 600 = 2000 alleles in the sample:

p = frequency of C^R allele = 1400/2000 = 0.7
q = frequency of C^W allele = 600/2000 = 0.3
$p + q$ = 0.7 + 0.3 = 1.0

[1]Genotype frequency = the number of individuals possessing a particular genotype divided by the total number of individuals in the sample.
[2]Total number of C^R or C^W alleles = the number of C^R or C^W alleles present in one individual with a particular genotype multiplied by the number of individuals with that genotype.

frequencies represent how the C^R and C^W alleles are distributed among individuals. In this example, examination of all of the plants revealed that 45% of individuals have the $C^R C^R$ genotype, 50% have the heterozygous $C^R C^R$ genotype, and the remaining 5% have the $C^W C^W$ genotype. Allele frequencies represent the commonness or rarity of each allele in the gene pool. As calculated in the table, 70% of the alleles in the population are C^R and 30% are C^W. Remember that for a gene locus with two alleles, there are three genotype frequencies but only two allele frequencies (p and q). The sum of the three genotype frequencies must equal 1; so must the sum of the two allele frequencies.

Once we have described the population, the next question might be, "is the population evolving?" or "is there evidence for evolution in the gene controlling flower colour?"

18.2b The Hardy–Weinberg Principle

When designing experiments, scientists often use control treatments to evaluate the effect of a particular factor. The control tells us what we would see if the experimental treatment had no effect. However, in studies using observational rather than **experimental data,** there is often no suitable control. In such cases, investigators develop conceptual models, called **null models,** that predict what they would see if that particular factor had no effect. Null models serve as theoretical reference points against which scientists can evaluate their observations.

Early in the twentieth century, geneticists were puzzled by the persistence of recessive traits because they assumed that natural selection replaced recessive or rare alleles with dominant or common ones. An English mathematician, G. H. Hardy, and a German physician, Wilhelm Weinberg, tackled this problem independently in 1908. Their analysis, now known as the **Hardy–Weinberg principle**, specifies the conditions under which a population of diploid organisms achieves **genetic equilibrium**, the point at which neither allele frequencies nor genotype frequencies change in succeeding generations. Their work also showed that dominant alleles need not replace recessive ones, and that the shuffling of genes in sexual reproduction does not in itself cause allele or genotype frequencies to change.

The Hardy–Weinberg principle is a mathematical model that describes how genotype frequencies are established in sexually reproducing organisms. According to this model, genetic equilibrium is possible only if *all* of the following conditions are met:

1. No mutations are occurring.
2. The population is closed to migration from other populations.
3. The population is infinite in size (i.e., there is no genetic drift; see Section 18.3c).
4. All genotypes in the population survive and reproduce equally well (selection is not acting on the trait being considered).
5. Individuals in the population mate randomly with respect to the trait being considered.

If all the conditions of the model are met, the allele frequencies of the population for an identified gene locus will never change, and the genotype frequencies will stop changing after one generation. The Hardy–Weinberg principle is thus a null model that serves as a reference point for evaluating the circumstances under which evolution *may* occur. If a population's genotype frequencies do not match the predictions of this null model, evolution may be occurring. If allele frequencies change over time, evolution is occurring. Determining which of the model's conditions are *not* met is a first step in understanding how and why the gene pool is changing (see "Using the Hardy–Weinberg Principle").

Using the Hardy–Weinberg Principle

Research Method Box

To see how the Hardy–Weinberg principle can be applied, we will analyze the snapdragon flower colour locus, using the hypothetical population of 1000 plants described in Table 18.1. This locus includes two alleles: C^R (with its frequency designated as p) and C^W (with its frequency designated as q), and three genotypes: homozygous $C^R C^R$, heterozygous $C^R C^W$, and homozygous $C^W C^W$. Table 18.1 lists the number of plants with each genotype and shows the calculation of both the genotype frequencies and the allele frequencies for the population.

Let's assume for simplicity that each individual produces only two gametes and that both gametes contribute to the production of offspring. This assumption is unrealistic, of course, but it meets the Hardy–Weinberg requirement that all individuals in the population contribute equally to the next generation. In each parent, the two alleles segregate and end up in different gametes:

450 $C^R C^R$ individuals produce → 900 C^R gametes

500 $C^R C^W$ individuals produce → 500 C^R gametes + 500 C^W gametes

50 $C^W C^W$ individuals produce → 100 C^W gametes

You can readily see that 1400 of the 2000 total gametes carry the C^R allele

(Continued)

Using the Hardy–Weinberg Principle (*Continued*)

and the other 600 carry the C^W allele. The frequency of C^R gametes is 1400/2000, or 0.7, which is equal to p; the frequency of C^W gametes is 600/2000, or 0.3, which is equal to q. Thus, the allele frequencies in the gametes are exactly the same as the allele frequencies in the parent generation. It could not be otherwise because each gamete carries one allele at each locus.

Now assume that these gametes, both sperm and eggs, encounter each other at random. In other words, individuals reproduce without regard to the genotype of a potential mate **(Figure 1).**

We can also describe the consequences of random mating—$(p + q)$ sperm fertilizing $(p + q)$ eggs—with an equation that predicts the genotype frequencies in the offspring generation:

$$(p + q) \times (p + q) = p^2 + 2pq + q^2$$

If the population is at genetic equilibrium for this locus, p^2 is the predicted frequency of the $C^R C^R$ genotype; $2pq$, the predicted frequency of the $C^R C^W$ genotype; and q^2, the predicted frequency of the $C^W C^W$ genotype. Using the gamete frequencies determined above, we can calculate the predicted genotype frequencies in the next generation:

frequency of $C^R C^R =$
$$p^2 = (0.7 \times 0.7) = 0.49$$
frequency of $C^R C^W =$
$$2pq = 2(0.7 \times 0.3) = 0.42$$
frequency of $C^W C^W =$
$$q^2 = (0.3 \times 0.3) = 0.09$$

Notice that the predicted genotype frequencies in the offspring generation have changed from the genotype frequencies in the parent generation: the frequency of heterozygous individuals has decreased, and the frequencies of both types of homozygous individuals have increased. This result occurred because the starting population was *not in equilibrium* at this gene locus. In other words, the distribution of parent genotypes did not conform to the predicted $p^2 + 2pq + q^2$ distribution.

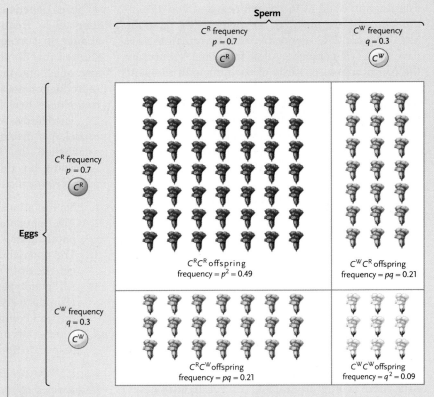

Figure 1
Mating table showing the process of random mating.

The 2000 gametes in our hypothetical population produced 1000 offspring. Using the genotype frequencies we just calculated, we can predict how many offspring will carry each genotype:

490 red ($C^R C^R$)
420 pink ($C^R C^W$)
90 white ($C^W C^W$)

In a real study, we would examine the offspring to see how well their numbers match these predictions.

What about the allele frequencies in the offspring? The Hardy–Weinberg principle predicts that they did not change. Let's calculate them and see. Using the method shown in Table 18.1 and the prime symbol (') to indicate offspring allele frequencies, we have

$p' = ([2 \times 490] + 420)/2000 =$
$$1400/2000 = 0.7$$
$q' = ([2 \times 90] + 420)/2000 =$
$$600/2000 = 0.3$$

You can see from this calculation that the allele frequencies did not change from one generation to the next, even though the alleles were rearranged to produce different proportions of the three genotypes. Thus, the population is now at genetic equilibrium for the flower colour locus. Neither the genotype frequencies nor the allele frequencies will change in succeeding generations as long as the population meets the conditions specified in the Hardy–Weinberg model.

To verify this, you can calculate the allele frequencies of the gametes for this offspring generation and predict the genotype frequencies and allele frequencies for a third generation. You could continue calculating until you ran out of either paper or patience, but these frequencies will not change.

Researchers use calculations such as these to determine whether an actual population is near its predicted genetic equilibrium for one or more gene loci. When they discover that a population is not at equilibrium, they infer that microevolution is occurring. They then investigate the factors that might be responsible.

1. What is the difference between the genotype frequencies and the allele frequencies in a population?
2. Why is the Hardy–Weinberg principle considered a null model of evolution?
3. If the five conditions of the Hardy–Weinberg principle are all met, when will genotype frequencies stop changing?

18.3 Evolutionary Agents

In the following sections, we describe the processes that produce genetic variation and foster microevolutionary change: mutation, gene flow, genetic drift, natural selection, and nonrandom mating. These are summarized in **Table 18.2.**

18.3a Mutations

A **mutation** is a spontaneous and heritable change in DNA. Mutations are rare events: roughly one gamete in 100 thousand to one in 1 million will include a new mutation at a particular gene locus. New mutations are so infrequent, in fact, that they exert little or no immediate effect on allele frequencies in most populations. But over evolutionary time scales, their numbers are significant; mutations have been accumulating in biological lineages for billions of years. And because it creates entirely new genetic variations, a *mutation can be a major source of heritable variation.*

For most animals, only mutations in the germ line (the cell lineage that produces gametes) are heritable; mutations in other cell lineages have no direct effect on the next generation. In plants, however, mutations may occur in meristem cells, which eventually produce flowers as well as nonreproductive structures (see Chapter 34); in such cases, a mutation may be passed to the next generation and ultimately influence the gene pool.

We classify mutations based on their effect on an organism's fitness, rather than on the underlying molecular changes and the mode of inheritance of the trait (e.g., dominant, recessive). It is not immediately known whether or not a new mutation will be advantageous, deleterious, or neutral.

Deleterious mutations alter an individual's structure, function, or behaviour in harmful ways. In mammals, for example, a protein called collagen is an essential component of most extracellular structures. Several simple mutations in humans cause forms of Ehlers–Danlos syndrome, a disruption of collagen synthesis that may result in loose skin; weak joints; or sudden death from the rupture of major blood vessels, the colon, or the uterus.

Lethal mutations can cause great harm to organisms carrying them. If a lethal allele is dominant, both homozygous and heterozygous carriers will, by definition, die from its effects; if recessive, it kills only homozygous recessive individuals. Manx cats have a recessive lethal allele for taillessness: homozygous dominant cats are normal, heterozygous cats have short tails and long legs, and homozygous recessive cats do not survive embryonic development.

Neutral mutations are neither harmful nor helpful. Because of the redundancy of the genetic code, several codons with different nucleotides in the third position may specify the same amino acid in the construction of a polypeptide chain (see Section 13.1). As a result, some DNA sequence changes—especially those in the third nucleotide of the codon—do not alter the amino acid sequence of the protein under construction. Not surprisingly, mutations at the third position appear to persist longer in populations than those at the first two positions. In other instances, mutations that change the amino acid sequence in a protein or even an organism's phenotype may have no influence on its survival and reproduction. A neutral mutation might even prove to be beneficial later if the environment changes.

Sometimes a change in DNA produces an advantageous mutation, which confers some benefit on an individual that carries it. However slight the advantage, natural selection may preserve the new allele and even increase its frequency over time. Once the mutation has been passed to a new generation, other agents of microevolution determine its long-term fate.

Table 18.2	Agents of Microevolutionary Change	
Agent	Definition	Effect on Genetic Variation
Mutation	Heritable change in DNA	Introduces new genetic variation into population; does not change allele frequencies quickly
Gene flow	Change in allele frequencies as individuals join a population and reproduce	May introduce genetic variation from another population
Genetic drift	Random changes in allele frequencies caused by chance events	Reduces genetic variation, especially in small populations; can eliminate rare alleles
Natural selection	Differential survivorship or reproduction of individuals with different phenotypes	One allele replacing another or allelic variation being preserved
Nonrandom mating	Choice of mates based on their their phenotypes and genotypes	Does not directly affect allele frequencies, but usually prevents genetic equilibrium

18.3b Gene Flow

Organisms or their genetic material (in the form of pollen, spores, or fertilized eggs) sometimes move from one population to another. If the immigrants reproduce, they may introduce novel alleles into a population, shifting its allele and genotype frequencies. This phenomenon is called **gene flow** and shows that populations are not completely closed, but can be open to migration.

Gene flow is common in some animal species. For example, young male baboons typically move from one local population to another after experiencing aggressive behaviour by older males. And many marine invertebrate eggs and larvae disperse long distances, carried by ocean currents.

Dispersal agents, such as pollen-carrying wind or seed-carrying animals, are responsible for gene flow in most plant populations. For example, blue jays foster gene flow among populations of oaks by carrying acorns from nut-bearing trees to their winter caches, which may be as much as a mile away **(Figure 18.6)**. Transported acorns that go uneaten may germinate and contribute to the gene pool of a neighbouring oak population.

The movement alone of individuals from one population to another is not sufficient to foster gene flow between two populations. The immigrants must also reproduce in the population they join, thereby contributing to its gene pool. In the San Francisco Bay area, for example, Bay checkerspot butterflies (*Euphydryas editha bayensis*) rarely move from one population to another because they are poor fliers. But when adult females do change populations, it is often late in the breeding season and their offspring have virtually no chance of finding enough food to mature. Thus, many immigrant females do not foster gene flow because they do not contribute to the gene pool of the population they join.

The evolutionary importance of gene flow depends on the degree of genetic differentiation between populations and the rate of gene flow between them. If two gene pools are very different, a little gene flow may increase genetic variability within the population that receives immigrants, and it will make the two populations more similar. But if populations are already genetically similar, even lots of gene flow will have little effect.

18.3c Genetic Drift

Chance events sometimes cause the allele frequencies in a population to change unpredictably. This phenomenon, known as **genetic drift**, has especially dramatic effects on small populations, clearly violating the Hardy–Weinberg assumption of an infinitely

Figure 18.6

Gene flow. Blue jays (*Cyanocitta cristata*) serve as agents of gene flow for oaks (genus *Quercus*) when the birds carry acorns from one oak population to another. An uneaten acorn may germinate and contribute to the gene pool of the population into which it was carried.

large population size. Two general circumstances, population bottlenecks and founder effects, often foster genetic drift.

Population Bottlenecks. On occasion, a stressful factor such as disease, starvation, or drought kills a large proportion of the individuals in a population, producing a **population bottleneck**, a dramatic reduction in population size. This cause of genetic drift greatly reduces genetic variation even if the population numbers later rebound **(Figure 18.7)**.

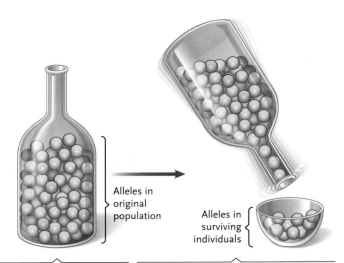

Alleles in original population

Alleles in surviving individuals

The gene pool of the original population, represented by a bottle filled with colored marbles, includes a locus with three alleles. Two of the alleles, represented by blue and green marbles, occur at high frequency; the third allele, represented by red marbles, occurs at low frequency.

If an environmental event randomly kills a large number of individuals in the population, the drastic reduction in population size is described as a population bottleneck. The process is analogous to shaking only a few of the marbles—the survivors—through the neck of the bottle. As a consequence of chance events associated with population bottlenecks, surviving individuals may not have the same allele frequencies as the original population. Rare alleles are inevitably lost.

Figure 18.7

Population bottlenecks, genetic drift, and the loss of genetic variability.

In the late nineteenth century, for example, hunters nearly wiped out northern elephant seals (*Mirounga angustirostris*) along the Pacific coast of North America. Since the 1880s, when the species received protected status, the population has increased to more than 30 000, all descended from a group of about 20 survivors. Today the population exhibits no variation in 24 proteins studied by gel electrophoresis. This low level of genetic variation, which is unique among seal species, is consistent with the hypothesis that genetic drift eliminated many alleles when the population experienced the bottleneck.

Founder Effect. When a few individuals colonize a distant locality and start a new population, they carry only a small sample of the parent population's genetic variation. By chance, some alleles may be totally missing from the new population, whereas other alleles that were rare "back home" might occur at relatively high frequencies. This change in the gene pool is called the **founder effect.**

The human medical literature provides some of the best-documented examples of the founder effect. The Old Order Amish, an essentially closed religious community in Lancaster County, Pennsylvania, have an exceptionally high incidence of Ellis–van Creveld syndrome, a genetic disorder caused by a recessive allele. In the homozygous state, the allele produces dwarfism, shortened limbs, and polydactyly (extra fingers). Genetic analysis suggests that although this syndrome affects fewer than 1% of the Amish in Lancaster County, as many as 13% may be heterozygous carriers of the allele. All of the individuals exhibiting the syndrome are descended from one couple, who helped found the community in the mid-1700s. Assuming Hardy–Weinberg, the allele frequency is 0.075 in the Amish population compared to about 0.003 in the entire population.

Small Population Implications. A simple analogy clarifies why genetic drift is more pronounced in small populations than in large ones. When individuals reproduce, male and female gametes often pair up randomly, as though the allele in any particular sperm or ovum was determined by a coin toss. Imagine that heads specifies the R allele and that tails specifies the r allele. If the two alleles are equally common (i.e., their frequencies, p and q, are both equal to 0.5), heads should be as likely an outcome as tails. But if you toss a coin 20 or 30 times to simulate random mating in a small population, you won't often see a 50:50 ratio of heads and tails. Sometimes heads will predominate and sometimes tails will—just by chance. Tossing the coin 500 times to simulate random mating in a somewhat larger population is more likely to produce a 50:50 ratio of heads and tails. And if you tossed the coin 5000 times, you would get even closer to a 50:50 ratio.

Genetic drift generally leads to the loss of alleles and reduced genetic variability; it therefore causes allele and genotype frequencies to differ from those predicted by the Hardy–Weinberg model.

Conservation Implications. Genetic drift has important implications for **conservation biology.** Because of their small population size, **endangered species** experience severe population bottlenecks, resulting in the loss of genetic variability. Moreover, the small number of individuals available for captive breeding programs may not fully represent a species' genetic diversity. Without such variation, no matter how large a population may become in the future, it will be less resistant to diseases or less able to cope with environmental change.

For example, scientists hypothesize that an environmental catastrophe produced a bottleneck in the African cheetah (*Acinonyx jubatus*) population about 10 000 years ago. Cheetahs today are remarkably uniform in genetic make-up. Their populations are highly susceptible to diseases; males also have a high proportion of sperm cell abnormalities and a reduced reproductive capacity. These observations support the hypothesis of a bottleneck resulting in a high frequency of deleterious alleles in the population. Thus, limited genetic variation, as well as small numbers, threatens populations of endangered species.

18.3d Genetic Variation in Populations

How much genetic variation actually exists within populations? In the 1960s, evolutionary biologists began to use gel electrophoresis (see Figure 15.7) to identify biochemical polymorphisms in diverse organisms. This technique separates two or more forms of a given protein if they differ significantly in shape, mass, or net electrical charge, as a result of mutation-induced changes in the underlying amino acid sequence. The identification of a protein polymorphism allowed researchers to infer genetic variation at the locus coding for that protein.

This approach revealed much more genetic variation in natural populations than anyone had imagined. For example, nearly half the loci surveyed in many populations of plants and invertebrates are polymorphic. Advances in molecular biology now allow scientists to survey genetic variation directly, and researchers have accumulated an astounding knowledge of the structure of DNA and its nucleotide sequences. In general, studies of chromosomal and mitochondrial DNA suggest that every locus exhibits some variability in its nucleotide sequence among individuals from a single population, between populations of the same species, and between related species. However, some variations detected in the protein-coding regions of DNA may not affect phenotypes because, as explained in Section 13.4a, they do not change the amino acid sequences of the proteins for which the genes code.

18.3e Natural Selection and Genetic Variability

The Hardy–Weinberg model requires all genotypes in a population to survive and reproduce equally well. But as you know, inherited traits can enable some individuals to survive better and produce more offspring than others. Natural selection is the process by which such traits become more common in subsequent generations. Thus, natural selection violates a requirement of the Hardy–Weinberg equilibrium and causes allele and genotype frequencies to differ from those predicted by the model.

Although natural selection can change allele frequencies in a population, it is the phenotype of an individual organism, rather than any particular allele, that is successful or not. When individuals survive and reproduce, their alleles, both favourable and unfavourable, are passed to the next generation. Of course, an organism with harmful or lethal dominant alleles will probably die before reproducing, and all the alleles it carries will share that unhappy fate, even those that are advantageous.

To evaluate reproductive success, evolutionary biologists consider **relative fitness**, the number of surviving offspring that an individual produces compared with the numbers left by others in the population. Thus, a particular allele will increase in frequency in the next generation if individuals carrying that allele leave *more* offspring than individuals carrying other alleles. Differences in the *relative* success of individuals are the essence of natural selection.

Natural selection tests fitness differences at nearly every stage of an organism's life cycle. One plant may be fitter than others in the population because its seeds survive colder conditions, because the arrangement of its leaves captures sunlight more efficiently, or because its flowers are more attractive to pollinators. However, natural selection exerts little or no effect on traits that appear during an individual's postreproductive life. For example, Huntington disease, a dominant-allele disorder that first strikes humans after the age of 40, is not subject to strong selection. Carriers of the disease-causing allele can reproduce before the onset of the condition, passing it to the next generation.

18.3f Nonrandom Mating

The Hardy–Weinberg model requires individuals to select mates randomly with respect to their genotypes. This requirement is, in fact, often met; humans, for example, generally marry one another in total ignorance of their genotypes for digestive enzymes or blood types.

Nevertheless, many organisms mate nonrandomly, selecting a mate with a particular phenotype. Snow geese, for example, usually select mates of their own colour (Figure 18.3), and human women are more likely to marry men who are taller than they are. If one phenotype is preferred by most potential mates, mating is not random. Hence the next generation will contain fewer heterozygous offspring—and more homozygous offspring—than the Hardy–Weinberg model predicts.

Inbreeding is a special form of nonrandom mating in which genetically related individuals mate with each other. **Self-fertilization** in plants (see Chapter 36) and a few animals (see Chapter 41) is an extreme example of inbreeding because offspring are produced from the gametes of a single parent. However, other organisms that live in small, relatively closed populations often mate with related individuals. Because relatives often carry the same alleles, inbreeding generally increases the frequency of homozygous genotypes and decreases the frequency of heterozygotes. Thus, recessive phenotypes are often expressed.

For example, the high incidence of Ellis–van Creveld syndrome among the Old Order Amish population, mentioned earlier, is partly caused by inbreeding. Although the founder effect originally established the disease-causing allele in this population, inbreeding in the small population increased the likelihood of its expression. Most human societies discourage matings between genetically close relatives, thereby reducing inbreeding and the inevitable production of recessive homozygotes that inbreeding causes.

STUDY BREAK

1. Which agents of microevolution tend to increase genetic variation within populations, and which ones tend to decrease it?
2. How does genetic drift cause the allele frequencies in a population to change?
3. In what way is sexual selection like directional selection?

18.4 Maintaining Genetic and Phenotypic Variation

Evolutionary biologists continue to discover extraordinary amounts of genetic and phenotypic variation in most natural populations. How can so much variation persist in the face of stabilizing selection and genetic drift?

18.4a Diploidy

The diploid condition reduces the effectiveness of natural selection on harmful recessive alleles. (Note that only a few recessive alleles are harmful.) Although such alleles are disadvantageous in the homozygous state, they may have little or no effect on heterozygotic individuals. Thus, recessive alleles can be protected from natural selection by the phenotypic expression of the dominant allele.

18.4b Balanced Polymorphisms

In a **balanced polymorphism**, two or more phenotypes are maintained in fairly stable proportions over many generations. Natural selection preserves balanced polymorphisms when heterozygotes have higher relative fitness, when different alleles are favoured in different environments, and when the rarity of a phenotype provides an advantage.

Heterozygote Advantage. A balanced polymorphism can be maintained by **heterozygote advantage**, when heterozygotes have higher relative fitness than either homozygote. As Darwin first discovered in his experiments on corn, the offspring of crosses between two homozygous strains of the same species often exhibit a robustness described as *hybrid vigour*. Apparently, being heterozygous at many gene loci provides some advantage, perhaps by allowing organisms to respond effectively to environmental variation.

The best-documented example of heterozygote advantage at a specific gene locus is the maintenance of the *HbS* (sickle cell) allele, which codes for a defective form of hemoglobin in humans. As you learned in Chapter 10, hemoglobin is an oxygen-transporting molecule in red blood cells. The hemoglobin produced by the *HbS* allele differs from normal hemoglobin (coded by the *HbA* allele) by just one amino acid. In *HbS/HbS* homozygotes, the faulty hemoglobin forms long, fibrous chains under low oxygen conditions, causing red blood cells to assume a sickle shape (as shown in Figure 10.1). Homozygous *HbS/HbS* individuals often die of sickle cell disease before reproducing. However, in tropical and subtropical Africa, *HbS/HbA* heterozygotes make up nearly 25% of many populations.

Why is the harmful allele maintained at such high frequency in some populations? It turns out that sickle cell disease is common in regions where malaria is prevalent **(Figure 18.8)**. Malaria is a disease transmitted by mosquitoes, in which parasites infect red blood cells. When heterozygous *HbA/HbS* individuals contract malaria, their infected red blood cells assume the same sickle shape as those of homozygous *HbS/HbS* individuals. The sickled cells lose potassium, killing the parasites, which limits their spread within the infected individual. Heterozygous individuals often survive malaria because the parasites do not multiply quickly inside them, their immune systems can effectively fight the infection, and they retain a large population of uninfected red blood cells. Homozygous *HbA/HbA* individuals are also subject to malarial infection, but because their infected cells do not sickle, the parasites multiply rapidly, causing a severe infection with a high mortality rate.

Therefore, *HbA/HbS* heterozygotes have greater resistance to malaria and are more likely to survive severe infections in areas where malaria is prevalent. Thus, natural selection preserves the *HbS* allele in these populations.

Selection in Different Environments. Genetic variability can also be maintained within a population when different alleles are favoured in different places or at different times (see "People behind Biology," p. 422). For example, the shells of European garden snails range in colour from nearly white to pink, yellow, or brown, and may be patterned by one to five coloured stripes (see Figure 18.1a). This polymorphism, which is relatively stable through time, is controlled by several

a. Distribution of *HbS* allele

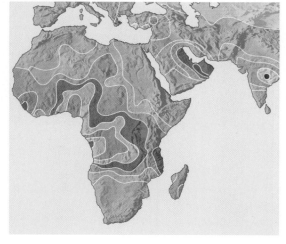

b. Distribution of malarial parasite

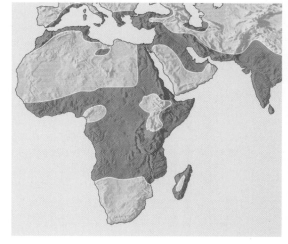

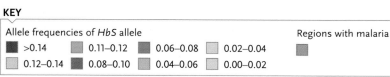

KEY

Allele frequencies of *HbS* allele

- ■ >0.14
- 0.11–0.12
- 0.06–0.08
- 0.02–0.04

- 0.12–0.14
- ■ 0.08–0.10
- 0.04–0.06
- 0.00–0.02

Regions with malaria
- ■

Figure 18.8

Heterozygote advantage. The distribution of the *HbS* allele **(a)**, which causes sickle cell disease in homozygotes, roughly matches the distribution of the malarial parasite *Plasmodium falciparum* **(b)** in southern Europe, Africa, the Middle East, and India. Gene flow among human populations has carried the *HbS* allele to some malaria-free regions.

PEOPLE BEHIND BIOLOGY 18.1
Dolf Schluter

A professor of evolutionary biology and a Canada Research Chair, Dolf Schluter is in the Department of Zoology at the University of British Columbia in Vancouver. Dolf and his students use experimental approaches to study how ecological situations affect the evolution of different wild populations. Are there genetic changes associated with populations of the same species living in different settings? Schluter and six colleagues looked at the genomes of populations of threespine sticklebacks (*Gasterosteus aculeatus*) from different locations.

Phenotypic changes in populations of sticklebacks living in fresh water differ from ancestral populations in salt water (see page 412). Schluter and his colleagues asked if large genetic effect changes (quantitative trait loci) differed among populations.

Three large-effect genes, *Ectodysplasin*, *Pitx1*, and *Kitlg*, underlie the changes in these fish between fresh and salt water. Bony lateral plate armour is controlled by *Ectodysplasin*, the presence of a pelvic girdle by *Pitx1*, and pigmentation by *Kitlg*. Using fish from four freshwater populations, Schluter and colleagues measured loci associated with these quantitative traits. The four freshwater lakes fell into two categories, one with predators (prickly sculpins, *Cottus asper*), the other without. Body armour occurred at three levels across the populations. The heaviest armour occurred in sticklebacks living in marine settings. Intermediate armour occurred in fish living in fresh water with prickly sculpins. The least armour occurred in freshwater settings without prickly sculpins **(Figure 1)**.

Using selective breeding, Schluter and his colleagues explored the genetics underlying the changes they observed. The results indicated that adaptive traits associated with new environments were evident in the genomes of wild populations. There also were obvious differences among the freshwater lakes without the sculpins **(Figure 2)**.

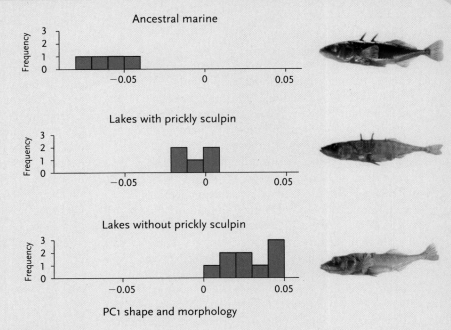

Figure 1

A comparison of the distinct phenotypic optima for threespine sticklebacks from ancestral marine populations, populations in freshwater with prickly sculpins, and those in freshwater lakes without the sculpins.

Sean M. Rogers, Patrick Tamkee, Brian Summers, Sarita Balabahadra, Melissa Marks, David M. Kingsley, and Dolph Schluter, "Genetic signature of adaptive peak shift in threespine stickleback," *Evolution*, Volume 66, Issue 8, pages 2439–2450, August 2012. John Wiley and Sons. © 2012 The Author(s). Evolution© 2012 The Society for the Study of Evolution.

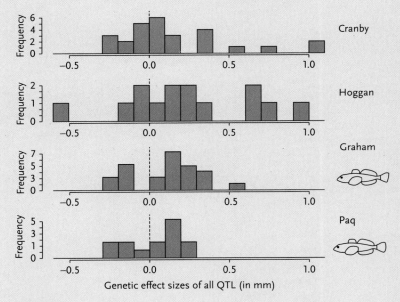

Figure 2

Variation in large-effect genetic changes in populations of threespine sticklebacks living in freshwater lakes with (Graham and Paq) and without (Cranby and Hoggan) sculpins.

Sean M. Rogers, Patrick Tamkee, Brian Summers, Sarita Balabahadra, Melissa Marks, David M. Kingsley, and Dolph Schluter, "Genetic signature of adaptive peak shift in threespine stickleback," *Evolution*, Volume 66, Issue 8, pages 2439–2450, August 2012. John Wiley and Sons. © 2012 The Author(s). Evolution© 2012 The Society for the Study of Evolution.

gene loci. The variability in colour and striping pattern can be partially explained by selection for camouflage in different habitats.

Predation by song thrushes (*Turdus ericetorum*) is a major agent of selection on the colour and pattern of these snails in England. When a thrush finds a snail, it smashes it against a rock, using it like an anvil, to break the shell. The bird eats the snail, but leaves the shell near its "anvil." Researchers collected the broken shells near an anvil and compared the phenotypes of captured snails to a random sample of the entire snail population. Their analyses indicated that thrushes are visual predators, usually capturing snails that are easy to find. Thus, well-camouflaged snails survive, and the alleles that specify their phenotypes increase in frequency.

The success of camouflage varies with habitat, however. Local subpopulations of the snail, which occupy different habitats, often differ markedly in shell colour and pattern. The predators eliminate the most conspicuous individuals in each habitat; thus, natural selection differs from place to place. In woods where the ground is covered with dead leaves, snails with unstriped pink or brown shells predominate. In hedges and fields, where the vegetation includes thin stems and grass, snails with striped yellow shells are the most common. In populations that span several habitats, selection preserves different alleles in different places, thus maintaining variability in the population as a whole.

STUDY BREAK

1. How does the diploid condition protect harmful recessive alleles from natural selection?
2. What is a balanced polymorphism?
3. Why is the allele that causes sickle cell disease very rare in human populations that are native to northern Europe?

18.5 Adaptation and Evolutionary Constraints

Although natural selection preserves alleles that confer high relative fitness on the individuals that carry them, researchers are cautious when they interpret the benefits that particular traits may provide.

18.5a The Evolution of Adaptive Traits

An **adaptive trait** is any product of natural selection that increases the relative fitness of an organism in its environment. **Adaptation** is the accumulation of adaptive traits over time, and examples range across all levels of biological organization, from the molecular to the ecological. For example, the water-retaining structures and special photosynthetic pathways of desert plants and the warning coloration of poisonous animals can be interpreted as adaptive traits. Note, however, that for adaptation to occur there must first be phenotypic variation for selection to act on.

In fact, we can concoct an adaptive explanation for almost any characteristic we observe in nature. But such explanations are just fanciful stories unless they are framed as testable hypotheses about the relative fitness of different phenotypes and genotypes. Unfortunately, evolutionary biologists cannot always conduct straightforward experiments because they sometimes study traits that do not vary much within a population or species. In such cases, they may compare variations of a trait in closely related species living in different environments. For example, one can test how the traits of desert plants are adaptive by comparing them to traits in related species from moister habitats.

When biologists try to unravel how and why a particular adaptive characteristic evolved, they must also remember that a trait they observe today may have had a different function in the past. For example, the structure of the shoulder joint in birds allows them to move their wings first upward and backward and then downward and forward during flapping flight. But analyses of the fossil record reveal that this adaptation, which is essential for flight, did not originate in birds. Some predatory nonflying dinosaurs, including the ancestors of birds, had similarly constructed shoulder joints. Researchers hypothesize that these fast-running predators may have struck at prey with a flapping motion similar to that used by modern birds. Thus, the structure of the shoulder may have evolved first as an adaptation for capturing prey, and only later proved useful for flapping flight. This hypothesis—however plausible it may be—cannot be tested by direct experimentation because the nonflying ancestors of birds have been extinct for millions of years. Instead, evolutionary biologists must use anatomical studies of birds and their ancestors, along with theoretical models about the mechanics of movement, to challenge and refine the hypothesis.

18.5b Factors Constraining Adaptive Evolution

We often marvel at how well adapted an organism is to its environment and mode of life. However, the adaptive traits of most organisms are compromises produced by competing selection pressures. Sea turtles, for example, must lay their eggs on beaches because their embryos cannot acquire oxygen under water. Although flippers allow the females to crawl to nesting sites on beaches, they are not ideally suited for terrestrial locomotion. Their structure reflects their primary function, swimming.

Moreover, no organism can be perfectly adapted to its environment because environments change over time. Natural selection preserves alleles that are successful under the prevailing environmental conditions. Thus, each generation is adapted to the environmental conditions under which its parents lived. If the environment changes from one generation to the next, adaptation will always lag behind.

Another constraint on the evolution of adaptive traits is historical. Natural selection is not an engineer that designs new organisms from scratch. Instead, it acts on new mutations and existing genetic variation. Because new mutations are fairly rare, natural selection works primarily with alleles that have been present for many generations. Thus, adaptive changes in the morphology of an organism are often based on small modifications of existing structures. The bipedal (two-footed) posture of humans, for example, evolved from the quadrupedal (four-footed) posture of our ancestors. Natural selection did not produce an entirely new skeletal design to accompany this radical behavioural shift. Instead, existing characteristics of the spinal column and the musculature of the legs and back were modified, albeit imperfectly, for an upright stance.

Evolution is the unifying theory in biology, drawing together virtually every aspect of the discipline. Microevolution focuses on variation, selection, selective advantages, and natural selection at the population level. Populations make up species, and speciation occurs when populations become diverse enough to be considered separate species. Macroevolution—evolution that occurs at or above the species level—is where we turn next.

STUDY BREAK

1. How can a biologist test whether a trait is adaptive or not?
2. Why are most organisms adapted to the environments in which their parents lived?

Review

18.1 Variation in Natural Populations

- Phenotypic traits exhibit either quantitative or qualitative variation within populations.

- Genetic variation, environmental factors, or an interaction between the two cause phenotypic variation within populations. Only genetically based phenotypic variation is heritable and subject to evolutionary change.

- Genetic variation arises within populations largely through mutation and genetic recombination. Artificial selection experiments and analyses of protein and DNA sequences reveal that most populations include significant genetic variation.

- Natural selection alters phenotypic variation in three ways. Directional selection increases or decreases the mean value of a trait, shifting it toward a phenotypic extreme. Stabilizing selection increases the frequency of the mean phenotype and reduces variability in the trait. Disruptive selection increases the frequencies of extreme phenotypes and decreases the frequency of intermediate phenotypes.

18.2 Population Genetics

- All the gene copies in a population make up its gene pool, which can be described in terms of allele frequencies and genotype frequencies.

- The Hardy–Weinberg principle of genetic equilibrium is a null model that describes the conditions under which microevolution, a change in allele frequencies through time, will not take place. Microevolution occurs in populations when the restrictive requirements of the model are not met.

18.3 Evolutionary Agents

- Several processes cause microevolution in populations (Table 18.2). Mutation introduces completely new genetic variation. Gene flow carries novel genetic variation into a population through the arrival and reproduction of immigrants. Genetic drift causes random changes in allele frequencies, especially in small populations. Natural selection occurs when the phenotypes of some individuals enable them to survive and reproduce more than others.

- Although nonrandom mating does not change allele frequencies, it can produce more homozygotic and fewer heterozygotic genotypes than the Hardy–Weinberg model predicts.

18.4 Maintaining Genetic and Phenotypic Variation

- Diploidy can maintain genetic variation in a population if recessive alleles are not expressed in heterozygotes and are thus hidden from natural selection.

- Polymorphisms are maintained in populations when heterozygotes have higher relative fitness than both homozygotes, when natural selection occurs in variable environments, or when the relative fitness of a phenotype varies with its frequency in the population.

18.5 Adaptation and Evolutionary Constraints

- Adaptive traits increase the relative fitness of individuals carrying them. Adaptive explanations of traits must be framed as testable hypotheses.

- Natural selection cannot result in perfectly adapted organisms because most adaptive traits represent compromises among conflicting needs, because most environments change constantly, and because natural selection can affect only existing genetic variation.

Questions

Self-Test Questions

1. Which of the following is an example of qualitative phenotypic variation?
   a. the lengths of people's toes
   b. the body sizes of pigeons
   c. human ABO blood types
   d. the birth weights of humans
   e. the number of leaves on oak trees

2. A population of mice is at Hardy–Weinberg equilibrium at a gene locus that controls fur colour. The locus has two alleles, M and m. A genetic analysis of one population reveals that 60% of its gametes carry the M allele. What percentage of mice contains both the M and m alleles?
   a. 60%
   b. 48%
   c. 40%
   d. 36%
   e. 16%

3. If the genotype frequencies in a population are 0.60 AA, 0.20 Aa, and 0.20 aa, and if the requirements of the Hardy–Weinberg principle apply, the genotype frequencies in the offspring generation will be
   a. 0.60 AA, 0.20 Aa, 0.20 aa
   b. 0.36 AA, 0.60 Aa, 0.04 aa
   c. 0.49 AA, 0.42 Aa, 0.09 aa
   d. 0.70 AA, 0.00 Aa, 0.30 aa
   e. 0.64 AA, 0.32 Aa, 0.04 aa

4. The reason spontaneous mutations do not have an immediate effect on allele frequencies in a large population is that
   a. mutations are random events, and mutations may be either beneficial or harmful
   b. mutations usually occur in males and have little effect on eggs
   c. many mutations exert their effects after an organism has stopped reproducing
   d. mutations are so rare that mutated alleles are greatly outnumbered by nonmutated alleles
   e. most mutations do not change the amino acid sequence of a protein

5. The phenomenon in which chance events cause unpredictable changes in allele frequencies is called
   a. gene flow
   b. genetic drift
   c. inbreeding
   d. balanced polymorphism
   e. stabilizing selection

6. An Eastern European immigrant carrying the allele for Tay–Sachs disease settled in a small village on the St. Lawrence River. Many generations later, the frequency of the allele in that village is statistically higher than it is in the immigrant's homeland. The high frequency of the allele in the village is probably an example of
   a. natural selection
   b. the concept of relative fitness
   c. Hardy–Weinberg genetic equilibrium
   d. phenotypic variation
   e. the founder effect

7. If a storm kills many small sparrows in a population, but only a few medium-sized and large ones, which type of selection is probably operating?
   a. directional selection
   b. stabilizing selection
   c. disruptive selection
   d. sexual selection
   e. artificial selection

8. Which of the following phenomena explains why the allele for sickle cell hemoglobin is common in some tropical and subtropical areas where the malaria parasite is prevalent?
   a. balanced polymorphism
   b. heterozygote advantage
   c. sexual dimorphism
   d. neutral selection
   e. stabilizing selection

9. The **neutral mutation hypothesis** proposes that
   a. complex structures in most organisms have not been fostered by natural selection
   b. most mutations have a strongly harmful effect
   c. some mutations are not affected by natural selection
   d. natural selection cannot counteract the action of gene flow
   e. large populations are subject to stronger natural selection than small populations

10. Phenotypic characteristics that increase the fitness of individuals are called
   a. mutations
   b. founder effects
   c. heterozygote advantages
   d. adaptive traits
   e. polymorphisms

Questions for Discussion

1. Many human diseases are caused by recessive alleles that are not expressed in heterozygotes. Some people think that eugenics—the selective breeding of humans to eliminate undesirable genetic traits—provides a way for us to rid our populations of such harmful alleles. Explain why eugenics cannot eliminate such genetic traits from human populations.

2. Using two types of beans to represent two alleles at the same gene locus, design an exercise to illustrate how population size affects genetic drift.

3. In what ways are the effects of sexual selection, disruptive selection, and nonrandom mating different? How are they similar?

4. Captive breeding programs for endangered species often have access to a limited supply of animals for a breeding stock. As a result, their offspring are at risk of being highly inbred. Why and how might zoos and conservation organizations avoid or minimize inbreeding?

19

szefei/Shutterstock.com

Birds of paradise. A male Count Raggi's bird of paradise (*Paradisaea raggiana*). There are 43 known birds of paradise species, 35 of them found only on the island of New Guinea.

Species and Macroevolution

WHY IT MATTERS

In 1927, nearly 100 years after Darwin boarded the *Beagle,* a young German naturalist named Ernst Mayr embarked on his own journey: to the highlands of New Guinea. He was searching for rare birds of paradise. These birds were known in Europe only through their ornate and colourful feathers, which were used to decorate women's hats. On his trek through the remote Arfak Mountains, Mayr identified 137 bird species (including many birds of paradise) based on differences in their size, plumage, colour, and other external characteristics.

To Mayr's surprise, the native Papuans—who were untrained in the ways of Western science, but who hunted these birds for food and feathers—had their own names for 136 of the 137 species he identified. The close match between the two lists confirmed Mayr's belief that the *species* is a fundamental level of organization in nature. Each species has a unique combination of genes underlying its distinctive appearance and habits. Thus, people who observe them closely—whether indigenous hunters or Western scientists—can often distinguish one species from another.

Mayr also discovered some remarkable patterns in the geographical distributions of the bird species in New Guinea. For example, each mountain range he explored was home to some species that lived nowhere else. Closely related species often lived on different mountaintops, separated by deep valleys of unsuitable habitat. In 1942, Mayr published the book *Systematics and the Origin of Species,* in which he described the role of geography in the evolution of new species; the book quickly became a cornerstone of the modern synthesis (see Section 17.4).

What mechanisms produce distinct species? As you discovered in Chapter 18, microevolutionary processes alter the pattern and extent of genetic and phenotypic variation within populations. When these processes differ between populations, the populations will diverge genetically, and they may eventually become so different that we recognize them as distinct species. These changes are considered macroevolution, evolution that occurs at or above the species level. Although Darwin's famous book was titled *On the Origin of Species,* he didn't dwell on the question of *how* new species arise. But the concept of **speciation**—the process of species formation—was implicit in his insight that similar species often share inherited characteristics and a common ancestry. Darwin also recognized that "descent with modification" had generated the amazing diversity of organisms on Earth.

Today, evolutionary biologists view speciation as a *process,* a series of events that occur through time. However, they usually study the *products* of speciation, species that are alive today. Because they can rarely witness the process of speciation from start to finish, scientists make inferences about it by studying organisms in various stages of species formation.

In this chapter, we consider four major topics: how biologists define and recognize species, how species maintain their genetic identity, how the geographical distributions of organisms influence speciation, and how different macroevolutionary genetic mechanisms produce new species.

19.1 What Is a Species?

A simple definition of a **species** (singular and plural, *species*) is a population of organisms capable of interbreeding and producing fertile offspring. The concept of species is based on our perception that Earth's biological diversity is packaged in discrete, recognizable units, and not as a continuum of forms grading into one another. As a group of organisms capable of interbreeding to produce fertile offspring, a species should be genetically distinct from other species. In reality, this definition does not work for organisms that reproduce asexually or those, such as some of Darwin's finches, that hybridize. The working definition of species depends upon the organisms to which it is applied. We should not be surprised that any one definition of *species* in biology is not uniformly used by all biologists. Organisms are the product of evolution, a dynamic process that does not easily accommodate rigid definitions. Our concepts of species appear to be most readily applicable to static situations but not to all species in all situations.

Although a species is a fundamental unit in biology, the diversity of living organisms makes it challenging to have one all-encompassing definition of the word *species*. In previous chapters we saw the importance of variation and selection to particular variants. Variation in reproductive patterns complicates the application of the species concept, as does **hybridization**, which is reproduction involving more than one species. Species are the products of evolution, an ongoing process, so we should not be surprised

at the diversity of biological species and of concepts about species.

19.1a Naming Species

The Swedish naturalist Carl von Linné (1707–1778), better known by his Latinized name, Carolus Linnaeus, was the first modern practitioner of **taxonomy**, the science that identifies, names, and classifies new species. A professor at the University of Uppsala, Linnaeus sent ill-prepared students around the world to gather specimens, losing perhaps a third of his followers to the rigours of their expeditions. Although he may not have been a commendable student adviser, Linnaeus developed the basic system of naming and classifying organisms that biologists have used for more than two centuries. The Linnaeus naming system holds so much information—just in an organism's name—that it's worthwhile to learn the rules and how they are applied.

19.1b Binomial Nomenclature

Linnaeus invented the system of **binomial nomenclature**, in which species are assigned a Latinized two-part name, a species name or **binomial**. The first part of the name identifies a **genus** (plural, *genera*), a group of species with similar characteristics. The second part is the **specific epithet**. The combination of the generic name and the specific epithet provides a unique name for every species. For example, *Ursus maritimus* is the binomial or species name for the polar bear and *Ursus arctos* is the brown bear. By convention, the first letter of a generic name is always capitalized, the species name is never capitalized, and the entire binomial is italicized (or underlined in handwritten work). In addition, the specific epithet is never used without the full or abbreviated generic name preceding it because the same species name is often given to species in different genera. For example, *Castor canadensis* is the North American beaver, *Papilio canadensis* is the Canadian tiger swallowtail butterfly, and *Cornus canadensis* is the Canadian bunchberry. The first use of an organism's binomial name is written in full, while subsequent mentions can use the first letter of the genus plus the specific epithet, for example, *U. maritimus*. Make sure everything is clear: *C. canadensis* refers equally to beaver and bunchberry.

Nonscientists often use different common names to identify a species. For example, *Bothrops asper,* a poisonous snake native to Central and South America, is called "barba amarilla" (in Spanish, meaning "yellow beard") in some places and "cola blanca" (meaning "white tail") in others. In fact, this species has about 50 local names. Adding to the confusion, the same common name is sometimes used for several different species. Binomials, however, allow people everywhere to discuss organisms unambiguously.

The naming of newly discovered species follows a formal process of publishing a description of the species in a scientific journal. International commissions meet periodically to settle disputes about scientific names.

Many binomials are descriptive of the organism or its habitat. *Asparagus horridus,* for example, is a spiny plant. Other species, such as the South American bird *Rhea darwinii,* are named for notable biologists. Some are named with a sense of humour; even Linnaeus called the praying mantis *Mantis religiosa. Wunderpus photogenicus* is a beautiful Indo-Malayan octopus, while *Gelae baen, Gelae belae, Gelae donut, Gelae fish,* and *Gelae rol* are a group of fungus beetles, all named by the same systematists. The submersibles used in deep-sea research—Alvin, Mir-1 and Mir-2, Nautil, and Shinkai-6500—have all had deep-sea species named after them.

19.1c The Taxonomic Hierarchy

Linnaeus described and named thousands of species on the basis of their morphological similarities and differences. Keeping track of so many species was no easy task, so he devised a **classification**, a conceptual filing system that arranges organisms into ever larger, more inclusive categories. Linnaeus' classification, called the **taxonomic hierarchy**, comprises a nested series of formal categories: **domain, kingdom, phylum, class, order, family**, genus, species, and subspecies **(Figure 19.1).** The organisms included within any category of the taxonomic hierarchy comprise a **taxon** (plural, *taxa*). Woodpeckers, for example, are a taxon (Picidae) at the family level, and pine trees are a taxon (*Pinus*) at the genus level.

19.2 Species Concepts

The concept of species is based on our observations that Earth's biological diversity is packaged in discrete, recognizable units, and not as a continuum of forms grading into one another. As biologists have learned more about evolutionary processes—and the dazzling biodiversity these processes have produced—they have developed about 23 complementary species concepts. We discuss the most used concepts here: morphological, biological, and phylogenetic.

19.2a The Morphological Species Concept

Biologists often describe new species on the basis of visible anatomical characteristics, a process that dates back to Linnaeus' classification system. This approach is based on

Domain *Eukarya*
Kingdom *Animalia*
Phylum *Chordata*
Class *Mammalia*

Order *Monotremata*	*Rodentia*	*Docodonta*
Family *Ornithorhynchidae*	*Castoridae*	*incertae sedis*
Genus *Ornithorhynchus*	*Castor*	*Castorocauda*
Species *Ornithorhynchus anatinus*	*Castor canadensis*	*Castorocauda lutrasimilis*

Figure 19.1

The taxonomic hierarchy. The classifications of the duck-billed platypus, the Canadian beaver, and the extinct Mesozoic beaver reflect their similarities to other species in the orders to which they belong: *Monotremata, Rodentia,* and *Docodonta,* respectively. *Incertae sedis* (Latin: of uncertain placement) indicates that we do not know in which family to place the Mesozoic beaver.

the **morphological species concept**, the idea that all individuals of a species share measurable traits that distinguish them from individuals of other species.

The morphological species concept has many practical applications. For example, paleobiologists use morphological criteria to identify the species of fossilized organisms (see Chapter 20). And because we can observe the external traits of organisms in nature, field guides to plants and animals list diagnostic (that is, distinguishing) physical characters that allow us to recognize them **(Figure 19.2).**

Nevertheless, relying exclusively on morphology to identify species can present problems. Consider the variation in the shells of the European garden snail (*Cepaea nemoralis,* Figure 18.1a). How could anyone imagine that such a variety of shells represents just one species of snail? Conversely, morphology does not help us distinguish some closely related species that are

Yellow-throated warbler
(Dendroica dominica)

Myrtle warbler
(Dendroica coronata)

Figure 19.2

A guide book shows the diagnostic characters of yellow-throated warblers and myrtle warblers, which can be distinguished by the colour of feathers on the throat and rump.

nearly identical in appearance (like some mice). Finally, morphological species definitions tell us little about the evolutionary processes that produce new species.

19.2b The Biological Species Concept

The **biological species concept** emphasizes the dynamic nature of species. Ernst Mayr defined biological species as "groups of ... interbreeding natural populations that are reproductively isolated from [do not produce fertile offspring with] other such groups." The concept is based on reproductive criteria and is easy to apply, at least in principle: if the members of two populations interbreed and produce fertile offspring *under natural conditions,* they belong to the same species; their fertile offspring will, in turn, produce the next generation of that species. If two populations do not interbreed in nature, or fail to produce fertile offspring when they do, they belong to different species.

The biological species concept defines species in terms of population genetics and evolutionary theory. The first half of Mayr's definition notes the genetic *cohesiveness* of species: populations of the same species experience gene flow, which mixes their genetic material. Thus, we can think of a species as one large gene pool, which may be subdivided into local populations.

The second part of the biological species concept emphasizes the genetic *distinctness* of each species. Because populations of different species are reproductively isolated, they cannot exchange genetic information. In fact, the process of speciation is frequently defined as the evolution of reproductive isolation between populations.

The biological species concept also explains why individuals of a species generally look alike: members of the same gene pool share genetic traits that determine their appearance. Individuals of different species generally do not resemble one another as closely because they share fewer genetic characteristics. In practice, biologists often still use similarities or differences in morphological traits as convenient markers of genetic similarity or reproductive isolation.

However, the biological species concept does not apply to the many forms of life that reproduce asexually, including most bacteria; some protists, fungi, and plants; and a few animals. In these species, individuals don't interbreed, so it is pointless to ask whether different populations do. Similarly, we cannot use the biological species concept to study extinct organisms, because we have little or no data on their specific reproductive habits. These species must all be defined using morphological or biochemical criteria. Yet, despite its limitations, the biological species concept currently provides one of the best evolutionary definitions of a sexually reproducing species.

The diversity of species and their lifestyles partly reflects the mechanisms underlying the processes of speciation. The biological species concept defines species in terms of population genetics and evolutionary theory in a static world. The definition alludes to the genetic cohesiveness of species. Populations of the same species are said to experience gene flow that mixes their genetic material and could be the "glue" holding a species together. The second part of this concept emphasizes the genetic distinctness of each species. Because populations of different species are reproductively isolated, they cannot exchange genetic information.

The biological species concept could explain why individuals of a species generally look alike. If phenotype reflects genotype, members of the same gene pool should share genetic traits (genotype) that determine phenotype. Individuals of different species generally do not resemble one another as closely because they share fewer genetic characteristics. In practice, biologists often use similarities or differences in morphological traits as convenient markers of genetic similarity or reproductive isolation. Recently, scientists have used a short DNA sequence in a particular gene to analyze relatedness (see "People behind Biology," p. 430).

19.2c The Phylogenetic Species Concept

Recognizing the limitations of the biological species concept, biologists have developed dozens of other ways to define a species. A widely accepted alternative is the **phylogenetic species concept.** Using both morphological and genetic sequence data, scientists first construct an evolutionary tree for the organisms of interest. They then define a phylogenetic species as a cluster of populations—the tiniest twigs on this part of the tree of life—that emerge from the same small branch. Thus, a phylogenetic species comprises populations that share a recent evolutionary history. We will consider this approach for understanding the evolutionary relationships of organisms in Chapter 20.

One advantage of the phylogenetic species concept is that biologists can apply it to any group of organisms, including species that have long been extinct, as well as living organisms that reproduce asexually. Proponents of this approach also argue that the morphological and genetic distinctions between organisms on different branches of the tree of life reflect the absence of gene flow between them—one of the key requirements of the biological species definition. Nevertheless, because detailed evolutionary histories have been described for relatively few groups of organisms, biologists are not yet able to apply the phylogenetic species concept to all forms of life. Continued research on the details of evolutionary relationships will increase this concept's usefulness in the future.

Barcode of Life Project

Today, most systematists working on living organisms include molecular characters as part of the data set when determining evolutionary relationships. Molecular data include nucleotide base sequences of DNA and RNA or the amino acid sequences of the proteins for which they code. Technological advances have automated many of the necessary laboratory techniques, and analytical software makes it easy to compare new data with information filed in data banks, for example, the Barcode of Life project.

In 2003, Paul Hebert, of the Biodiversity Institute of Ontario at the University of Guelph, proposed a new system of species identification using a sequence of approximately 600 DNA base pairs of the mitochondrial gene cytochrome oxidase I. He anticipated a database that would "provide a new master key for identifying species, one whose power will rise with increased taxon coverage and with faster, cheaper sequencing." He compared the system to the barcode system for retail products such as grocery items, calling it the Barcode of Life **(Figure 1).**

Molecular sequences have some practical advantages over organismal characters. First, they provide abundant data because every base in a nucleic acid can serve as a separate, independent character for analysis. Moreover, because many genes have been conserved by evolution, molecular sequences can be compared between distantly related organisms that share

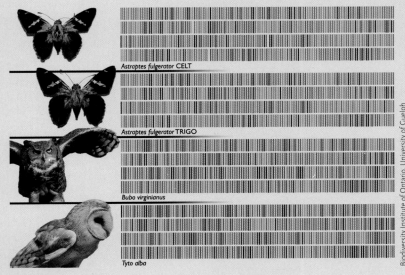

Astraptes fulgerator CELT

Astraptes fulgerator TRIGO

Bubo virginianus

Tyto alba

Figure 1

no organismal characters. Molecular characters can also be used to study closely related species that have only minor morphological differences. Finally, many nucleic acids are not directly affected by the developmental or environmental factors that cause nongenetic morphological variations (see Chapter 18).

But there are drawbacks to using molecular characters in taxonomy. There are only four alternative character states (the four nucleotide bases) at each position in a DNA or RNA sequence (see Chapter 11). If two species have the same nucleotide base substitution at a given position in a DNA segment, their similarity may have evolved independently. As a

result, systematists often find it difficult to verify that molecular similarities were inherited from a common ancestor.

For organismal characters, biologists can establish that similarities are homologous by analyzing the characters' embryonic development or details of their function. But molecular characters have no embryonic development, and biologists still do not understand the functional significance of most molecular differences. Despite these disadvantages, molecular characters represent the genome directly, and researchers use them with great success in phylogenetic analyses and in embryology.

STUDY BREAK

1. Why do scientists need a nomenclature system for organisms?
2. Describe the Linnean binomial system.
3. Define the morphological, biological, and phylogenetic species concepts. Compare and contrast these species concepts.

19.3 Maintaining Reproductive Isolation

Reproductive isolation is central to the biological species concept. A **reproductive isolating mechanism** is any biological characteristic that prevents the gene pools of two species from mixing. Biologists classify reproductive isolating mechanisms into two categories:

Biodiversity Institute of Ontario, University of Guelph

prezygotic isolating mechanisms, which exert their effects before the production of a zygote (fertilized egg), and **postzygotic isolating mechanisms**, which operate after zygote formation **(Table 19.1)**. These isolating mechanisms are not mutually exclusive; two or more of them may operate simultaneously. These mechanisms are considered as macroevolution.

19.3a Prezygotic Isolating Mechanisms

Biologists have identified five mechanisms that can prevent interspecific (between species) matings or fertilizations and thus prevent the production of hybrid offspring. These five prezygotic mechanisms are *ecological, temporal, behavioural, mechanical,* and *gametic isolation.*

Species living in the same geographical region may experience **ecological isolation** if they live in different habitats. For example, lions and tigers were both common in India until the mid-nineteenth century, when hunters virtually exterminated the Asian lions. However, because lions live in open grasslands and tigers in dense forests, the two species did not encounter one another and did not interbreed. Lion–tiger hybrids are sometimes born in captivity, but they do not occur under natural conditions.

Species living in the same habitat can experience **temporal isolation** if they mate at different times of day or different times of year. For example, the fruit flies *Drosophila persimilis* and *Drosophila pseudoobscura* overlap extensively in their geographical distributions, but they do not interbreed, in part because *D. persimilis* mates in the morning and *D. pseudoobscura* in the afternoon. Similarly, two species of pine in California are reproductively isolated where their geographical distributions overlap: even though both rely on the wind to carry male gametes (within pollen grains) to female gametes (ova) in other cones, *Pinus radiata* releases pollen in February and *Pinus muricata* releases pollen in April.

Many animals rely on specific signals, which may differ dramatically between species, to identify the species of a potential mate. **Behavioural isolation** results when the signals used by one species are not recognized by another. For example, female birds rely on the song, colour, and displays of males to identify members of their own species. Similarly, female fireflies identify males by their flashing patterns **(Figure 19.3)**. These behaviours (collectively called **courtship displays**) are often so complicated that signals sent by one species are like a foreign language that another species simply does not understand.

Mate choice by females and sexual selection (see Section 17.5c) generally drive the evolution of mate recognition signals. Females often spend substantial energy in reproduction, and choosing an appropriate mate—that is, a male of her own species—is critically important for the production

Table 19.1	Isolating Mechanisms	
Timing Relative to Fertilization	Mechanism	Mode of Action
Prezygotic ("premating") mechanisms	Ecological isolation	Species live in different habitats
	Temporal isolation	Species breed at different times
	Behavioural isolation	Species cannot communicate
	Mechanical isolation	Species cannot physically mate
	Gametic isolation	Species have nonmatching receptors on gametes
Postzygotic ("postmating") mechanisms	Hybrid inviability	Hybrid offspring do not complete development
	Hybrid sterility	Hybrid offspring cannot produce gametes
	Hybrid breakdown	Hybrid offspring have reduced survival or fertility

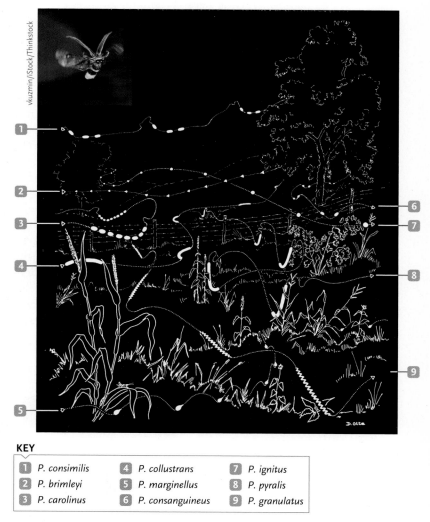

vkuzmin/iStock/Thinkstock

KEY

1 *P. consimilis*	4 *P. collustrans*	7 *P. ignitus*
2 *P. brimleyi*	5 *P. marginellus*	8 *P. pyralis*
3 *P. carolinus*	6 *P. consanguineus*	9 *P. granulatus*

Figure 19.3

Behavioural reproductive isolation. Male fireflies use bioluminescent signals to attract potential mates. The different flight paths and flashing patterns of males in nine North American *Photinus* species are represented here. Females respond only to the display given by males of their own species. The inset photo shows *P. pyralis*.

Illustration courtesy of James E. Lloyd. Miscellaneous Publications of the Museum of Zoology of the University of Michigan, 130:1–195, 1966.

of successful young. By contrast, a female that mates with a male from a different species is unlikely to leave any surviving offspring at all. Over time, the number of males with recognizable traits, as well as the number of females able to recognize the traits, increases in the population.

Differences in the structure of reproductive organs or other body parts—**mechanical isolation**—may prevent individuals of different species from interbreeding. In particular, many plants have anatomical features that allow only certain pollinators, usually particular bird or insect species, to collect and distribute pollen (see Chapter 36). For example, the flowers and nectar of two native California plants, the purple monkey-flower (*Mimulus lewisii*) and the scarlet monkey-flower (*Mimulus cardinalis*), attract different animal pollinators **(Figure 19.4)**. *Mimulus lewisii* is pollinated by bumblebees. The broad petals of its shallow purple flowers provide a landing platform for the bees. Bright yellow streaks on the petals serve as "nectar guides," directing bumblebees to the short nectar tube and reproductive parts, which are located among the petals. Bees enter the flowers to drink their concentrated nectar, and they pick up and deliver pollen as their legs and bodies brush against the reproductive parts of the flowers. *Mimulus cardinalis,* by contrast, is pollinated by hummingbirds. It has long red flowers with no yellow streaks, and the reproductive parts extend above the petals. The red colour attracts hummingbirds but lies outside the colour range detected by bumblebees. The nectar of *M. cardinalis* is more dilute than that of *M. lewisii* but is produced in much greater quantity, making it easier for hummingbirds to ingest. When a hummingbird visits *M. cardinalis* flowers, it pushes its long bill down the nectar tube, and its forehead touches the reproductive parts, picking up and delivering pollen. Recent research has demonstrated that where the two monkey-flower species grow side by side, animal pollinators restrict their visits to either one species or

the other 98% of the time, providing nearly complete reproductive isolation.

Even when individuals of different species mate, **gametic isolation**, an incompatibility between the sperm of one species and the eggs of another, may prevent fertilization. Many marine invertebrates release gametes into the environment for external fertilization. The sperm and eggs of each species recognize one another's complementary surface proteins (see Chapter 42), but the surface proteins on the gametes of different species don't match. In animals with internal fertilization, sperm of one species may not survive within the reproductive tract of another. Interspecific matings between some *Drosophila* species, for example, induce a reaction in the female's reproductive tract that blocks "foreign" sperm from reaching eggs. Parallel physiological incompatibilities between a pollen tube and a stigma prevent interspecific fertilization in some plants.

19.3b Postzygotic Isolating Mechanisms

If prezygotic isolating mechanisms between two closely related species are incomplete or ineffective, sperm from one species may fertilize an egg of the other species. In such cases the two species will be reproductively isolated if their offspring, called hybrids, have lower fitness than those produced by intraspecific matings. Three postzygotic isolating mechanisms—*hybrid inviability, hybrid sterility,* and *hybrid breakdown*—can reduce the fitness of hybrid individuals.

Many genes govern the complex processes that transform a zygote into a mature organism. Hybrid individuals have two sets of developmental instructions, one from each parent species, which may not interact properly for the successful completion of embryonic development. As a result, hybrid organisms frequently die as embryos or at an early age, a phenomenon called **hybrid inviability**. For example, domestic sheep and goats can mate and fertilize one another's ova, but the hybrid embryos always die before coming to term, presumably because the developmental programs of the two parent species are incompatible.

Although some hybrids between closely related species develop into healthy and vigorous adults, they may not produce functional gametes. This **hybrid sterility** often results when the parent species differ in the number or structure of their chromosomes, which cannot pair properly during meiosis. Such hybrids have zero fitness because they leave no descendants. The most familiar example is a mule, the product of mating between a female horse ($2n = 64$) and a male donkey ($2n = 62$). Zebroids, the offspring of matings between horses and zebras, are also usually sterile **(Figure 19.5)**.

Purple monkey-flower
(Mimulus lewisii)

Scarlet monkey-flower
(Mimulus cardinalis)

TIM LAMAN/National Geographic Creative

South12th Photography/Shutterstock.com

Figure 19.4
Mechanical reproductive isolation. Because of differences in floral structure, two species of monkey-flower attract different animal pollinators. *Mimulus lewisii* attracts bumblebees, and *Mimulus cardinalis* attracts hummingbirds.

Some first-generation hybrids (F₁; see Section 10.1) are healthy and fully fertile. They can breed with other hybrids and with both parental species. However, the second generation (F₂), produced by matings between F₁ hybrids, or between F₁ hybrids and either parental species, may exhibit reduced survival or fertility, a phenomenon known as **hybrid breakdown.** For example, experimental crosses between *Drosophila* species may produce functional hybrids, but their offspring experience a high rate of chromosomal abnormalities and harmful types of genetic recombination. Thus, reproductive isolation is maintained between the species because there is little long-term mixing of their gene pools. In cases where hybrids have lower fitness, there is strong selection for mechanisms that promote assertive mating or prevent hybridization prior to mating or fertilization.

Figure 19.5

Interspecific hybrids. Horses and zebroids (hybrid offspring of horses and zebras) run in a mixed herd. Zebroids are usually sterile.

STUDY BREAK

1. What is the difference between prezygotic and postzygotic isolating mechanisms?
2. When a male duck of one species performed a courtship display to a female of another species, she interpreted his behaviour as aggressive rather than amorous. What type of reproductive isolating mechanism does this scenario illustrate?

19.4 The Geography of Speciation

Just as individuals within populations exhibit genotypic and phenotypic variation (see Figure 18.1), populations within species also differ both genetically and phenotypically. Neighbouring populations often have shared characteristics because they live in similar environments, exchange individuals, and experience comparable patterns of natural selection. Widely separated populations, by contrast, may live under different conditions and experience different patterns of selection. Because gene flow is less likely to occur between distant populations, their gene pools and phenotypes often differ.

When geographically separated populations of a species exhibit dramatic, easily recognized phenotypic variation, biologists may identify them as different subspecies **(Figure 19.6). Subspecies** are local variants of a species. Individuals from different subspecies usually interbreed where their geographical distributions meet, and their offspring often exhibit intermediate phenotypes. Various patterns of geographical variation—as well as analyses of how the variation may relate to climatic or habitat variation—have provided great insight into the speciation process. Two of the best-studied patterns are *ring species* and *clinal variation.*

19.4a Ring Species

Some plant and animal species have a ring-shaped geographical distribution that surrounds uninhabitable terrain. Adjacent populations of these **ring species** can exchange genetic material directly, but gene flow between distant populations occurs only through the intermediary populations.

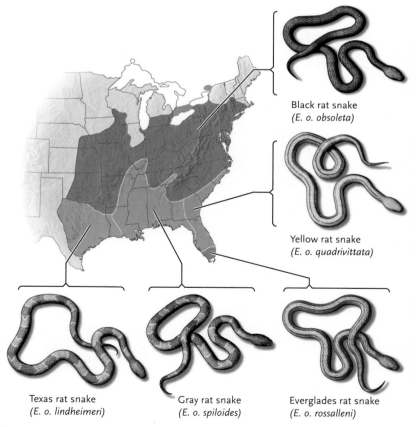

Black rat snake
(*E. o. obsoleta*)

Yellow rat snake
(*E. o. quadrivittata*)

Texas rat snake
(*E. o. lindheimeri*)

Gray rat snake
(*E. o. spiloides*)

Everglades rat snake
(*E. o. rossalleni*)

Figure 19.6

Subspecies. Five subspecies of rat snake (*Elaphe obsoleta*) in eastern North America differ in colour and in the presence or absence of stripes or blotches.

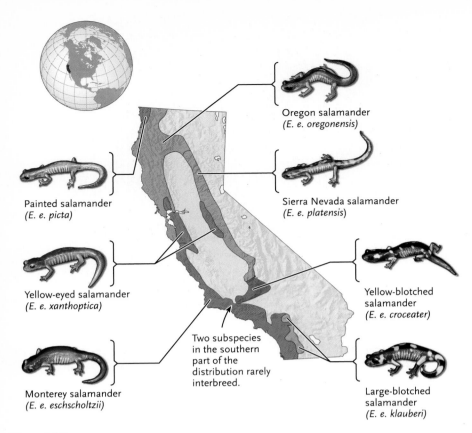

Figure 19.7
Ring species. Six of the seven subspecies of the salamander *Ensatina eschscholtzii* are distributed in a ring around California's Central Valley. Subspecies often interbreed where their geographical distributions overlap. However, the two subspecies that nearly close the ring in the south (marked with an arrow), the Monterey salamander and the yellow-blotched salamander, rarely interbreed.

Oregon salamander
(*E. e. oregonensis*)

Painted salamander
(*E. e. picta*)

Sierra Nevada salamander
(*E. e. platensis*)

Yellow-eyed salamander
(*E. e. xanthoptica*)

Yellow-blotched salamander
(*E. e. croceater*)

Two subspecies in the southern part of the distribution rarely interbreed.

Monterey salamander
(*E. e. eschscholtzii*)

Large-blotched salamander
(*E. e. klauberi*)

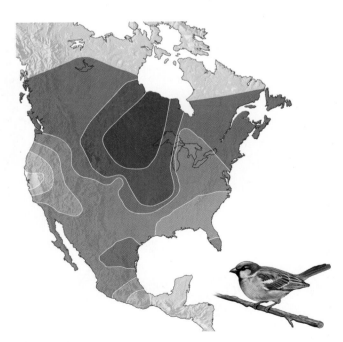

Figure 19.8
Clinal variation. House sparrows (*Passer domesticus*) exhibit clinal variation in overall body size, which was summarized from measurements of 16 skeletal features. Darker shading in the map indicates larger size.

The lungless salamander *Ensatina eschscholtzii,* an example of a ring species, is widely distributed in the coastal mountains and the Sierra Nevada of California, but it cannot survive in the hot, dry Central Valley **(Figure 19.7).** Seven subspecies differ in biochemical traits, colour, size, and ecology. Individuals from adjacent subspecies often interbreed where their geographical distributions overlap, and intermediate phenotypes are fairly common. But at the southern end of the Central Valley, adjacent subspecies rarely interbreed. Apparently, they have differentiated to such an extent that they can no longer exchange genetic material directly.

Are the southernmost populations of this salamander subspecies or different species? A biologist who saw *only* the southern populations, which coexist without interbreeding, might define them as separate species. However, they can still exchange genetic material through the intervening populations that form the ring. Hence, biologists recognize these populations as belonging to the same species. Most likely, the southern subspecies are in an intermediate stage of species formation.

Another example is the Greenish warblers (*Phylloscopus trochiloides*) from Asia, which live surrounding the Tibetan Plateau. In these birds, courtship songs function in species recognition. If the birds do not respond to playbacks of courtship songs, they do not recognize a potential mate or competitor. Responses to playback presentations indicated that the birds recognized the songs as signals from their own species.

19.4b Clinal Variation

When a species is distributed over a large, environmentally diverse area, some traits may exhibit a **cline**, a pattern of smooth variation along a geographical gradient. For example, many birds and mammals in the northern hemisphere show clinal variation in body size **(Figure 19.8)** and the relative length of their appendages. In general, populations living in colder environments have larger bodies and shorter appendages, a pattern that is usually interpreted as a mechanism to conserve body heat (see Chapter 50).

Clinal variation usually results from gene flow between adjacent populations that are each adapting to slightly different conditions. However, if populations at opposite ends of a cline are separated by great distances, they may exchange very little genetic material through reproduction. Thus, when a cline extends over a large geographical gradient, distant populations may be genetically and morphologically distinct.

Despite the geographical variation that many species exhibit, even closely related species are genetically and morphologically different from each other. In the next section, we consider the mechanisms that maintain the genetic distinctness of closely related species by preventing their gene pools from mixing.

19.4c Allopatric Speciation

Biologists define three modes of speciation, based on the geographical relationship of populations as they become reproductively isolated: *allopatric speciation* (*allo* = different, *patria* = homeland), *parapatric speciation* (*para* = beside), and *sympatric speciation* (*sym* = together). **Allopatric speciation** may take place when a physical barrier subdivides a large population or when a small population becomes separated from a species' main geographical distribution. Allopatric speciation occurs in two stages. First, two populations become *geographically* separated, preventing gene flow between them. Then, as the populations experience distinct mutations as well as different patterns of natural selection and genetic drift, they may accumulate genetic differences that isolate them *reproductively*. Allopatric speciation is probably the most common mode of speciation in large animals.

Geographical separation sometimes occurs when a barrier divides a large population into two or more units **(Figure 19.9)**. For example, hurricanes may create new channels that divide low coastal islands and the populations inhabiting them. Uplifting mountains or landmasses as well as rivers or advancing glaciers can also produce barriers that subdivide populations. The uplift of the Isthmus of Panama, caused by movements of Earth's crust about 5 million years ago, separated a once-continuous shallow sea into the eastern tropical Pacific Ocean and the western tropical Atlantic Ocean. Populations of marine organisms were subdivided by this event. In the tropical Atlantic Ocean, populations experienced patterns of mutation, natural selection, and genetic drift that were different from those experienced by populations in the tropical Pacific Ocean. As a result, the populations diverged genetically, and pairs of closely related species now live on either side of this divide **(Figure 19.10)**.

In other cases, small populations may become isolated at the edge of a species' geographical distribution. Such peripheral populations often differ genetically from the central population because they are adapted to somewhat different environments. Once a small population is isolated, founder effects and small population size may promote genetic drift, and natural selection may favour the evolution of distinctive traits. If the isolated population experiences

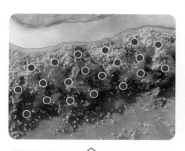  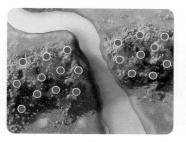 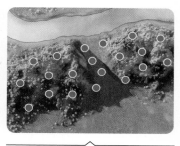

1 At first, a population is distributed over a large geographical area. A river flows along one edge of the population's geographical range.

2 A geographical change, such as a change in the river's course, separates the original population, creating a barrier to gene flow.

3 In the absence of gene flow, the separated populations evolve independently and diverge into different species.

4 When the river later changes course again, allowing individuals of the two species to come into secondary contact, they do not interbreed.

Figure 19.9
The model of allopatric speciation and secondary contact.

Isthmus of Panama

Cortez rainbow wrasse *(Thalassoma lucasanum)*

Blue-headed wrasse *(Thalassoma bifasciatum)*

Javontaevious. This work is licensed under the Creative Commons Attribution-ShareAlike 3.0 License, http://creativecommons.org/licenses/by-sa/3.0/

NASA

Fred McConnaughey/Science Source

Figure 19.10
Geographical separation. The uplift of the Isthmus of Panama divided an ancestral wrasse population. The Cortez rainbow wrasse now occupies the eastern Pacific Ocean, and the blue-headed wrasse now occupies the western Atlantic Ocean.

limited gene flow from the parent population, these agents of evolution will foster genetic differentiation between them. In time, the accumulated genetic differences may lead to reproductive isolation.

Populations established by colonization of oceanic islands represent extreme examples of this phenomenon. The founder effect makes the populations genetically distinct. And on oceanic archipelagos, such as the Galápagos and Hawaiian islands, individuals from one island may colonize nearby islands, founding populations that differentiate into distinct species. Each island may experience multiple invasions, and the process may be repeated many times within the archipelago, leading to the evolution of a **species cluster**, a group of closely related species recently descended from a common ancestor **(Figure 19.11)**. Sometimes a species cluster can evolve relatively quickly; for example, the nearly 800 species of fruit flies now living on the Hawaiian Islands (see "Speciation in Hawaiian Fruit Flies") evolved in less than 5 million years.

Sometimes, allopatric populations reestablish contact when a geographical barrier is eliminated or breached (see Figure 19.9, step 4). This *secondary contact* provides a test of whether or not the populations have diverged into separate species. If their gene pools did not differentiate much during geographical separation, the populations will interbreed and merge. But if the populations have differentiated enough to be reproductively isolated, they have become separate species.

During the early stages of secondary contact, prezygotic reproductive isolation may be incomplete. Some members of each population may mate with individuals from the other, producing viable, fertile offspring in areas called **hybrid zones**. Although some hybrid zones have persisted for hundreds or thousands of years **(Figure 19.12, p. 438)**, they are generally narrow, and ecological or geographical factors maintain the separation of the gene pools for the majority of individuals in both species.

If postzygotic isolating mechanisms cause hybrid offspring to have lower fitness than those produced within each population, natural selection will promote the evolution of prezygotic isolating mechanisms, favouring individuals that mate only with members of their own population. Recent studies of *Drosophila* suggest that this phenomenon, called **reinforcement**, enhances reproductive isolation that had begun to develop while the populations were geographically separated.

19.4d Parapatric Speciation

Sometimes a single species is distributed across a discontinuity in environmental conditions, such as a major change in soil type. Although organisms on one side of the discontinuity may interbreed freely with

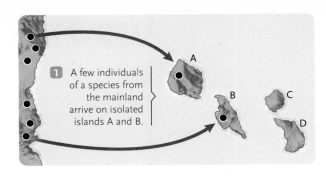
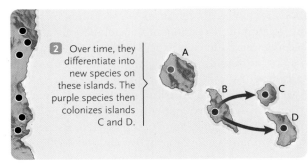
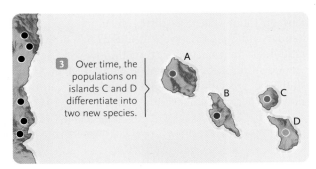
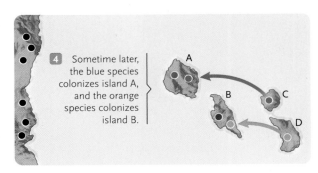

Figure 19.11

Evolution of a species cluster on an archipelago. Letters identify four islands in a hypothetical archipelago, and coloured dots represent different species. The ancestor of all the species is represented by black dots on the mainland. At the end of the process, islands A and B are each occupied by two species, and islands C and D are each occupied by one species, all of which evolved on the islands.

those on the other side, natural selection may favour different alleles on either side, limiting gene flow. In such cases, **parapatric speciation**—speciation arising between adjacent populations—may occur if hybrid offspring have low relative fitness.

Some strains of bent grass (Agrostis tenuis), a common pasture plant in the United Kingdom, have the

Speciation in Hawaiian Fruit Flies

The islands of the Hawaiian archipelago have been geographically isolated throughout their history, lying at least 3200 km from the nearest continents or other islands (Figure 1). Built by undersea volcanic eruptions over millions of years, they emerged from northwest to southeast: Kauai is at least 5 million years old, and Hawaii, the "Big Island," is less than 1 million years old. Individual islands differ in maximum elevation and include diverse habitats, from sparse, dry vegetation to lush, wet forests.

Resident species must have arrived from distant mainland localities or evolved on the islands from colonizing ancestors. The islands' isolation, different ages, and geographical and ecological complexity allowed repeated interisland colonizations followed by allopatric speciation events. Thus, it is not surprising that species clusters have evolved in several groups of organisms (including flowering plants, insects, and birds).

Nearly 800 species of Hawaiian fruit flies have been discovered, most of which live on only one island. Biologists used many characters to identify the different species, including external and internal anatomy, cell structure, chromosome structure, ecology, and mating behaviour. Their data suggest that the vast majority of native Hawaiian species arose from a single ancestral species that colonized the archipelago long ago, probably from eastern Asia. The fruit flies of the Hawaiian Islands now represent more than 25% of all known fruit fly species.

Hampton Carson, of the University of Hawaii, spearheaded studies on the evolutionary relationships of Hawaiian fruit flies. He and his colleagues gathered data on hundreds of fly species—a daunting task. Most species of fruit flies are sexually dimorphic. The females of different species may be similar in appearance, but the males of even closely related species differ in virtually every aspect of their external anatomy: body size; head shape; and the structure of their eyes, antennae, mouthparts, bristles, legs, and wings. Their mating behaviour and choice of mating sites also vary dramatically.

Nevertheless, closely related species on different islands occupy comparable habitats and associate with related plant species. Carson suggested that speciation in these flies resulted from the evolution of different genetically determined *mating systems*, the behaviours and morphological characteristics that males display when seeking a mate. The mating systems serve as prezygotic isolating mechanisms.

The 100 or more species of "picture-wing" *Drosophila*, relatively large flies with patterns on their wings, illustrate the evolution of a species cluster. Carson and his colleagues used similarities and differences in the banding patterns on the giant chromosomes in the flies' salivary gland cells to trace the evolutionary origin of species on the younger islands by identifying their closest relatives on the older islands. Their analysis of 26 species on Hawaii, the youngest island, suggested that flies from the older islands colonized Hawaii at least 19 different times, and each founder population evolved into a new species there. Additional species apparently evolved when lava flows on Hawaii subdivided existing populations.

Among the picture-wing fruit flies, some interspecies matings result in hybrid sterility or hybrid breakdown. But for most species, prezygotic reproductive isolation is maintained by differences in their mating systems. For example, *Drosophila silvestris* and *D. heteroneura*, which produce healthy and fertile hybrids in the laboratory, have similar geographical distributions; however, differences in courtship behaviour and in the shape of the males' heads, a characteristic that females use to recognize males of their own species (Figure 2), keep these two species reproductively isolated. In nature, they hybridize only in one small geographical area.

The work of Carson and his colleagues suggests that most speciation in Hawaiian *Drosophila* has

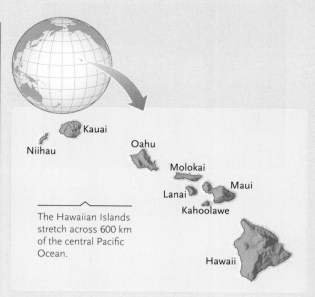

Figure 1
Geographic isolation of the Hawaiian Islands.

resulted from founder effects and genetic drift. When a fertile female—or a small group of males and females—moves to a new island, this founding population responds to novel selection pressures in its new environment. Sexual selection then exaggerates distinctive morphological and behavioural characteristics, maintaining the population's reproductive isolation from its new neighbours. The tremendous variety of Hawaiian fruit flies has undoubtedly been produced by repeated colonizations of newer islands by flies from older islands and by the back-colonization of older islands by newly evolved species. Thus, they represent what evolutionary biologists describe as an *adaptive radiation,* a cluster of closely related species that are ecologically different.

Drosophila heteroneura   *Drosophila silvestris*

Figure 2
Two Drosophila *species in which the males' head shapes differ.*

Bullock's oriole (*Icterus bullockii*)

Baltimore oriole (*Icterus galbula*)

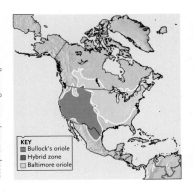

KEY
■ Bullock's oriole
■ Hybrid zone
□ Baltimore oriole

Figure 19.12

Hybrid zones. Males of the Bullock's oriole and Baltimore oriole differ in colour and courtship song. Populations of these species have maintained a hybrid zone for hundreds of years, studied by Dr. James Rising of the University of Toronto. The two oriole species now hybridize less frequently than they once did, leading some researchers to suggest that their reproductive isolation evolved recently.

physiological ability to grow on mine tailings, where the soil is heavily polluted by copper or other metals. Plants of the copper-tolerant strains grow well on polluted soils, but plants of the pasture strain do not. Conversely, copper-tolerant plants don't survive as well as pasture plants on unpolluted soils. These strains often grow within a few metres of each other where polluted and unpolluted soils form an intricate environmental mosaic. Because bent grass is wind pollinated, pollen is readily transferred from one strain to another. Laboratory tests have shown that the strains are fully interfertile. However, copper-tolerant plants flower about one week earlier than the pasture plants, which promotes prezygotic (temporal) isolation of the two strains. If the flowering times become further separated, the two strains may attain complete reproductive isolation and become separate species.

Some biologists argue that the places where parapatric populations of bent grass interbreed are really hybrid zones where previously allopatric populations have established secondary contact. Unfortunately, there is no way to determine whether the hybridizing populations were parapatric or allopatric in the past. Thus, a thorough evaluation of the parapatric speciation hypothesis must await the development of techniques that enable biologists to distinguish clearly between the products of allopatric and parapatric speciation.

19.4e Sympatric Speciation

In **sympatric speciation**, reproductive isolation evolves between distinct subgroups that arise within one population. Models of sympatric speciation do not require that the populations be either geographically or environmentally separated as their gene pools diverge. We examine below general models of sympatric speciation in animals and plants; the genetic basis of sympatric speciation is one of the topics we consider in the next section.

Insects that feed on just one or two plant species are among the animals most likely to evolve by sympatric speciation. These insects generally carry out most important life cycle activities on or near their "host" plants. Adults mate on the host plant; females lay their eggs on it; and larvae feed on the host plant's tissues, eventually developing into adults, which initiate another round of the life cycle. Host-plant choice is genetically determined in many insect species. In others, individuals associate with the host-plant species they ate as larvae.

Theoretically, a genetic mutation could suddenly change some insects' choice of host plant. Mutant individuals would shift their life cycle activities to the new host and then interact primarily with others preferring the same new host, an example of ecological isolation. These individuals would collectively form a separate subpopulation, called a **host race.** Reproductive isolation could evolve between different host races if the individuals of each host race are more likely to mate with members of their own host race than with members of another. Some biologists criticize this model, however, because it assumes that the genes controlling two traits, the insects' host-plant choice and their mating preferences, change simultaneously. Moreover, host-plant choice is controlled by multiple gene loci in some insect species, and it is clearly influenced by prior experience in others.

The apple maggot (*Rhagoletis pomonella*) is one of the most thoroughly studied examples of possible sympatric speciation in animals **(Figure 19.13).** This fly's natural host plant in eastern North America is the hawthorn (*Crataegus* sp.), but at least two host races have appeared in the past 150 years. The larvae of a new host race were first discovered feeding on apples in New York State in the 1860s. In the 1960s, a cherry-feeding host race appeared in Wisconsin. This is also an example of disruptive selection (see Section 18.1c).

Genetic analyses have shown that variations at just a few gene loci underlie differences in the feeding preferences of *Rhagoletis* host races; other genetic differences cause the host races to develop at different rates. Moreover, adults of the three races mate during

Figure 19.13

Sympatric speciation in animals. Male and female apple maggots (*Rhagoletis pomonella*) court on a hawthorn leaf. The female will later lay her eggs on the fruit, and the offspring will feed, mate, and lay their eggs on hawthorns as well.

different summer months. Nevertheless, individuals show no particular preference for mates of their own host race, at least under simplified laboratory conditions. Thus, although behavioural isolation has not developed between races, ecological and temporal isolation may separate adults in nature. Researchers are still not certain that the different host races are reproductively isolated under natural conditions.

In 2010, Andrew P. Michel and colleagues in the United States and Germany published a genomic analysis of the apple- and hawthorn-feeding races of *Rhagoletis*. Their results suggest that over the past 150 years, the two races have diverged at many loci in their genomes—not just at the loci that influence food choice and developmental rate—and that the divergence has largely been driven by natural selection. Ongoing genetic divergence may prevent them from interbreeding in the future.

Sympatric speciation often occurs in plants through a genetic phenomenon, **polyploidy**, in which an individual has one or more *extra* copies of the entire haploid complement of chromosomes (see Section 11.3). Polyploidy can lead to speciation because these large-scale genetic changes may prevent polyploid individuals from breeding with individuals of the parent species. Nearly half of all flowering plant species are polyploid, including many important crops and ornamental species. The genetic mechanisms that produce polyploid individuals in plant populations are well understood; we describe them in the next section as part of a larger discussion of the genetics of speciation.

STUDY BREAK

1. What are the two stages required for allopatric speciation?
2. How do the conditions leading to parapatric and sympatric speciation differ?
3. Why might insects from different host races be unlikely to mate with each other?

19.5 Genetic Mechanisms of Speciation

In this section we examine three macroevolutionary genetic mechanisms that can lead to reproductive isolation between populations. Two are related to geographical distribution: *genetic divergence* between allopatric populations and *polyploidy* in sympatric populations. The third, **chromosome alterations**, occur independently of the geographical distributions of populations.

19.5a Genetic Divergence in Allopatric Populations

In the absence of gene flow, geographically separated populations inevitably accumulate genetic differences through the actions of mutation, genetic drift, and natural selection. How much genetic divergence is necessary for speciation to occur? To understand the genetic basis of speciation in closely related species, researchers first identify the specific causes of reproductive isolation. They then use standard techniques of genetic analysis, along with new molecular approaches such as gene mapping and sequencing, to analyze the genetic mechanisms that establish reproductive isolation. These techniques now allow researchers to determine the minimum number of genes responsible for reproductive isolation in particular pairs of species.

In cases of postzygotic reproductive isolation, mutations in just a few gene loci can establish reproductive isolation. For example, if two common aquarium fishes, swordtails (*Xiphophorus helleri*) and platys (*Xiphophorus maculatus*), mate, two genes induce the development of lethal tumours in their hybrid offspring. When hybrid sterility is the primary cause of reproductive isolation between *Drosophila* spp., at least five gene loci are responsible. About 55 gene loci contribute to postzygotic reproductive isolation between the European fire-bellied toad (*Bombina bombina*) and the yellow-bellied toad (*Bombina variegata*).

In cases of prezygotic reproductive isolation, some mechanisms have a surprisingly simple genetic basis. For example, a single mutation reverses the direction of coiling in the shells of some snails (*Bradybaena* spp.). Snails with shells that coil in opposite directions cannot approach each other close enough to mate, making reproduction between them mechanically impossible.

Many traits that now function as prezygotic isolating mechanisms may originally have evolved in response to sexual selection (see Section 17.5c). In sexually dimorphic species, this evolutionary process exaggerates showy structures and courtship behaviours in males, the traits that females use to select appropriate mates. When two populations encounter one another on secondary contact, these traits may also prevent

Molecular Investigation Techniques: Monkey-Flower Speciation

Reproductive isolation is the primary criterion that biologists use to distinguish species. As noted earlier, the monkey-flower species *Mimulus lewisii* and *Mimulus cardinalis* are mechanically reproductively isolated in nature because differences in flower structure keep bumblebees or hummingbirds from carrying pollen from one species to the other (see Figure 19.4). However, the two species are easily crossed in the laboratory and produce fertile F_1 hybrids. The F_2 offspring have flowers with various forms intermediate between the two parental types, suggesting that several gene loci control the traits separating the species.

Because little was known about the genetics of the two monkey-flower species, it was not possible to use a direct genetic analysis to identify and map the flower trait genes to the chromosomes. Instead, H. D. Bradshaw and other researchers at the University of Washington used an indirect molecular approach that identified DNA sequence variations (analogous to those used in DNA fingerprinting; see Chapter 15) at various loci in the genome. Just as morphological and biochemical traits vary within a population, DNA sequences vary at particular sites in the genome. The different DNA sequence alleles can be distinguished using a polymerase chain reaction (PCR).

The researchers used a random set of 153 DNA sequence variations. They correlated the segregation of these variations with the segregation of flower traits in 93 plants of the F_2 generation. Some of the DNA sequences segregated closely with a particular flower trait. This result indicated that the particular DNA sequence variation locus was very near the gene for that flower trait on the chromosome. In other words, the flower trait locus was identified indirectly through the close linkage between the DNA variation locus and the flower trait locus.

But where are the genes? To answer this question, the researchers used the DNA sequences linked to flower trait loci as probes to find the sites on the chromosomes where they originated. For any given DNA variation locus, once its position on the chromosomes was determined, the investigators knew that the flower trait locus correlated with it must be nearby.

The investigation showed that reproductive isolation of *M. lewisii* and *M. cardinalis* results from differences in eight floral traits: the amount of (1) anthocyanin pigments and (2) carotenoid pigments in petals, (3) flower width, (4) petal width, (5) nectar volume, (6) nectar concentration, and the lengths of the stalks supporting the (7) male and (8) female reproductive parts. Although the investigators could not directly determine the number of genes controlling each trait, the characteristics of the traits, their locations at eight sites on six of the eight chromosomes, and their pattern of inheritance make it most likely that each trait is controlled by a single gene, giving a likely minimum of eight genes.

Mutations in as few as eight genes may have established reproductive isolation between these two species of monkey-flower. Thus, it appears that in some cases surprisingly little genetic change is required for the evolution of a new species.

---

interspecific mating. For example, many closely related duck species exhibit dramatic variation in the appearance of males, but not females **(Figure 19.14)**, an almost certain sign of sexual selection. Yet these species hybridize readily in captivity, producing offspring that are both viable and fertile.

Reproductive isolation and speciation in ducks and other sexually dimorphic birds probably result from geographical isolation and sexual selection on just a few morphological and behavioural characteristics that influence their mating behaviour. Thus, sometimes the evolution of reproductive isolation may not require much genetic change at all. Indeed, sexual selection appears to increase the rate at which new species arise: bird lineages that are sexually dimorphic generally include more species than do related lineages in which males and females have a similar appearance.

19.5b Polyploidy

Common among plants, polyploidy may also be an important factor in the evolution of some fish, amphibian, and reptile species. Polyploid individuals can arise from chromosome duplications within a single species (autopolyploidy) or through hybridization of different species (allopolyploidy).

Mallard ducks (*Anas platyrhynchos*)

Pintail ducks (*Anas acuta*)

Figure 19.14
In closely related species, such as mallard and pintail ducks, males have much more distinctive coloration than females, a sure sign of sexual selection at work.

a. Speciation by autopolyploidy in plants

A spontaneous doubling of chromosomes during meiosis produces diploid gametes. If the plant fertilizes itself, a tetraploid zygote will be produced.

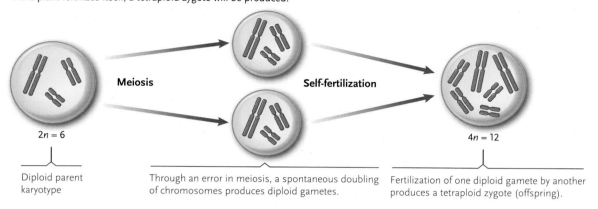

Diploid parent karyotype
$2n = 6$

Through an error in meiosis, a spontaneous doubling of chromosomes produces diploid gametes.

Fertilization of one diploid gamete by another produces a tetraploid zygote (offspring).
$4n = 12$

Meiosis **Self-fertilization**

b. Speciation by hybridization and allopolyploidy in plants

A hybrid mating between two species followed by a doubling of chromosomes during mitosis in gametes of the hybrid can instantly create sets of homologous chromosomes. Self-fertilization can then generate polyploid individuals that are reproductively isolated from both parent species.

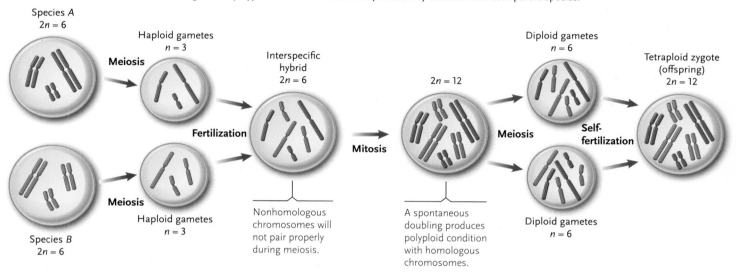

Species A
$2n = 6$

Haploid gametes
$n = 3$

Interspecific hybrid
$2n = 6$

Diploid gametes
$n = 6$

Tetraploid zygote (offspring)
$2n = 12$

$2n = 12$

Meiosis

Fertilization

Mitosis

Meiosis

Self-fertilization

Meiosis

Haploid gametes
$n = 3$

Species B
$2n = 6$

Nonhomologous chromosomes will not pair properly during meiosis.

A spontaneous doubling produces polyploid condition with homologous chromosomes.

Diploid gametes
$n = 6$

Figure 19.15

Polyploidy in plants. **(a)** Speciation by autopolyploidy in plants can occur by a spontaneous doubling of chromosomes during meiosis, producing diploid gametes. If the plant fertilizes itself, a tetraploid zygote will result. **(b)** Speciation by hybridization and allopolyploidy in plants can occur when two species mate, producing a hybrid. If chromosomes are doubled during mitosis in gametes of the hybrid, sets of homologous chromosomes are instantly created. Self-fertilization can then generate polyploid individuals that are reproductively isolated from both parent species.

In **autopolyploidy (Figure 19.15a)**, a diploid ($2n$) individual may produce, for example, tetraploid ($4n$) offspring, each of which has four complete chromosome sets. Autopolyploidy often results through an error in either mitosis or meiosis, when gametes spontaneously receive the same number of chromosomes as a somatic cell. Such gametes are called **unreduced gametes** because their chromosome number has not been halved.

Diploid pollen can fertilize the diploid ovules of a self-fertilizing individual, or it may fertilize diploid ovules on another plant with unreduced gametes. The resulting tetraploid offspring can reproduce either by self-pollination or by breeding with other tetraploid individuals. However, a tetraploid plant cannot produce fertile offspring by hybridizing with its diploid parents. The fusion of a diploid gamete with a normal haploid gamete produces a triploid ($3n$) offspring, which is usually sterile because its odd number of chromosomes cannot segregate properly during meiosis. Thus, the tetraploid is reproductively isolated from the original diploid population. Many species of grasses, shrubs, and ornamental plants—including violets, chrysanthemums, and nasturtiums—are autopolyploids, having anywhere from 4 to 20 complete chromosome sets.

In **allopolyploidy** (Figure 19.15b), two closely related species hybridize and subsequently form polyploid offspring. Hybrid offspring are sterile if the two parent species have diverged enough that their chromosomes do not pair properly during meiosis. However, if the hybrid's chromosome number is doubled, the chromosome complement of the gametes is also doubled, producing homologous chromosomes that *can* pair during meiosis. The hybrid can then produce polyploid gametes and, through self-fertilization or fertilization with other doubled hybrids, establish a population of a new polyploid species. Compared with speciation by genetic divergence, speciation by allopolyploidy is extremely rapid, causing a new species to arise in one generation without geographical isolation.

Even when sterile, polyploids are often robust, growing larger than either parent species. For that reason, both autopolyploids and allopolyploids have been important to agriculture. For example, the wheat used to make flour (*Triticum aestivum*) has six sets of chromosomes **(Figure 19.16)**. Other polyploid crop plants are apples, coffee, strawberries, potatoes, oats, and tobacco.

Plant breeders often try to increase the probability of forming an allopolyploid by using chemicals that foster nondisjunction of chromosomes during mitosis. In the first such experiment, undertaken in the 1920s, scientists crossed a radish and a cabbage, hoping to develop a plant with both edible roots and leaves. Instead, the new species, *Raphanobrassica,* combined the least desirable characteristics of each parent, growing a cabbagelike root and radishlike leaves. Recent experiments have been more successful. For example, plant scientists have produced an allopolyploid grain, triticale, that has the disease resistance of its rye parent and the high productivity of its wheat parent.

In both autopolyploidy and allopolyploidy, a spontaneous doubling of chromosome number produces gametes with twice the original number of chromosomes, but the timing of doubling is different. In autopolyploidy, the doubling occurs during a meiotic cell division that produces $2n$ gametes in the parent. In allopolyploidy, the doubling occurs after a hybrid offspring is produced, when some of its cells are undergoing mitosis; meiosis in the polyploid hybrid then produces polyploid gametes.

19.5c Speciation from Chromosome Alterations

Other changes in chromosome structure or number may also foster speciation. Closely related species often have a substantial number of chromosome differences between them, including inversions, translocations, deletions, and duplications (described in Section 11.3). These differences, which may foster postzygotic isolation, can often be identified by comparing the *banding patterns* in stained chromosome preparations from the different species. In all species, banding patterns vary from one chromosome segment to another. When researchers find identical banding patterns in chromosome segments from two or more related species, they know that they are examining comparable portions of the species' genomes. Thus, the banding patterns allow scientists to identify specific chromosome segments and compare their positions in the chromosomes of different species.

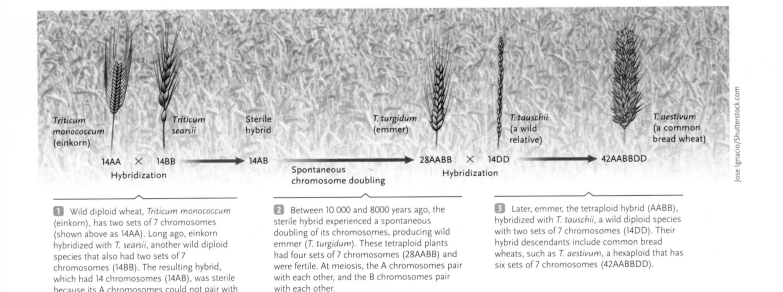

1 Wild diploid wheat, *Triticum monococcum* (einkorn), has two sets of 7 chromosomes (shown above as 14AA). Long ago, einkorn hybridized with *T. searsii*, another wild diploid species that also had two sets of 7 chromosomes (14BB). The resulting hybrid, which had 14 chromosomes (14AB), was sterile because its A chromosomes could not pair with its B chromosomes during meiosis.

2 Between 10 000 and 8000 years ago, the sterile hybrid experienced a spontaneous doubling of its chromosomes, producing wild emmer (*T. turgidum*). These tetraploid plants had four sets of 7 chromosomes (28AABB) and were fertile. At meiosis, the A chromosomes pair with each other, and the B chromosomes pair with each other.

3 Later, emmer, the tetraploid hybrid (AABB), hybridized with *T. tauschii*, a wild diploid species with two sets of 7 chromosomes (14DD). Their hybrid descendants include common bread wheats, such as *T. aestivum*, a hexaploid that has six sets of 7 chromosomes (42AABBDD).

Figure 19.16
The evolution of wheat. Researchers believe the evolution of common bread wheat resulted from a hybridization between two diploid species, followed by a spontaneous doubling of the hybrid's chromosomes, and a second hybridization between the polyploid hybrid and a third diploid species.

STUDY BREAK

1. How can natural selection promote reproductive isolation in allopatric populations?
2. How does polyploidy promote speciation in plants?

It is ironic that species are so fundamental to biology and evolution and yet present such a challenge when it comes to a clear articulation of an underlying concept. But this reality is hardly surprising given the diversity of living organisms and the range of situations that have led to their evolution. The study of speciation—macroevolution—is the study of evolution in action. Naming species is thus the use of fixed features of an organism to allow it to be identified. However, the application of a fixed system to a dynamic one is rarely successful. If we remember this, then topics from Hardy–Weinberg to gene flow, as well as phenotypic and genotypic variation, should all come into perspective.

Review

aplia™

To access course materials such as Aplia and other companion resources, please visit www.NELSONbrain.com.

19.1 What Is a Species?

• Most biologists define a species as a population of organisms capable of interbreeding and producing fertile offspring.

• Linnaeus invented a system of binomial nomenclature in which each species is given a unique two-part name, called a binomial.

• Species of organisms are given Latinized binomial names that are presented in italics. *Castor canadensis* is the scientific name of the Canadian beaver. *Castor* is the genus, *canadensis* is the trivial name, and the two names together, *Castor canadensis,* are the binomial or species name.

• Species are organized into a taxonomic hierarchy, comprising a nested series of formal categories: domain, kingdom, phylum, class, order, family, genus, species, and subspecies. The organisms included within any category of the taxonomic hierarchy make up a taxon (plural, *taxa*).

19.2 Species Concepts

• The morphological species concept is based on the idea that all individuals of a species share measurable traits that distinguish them from individuals of other species. This concept dates to Linnaeus' classification system.

• The biological species concept defines species as groups of interbreeding populations that are reproductively isolated from populations of other species in nature. A biological species thus represents a gene pool within which genetic material is potentially shared among populations. The biological species concept cannot be applied to organisms that reproduce only asexually, to those that are extinct, or to geographically separated populations.

• The phylogenetic species concept defines a species as a group of populations with a recently shared evolutionary history. Using both morphological and genetic sequence data, scientists first reconstruct an evolutionary tree for the organisms of interest. They then define a phylogenetic species as a cluster of populations that emerge from the same small branch.

19.3 Maintaining Reproductive Isolation

• Reproductive isolating mechanisms are biological characteristics that prevent two species from interbreeding.

• Prezygotic isolating mechanisms either prevent individuals of different species from mating or prevent fertilization between their gametes. Prezygotic isolation occurs because species live in different habitats, breed at different times, use different courtship behaviour, or differ anatomically. Prezygotic isolation can also result from genetic and physiological incompatibilities between male and female gametes.

• Postzygotic isolating mechanisms reduce the fitness of interspecific hybrids through hybrid inviability, hybrid sterility, or hybrid breakdown.

19.4 The Geography of Speciation

• Most species exhibit geographical variation of phenotypic and genetic traits. In ring species, populations are distributed in a ring around unsuitable habitat. Many species exhibit clinal variation of characteristics, which change smoothly over a geographical gradient.

• The model of allopatric speciation proposes that speciation results from divergent evolution in geographically separated populations. If allopatric populations accumulate enough genetic differences, they will be reproductively isolated upon secondary contact. Nevertheless, some species hybridize over small areas of secondary contact.

• The model of parapatric speciation suggests that reproductive isolation can evolve between parts of a population that occupy opposite sides of an environmental discontinuity.

• A model of sympatric speciation suggests that reproductive isolation may evolve between host races that rarely contact one another under natural conditions. Sympatric speciation commonly occurs in flowering plants by allopolyploidy.

19.5 Genetic Mechanisms of Speciation

- Allopatric populations inevitably accumulate genetic differences, some of which contribute to their reproductive isolation. Reproductive isolating mechanisms evolve as by-products of genetic changes that occur during divergence. Prezygotic isolating mechanisms may evolve in populations experiencing secondary contact.
- Speciation by polyploidy in flowering plants involves the duplication of an entire chromosome complement through nondisjunction of chromosomes during meiosis or mitosis. Polyploids can arise among the offspring of a single species (autopolyploidy) or, more commonly, after hybridization between closely related species (allopolyploidy).
- Chromosome alterations can promote speciation by fostering the genetic divergence of, and reproductive isolation between, populations with different numbers of chromosomes or different chromosome structure.

Questions

Self-Test Questions

1. Who is the "father" of taxonomy?
   a. Charles Darwin
   b. Charles Lyell
   c. Alfred Wallace
   d. Carolus Linnaeus
   e. Jean Baptiste de Lamarck

2. In the Linnaean hierarchy, what are the organisms classified within the same taxonomic category called?
   a. a phylum
   b. a taxon
   c. a genus
   d. a binomial
   e. an epithet

3. On what basis does the biological species concept define species?
   a. reproductive characteristics
   b. biochemical characteristics
   c. morphological characteristics
   d. behavioural characteristics
   e. all of the above

4. What is a characteristic that exhibits smooth changes in populations distributed along a geographical gradient called?
   a. ring species
   b. hybrid
   c. cline
   d. hybrid breakdown
   e. subspecies

5. If two species of holly (genus Ilex) flower during different months, how might their gene pools be kept separate?
   a. mechanical isolation
   b. ecological isolation
   c. gametic isolation
   d. temporal isolation
   e. behavioural isolation

6. Which of the following is true about prezygotic isolating mechanisms?
   a. They reduce the fitness of hybrid offspring.
   b. They generally prevent individuals of different species from producing zygotes.
   c. They are found only in animals.
   d. They are found only in plants.
   e. They are observed only in organisms that reproduce asexually.

7. In the model of allopatric speciation, which is true of the geographical separation of two populations?
   a. It is sufficient for speciation to occur.
   b. It occurs only after speciation is complete.
   c. It allows gene flow between them.
   d. It reduces the relative fitness of hybrid offspring.
   e. It inhibits gene flow between them.

8. Adjacent populations that produce hybrid offspring with low relative fitness may be undergoing which of the following?
   a. clinal isolation
   b. parapatric speciation
   c. allopatric speciation
   d. sympatric speciation
   e. geographical isolation

9. An animal breeder, attempting to cross a llama with an alpaca for finer wool, found that the hybrid offspring rarely lived more than a few weeks. What did this outcome probably result from?
   a. genetic drift
   b. prezygotic reproductive isolation
   c. postzygotic reproductive isolation
   d. sympatric speciation
   e. polyploidy

10. Which of the following could be an example of allopolyploidy?
    a. One parent has 8 chromosomes, the other has 10, and their offspring have 36.
    b. Gametes and somatic cells have the same number of chromosomes.
    c. Chromosome number increases by one in a gamete and in the offspring it produces.
    d. Chromosome number decreases by one in a gamete and in the offspring it produces.
    e. Chromosome number in the offspring is exactly half of what it is in the parents.

Questions for Discussion

1. All domestic dogs are classified as members of the species *Canis familiaris*. But it is hard to imagine how a tiny Chihuahua could breed with a gigantic Great Dane. Do you think that artificial selection for different breeds of dogs will eventually create different dog species?

2. Human populations often differ dramatically in external morphological characteristics. On what basis are all human populations classified as a single species?

3. If intermediate populations in a ring species go extinct, eliminating the possibility of gene flow between populations at the two ends of the ring, would you now identify the remaining populations as full species? Explain your answer.

4. How do human activities (such as destruction of natural habitats, diversion of rivers, and construction of buildings) influence the chances that new species of plants and animals will evolve in the future? Frame your answer in terms of the geographical and genetic factors that foster speciation.

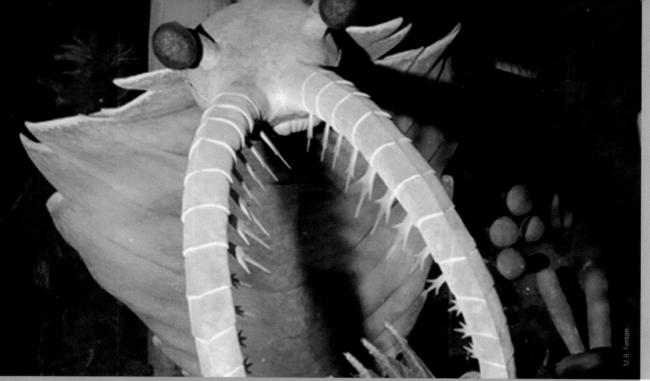

A reconstruction of a 1-m-long *Anomalocaris* that has just caught a prey gives an idea of its bizarre appearance.

Understanding the History of Life on Earth

WHY IT MATTERS

The Burgess Shale biota, in British Columbia, is an assemblage of organisms fossilized over 500 million years ago (mya) in the Cambrian period. The organisms found there present challenges for paleontologists. Many of the specimens are incomplete and most were softbodied organisms that normally did not fossilize well. As well, many of the organisms are hard to imagine for people who grew up knowing dinosaur fossils. The Burgess Shale biota is an example of a treasure trove of fossils, or lagerstätten (German, from *Lager*, "lair or den," and *Stätte*, "place"). Such treasure troves are preserved in deposits of sedimentary rock that contain large numbers of exceptionally well-preserved fossils. The Burgess Shale is one of about 60 lagerstätten known worldwide.

Anomalocaris (meaning "abnormal shrimp," opening photograph) is one of the most spectacular animals of the Burgess Shale. Parts of *Anomalocaris* were originally described as different animals **(Figure 20.1, p. 446).** When it was discovered in 1892, the appendages of *Anomalocaris* were described as a section of the abdomen of a large shrimp. Then *Peytoia* was described from what appeared to be 32 subumbrel lobes thought to belong to a jellyfish. Another fossil, *Tuzoia,* was described as the carapace or shell of a crustacean. The last component was *Laggania,* originally thought to be part of a sea cucumber, but later recognized as another specimen with smaller frontal appendages. Over time, the discovery of more and better-preserved specimens changed our view of this animal. *Anomalocaris* and other arthropods with large frontal ("great") appendages are now classified in the order Megacheira (phylum Arthropoda). Their great appendages are modified mouthparts.

As recently as 2013, most paleontologists agreed that the parts originally described as *Anomalocaris* were the remains of two great appendages. *Peytoia* were plates around the mouth of *Anomalocaris,* while *Tuzoia* was its carapace or shell. *Laggania* was a

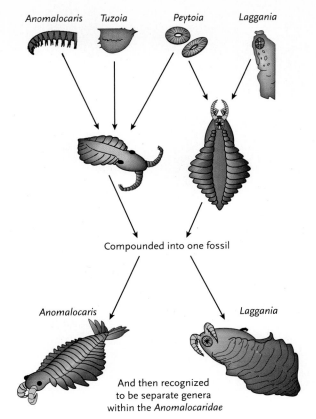

Anomalocaris   Tuzoia   Peytoia   Laggania

Compounded into one fossil

Anomalocaris   Laggania

And then recognized
to be separate genera
within the *Anomalocaridae*

Figure 20.1
How our knowledge of *Anomalocaris* developed and emerged as more fossils were discovered. In the end, *Anomalocaris* is a composite, including parts originally described as a sea cucumber (*Laggania*), a schyphozoan medusa (jellyfish, *Peytoia*), and a bivalve arthropod (*Tuzoia*).

From FENTON/DUMONT/OWEN. *Integrative Animal Biology*, 1E. © 2014 Nelson Education Ltd. Reproduced by permission. www.cengage.com/permissions

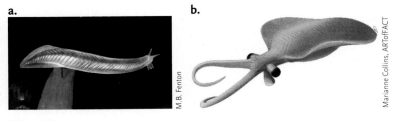

a.   **b.**

Figure 20.2
Two reconstructions of *Nectocaris*. **(a)** The earlier one presented it as a primitive chordate; **(b)** the later one as a cephalopod mollusc.

poorly preserved remains of another megacheirid, one with smaller great appendages. We presume that the two great appendages were used to seize and break the shells of prey such as trilobites. Megacheirans are known from the Cambrian through the Ordovician, but the lineage does not survive today.

Our views of other animals from the Burgess Shale have also changed considerably. Originally described as a distant relative of vertebrates **(Figure 20.2a)**, *Nectocaris* was more recently recognized as a cephalopod mollusc (Figure 20.2b). Still other animals are simply astonishing, such as *Opabinia* **(Figure 20.3)**.

The Burgess Shale and other Cambrian lagerstätten have provided the world with a glimpse of a long-extinct, highly diverse assemblage of animals. These fossils

Figure 20.3
A reconstruction of the five-eyed *Opabinia* from the Burgess Shale.

have intrigued biologists and paleontologists for over 100 years. This assemblage reinforces at least three lessons:

1. We must keep an open mind when looking at life. The fossil you are looking at may be a small part of something rather than the whole thing. Some organisms, both living and fossil, are strikingly different from anything with which we are familiar.

2. Many new and astonishing organisms remain to be discovered. Some are fossils, while others are still with us.

3. How could life have diversified so much by about 600 mya? Where were all of these organisms before the Cambrian? And what happened in the millions of years since they lived?

Setting the Stage for Fossils and Phylogeny. Imagine our world without fossils. No Tyrannosaurus rex to frighten you as a child. No insects trapped in amber for hundreds of millions of years. No footprints across sandstone (Figure 21.1). Without fossils, we would understand very little about the history of life on Earth. But how do we relate these species to those that live today?

Millions of species either live or have lived on Earth. You have learned that each was formed from another species by speciation. You also know from experience that dogs and cats are more related than either is to cows or lizards. Systematists keep the binomial classification system up to date and determine the relatedness of all species. One way they do this is by drawing phylogenetic trees similar to Darwin's Tree of Life (Figure 17.9).

In this chapter, we will look at the geological history and the history of life on the planet. Then, we will use what we know about extinct animals (from the fossil record) and modern-day organisms to draw phylogenies to understand how today's species have descended from yesterday's.

20.1 The Geological Time Scale

Many fossils are found in sedimentary rock. Sediments found in any one place form distinctive strata (layers) that usually differ in colour, mineral composition, particle size, and thickness **(Figure 20.4)**. If they have not been disturbed, the strata are arranged in the order in which they formed, with the youngest layers on top. But strata have sometimes been uplifted, warped, or even inverted by geologic processes.

Geologists of the nineteenth century deduced that the fossils discovered in a particular sedimentary stratum,

a. Sedimentation

Highest strata contain the most recent fossils.

Lowest strata contain the oldest fossils.

b. Geological strata in the Painted Desert, Arizona

Nick Greaves/Alamy

Figure 20.4

Sedimentation and geological strata. **(a)** Sedimentation deposits successive layers at the bottom of a lake or sea. **(b)** Over millions of years, the upper layers compress those below them into rock. When the rocks are later exposed by uplifting or erosion, the different layers are evident as geological strata.

no matter where it was found, represent organisms that lived and died at roughly the same time in the past (see Chapter 17). Because each stratum was formed at a specific time, the sequence of fossils from lowest (oldest) to highest (newest) strata reveals their relative ages. Geologists used the sequence of strata and their distinctive fossil assemblages to establish the geologic time scale that diagrams the history of life on Earth **(Table 20.1).**

20.2 The Fossil Record

20.2a How Fossils Were Formed

We often see spectacular mounted fossil skeletons in museums **(Figure 20.5, p. 449).** These were formed when dissolved minerals entered the spaces within the bones and then solidified. Some fossils, such as those preserved in amber (tree resin, **Figure 20.6a, p. 449**), show fine

Table 20.1	**The Geological Time Scale and Major Evolutionary Events**					
Eon	Era	Period	Epoch	Millions of Years Ago	Major Evolutionary Events	
Phanerozoic	Cenozoic	Quaternary	Holocene	0.01		
			Pleistocene	2.6	Origin of humans; major glaciations	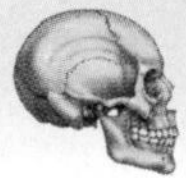
		Neogene	Pliocene	5.3	Origin of apelike human ancestors	
			Miocene	23.0	Angiosperms and mammals further diversify and dominate terrestrial habitats	 
		Paleogene	Oligocene	33.9	Primates diversify; origin of apes	
			Eocene	55.8	Angiosperms and insects diversify; modern orders of mammals differentiate	 
			Paleocene	65.5	Grasslands and deciduous woodlands spread; modern birds, mammals, snakes, pollinating insects diversify; continents approach current positions	 

(Continued)

Eon	Era	Period	Epoch	Millions of Years Ago	Major Evolutionary Events	
Phanerozoic	Mesozoic	Cretaceous		145.5	Angiosperms, insects, marine invertebrates, fishes, dinosaurs diversify; asteroid impact causes mass extinction at end of period, eliminating dinosaurs and many other groups	
		Jurassic		201.6	Gymnosperms abundant in terrestrial habitats; modern fishes diversify; dinosaurs diversify and dominate terrestrial habitats; frogs, salamanders, lizards, and birds appear; continents continue to separate	
		Triassic		251.0	Predatory fishes and reptiles dominate oceans; gymnosperms dominate terrestrial habitats; radiation of dinosaurs; early mammals; Pangaea starts to break up; mass extinction at end of period	
	Paleozoic	Permian		299.0	Insects and reptiles abundant and diverse in swamp forests; some reptiles colonize oceans; fishes colonize freshwater habitats; continents coalesce into Pangaea, causing glaciation and decline in sea level; mass extinction at end of period eliminates 85% of species	
		Carboniferous		359.0	Vascular plants form large swamp forests; first flying insects; amphibians diversify; first reptiles appear	
		Devonian		416.0	Terrestrial vascular plants diversify; fungi, invertebrates, amphibians colonize land; first insects and seed plants; major glaciation at end of period; mass extinction, mostly of marine life	
		Silurian		444.0	Jawless fishes diversify; first jawed fishes, arthropods, terrestrial vascular plants	
		Ordovician		488.0	Major radiations of marine invertebrates and jawless fishes; major glaciation at end of period causes mass extinction of marine life	
		Cambrian		542.0	Appearance of modern animal phyla, including earliest vertebrates (Cambrian explosion); simple marine communities	
Proterozoic				2500	High concentration of oxygen in atmosphere; origin of eukaryotic cells; evolution and diversification of "protists," fungi, softbodied animals	
Archean				3850	Evolution of prokaryotes, including anaerobic and photosynthetic bacteria; oxygen starts to accumulate in atmosphere; origin of aerobic respiration	
Hadean				4600	Formation of Earth, including crust, atmosphere, and oceans; origin of life	

Figure 20.5

Camarasaurus supremus, a herbivorous dinosaur, in the Royal Tyrrell Museum.

M.B. Fenton

details of the organisms. Some fossils, particularly plant fossils, are moulds or impressions (Figure 20.6b). These organisms may have been compressed shortly after death, so three-dimensional analysis may be difficult. Footprints may be fossilized when an animal walks across mud (Figure 20.6b and Figure 21.1). Even droppings (coprolites; Figure 20.6d) and gastroliths, stones in the stomach used to grind food, have been preserved as fossils (Figure 20.6e). Other fossilized remains, including some early humans, are frozen or mummified (Figure 20.6f). Petrified forests are trees fossilized by minerals.

Unfortunately, however, the fossil record is incomplete because few organisms fossilize completely, because some organisms are more likely

a.

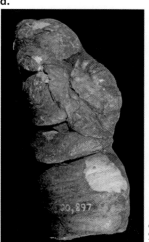

d.

b.

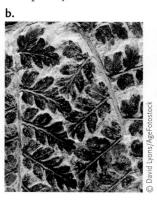

© David Lyons/AgeFotostock

c.

M.B. Fenton

e.

M.B. Fenton

M.B. Fenton

Figure 20.6

A sampling of fossils. **(a)** Insects in amber; **(b)** an impression of a fern (*Sphenopteris*) from the Carboniferous period, preserved in coal; **(c)** dinosaur footprint; **(d)** coprolite; **(e)** gastroliths (the rounded, polished stones); and **(f)** a frozen baby mammoth.

f.

Novosti/Science Source

X-Ray Tomography and 3-D Structure of Fossils

Tomography is the use of thin sections cut from a fossil to reconstruct the organism's appearance, for example, the structures of the wormlike mollusc Acaenoplax **(Figure 1).** X-rays allow researchers to observe fine details of specimens and otherwise invisible specimens **(Figure 2).** Applying tomographic techniques to synchrotron X-ray views allows reconstruction of the fine details of structures such as the surfaces of eggs or the nuclei **(Figure 3).**

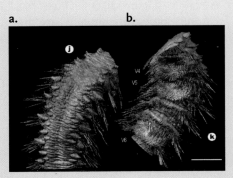

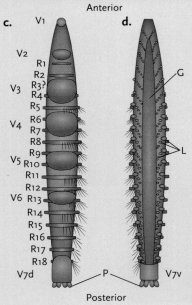

Figure 1

Dorsal (a, c) and ventral (b, d) views of reconstructions of a worm-shaped mollusc, Acaenoplax. (a) and (b) show the anterior 7 mm based on 243 slices at 30 μm intervals. (c) and (d) are overviews of the animals.

SUTTON, M. D., D. E. G. BRIGGS, et al. (2004). "Computer reconstruction and analysis of the vermiform mollusc Acaenoplax hayae from the Herefordshire Lagerstätte (Silurian, England), and implications for molluscan phylogeny," *Palaeontology* 47(2): 293–318. John Wiley and Sons. The Palaeontological Association.

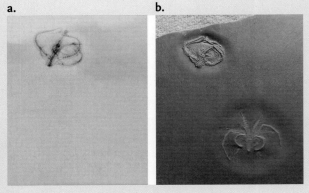

Figure 2

A Photo (a) and X-ray (b) of a 3 to 4 mm thick rock slab from the Lower Devonian. The top animal is an echinoderm (asteroid, Taeniaster beneckei), *while the bottom is an arthropod* (Mimetaster hexagonalis), *visible only in the X-ray.*

Republished with permission of British Institute of Radiology, from P. Hohenstein, "X-ray imaging for palaeontology," *British Journal of Radiology* (2004) 77, 420–425.

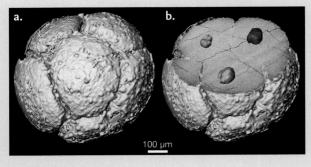

Figure 3

Synchrotron X-ray tomographic reconstructions of embryos of Tianzhushaia from the Ediacaran in China. (a) is a rendering of the surface showing six cells, while (b) shows three nuclei in a slice through the embryo.

From Huldtgren, T., J.A. Cunningham, C. Yin, M. Stampanoni, F. Marone, P.C.J. Donoghue and S. Bengtson, "Fossilized nuclei and germination structures identify Ediacaran "animal embryos" as encysting protists," *Science* 23 December 2011: Vol. 334 no. 6063 pp. 1696–1699. Reprinted with permission from AAAS.

---

to fossilize than others, and because natural processes destroy many fossils.

20.2b Early Fossils

Stromatolites, the first fossil evidence of life, date to about 3.5 billion years ago (bya). Oxygenic photosynthesis, by blue-green prokaryotes and plants (later), resulted in an increase in atmospheric oxygen from about 2.5 bya. The earliest unicellular eukaryotes date from just over 2 bya, and multicellular eukaryotes appeared by 1.2 bya (see Section 3.4 and Figure 3.20).

The history of life on Earth was not gradual. An "explosion" in the diversity of groups of organisms, about 600 mya, marked the beginning of the Cambrian

a.

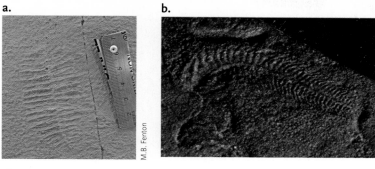

b.

Figure 20.7
Two fossils from the Ediacaran, a rangeomorph **(a)** from Newfoundland and a *Spriggina* **(b)** from Australia.

(Table 20.1). Complex multicellular animals such as those discussed in "Why It Matters" first appeared at this time. It is often difficult to assign some of these fossil animals to any living group, for example, rangeomorphs **(Figure 20.7a)**, which may have been suspension-feeding animals, and *Spriggina* (Figure 20.7b), which appears to have been a polychaete annelid. In each case, the fossils are moulds on a rock surface. Two other explosions have occurred: land plants in the Devonian period and flowering plants in the Cretaceous.

Evolution of Compound Eyes

The eyes of *Anomalocaris* from the Burges Shales revealed previously unknown details of the visual systems of these animals. This evidence indicates an early evolutionary origin of compound eyes **(Figure 1;** see also page 445) and supports the view that *Anomalocaris* was a predator that oriented itself visually in open-water marine environments. Fossils preserved as iron oxide and as calcium phosphate both provide the same detailed picture of the surfaces of the ommatidia making up the compound eyes. Each *Anomalocaris* eye consisted of more than 16 thousand packed ommatidia, approaching the structure of modern insect eyes (Figure 1b).

Several mass extinctions have also occurred, the best known being the Cretaceous–Paleogene event, about 66 mya, in which an asteroid hit Earth near the Yucatán Peninsula, killing most of the dinosaurs along with 75% of all species. Other major mass extinction events occurred at 200, 251, 360, and 450–440 mya. In each event, at least 70% of all species went extinct. Mass extinctions have a variety of causes, such as volcanic eruptions and climate change.

After each mass extinction, many niches and other opportunities opened up for the surviving species, which underwent a period of **adaptive radiation**. Each lineage could diversify rapidly, taking advantage of the newly available niches and the reduction in competition.

Figure 1
Ommatidia making up compound eyes in a fossil trilobite (a) and an extant deer fly (b). The ommatidia of Anomalocaris *are more like those of the fly than of a trilobite. A combination of scanning electron micrography and dispersive spectrometry was used to obtain structural details of the fossils.*

David Evans

David Evans is a young paleontologist at the Royal Ontario Museum in Toronto. Not only does he search for dinosaur fossils in places as diverse as Alberta and Mongolia, he also tries to determine the function(s) of certain structures of the extinct reptiles. With colleagues in Ohio, Dr. Evans appears to have solved a long-standing puzzle about lambeosaurines, a group of huge duck-billed dinosaurs that lived in the swampy habitats of western North America in the late Cretaceous period. Lambeosaurine heads sported bony crests with gigantic nasal passages **(Figure 1)**. Paleobiologists proposed several functions of the crests: as weapons in male combat, adornments that attracted mates, snorkels that facilitated breathing underwater, radiators that cooled the dinosaurs' bodies, structures that enhanced the sense of smell, or resonating chambers that produced honking vocalizations. Based on anatomical analyses, researchers accepted the vocalization hypothesis as the most probable: computerized acoustic models predicted that air flowing through the nasal passages would have produced low-frequency sounds (30–375 Hz).

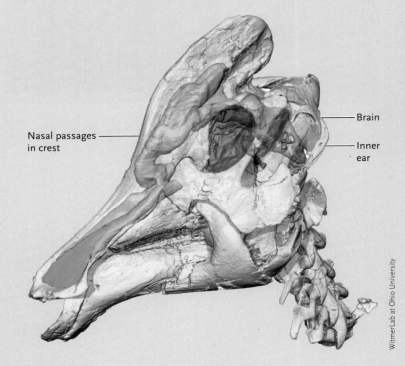

Figure 1

Honking dinosaurs. Analysis of the sinuses and braincase of lambeosaurines (Corythosaurus spp.) *revealed that the nasal passages in their crests served as resonating chambers for the production of low-frequency sounds that their inner ears could detect.*

But could lambeosaurines *hear* sounds in that frequency range? In 2009, Dr. Evans and his colleagues used computed tomography and 3-D visualization software to scan and reconstruct the interior anatomy of the skulls of several lambeosaurine species. Their findings show that the inner ears of lambeosaurines were attuned to hear low-frequency sounds that matched those predicted by the earlier research. They concluded that the elaborate nasal passages in lambeosaurine crests, along with the structure of their inner ears, facilitated vocal communication.

20.2c   The Importance of Skeletons

Not surprisingly, the apparent explosion of life marking the beginning of the Cambrian coincides with the appearance of skeletons—structures that support organisms. The early fossils of bacteria and plants are recognizable due to the presence of cell walls that provided physical support and maintained the integrity of cell shape and structure. The absence of cell walls and analogous structures in invertebrate animals led to fewer fossils. However, some early animal fossils from the Ediacaran (e.g., *Spriggina,* Figure 20.7) suggest that they had supporting systems such as cuticles (see "Evolution of Compound Eyes," p. 451). Insects and other arthropods fossilized due to their exoskeleton.

When teeth and bones appear in fossil finds, it marks a sharp increase in the abundance of fossils because hard tissues lend themselves to mineral fossilization, while soft tissues are usually fossilized as moulds. Dinosaurs are particularly well represented in the fossil record because of their size and their well-developed bones (see "People behind Biology").

STUDY BREAK

1. How old are the first fossils? How old are the first fossil multicellular organisms?
2. What is a mass extinction event? Describe the Cambrian explosion.
3. Why are skeletons important to the fossil record?

20.3 Lives of Prehistoric Organisms

20.3a The Move onto Land

The movement of organisms out of the water and onto land was a momentous event in diversification. Fresh-water algae existed before the first land plants appeared, during the Ordovician Period. These land plants were similar to today's liverworts and mosses. By the Silurian, lycopsids, the oldest of the vascular plants, had diverged from the main lineage that led to ferns, horsetails, and seed plants. By the middle of the Devonian, treelike lycopsids grew to at least 8 m tall, constituting the first forests and adding a vertical dimension to terrestrial ecosystems. In the Permian, some lycopsid trees were 40 m tall. These trees probably grew rapidly, perhaps achieving maturity in a few years. At maturity, lycopsids reproduced with spores and then died. Later in the Permian, tree ferns (still seen today in New Zealand and parts of Australia) replaced lycopsids as the trees of swamp forests. Today lycopsids survive as quillworts and clubmosses.

For animals to move onto land required several adaptations, such as sufficient support to allow them to maintain their integrity, and some form of protection from ultraviolet radiation. Adaptations to ensure a waterproof coating, as well as necessary changes in mechanisms of gaseous exchange, excretion, and locomotion, were also needed. Terrestrial plants provided some of the necessities for animals, such as food, oxygen, and shelter.

A variety of modern bony fishes, such as mudskippers, climbing perch, and some catfishes, regularly move onto land. The same is true of many annelids, molluscs, and arthropods. This suggests that the earliest movements of animals onto land occurred as short visits.

By the Devonian, the diversification of insects reflects the appearance of lycopsid forests. Living insects can generally be sorted into two groups. Hemimetabolous insects (e.g., dragonflies, cockroaches) have three-stage life cycles, with eggs, nymphs, and adults. Holometabolous insects have four-stage life cycles, with eggs, larvae, pupae, and adults. The latter group is the more diverse, including beetles, lepidopterans (moths and butterflies), hymenopterans (ants, bees, and wasps), and flies. In 2013, the description of previously unknown fossil insects revealed that holometabolous insects had appeared by the middle of the Carboniferous, about 300 mya, but underwent a striking diversification after the mass extinction event that marked the end of the Permian.

By the Carboniferous, terrestrial insects were diverse and included winged forms. Of the ~16 orders of insects in the Carboniferous, 11 did not survive into the Mesozoic. The wings of mid-Carboniferous dragonflies (Odonata) showed specializations that persist in living dragonflies. Coevolutionary relationships between plants and insects developed by the Late Carboniferous when some insect larvae formed galls in the internal tissue of tree fern fronds. Fossilized plant tissues show diagnostic histological and cellular details **(Figure 20.8)** typical of modern gall-forming plant–insect interactions. Amber (Figure 20.6a) had appeared in the Carboniferous, demonstrating that some trees had the biosynthetic mechanisms necessary to produce resins in defence against insects. Dissection of insects, plant parts, and other organisms in amber have revealed a great deal about their evolution.

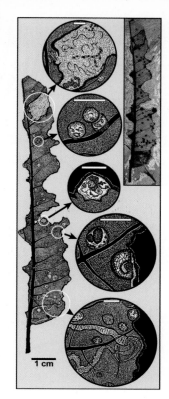

Figure 20.8
This leaf shows evidence of attack by insects, including mining, galling, and external feeding. Photographed specimen (inset upper right) is shown in detail on the left. From top to bottom: a blotch mine with insect feces (frass). Then three galls, one with an exit hole along a secondary vein, another with its margin consumed by an external feeder. Bottom, two galls, one partly consumed, along with two linear mines.
Wilf et al. 2005. "Richness of plant–insect associations in Eocene Patagonia: A legacy for South American biodiversity," *PNAS*, vol. 102 no. 25 8944–8948. Copyright 2005 National Academy of Sciences, U.S.A.

STUDY BREAK

1. Why was the movement of plants onto land fundamentally important for animals?
2. What are lycopsids and tree ferns?

20.3b The Move Back to Water

The first mammals appeared about 165 mya as small furred land animals. After the Cretaceous–Paleogenic mass extinction, the surviving mammals expanded, filling many of the now-vacant niches. Some mammals, such as otters and beavers, became semi-aquatic, bearing their young on land while living much of their lives in water.

Several lineages of mammals are marine, some of which never leave the water (Sirenia: manatees and dugongs; Cetacea: whales). Whales and toothed whales are a distinctive group of mammals whose fossil record extends back more than 50 Ma. One of the Eocene whales, *Basilosaurus cetoides,* had a pelvic girdle and leg and foot bones **(Figure 20.9, p. 454)**, features lacking in modern cetaceans. The discovery of an Eocene fossil *Indohyus*, a raccoon-sized mammal known from fossils in Pakistan, has had interesting repercussions for the classification of mammals. *Indohyus* had a groovelike structure (the involucrum; **Figure 20.10, p. 454**) at the back of its skull, a feature now known to occur in both cetaceans and even-toed ungulates or artiodactyls (pigs,

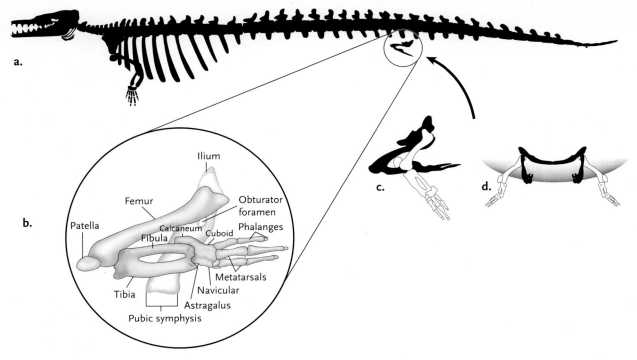

Figure 20.9

Unlike modern whales (Cetacea), the Eocene whale *Basilosaurus cetoides* from Egypt had a functional pelvic girdle and leg and foot bones. **(a)** The skeleton of the Eocene whale. **(b)** Skeletal details of the hind limb. **(c)** The pelvic girdle in resting posture; **(d)** in extended posture.

Based on Gingerich, P.D., B.H. Smith and E.L. Simons. 1990. "Hind limbs of Eocene Basilosaurus: evidence of feet in whales" *Science* 249:154–157. Gingerich, P.D., M. ul-Haq, W. von Koenigswald, W.J. Saunders, B.H. Smith and I.S. Zalmout. 2009. "New protocetid whale from the Middle Eocene of Pakistan: birth on land, precocial development, and sexual dimorphism," PLoS ONE, 4:e4366. From FENTON/DUMONT/OWEN. Integrative Animal Biology, 1E. © 2014 Nelson Education Ltd. Reproduced by permission. www.cengage.com/permissions

deer, camels, and hippopotamuses). The involucrum indicated that cetaceans and artiodactyls are more closely related to one another than either is to any other mammals. This discovery is the grounds for placing whales and artiodactyls in one order, the Cetartiodactyla.

Isotopic analysis of the teeth of *Indohyus* illustrated that over time, these animals showed a change in diet, from eating plants to eating fish. *Indohyus* apparently lived a hippopotamus-like existence, originally going ashore to graze and returning to the water, probably to avoid predators. This information provides clues to the origin of cetaceans and the fundamental changes that

occurred as they became more and more aquatic. Modern whales never leave the water, unlike seals and their relatives, which haul out at least to give birth. They are divided into two main groups: toothed whales (odontocetes) that eat mainly fish and other larger marine animals, and baleen whales (mystacetes) that eat plankton (small marine organisms). The initial diversification of cetaceans involved a switch in diet from plants to fish.

The sea cows (manatees or dugongs, Sirenia) are fully aquatic herbivores apparently belonging to an African lineage of mammals, the Afrotheria, that also

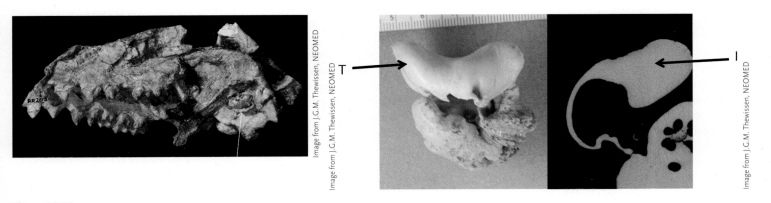

Figure 20.10

(a) The skull of an *Indohyus,* an even-toed ungulate (Artiodactyla), has a well-developed involucrum (white arrow). This feature indicates that whales and even-toed ungulates belong in a single order, Cetartiodactyla. The skull is about 15 cm long. **(b)** A lateral view of the bony part of the right ear of a modern whale (beluga, *Delphinapterus leucas*) showing the lateral wall of the tympanic plate (T). In **(c),** a CT scan (bone in white) illustrates the thin wall of the tympanic plate and the thicker involucrum (I). Scale upper left is in millimetres. Dorsal is top of images.

includes elephants. Sirenians may be older (Palaeocene) than cetaceans and apparently originated in fresh water. The earliest complete fossil sirenians are from Jamaica. The bones indicate that the animals could move onto land but probably spent most of their lives in the water. The sirenians remained herbivores, unlike cetaceans.

A return to an aquatic existence also evolved several times in reptiles. In terms of body form, ichthyosaurs **(Figure 20.11)** were most similar to whales and more fishlike than the other marine reptiles. Sea turtles, mosasaurs, and plesiosaurs are other marine reptiles. The flipperlike limbs of marine mammals and reptiles (Figure 20.11) are clear evidence of anatomical convergence. The similarity extends to the flipperlike wings of penguins, marine birds that "fly" underwater.

Once again, the fossil record provides details about morphological adaptations and examples of convergence in form and function, as well as sometimes surprising information about evolutionary relationships among organisms.

STUDY BREAK

1. What is the involucrum? What does it tell us about relatedness among mammals?
2. How did the return to an aquatic existence differ between cetaceans and sirenians?

20.4 Phylogeny

Earlier, you saw Darwin's conception of branching evolution—a tree of life (Figure 17.9 and **Figure 20.12, p. 456**), as well as an evolutionary tree showing the three domains of life (Figure 3.20). Biologists call this type of diagram a phylogeny. Just like a family tree, **phylogenies** show the evolutionary history of a group of organisms. Phylogenies are presented as **phylogenetic trees**, which are formal hypotheses identifying likely relationships among groups of organisms (Figure 3.20 and **Figure 20.13, p. 457**). Like all hypotheses, they can be tested with data and are often revised as scientists gather new data. Some

a.

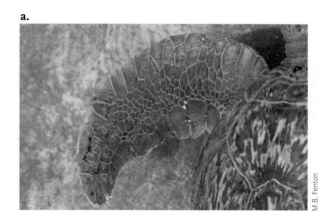

b.

c.

d.

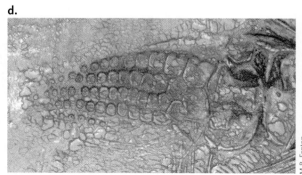

Figure 20.11
Pectoral flippers from marine reptiles. **(a)** An extant hawksbill turtle (*Eretmochelys imbricata*), **(b)** the Cretaceous turtle *Toxochelys*, **(c)** a plesiosaur, and **(d)** an ichthyosaur.

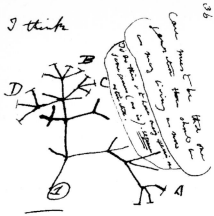

Figure 20.12

Darwin and branching evolution. This entry from his notebook on the "transmutation of species" demonstrates that Charles Darwin first thought about the branching pattern of evolution in 1837, more than 20 years before he published *On the Origin of Species*.

Charles Darwin's First Notebook on Transmutation of Species (1837)

phylogenetic trees show relatedness of large groups of organisms (phyla or classes); others show how genera or species are related. Many phylogenies include prehistoric organisms along with modern-day ones.

From before Linnaeus (see Section 19.1) until the mid-1900s, systematists classified organisms and developed phylogenies based mainly on morphological characters. Over the past 50 years, along with morphology, they considered patterns of behaviour and traits such as chromosomal anatomy, details of physiology, morphology of subcellular structures, cells, and organ systems. Modern systematists also use molecular sequences of nucleic acids and proteins as additional characters.

Accurate phylogenetic trees are essential components of the comparative method that biologists use to analyze evolutionary processes. Robust phylogenetic hypotheses allow us to distinguish similarities inherited from a common ancestor from those that evolved independently in response to similar environments.

Data collected and organized by systematists also allow biologists to select appropriate organisms for their work. Many biological experiments are first conducted with individuals of a single species (see Chapter 18), particularly a species that is a closed genetic system, where individuals do not hybridize with members of related species. If a researcher inadvertently used two species that responded differently, the mixed results would probably confuse the underlying picture.

20.4a Evaluating Systematic Characters

Systematists use guidelines to select characters for study. As we saw previously, there is more to being a beaver than being a mammal with a flattened tail. Systematists seek characters that are independent markers of underlying genetic similarity and differentiation. Ideally, systematists create phylogenetic

hypotheses and classifications by analyzing the genetic changes that caused speciation and differentiation. But the fossil record is not complete, so systematists often rely on phenotypic traits as indicators of genetic similarity or divergence. Systematists study traits in which phenotypic variation reflects genetic differences, while trying to exclude differences caused by environmental conditions.

Useful systematic characters must be genetically independent, reflecting different parts of organisms' genomes. This precaution is necessary because different organismal characters can have the same genetic basis. We want to use each genetic variation only once in an analysis. For example, tropical lizards in the genus *Anolis* can climb trees because they grip the bark with small adhesive pads on the undersides of their toes. The number of pads varies from species to species and toe to toe. Researchers have used the number of pads on the fourth toe of the left hind foot as a systematic character. They do not use the number of pads on the fourth toe of the right hind foot as a *separate* character because the same genes almost certainly control the number of pads on the toes of both hind feet. The point here is not the fine-grained detail about toes, but rather the kinds of characters that can be used when assembling a picture of adaptive radiation.

Homologous Characters. The limbs of tetrapod vertebrates are homologous characters that are similar in their evolutionary history but not necessarily their function. Homologous characters are useful in preparing phylogenies because they reflect underlying genetic similarities, and the comparison of homologous characters can indicate common ancestry and genetic relatedness.

Where their functions have changed, homologous structures (inherited from a common ancestor) can differ considerably among species. The stapes, a bone in the middle ear of tetrapod vertebrates, evolved from (is homologous to) the hyomandibula, a bone that supports the jaw joint of most fishes. The structure, position, and function of the hyomandibula are different in tetrapods than they are in fishes **(Figure 20.14)**.

Homologous characters emerge from comparable embryonic structures and grow in similar ways during development. Systematists have put great stock in embryological indications of homology on the assumption that evolution has conserved the pattern of embryonic development in related organisms. Indeed, recent discoveries in **evolutionary developmental biology** have revealed that some genetic controls of developmental pathways can be very similar across a wide variety of organisms (e.g., *Pax-6* genes control the development of eyes; see Chapters 1 and 42). The same situation applies to *HoxD* genes that control the development of limbs.

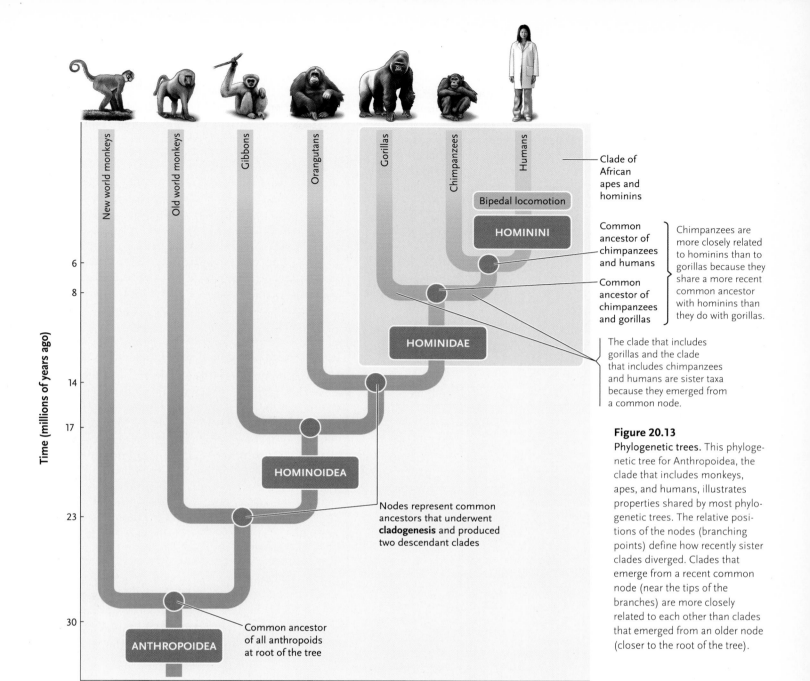

New world monkeys

Old world monkeys

Gibbons

Orangutans

Gorillas

Chimpanzees

Humans

Clade of
African
apes and
hominins

Bipedal locomotion

HOMININI

Common
ancestor of
chimpanzees
and humans

Chimpanzees are
more closely related
to hominins than to
gorillas because they
share a more recent
common ancestor
with hominins than
they do with gorillas.

Common
ancestor of
chimpanzees
and gorillas

HOMINIDAE

The clade that includes
gorillas and the clade
that includes chimpanzees
and humans are sister taxa
because they emerged from
a common node.

HOMINOIDEA

Nodes represent common
ancestors that underwent
cladogenesis and produced
two descendant clades

Time (millions of years ago)

6

8

14

17

23

30

ANTHROPOIDEA

Common ancestor
of all anthropoids
at root of the tree

Figure 20.13
Phylogenetic trees. This phylogenetic tree for Anthropoidea, the
clade that includes monkeys,
apes, and humans, illustrates
properties shared by most phylogenetic trees. The relative positions of the nodes (branching
points) define how recently sister
clades diverged. Clades that
emerge from a recent common
node (near the tips of the
branches) are more closely
related to each other than clades
that emerged from an older node
(closer to the root of the tree).

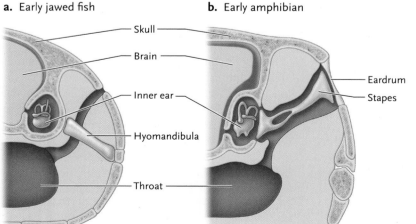

a. Early jawed fish

b. Early amphibian

Skull

Brain

Inner ear

Hyomandibula

Throat

Eardrum

Stapes

Figure 20.14
Homologous bones, different structures and functions.
(a) The hyomandibula braced the jaw joint against the skull in early
jawed fishes. **(b)** The hyomandibula is homologous to the stapes,
which transmits sound to the inner ear in tetrapods, exemplified
here by an early amphibian. Both diagrams show a cross-section
through the head just behind the jaw joint.

Analogous Characters. Analogous characters are those in different animals that serve the same function. They are homoplasious (plural noun form **homoplasies**), phenotypic similarities that evolved independently in different lineages. For example, the flattened tails of aquatic mammals such as beavers and platypuses appear to be homoplasious. Systematists exclude homoplasies from their analyses because they provide no information about shared (genetic) ancestry.

The situation can be complex. For example, flight in animals has evolved at least four times (bats, birds, insects, pterosaurs). Bones in the wings of flying vertebrates (bats, birds, and pterosaurs) are homologous **(Figure 20.15).** They have the same basic structural elements (arm, wrist, and hand), with similar spatial relationships to each other and to the bones that attach the wing to the rest of the skeleton (shoulder girdle). In the details of the bones, however, the wings of bats, birds, and pterosaurs are quite different (Figure 20.15). Wing bones of bats, birds, and pterosaurs are homologous to the forelimbs of other tetrapods.

But the large flat surfaces of bird wings are homoplasious with those of bats and pterosaurs. Feathers form the flight surfaces of birds, whereas those of bats and pterosaurs are made of skin. Therefore, one could assert that in their flight membranes, birds are convergent with bats and pterosaurs. When we extend the comparison, the wings of insects are convergent with those of vertebrates (bats, birds, pterosaurs) (Figure 20.15). In this situation, the wings of vertebrates are examples of parallel evolution. When you consider the fine details, the basic elements supporting the wings of bats, birds, and pterosaurs are homologous. However, the details of the forearm, hand, and finger bones differ substantially among these three groups of animals.

The example of wings illustrates that the distinction between parallel and convergent evolution is based on closeness of relationships, and the groups included in the comparison.

Ancestral and Derived Characters. **Mosaic evolution** refers to the reality that in all evolutionary lineages, some characters evolve slowly, while others evolve rapidly. Mosaic evolution is pervasive. Every species displays a mixture of **ancestral characters** (old forms of traits) and **derived characters** (new forms of traits). Derived characters provide the most useful information about evolutionary relationships because once a derived character is established, it usually persists in all of that species' descendants. Thus, unless derived characters are lost or replaced by newer characters over evolutionary time, they can serve as markers for entire evolutionary lineages.

Systematists score characters as either ancestral or derived only when comparing them among organisms. Thus, any particular character is derived *only in relation to* what occurs in other organisms. The comparison may be with an older version of the same character or, sometimes, its absence and the appearance of a new trait.

Most species of animals are invertebrates, by definition lacking a vertebral column. Backbones are a defining feature of vertebrates, the animal lineage that includes fishes, amphibians, reptiles, birds, and mammals. When systematists compare vertebrates with all animals lacking a vertebral column, they score the absence of a vertebral column as the ancestral condition. The presence of a backbone is a derived character.

Systematists also distinguish between ancestral and derived characters to ascertain in which direction a character has evolved. In some cases, the fossil record is detailed enough to provide unambiguous information about the direction of evolution. Biologists are confident that the presence of a vertebral column is a derived character because the earliest fossil animals were invertebrates.

Systematists use **outgroup comparison** to distinguish ancestral from derived characters. This involves comparing the group under study with more distantly related species constituting a group not otherwise included in the analysis. For example, most modern butterflies have six walking legs, but some species in two families (Nymphalidae and Papillionidae) have four walking legs and two small, nonwalking legs **(Figure 20.16).** Which is the ancestral character state? Which is derived? Outgroup comparison with other insects shows that six walking legs is the prevalent condition, representing an ancestral character. Four walking legs is a derived character. The same would apply when trying to understand the almost legless condition of female bagworms (Psychidae), another group of butterflies.

Figure 20.15
Arm, hand, and finger bones support the wings of a bat **(a)**, a bird **(b)**, and a pterosaur **(c)**. The wrist position (arrow) is more similar between bats and birds than either is to the pterosaur.

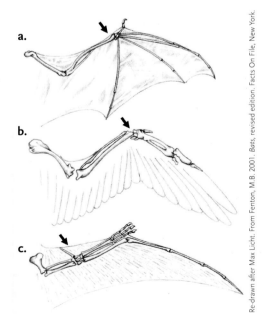

Re-drawn after Max Licht. From Fenton, M.B. 2001. *Bats*, revised edition. Facts On File, New York.

a. Caddis fly

b. Orange palm dart butterfly

c. Monarch butterfly

Figure 20.16
In preparing a phylogeny, it is vital to include in the comparison a species that is an outgroup. In this case, **(a)** the caddis fly (order Trichoptera, family Limnephilidae) is not as closely related as the two butterflies: **(b)** orange palm dart butterfly (*Cephrenes auglades,* family Hesperiidae) and **(c)** monarch butterfly (*Danaeus plexippus,* family Nymphalidae). The comparison suggests that six walking legs (a and b) is ancestral in insects, and four walking legs (c) is a derived character state.

20.4b Phylogenetic Inference and Classification

Phylogenetic trees portray the evolutionary diversification of lineages as a hierarchy that reflects the branching pattern of evolution. Each branch represents the descendants of a single ancestral species. When converting the phylogenetic tree into a classification, systematists use the **principle of monophyly**. They try to identify **monophyletic taxa**, those derived from a single ancestral species **(Figure 20.17)**. **Polyphyletic taxa** include species from separate evolutionary lineages. If, based on the presence of wings, we placed bats, birds, pterosaurs, and insects in one taxonomic group (flying animals), it would be polyphyletic. A **paraphyletic taxon** includes an ancestor and some, but not all, of its descendants. The traditional taxon class Reptilia is paraphyletic because it includes some obvious reptiles, such as turtles, lizards, dinosaurs, and crocodiles, but not other descendants, such as mammals and birds.

Many systematists also strive to create parsimonious phylogenetic hypotheses. According to the **assumption of parsimony** (also known as Occam's Razor), the simplest explanation of an issue is usually the most accurate. Systematists assume that any particular evolutionary change is an unlikely event and presumably happened only once in any evolutionary lineage. Phylogenetic trees illustrate hypotheses that place all organisms on a single branch. Birds are portrayed as a single evolutionary branch, implying that feathered wings evolved once in their common ancestor. This hypothesis is more parsimonious than one proposing that feathered wings evolved independently in two or more vertebrate lineages. The monophyly of birds is not contradicted by the repeated evolution of flightlessness in this group. Recently discovered fossils have changed our appreciation of the evolution of birds (see Section 20.2).

20.4c Phenotypic Similarities and Differences: Traditional Evolutionary Systematics

For a century after the publication of Charles Darwin's theory of evolution by natural selection, most systematists followed Linnaeus' practice of inferring

Monophyletic taxon

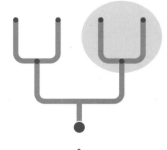

A monophyletic taxon includes an ancestral species and all of its descendants.

Polyphyletic taxon

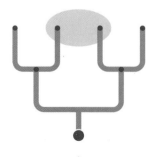

A polyphyletic taxon includes species from different evolutionary lineages.

Paraphyletic taxon

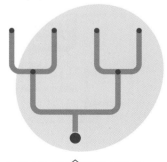

A paraphyletic taxon includes an ancestral species and only some of its descendants.

Figure 20.17
Defining taxa in a classification. Systematists can create different classifications from the same phylogenetic tree by identifying different groups of species as a single taxon (shaded).

evolutionary relationships from phenotypic similarities and differences. This represents **traditional evolutionary systematics,** which places together groups of species sharing ancestral and derived characters. Mammals are defined by their internal skeleton, vertebral column, and four limbs—ancestral characters among tetrapod vertebrates. But mammals also have derived characters such as hair, mammary glands, and a four-chambered heart (see Chapter 28). The four-chambered heart (see Section 40.1) also occurs in birds and some reptiles.

Classifications produced by traditional systematics reflect evolutionary branching and morphological divergence **(Figure 20.18a).** Among tetrapod vertebrates, the amphibian and mammalian lineages diverged early, followed shortly by the divergence of the turtle lineage and then that of other reptiles. After this, subsequent divergences produced two groups: lepidosaurs (lizards and snakes) and archosaurs (crocodilians, dinosaurs, and birds). Although crocodilians outwardly resemble lizards, they share a more recent common ancestor with birds. Yet birds differ from crocodilians in many morphological characters, including feathers and wings.

Even though the phylogenetic tree of tetrapod vertebrates shows six living groups, the traditional classification recognizes four classes: Amphibia, Mammalia, Reptilia, and Aves (birds). These groups (classes in classification) are given equal ranking because each represents a distinctive body plan and way of life. The class Reptilia, however, is a paraphyletic taxon because it includes some descendants of the common ancestor, namely turtles, lizards, snakes, and crocodilians (Figure 20.18a). But class Reptilia does not include other descendants of archosaurs, namely birds.

Traditional evolutionary systematists justify this definition of Reptilia because it includes morphologically similar animals with close evolutionary relationships. Crocodilians are classified with lizards, snakes, and turtles because they have a common ancestry and are covered with dry, scaly skin. Traditional systematists also argue that the key innovations thought to have initiated the adaptive radiation of

a. Traditional phylogenetic tree with classification

b. Cladogram with classification

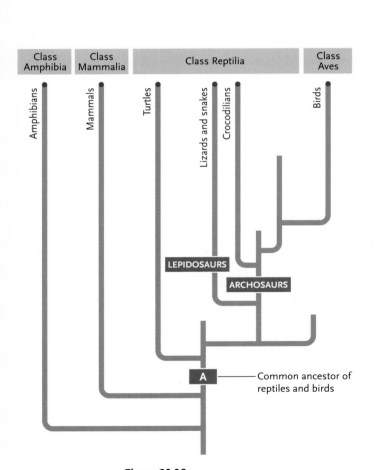

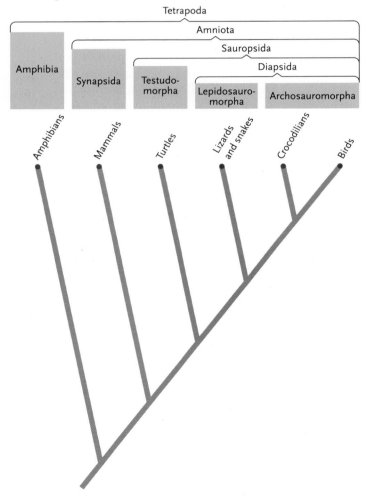

Figure 20.18

Phylogenetic trees and classifications for tetrapod vertebrates. **(a)** Traditional and **(b)** cladistic phylogenies produce different phylogenetic trees and classifications. Classifications are presented above the trees.

birds—wings, feathers, high metabolic rates, and flight—are extreme divergences from the ancestral morphology. Therefore, birds merit recognition as a separate class. The question is where to draw the line between birds and reptiles.

20.4d Cladistics: Analytical Classifications

Cladistics emerged in the 1950s and 1960s when some researchers criticized the inherent lack of clarity in classifications based on two distinct phenomena, branching evolution and morphological divergence. After all, how can we tell why two groups or organisms are classified in the same higher taxon? In some cases they have a recent common ancestor (e.g., lizards and snakes), but in other cases they do not (e.g., lizards and crocodilians).

To minimize such confusion, many systematists followed the philosophical and analytical lead of Willi Hennig, the German entomologist who wrote *Phylogenetic Systematics,* published in 1966. Hennig and his followers argued that classifications should be based solely on evolutionary relationships. **Cladistics** ignores morphological divergence, producing phylogenetic hypotheses and classifications that reflect only the branching pattern of evolution.

Cladists place species that share derived characters in one group. They argue that mammals form a monophyletic lineage, a **clade,** because they have a unique set of derived characters, including hair, mammary glands, reduction of bones in the lower jaw, and a four-chambered heart. The ancestral characters of mammals, such as an internal skeleton, a vertebral column, and four legs, do not distinguish them from other tetrapod vertebrates, so these traits are excluded from analysis.

Phylogenetic trees produced by cladists **(cladograms)** illustrate the hypothesized sequence of evolutionary branchings, with a hypothetical ancestor at each branching point (Figure 20.18b). Cladograms portray strictly monophyletic groups and are usually constructed using the assumption of parsimony. Once a researcher identifies derived, homologous characters, constructing a cladogram is straightforward (see "Constructing a Cladogram," p. 462).

Classifications produced by cladistic analysis often differ radically from those of traditional evolutionary systematics. Pairs of higher taxa are defined directly from the two-way branching pattern of the cladogram. Thus, the clade **Tetrapoda** (the traditional amphibians, reptiles, birds, and mammals) is divided into two taxa, Amphibia (tetrapods with no amnion (they lay their eggs in water); see Chapters 28 and 42) and Amniota (tetrapods with an amnion). Amniota is subdivided into two taxa on the basis of skull morphology and other characters, namely Synapsida (mammals) and Sauropsida (turtles, lizards, snakes, crocodilians, and birds). Based on cranial structure, Sauropsida is

further divided into Testudomorpha (turtles) and Diapsida (lizards and snakes, crocodilians, and birds). Finally, based on anatomical details, Diapsida is subdivided into two more recently evolved taxa, Lepidosauromorpha (lizards and snakes) and Archosauromorpha (crocodilians and birds) (Figure 20.18b). The strictly cladistic classification parallels the pattern of branching evolution that produced the organisms included in the classification. These parallels are the essence and strength of the cladistic method.

Today most biologists use the cladistic approach because of its evolutionary focus, clear goals, and precise methods. Some systematists advocate abandoning the Linnaean hierarchy for classifying and naming organisms. They propose using a strictly cladistic system, called **PhyloCode**, that identifies and names clades instead of placing organisms into the familiar taxonomic groups. However, traditional evolutionary systematics has guided most laypeople's understanding of biological diversity.

20.5 Using Phylogenetics: The Evolution of Birds

The changes in the phylogeny of birds and their theropod relatives raise the obvious question, "What is a bird?" Twenty years ago, feathers were an obvious answer. The diversity of modern birds (Section 28.12, Chapter 28) makes it clear that all of them are feathered. But many dinosaurs also had feathers; indeed feathers probably evolved first as insulation for warm-blooded dinosaurs. Morphological and skeletal features associated with flight are also not the answer to "what is a bird?": not all living birds fly.

Most paleontologists accept that birds evolved from a lineage of theropod dinosaurs **(Figure 20.19, p. 463).** The lineage leading to theropods had appeared in the early Triassic and includes well-known species such as allosaurs, tyrannosaurs, and their relatives. The following numbers correspond to numbers in Figure 20.19:

1. The long bones are hollow, and the first digit of the foot plays little role in weight support.
2. The development of a rotary wrist joint is associated with a grasping hand.
3. Bones in the shoulder girdle and the sternum (breastbone) are expanded to support chest muscles. As well, feathers appear for insulation.
4. Vaned feathers appear.
5. The trunk is shortened and the tail stiffened, resulting in better balance and maneuverability.
6. Basic perching and flight behaviour appears by the end of the Jurassic. The last three modifications are associated with powered flight.
7. A deep thorax develops.
8. A canal houses the main wing rotation muscles.
9. An elastic wishbone and a strongly keeled sternum appear.

Constructing a Cladogram

Research Method
Box

Cladograms allow systematists (and others) to visualize hypothesized evolutionary relationships by grouping organisms that share derived characters. The cladogram also indicates where derived characters evolved.

Here we develop a cladogram for the nine extant groups of chordate vertebrates: lampreys (Agnatha), sharks (Chondrichthyes), bony fishes (Osteichthyes), amphibians (Amphibia), reptiles (turtles, lizards and snakes, crocodilians), birds, and mammals (see also Chapter 28). We also include lancelets (marine organisms in the subphylum Cephalochordata). Lancelets serve as the outgroup in our comparison.

We have chosen characters on which to base the cladogram **(Table 1),** noting the presence (+) or absence (−) of 10 different characters. The characters are ancestral or derived in each group, but the outgroup (the lancelets) lacks all of these traits. We construct the cladogram from the information in the table, grouping organisms that share derived characters (right branch, **Figure 1a**), whereas the lancelets form the left branch because they lack the derived characters.

The remaining organisms except lancelets and lampreys have jaws. Now the right branch (Figure 1b) includes all living vertebrates sharing derived characters, separating them from lancelets and lampreys. The selection of different characters might give different outcomes.

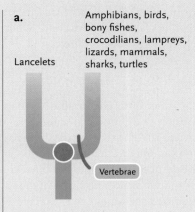

a.

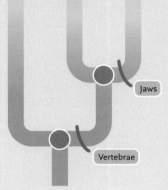

b.

Figure 1

(a) A cladogram showing the separation of lancelets from living chordates. *(b)* A cladogram showing the separation of lancelets and lampreys from most living chordates. Lampreys are chordates, but the cladogram suggests that they are the earliest chordates. These cladograms were prepared from the data in Table 1.

Table 1

	Vertebrae	Jaws	Swim Bladder or Lungs	Paired Limbs	Extraembryonic Membranes	Mammary Glands	Dry, Scaly Skin	Two Openings at Back of Skull	One Opening in Front of Eye	Feathers
Lancelets	−	−	−	−	−	−	−	−	−	−
Lampreys	+	−	−	−	−	−	−	−	−	−
Sharks	+	+	−	−	−	−	−	−	−	−
Bony fishes	+	+	+	−	−	−	−	−	−	−
Amphibians	+	+	+	+	−	−	−	−	−	−
Mammals	+	+	+	+	+	+	−	−	−	−
Turtles	+	+	+	+	+	−	+	−	−	−
Lizards	+	+	+	+	+	−	+	+	−	−
Crocodilians	+	+	+	+	+	−	+	+	+	−
Birds	+	+	+	+	+	−	+	+	+	+

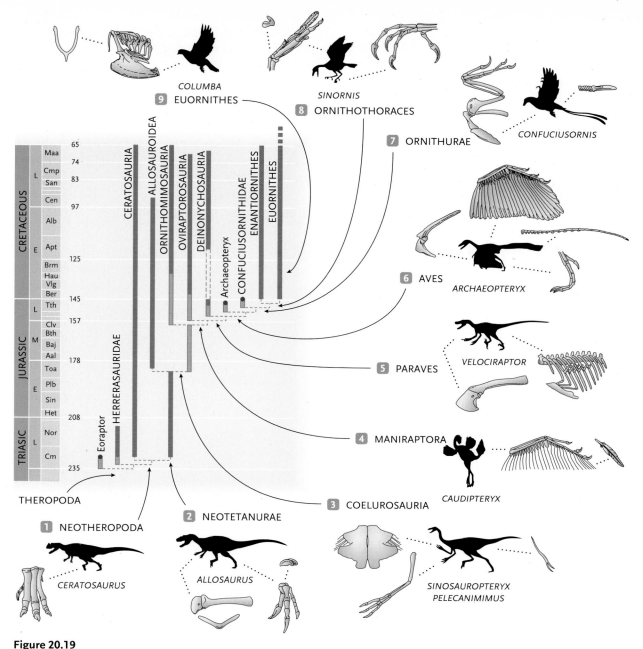

Figure 20.19

Nine morphological changes in the development of a bird from its theropod ancestors. Details in text.

From Paul C. Sereno, "The Evolution of Dinosaurs," *Science* 25 June 1999: Vol. 284 no. 5423 pp. 2137–2147. Reprinted with permission from AAAS.

In 2013, the theropod lineage Eumaniraptora was described as having three families: Dromaeosauridae, Troodontidae, and Avialae. The latter two make up the clade Paraves. Even in 2011, *Archaeopteryx* had been placed in the Avialae, arguably with other birds. Described in 1861, *Archaeopteryx lithographica* remains one of the best known fossil "birds," and there is ongoing disagreement about this species' ability to fly. Animals in the Deinonychosauria (4 and 5, Figure 20.19) would probably have been unable to fly, while *Archaeopteryx* would have had limited flight ability. In Figure 20.19, only Euornithes are "real" birds.

The arrangement of species in families changed with the discovery of fossils from a Late Jurassic lagerstätten (~150 mya) in northeast China. These deposits contain thousands of specimens preserved in lake deposits rich in volcanic ash. Fine details of soft tissues, including feathers, are preserved in these specimens. Two species are particularly important, *Xiaotingia zhengi* and *Aurornis xui,* described in 2011 and 2013, respectively. The anatomical details of these species and those of *Archaeopteryx* place all three genera in Paraves, but now *Archaeopteryx* falls within the Troodontidae; *Xiaotingia* and *Aurornis* in Avialae. If *Archaeopteryx* and other troodontids were capable of limited flight, it would appear that flight evolved once in this Paraves clade.

The story about the origin and evolution of birds is ongoing. It demonstrates the importance of fossils and modern techniques for examining them. More important, the story illustrates that phylogenies are hypotheses about evolutionary history. The hypotheses change as more data are brought to bear on the situation. In the next chapter, we will see how the same situation emerges about the history of our own species.

STUDY BREAK

1. What are the diagnostic characteristics of birds?
2. What explains the changes in the position of *Archaeopteryx* in the classification of birds?

20.6 Parallelism and Convergence Can Complicate the Scene

Understanding and documenting the diversity of life is a major challenge to biologists. The evolution of flight (above) illustrates the potential difficulty in recognizing parallel and convergent evolution. This exercise means determining evolutionary relationships among organisms and deciding if similar structures or similar-looking structures are grounds for grouping species together. In other words, if they look the same, are they closely related in an evolutionary sense?

When first discovered, ichthyosaurs **(Figure 20.20)** were thought to be fish. Later, details of their skeletons confirmed that they were not fish, but reptiles. The initial confusion is understandable because ichthyosaurs have fishlike bodies, as do mammals such dolphins and whales. Many aquatic vertebrates have fishlike bodies— think of sharks (cartilaginous fishes) and tunas (bony fishes). But these are two very different kinds of "fish." Other fish have very different bodies—think of eels, flatfish, anglerfish, and sea horses.

There is a tendency among organisms living under the same conditions to develop similar body forms. This can be called parallel or convergent evolution, depending on the evolutionary relatedness of the organisms involved. Convergent evolution is used when referring to phylogenetically more distantly related organisms; parallel evolution when referring to more closely related ones. How can you tell?

One of the most exciting and satisfying things about biology is learning to look at something, recognize it, and understand just what you are seeing. Below are two cases, examples of where even an experienced biologist might be fooled. At first, this may seem strange or even preposterous when you know what a flower looks like; surely, you will always recognize one. Read on to see that things are not always as they appear.

20.6a Case 1: What Looks Like a Flower May Not Be One

You may be astonished to realize that what you thought was a flower **(Figure 20.21a, b)** is actually a leaf modified by a fungus. The fungus, the rust *Puccinia monoica*, affects the growth of the leaves, changing their appearance and odour. The fungus-induced "flowers" have nectaries (glands that produce nectar; Figure 20.21c). Just as many biologists are fooled by the flowerlike leaves, so are insects that come to pollinate the flowers. In this way, the rust interferes with fertilization of the plant. The rust also inhibits the formation of the plant's own flowers, minimizing confusion among pollinators. So when is a flower not a flower?

20.6b Case 2: Some Plants Are Carnivorous

In some places with an abundance of water and sunlight, nitrogen for plants can be in short supply. Here we find a diversity of ways that plants trap insects to directly (or indirectly) obtain nitrogen (**Figure 20.22**; see Figure 1, "Pitcher Plant Ecosystems," Chapter 31). Like people trying to catch insects (but perhaps not for their nitrogen), plants use different methods. Insectivorous plants catch insects in sticky traps (flypaper), snap traps, and pitfall traps (pitchers). Flypaper traps have appeared in at least five evolutionary lines of plants and pitchers at least three times. For the most part, carnivorous plants do not share a close common ancestor.

20.7 Putting Fossils and Phylogeny in Perspective

The fossil record is the history book for life on Earth, and the study of extant organisms is not complete without the study of extinct ones. This chapter illustrates the richness, completeness—and incompleteness—of the fossil record. Systematists cannot complete phylogenetic trees or cladograms without considering early organisms. These phylogenetic diagrams indicate the evolutionary relatedness of organisms.

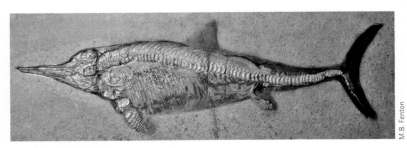

Figure 20.20
An ichthyosaur (*Stenopterygius* sp.) from Germany on display at the Royal Tyrrell Museum in Drumheller, Alberta. Although this specimen is about 2 m long, the largest ichthyosaurs were over 20 m in length.

a.

b.

c.

Figure 20.21
(a) There is an obvious difference between an uninfected *Boerchera* species (left) and an infected one (right).
(b) When infected by the fungus *Puccinia monoica*, a rust, the leaves of *Boerchera* become flowerlike and **(c)** appear to produce nectar. The rust inhibits flowering so that insects visiting the "flowers" to collect nectar fertilize the rust. The insects do not visit the host plant's flowers.

a.

b.

c.

d.

e.

f.

g.

h.

Figure 20.22
Plants that catch insects in pitchers: **(a)** *Cephalotus follicularis*, **(b)** *Sarracenia purpurea*, **(c)** *Darlingtonia california*, and **(d)** *Nepenthes*; on flypaper: **(e)** *Drosera capensis*, **(f)** *Pinguicula* spp., and **(g)** *Brocchinia reducta*; or in a snap trap: **(h)** *Dioneae muscipula*.

Review

To access course materials such as Aplia and other companion resources, please visit www.NELSONbrain.com.

20.1 The Geological Time Scale

- Sediments found in any one place form distinctive strata (layers) that usually differ in colour, mineral composition, particle size, and thickness.
- Geologists used the sequence of strata and their distinctive fossil assemblages to establish the geologic time scale that diagrams the history of life on Earth.

20.2 The Fossil Record

- Fossils are the parts of organisms preserved in sedimentary rocks or in oxygen-poor environments.
- Fossils can include vertebrate bones, insects in amber, impression of plants and softbodied animals, footprints, coprolites, gastroliths, and frozen or mummified organisms.
- An "explosion" in the diversity of groups of organisms, about 600 mya, marked the beginning of the Cambrian. Complex multicellular animals first appeared at this time.
- Mass extinctions have occurred at least five times in the history of life, with possible causes being volcanic activity, climate change, and asteroid strikes. Following a mass extinction, the surviving organisms undergo adaptive radiation.

20.3 Lives of Prehistoric Organisms

- The first land plants, similar to today's liverworts and mosses, appeared during the Ordovician Period.
- For animals to move onto land required several adaptations, such as sufficient support to allow them to maintain their integrity, and some form of protection from ultraviolet radiation. Adaptations to mechanisms of gaseous exchange, excretion, and locomotion were also needed.
- Whales and toothed whales are a distinctive group of mammals whose fossil record extends back more than 50 mya.

20.4 Phylogeny

- Phylogenetic trees are hypotheses that portray the branching pattern of evolution. Most phylogenetic trees have an implicit or explicit time line that indicates the relative times for cladogenesis. Branch points are described as nodes, and monophyletic lineages are called clades. Two lineages that share a node are called sister clades. Clades can be rotated at nodes without changing the meaning of the tree.
- In preparing a phylogeny, systematists seek traits in which phenotypic variation reflects genetic differences rather than environmental variation. Systematists also use genetically independent traits, which reflect different parts of an organism's genome.

- Homologous characters have been inherited from a common ancestor, so phenotypic similarities between organisms reflect underlying genetic similarities. Homologous characters can differ considerably among species. Analogous (homoplasious) characters are phenotypically similar and have similar functions but evolved independently in different lineages.
- Systematists compare homologous characters to determine common ancestry and genetic relatedness. They exclude analogous structures because they provide no information about shared ancestry or genetic relatedness.
- Mosaic evolution refers to the reality that some characters evolve more slowly or more quickly than others. Ancestral characters are old forms of traits, and derived characters are new forms. Once a derived character becomes established, it occurs in all the species' descendants and is a marker for evolutionary lineages.
- An outgroup comparison can be used to identify ancestral and derived traits because it compares the group under study with more distantly related species not otherwise included in the analysis.
- A monophyletic taxon is a group of species derived from a single ancestral species. A polyphyletic taxon includes species from separate evolutionary lineages. A paraphyletic taxon includes an ancestor and some, but not all, of its descendants.
- According to the assumption of parsimony, any particular evolutionary change is a rare event, unlikely to have occurred twice in one lineage. Therefore, the fewest possible evolutionary changes should be used to account for within-lineage diversity.
- Derived character states can serve as markers of clades.
- Systematists use evidence from the fossil record as well as outgroup comparison to identify which character states are derived and which are ancestral.
- Cladistic analyses use synapomorphies (derived character states) to construct phylogenetic hypotheses.

20.5 Using Phylogenetics: The Evolution of Birds

- Most paleontologists accept that birds evolved from a lineage of theropod dinosaurs.

20.6 Parallelism and Convergence Can Complicate the Scene

- Convergent evolution refers more to distantly related organisms and parallel evolution to more closely related ones. The evolution of flight is convergent between insects (phylum Arthropoda) and vertebrates (phylum Chordata: birds, pterosaurs, and bats) and parallel within the vertebrates.

Questions

Self-Test Questions

1. The fossil record does which of the following?
   a. provides direct and indirect evidence about life in the past
   b. shows that all morphological novelties arise rapidly
   c. provides abundant data about rare species with local distributions
   d. is equally good for all organisms that ever lived
   e. provides no evidence about the physiology or behaviour of ancient organisms

2. What is the absolute age of a geological stratum determined by?
   a. the thickness of its rocks
   b. the particle size in its rocks
   c. the types of fossils found within it
   d. its position relative to other layers
   e. radiometric dating techniques

3. Fossils can be which of the following?
   a. mounted skeletons of mineralized bone
   b. spiders preserved in amber
   c. moulds or impressions of Anomalocaris
   d. coprolites or gastroliths
   e. all of the above

4. Why do adaptive radiations often follow mass extinctions?
   a. Mass extinctions limit the impact of paedomorphosis.
   b. Mass extinctions foster allometry and heterochrony.
   c. Mass extinctions decimate all forms of life on Earth.
   d. Species that form transitional fossils often survive mass extinctions.
   e. Extinctions open adaptive zones that had been previously occupied.

5. Fill in the blank. A phylogenetic tree portrays the ___ of a group of organisms.
   a. classification
   b. evolutionary history
   c. domain
   d. distribution

6. When systematists use morphological or behavioural traits to reconstruct the evolutionary history of a group of animals, what are they assuming?
   a. Phenotypic characters reflect underlying genetic similarities and differences.
   b. The animals use exactly the same traits to identify appropriate mates.
   c The adaptive value of these traits can be explained.
   d. Variations are produced by environmental effects during development.

7. Which of the following pairs of structures are homoplasious?
   a. the wing skeleton of a bird and the wing skeleton of a bat
   b. the wing of a bird and the wing of a fly
   c. the eye of a fish and the eye of a human
   d. the wing structures of a pterosaur and those of a bird

8. Which of the following does NOT help systematists determine which version of a morphological character is ancestral and which is derived?
   a. outgroup comparison
   b. patterns of embryonic development
   c. studies of the character in more related species
   d. dating of the character by molecular clocks

9. In a cladistic analysis, a systematist groups together organisms that share which of the following?
   a. derived homologous traits
   b. derived homoplasious traits
   c. ancestral homologous traits
   d. ancestral homoplasious traits

10. How would one construct a cladogram by applying the parsimony assumption to molecular sequence data?
    a. Start by making assumptions about variations in the rates at which different DNA segments evolve.
    b. Group organisms sharing the largest number of ancestral sequences.
    c. Group organisms that share derived sequences, matching the groups to those defined by morphological characters.
    d. Group organisms sharing derived sequences, minimizing the number of hypothesized evolutionary changes.

Questions for Discussion

1. Traditional evolutionary systematists identify the Reptilia as one class of vertebrates, even though this taxon is paraphyletic. What are the advantages and disadvantages of defining paraphyletic taxa in a classification?

2. Create an imaginary phylogenetic tree for an ancestral species and its 10 descendants. Circle a monophyletic group, a polyphyletic group, and a paraphyletic group on the tree. Explain why the groups you identify match the definitions of the three types of groups.

3. Imagine that you are trying to determine the evolutionary relationships among six groups of animals that look very much alike because they have few measurable morphological characters. What data would you collect to reconstruct their phylogenetic history?

4. The geological evolution of Earth has had an obvious effect on biological evolution. You have read about how the release of oxygen by photosynthetic organisms increased atmospheric oxygen concentration. How are human activities changing the physical environment on Earth? What new selection pressures do these environmental alterations establish?

© Malcolm Park editorial/Alamy

Neanderthal Museum

Neandertal (left) and *Homo sapiens* (right). DNA techniques, as well as fossils, are advancing our knowledge about the evolutionary history of humans.

21

Humans and Evolution

WHY IT MATTERS

From about 500 thousand to 30 thousand years ago, neandertals (*Homo neandertalensis*) roamed much of Europe and eastern and central Asia. Humans (*Homo sapiens*) arrived in the area about 40 thousand years ago. Neandertals were shorter, more heavily built, and stronger than humans, and their brains were larger as well. But the neandertals disappeared about 30 thousand years ago, and paleontologists have often wondered why humans won out. Hypotheses include interbreeding between the two species (or subspecies), larger eyes in the neandertal, and climate change. Only about 2% of the DNA of modern Europeans is neandertal, so why didn't the two species simply interbreed and blend together? Also, surprisingly, no neandertal mitochondrial DNA (mtDNA) has shown up in human mDNA. Since mDNA is passed only from mother to child, this suggests that while neandertal fathers and human mothers might have produced viable offspring, neandertal mothers and human fathers could not. Was this the only thing that separated *H. sapiens* from *H. neandertalensis*?

Recently, researchers in Seattle have developed the "Brainscan Atlas," a genetic reference about how the human brain is constructed and how it develops embryonically. Drs. Mohammed Uddin and Stephen Scherer, at Toronto's Hospital for Sick Children, used the Atlas and the Exome Variant Server (a database of all of a human's exomes) to determine which genes were responsible for autism. Uddin and Scherer determined that about 1700 genes were related only to brain development, not to any other function in the body. They then discovered that at least one-third of these 1700 genes had been implicated by other researchers in other brain and cognitive disorders.

Now, back to the neandertal conundrum: Uddin and Scherer suddenly thought that the 1700 genes could somehow be related to what makes humans uniquely human. People with autism have difficulties with communication and socialization; could these abilities also be what sets us apart from neandertals?

Dr. Ajit Varki at the Univerity of California, San Diego, thinks that despite the physical similarities between humans and neandertals, they could have been very different in **social behaviour** and in their abilities to communicate. Consequently, children with one *H. sapiens* parent and one *H. neandertalensis* parent could have been "cognitively sterile," with difficulties in communication making them likely to reproduce successfully.

This research was published in 2014, not long before this book was printed. By the time you read this, paleontologists may know a lot more about the *H. sapiens*–*H. neandertalensis* relationship.

DNA techniques as well as fossils are advancing our knowledge about the evolutionary history of humans. In this chapter we focus on some of the most important changes in our ancestry. We consider the implications of **bipedalism**, showing that it is much more than walking erect on two legs. We also present some of the recently discovered fossils and consider how the biological species concept applies to our own species.

21.1 The Fossil Record of Hominins

A combination of genetic and morphological analyses of living and fossil species indicates that between 5 and 10 mya in Africa, **hominoids** (superfamily Hominoidea, including apes and humans) had diverged into several lineages. One lineage, the **hominins** (family Hominidae, subfamily Homininae), includes modern humans and our bipedal ancestors (see Figure 20.13 and "The Cast of Characters: Fossil Hominins"). Where only one species of hominin (*Homo sapiens*) exists today, several species lived in the past.

The Cast of Characters: Fossil Hominins

Most of our ancestors' fossils have been found in Africa **(Figure 1)**. Brain capacity varies with overall body size, so that large individuals (typically males) have more brain capacity than smaller ones (typically females). Across the species presented below, brain capacity ranges from the size of chimpanzee brain capacity (275–500 cm³, *Orrorin tugensis*) to that of *Homo sapiens* (1000–1900 cm³). Species of *Australopithecus* have brain capacities about 400–500 cm³; *Homo habilis* about 640 cm³. *Homo erectus* brain capacity ranges from 930 to 1030 cm³; *Homo neandertalensis* from 1300 to 1600 cm³.

Orrorin tugensis: In 2000, researchers found 13 fossils of *O. tugensis* ("first man" in a local African language), a species that lived in the forests of eastern Africa about 6 mya. The thigh bones and pelvis indicate that it was bipedal.

Ardipithecus ramidus: In 1994, *Ardipithecus ramidus* was described from bone fragments (teeth and jaw fragments) collected in South Africa. These hominids stood 120 cm tall and had apelike teeth. The October 2, 2009, issue of *Science* included 11 papers about *A. ramidus* by an international team of researchers who reported data from 110 specimens. This species lived from about 4 to 6 mya, and many of its features

overturned ideas about the evolution of our own species. The structure of its pelvis and feet suggested that it was bipedal. Both males and females had small canine teeth (compared to those of other primates), which imply reduced competition between males, presumably for females in oestrus, in turn suggesting concealed ovulation (see Chapter 42). Bipedal locomotion could have enabled the hominins to exploit both land surface and trees in the search for food and shelter. Bipedal animals can carry food and be more effective provisioners. Many features of *A. ramidus* demonstrate that our ancestors showed an earlier than expected departure from a chimpanzee-like existence.

Australopithecus africanus: The first australopith to be described, *Australopithecus africanus*, was discovered by Raymond Dart in 1924. With its relatively small brain, this bipedal species was not immediately recognized as a hominin.

Australopithecus afarensis: Specimens of more than 60 individuals have been found in northern Ethiopia. The sample includes about 40% of a female's skeleton, named "Lucy" (Figure 21.5, p. 474; apparently the Beatles' song "Lucy in the Sky with Diamonds" was playing on the radio when the skeleton was first uncovered). *A. afarensis* lived

3 to 3.5 mya. This species retained several ancestral characters, including moderately large and pointed canine teeth and a relatively small brain. Males and females were 150 cm and 120 cm tall, respectively. Skeletal analyses suggest that Lucy was fully bipedal, a conclusion supported by fossilized footprints preserved in a layer of volcanic ash (Figure 21.1, p. 473). In 2010, the description of a male specimen of *Australopithecus afarensis* ("big man") provided further evidence of bipedalism, specifically details of the pelvic girdle and sacrum not preserved in Lucy. Furthermore, a well-preserved scapula (shoulder blade) provided no evidence of suspensory climbing evident in the scapulae of great apes.

Australopithecus anamensis: One of the oldest known species in the genus, *Australopithecus anamensis* lived in eastern Africa around 4 mya. Its teeth had thick enamel, which is typically a derived hominin character. A fossilized thigh bone suggests that it was bipedal.

Australopithecus sediba: In September 2011, the journal *Science* published a series of papers describing the newly discovered *Australopithecus sediba*, an "early" (2 mya) version of Lucy (*A. afarensis*). The fossils, from the Malapa site in South Africa, provided a new perspective on the evolution of humans. The ankles and feet suggest a

(Continued)

The Cast of Characters: Fossil Hominins (*Continued*)

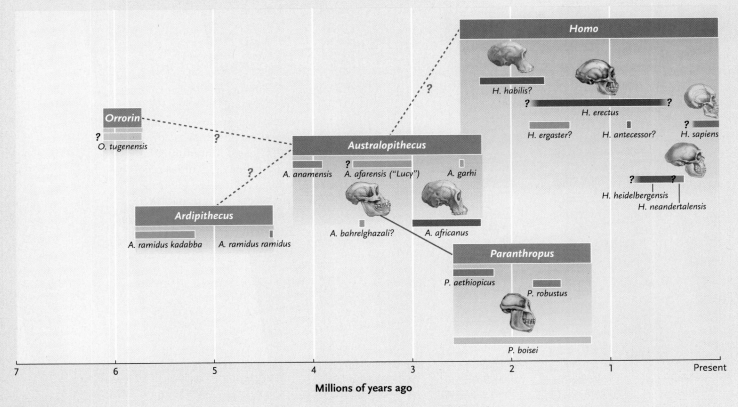

Figure 1

Hominin timeline showing several of the species described in the text. These species lived at the same place and time in eastern and southern Africa. The timeline is shown for each species, and question marks indicate uncertainty about classification and/or ages of fossils. Some skulls are reconstructions from fragments.

combination of climbing ability (arboreal life style) and bipedalism that differed from that of Lucy (see page 474). The pelvis of *A. sediba* shared features, such as the sacral and pubic areas, with those of Lucy and other australopithecines. But in the shape of its ilium, *A. sediba* resembled species of *Homo*. These features in *A. sediba* suggest that giving birth to offspring with large brains had not appeared at this stage of australopithecine evolution.

Some of the fossil *A. sediba* had nearly complete wrists and hands, indicating the capacity for strong flexion typically associated with tree climbing. But the long thumb and short fingers of *A. sediba* imply a capacity for precision gripping (see Figure 21.8, p. 476), suggesting the potential for use of tools. Construction of a virtual endocast of the skull of *A. sediba* indicates that the frontal lobes of the brain were generally like those of other australopithecines. But the features of the brain suggest its

gradual reorganization toward the appearance in species of *Homo*.

This new fossil demonstrates, once again, how such a find can influence our view of evolutionary history. This is not a "missing link" but rather a species whose features reflect the general state of evolutionary development of species in the genus *Australopithecus*. It also presages the transitions that later occur with the emergence of species in the genus *Homo*.

Homo habilis: Pliocene fossils of the earliest *Homo* are fragmentary and widely distributed in space and time. They are thought to have belonged to *Homo habilis* (meaning "handy man"). From 1.7 to 2.3 mya, *H. habilis* occupied the woodlands and savannas of eastern and southern Africa, sharing these habitats with various species of *Paranthropus*. The two genera are easy to distinguish because the brains of *H. habilis* were at least 20% larger, and their **incisors** were larger and **molars** smaller than those of *Paranthropus* spp. They ate hard-shelled nuts and

seeds, as well as soft fruits, tubers, leaves, and insects. They may also have hunted small prey or scavenged carcasses left by larger predators.

Researchers have found numerous tools dating to the time of *H. habilis* but are not sure which species made them. Many hominid species of that time probably cracked marrowbones with rocks or scraped flesh from bones with sharp stones. Paleoanthropologist Louis Leakey was the first to discover evidence of tool-*making* at eastern Africa's Olduvai Gorge, which cuts through a great sequence of sedimentary rock layers. The oldest tools at this site are crudely chipped pebbles, probably manufactured by *H. habilis*. However, humans are not the only animals to use tools (see Chapter 47).

Homo erectus: Early in the Pleistocene, about 1.8 mya, a new species of humans, *Homo erectus* ("upright man"), appeared in eastern Africa **(Figure 2).** One nearly complete skeleton suggests that *H. erectus* was taller than its ancestors and had a much larger brain, a thicker skull, and protruding brow ridges. *H. erectus* made fairly sophisticated tools, such as hand axes **(Figure 3)** used to cut food and other materials, to scrape meat from bones, and to dig for roots. *H. erectus* probably ate both plants and animals and may have hunted and scavenged animal prey. Archaeological data point to their use of fire to cook food and to keep warm. Near Lake Turkana in Kenya, fossils identified as *Homo* and dating from 1.45 to 1.55 mya were described in 2007. These suggested that *H. erectus* and *H. habilis* lived together in the same habitats for a considerable time, much as chimps and gorillas do today. Adult male *H. erectus* were much larger than adult females, suggesting a polygynous lifestyle, one male with several females (see Chapter 47).

About 1.5 mya, the pressure of growing populations apparently forced groups of *H. erectus* out of Africa. They dispersed northward from eastern Africa into both northwestern Africa and Eurasia. Some moved eastward through Asia as far as the island of Java. Judging from its geographic distribution, *H. erectus* was successful in many environments. It produced several descendant species, of which modern humans (*H. sapiens*, meaning "wise man") are the only survivors. Now-extinct descendants of *H. erectus*, archaic humans, first appeared at least 400 thousand years ago. They generally had larger brains, rounder skulls, and smaller molars than *H. erectus*.

Homo floresiensis: *H. floresiensis* was described in 2004 from Flores Island in Indonesia. Although first proposed as a distinct species, its small size was used to support the view that it was just a small individual. In 2013, analyses of various aspects of the morphology of *H. floresiensis* indicated that it was not a dwarf or microcephalic, rather a distinct species most closely related to *H. erectus*.

Homo neandertalensis: Neandertals lived in Europe and western Asia from 28 thousand to 150 thousand years ago. They are the best known of the archaic humans and sometimes have been treated as a subspecies of *Homo sapiens*. Compared with modern humans, they had a heavier build,

a. *Homo erectus*

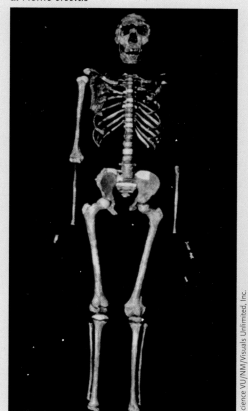

Figure 2
Homo erectus, *a nearly complete skeleton from Kenya.*

b. Hand axe

Figure 3
A hand axe found at a site used by Homo erectus.

more pronounced brow ridges, and slightly larger brains. Neandertals were culturally and technologically sophisticated. They made complex tools, including wooden spears, stone axes, flint scrapers, and knives. At some sites, they built shelters of stones, branches, and animal hides, and they routinely used fire. They were successful hunters and probably ate nuts, berries, fishes, and bird eggs. Some groups buried their dead, and they may have had rudimentary speech. There is evidence that some were cannibals.

In 1997, two teams of researchers independently analyzed short segments of mtDNA extracted from the fossilized arm bone of a neandertal. Unlike nuclear DNA, which individuals inherit from both parents, only mothers pass mtDNA to offspring. mtDNA does not undergo genetic recombination (see Chapter 9) and has a high mutation rate, making it useful for phylogenetic analyses. If mutation rates in mtDNA are fairly constant, this molecule can serve as a **molecular clock.** Comparing the

neandertal sequence with mtDNA from 986 living humans revealed three times as many differences between the neandertals and modern humans as between pairs of modern humans in their sample. These data suggest that neandertals and modern humans are different species that diverged from a common ancestor 550 thousand to 690 thousand years ago.

Homo sapiens: Modern humans differ from neandertals and other archaic humans in having a slighter build, less protruding brow ridges, and a more prominent chin. The earliest fossils of modern humans found in Africa and Asia are 150 thousand years old; those from the Middle East are 100 thousand years old. Fossils from about 20 thousand years ago are known from western Europe, the most famous being those of the Cro-Magnon deposits in southern France. The widespread appearance of modern humans roughly coincided with the demise of neandertals in western Europe and the Middle East, 28 thousand to 40 thousand years ago.

Most of the hominins that lived in eastern and southern Africa from 1 to 6 mya are currently classified in the genera *Australopithecus* (*australo* = southern; *pithecus* = ape) and *Paranthropus* (*para* = beside; *anthropus* = man). With large faces, protruding jaws, and small skulls and brains, these hominins resembled apes. Between 1 and 3.7 mya, several other species of hominins occurred in eastern and southern Africa.

These adult males ranged from 40 to 50 kg in mass and from 130 to 150 cm in height; females were smaller. Most of these species had deep jaws and large molars, suggesting a diet of hard food, such as nuts, seeds, and other vegetable products. *Australopithecus africanus,* known only from southern Africa, had small jaws and teeth, suggesting a diet of softer food. The phylogenetic relationships among species in the

Chromosomal Similarities and Differences among the Great Apes

The banding patterns of humans and their closest relatives among the apes—chimpanzees, gorillas, and orangutans—reveal that whole sections of chromosomes have been rearranged over evolutionary time. For example, humans have a diploid complement of 46 chromosomes, whereas chimpanzees, gorillas, and orangutans have 48 chromosomes. The difference can be traced to the fusion (i.e., the joining together) of two ancestral chromosomes into chromosome 2 of humans; the ancestral chromosomes are separate in the other three species **(Figure 1).**

Jorge J. Yunis and Om Prakash of the University of Minnesota Medical School analyzed the banding patterns on metaphase chromosome preparations from humans, chimpanzees, gorillas, and orangutans. They identified about 1000 bands that are present in the four species. By matching the banding patterns on the chromosomes, the researchers verified that they were comparing the same segments of the genomes in the four species. They then searched for similarities and differences in the structure of the chromosomes.

Analysis of human chromosome 2 reveals that it was produced by the fusion of two smaller chromosomes that are still present in the other three species. Although the position of the centromere in human chromosome 2 matches that of the centromere in one of the chimpanzee chromosomes, in gorillas and orangutans it falls within an inverted segment of the chromosome. (Recall from Section 11.3 that an inverted chromosome segment has a reversed orientation, so the order of genes on it is reversed relative to the order in a segment that is not inverted.) Humans and chimps also differ from each other in centromeric inversions in six other chromosomes.

How might such chromosome rearrangements promote speciation? In a paper published in 2003, Arcadi Navarro of the Universitat Pompeu Fabra in Spain and Nick H. Barton of the University of Edinburgh in Scotland compared the rates of evolution in protein-coding genes that lie within rearranged chromosome segments of humans and chimpanzees to those in genes outside the rearranged segments. They discovered that proteins evolved more than twice as quickly in the rearranged chromosome segments. Navarro and Barton reasoned that because chromosome rearrangements inhibit chromosome pairing and recombination during meiosis, new genetic variations favoured by natural selection would be conserved within the rearranged segments. These variations accumulate over time, contributing to genetic divergence between populations with the rearrangement and those without it. Thus, chromosome rearrangements can be a trigger for speciation: once a chromosome rearrangement becomes established within a population, that population will diverge more rapidly from populations lacking the rearrangement. The genetic divergence eventually causes reproductive isolation.

Figure 1

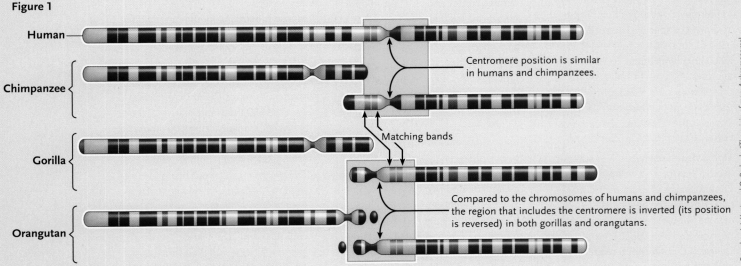

Centromere position is similar in humans and chimpanzees.

Matching bands

Compared to the chromosomes of humans and chimpanzees, the region that includes the centromere is inverted (its position is reversed) in both gorillas and orangutans.

Based on J. J. Yunis and O. Prakash, "The origin of man: A chromosomal pictorial legacy," *Science* 19 March 1982: Vol. 215 no. 4539 pp. 1525–1530

genera *Australopithecus* and *Paranthropus,* and their exact relationships to later hominids, are not yet fully understood (see Figure 20.13 and "The Cast of Characters: Fossil Hominins"). *Australopithecus* was likely ancestral to humans (various species in the genus *Homo*).

Many people believe that evolutionary biologists say that our species evolved *from* apes. But while the fossil record clearly demonstrates that the evolutionary lineage to which *H. sapiens* belongs includes chimps and gorillas (Figure 20.13), the lineage leading to humans has been distinct from the one leading to gorillas and chimps for over 6 million years. Belonging to an evolutionary lineage does not mean one species in the lineage gives rise to another. Evolutionary biologists are not proposing that humans evolved from apes, but that apes are our closest living relatives (see "Chromosomal Similarities and Differences among the Great Apes").

STUDY BREAK

1. What are three adaptations of humans that separate them from chimpanzees?
2. How old are the oldest fossils of humans? Of hominoids?
3. When did the lineage leading to chimps and orangutans diverge from the hominin lineage?

21.2 Morphology and Bipedalism

Upright posture and bipedal locomotion distinguish hominins from apes. Bipedal locomotion largely meant that the hands were not used in locomotion, allowing them to become specialized for other activities, such as carrying things and using and making tools. Sometimes paleontologists find fossilized hominin footprints, indicating bipedalism **(Figure 21.1).** But usually the fossil record of mammals such as *Homo sapiens* consists mainly of bones and teeth, the body parts most often fossilized. Bipedalism is obvious from the feet, thighs, pelvis, shoulders, and arms of these fossil and human skeletons **(Figure 21.2).** When appropriate fossils are available, it may be possible to learn when the skeletal features associated with bipedalism appeared over time.

Bipedalism in hominins involves a suite of anatomical features, not just the pattern of footfall (opening photograph) and walking and running behaviour.

21.2a Feet, Legs, and Pelvis

Paleontologists have long inferred bipedalism from the structure of the thigh bones (femora) and pelvis **(Figure 21.3, p. 474).** Both ends of the femur, at the hip and knee joints, are larger in the human than in the chimpanzee because more weight is directed through the human joints. Humans have a smaller angle at the hip end of the femur (the ball and socket joint) because

Figure 21.1
Mary Leakey discovered these fossilized footprints of an australopithecine made in soft, damp volcanic ash about 3.7 mya. The footprints appear to have been made by an adult and a young.

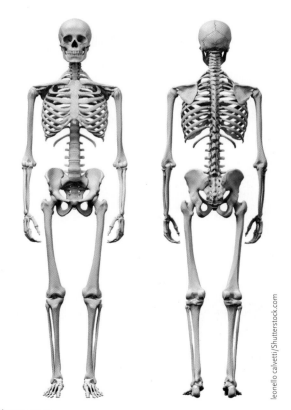

Figure 21.2
Human skeleton, front and back views.

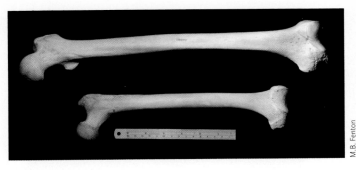

Figure 21.3
A comparison of the thigh bones (femora) of a chimpanzee and a human (top).

Australopithecus sediba from the Malapa site in South Africa was an earlier (about 2 mya) version of Lucy (*Australopithecus afarensis*, **Figure 21.5**). The ankles and feet of *A. sediba* suggest a combination of climbing ability (arboreal lifestyle) and bipedalism that differed from that of Lucy. The sacral and pubic areas of the pelvis of *A. sediba* resemble those of Lucy and other australopithecines. But the shape of the ilium of *A. sediba* resembled that of species of *Homo*.

In 2007, S. K. S. Thorpe, R. L. Holder, and R. H. Crompton proposed that bipedalism arose in an arboreal setting. Specifically, they asserted that hand-assisted bipedalism allowed the ancestors of humans and great apes to move on flexible supports (branches) that would otherwise have been too small. Thorpe et al. compared human and orangutan (*Pongo abelii*) locomotion and found that orangutans walking on flexible branches increase knee and hip extension, just as humans do when running on a springy track. In a bipedal gait, humans and orangutans flex the hind limbs in a manner that differs from the gait of gorillas and chimps.

of their upright stance. Also, human leg bones (both the upper leg and lower leg) are longer than in chimpanzees, while chimps have longer foreleg bones than humans do.

More recently, the metatarsals (the long foot bones) provided additional features for recognizing bipedalism in hominins. For example, the fourth metatarsals of *Australopithecus afarensis* were more like those of humans than those of either chimps or gorillas **(Figure 21.4)**. Comparable features also appeared in *Ardipithecus ramidus* from 3.4 mya, indicating a longer than expected history of bipedalism.

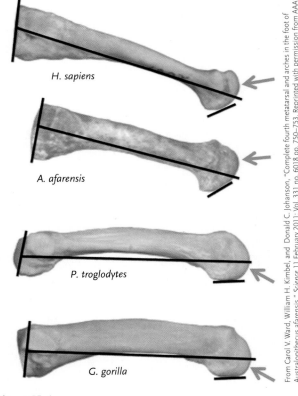

H. sapiens

A. afarensis

P. troglodytes

G. gorilla

From Carol V. Ward, William H. Kimbel, and Donald C. Johanson, "Complete fourth metatarsal and arches in the foot of *Australopithecus afarensis*," *Science* 11 February 2011: Vol. 331 no. 6018 pp. 750–753. Reprinted with permission from AAAS.

Figure 21.4
The black lines indicate the angles between the proximal (on the left, the ankle end) and distal ends of the fourth metatarsal bone (arrows) of a human, *Australopithecus afarensis*, a chimpanzee (*Pan troglodytes*), and a gorilla (*Gorilla gorilla*). Note the angle in the hominins compared to the parallel lines in the ape species.

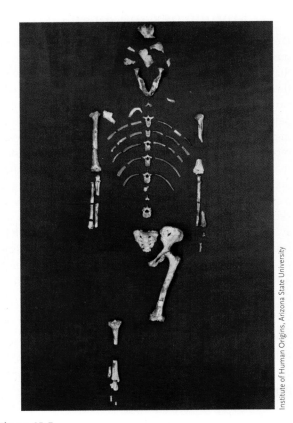

Institute of Human Origins, Arizona State University

Figure 21.5
The fossil skeleton of *Australopithecus afarensis*, popularly known as "Lucy."

21.2b Shoulders and Arms

The importance of climbing for australopiths is supported by the appearance of their shoulder blades (scapulae). The angle of the socket (glenoid fossa, **Figure 21.6**) that receives the head of the upper arm bone (humerus) faces cranially (toward the head), as it does in apes that hang from their arms. In humans, the glenoid fossa faces laterally but changes with age. A lateral-facing glenoid fossa also contributes to humans' ability to throw projectiles such as spears or stones at high speeds.

Species in the genus *Homo* have three specializations associated with throwing ability:

1. They have a "long waist" because of an increase in the number of lumbar vertebrae combined with longer individual vertebrae. The long waist allows the movement of hips and thorax to be decoupled, resulting in a large range of motion of the shoulders and the development of torque.
2. Torsion of the humerus (upper arm bone) between the orientation of its head and the axis of the elbow extends the range of motion during rotation.
3. The laterally directed glenoid fossa aligns the moment generated by flexion of a muscle, the pectoralis major, with the rotation of the torso.

This set of specializations appeared more than 2 mya in *Homo erectus*. It permits elastic storage of energy, which contributes to accurate spear throwing. Effective throwing increased the hunting potential of species in the genus *Homo*.

21.2c Hands

Species in the genus *Homo* have hands that are quite distinct from those of apes. The palms and fingers are short, while the thumbs are long, strong, and mobile **(Figure 21.7)**. This results in our ability to use two different grips **(Figure 21.8, p. 476)**, a power grip and a precision grip, allowing manipulative skills and a capacity to make precise tools. The proportions of hominin hands also deliver a performance advantage when striking with a fist. Buttressing of the elements in the hand increases the stiffness of the joint between the second metacarpal (a hand bone) and phalanges (finger

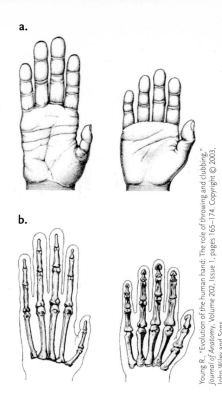

Young R., "Evolution of the human hand: The role of throwing and clubbing," *Journal of Anatomy*, Volume 202, Issue 1, pages 165–174. Copyright © 2003, John Wiley and Sons

Figure 21.7
The chimp's hands (left) are adapted for grasping branches, while human hands (right) are adapted for precision and power grips. The relatively longer thumbs of humans also contribute to our precision grip.

bones), enabling hominins to punch with more force. The ability to present our hands palms up (supination) or palms down (pronation) also increases their versatility.

Increasing the force that can be delivered can also be achieved by the leverage associated with attaching a handle or strap (haft) to a projectile point, such as a spear or an arrow. At Kathu Pan in South Africa, hafted tools date from 500 thousand years ago. Hafted tools such as spears further enhance the impact of high-velocity throwing, while shorter stabbing blows reflect the importance of the power grip. Other tools found among hominin fossils include stone cutting tools used to take meat off bones and stone axes used for chopping trees. Through time, the tools found show improvement in the technique and the fine motor skills required to make them.

21.2d Pelvis and Birth

Female hominins suffer at least one consequence of being bipedal, namely the shift in the body's centre of mass during pregnancy. A marked posterior concavity of individual lower back (lumbar) vertebrae stabilizes the centre of mass of the upper body over the hips. Females have a derived curvature of the lumbar area and reinforcement of those vertebrae to compensate for the additional load associated with pregnancy. The anatomy of the pelvis and lower back of *Australopithecus* indicates that

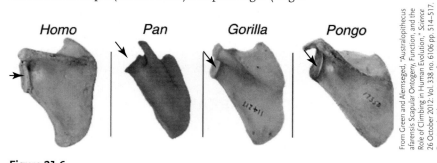

From Green and Alemseged, "*Australopithecus afarensis* Scapular Ontogeny, Function, and the Role of Climbing in Human Evolution," *Science* 26 October 2012: Vol. 338 no. 6106 pp. 514–517. Reprinted with permission from AAAS.

Homo   *Pan*   *Gorilla*   *Pongo*

Figure 21.6
A comparison of the shoulder blades (scapulae) of a human, a chimp (*Pan*), a gorilla, and an orangutan (*Pongo*) showing the orientation of the glenoid fossa (arrows).

Figure 21.8

Power grip versus precision grip. Hominins grasp objects in two distinct ways. The power grip **(a)** allows us to grasp an object firmly, whereas the precision grip **(b)** allows us to manipulate objects by fine movements.

these adaptations to bipedalism preceded the evolution of species in the genus *Homo*. Compared to modern humans, the birth canals of neandertals were not as specialized. The birth process in our species is specialized and may be a relatively recent development.

STUDY BREAK

1. Name three morphological specializations associated with bipedalism.
2. What effect do these specializations have on hominins, and how do they distinguish hominins from chimps?

21.3 Human Features That Do Not Fossilize

Some features characteristic of humans are unlikely to fossilize, such as behavioural and soft tissue features associated with social organization and language. The structure of jaws and teeth as well as fossilized dung (coprolites) can be used to infer (jaws and teeth) or reveal (coprolites) what fossil animals ate. Although jaw and tooth structure can suggest the ability to eat hard food, our ancestors may have used tools to break up hard foodstuffs and fire to soften and cook them. Meanwhile, early hominins exploited a range of habitats and thrived on a diversity of food (see "Molecule behind Biology").

Humans show a great capacity for making friends—not genetically related individuals—with whom they have long-term, nonreproductive relationships that involve cooperation and mutual influence. These relationships underlie social networks, a feature of many social species. In humans, social networks may include individuals of other species, such as dogs and cats. Social networks and associated cooperative behaviour are well known from human hunter–gatherer societies, such as the Hadza of Tanzania.

One apparently unique feature of human social networks is the common use of some land areas by different groups. This leads to a pattern of movement among groups (dispersal), which enhances social learning and a cumulative culture.

Effective communication among individuals is an essential part of social networks. One aspect is an individual's ability to read and interpret the body language (see Chapter 47, page 1166) of another. Humans use both body cues and facial expressions to distinguish between intense positive and negative emotions **(Figure 21.9)**.

Language is a means of communication that involves symbolism and syntax. Although language is sometimes considered a hallmark of *Homo sapiens,* other animals also use a combination of symbolism and syntax in communication. For example, when a vervet monkey (*Chlorocebus pygerythrus*) gives the "eagle" alarm call, its fellows look skyward and move closer to the trunks of trees. When the same monkey gives a "leopard" alarm call, other monkeys in trees look down and those on the ground climb trees.

Language is not unique to *H. sapiens,* and it is not clear when language appeared in human evolution. The *FOXP2* gene is associated with speech and language in *H. sapiens,* and genomic analysis reveals that *H. sapiens* and *Homo neandertalensis* have similar *FOXP2* genes, implying that neandertals had language. A trait of many (all?) present-day human languages is a combination of simple categories (to minimize details) and a high level of informativeness to maximize communication efficiency.

STUDY BREAK

1. How does communication figure in the evolution of humans?
2. What are friends? How do they distinguish humans from other animals?

Lactose Intolerance

Lactose, a sugar in milk, is broken down by the enzyme lactase. Before young mammals are weaned, they ingest milk and digest lactose. By adulthood, many mammals no longer produce lactase and therefore cannot digest lactose. In humans, a single cytosine to thymine mutation in the regulatory region of the lactase allele (LP) results in lactase persistence. Possession of the LP allele allows adults to drink milk. About 66% of adult humans are lactose intolerant (do not have the LP allele), and drinking milk results in severe intestinal distress.

Before the appearance of the LP allele, our ancestors were still able to get some nutritional value from milk. They ate fermented cheeses and yogurt, which have much lower levels of lactose than milk does. Hard cheeses such as Parmesan have very little lactose.

The LP allele appeared about 7500 years ago in what is now Hungary. It rapidly spread through the population because individuals carrying it produced about 19% more fertile offspring than those lacking the allele, a selective advantage. The LP allele illustrates how changes at the gene level can result in changes at the population level.

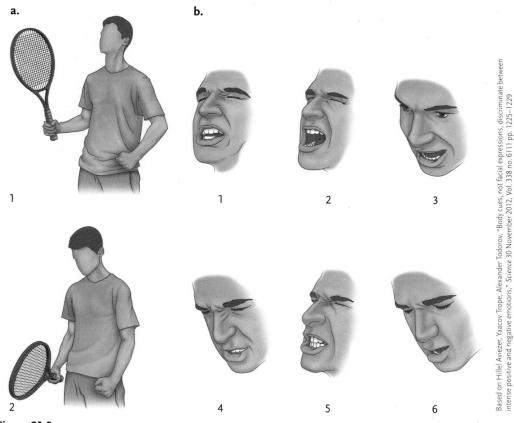

a. b.

1 2 3

1

2 4 5 6

Based on Hillel Aviezer, Yaacov Trope, Alexander Todorov, "Body cues, not facial expressions, discriminate between intense positive and negative emotions," *Science* 30 November 2012, Vol. 338 no. 6111 pp. 1225–1229

Figure 21.9
(a) Body language in response to winning (1) and losing (2) a point. **(b)** Facial expressions presented as isolated views in response to winning (2, 3, 5) and losing (1, 4, 6) a point.

21.4 Dispersal of Early Humans

There are two main theories about the dispersal of our ancestors from Africa. The **African emergence hypothesis** proposes that early hominin descendants (archaic humans) left Africa and established populations in the Middle East, Asia, and Europe. Some time later, 100 thousand to 200 thousand years ago, *H. sapiens* arose in Africa and also migrated into Europe and Asia. Perhaps through competition, *H. sapiens* eventually drove archaic humans to extinction. This hypothesis suggests that all modern humans are descended from a fairly recent African ancestor.

The **multiregional hypothesis** suggests that populations of *H. erectus* and archaic humans had spread through much of Europe and Asia by 500 thousand years ago and modern humans (*H. sapiens*) evolved from descendants of these earlier dispersals. Although these geographically separated populations may have experienced some evolutionary differentiation, gene flow between them prevented reproductive isolation and maintained them as a single but variable species, *H. sapiens*.

Paleontological data do not clearly support either hypothesis, but as of 2011, genetic data **(Figure 21.10)** generally supported the African emergence hypothesis.

Further work on the Y chromosomes of thousands of men from Africa, Europe, Asia, Australia, and the Americas has confirmed that all modern humans are the descendants of a single migration out of Africa.

A rapid exodus of anatomically modern humans out of Africa may have occurred along the coast of the Indian Ocean. Archaeological material from the United Arab Emirates suggests that early emigrants may have taken advantage of lower sea levels to move along the Arabian coast around 60 thousand years ago.

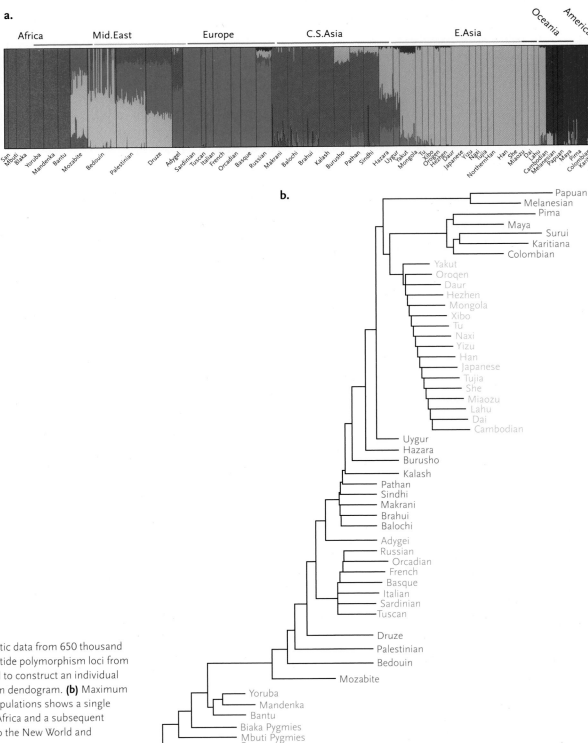

Figure 21.10

Out of Africa. **(a)** Genetic data from 650 thousand common single-nucleotide polymorphism loci from 928 humans were used to construct an individual ancestry and population dendogram. **(b)** Maximum likelihood tree of 51 populations shows a single origin in sub-Saharan Africa and a subsequent radiation across Asia to the New World and Polynesia.

From Jun Z. Li, Devin M. Absher, Hua Tang, Audrey M. Southwick, Amanda M. Casto, Sohini Ramachandran, Howard M. Cann, Gregory S. Barsh, Marcus Feldman, Luigi L. Cavalli-Sforza, and Richard M. Myers, "Worldwide Human Relationships Inferred from Genome-Wide Patterns of Variation," *Science*, February 22, 2008, vol. 319, pp. 1100–1104. Reprinted with permission from AAAS.

Earlier dispersal is clear from material found at Attirampakkam in India. These fossils indicate that Acheulian humans (probably *Homo erectus*) had occupied this site by about 1.5 mya. Acheulian cultures are typified by large cutting tools with bifaces. Other records indicate that hominins (*Homo floresiensis*) were on Flores Island (Indonesia) by 1 mya. Humans occupied sites on the highlands in New Guinea by about 49 thousand years ago. These humans exploited endemic nuts (*Pandanus* spp.) and appeared to have cleared forests to promote growth of their preferred plants. The timing of the arrival of humans in the New World is less well known. Dating of sites at caves in Oregon indicates human occupancy by about 12 thousand years ago.

21.4a The Denisovans

Genetic information in the form of DNA recovered from a finger bone of a girl who lived over 50 thousand years ago indicates that she had dark skin, brown hair, and brown eyes. The girl's fossilized bone fragments were found in Denisova Cave in Siberia. The name of the cave has been applied to the people, Denisovans, who apparently were a sister group to the neandertals. Subsequent genomic analysis revealed that the Denisovans lived in southeast Asia and interbred with the ancestors of today's Melanesians.

Vital components of our immune system (HLA class I) were acquired through the *HLA-B*73* allele inherited from Denisovans in west Asia. Genome analysis also indicates that some *HLA* haplotypes entered modern European and Oceanian human populations from both neandertals and Denisovans.

STUDY BREAK

1. What evidence supports the African emergence hypothesis?
2. By when had humans arrived in India? How do we know?

21.5 Hominins and the Species Concepts

The history of hominins clearly demonstrates the challenges inherent in recognizing species and the boundaries between them. This example is particularly illuminating because it involves paleontological, archaeological, and genomic evidence. The genomic analysis shows that the ancestors of modern humans interbred with both neandertals and Denisovans. If we apply the biological species concept (see Chapter 19), these three groups are not separate species. If we apply the phylogenetic species concept, the distinction is less clear. The morphological evidence from fossils is incomplete and does not necessarily settle the matter.

In 2013, the description of fossil hominins from Georgia (Dmanisi) changed our view of our ancestors. In 2000, these fossils from 1.7 mya were identified as *Homo georgicus,* but the 2013 presentation of data from five skulls shows the same amount of variation in morphology that we know from living populations of humans and chimpanzees. This discovery suggests a highly variable lineage and obliges us to reconsider which of the named species of *Homo* are valid.

STUDY BREAK

1. What is the impact of the fossils from Dmanisi on our view of the species diversity in the genus *Homo*?
2. How do the "species" of hominins fit the biological species concept?

With recent advances in molecular technology, we are learning a great deal more about our evolutionary history. The story continues about what sets humans apart from other animals, whether it is communication abilities, social cooperation, or even an understanding of the future. As we will see later in the book, our animal ancestry is still clear, whether the topic is population ecology (Chapter 29) or animal behaviour (Chapter 47).

Review

To access course materials such as Aplia and other companion resources, please visit www.NELSONbrain.com.

21.1 The Fossil Record of Hominins

- Upright posture and bipedal locomotion are key adaptations distinguishing hominins from apes.
- Evolutionary biologists are not proposing that humans evolved from apes, but that apes are our closest living relatives (see "Chromosomal Similarities and Differences among the Great Apes").

21.2 Morphology and Bipedalism

- Bipedalism may have arisen in an arboreal setting that allowed the ancestors of humans to move on flexible supports (branches). Bipedal locomotion freed the hands from locomotor functions, allowing them to become adapted for other activities, such as tool use.
- The structure and length of the femur, and the structure of the scapula (shoulder blade) can indicate whether a hominin was bipedal or not.

- Bipedal hominins were able to use their arms for carrying things as well as making and using tools.
- The hands of hominins are quite distinct from those of apes. The palms and fingers are short, while the thumbs are long, strong, and mobile. This results in our ability to use a power grip and a precision grip, allowing manipulative skills and a capacity to make precise tools.

21.3 Human Features That Do Not Fossilize

- Humans are able to develop social networks among friends and family groups.
- Effective communication among individuals is an essential part of social networks.

21.4 Dispersal of Early Humans

- The African emergence hypothesis suggests that between 0.5 and 1.5 mya, a population of *H. erectus* gave rise to several descendant species that left Africa and established populations in the Middle East, Asia, and Europe. Some time later, *H. sapiens* arose in Africa. These modern humans also migrated into Europe and Asia and eventually drove archaic humans to extinction.
- The multiregional hypothesis proposes that populations of *H. erectus* and archaic humans had spread through much of Europe and Asia by 0.5 mya. Modern humans (*H. sapiens*) then evolved from archaic humans in many regions simultaneously. Although they were geographically separated, these populations may have experienced some evolutionary differentiation; gene flow between them prevented reproductive isolation and maintained them as a single species, *H. sapiens*.

21.5 Hominins and the Species Concepts

- The history of hominins shows the challenges inherent in recognizing species and the boundaries between them. This example involves paleontological, archaeological, and genomic evidence.

Questions

Self-Test Questions

1. Which of the following is true about neandertals (*Homo neandertalensis*)?
   a. They did not occur in the same places and times as *Homo sapiens*.
   b. They did not interbreed with *Homo sapiens*.
   c. They were not behaviourally distinct from *Homo sapiens*.
   d. They were extinct by 20 thousand years ago.

2. Which describes the Hominidae?
   a. the group that includes gorillas and chimps
   b. first appear in the fossil record about 1 mya
   c. include species in the genera *Homo*, *Australopithecus*, and *Paranthropus*
   d. include only species in the genus *Homo*

3. In hominids, bipedalism involves specializations of which of the following?
   a. feet
   b. knees
   c. ankles
   d. pelvis and legs

4. In humans, these are specializations for throwing.
   a. feet
   b. hands
   c. shoulders
   d. waist, shoulders, and arms

5. This is true about the ability to make friends.
   a. It occurs in social organisms.
   b. It is unique to humans.
   c. It depends upon individual recognition.
   d. It occurs only in some people.

6. A variety of evidence indicates that humans evolved in which place?
   a. Africa
   b. South America
   c. Europe
   d. Australia

7. The Denisovans were which of the following?
   a. another species in the genus *Homo*
   b. larger than *Homo sapiens*
   c. originally discovered in Siberia from genetic analysis of one finger bone
   d. a variety of neandertal

8. Which of the following species of fossil hominin is the oldest?
   a. *Homo neanderetalensis*
   b. *Homo sapiens*
   c. *Ardipithecus ramidus*
   d. *Australopithecus sediba*

9. What does genetic similarity among humans, chimps, gorillas, and orangutans mean?
   a. They all belong to the same species.
   b. They have a relatively recent common ancestor compared to other primates.
   c. They are evidence of creation.
   d. They can interbreed and produce fertile offspring.

10. When did hominins move out of Africa?
    a. 2 mya
    b. 3 mya
    c. 1.5 mya
    d. 100 thousand years ago

Questions for Discussion

1. Which of the species concepts presented in Chapter 19 best fits the species of hominins currently known to us? Why?

2. What does bipedal mean? Name three groups of tetrapods that are bipedal. How do the modes of locomotion differ among these groups?

3. What is language? Is language unique to humans? What evidence would you use to support your definition of language and its appearance in humans?

4. What does lactose intolerance tell us about the evolution of humans?

5. Is tool use a characteristic of humans? Explain your answer. What advantage(s) can use of tools confer on an individual?

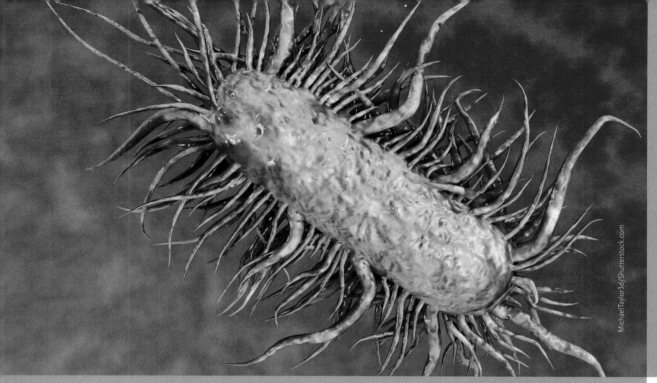

The bacterium *Escherichia coli*.

Bacteria and Archaea

WHY IT MATTERS

Who are you? What makes you "you"? Would you feel less like "you" if you knew that most of the cells in your body weren't human cells at all? The bacterial cells on and in our body outnumber our cells by ten to one. And given that, as Princeton University scientist Bonnie Bassler points out, the average person has about 30 000 human genes but more than 3 million bacterial genes, we are at most 1% human! But these bacteria aren't alien invaders—many of them may be crucial for making us unique individuals. There are about a hundred trillion bacteria of hundreds (or thousands) of different species lining your large intestine. When you were born, your gut was sterile, but immediately after birth, your intestines started to be colonized—the exact composition of these "pioneers" depends on where you were born and whether you were breastfed, among other factors. The early colonists were essential for the normal development of your gut as an infant, and throughout your life, your gut bacteria have continued to help you in many ways: they help digest your food, synthesize vitamin K for you, and produce antimicrobial factors to protect you from pathogens. The composition of your gut bacteria is more similar to that of your family than to people not related to you, but it is still unique to you; even identical twins, who share so much else, have different sets of gut bacteria. Recent research has revealed that the diversity of your gut bacteria plays a role in your odds of developing metabolic diseases and becoming obese, and may even be involved in your mental health.

CONCEPT FIX As you can tell from this introduction, not all or even most prokaryotic organisms are harmful! The idea that all bacteria cause disease is one of the major misconceptions about these organisms, but nothing could be farther from the truth: most known bacteria and members of the other group of prokaryotic organisms, archaea, play a crucial role in ecosystems, recycling nutrients and breaking down compounds that no other organisms can. Others carry out reactions important in food production, in industry (e.g., production of pharmaceutical products), and in **bioremediation** of polluted sites.

In this chapter, we first look at the structure and function of prokaryotic organisms, emphasizing the features that differentiate them from other organisms, and conclude with a look at the diversity of these fascinating organisms.

22.1 The Full Extent of the Diversity of Bacteria and Archaea Is Unknown

While reading this chapter, keep in mind that everything we know so far about bacteria and archaea is based on a tiny fraction of the total number of species. We have isolated and identified only about 6000 species, which may be as low as 1% of the total number. We know almost nothing about the prokaryotic organisms of entire habitats, such as the oceans, which make up 70% of Earth's surface. Why have we only identified so few, and why are we not even sure how many prokaryotic organisms there might be? In the past, we identified and classified bacteria and archaea based on external features (e.g., cell wall structure) and physiological differences, which meant that we had to be able to grow the organisms in culture. We have learned a great deal about the biology of some bacteria and archaea but have been unable to learn much about the majority of prokaryotic organisms, since they cannot be grown in culture (e.g., those that require extreme physicochemical conditions). Recently, molecular techniques have been developed that allow us to isolate and clone DNA from an environment and then analyze gene sequences; this means that we can now identify and characterize bacteria and archaea without having to culture them. This approach, known as **metagenomics**, now enables us to investigate the diversity of prokaryotic organisms in a wide range of environments. However, our understanding of the full extent of microbial diversity still faces other challenges, such as the fact that many environments (e.g., the deep ocean) are remote and thus very difficult and/or costly to sample.

22.1a Prokaryotic Organisms Make Up Two of the Three Domains of Life

Two of the three domains of living organisms, **Archaea** and **Bacteria**, consist of prokaryotic organisms (the third domain, **Eukarya**, includes all eukaryotes). Bacteria are the prokaryotic organisms most familiar to us, including those responsible for diseases of humans and other animals as well as those that we rely on for production of cheese, yogurt, chocolate, and other foods. Archaea are not as well known, as they were only discovered about 40 years ago. As you will see in this chapter, archaea share some cellular features with eukaryotes and some with bacteria but have still other features that are unique. Many of the archaea live under very extreme conditions that no other organisms, including bacteria, can survive.

22.2 Prokaryotic Structure and Function

We begin our survey by examining prokaryotic cellular structures and modes of reproduction, and how they obtain energy and nutrients.

In general, prokaryotic organisms are the smallest in the world **(Figure 22.1)**. Few species are more than 1 to 2 μm long (although the longest is 600 μm long, which is larger than some eukaryotes!); from 500 to 1000 of them would fit side by side across the dot on this letter "i." Despite the small size of bacteria and archaea, they dominate life on Earth: current estimates of total prokaryotic diversity are in the billions of species, and their total collective mass (their **biomass**) on Earth exceeds that of animals and may be greater than that of all plant life. Prokaryotic organisms colonize every niche on Earth that supports life, and even occur deep in the Earth's crust. They also colonize other organisms—for example, huge numbers of bacteria inhabit the surfaces and cavities of a healthy human body, including the skin, mouth and nasal passages, and large intestine. As mentioned in "Why It Matters," collectively, the bacteria in and on your body outnumber all the other cells in your body. It is not surprising that the diversity of bacteria and archaea should be so much greater than that of eukaryotes because for about 3 billion years they were the only forms of life on Earth and so had time to diversify and expand into every habitat on Earth before the first eukaryotes appeared on the scene (see Chapter 2).

22.2a Prokaryotic Cells Appear Simple in Structure Compared with Eukaryotic Cells

Three cell shapes are common among prokaryotes: spiral, spherical (or **coccoid**; *coccus* = berry), and cylindrical (known as **rods**), but some archaea even have square cells **(Figure 22.2)**.

At first glance, a typical prokaryotic cell seems much simpler than a eukaryotic cell **(Figure 22.3, p. 484)**: images taken with standard electron microscopy typically reveal little more than a cell wall and plasma membrane surrounding cytoplasm with DNA concentrated in one region and ribosomes scattered throughout. The chromosome is not contained in a membrane-bound nucleus but is packed into an area of the cell called the **nucleoid.** Prokaryotic cells have no cytoplasmic organelles equivalent to the endoplasmic reticulum (ER) or Golgi complex of eukaryotic cells (see Chapter 2). With few exceptions, the reactions carried out by organelles in eukaryotes are distributed between the plasma membrane and the cytoplasmic solution of prokaryotic cells; this means that macromolecules such as proteins are very concentrated in the cytoplasm of these cells, making the cytoplasm quite viscous. This evident simplicity of prokaryotic cells led

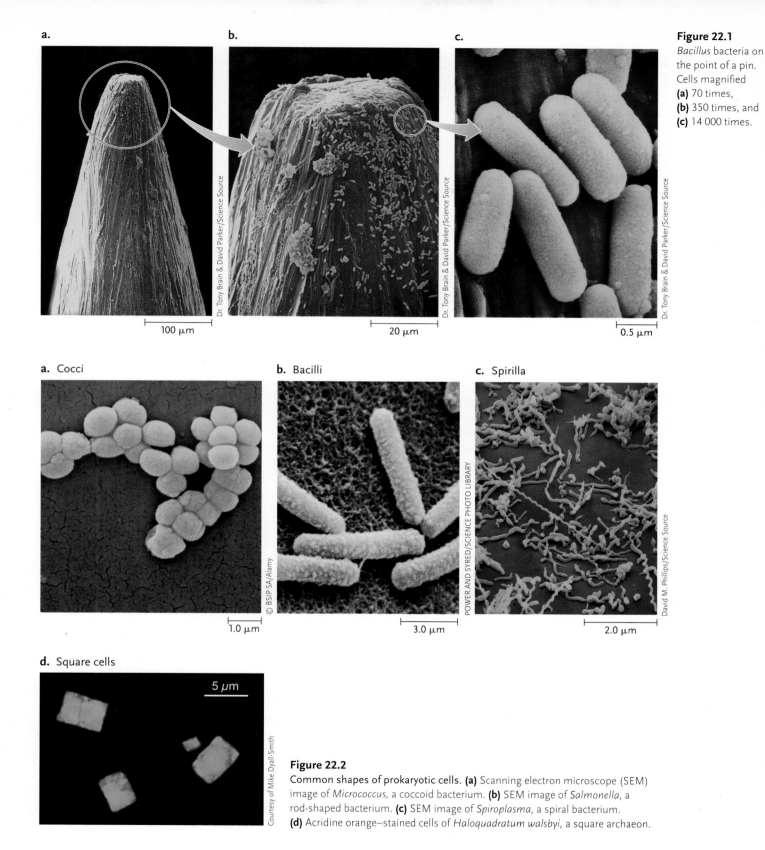

Figure 22.1
Bacillus bacteria on the point of a pin. Cells magnified **(a)** 70 times, **(b)** 350 times, and **(c)** 14 000 times.

a. Cocci

b. Bacilli

c. Spirilla

d. Square cells

5 μm

Figure 22.2
Common shapes of prokaryotic cells. **(a)** Scanning electron microscope (SEM) image of *Micrococcus*, a coccoid bacterium. **(b)** SEM image of *Salmonella*, a rod-shaped bacterium. **(c)** SEM image of *Spiroplasma*, a spiral bacterium. **(d)** Acridine orange–stained cells of *Haloquadratum walsbyi*, a square archaeon.

people to regard these cells as featureless and disorganized. However, the apparent simplicity of these cells is misleading. New microscopic techniques reveal that prokaryotic cells do have a cytoskeleton—not homologous to that of a eukaryote but serving some of the same functions—and have more sophisticated organization than was previously thought. In fact, recent research carried out by Laura van Niftrik of the Netherlands and her colleagues has identified a prokaryotic organelle! Certain bacteria that obtain energy by oxidizing ammonia have an internal membrane-bound compartment, where ammonia oxidation

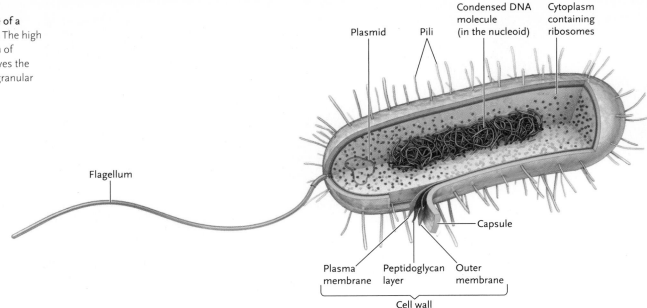

occurs. It was hypothesized that as ammonia oxidation proceeds inside this compartment, a proton-motive force could be generated across the membrane, generating ATP. Van Niftrik and her colleagues found that the membrane around this compartment does contain ATP synthase, supporting the above hypothesis. Thus, it appears that some prokaryotic cells have organelles with specialized functions.

Internal Structures. The genome of most prokaryotic cells consists of a single, circular DNA molecule, although some, such as the causative agent of Lyme disease (*Borrelia burgdorferi*), have a linear chromosome. Many prokaryotic cells also contain small circles of DNA called **plasmids (Figure 22.4),** which generally contain genes for nonessential but beneficial functions such as antibiotic resistance. Plasmids replicate independently of the cell's chromosome and can be transferred from one cell to another, meaning that genes for antibiotic resistance are readily shared among prokaryotic cells, even among cells of different species. This *horizontal gene transfer* allows antibiotic resistance and other traits to spread very quickly in bacterial populations. Horizontal gene transfer also occurs when bacterial cells take up DNA from their environment (e.g., from other cells that have lysed) or when viruses transfer DNA from one bacterium to another (see Chapter 23). Evidence indicates that a virus transferred toxin-encoding genes from *Shigella dysenteriae* (which causes bloody diarrhea) to *E. coli,* resulting in the deadly O157:H7 strain responsible for serious illness or even death of people eating beef and other food contaminated with this bacterium.

Like eukaryotic cells, prokaryotic cells contain ribosomes. Bacterial ribosomes are smaller than eukaryotic ribosomes but carry out protein synthesis by essentially the same mechanisms as those of eukaryotes (see Chapter 13). Archaeal ribosomes resemble those of bacteria in size but differ in structure; protein synthesis in Archaea is a combination of bacterial and eukaryotic processes, with some unique archaeal features. As a result, antibiotics that stop bacterial infections by targeting ribosome activity do not interfere with archaeal protein synthesis.

Prokaryotic Cell Walls. Most prokaryotic cells have a cell wall that lies outside their plasma membrane and protects the cell from lysing if subjected to hypotonic conditions or exposed to membrane-disrupting compounds such as detergents. The primary component of bacterial cell walls is **peptidoglycan,** a polymer of sugars and amino acids that forms linear chains. Peptide cross-linkages between the chains give the cell wall great strength and rigidity. The antibiotic penicillin prevents the formation of these cross-linkages, resulting in a weak cell wall that is easily ruptured, killing the cell **(Figure 22.5).**

Bacteria can be divided into two broad groups, Gram-positive and Gram-negative cells, based on their reaction to the **Gram stain procedure,** traditionally used as the first step in identifying an unknown bacterium.

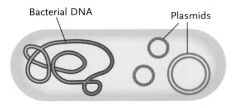

Figure 22.4
Plasmids inside a prokaryotic cell.

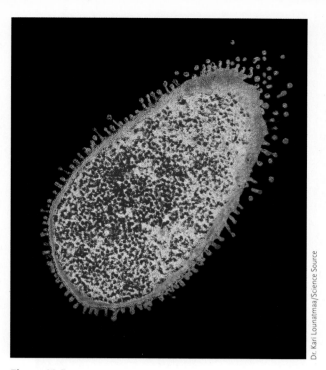

Figure 22.5

Image showing degradation of the cell wall following antibiotic treatment. The cell will eventually lyse, killing the bacterium.

Dr. Kari Lounatmaa/Science Source

Cells are first stained with crystal violet and then treated with iodine, which forms a complex with crystal violet. The cells are then rinsed with ethanol and counterstained with safranin. Some cells retain the crystal violet–iodine complex and thus appear purple when viewed under the microscope; these are termed Gram-positive cells. In other bacteria, ethanol washes the crystal violet–iodine complex out of the cells, which are colourless until counterstained with safranin; these Gram-negative cells appear pink under the microscope. The differential response to staining is related to

differences in cell wall structure: **Gram-positive** bacteria have cell walls composed almost entirely of a single, relatively thick layer of peptidoglycan (**Figure 22.6a**). This thick peptidoglycan layer retains the crystal violet–iodine complex inside the cell. **Gram-negative** cells have only a thin peptidoglycan layer in their walls, and the crystal violet–iodine complex is washed out. In contrast, the cell wall of Gram-negative bacteria has two distinct layers (Figure 22.6b), a thin peptidoglycan layer just outside the plasma membrane and an **outer membrane** external to the peptidoglycan layer. This outer membrane contains **lipopolysaccharides (LPSs)** and thus is very different from the plasma membrane. The outer membrane protects Gram-negative bacteria from potentially harmful substances in the environment; for example, it inhibits entry of penicillin. Therefore, Gram-negative cells are less sensitive to penicillin than are Gram-positive cells.

The cell walls of some archaea are assembled from a molecule related to peptidoglycan but with different molecular components and bonding structure. Others have walls assembled from proteins or polysaccharides instead of peptidoglycan. Archaea have a variable response to the Gram stain, so this procedure is not useful in identifying archaea.

The cell wall of many prokaryotic cells is surrounded by a layer of polysaccharides known as a **capsule** (**Figure 22.7, p. 486;** see also Figure 22.6). Capsules are "sticky" and play important roles in protecting cells in different environments. Cells with capsules are protected to some extent from desiccation, extreme temperatures, bacterial viruses, and harmful molecules such as antibiotics and antibodies. In many pathogenic bacteria, the presence or absence of the protective capsule differentiates infective from noninfective forms. For example, normal *Streptococcus pneumoniae* bacteria are capsulated and virulent, causing severe pneumonia in humans and other mammals. Mutant *S. pneumoniae* without capsules are nonvirulent and can easily be

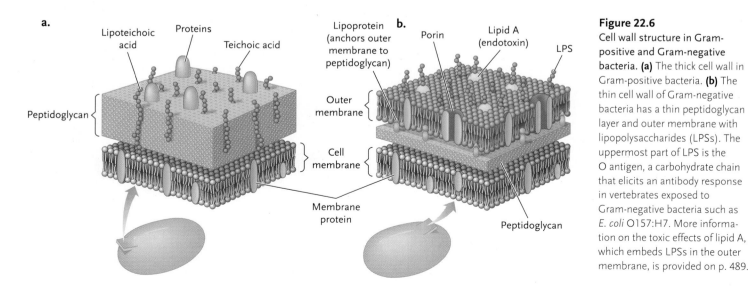

a. Lipoteichoic acid · Proteins · Teichoic acid · Peptidoglycan · Cell membrane

b. Lipoprotein (anchors outer membrane to peptidoglycan) · Porin · Lipid A (endotoxin) · LPS · Outer membrane · Membrane protein · Peptidoglycan

Figure 22.6

Cell wall structure in Gram-positive and Gram-negative bacteria. **(a)** The thick cell wall in Gram-positive bacteria. **(b)** The thin cell wall of Gram-negative bacteria has a thin peptidoglycan layer and outer membrane with lipopolysaccharides (LPSs). The uppermost part of LPS is the O antigen, a carbohydrate chain that elicits an antibody response in vertebrates exposed to Gram-negative bacteria such as *E. coli* O157:H7. More information on the toxic effects of lipid A, which embeds LPSs in the outer membrane, is provided on p. 489.

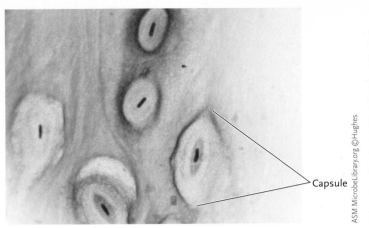

Figure 22.7
Capsules surrounding the cell wall of *Rhizobium*, a Gram-negative soil bacterium.

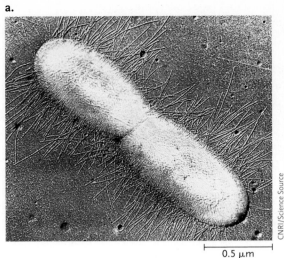

0.5 μm

a.

b.

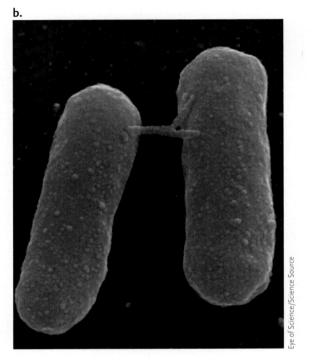

c.

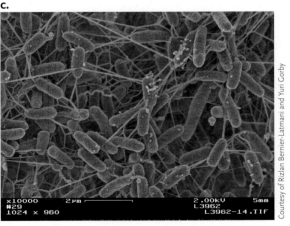

Figure 22.8
(a) Pili extending from the surface of a dividing *E. coli* bacterium.
(b) Sex pilus connecting two bacterial cells. **(c)** Nanowires (pili that conduct electricity) on *Shewanella oneidensis*. Note that these nanowires are much longer than the cells.

eliminated by the body's immune system if they are injected into mice or other animals.

Flagella and Pili. Many prokaryotic cells can move actively through liquids and even through films of liquid on a surface, most commonly via **flagella** (singular, *flagellum* = whip) extending from the cell wall (see Figure 22.3, p. 484). As outlined in Chapter 2, prokaryotic flagella are very different from eukaryotic flagella in both structure and pattern of movement. Prokaryotic flagella are made of rigid helical proteins, some of which act as a motor, rotating the flagellum much like the propeller of a boat. Archaeal flagella are superficially similar to bacterial flagella and carry out the same function, but the two types of flagella contain different components, develop differently, and are coded for by different genes.

Some prokaryotic cells have rigid shafts of protein called **pili** (singular, *pilus* = "hair") extending from their cell walls **(Figure 22.8a),** which enable them to adhere to or move along a surface. One type, called a *sex pilus,* not only allows bacterial cells to adhere to each other but also acts as a conduit for the transfer of plasmids from one cell to another (Figure 22.8b). Other types of pili enable bacteria to bind to animal cells. The bacterium that causes gonorrhea (*Neisseria gonorrhoeae*) uses pili to adhere to cells of the throat, eye, urogenital tract, or rectum in humans. In 2005, it was discovered that the pili of some bacteria (e.g., species of *Geobacter* and *Shewanella*) conduct electricity; these "nanowires" transfer electrons out of the cell onto minerals such as iron oxides in their environment (Figure 22.8c). Such electricity-generating bacteria hold promise for the development of microbial fuel cells as an alternative energy source.

Even though prokaryotes are simpler and less structurally diverse than eukaryotic cells, they are much more diverse metabolically, as we will now explore.

1. What features differentiate a prokaryotic cell from a eukaryotic cell? What features do both kinds of cells have?
2. How does the presence of a capsule affect the ability of the human body to mount an immune response to those bacteria?
3. How is a pilus similar to a flagellum? How is it different?
4. How does the amount of peptidoglycan in a bacterial cell wall relate to its Gram-stain reaction?

Table 22.1	Modes of Nutrition Used by Living Organisms		
Energy Source		Oxidation of Molecules*	Light
Carbon source	**CO_2**	**Chemoautotroph** Some bacteria and archaea; no eukaryotes	**Photoautotroph** Some bacteria, some protists, and most plants
	Organic molecules	**Chemoheterotroph** Some bacteria, archaea, and protists; also fungi, animals, and even some plants	**Photoheterotroph** Some bacteria

*Inorganic molecules for chemoautotrophs and organic molecules for chemoheterotrophs.

22.2b Prokaryotic Organisms Have the Greatest Metabolic Diversity of All Organisms

Organisms can be grouped into four modes of nutrition based on sources of energy and carbon (see **Table 22.1**).

In this approach to classification, we focus on carbon rather than other nutrients because carbon is the backbone of all organic molecules synthesized by an organism. Organisms such as plants that synthesize organic carbon molecules using inorganic carbon (CO_2) are **autotrophs** (*auto* = self; *troph* = nourishment). (Note that although CO_2 contains a carbon atom, oxides containing carbon are considered inorganic molecules.) All animals are **heterotrophs,** meaning that they obtain carbon from organic molecules, either from living hosts or from organic molecules in the products, wastes, or remains of dead organisms.

Organisms are also divided according to the source of the energy they use to drive biological activities. **Chemotrophs** (*chemo* = chemical) obtain energy by oxidizing inorganic or organic substances, whereas **phototrophs** obtain energy from light. Combining the carbon and energy sources allows us to group living organisms into four categories (Table 22.1).

Prokaryotic organisms (bacteria and archaea) show the greatest diversity in their modes of securing carbon and energy; they are the only representatives of two of the categories, chemoautotrophs and photoheterotrophs. **Photoheterotrophs** use light as an energy source and obtain carbon from organic molecules rather than from CO_2. **Chemoautotrophs** are commonly referred to as "lithotrophs" (*lithos* = rock, thus "rock-eaters"). As this name suggests, chemoautotrophs obtain energy by oxidizing inorganic substances such as hydrogen, iron, sulfur, ammonia, and nitrites and use CO_2 as their carbon source. Chemolithotrophs thrive in habitats such as deep-sea hydrothermal vents **(Figure 22.9),** where reduced inorganic compounds are abundant. The ability of these organisms to harness energy from these compounds makes them the foundation upon which the rest of the vent community ultimately depends, just as terrestrial organisms rely on the ability of plants and other photoautotrophs to capture light energy.

We breathe oxygen to provide the final electron acceptor for the electrons we remove from our food and pass down an electron transport chain (ETC) to make ATP via aerobic respiration (Chapter 6). Some prokaryotic organisms also use oxygen as a final electron acceptor; like us, these are aerobic organisms or **aerobes.** Aerobes may be **obligate aerobes;** that is, they cannot survive without oxygen. But some prokaryotic organisms "breathe" metals, using metals as the final electron acceptor for electrons; these organisms obtain energy via anaerobic respiration. **Anaerobic respiration** can also involve other inorganic molecules, such as nitrate or sulfate, as the final electron acceptors. Only prokaryotic organisms are capable of this type of respiration. **Obligate anaerobes** are poisoned by oxygen and survive either by fermentation, in which organic molecules are the final electron acceptors, or by anaerobic respiration. **Facultative anaerobes** use O_2 when it is present, but under anaerobic conditions, they live by fermentation or anaerobic respiration. As you learned in Chapter 6, prokaryotic organisms carry out a wider

Figure 22.9
Hydrothermal vents on the ocean floor.

Dr. Ken Macdonald/Science Source

range of fermentation reactions than do eukaryotes; many of these fermentations are economically important to humans, for example, in the production of foods such as cheese, yogurt, and chocolate.

22.2c Bacteria and Archaea Play Key Roles in Biogeochemical Cycles

The ability of prokaryotic organisms to metabolize such a wide range of substrates makes them key players in the life-sustaining recycling of elements such as carbon, oxygen, and nitrogen. The pathway by which a chemical element moves through an ecosystem is known as a **biogeochemical cycle.** As an element flows through its cycle, it is transformed from one form to another; prokaryotic organisms are crucial in many of these transformations. We will look at the nitrogen cycle as an example of the key role prokaryotic organisms play in biogeochemical cycles.

Nitrogen is a component of proteins and nucleotides and so is of vital importance for all organisms. The largest source of nitrogen on Earth is the atmosphere, which is almost 80% nitrogen. Why can't we just use this abundant atmospheric nitrogen? Most organisms cannot make use of this nitrogen because they cannot break the strong triple bond between the two nitrogen atoms. Only some bacteria and archaea can break this bond, using the enzyme nitrogenase, and convert N_2 into forms that can be used by other organisms. In this conversion process, known as **nitrogen fixation**, N_2 is reduced to ammonia (NH_3). Ammonia is quickly ionized to ammonium (NH_4^+), which prokaryotic cells then use to produce nitrogen-containing molecules such as amino acids and nucleic acids. Nitrogen fixation is the only means of replenishing the nitrogen sources used by most organisms—in other words, all organisms rely on nitrogen fixed by bacteria. Examples of nitrogen-fixing bacteria are cyanobacteria and *Rhizobium* (which is symbiotic with plants; see Chapter 38).

Other prokaryotic organisms carry out **nitrification**, the oxidation of ammonium (NH_4^+) to nitrate (NO_3^-). This oxidation process is carried out in two steps by two types of nitrifiers, ammonia oxidizers and nitrate oxidizers, present in soil and water. Ammonium oxidizers convert ammonium to nitrite (NO_2^-), whereas nitrite oxidizers convert nitrite to nitrate. Nitrate is then taken up by plants and fungi and incorporated into their organic molecules. Animals obtain nitrogen in organic form by eating other organisms or each other.

In sum, nitrification makes nitrogen available to many other organisms, including plants, animals, and bacteria that cannot metabolize ammonia. The metabolic versatility of bacteria and archaea is one factor that accounts for their abundance and persistence on the planet; another factor is their impressive reproductive capacity.

22.2d Asexual Reproduction Can Result in Rapid Population Growth

In prokaryotic organisms, asexual reproduction is the normal mode of reproduction. In this process, a parent cell divides by binary fission into two daughter cells that are exact genetic copies of the parent **(Figure 22.10).** Reproducing by binary fission means that under favourable conditions, populations of prokaryotic organisms can have very rapid exponential growth as one cell becomes two, two become four, and so on. Some prokaryotic cells can double their population size in only 20 minutes and will even begin a second round of cell division before the first round is complete; thus, one cell, given ideal conditions, can produce millions of cells in only a few hours.

These short generation times, combined with the small genomes (roughly one-thousandth the size of the genome of an average eukaryote), mean that prokaryotic organisms have higher mutation rates than do eukaryotic organisms. This translates to roughly 1000 times as many mutations per gene, per unit time, per individual as for eukaryotes. Genetic variability in prokaryotic populations, the basis for their diversity, derives largely from mutation and to a lesser degree from horizontal gene transfer (see Chapter 9). Further, the typically much larger populations of prokaryotic organisms compared with eukaryotes contribute to the much greater genetic variability in bacteria and archaea. In short, prokaryotic organisms have an enormous capacity to adapt, which is one reason for their evolutionary success.

As we have seen, the success of bacteria is beneficial to humans in many ways but can also be detrimental to us when dealing with successful pathogenic

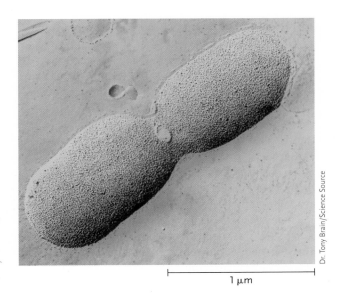

Dr. Tony Brain/Science Source

1 μm

Figure 22.10

E. coli cell dividing by binary fission. Note that a septum is forming between the two parent cells.

bacteria. In the next section, we investigate how some bacteria cause disease and how they are able to resist treatment with antibiotics.

22.2e Pathogenic Bacteria Cause Diseases by Different Mechanisms

Some bacteria produce **exotoxins**, toxic proteins that leak from or are secreted from the bacterium. For example, botulism food poisoning is caused by the exotoxin of the Gram-positive bacterium *Clostridium botulinum*, which grows in poorly preserved foods **(Figure 22.11).** The botulism exotoxin, botulin, is one of the most poisonous substances known: just a few nanograms can cause severe illness. What makes botulin so toxic? It produces muscle paralysis that can be fatal if the muscles that control breathing are affected. Interestingly, botulin is used under the brand name Botox for the cosmetic removal of wrinkles and in the treatment of migraine headaches and some other medical conditions. Exotoxins produced by certain strains of *Streptococcus pyogenes* have "superantigen properties" (i.e., overactivation of the immune system) that cause necrotizing fasciitis ("flesh-eating disease"). In 1994, Lucien Bouchard, who was then premier of Quebec, lost a leg to this disease.

Other bacteria cause disease through **endotoxins**. Endotoxins are the lipid A portion of the LPS molecule of the outer membrane of all Gram-negative bacteria, such as *E. coli, Salmonella,* and *Shigella*. When a Gram-negative cell lyses, the LPSs of the outer membrane are released; exposure to a specific component of this layer, known as lipid A, causes endotoxic shock. When Gram-negative bacteria enter the bloodstream, endotoxin overstimulates the host's immune system, triggering inflammation and an often lethal immune response. Endotoxins have different effects depending on the bacterial species and the site of infection.

22.2f Pathogenic Bacteria Commonly Develop Resistance to Antibiotics

An **antibiotic** is a natural or synthetic substance that kills or inhibits the growth of bacteria and other microorganisms. Prokaryotic organisms and fungi produce these substances naturally as defensive molecules, and we have also developed ways to synthesize several types of antibiotics. Different types of antibiotics have different modes of action: for example, streptomycins, produced by soil bacteria, block protein synthesis in their targets, whereas penicillins, produced by fungi, target the peptide cross-linkages in peptidoglycan, as described above.

How are bacteria able to block the actions of antibiotics? There are various mechanisms by which bacteria resist antibiotics **(Figure 22.12, p. 490).** For example, some bacteria are able to pump antibiotics out of the cell using membrane-bound pumps. They can also produce molecules that bind to the antibiotic or enzymes that break down the antibiotic, rendering it ineffective against its target. Alternatively, a simple mutation can result in a change in the structure of the antibiotic's target, so that the antibiotic cannot bind to it. Finally, bacteria can develop new enzymes or pathways that are not inhibited by the antibiotic.

Bacteria can develop resistance through mutations, but they can also acquire resistance via horizontal gene transfer (e.g., plasmid transfer). Taking antibiotics routinely in mild doses, or failing to complete a prescribed dosage, contributes to the spread of resistance by selecting strains that can survive in the presence of the drug. Prescription of antibiotics for

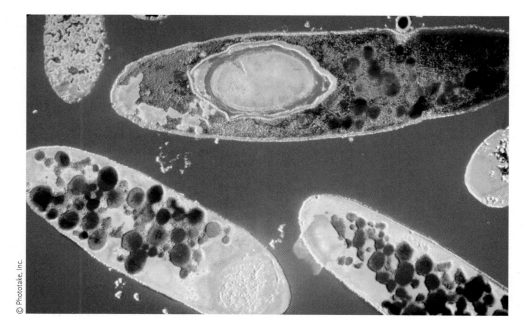

© Phototake, Inc.

Figure 22.11
The bacterium *Clostridium butyricum*, one of the *Clostridium* species that produces the toxin botulin (colourized TEM). The large stained structure in the cells is a spore (a survival structure).

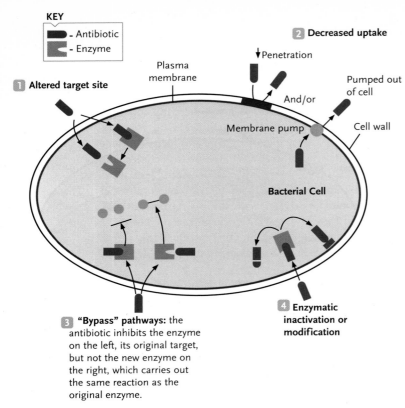

KEY

▮ - Antibiotic
▯ - Enzyme

1 Altered target site

2 Decreased uptake

Plasma membrane

↓Penetration

And/or

Pumped out of cell

Membrane pump

Cell wall

Bacterial Cell

3 "Bypass" pathways: the antibiotic inhibits the enzyme on the left, its original target, but not the new enzyme on the right, which carries out the same reaction as the original enzyme.

4 Enzymatic inactivation or modification

Figure 22.12
Four major mechanisms of antibiotic resistance. See text for further explanation of each mechanism.

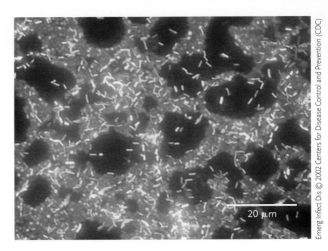

Emerg Infect Dis © 2002 Centers for Disease Control and Prevention (CDC)

20 μm

Figure 22.13
Biofilm grown on a stainless steel surface.

colds and other virus-caused diseases can also promote bacterial resistance because viruses are unaffected by antibiotics, but the presence of antibiotics in your system can lead to resistance. Antibacterial agents that may promote resistance are also commonly included in such commercial products as soaps, detergents, and deodorants. Resistance is a form of evolutionary adaptation; antibiotics alter the bacterium's environment, conferring a reproductive advantage on those strains best adapted to the altered conditions.

The development of resistant strains has made tuberculosis, cholera, typhoid fever, gonorrhea, and other bacterial diseases difficult to treat with antibiotics. For example, as recently as 1988, drug-resistant strains of *Streptococcus pneumoniae*, which causes pneumonia, meningitis, and middle-ear infections, were practically unknown. Now, resistant strains of *S. pneumoniae* are common and increasingly difficult to treat.

22.2g In Nature, Prokaryotic Organisms May Live in Communities Attached to a Surface

Often, researchers grow bacteria and archaea as individuals in pure cultures. We have learned a lot about prokaryotic organisms from these pure cultures, but in nature, prokaryotic organisms rarely exist as individuals or as pure cultures. Instead, bacteria and archaea live in communities where they interact in a variety of ways. One important type of community is known as a **biofilm**, which consists of a complex aggregation of microorganisms attached to a surface and surrounded by a film of polymers **(Figure 22.13)**. Life in a biofilm offers several benefits: organisms can adhere to hospitable surfaces, they can live on the products of other cells, conditions within the biofilm promote gene transfer between species, and the biofilm protects cells from harmful environmental conditions (see "Life on the Edge" 22.3). Biofilms form on any surface with sufficient water and nutrients. For example, you're probably familiar with how slippery rocks in a stream can be when you try to step from one to the next; the slipperiness is due to biofilms on the rocks. Dental plaque is also a biofilm; if this biofilm spreads below the gumline, it causes inflammation of the gums (gingivitis). Regular removal of plaque by brushing, flossing, and dental checkups helps prevent gingivitis.

Biofilms have practical consequences for humans, both beneficial and detrimental. On the beneficial side, for example, are the health effects each of us gains from the bacteria that live in biofilms in our gastrointestinal tracts. We also make use of biofilms in commercial applications: biofilms on solid supports are used in sewage treatment plants to process organic matter before the water is discharged, and they can be effective in bioremediating toxic organic molecules contaminating groundwater. But biofilms can also be harmful to human health. Biofilms adhere to many kinds of surgical equipment and supplies, including catheters, pacemakers, and artificial joints. Even if the bacteria colonizing these devices are not pathogenic, their presence is obviously not desirable given that these devices should be sterile. The presence of any Gram-negative bacteria is a concern, given their nature. As well, many heterotrophic bacteria will become opportunistic pathogens, given the right conditions. Biofilm infections are difficult to treat because bacteria in a biofilm

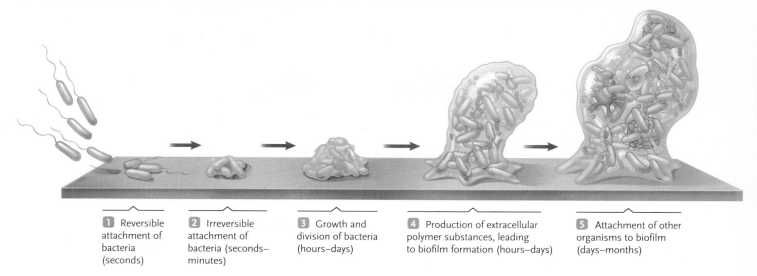

| 1 Reversible attachment of bacteria (seconds) | 2 Irreversible attachment of bacteria (seconds–minutes) | 3 Growth and division of bacteria (hours–days) | 4 Production of extracellular polymer substances, leading to biofilm formation (hours–days) | 5 Attachment of other organisms to biofilm (days–months) |

Figure 22.14
Steps in the formation of a biofilm.

are up to 1000 times as resistant to antibiotics as are the same bacteria in liquid cultures. For example, outbreaks of the disease caused by *E. coli* O157:H7 have been caused by biofilms that are very difficult to wash off spinach, lettuce, and other produce.

How does a biofilm form? Imagine a surface, such as a rock in a stream, over which water is flowing **(Figure 22.14)**. Due to the nutrients in the water, the surface rapidly becomes coated with polymeric organic molecules, such as polysaccharides or glycoproteins. Once the surface is conditioned with organic molecules, free cells attach in a reversible manner in a matter of seconds (see Figure 22.14, step 1). If the cells remain attached, the association may become irreversible (step 2), at which point the cells grow and divide on the surface (step 3). Next, the physiology of the cells changes, and they begin to secrete *extracellular polymeric substances* (EPSs), slimy, gluelike substances similar to the molecules found in bacterial capsules. EPS extends between cells in the mixture, forming a matrix that binds cells to each other and anchors the complex to the surface, thereby establishing the biofilm (step 4). The slime layer entraps a variety of materials, such as dead cells and insoluble minerals. The physiological change accompanying the formation of a biofilm results from marked changes in a prokaryotic organism's gene expression pattern—in effect, the prokaryotic cells in a biofilm become very different organisms. Over time, other organisms are attracted to and join the biofilm; depending on the environment, these may include other bacterial species, algae, fungi, or protozoa producing diverse microbial communities (step 5). As described in "Molecule behind Biology" 22.1, on p. 492, prokaryotic organisms in a biofilm communicate with each other via **quorum sensing;** in fact, this communication is part of biofilm

formation—it allows cells to start secreting EPS when a high enough cell density is reached.

Much remains to be learned about how organisms form and interact within a biofilm and how changes in gene expression during the transition are regulated.

In the next two sections, we describe the major groups of prokaryotic organisms.

STUDY BREAK

1. What is the difference between a chemoheterotroph and a photoautotroph?
2. What is the difference between an obligate anaerobe and a facultative anaerobe?
3. What is the difference between nitrogen fixation and nitrification? Why are nitrogen-fixing prokaryotic organisms important?
4. What is binary fission?
5. What is the difference between an endotoxin and an exotoxin? Explain how they differ with respect to how they cause disease.
6. Explain four mechanisms by which bacteria protect themselves from antibiotics.
7. What is a biofilm? Give an example of a biofilm that is beneficial to humans and one that is harmful. What advantages do prokaryotic cells in a biofilm gain?
8. What is quorum sensing?

22.3 The Domain Bacteria

As for other organisms, classification of bacteria and archaeans has been revolutionized by molecular techniques that allow researchers to compare nucleic acid

N-Acyl-l-Homoserine Lactone

Most bacteria are social organisms that interact in many ways and display social behaviours, such as hunting for food in swarms, bioluminescence (see Chapter 1), biofilm formation, and virulence in pathogenic bacteria, such as the strain of *E. coli* O157:H7 responsible for "hamburger disease." These behaviours happen only when a critical population density is reached, meaning that bacteria must be able to sense the presence of other cells. How does a bacterial cell know that it is not alone? Bacteria use quorum sensing to communicate; this mechanism involves the release of signalling molecules into the environment. Accumulation of signalling molecules enables the cell to determine the density of other cells around it and respond accordingly; the response occurs after the signalling molecule is perceived by specific receptors on the cell's membrane and triggers activation of specific genes. Different bacterial species use different signalling molecules; for example, many Gram-negative bacteria use *N*-acyl-l-homoserine lactones (a lactone is a type of cyclic ester) such as that shown in **Figure 1.** Gram-positive cells also signal each other but use small peptides rather than lactones. If we can learn to "speak" or "translate" these bacterial languages, could we interfere with the social behaviours they control? The possibility has important implications for medical science given the role of these signals in processes such as the onset of virulence in pathogenic bacteria and communication within biofilms such as those that form on medical devices implanted in patients.

Figure 1
N-Acyl-l-homoserine lactone from Vibrio fischeri.

and protein sequences as tests of evolutionary relatedness. Ribosomal RNA (rRNA) sequences have been most widely used in the evolutionary studies of prokaryotic organisms. Researchers have identified several evolutionary branches within each prokaryotic domain **(Figure 22.15)**, but these classifications will likely change in the future when full genomic sequences can be compared. We discuss the major groups of the domain Bacteria, which is much better characterized than the domain Archaea, in this section, and those of the domain Archaea in the next section.

22.3a Molecular Studies Reveal Numerous Evolutionary Branches in the Bacteria

Bacteria as a domain is much better characterized than Archaea: sequencing studies reveal that bacteria have several distinct and separate evolutionary branches. We restrict our discussion to six particularly important groups: proteobacteria, green bacteria, cyanobacteria, Gram-positive bacteria, spirochetes, and chlamydias (see Figure 22.15).

Proteobacteria: The Purple Bacteria and Their Relatives. This highly diverse group of Gram-negative bacteria likely evolved from a purple, photosynthetic ancestor. Their purple colour comes from their photosynthetic pigment, a type of chlorophyll distinct from that of plants. Many present-day species are either photoautotrophs (the purple sulfur bacteria) or photoheterotrophs (the purple nonsulfur bacteria); both groups carry out a type of photosynthesis that does not use water as an electron donor and does not release oxygen as a by-product.

Other present-day proteobacteria are **chemoheterotrophs** that are thought to have evolved as an evolutionary branch following the loss of photosynthetic capabilities in an early proteobacterium. The evolutionary ancestors of mitochondria are considered likely to have been ancient nonphotosynthetic proteobacteria.

Among the chemoheterotrophs classified with the proteobacteria are *E. coli*; plant pathogenic bacteria; and bacteria that cause human diseases such as bubonic plague, gonorrhea, and various forms of gastroenteritis and dysentery. The proteobacteria also include both free-living and symbiotic nitrogen-fixing bacteria.

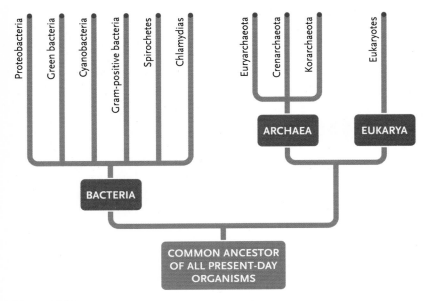

Figure 22.15
An abbreviated phylogenetic tree of Bacteria and Archaea.

Myxobacteria are an unusual group of nonphotosynthetic proteobacteria that form colonies held together by the slime they produce. Enzymes secreted by the colonies digest "prey"—other bacteria, primarily—that become stuck in the slime. When environmental conditions become unfavourable, as when soil nutrients or water are depleted, myxobacteria form a fruiting body, a differentiated multicellular stage large enough to be visible to the naked eye **(Figure 22.16)**. The fruiting body contains clusters of spores that are dispersed to form new colonies when the fruiting body bursts. Quorum sensing is involved in spore formation.

Helicobacter pylori, the cause of many gastric ulcers (see "People behind Biology" 22.2, p. 494), is also a proteobacterium.

Green Bacteria. This diverse group of photosynthetic Gram-negative bacteria is named for the chlorophyll pigments that give the cells their green colour (a different form of chlorophyll than that found in plants). Like the purple bacteria, they do not release oxygen as a by-product of photosynthesis. Also like the purple bacteria, some are photoautotrophs, whereas others are photoheterotrophs. The photoautotrophic green bacteria are fairly closely related to the Archaea and are usually found in hot springs, whereas the photoheterotrophic type is typically found in marine and high-salt environments.

Cyanobacteria. These Gram-negative photoautotrophs are blue-green in colour **(Figure 22.17)** and carry

a.

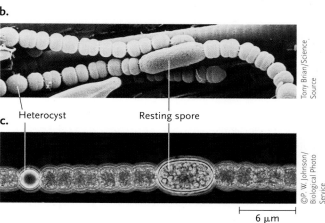
b.

Heterocyst   Resting spore

c.

6 μm

Figure 22.17
Cyanobacteria. **(a)** A population of cyanobacteria covering the surface of a pond. **(b)** and **(c)** Chains of cyanobacterial cells. Some cells in the chains form spores. The heterocyst is a specialized cell that fixes nitrogen.

out photosynthesis by the same pathways and using the same chlorophyll as eukaryotic algae and plants. Like plants and algae, they release oxygen as a by-product of photosynthesis.

The direct ancestors of present-day cyanobacteria were the first organisms to use the water-splitting reactions of photosynthesis. As such, they were critical to the accumulation of oxygen in the atmosphere, which allowed the evolutionary development of aerobic organisms. Chloroplasts probably evolved from early cyanobacteria that were incorporated into the cytoplasm of primitive eukaryotes, which eventually gave rise to the algae and higher plants, as discussed in Chapter 26. Besides releasing oxygen, present-day cyanobacteria help fix nitrogen into organic compounds in aquatic habitats and act as symbiotic partners with fungi in lichens (see Chapter 25).

Gram-Positive Bacteria. This large group contains many species that live primarily as chemoheterotrophs. Some cause human diseases, including *Bacillus anthracis,* the causal agent of anthrax;

100 μm

Figure 22.16
The fruiting body of *Chondromyces crocatus,* a myxobacterium. Cells of this species collect together to form the fruiting body.

Barry Marshall, *University of Western Australia;* Robin Warren, *Royal Perth Hospital (retired)*

A few hours after you eat, you go to your doctor complaining of stomach pain, abdominal bloating, and nausea; most worryingly, you have started to vomit blood. Your doctor tells you that you have a gastric ulcer, a lesion in your stomach lining. If this visit to your doctor had occurred before the mid-1980s, your doctor would have explained that ulcers are caused by increased stomach acidity due to stress. The treatment? Drink lots of milk, take antacids, and give up alcohol and your favourite spicy foods—no more curries or chili that would aggravate your ulcer. This view of ulcers was accepted for years until two Australian physicians, Barry Marshall and Robin Warren, of the University of Western Australia, demonstrated that most ulcers are caused by a bacterial infection. Marshall and Warren observed that biopsies from patients with ulcers revealed large numbers of spiral-shaped bacterial cells in inflamed tissues. Together, the two physicians carried out a series of studies that demonstrated the link between ulcers and the presence of the bacterium; as you can see in **Table 1,** the bacterium was associated with almost all gastric and duodenal (part of the small intestine) ulcers. The bacterium was not known at the time but was later named *Helicobacter pylori*

Table 1	Association of Bacteria with Biopsy Samples		
Biopsy Appearance	Total Samples	Number (%) Associated with Bacteria
Gastric ulcer	22	18 (77%)
Duodenal ulcer	13	13 (100%)
Total	35	31 (89%)

(Figure 1). But despite Marshall and Warren having research published in respected medical journals, the medical community did not believe their findings—how could bacteria possibly survive in the very acidic conditions of the stomach? Out of frustration, and anxious to get proper treatment for his patients, Marshall drank a culture of *H. pylori*! After about a week, he developed severe abdominal pain and vomiting, and endoscopic examination of his stomach showed regions of inflammation teeming with *H. pylori*. Much to his disappointment, he did not develop ulcers, but he had made the point that *H. pylori* is pathogenic. Marshall and Warren also showed that antibiotics were effective in treating ulcers, and in 2005, they were awarded the Nobel Prize in Medicine. So how is *H. pylori* able to survive in the stomach? It is able to burrow deep into the mucus lining the stomach by means of its numerous flagella, and it produces urease, which converts urea into CO_2 and ammonia, making the region around its cells more basic.

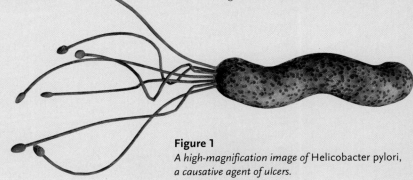

Figure 1
A high-magnification image of Helicobacter pylori, *a causative agent of ulcers.*

Figure 22.18
Streptococcus bacteria forming the long chains of cells typical of many species in this genus.

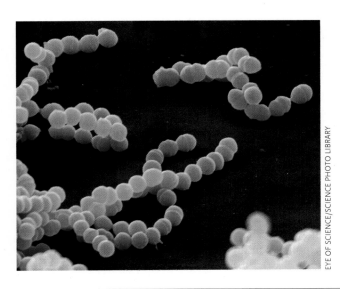

EYE OF SCIENCE/SCIENCE PHOTO LIBRARY

Staphylococcus, which causes some forms of food poisoning, toxic shock syndrome, pneumonia, and meningitis; and *Streptococcus* **(Figure 22.18),** which causes strep throat, necrotizing fasciitis, and some forms of pneumonia. However, some Gram-positive bacteria are beneficial to humans; *Lactobacillus,* for example, carries out the lactic acid fermentation used in the production of pickles, sauerkraut, and yogurt. One unusual group of bacteria, the mycoplasmas, is placed among the Gram-positive bacteria by molecular studies even though they show a Gram-negative staining reaction. This staining reaction results because they are naked cells that secondarily lost their cell walls in evolution. Some mycoplasmas, with diameters from 0.1 to 0.2 μm, are the smallest known cells.

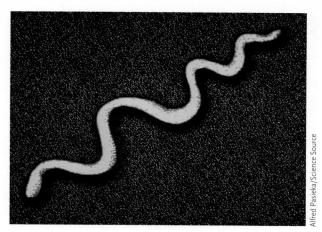

Figure 22.19
Treponema pallidum, a spirochete bacterium that causes syphilis (scanning electron microscope image).

Spirochetes. These organisms have helically spiralled flagella embedded in their cytoplasm, causing the cells to move in a twisting, corkscrew pattern **(Figure 22.19).** Their corkscrew movements enable them to move in viscous environments such as mud and sewage, where they are common. Some spirochetes are harmless inhabitants of the human mouth; another species, *Treponema pallidum*, is the cause of syphilis. Termites have symbiotic spirochetes in their intestines that enable them to digest cellulose.

Chlamydias. These bacteria are unusual because although they are Gram-negative and have cell walls with an outer membrane, they lack peptidoglycan. All the known chlamydias are intracellular parasites that cause various diseases in animals. One bacterium of this group, *Chlamydia trachomatis*, is responsible for one of the most common sexually transmitted infections of the urinary and reproductive tracts of humans and also causes trachoma, an infection of the cornea that is the leading cause of blindness in humans.

In this section, you have seen that bacteria thrive in nearly every habitat on Earth. However, some members of the second prokaryotic domain, the Archaea, the subject of the next section, live in habitats that are too forbidding even for bacteria.

STUDY BREAK

1. What methodologies have been used to classify prokaryotic organisms?
2. What were the likely characteristics of the evolutionary ancestor of present-day proteobacteria?
3. How does photosynthesis in photosynthetic proteobacteria differ from photosynthesis in cyanobacteria?

22.4 The Domain Archaea

The first Archaea were isolated from extreme environments, such as hot springs, hydrothermal vents on the ocean floor, and salt lakes **(Figure 22.20).** For that reason, these prokaryotes were called *extremophiles* (organisms that live in extreme environments). Subsequently, archaea have also been found living in less extreme environments.

Archaea share some cellular features with eukaryotes and some with bacteria and have other features that are unique **(Table 22.2, p. 496).**

22.4a Unique Characteristics of Archaea

Among their unique characteristics are certain features of their plasma membranes and cell walls. The lipid molecules in archaeal plasma membranes are unlike those in the plasma membranes of the majority of bacteria: there is a different linkage between glycerol and the hydrophobic tails, and the

a.

b.

Figure 22.20
Typically extreme archaeal habitats. (a) Highly saline water in Great Salt Lake, Utah, coloured red purple by archaeans. **(b)** Hot, sulfur-rich water in Emerald Pool, Yellowstone National Park, coloured brightly by the oxidative activity of archaea, which convert H_2S to elemental sulfur.

LIFE ON THE EDGE 22.3

Snottite Bacteria

Some of the most extreme and inhospitable environments on Earth are deep caves that have formed in sulfur-rich rocks. As water flows through these rocks, toxic H_2S gas is released at concentrations that can make the cave atmosphere toxic to humans, who cannot survive in the caves without gas masks. But extremophile bacteria and archaea thrive in these caves, including bacteria that grow in biofilms to form *snottites* **(Figure 1),** mucous stalactites that hang from the walls and ceiling of the cave. These bacteria obtain energy from H_2S and other sulfur compounds, producing sulfuric acid as a waste product that drips from the snottites. The biofilm that surrounds the bacteria protects them from the extremely acid environment (pH < 2) they have helped to create, but the acid eats away at the surrounding rock,

enlarging the cave. In addition to actively contributing to cave formation, these extremophile bacteria are the foundation of the cave ecosystem. Their ability to convert inorganic

chemicals such as H_2S into the organic molecules that make up their cells provides a source of organic carbon to other organisms, making them the base of the food web in such caves.

Figure 1
Snottites, Cueva de Villa Luz, Mexico.

Peter Lane Taylor, VISUALS UNLIMITED/SCIENCE PHOTO LIBRARY

Table 22.2	Characteristics of the Bacteria, Archaea, and Eukarya		
Characteristic	Bacteria	Archaea*	Eukarya
DNA arrangement	Single, circular in most, but some linear and/or multiple	Single, circular	Multiple linear molecules
Chromosomal proteins	Prokaryotic histonelike proteins	Five eukaryotic histones	Five eukaryotic histones
Genes arranged in operons	Yes	Yes	No
Nuclear envelope	No	No	Yes
Mitochondria	No	No	Yes
Chloroplasts	No	No	Yes
Peptidoglycan in cell wall	Present	Absent; some have pseudopeptidoglycan	Absent
Membrane lipids	Unbranched; linked by ester linkages	Branched; linked by ether linkage; may have polar heads at both ends	Unbranched; linked by ester linkages
RNA polymerase	Limited variations	Multiple types	Multiple types
Ribosomal proteins	Prokaryotic	Some prokaryotic, some eukaryotic	Eukaryotic
First amino acid placed in proteins	Formylmethionine	Methionine	Methionine
Aminoacyl–tRNA synthetases	Prokaryotic	Eukaryotic	Eukaryotic
Cell division proteins	Prokaryotic	Prokaryotic	Eukaryotic
Proteins of energy metabolism	Prokaryotic	Prokaryotic	Eukaryotic

*Given that very few Archaea have been identified or cultured, the information in this table is based on an extremely small data set.

tails are isoprenes rather than fatty acids (see Chapter 5). Also, some lipids have polar head groups at both ends. Why would such seemingly minor differences be significant? These unique lipids are more resistant to disruption, making the plasma membranes better suited to extreme environments. Similarly, the unique cell walls of archaea are more resistant to extremes than those of bacteria; some archaea can even survive being boiled in strong detergents!

Many archaea are chemoautotrophs, whereas others are chemoheterotrophs. Interestingly, no known member of the Archaea has been shown to be pathogenic.

22.4b Molecular Studies Reveal Three Evolutionary Branches in the Archaea

The phylogeny of Archaea is poorly developed relative to Bacteria and in quite a state of flux because a tremendous number of archaea have not been cultured, meaning that we have only metagenomic data for most of these organisms. Based on differences in rRNA sequence data, the domain Archaea is divided into three groups (see Figure 22.15, p. 492). Two major groups, the **Euryarchaeota** and the **Crenarchaeota**, contain archaea that have been cultured in the laboratory. The third group, the **Korarchaeota**, has been recognized solely on the basis of DNA taken from environmental samples.

Euryarchaeota. These organisms are found in various extreme environments. They include methanogens, extreme halophiles, and some extreme thermophiles, as described below.

Methanogens (methane generators) live in low-oxygen environments **(Figure 22.21)** and represent about one-half of all known species of Archaea. Methanogens are obligate anaerobes that live in the anoxic (oxygen-lacking) sediments of swamps, lakes,

RALPH ROBINSON, VISUALS UNLIMITED/SCIENCE PHOTO LIBRARY

5 μm

Figure 22.21
A colony of the methanogenic archaeon *Methanosarcina*, which lives in the sulfurous, waterlogged soils of marshes and swamps.

marshes, and sewage works, as well as in more moderate environments, such as the rumen of cattle and sheep, the large intestine of dogs and humans, and the hindgut of insects such as termites and cockroaches. Methanogens generate energy by converting various substrates such as carbon dioxide and hydrogen gas or acetate into methane gas, which is released into the atmosphere.

Halophiles are salt-loving organisms. Extreme halophilic Archaea live in highly saline environments such as the Dead Sea and on foods preserved by salting. They require a minimum NaCl concentration of about 1.5 M (about 9% solution) to survive and can live in a fully saturated solution (5.5 M, or 32%). Most are aerobic chemoheterotrophs, which obtain energy from sugars, alcohols, and amino acids using pathways similar to those of bacteria. Many extreme halophiles use light as a secondary energy source, supplementing the oxidations that are their primary source of energy.

Extreme thermophiles live in extremely hot environments such as hot springs and ocean floor hydrothermal vents. Their optimal temperature range for growth is 70 to 95°C, close to the boiling point of water. By comparison, no eukaryotic organism is known to live at a temperature higher than 60°C. Some extreme thermophiles are members of the Euryarchaeota, but most belong to the **Crenarchaeota**, the next group that we discuss.

Crenarchaeota. This group includes most of the extreme thermophiles, which have a higher optimal temperature range than those belonging to the Euryarchaeota. For example, the most thermophilic member of this group, *Pyrobolus*, dies below 90°C, grows optimally at 106°C, and can survive an hour of autoclaving at 121°C! *Pyrobolus* lives in ocean floor hydrothermal vents, where the pressure creates water temperatures greater than the boiling point of water on Earth's surface.

Also in this group are **psychrophiles** ("cold loving"), organisms that grow optimally in cold temperatures in the range from –10 to –20°C. These organisms are found mostly in the Antarctic and Arctic oceans, which are frozen most of the year, and in the intense cold at ocean depths.

Mesophilic members of the Crenarchaeota make up a large part of plankton found in cool, marine waters, where they are food sources for other marine organisms.

Korarchaeota. This group has been recognized solely on the basis of DNA samples obtained from marine and terrestrial hydrothermal environments. To date, no members of this group have been isolated and cultivated in the lab, and nothing is known about their physiology. Molecular data indicate that they are the oldest archaeal lineage.

Thermophilic archaea are important commercially. For example, they are very important in biotechnological applications as sources of enzymes that function under extreme physicochemical conditions (e.g., high temperature, high salinity).

In this chapter, we have focused on bacteria and archaea, whose metabolic diversity and environmental range and ecological importance belie their structural simplicity. In the next chapter, we look at still simpler entities: viruses, viroids, and prions, which are derived from living organisms and retain only some of the properties of life.

STUDY BREAK

1. What distinguishes members of the Archaea from members of the Bacteria and Eukarya?
2. How does a methanogen obtain energy? In which group or groups of Archaea are methanogens found?
3. Where do extreme halophilic Archaea live? How do they obtain energy? In which group or groups of Archaea are the extreme halophiles found?
4. What are extreme thermophiles and psychrophiles?

Review

aplia

To access course materials such as Aplia and other companion resources, please visit www.NELSONbrain.com.

22.1 The Full Extent of the Diversity of Bacteria and Archaea Is Unknown

- Bacteria and Archaea are the most abundant and diverse organisms on Earth; however, the vast majority of prokaryotic organisms have not been described because they cannot be cultured using standard techniques and because many environments are very difficult and/or expensive to access.
- Prokaryotic organisms make up two of the three domains of life, the Archaea and the Bacteria.

22.2 Prokaryotic Structure and Function

- Prokaryotic genomes typically consist of a single, circular DNA molecule packaged into the nucleoid (Figure 22.4). Many prokaryotic cells also contain plasmids, which replicate independently of the chromosome and can be passed to other cells.
- Gram-positive bacterial cell walls consist of a single, relatively thick peptidoglycan layer. Gram-negative bacteria have walls consisting of a relatively thin peptidoglycan sheath surrounded by an outer lipopolysaccharide (LPS) membrane (Figure 22.6).
- A polysaccharide capsule (Figures 22.3, and 22.7) surrounds many bacteria, protecting them and helping them adhere to surfaces.
- Archaea and bacteria show great diversity in their modes of obtaining energy and carbon. Two of the modes of nutrition found among eukaryotic organisms are also found in prokaryotic organisms (chemoheterotrophy and photoautotrophy), but two other modes are unique to prokaryotic organisms: chemoautotrophs obtain energy by oxidizing inorganic substrates and use carbon dioxide as their carbon source, and photoheterotrophs use light as a source of energy and obtain their carbon from organic molecules.

- Bacteria and Archaea use a range of pathways to transform energy: aerobic respiration, anaerobic respiration, and/or various forms of fermentation.
- Some bacteria and archaea are capable of nitrogen fixation, the conversion of atmospheric nitrogen to ammonia; others are responsible for nitrification, the conversion of ammonium to nitrate.
- Prokaryotic cells normally reproduce asexually by binary fission (Figure 22.10), which can result in very rapid population growth under favourable conditions.
- In nature, bacteria and archaea live in complex communities, such as biofilms (Figure 22.13).
- Pathogenic bacteria cause disease via exotoxins and endotoxins.
- Bacteria may develop resistance to antibiotics through mutation of their own genes or by acquiring resistance genes from other bacteria.

22.3 The Domain Bacteria

- Bacteria are divided into more than a dozen evolutionary branches, including Gram-negative proteobacteria, Gram-negative green bacteria, cyanobacteria, Gram-positive bacteria, spirochetes, and chlamydias.

22.4 The Domain Archaea

- A very large number of archaea have not been cultured, but we know that archaea have some features that are like those of bacteria, other features that are eukaryotic, and some that are unique (see Table 22.2).
- Archaea are classified into three groups: the Euryarchaeota (methanogens, extreme halophiles, and some extreme thermophiles); the Crenarchaeota (which includes most of the extreme thermophiles, but also some psychrophiles and mesophiles); and the Korarchaeota, known only from DNA samples.

Questions

Self-Test Questions

1. Which of the following structures is found in prokaryotic cells?
   a. cellulose cell wall
   b. ribosome
   c. mitochondria
   d. nuclear membrane

2. Which of the following statements about archaea is correct?
   a. Their cell walls contain peptidoglycan.
   b. Most are pathogens.
   c. Many are extremophiles.
   d. They have no traits in common with eukaryotic cells.

3. Which of the following statements accurately describes a plasmid?
   a. It can only replicate when the cell's chromosome replicates.
   b. It is a small circular piece of RNA outside a cell's chromosome.
   c. It is a small circular piece of DNA outside a cell's chromosome.
   d. It refers to a piece of DNA taken up from the environment by a prokaryotic cell.

4. You have isolated an unknown bacterium that produces a toxin and you are trying to determine if this is an endotoxin or an exotoxin. Which of the following features would be associated with the toxin, if it were an endotoxin?
   a. It would be secreted from the cell.
   b. It would be part of the cell wall.
   c. It would be part of the plasma membrane.
   d. It would be produced by an archaeon.

5. Place the following steps by which prokaryotic cells form a biofilm in the correct order:
   1. Cells grow and divide.
   2. The cells' physiology changes.
   3. Cells attach to a surface that is covered in organic polymers.
   4. Cells secrete extracellular polymers that "glue" the cells to the surface and to each other.
      a. 1, 2, 3, 4
      b. 2, 1,4, 3
      c. 4, 3, 1, 2
      d. 3, 1, 2, 4

6. You are growing a facultative anaerobic archaeon in culture under two conditions: one culture is in anaerobic conditions, and the other is in aerobic conditions. How would you expect the growth of the cells to compare between the two cultures?
   a. Growth would be greater in the culture in aerobic conditions.
   b. Growth would be greater in the culture in anaerobic conditions.
   c. Growth would be the same in both conditions.

7. A bacterium that oxidizes nitrite as its only energy source was found deep in a cave. How would you classify this bacterium, based on its carbon and energy source?
   a. as a chemolithotroph
   b. as a chemoheterotroph
   c. as a photoautotroph
   d. as a photoheterotroph

8. Which of the following processes converts ammonium (NH_4^+) into nitrate (NO_3^-)?
   a. nitrogen fixation
   b. ammonification
   c. nitrification
   d. denitrification

9. Which of the following groups of bacteria are all oxygen-producing photoautotrophs?
   a. spirochetes
   b. cyanobacteria
   c. proteobacteria
   d. green bacteria

10. Which of the following statements about chlamydias is correct?
    a. They lack peptidoglycan.
    b. They are Gram-positive.
    c. They are not pathogenic.
    d. They have no outer membrane in the cell wall.

Questions for Discussion

1. In the lab, you have isolated some prokaryotic cells that belong either to a Gram-positive bacterium or to an archaeon. What cellular (structural) features could you look for to determine which type of organism you have isolated? Indicate how that feature would differ between the two kinds of organism (assume that you have the necessary equipment to test for any cellular feature you want).

2. List several functions of the outer wall layer in Gram-negative bacteria.

3. You are doing research to develop new drugs and have developed a new antibiotic. This drug acts by inhibiting ribosome function in prokaryotic cells. When you test it on animal cells, however, you find that the growth of animal cells is inhibited. Explain why this drug inhibits the growth of animal cells.

4. How do bacteria resist antibiotics? Why do antibiotics lose their effectiveness so quickly?

5. In which nutritional class would you place a prokaryotic organism that uses glucose as its only energy and carbon source? What about an organism that uses elemental sulfur as an energy source and carbon dioxide as a carbon source? What is the energy source for phototrophic organisms?

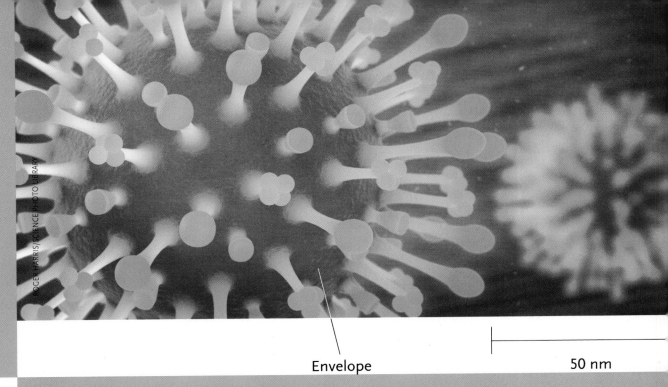

ROGER HARRIS/SCIENCE PHOTO LIBRARY

Envelope

50 nm

Influenza virus.

Viruses, Viroids, and Prions: Infectious Biological Particles

WHY IT MATTERS

Imagine yourself sitting in a crowded airplane bound from London, United Kingdom, to Vancouver. The person sitting beside you has a runny nose; is sneezing, coughing, and sucking on lozenges; and appears to have a fever. Recognizing that your seatmate is exhibiting many of the symptoms of influenza, a respiratory illness caused by the influenza virus shown in the micrograph above, you worry that the virus will spread to you through your seatmate's coughing and sneezing. You have just seen how influenza ("the flu") can affect people and how air travel can help it spread around the world.

At any given time, 5 to 15% of the global population of people exhibits the symptoms of influenza, and each year, about 500 000 people are killed by influenza A, one type of influenza virus. Recent research has shown that new strains of influenza A arise each year from just a few initial sources in East and Southeast Asia and then spread around the world. Colin Russell of Cambridge University in the United Kingdom and his colleagues analyzed 13 000 human influenza A viruses of the H3N2 subtype by collecting viral material from infected people on six continents between 2002 and 2007. Their analysis revealed almost continuous circulation of H3N2 in East and Southeast Asia. This regional network of overlapping epidemics appeared to be the source of influenza outbreaks elsewhere in the world. The epidemics then spread to Australia and other islands in the central and south Pacific, North America, and Europe, finally arriving in South America. As influenza viruses travel through populations around the world, they evolve, changing so much that the vaccines we developed in previous years are no longer effective, and new vaccines must be developed.

The good news is that understanding the global pattern of influenza migration will help the World Health Organization develop effective vaccines. Knowing which strains cause the initial outbreak in Asia allows scientists to formulate vaccines to target these strains, offering people in other regions some protection from the illness. Flu vaccines are commonly prepared using killed viruses (viruses that have been inactivated, for example, by chemical treatment, so that they are no longer infective) meaning that when you get your flu shot, you won't develop influenza, but your body will produce antibodies against the virus, protecting you against subsequent infection by any of those specific strains (antibodies are highly specific protein molecules produced by the immune system that recognize and bind to specific proteins of a pathogen, such as the proteins in a virus's coat).

Each winter in Canada and other countries, people line up for their annual flu shot. For many of us, this shot represents a gamble that we will be protected against the strains of the influenza virus making the rounds that winter.

For most people, getting the flu means feeling awful for a few days, but for the very young, the elderly, and people with weakened immune systems, the stakes are higher: a bout of flu can be fatal. Some flu outbreaks have been devastatingly lethal to a very large proportion of the population. The worst recorded example is the flu *pandemic* (an outbreak or epidemic that spreads around the world) of 1918. A strain of influenza virus known as the Spanish flu infected almost half of the world's population, killing about 1 in every 20 people.

Why was the Spanish flu so deadly? And why do we need to develop new flu vaccines so often? We investigate these questions later in this chapter. We also look at the beneficial roles played by viruses—not all are pathogenic—and investigate ways in which we may be able to harness the infective abilities of viruses for our own uses. For example, can we use viruses as vectors for gene therapy to fight diseases? We start with a look at the defining characteristics of viruses: how they are able to enter cells and take over the cell's machinery to make more copies of themselves. We also compare viruses with viroids and prions, other infectious particles.

23.1 What Is a Virus? Characteristics of Viruses

If you look back at the tree of life (Figure 3.20), you'll notice that viruses are not shown. That is because they lack many of the properties of life shared by all organisms (Section 3.1a) and so are not considered to be living organisms. For example, viruses cannot reproduce on their own and lack a metabolic system to provide energy for their life cycles; instead, they depend on the host cells that they infect for these functions. For this reason, viruses are infectious biological particles rather than organisms. The structure of a virus is reduced to the minimum necessary to transmit its genome from one host cell to another. A virus is simply one or more nucleic acid molecules surrounded by a protein **coat** or **capsid (Figure 23.1a, b).** Some capsids may be enclosed within a membrane or **envelope** derived from their host cell's membrane (Figure 23.1c). So a virus is not a cell—it does not have cytoplasm enclosed by a plasma membrane, as do all known living organisms.

The nucleic acid genome of a virus may be either DNA or RNA and can be composed of either a single strand or a double strand of RNA or DNA. Viral genomes range from just a few genes to over a hundred genes; all viruses have genes that encode at least their coat proteins, as well as proteins involved

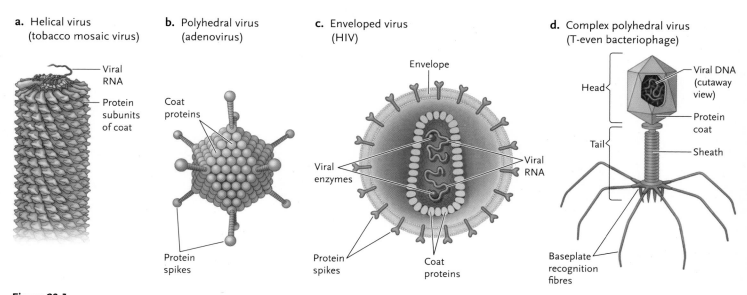

Figure 23.1

Viral structure. All viruses consist of nucleic acid surrounded by a protein coat, but they can have a very wide range of sizes and shapes.

in regulation of transcription. Genomes of **enveloped viruses** also include genes required for the synthesis of envelope proteins. Some viral genomes also include virus-specific enzymes for nucleic acid replication.

Most viruses take one of two basic structural forms, helical and polyhedral. In **helical viruses** the protein subunits assemble in a rodlike spiral around the genome (see Figure 23.1a). A number of viruses that infect plant cells are helical. In **polyhedral viruses**, the coat proteins form triangular units that fit together like the parts of a soccer ball (see Figure 23.1b). The polyhedral viruses include forms that infect animals, plants, and bacteria. In some polyhedral viruses, protein spikes that provide host cell recognition extend from the corners, where the facets fit together. Both helical and polyhedral viruses can be enveloped in a membrane derived from the host's membrane (see Figure 23.1c and **Figure 23.2**). In enveloped viruses, proteins synthesized from the viral genome in the host cell are transported to and embedded in the membrane before the virus particle buds through the host cell. These proteins allow the virus to recognize and bind to host cells.

Although they are not considered to be alive, viruses are classified into orders, families, genera, and species using several criteria, including virus size and structure, genome structure (RNA or DNA, single stranded or double stranded), and how their nucleic acid is replicated. More than 4000 species of viruses have been classified into more than 80 families. The family names end in *-viridae* and may refer either to the geographic region where the virus was first discovered or to the structure of the virus. For example, Coronaviridae, the family to which influenza virus belongs, is named for the "crown" of protein spikes on the capsid, as shown in the photomicrograph at the start of this chapter (*corona* = crown). Like some bacteria, some viruses are named for the disease they cause; these names can be one or two words, for example, herpesvirus or Ebola virus. Each type of virus is made up of many strains, differentiated by their virulence.

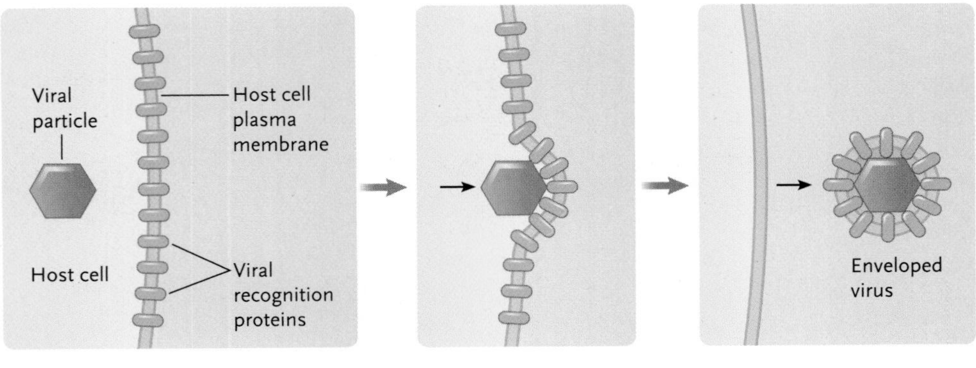

As was the case for our look at prokaryotic organisms in the previous chapter, we will just scratch the surface of viral diversity in this chapter; for example, there are millions of viruses in every millilitre of ocean water, most of which have not been identified. As we learn more about viruses, their classification will likely change.

Every living organism is likely permanently infected by one or more kinds of viruses. Usually, a virus infects only a single species or a few closely related species. A virus may even infect only one organ system or a single tissue or cell type in its host. However, some viruses are able to infect unrelated species, either naturally or after mutating.

Of the roughly 80 viral families described to date, 21 include viruses that cause human diseases. Viruses also cause diseases of wild and domestic animals. Plant viruses cause annual losses of millions of tonnes of crops, especially cereals, potatoes, sugar beets, and sugar cane. (**Table 23.1** lists some important families of viruses that infect animals.) The effects of viruses on the organisms they infect range from undetectable, to merely bothersome, to seriously debilitating or lethal. For instance, some viral infections of humans, such as those causing cold sores, chickenpox, and the common cold, are usually little more than a nuisance to healthy adults. Others cause some of the most severe and deadly human diseases, including AIDS, encephalitis, and Ebola hemorrhagic fever.

However, not all viruses are pathogens. Many viruses actually benefit their hosts; for example, infection by certain nonpathogenic viruses protects human hosts against pathogenic viruses. The "protective" viruses interfere with replication or other functions of the pathogenic viruses. Some viruses also act to defend their host cells. For example, one of the primary reasons that bacteria do not completely overrun this planet is that they are destroyed in incredibly huge numbers by viruses known as **bacteriophages**, or **phages** for short (*phagein* = to eat) (see Figure 23.1d). Viruses also provide a natural means to control some insect pests, such as spruce budworm.

Viruses are vital components of ecosystems and may be the dominant entity in some ecosystems, such as the oceans. We don't yet fully understand their roles in these ecosystems, but it is clear that they affect nutrient cycling through their effects on prokaryotic organisms. For example, in certain regions of the ocean, a few genera of cyanobacteria dominate the marine phytoplankton, making major contributions to global photosynthesis. Bacteriophages infect these cyanobacteria, causing high levels of mortality, thus influencing cyanobacterial population dynamics as well as the release of nutrients from bacterial cells. But these viruses also help keep photosynthesis

Figure 23.2
How enveloped viruses acquire their envelope.

Table 23.1 | Major Animal Viruses

Viral Family	Envelope	Nucleic Acid	Diseases
Adenovirus	No	ds DNA	Respiratory infections, tumours
Flavivirus	Yes	ss RNA	Yellow fever, dengue, hepatitis C
Hepadnavirus	Yes	ds DNA	Hepatitis B
Human herpesvirus	Yes	ds DNA	
Herpes simplex I			Oral herpes, cold sores
Herpes simplex II			Genital herpes
Varicella-zoster virus			Chickenpox, shingles
Herpesvirus 4 (Epstein–Barr virus)			Infectious mononucleosis
Orthomyxovirus	Yes	ss RNA	Influenza
Papovavirus	No	ds DNA	Benign and malignant warts
Papillomavirus	No	ds DNA	Human papillomavirus (genital warts)
Paramyxovirus	Yes	ss RNA	Measles, mumps, pneumonia
Picornavirus	No	ss RNA	
Enterovirus			Polio, hemorrhagic eye disease, gastroenteritis
Rhinovirus			Common cold
Hepatitis A virus			Hepatitis A
Apthovirus			Foot-and-mouth disease in livestock
Poxvirus	Yes	ds DNA	Smallpox, cowpox
Retrovirus	Yes	ss RNA	
HTLV I, II			T-cell leukemia
HIV			AIDS
Rhabdovirus	Yes	ss RNA	Rabies, other animal diseases

ds = double-stranded; HTLV = human T lymphotropic virus; ss = single-stranded.

going in their cyanobacterial hosts, as recently discovered by Nicholas Mann and colleagues at the Univeristy of Warwick. As you read in Chapter 7, one of the proteins that make up photosystem II is very susceptible to light-induced damage and so is constantly being replaced by newly synthesized molecules. As long as the cell can make new protein quickly enough to keep up with damage, photosynthesis can continue, but if the rate of damage to photosystem II exceeds the repair rate, the rate of photosynthesis will drop. When these bacteriophages infect cyanobacteria, they shut down their host's protein synthesis. Without continued synthesis of the photosystem protein, photosynthesis should slow down following infection—but it doesn't. How is the photosynthetic rate maintained? Mann and his colleagues found that the virus's genome includes genes for this protein; expression of these viral proteins enables the repair rate to keep up with light-induced damage, allowing the cell to photosynthesize. Although the virus is doing this for "selfish" reasons (i.e., to ensure that its host has sufficient resources for the virus to complete its life cycle), the outcome of this association is that much of the carbon fixed on Earth may be facilitated by virus-controlled photosynthesis.

STUDY BREAK

1. What is a virus?
2. List three features of viruses that distinguish them from living organisms.

23.2 Viruses Infect Bacterial, Animal, and Plant Cells by Similar Pathways

Viral particles move by random molecular motions until they contact the surface of a host cell. For infection to occur, the virus or the viral genome must then enter the cell. Inside the cell, the viral genes are expressed, leading to replication of the viral genome and assembly of progeny viruses. The new viral particles or **virions**, as the extracellular form of a virus is known, are then released from the host cell, a process that often ruptures the host cell, killing it.

23.2a Bacteriophages Are Viruses That Infect Bacteria

We have learned a great deal about the infective cycles of viruses, as well as the genetics of both viruses and bacteria, from studies of the bacteriophages infecting *Escherichia coli* (*E. coli*). Some of these are **virulent bacteriophages**, which kill their host cells during each cycle of infection, whereas others are **temperate bacteriophages.** Temperate bacteriophages enter an inactive phase inside the host cell and can be passed on to several generations of daughter cells before becoming active and killing their host.

Virulent Bacteriophages. Among the virulent bacteriophages infecting *E. coli*, the **T-even bacteriophages** T2, T4, and T6 have been the most valuable in genetic studies. The coats of these phages are divided into a *head* and a *tail* (see Figure 23.1d, p. 501). A double-stranded linear molecule of DNA is packed into the head. The tail, assembled from several different proteins, has **recognition proteins** at its tip that can bind to the surface of the host cell. Once the tail is attached, it functions as a sort of syringe that injects the DNA genome into the cell **(Figure 23.3).**

Infection begins when a T-even phage collides randomly with the surface of an *E. coli* cell and the tail attaches to the host cell wall (**Figure 23.4,** step 1). An enzyme present in the viral coat, *lysozyme,* then digests a hole in the cell wall through which the tail injects the DNA of the phage (step 2). The proteins of the viral coat remain outside. Throughout its life cycle within the bacterial cell, the phage uses host cell machinery to express its genes. One of the proteins produced early in the infection is an enzyme that breaks down the bacterial chromosome. The phage gene for a DNA polymerase that replicates the phage's DNA is also expressed early on. Eventually, 100 to 200 new viral DNA molecules are synthesized (step 3). Later in the infection, the host cell machinery transcribes the phage genes for the viral coat proteins (step 4). As the head and tail proteins assemble, the replicated viral DNA is packed into the heads (step 5).

When viral assembly is complete, the cell synthesizes a phage-encoded lysozyme that lyses the bacterial cell wall, causing the cell to rupture and releasing viral particles that can infect other *E. coli* cells (step 6). This whole series of events, from infection of a cell through to the release of progeny phages from the ruptured (or **lysed**) cell, is called the **lytic cycle.**

Some virulent phages (although not T-even phages) may package fragments of the host cell's DNA in the heads as the viral particles assemble. This transfer of bacterial genes from one bacterium to another via a virus is known as transduction. In the type of transduction described above, bacterial genes from essentially any DNA fragment can be randomly incorporated into phage particles; thus, gene transfer by this mechanism is termed generalized transduction.

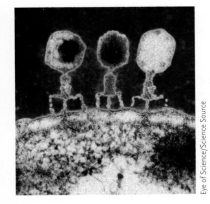

Figure 23.3

Bacteriophages injecting their DNA into *E. coli*.

Eye of Science/Science Source

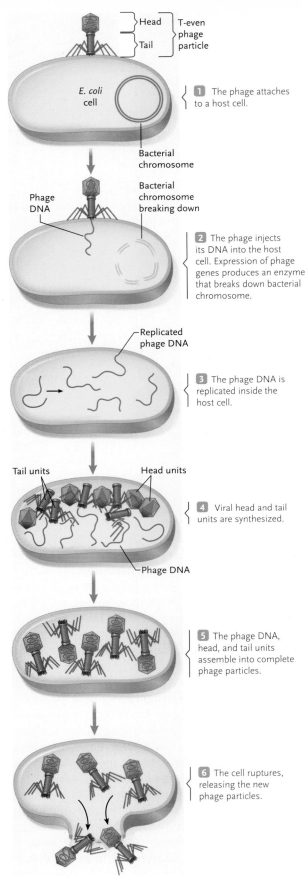

Figure 23.4

The infective cycle of a T-even bacteriophage, an example of a virulent phage.

A Scientist's Favourite Temperate *E. coli* Bacteriophage, Lambda. The infective cycle of the bacteriophage lambda (λ), an *E. coli* phage used extensively in research, is typical of temperate phages. Phage lambda infects *E. coli* in much the same way as the T-even phages. The phage injects its double-stranded linear DNA chromosome into the bacterium (**Figure 23.5,** step 1). Once inside, the linear chromosome forms a circle and then follows one of two paths. Sophisticated molecular switches govern which path is followed at the time of infection.

One path is the lytic cycle, which is like the lytic cycles of virulent phages. The lytic cycle (see Figure 23.5, left side) starts with steps 1 and 2 (infection) and then goes directly to steps 7 through 9 (production and release of progeny virus) and back to step 1. A second and more common path is the **lysogenic cycle** (see Figure 23.5, right side). This cycle begins when the viral chromosome integrates into the host cell's DNA by recombination (see Figure 23.5, steps 1 through 3). The DNA of a temperate phage typically inserts at one or possibly a few specific sites in the bacterial chromosome through the action of a phage-encoded enzyme that recognizes certain sequences in the host DNA. Once integrated, the lambda genes are mostly inactive, so no structural components of the phage are made. While inserted in the host cell DNA, the virus is known as a **prophage** (*pro* = before). When the host cell DNA replicates, so does the integrated viral DNA, which is passed on to daughter cells along with the host cell DNA (see Figure 23.5, steps 4 and 5).

What triggers the integrated prophage to become active (step 6)? Certain environmental signals, such as nutrient availability and ultraviolet irradiation, stimulate this change, causing the prophage to enter the lytic cycle (see Figure 23.5, steps 6 through 9). Genes that were inactive in the prophage are now transcribed. Among the first viral proteins synthesized are enzymes that excise the lambda chromosome from the host chromosome. The result is a circular lambda chromosome that replicates itself and directs the production of linear viral DNA and coat proteins. This active stage culminates in the lysis of the host cell and the release of infective viral particles.

The excision of the prophage from its host's DNA is not always precise, resulting in the inclusion of one or more host cell genes with the viral DNA. These genes are replicated with the viral DNA and packed into the coats and may be carried to a new host cell in the next cycle of infection. Clearly, only genes that are adjacent to the integration site(s) of a temperate phage can be cut out with the viral DNA, can be included in

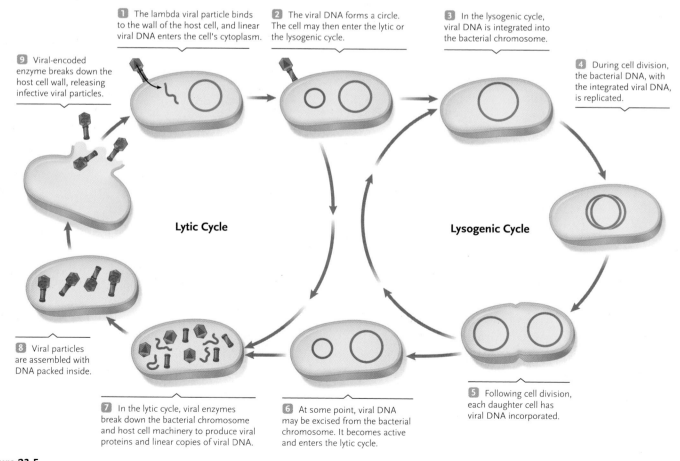

1 The lambda viral particle binds to the wall of the host cell, and linear viral DNA enters the cell's cytoplasm.

2 The viral DNA forms a circle. The cell may then enter the lytic or the lysogenic cycle.

3 In the lysogenic cycle, viral DNA is integrated into the bacterial chromosome.

9 Viral-encoded enzyme breaks down the host cell wall, releasing infective viral particles.

4 During cell division, the bacterial DNA, with the integrated viral DNA, is replicated.

Lytic Cycle

Lysogenic Cycle

8 Viral particles are assembled with DNA packed inside.

7 In the lytic cycle, viral enzymes break down the bacterial chromosome and host cell machinery to produce viral proteins and linear copies of viral DNA.

6 At some point, viral DNA may be excised from the bacterial chromosome. It becomes active and enters the lytic cycle.

5 Following cell division, each daughter cell has viral DNA incorporated.

Figure 23.5
The infective cycle of lambda, an example of a temperate phage, which can go through the lytic cycle or the lysogenic cycle.

phage particles during the lytic stage, and can undergo transduction. Accordingly, this mechanism of gene transfer is termed **specialized transduction.**

Infection of Animal Cells. Viruses infecting animal cells follow a pattern similar to that for bacterial cells, except that both the viral coat and the genome enter a host cell. Depending on the virus, removal of the coat to release the genome occurs during or after cell entry; the envelope does not enter the cell.

Viruses without an envelope, such as poliovirus, bind by their recognition proteins to the plasma membrane and are then taken into the host cell by endocytosis. The virus coat and genome of some enveloped viruses, such as herpesvirus, HIV, and the virus causing rabies, enter the host cell by fusion of their envelope with the host cell plasma membrane. Other enveloped viruses, such as influenza virus, enter host cells by endocytosis.

Once inside the host cell, the genome directs the synthesis of additional viral particles by basically the same pathways as bacterial viruses. Some animal viruses, however, replicate themselves in very complex ways; one example is HIV, the virus that causes AIDS (see "Molecule behind Biology" 23.1). Newly completed viruses that do not acquire an envelope are released by rupture of the host cell's plasma membrane, typically killing the cell. In contrast, most enveloped viruses receive their envelope as they pass through the plasma membrane, usually without breaking the membrane (see Figure 23.2, p. 502). This pattern of viral release typically does not cause immediate damage to the host cell unless very high numbers of virus particles are released.

The vast majority of animal virus infections are asymptomatic because causing disease is of no benefit to the virus. However, a number of pathogenic viruses cause diseases in a variety of ways. Some viruses (e.g., herpesvirus) cause cell death when progeny viruses are released from the cell. This can lead to massive cell death, destroying vital tissues such as nervous tissue or white or red blood cells, or causing lesions in skin and mucous membranes. Other viruses release cellular molecules when infected cells break down, which can induce fever and inflammation (e.g., influenza virus). Yet other viruses alter gene function when they insert into the host cell DNA, leading to cancer and other abnormalities.

Some animal viruses enter a **latent phase,** similar to the lysogenic cycle for bacteriophages, in which the virus remains in the cell in an inactive form. The herpesviruses that cause oral and genital ulcers in humans remain in a latent phase in the cytoplasm of some body cells for the life of the individual. At times, particularly during periods of stress, the virus becomes active in some cells, directing viral replication and causing ulcers to form as cells break down during viral release.

Plant Viruses. Plant viruses may be rodlike or polyhedral. Although most include RNA as their nucleic acid, some contain DNA. None of the known plant viruses have envelopes. They enter cells through mechanical injuries to leaves and stems; they can also be transmitted from one plant to another during pollination or via herbivorous animals such as leafhoppers, aphids, and nematodes. Plant viruses can also be transmitted from one generation to the next in seeds. Once inside a cell, plant viruses replicate via the same processes as animal viruses. However, within plants, virus particles can pass from infected to healthy cells through plasmodesmata, the openings in cell walls that interconnect the cytoplasm of plant cells, and through the vascular system.

Plant viruses are generally named and classified by the type of plant they infect and their most visible effects. *Tomato bushy stunt virus*, for example, causes dwarfing and overgrowth of leaves and stems of tomato plants, and *tobacco mosaic virus* causes a mosaic-like pattern of spots on the leaves of tobacco plants. Most species of crop plants can be infected by at least one destructive virus.

The tobacco mosaic virus was the first virus to be isolated, disassembled, and reassembled in a test tube (see Figure 23.1a, p. 501).

STUDY BREAK

1. What is the difference between a virulent phage and a temperate phage?
2. What are the two types of transduction? How do they differ from each other?
3. How do plant viruses differ from animal viruses?

23.3 It Is Typically Difficult to Treat and Prevent Viral Infections

Viral infections are typically difficult to treat because viruses are, for much of the infection, "hidden" inside host cells and use host cell machinery to replicate. Thus, there often are no obvious viral products to be targeted by drugs. Viral infections are unaffected by antibiotics and other treatment methods used for bacterial infections. As a result, many viral infections are allowed to run their course, with treatment limited to relieving the symptoms while the natural immune defences of the patient attack the virus. Some viruses, however, cause serious and sometimes deadly symptoms on infection; for these, the focus has often been on prevention through vaccine development (e.g., measles, polio). Viruses that use their own polymerases (e.g., RNA viruses such as influenza) provide more obvious targets, so researchers have spent considerable effort developing antiviral drugs to treat them. Many of

Reverse Transcriptase

Acquired immune deficiency syndrome (AIDS) is a disease caused by the human immunodeficiency virus (HIV). This disease has likely already killed about 75 million people worldwide, and the epidemic continues to grow, with infection rates in some areas of Africa as high as one in three adults. Even more concerning, infection rates are increasing in south and east Asia, some of the most densely populated regions of the world. If the epidemic continues to spread at current rates, the World Health Organization has

projected AIDS as the fourth-leading cause of death by 2030 (behind heart disease, other chronic diseases, and car accidents). Although drug treatments to hold AIDS in check do exist, they are very expensive, and most people in developing countries cannot afford them. There is no cure for AIDS, so the millions of people currently infected will die prematurely.

HIV is a retrovirus that contains two copies of single-stranded RNA. It also carries several molecules of an enzyme, reverse transcriptase, in its

capsid. Replication of retroviruses is unusual: the virus's genome enters the host cell along with reverse transcriptase, which copies the viral RNA onto a complementary strand of DNA **(Figure 1).** A second strand of DNA is then synthesized, using the first strand as a template. The resulting double-stranded DNA integrates into the host cell's DNA as a provirus (comparable to the prophage described above). This DNA is transcribed by the host cell into mRNA, which is translated to produce viral

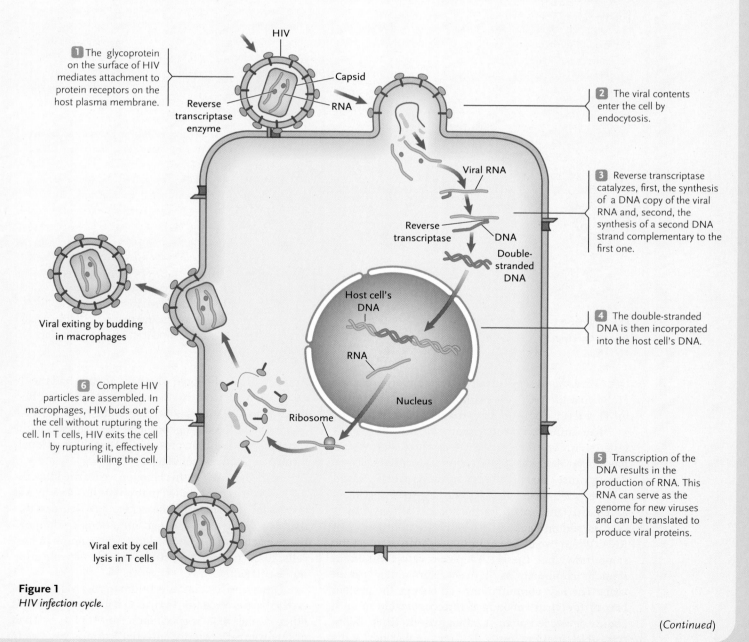

1 The glycoprotein on the surface of HIV mediates attachment to protein receptors on the host plasma membrane.

HIV

Capsid

Reverse transcriptase enzyme

RNA

2 The viral contents enter the cell by endocytosis.

Viral RNA

Reverse transcriptase

DNA

Double-stranded DNA

3 Reverse transcriptase catalyzes, first, the synthesis of a DNA copy of the viral RNA and, second, the synthesis of a second DNA strand complementary to the first one.

Host cell's DNA

RNA

Nucleus

4 The double-stranded DNA is then incorporated into the host cell's DNA.

Viral exiting by budding in macrophages

6 Complete HIV particles are assembled. In macrophages, HIV buds out of the cell without rupturing the cell. In T cells, HIV exits the cell by rupturing it, effectively killing the cell.

Ribosome

5 Transcription of the DNA results in the production of RNA. This RNA can serve as the genome for new viruses and can be translated to produce viral proteins.

Viral exit by cell lysis in T cells

Figure 1
HIV infection cycle.

(Continued)

Reverse Transcriptase (*Continued*)

proteins, including capsid proteins and reverse transcriptase molecules. New virus particles are released from the cell to infect other cells or be passed to new hosts.

Why is HIV so lethal? It targets cells of the human immune system. Obviously, infection of these cells compromises the body's ability to fight off the virus. In addition, some of the immune system cells are not killed by the virus but instead act as a continuing source of infection.

Because reverse transcriptase **(Figure 2)** is a unique feature of HIV, it makes a good target for drug treatment (if the drugs affect only this enzyme, they will not harm the human host). Several antiretroviral drugs have been developed, although HIV has become resistant to some of these drugs. The search continues for a vaccine that would prevent HIV infection, but despite years of research, no vaccine exists yet.

Why is there no vaccine, and how does HIV become resistant so quickly to drugs? The answer to both questions is that HIV mutates quickly and extensively. In a cell's normal DNA replication process, DNA polymerase has proofreading capabilities, so the replicated DNA contains few errors. Reverse transcriptase does not have any proofreading ability, so any errors made when it catalyzes the synthesis of DNA from RNA (and there are many such errors) persist. Proteins encoded by this mutated DNA will be different from those of the original virus; for example, the proteins of the viral coat will be different and so will not be recognized by existing antibodies.

However, reverse transcriptase has also made important positive contributions to biomedical research. For example, retroviruses play an important role in gene therapy, in which new diseases are treated by the introduction of new genes into the body. Viruses are very effective vectors for introducing genes into cells. The desired genes are cloned into the viral genome, and once the virus is taken up by the cell, those genes are introduced into all cells infected by the virus. Retroviruses are particularly useful in gene therapy since the genetic material they carry is integrated into the host cell genome. Reverse transcriptase is also an important tool in molecular biology, as it can be used to synthesize complementary DNA (cDNA) from mRNA, allowing for the cloning of actively expressed genes. It can also be used in genetic engineering; for example, reverse transcriptase can be used to make cDNA out of the mRNA for insulin. This cDNA does not have introns because it is synthesized from an mRNA template and thus can be expressed in a bacterial host (which lacks the enzymes to process DNA that contains introns), allowing insulin to be produced in large quantities.

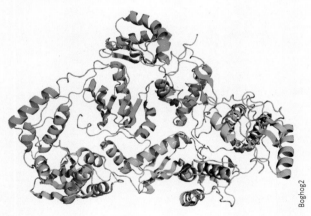

Boghog2

Figure 2
Reverse transcriptase.

these drugs fight the virus directly by targeting a stage of the viral life cycle; for example, the drug zanamivir inhibits release of influenza virus particles from cells.

The influenza virus illustrates the difficulties inherent in controlling or preventing viral diseases. As mentioned at the start of the chapter, the influenza type A virus causes flu epidemics that sweep over the world each year. Why does a new vaccine have to be developed each year? One reason for the success of this virus is that its genome consists of eight separate pieces of RNA. When two different influenza viruses infect the same individual, these RNA pieces can assemble in random combinations derived from either parent virus. The new combinations can change the protein coat of the virus, making it unrecognizable to antibodies developed against either parent virus. Being "invisible" to these antibodies means that new virus strains can infect people who have already had the flu caused by a different strain or who had flu shots effective only against the parent strains of the virus. Random mutations in the RNA genome of the virus add to the variations in the coat proteins that make previously formed antibodies ineffective.

In the opening to this chapter, we learned that the 1918 influenza virus killed many of its hosts. Why was this strain so virulent? Researchers have learned that the 1918 influenza virus had mutations in the polymerase genes that replicated the viral genome in host cells, likely making this strain capable of replicating more efficiently.

Other viruses that infect humans are also considered to have evolved from a virus that previously infected other animals. HIV is one of these; until the second half of the twentieth century, infections of this virus were apparently restricted almost entirely to chimpanzees and gorillas in Africa. Now the virus infects nearly 36 million people worldwide, with the greatest concentration of infected individuals in sub-Saharan Africa.

As illustrated by this example, our efforts to control or eliminate human diseases caused by viral pathogens are complicated when dealing with viruses that have broad host specificity and can infect other animals besides humans. Because other animals can harbour these viruses, we can never successfully eradicate the diseases they cause. For example, the influenza virus can infect birds, swine, and other animals in addition to humans.

Also, as human encroachment on wildlife habitats increases, we create the potential for the evolution of new human viruses, as strains that infect other animals mutate to infect humans. These factors, together with increasing global travel and trade, create the potential for a new human pathogenic virus to readily become a global problem, as we have experienced with HIV. A better understanding of the evolution and life cycles of viruses is crucial if we are to prevent or treat emerging viral diseases.

STUDY BREAK

What can make a viral infection more difficult to treat than a bacterial infection?

23.4 Viruses May Have Evolved from Fragments of Cellular DNA or RNA

Where did viruses come from? Several different hypotheses have been proposed to explain the origin of viruses. Some biologists have suggested that because viruses can duplicate only by infecting a host cell, they probably evolved after cells appeared. They may represent "escaped" fragments of DNA molecules that once formed part of the genetic material of living cells or an RNA copy of such a fragment. The fragments first became surrounded by a protective layer of protein with recognition functions, and then these fragments escaped from their parent cells. As viruses evolved, the information encoded in the core of the virus became reduced to a set of directions for producing more viral particles of the same kind.

More recent hypotheses suggest that viruses are very ancient, with virus like particles predating the first cells. The first viruses originated from the "primordial gene pool"—the pool of RNA that is thought to have been the first genetic material.

Regardless of when viruses originated, they do not share a common evolutionary origin. Thus, unlike cellular life, there is no common ancestor for all viruses and we cannot draw a phylogenetic tree for all viruses. However, viruses have played an important role in the evolution of cellular life because of their ability to integrate their genes into their hosts and to acquire genes from their hosts, as described above. In this way, viruses can be a source of new cellular genetic material, providing new enzymes and other proteins to a cell. Viruses may also have played a more direct role in the evolution of eukaryotic cells: some biologists have suggested that the nucleus originated from a large, double-stranded DNA virus that infected prokaryotic cells, resulting in the first eukaryotic cell.

STUDY BREAK

Why do some biologists think viruses must have originated after cells evolved, rather than predating cells?

23.5 Viroids and Prions Are Infective Agents Even Simpler in Structure than Viruses

Viroids, first discovered in 1971, are small, infectious pieces of RNA. Although the RNA is single stranded, bonding within the molecule causes it to become circular. Viroids are smaller than any virus and lack a protein coat. They also differ from viruses in that their RNA genome does not code for any proteins. Viroids are plant pathogens that can rapidly destroy entire fields of citrus, potatoes, tomatoes, coconut palms, and other crop plants. How do viroids cause such devastating diseases without synthesizing any proteins?

The manner in which viroids cause disease remains unknown. In fact, researchers believe that there is more than one mechanism. Recent research indicates that the viroid may cause disease when its RNA interacts with molecules in the cell; for example, it may disrupt normal RNA processing of the host cell: if the viroid's RNA sequence is complementary to the mRNA of the host cell, it can bind to the host's mRNA, thus preventing normal protein synthesis and causing disease.

Like viruses and virions, **prions** are small infectious particles, but they are not based on nucleic acids; instead, they are infectious protein molecules (the term "prion" is a loose acronym for *proteinaceous infectious particle*). Prions cause spongiform encephalopathies (SEs), degenerate diseases of the nervous system in mammals characterized by loss of motor control and erratic behaviour. The brains of affected animals are full of spongy holes (**Figure 23.6, p. 510**) (hence the "spongiform" designation) and deposits of proteinaceous material. Under the microscope, aggregates of misfolded proteins, called amyloid fibres, are seen in brain tissues; the accumulation of these proteins is the likely cause of the brain damage. SEs progress slowly, meaning that animals may be sick for a long time before their symptoms become obvious, but death is inevitable.

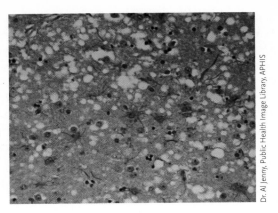

Figure 23.6

Bovine spongiform encephalopathy (BSE). The light-coloured patches in this thin section from a brain damaged by BSE are areas where tissue has been destroyed.

One SE disease is *scrapie,* a brain disease that causes sheep to rub against fences, rocks, or trees until they scrape off most of their wool. In cattle, a similar disease is bovine spongiform encephalopathy (BSE), also known as "mad cow disease." Humans also have SE diseases, such as *kuru,* found among cannibals in New Guinea, who became infected by eating raw human brain during ritual feasts following the death of an individual. *Creutzfeldt–Jakob disease (CJD)* is a very rare SE disease that affects about one person in a million per year, globally. The symptoms of CJD include rapid mental deterioration, loss of vision and speech, and paralysis; autopsies show spongy holes and deposits in brain tissue similar to those of cattle with BSE. We don't know how CJD is transmitted naturally, but we know it can be transmitted inadvertently, for example, with corneal transplants.

SE diseases hit the headlines worldwide in the late 1980s when farmers in the United Kingdom reported a new disease, later determined to be BSE, spreading among their cattle. It is estimated that over 900 000 cows in the United Kingdom were affected, many of which entered the human food chain before they developed symptoms. Where did BSE come from? The source was determined to be meat and bone meal fed to the cows; this meal came from the carcasses of sheep and cattle. The practice of feeding animal meal to cattle had been followed for years, but a money-saving change in processing in the early 1980s (a reduction in how long rendered material was held at high temperature) allowed the infectious agent—maybe from scrapie-infected sheep—to survive in the meat and bone meal. Worse was to come when it became evident that BSE had spread to humans who had eaten contaminated beef. This new human disease, known as variant CJD, is linked to eating meat products from cattle with BSE. Between 1996, when variant CJD was first described, and 2007, there were 208 cases from 11 countries, with the vast majority of these in the United Kingdom. A 12-year study of human tissue samples removed during appendix operations in the United Kingdom suggests that about 1 in every 2000 people in the United Kingdom is a carrier for variant CJD. Will these people actually develop the disease? Evidence from studies of kuru suggests that it may take more than 50 years for prion diseases to develop, so there is some concern that a spike in variant CJD cases is still to come.

Concern about variant CJD explains why the discovery of even one cow with BSE can wreak havoc on a country's beef exports, as happened in Canada when an infected cow was found in Alberta in 2003. The United States closed its border to all beef from Canada within a day, followed shortly by border closings of 40 other countries. Loss of these markets caused serious economic hardship for Canadian ranchers and farmers.

What is the cause of BSE and other SE diseases, and how does this causative agent spread? As explained in "People behind Biology," Stanley Prusiner demonstrated that infectious proteins cause these diseases. Prions are the only known infectious agents that do not include a nucleic acid molecule, and their discovery changed some fundamental views of biology.

Our current understanding of prion infection is that prion proteins are able to survive passage through the stomach of an animal consuming them; they then enter that animal's bloodstream and proceed to the brain, where they somehow interact with normal prion proteins, causing these proteins to change shape to become abnormal and infectious. Prion proteins and the normal precursor proteins share the same amino acid sequences but differ in how they are folded. Prions are somehow able to impose their folding on normal proteins, thus "infecting" the normal proteins. As the infection spreads, neural functioning is impaired and protein fibrils accumulate, producing aggregations of fibrils that trigger apoptosis of infected cells, leading to the SE characteristic of these diseases.

What is the function of "normal" prion proteins? We don't know yet, but evidence suggests that normal prions may regulate the protein synthesis required for growth, development, and protection of brain cells. Mice lacking normal prion proteins have subtle impairments in memory and cognition. Perhaps the inability of the misfolded prion proteins to carry out their normal functions results in dementia and the other symptoms of BSE.

In this chapter, we focused on the simplest biological entities: viruses, viroids, and prions, which possess only some of the properties of life. In the next five chapters, we investigate more structurally complex organisms: the eukaryotic protists, fungi, plants, and animals.

STUDY BREAK

How do viroids and prions differ from viruses? How do they differ from each other?

PEOPLE BEHIND BIOLOGY 23.2
Stanley Prusiner, *University of California*

For several decades, scientists had hypothesized that a slow virus—a disease-causing virus with a long incubation period and gradual onset of pathogenicity—was responsible for scrapie and other spongiform encephalopathies. However, scientists had repeatedly examined the brains of infected animals and not found any evidence of viral infection. In 1982, Stanley Prusiner, a researcher at University of California, San Francisco, determined that the infectious agent was a protein. He pointed to the accumulation of protein fibrils in the brains of infected animals and termed this protein the prion protein (PrP). The research community mostly rejected this hypothesis because it went against all the accepted dogma of biology—genes in the form of DNA or RNA were necessary to cause disease. How could a protein make copies of itself? Prusiner located the gene for PrP and then found that prion proteins are naturally occurring membrane proteins in many types of cells, including neurons. In sheep infected with scrapie, Prusiner found "rogue" forms of the prion proteins that were abnormally folded. He proposed that these infectious prion proteins somehow interacted with "normal" prion proteins to cause misfolding of these proteins; thus, the abnormal protein structure is "infectious." The misfolded prion proteins aggregate, forming the masses of fibrils characteristic of SE diseases. In 1997, Prusiner received a Nobel Prize for his discovery of prions.

Review

aplia™

To access course materials such as Aplia and other companion resources, please visit www.NELSONbrain.com.

23.1 What Is a Virus? Characteristics of Viruses

- Viruses are nonliving infective agents. A free virus particle consists of a nucleic acid genome enclosed in a protein coat (Figure 23.1). Recognition proteins enabling the virus to attach to host cells extend from the surface of infectious viruses.

23.2 Viruses Infect Bacterial, Animal, and Plant Cells by Similar Pathways

- Viruses reproduce by entering a host cell and directing the cellular machinery to make new particles of the same kind (Figures 23.2, 23.4).

23.3 It Is Typically Difficult to Treat and Prevent Viral Infections

- Viruses are unaffected by antibiotics and most other treatment methods. As well, many viruses have great genetic variability and are located inside cells for much of the infection.

  For these reasons, viral infections are difficult to treat, which is why efforts have focused development of vaccines on preventing infection by those viruses that cause serious or fatal diseases.

23.4 Viruses May Have Evolved from Fragments of Cellular DNA or RNA

- There are several hypotheses about the origin of viruses. Viruses may have evolved after cells did and may have descended from nucleic acid fragments that "escaped" from a cell. Evidence for this hypothesis comes from the fact that viruses can duplicate only by infecting a host cell. On the other hand, a competing hypothesis suggests that viruses evolved before the first cells, with the first virus like particles originating from the pool of RNA that was the first genetic material.

- Viruses have different evolutionary origins; i.e., they do not share a common ancestor.

23.5 Viroids and Prions Are Infective Agents Even Simpler in Structure Than Viruses

- Viroids, which infect crop plants, consist of only a very small, single-stranded RNA molecule. Prions, which cause brain diseases in some animals, are infectious proteins with no associated nucleic acid. Prions are misfolded versions of normal cellular proteins, which can induce other normal proteins to misfold.

Questions

Self-Test Questions

1. Which of the following best defines a virus?
   a. a naked fragment of nucleic acid
   b. a disease-causing group of proteins
   c. an entity composed of proteins and nucleic acids
   d. an entity composed of proteins, nucleic acids, and ribosomes

2. Viruses form a capsid around their nucleic acid core. What is this capsid composed of?
   a. protein
   b. lipoprotein
   c. glycoprotein
   d. polysaccharides

3. Which of the following statements about viral envelopes is correct?
   a. They contain glycoproteins of viral origin.
   b. They are located between the virus's capsid and its nucleic acid.
   c. They are composed of a lipid bilayer derived from the viral membrane.
   d. They are composed of peptidoglycan, the same material as bacterial cell walls.

4. Which of the following characteristics distinguishes plant viruses from animal viruses?
   a. Plant viruses are easily curable.
   b. Plant viruses are covered by a membrane envelope.
   c. Plant viruses lack the ability to actively infect a host cell.
   d. Plant viruses lack the ability to replicate their RNA genome.

5. Which of the following describes what happens when a bacteriophage enters the lysogenic stage?
   a. It enters the host cell and kills it immediately.
   b. It enters the host cell, picks up host DNA, and leaves the cell unharmed.
   c. It merges with the host cell plasma membrane, forming an envelope, and then exits the cell.
   d. It injects its DNA into the host cell DNA, and the host DNA integrates viral DNA into the host genome.

6. Which of the following does reverse transcriptase synthesize?
   a. RNA from DNA
   b. DNA from RNA
   c. proteins from DNA
   d. proteins from RNA

7. Which of the following correctly describes a viroid?
   a. the smallest type of virus
   b. small infectious pieces of DNA
   c. small infectious pieces of RNA
   d. infectious pieces of RNA wrapped in a protein coat

8. Which of the following statements about temperate phages is correct?
   a. They never lyse host cells.
   b. They turn their host cell into a prophage.
   c. They integrate their DNA into the host cell chromosome.
   d. They break down the host cell's chromosome when their DNA enters the cell.

9. There are many similarities in how animal and bacterial viruses infect their host cells. Which of the following correctly states one such similarity?
   a. Both animal and bacterial viruses commonly have envelopes.
   b. For bacterial and animal viruses, only their nucleic acid enters the host cell.
   c. Both bacterial and animal viruses have a capsid divided into a head and a tail.
   d. Both animal and bacterial viruses bind to specific receptors on the host cell.

10. Which of the following statements about prions is correct?
    a. Prions can only be transmitted from animals to humans.
    b. The diseases caused by prions progress very rapidly.
    c. Prions have a different amino acid sequence from the normal protein.
    d. Prion proteins reproduce by misfolding normal proteins.

Questions for Discussion

1. From what you have read in this chapter, would you consider viruses to be alive? Why or why not?

2. Why do animal viruses have envelopes, whereas bacteriophages do not?

3. Why is it difficult to design an effective, long-lasting vaccine for the flu virus and the HIV virus?

4. What is a retrovirus, and why have these viruses been given this name?

5. From an evolutionary standpoint, why would the lysogenic state be favoured?

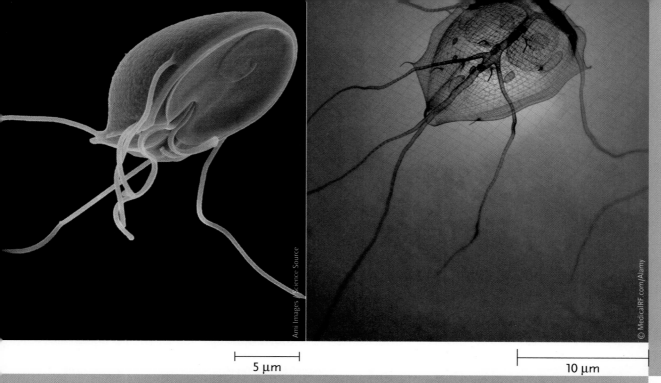

24

Giardia lamblia. (left) Scanning electron microscope image. (right) Light microscope image.

Protists

WHY IT MATTERS

You are on a backpacking trip in your favourite wilderness area on a hot and sunny day. You pause to take a drink of water from your water bottle but discover it is almost empty. You are very thirsty, so you refill your bottle from a nearby stream. The water is clear and cold and looks clean, and, besides, you're out in the middle of nowhere, so it must be safe to drink, right? You continue on the hike and feel fine. But a few days after you get home, you don't feel so great: you have abdominal pain, cramps, and diarrhea. Your doctor says that you have giardiasis, or "beaver fever," caused by *Giardia lamblia,* the most common intestinal parasite in North America (it is very prevalent in water bodies formed by beaver dams). What is *Giardia,* and how does it make you sick?

    Giardia is a single-celled eukaryote that can exist in two forms: a dormant cyst and a motile feeding stage. When you drank from that seemingly clean stream, you ingested some cysts. The cysts can survive for months, so it is important to boil or filter water when you are out hiking or camping. As the swallowed cysts moved from your stomach into your small intestine, the cysts released the motile feeding stage, **trophozoites** (*troph* = food; *zoon* = animal), shown in the photographs at the top of the previous page. Using their multiple flagella, the trophozoites were able to swim about in your intestinal space and attach themselves to the epithelial cells of your intestine. Infection with *Giardia* can become chronic, causing inflammation and reduction of the absorptive capacity of the gut. So why doesn't your immune system detect the presence of *Giardia* and get rid of the parasite? *Giardia* can alter the proteins on its surface that your immune system relies on to recognize an invader, and so it escapes recognition; thus, *Giardia* infections can be persistent or recur.

    Giardia is a **protist** (Greek, *protistos* = the very first). Protists are a very heterogeneous collection of about 200 000 eukaryotes. Most are unicellular and microscopic, but some are large, multicellular organisms. Like their most ancient

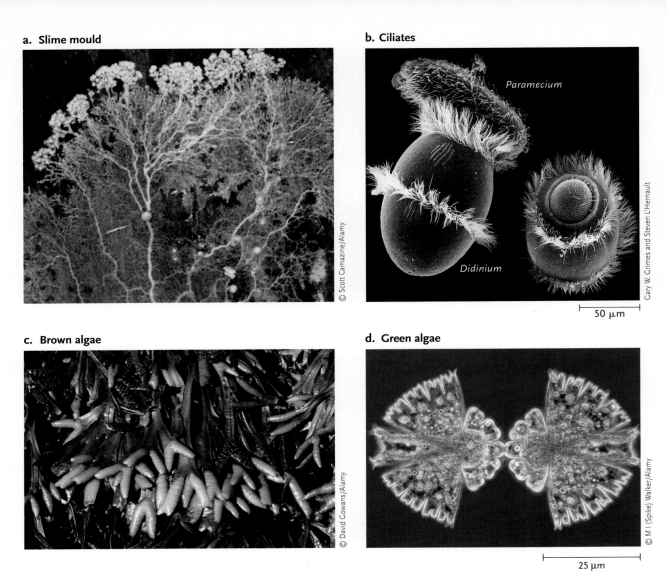

a. Slime mould

b. Ciliates

Paramecium

Didinium

50 μm

c. Brown algae

d. Green algae

25 μm

Figure 24.1

A sampling of protist diversity. **(a)** *Physarum*, a plasmodial slime mould. **(b)** *Didinium*, a ciliate, consuming another ciliate, *Paramecium*. **(c)** *Fucus gardneri* (common rockweed), a brown alga growing in rocky intertidal zones. **(d)** *Micrasterias*, a single-celled green alga, here shown dividing in two.

ancestors, almost all of these eukaryotic species are aquatic. **Figure 24.1** shows a number of protists, illustrating their great diversity.

24.1 The Vast Majority of Eukaryotes Are Protists

The diversity among protists makes it very difficult to define what a protist is. The simplest definition, and the one we will use in this book, is that a protist is any eukaryotic organism that is not an animal, a land plant, or a fungus. Earlier classifications grouped all of these "other" eukaryotes together in one kingdom, Protista. This oversimplified classification reflected our earlier understanding of eukaryote biology, which traditionally has been almost entirely based on the study of animals, land plants, and fungi—the multicellular eukaryotes. But these groups are only three branches of the very

large and diverse tree of living eukaryotes **(Figure 24.2)**. This evolutionary tree is based on molecular data, which are considered the most informative data for determining evolutionary relationships. The tree shows that eukaryotic organisms are divided into approximately five "supergroups," a taxonomic level above that of "kingdom." As you can see by looking at Figure 24.2, the vast majority of eukaryotes are not land plants, animals, or fungi but protists. The tree shown here represents our current understanding of the relationships among eukaryotic organisms, which is actively changing as researchers continue to investigate the evolutionary history of eukaryotes; the actual number of supergroups is still being debated, with some researchers dividing eukaryotes into additional supergroups to those shown in Fig 24.2. You may notice that the "root" of the tree— the last eukaryote common ancestor (LECA)—is not identified. The identity of this ancestral group is a major mystery that researchers are actively working to unravel.

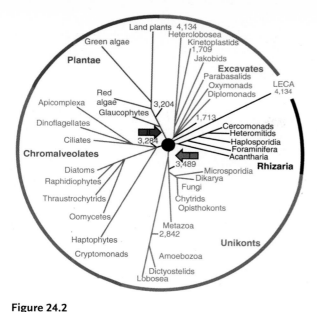

Figure 24.2

Major lineages of protists within the supergroups of eukaryotic organisms. The current evolutionary tree for eukaryotes divides these organisms among approximately five supergroups: Excavates, Unikonts (including animals and fungi), Plantae (including land plants and their algal relatives), Chromalveolates, Excavates and Rhizaria. Selected lineages of protists in each supergroup are discussed in this chapter.

The first eukaryotes likely evolved about 1.5–2 billion years ago (Ba) ago. While we don't fully understand how they evolved, we know that endosymbiosis played an important role in the process. Eukaryotes contain mitochondria (although some have very reduced versions of this organelle) and many also contain chloroplasts. As outlined in Chapter 2, mitochondria and chloroplasts are the descendants of free-living prokaryotes that, over evolutionary time, became organelles. All mitochondria are thought to have arisen from a single endosymbiotic event, but the history of chloroplasts is more complex. We will return to the evolution of chloroplasts at the end of this chapter, once you have had a chance to become familiar with the various groups of protists.

In this chapter, we will start with an overview of features of protists and then focus on key protist lineages in each of the eukaryotic supergroups. In this way, you will gain an understanding of how diverse protists are morphologically, functionally, and ecologically. As you read about the various groups of protists, think about how they differ from animals and plants, and how learning about these "other" eukaryotes changes your understanding of eukaryote biology. Protists are sometimes called the "rule-breakers" of the eukaryotic world: many of the general rules or "facts" we think we know about eukaryotic organisms are revealed as not being generally true at all once protists are considered, forcing us to rethink what is "typical" or "normal" in eukaryote biology.

STUDY BREAK

By what process did eukaryotes such as protists acquire mitochondria and chloroplasts?

24.2 Characteristics of Protists

Because protists are eukaryotes, the boundary between them and prokaryotic organisms is clear and obvious. Unlike bacteria and archaea, protists have a membrane-bound nucleus, with multiple, linear chromosomes. In addition to cytoplasmic organelles, including mitochondria and chloroplasts (in some species), protists have microtubules and microfilaments, which provide motility and cytoskeletal support. As well, they share characteristics of transcription and translation with other eukaryotes.

The phylogenetic relationship between protists and other eukaryotes is more complex (Figure 24.2). Over evolutionary time, the eukaryotic family tree branched out in many directions. All of the organisms in the eukaryotic lineages consist of protists except for three groups, the animals, land plants, and fungi, which arose from protist ancestors. Although some protists have features that resemble those of the fungi, plants, or animals, several characteristics are distinctive. In contrast to fungi, most protists are motile or have motile stages in their life cycles, and their cell walls are made of cellulose, not chitin.

How do photosynthesizing protists differ from plants? Unlike plants, many photoautotrophic protists can also live as heterotrophs, and some regularly combine both modes of nutrition. Protists do not retain developing embryos in parental tissue, as plants do, nor do they have highly differentiated structures equivalent to roots, stems, and leaves. Photosynthetic protists are sometimes referred to as *algae;* these protists are generally aquatic and often unicellular and microscopic (although many are multicellular). However, the different groups of algae are not closely related to each other (see Figure 24.2), so the term *algae* does not indicate any sort of relatedness among organisms referred to by that term.

How do protists differ from animals? Unlike protists, all animals are multicellular and have features such as an internal digestive tract and complex developmental stages. Protists also lack features that characterize many animals, including nerve cells; highly differentiated structures such as limbs and a heart; and collagen, an extracellular support protein.

STUDY BREAK

What features distinguish protists from prokaryotic organisms? What features distinguish them from fungi, plants, and animals?

24.3 Protists' Diversity Is Reflected in Their Metabolism, Reproduction, Structure, and Habitat

As you might expect from looking at Figure 24.2, protists are highly diverse in metabolism, reproduction, structure, and habitat.

Habitat. Protists live in aqueous habitats, including aquatic or moist terrestrial locations, such as oceans, freshwater lakes, ponds, streams, and moist soils, and within host organisms. In bodies of water, small photosynthetic protists collectively make up the **phytoplankton** (*phytos* = plant; *planktos* = drifting), the organisms that capture the energy of sunlight in nearly all aquatic habitats. These phototrophs provide organic substances and oxygen for heterotrophic bacteria, other protists, and the small crustaceans and animal larvae that are the primary constituents of **zooplankton** (*zoe* = life, usually meaning animal life). Although protists are not animals, biologists often include them among the zooplankton. Phytoplankton and larger multicellular protists forming seaweeds collectively account for about half of the total organic matter produced by photosynthesis.

In the moist soils of terrestrial environments, protists play important roles among the detritus feeders that recycle matter from organic back to inorganic form. In their roles in phytoplankton, in zooplankton, and as detritus feeders, protists are enormously important in world ecosystems.

Protists that live in host organisms are **parasites**, obtaining nutrients from the host. Indeed, many of the parasites that have significant effects on human health are protists, causing diseases such as malaria, sleeping sickness, and amoebic dysentery.

Structure. Whereas most protists are single cells, others live as **colonies (Figure 24.3)** in which individual cells show little or no differentiation and are potentially independent. Within colonies, individuals use cell signalling to cooperate on tasks such as feeding and movement. Some protists are large multicellular organisms; for example, the giant kelp of coastal waters can rival forest trees in size.

Many single-celled and colonial protists have complex intracellular structures, some found nowhere else among living organisms **(Figure 24.4).** These unique structures reflect key aspects of the habitats in which protists live. For example, consider a single-celled protist living in a freshwater pond. Its cytoplasm is

Figure 24.3
Colonial protist (*Dinobryon*).

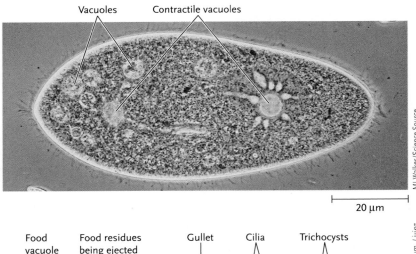

Vacuoles   Contractile vacuoles

20 μm

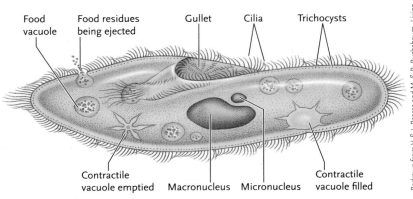

Food vacuole   Food residues being ejected   Gullet   Cilia   Trichocysts

Contractile vacuole emptied   Macronucleus   Micronucleus   Contractile vacuole filled

Figure 24.4
A ciliate, *Paramecium*, showing the cytoplasmic structures typical of many protists.

hypertonic to the water surrounding it, meaning that water flows into the cell by osmosis (see Section 5.5). How can the protist stop itself from bursting? A specialized cytoplasmic organelle, the **contractile vacuole**, gradually fills with fluid. When this vacuole reaches its maximum size, it moves to the plasma membrane and forcibly contracts, expelling the fluid to the outside through a pore in the membrane.

The cells of some protists are supported by an external cell wall or by an internal or external shell built up from organic or mineral matter; in some, the shell takes on highly elaborate forms. Instead of a cell wall, other protists have a **pellicle**, a layer of supportive protein fibres located inside the cell just under the plasma membrane, providing strength and flexibility **(Figure 24.5)**.

At some time during their lives, almost all protists move. Some move by amoeboid motion, in which the cell extends one or more lobes of cytoplasm called **pseudopodia** ("false feet"; see **Figure 24.6**). The rest of the cytoplasm and the nucleus then flow into the pseudopodium, completing the movement. Other protists move by the beating of flagella or cilia. In some protists, cilia are arranged in complex patterns, with an

equally complex network of microtubules and other cytoskeletal fibres supporting the cilia under the plasma membrane.

Many protists can exist in more than one form, for example, as a motile form and as a nonmotile cyst that can survive unfavourable conditions. This morphological variability allows the species to live in different habitats at different stages in its life.

Metabolism. Almost all protists are aerobic organisms that live either as heterotrophs—obtaining carbon from organic molecules produced by other organisms—or as photoautotrophs, by producing organic molecules for themselves by photosynthesis (see Chapter 7). Some heterotrophic protists obtain organic molecules by engulfing part or all of other organisms (*phagocytosis*) and digesting them internally. Others absorb small organic molecules from their environment by diffusion. Some protists can live as either heterotrophs or autotrophs.

Reproduction. Reproduction may be asexual, by mitosis, or sexual, through meiotic cell division and formation of gametes. In protists that reproduce by both mitosis and meiosis, the two modes of cell division are often combined into a **life cycle** that is highly distinctive among the different protist groups. We do not yet have a complete understanding of the reproductive biology of many protists.

STUDY BREAK

Define each of the following terms in your own words, and indicate the role that each plays in the life of a protist: *pellicle, pseudopodia, contractile vacuole*.

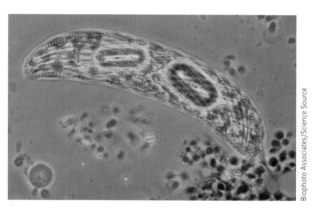

Figure 24.5
Euglena spirogyra, showing pellicle (strips of protein fibres).

Biophoto Associates/Science Source

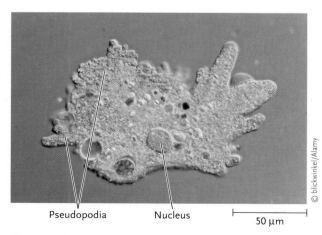

Pseudopodia   Nucleus
50 μm

© blickwinkel/Alamy

Figure 24.6
Amoeba proteus of the Amoebozoa is perhaps the most familiar protist of all.

24.4 The Eukaryotic Supergroups and Their Key Protist Lineages

In this section, we look at the biological features of the major protist lineages in each eukaryote supergroup shown in Figure 24.2, p. 515. Our focus is the ecological or economic importance of each lineage, the habitats in which you would find these organisms, and key features that differentiate the group from other protists. As you read through the information on each lineage, think about how the structural features of that group relate to its habitat and lifestyle.

24.4a Excavata Are Unicellular, Flagellated Protists, Many of Which Lack Mitochondria

This supergroup takes its name from the hollow (excavated) ventral feeding groove found in most members. Protists of this supergroup are sometimes referred to as protozoa (*proto* = first; *zoon* = animal) because, like animals, they ingest their food and move by themselves. We will consider four lineages of Excavates: Diplomonads, Parabasalids, Euglenoids, and Kinetoplastids.

Euglenoids. You have probably seen an example of one genus of euglenoids, *Euglena*, in your earlier biology classes **(Figure 24.7),** as they are often used to illustrate how some protists have plantlike features (photosynthesis) combined with features that we consider animal-like (movement). Euglenoids are important primary producers in freshwater ponds, streams, and lakes, and even some marine habitats. Most are autotrophs that carry out photosynthesis using the same photosynthetic pigments and mechanisms as plants. If light is not available, many of the photosynthetic euglenoids can also live as heterotrophs by absorbing organic molecules through the plasma membrane or by engulfing small particles. Other euglenoids lack chloroplasts and live entirely as heterotrophs.

The name *Euglena* roughly translates as "eyeball organism," a reference to the large *eyespot* that is an obvious feature of photosynthetic euglenoids (see Figure 24.7). The eyespot contains pigment granules in association with a light-sensitive structure and is part of a sensory mechanism that stimulates cells to swim toward moderately bright light or away from intensely bright light so that the organism finds optimal conditions for photosynthetic activity. In addition to an eyespot, euglenoids contain numerous organelles, including a contractile vacuole.

Rather than an external cell wall, euglenoids have a spirally grooved pellicle formed from strips of transparent, protein-rich material underneath the membrane (see Figure 24.5, p. 517). In some euglenoids, the strips are arranged in a spiral pattern, allowing the cell to change its shape in a wriggling sort of motion (known as euglenoid movement) that allows the cell to change direction. Euglenoids can also swim by whiplike movements of flagella that extend from one end of the cell. Most have two flagella: one rudimentary and short, the other long.

Kinetoplastids. Sleeping sickness is a fatal disease endemic to sub-Saharan Africa. Although the disease was almost eradicated about 40 years ago, it has been making a comeback due to wars and the subsequent refugee movement and damage to healthcare systems. Sleeping sickness is caused by various subspecies of *Trypanosoma brucei* **(Figure 24.8)** that are transmitted from one host to another by bites of the tsetse fly. Early symptoms include fever, headaches, rashes, and anemia. Untreated, the disease damages the central nervous system, leading to a sleeplike coma and eventual death. The disease has proved difficult to control because the same trypanosomes infect wild mammals, providing an inexhaustible reservoir for the parasite. Other trypanosomes, also transmitted by insects, cause Chagas disease in Central and South America and leishmaniasis in many tropical countries. Humans with Chagas disease have an enlarged liver and spleen and may experience severe brain and heart damage; leishmaniasis causes skin sores and ulcers, as well as liver and spleen damage.

Like trypanosomes, other kinetoplastids are heterotrophs that live as animal parasites. Kinetoplastid cells are characterized by a single mitochondrion that contains a large DNA-protein deposit called a *kinetoplast* (see Figure 24.8). Most kinetoplastids also have a leading and a trailing flagellum, which are used for movement. In some cases, the trailing flagellum is attached to the side of the cell, forming an undulating membrane that allows the organism to glide along or attach to surfaces.

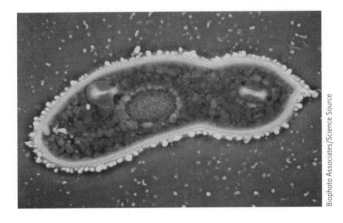

Biophoto Associates/Science Source

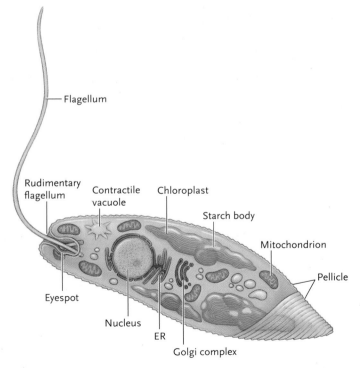

— Flagellum

Rudimentary flagellum
Contractile vacuole
Chloroplast
Starch body
Mitochondrion
Eyespot
Pellicle
Nucleus
ER
Golgi complex

Figure 24.7
Body plan and a colour photo *of Euglena gracilis.*

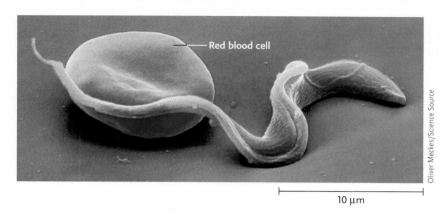

Red blood cell

10 μm

Oliver Meckes/Science Source

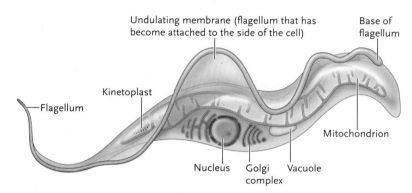

Undulating membrane (flagellum that has become attached to the side of the cell)

Base of flagellum

Kinetoplast

Flagellum

Mitochondrion

Nucleus Golgi Vacuole
complex

Figure 24.8

Trypanosoma brucei, the parasitic kinetoplastid that causes African sleeping sickness.

group did have mitochondria. The nuclei of Excavata that lack mitochondria contain genes derived from mitochondria, and they also have organelles that likely evolved from mitochondria. These Excavata may have lost their mitochondria as an adaptation to the parasitic way of life, in which oxygen is in short supply.

Diplomonads. Diplomonad means *double cell*, and these organisms do look like two cells together (see the figure at the beginning of the chapter), with their two apparently identical, functional nuclei and multiple flagella arranged symmetrically around the cell's longitudinal axis. The best-known diplomonad is *Giardia lamblia,* profiled at the beginning of this chapter. Some are free living, but many live in animal intestines; some diplomonads do not cause harm to the host, whereas others, like *Giardia,* live as parasites.

Like many Excavata, Diplomonads and Parabasalids are single-celled animal parasites that lack mitochondria and move by means of flagella. Because they lack mitochondria, these organisms are limited to producing ATP via glycolysis (see Chapter 6). Originally, the lack of mitochondria in many Excavata led biologists to consider this group as the most ancient line of protists; however, it now appears that the ancestor of this

Parabasalids. The sexually transmitted disease trichomoniasis is caused by the parabasalid *Trichomonas vaginalis* **(Figure 24.9a).** The infection is usually symptomless in men, but in women, *T. vaginalis* can cause severe inflammation and irritation of the vagina and vulva. If untreated, trichomoniasis can cause infection of the uterus and fallopian tubes that can result in infertility. Luckily, drugs can easily cure the infection.

a. *Trichomonas vaginalis*

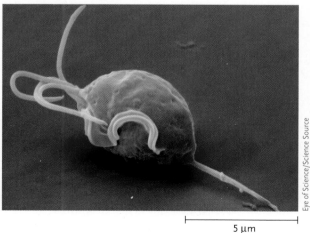

Eye of Science/Science Source

5 μm

b. *Trichonympha*

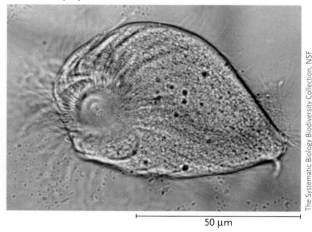

The Systematic Biology Biodiversity Collection, NSF

50 μm

Figure 24.9

Examples of parabasalids (Excavata). **(a)** A parabasalid, *Trichomonas vaginalis,* that causes a sexually transmitted disease, trichomoniasis. **(b)** *Trichonympha,* a parabasalid that lives in the guts of termites.

Parabasalids take their names from cytoplasmic structures associated with the nucleus, *parabasal bodies;* some biologists consider these structures to be the Golgi apparatus of these cells. Parabasalids are also characterized by a sort of fin called an **undulating membrane**, formed by a flagellum buried in a fold of the cytoplasm, in addition to freely beating flagella. The buried flagellum allows parabasalids to move through thick, viscous fluids, such as those lining human reproductive tracts.

Other parabasalids (e.g., *Trichonympha;* Figure 24.9b) are symbionts that live in the guts of termites and other wood-eating insects, digesting the cellulose in the wood for their hosts. As if this endosymbiotic relationship were not complex enough, biologists recently discovered that the protists themselves cannot produce the enzymes necessary to break down cellulose but instead rely on bacterial symbionts to do it.

24.4b Chromalveolates Have Complex Cytoplasmic Structures and Move via Flagella or Cilia

This group is named for the small, membrane-bound vesicles called *alveoli* (*alvus* = belly) in a layer just under the plasma membrane. The Chromalveolate supergroup includes two motile, free-living lineages as well as a motile parasitic group. We will take a closer look at some representative lineages over the next few pages.

Ciliates. This group of protists has helped us understand key aspects of eukaryotic cells, such as the existence of telomeres at the ends of eukaryotic chromosomes and the function of telomerase. These protists are examples of model organisms—organisms that are easily manipulated and easily raised in the lab and for which we have abundant data, for example, genome sequences (see *The Purple Pages*). Several protists are ideal model organisms because, even though they are single celled, the complexity of their structures and functions is comparable to that of humans and other animals. One ciliate, *Tetrahymena* **(Figure 24.10)**, was the organism in which telomeres and telomerase were discovered; it was also the cell in which the first motor protein was identified, cell cycle control mechanisms were first described, and ribozymes were discovered. The involvement of ciliates with scientific research dates back several centuries—they were among the first organisms observed in the seventeenth century by the pioneering microscopist Anton van Leeuwenhoek.

The ciliates are a large group, with nearly 10 000 known species of primarily single-celled but highly complex heterotrophic organisms that swim by means of cilia (see Figures 24.4, p. 516, and 24.10). Any sample of pond water or bottom mud contains a wealth of

a. Ciliate

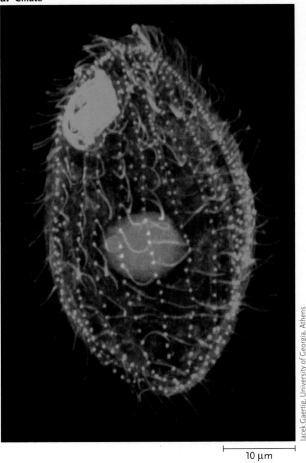

10 μm

b. Cilia

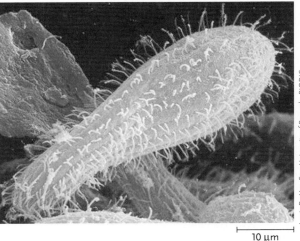

10 μm

Figure 24.10
Tetrahymena, a ciliate: **(a)** stained with fluorescent dye to show cilia and microtubules; **(b)** SEM image showing cilia.

these creatures. Some ciliates live individually, whereas others are colonial. Certain ciliates are animal parasites; others live and reproduce in their hosts as mutually beneficial symbionts. A compartment of the stomach of cattle and other grazing animals contains large numbers of symbiotic ciliates that digest the cellulose in their hosts' plant diet. The host animals then digest the excess ciliates.

Ciliates have many highly developed organelles, including a mouthlike gullet lined with cilia, structures that exude toxins and other defensive materials from the cell surface, contractile vacuoles, and a complex system of food vacuoles. A pellicle reinforces the cell's shape. A complex cytoskeleton anchors the cilia just below the pellicle and coordinates the ciliary beating. The cilia can stop and reverse their beating in synchrony, allowing ciliates to stop, back up, and turn if they encounter negative stimuli.

Ciliates are the only eukaryotes that have two types of nuclei in each cell: one or more small nuclei called *micronuclei* and a single larger *macronucleus* (see Figure 24.4, p. 516). A **micronucleus** is a diploid nucleus that contains a complete complement of genes. It functions primarily in cellular reproduction, which may be asexual or sexual. The number of micronuclei present depends on the species. The **macronucleus** develops from a micronucleus but loses all genes except those required for basic functions (e.g., feeding, metabolism) of the cell and for synthesis of ribosomal RNA. The macronucleus contains numerous copies of these genes, allowing it to synthesize large quantities of proteins and rRNA.

Ciliates abound in freshwater and marine habitats, where they feed voraciously on bacteria, algae, and each other. *Paramecium* and *Tetrahymena* are typical of the group (see Figures 24.4, p. 516 and 24.10). Their rows of cilia drive them through their watery habitat, rotating the cell on its long axis while it moves forward or back and turns. The cilia also sweep water laden with prey and food particles into the gullet, where food vacuoles form. The ciliate digests food in the vacuoles and eliminates indigestible material through an anal pore. Contractile vacuoles with elaborate, raylike extensions remove excess water from the cytoplasm and expel it to the outside. When under attack or otherwise stressed, *Paramecium* discharges many dartlike protein threads from surface organelles called **trichocysts**.

Dinoflagellates. In spring and summer, the coastal waters of Canada sometimes turn reddish in colour **(Figure 24.11a).** These **red tides** are caused by a population explosion, or *bloom,* of certain dinoflagellates that make up a large proportion of marine phytoplankton. These protists typically have a shell formed from cellulose plates (Figure 24.11b). The beating of flagella, which fit into grooves in the plates, makes dinoflagellates spin like a top (*dinos* = spinning) as they swim.

Red tides are caused by conditions such as increased nutrient runoff into coastal waters (particularly from farms and industrial areas), warm ocean surface temperatures, and calm water. Red tides occur in the waters of many other countries besides Canada and are more common in warmer waters. Some red tide

dinoflagellates produce a toxin that interferes with nerve function in animals that ingest them (see "Molecule behind Biology" 24.1).

More than 4000 dinoflagellate species are known, and most, like those that cause red tides, are single-celled organisms in marine phytoplankton. Their abundance in phytoplankton makes dinoflagellates a major primary producer of ocean ecosystems. You can sometimes see their abundance because some are **bioluminescent,** that is, they glow or release a flash of light, particularly when disturbed. Dinoflagellate luminescence can make the sea glow in the wake of a boat at night and coat nocturnal surfers and swimmers with a ghostly light **(Figure 24.12).** Why do these organisms emit light? One explanation is that this burst of light would be likely to scare off predators. The production of light depends on the enzyme *luciferase* and its substrate *luciferin* in forms similar to the system that produces light in fireflies.

Dinoflagellates live as heterotrophs or autotrophs; many can carry out both modes of nutrition. Some dinoflagellates live as symbionts in the tissues of other marine organisms, such as jellyfish, sea anemones, corals, and molluscs, and give these organisms their distinctive colours. Dinoflagellates in coral use the coral's carbon dioxide and nitrogenous waste while supplying 90% of the coral's carbon. The vast numbers of dinoflagellates living as photosynthetic symbionts in tropical coral reefs allow the

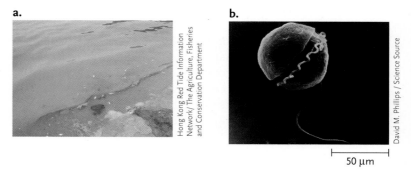

a.

Hong Kong Red Tide Information Network/ The Agriculture, Fisheries and Conservation Department

b.

David M. Phillips / Science Source

50 μm

Figure 24.11
(a) Red tide caused by dinoflagellate bloom. **(b)** *Karenia brevis,* a toxin-producing dinoflagellate.

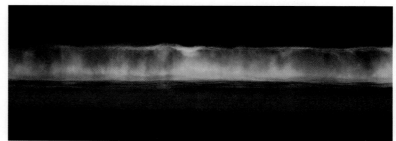

Mike Sauder. This file is licensed under the Creative Commons Attribution-Share Alike 2.0 Generic license, https://creativecommons.org/licenses/by-sa/2.0/

Figure 24.12
Bioluminescent dinoflagellates (*Lingulodinium polyedrum*) lighting a breaking wave at midnight.

MOLECULE BEHIND BIOLOGY 24.1

Saxitoxin

Some dinoflagellates that cause red tides also produce neurotoxins. Fish that feed on the dinoflagellates and birds that feed on the fish may be killed in huge numbers by the toxins. Dinoflagellate toxins do not noticeably affect clams, oysters, and other molluscs but become concentrated in their tissues. Eating the tainted molluscs can cause paralytic shellfish poisoning in humans and other animals, characterized by nausea, vomiting, shortness of breath, and a choking feeling. The main toxin responsible is saxitoxin **(Figure 1),** a neurotoxic alkaloid that is the most lethal nonprotein toxin known—a dose of just 0.2 mg is

Figure 1

Saxitoxin, one of the neurotoxins produced by dinoflagellates.

enough to kill an average-weight person. Saxitoxin acts by binding to sodium channels of nerve cells, thus preventing the normal movement of sodium ions through the channel and blocking the transmission of nerve impulses.

Saxitoxin is especially deadly for mammals because it paralyzes the diaphragm and other muscles required for breathing. There is no cure, and death can occur within minutes if the person is not treated quickly; treatment involves artificial respiration to support breathing. Saxitoxin has been experimented with as a chemical weapon but also has more constructive uses. For example, it has been used to determine the components of sodium channels in cell membranes and in studies of various nerve disorders.

Other photosynthetic protists also produce blooms, and some of these also produce toxins.

---

reefs to reach massive sizes; without dinoflagellates, many coral species would die. When stressed, corals eject their endosymbionts, a phenomenon known as coral bleaching because the absence of the pigmented dinoflagellates allows the coral's calcareous skeleton to be visible **(Figure 24.13).** What causes the coral to become stressed? Increased water temperatures appear to be the main cause, although exposure to contaminants such as oil can also cause bleaching. If the stress causing the bleaching is transient, the coral usually regains its endosymbionts, but if the stress persists, the coral will die. The severity and spatial extent of coral bleaching has been increasing over the past few decades such that it is now a global problem. In 1998, a serious bleaching event destroyed 16% of the world's reefs. Localized high ocean temperatures

in the Caribbean in 2005 resulted in more than 80% of corals bleaching, with more than 40% of these being killed.

Apicomplexans. Apicomplexans are nonmotile parasites of animals. They take their name from the *apical complex,* a group of organelles at one end of a cell, which helps the cell attach to and invade host cells. Apicomplexans absorb nutrients through their plasma membranes (rather than by engulfing food particles) and lack food vacuoles. One genus, *Plasmodium,* is responsible for malaria, one of the most widespread and debilitating human diseases. About 500 million people are infected with malaria in tropical regions, including Africa, India, Southeast Asia, the Middle East, Oceania, and Central and South America. In 2012, malaria killed an estimated 627 000 people, about half as many as were killed by AIDS that year. It is particularly deadly for children younger than six. In many countries where malaria is common, people are often infected repeatedly, with new infections occurring alongside preexisting infections.

Plasmodium is transmitted by 60 different species of mosquitoes, all members of the genus *Anopheles.* Infective cells develop inside the female mosquito, which transfers the cells to human or bird hosts **(Figure 24.14).** The infecting parasites divide repeatedly by asexual reproduction in their hosts, initially in liver cells and then in red blood cells. Their growth causes red blood cells to rupture in regular cycles every 48 or 72 hours, depending on the *Plasmodium* species. The ruptured red blood cells clog vessels and release the parasite's metabolic wastes, causing cycles of chills and fever.

Figure 24.13

Bleached elkhorn coral (*Acropora palmata*).

NOAA

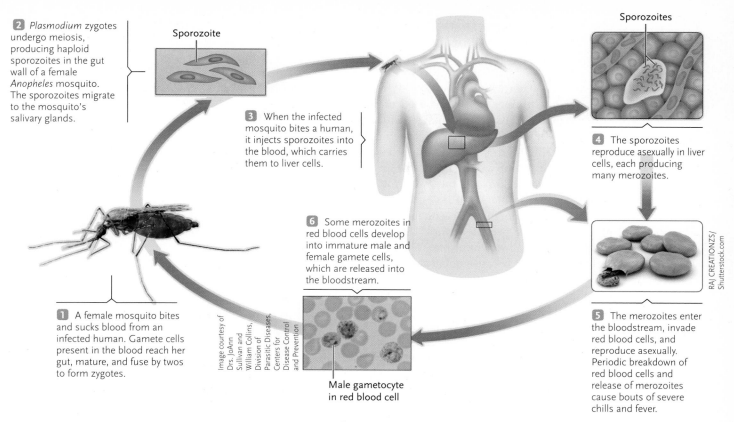

2 *Plasmodium* zygotes undergo meiosis, producing haploid sporozoites in the gut wall of a female *Anopheles* mosquito. The sporozoites migrate to the mosquito's salivary glands.

Sporozoite

3 When the infected mosquito bites a human, it injects sporozoites into the blood, which carries them to liver cells.

Sporozoites

4 The sporozoites reproduce asexually in liver cells, each producing many merozoites.

RAJ CREATIONZS/ Shutterstock.com

6 Some merozoites in red blood cells develop into immature male and female gamete cells, which are released into the bloodstream.

5 The merozoites enter the bloodstream, invade red blood cells, and reproduce asexually. Periodic breakdown of red blood cells and release of merozoites cause bouts of severe chills and fever.

1 A female mosquito bites and sucks blood from an infected human. Gamete cells present in the blood reach her gut, mature, and fuse by twos to form zygotes.

Image courtesy of Drs. JoAnn Sullivan and William Collins, Division of Parasitic Diseases, Centers for Disease Control and Prevention

Male gametocyte in red blood cell

Figure 24.14
Life cycle of a *Plasmodium* species that causes malaria.

The victim's immune system is ineffective because during most of the infective cycle, the parasite is inside body cells and thus "hidden" from antibodies. Furthermore, like *Giardia*, *Plasmodium* regularly changes its surface molecules, continuously producing new forms that are not recognized by antibodies developed against a previous form. In this way, the parasite keeps one step ahead of the immune system, often making malarial infections essentially permanent. For a time, malaria was controlled in many countries by insecticides such as DDT. However, the mosquitoes developed resistance to the insecticides and have returned in even greater numbers than before the spraying began.

In addition to the asexual reproduction described above for *Plasmodium*, apicomplexans also reproduce sexually, forming gametes that fuse to produce cysts. As in *Giardia*, when a host organism ingests the cysts, they divide to produce infective cells. Many apicomplexans use more than one host species for different stages of their life cycle. For example, another organism in this group, *Toxoplasma*, has the sexual phase of its life cycle in cats and the asexual phases in humans, cattle, pigs, and other animals. Feces of infected cats contain cysts; humans ingesting or inhaling the cysts develop toxoplasmosis, a disease that is usually mild in adults but can cause severe brain damage or even death to a fetus. Because of the danger of toxoplasmosis, pregnant women should avoid emptying litter boxes or otherwise cleaning up after a cat.

The groups of Chromalveolates discussed below all share a distinctive arrangement of flagella at some stage of their life cycles. As indicated in Figure 24.15, motile cells in these organisms have two different flagella: one smooth and a second covered with bristles, giving it a "hairy" appearance **(Figure 24.15)**. In many of these chromalveolates, the flagella occur only on reproductive cells such as eggs and sperm. This group of Chromalveoloates includes the oomycetes (water moulds), Diatoms, Golden algae, and Brown algae.

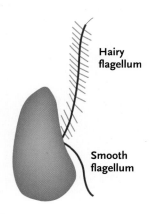

Hairy flagellum

Smooth flagellum

Figure 24.15
Stramenopile protist, with "smooth" and "hairy" flagella.

Recall that algae is a general term for photosynthetic protists, but the different groups of algae are not closely related to each other, so the term does not imply a phylogenetic grouping.

Oomycetes: Water Moulds and Downy Mildews. In Ireland, the summer of 1846 started off warm and sunny. This was a welcome change, as the previous summer had been cool and damp, causing the potato crop to fail. But then the weather turned wet and cold again and within one week at the end of July, the entire potato crop was destroyed—the leaves rotting and the tubers turning to black, putrid mush **(Figure 24.16).** Worse was to come: the unseasonably cool and damp growing seasons persisted until 1860, causing the potato crops to fail year after year. These crop failures were catastrophic because potatoes were virtually the only food source for most people. Altogether, about one-third of the Irish population died or emigrated (to Canada and the United States, among other countries) due to the potato famines.

In 1861, the organism that caused the blight was identified as a water mould, *Phytophthora infestans*. Originally thought to be a fungus, *P. infestans* produces infective cells that are easily dispersed by wind and water. The blight caused by this organism has recently re-emerged as a serious disease in potato-growing regions of Canada and the United States due to the migration of new strains from Mexico that are resistant to existing pesticides.

Water moulds are not fungi; they are oomycetes **(Figure 24.17a),** but they do share some features with fungi. Like fungi, oomycetes grow as microscopic, nonmotile filaments called **hyphae** (singular, *hypha*), forming a network called a **mycelium** (Figure 24.17b). Also like fungi, they are heterotrophs, which secrete enzymes that digest the complex molecules of surrounding organic matter or living tissue into simpler molecules that are small enough to be absorbed into their cells. Other features, however, set the Oomycota apart from the fungi; chief among them are differences in nucleotide sequence, which clearly indicate close evolutionary relationships to other heterokonts rather than to the fungi.

The water moulds live almost exclusively in freshwater lakes and streams or moist terrestrial habitats, where they are key decomposers. Dead animal or plant material immersed in water commonly becomes coated with cottony water moulds. Other water moulds, such as the mould growing on the fish shown in Figure 24.17b, parasitize living aquatic animals. The downy mildews are parasites of land plants (Figure 24.17c). Oomycetes may reproduce asexually or sexually.

Diatoms. The organisms shown in **Figure 24.18** may not look like living organisms at all but instead like artwork or jewels. These are **diatoms**, single-celled

a. Water mould

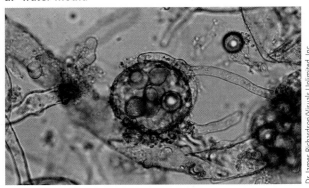

50 μm

b. Water mould infecting fish

c. Downy mildew

Figure 24.17
Oomycetes. **(a)** The water mould *Saprolegnia parasitica*. **(b)** *S. parasitica* growing as cottony white fibres on the tail of an aquarium fish. **(c)** Downy mildew, *Plasmopara viticola*, growing on grapes. At times, it has nearly destroyed vineyards in Europe and North America.

Figure 24.16
Blight caused by *Phytophthora infestans* in a potato crop.

Figure 24.18

Diatoms. Depending on the species, the shells are either radially or bilaterally symmetrical, as seen in this sample.

50 μm

Ian Hinsch/Science Source

organisms with a glassy silica shell, which is intricately formed and beautiful in many species. The two halves of the shell fit together like the top and bottom of a Petri dish or box of chocolates (see Figure 24.18). Substances move in and out of the cell through elaborately patterned perforations in the shell. Diatom shells are common in fossil deposits. In fact, more diatoms are known as fossils than as living species—some 35 thousand extinct species have been described compared with 7000 living species. For about 180 million years, diatom shells have been accumulating into thick layers of sediment at the bottom of lakes and seas.

In fact, you probably use diatoms—or their remnants—a couple of times a day when you brush your teeth. Most toothpaste contains a mild abrasive to assist in removing plaque, a bacterial biofilm that forms on your teeth. This abrasive is commonly made from grinding the fossilized shells of diatoms into a fine powder, called *diatomaceous earth*. In addition to toothpaste, diatomaceous earth is used in filters, as an insulating material, and as a pesticide. Diatomaceous earth kills crawling insects and insect larvae by abrading their exoskeleton, causing them to dehydrate and die. Insects also die when they eat the powder, but larger animals, including humans, are unaffected by it.

Diatoms are photoautotrophs that carry out photosynthesis by pathways similar to those of plants. They are among the primary photosynthetic organisms in marine plankton and are also abundant in freshwater habitats as both phytoplankton and bottom-dwelling species. Although most diatoms are free living, some are symbionts inside other marine protists. One diatom, *Pseudonitzschia,* produces a toxic amino acid that can accumulate in shellfish. The amino acid, which acts as a nerve poison, causes amnesic shellfish poisoning when ingested by humans; the poisoning can be fatal.

Asexual reproduction in diatoms occurs by mitosis followed by a form of cytoplasmic division in which each daughter cell receives either the top or the bottom half of the parent shell. The daughter cell then secretes the missing half, which becomes the smaller, inside shell of the box. The daughter cell receiving the larger top half grows to the same size as the parent shell, but the cell receiving the smaller bottom half is limited to the size of this shell. As asexual divisions continue, the cells receiving bottom halves become progressively smaller. When a minimum size is reached, sexual reproduction is triggered. The cells produce flagellated gametes, which fuse to form a zygote. The zygote grows to normal size before secreting a completely new shell with full-sized top and bottom halves.

Although flagella are present only in gametes, many diatoms move by an unusual mechanism in which a secretion released through grooves in the shell propels them in a gliding motion.

Golden Algae. Nearly all golden algae are autotrophs and carry out photosynthesis using pathways similar to those of plants. Their colour is due to a brownish carotenoid pigment, fucoxanthin, which masks the green colour of the chlorophylls **(Figure 24.19a, p. 526)**. However, most of these organisms can also live as heterotrophs if there is insufficient light for photosynthesis. They switch to feeding on dissolved organic molecules or preying on bacteria and diatoms. Golden algae are important in freshwater habitats and in *nanoplankton,* a community of marine phytoplankton composed of huge numbers of extremely small cells. During the spring and fall, blooms of golden algae are responsible for the fishy taste of many cities' drinking water.

Most golden algae are colonial forms (see Figures 24.3, p. 516, and 24.19a) in which each cell of the colony bears a pair of flagella. The golden algae have glassy shells, but in the form of plates or scales rather than in the Petri dish form of the diatoms.

Brown Algae. If you were asked where in Canada you'd find forests of giant trees, you'd likely think of the **temperate rain forests** in British Columbia. But there are also vast underwater forests in the waters off the British Columbia coast, formed not by trees but by a type of brown algae known as kelp (*Macrocystis integrifolia*), which can grow to lengths of 30 m. A related species, giant kelp (*M. pyrifera*) (Figure 24.19b–d), can grow up to 60 m long. Kelps are the largest and most complex of all protists. Their tissues are differentiated into leaflike *blades,* stalklike *stipes,* and rootlike *holdfasts* that anchor them to the bottom. Hollow, gas-filled bladders give buoyancy to the stipes and blades and help keep them upright and oriented toward the sunlit upper layers of water (Figure 24.19b). The stipes of some kelps contain tubelike vessels, similar to the vascular elements of plants, which rapidly distribute the products of photosynthesis throughout the body of the alga.

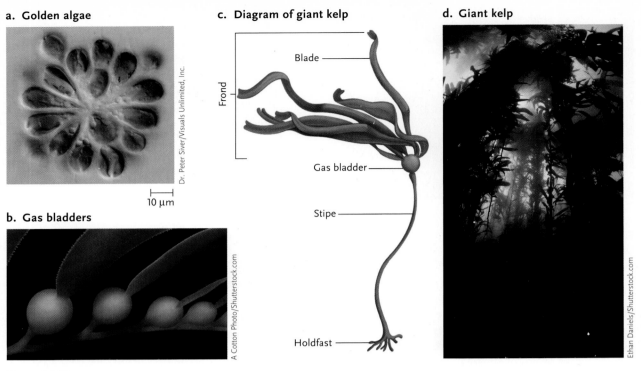

a. Golden algae

Dr. Peter Siver/Visuals Unlimited, Inc.

10 μm

b. Gas bladders

A Cotton Photo/Shutterstock.com

c. Diagram of giant kelp

Frond

Blade

Gas bladder

Stipe

Holdfast

d. Giant kelp

Ethan Daniels/Shutterstock.com

Figure 24.19
Golden and brown algae. **(a)** A microscopic swimming colony of Synura. Each cell bears two flagella, which are not visible in this light micrograph. **(b)** Gas bladders connect kelp's stipes ("stems") to its blades ("leaves"). **(c)** The fronds of giant kelp are borne on stalks known as stipes, which are anchored to the substrate by holdfasts. **(d)** A forest of *Macrocystis pyrifera* (giant kelp).

Kelps have an astonishingly fast growth rate—giant kelp can grow up to 30 cm per day!

Just as for terrestrial forests, kelp forests provide food and habitat for many marine organisms. Herds of sea otters (*Enhydra lutris*), for example, tend to live in and near kelp forests. When sea otters sleep at sea, they wrap kelp around themselves to keep from drifting away **(Figure 24.20)**. Although the forest is an important habitat for the sea otters, the otters, in turn, are critical for the survival of these forests. Sea otters are one of the few predators of sea urchins, which graze on the kelp and can cause deforestation if their populations get very large. Predation by sea otters keeps sea urchin populations in control, preventing destruction of kelp forests.

All brown algae are photoautotrophs, but not all are as large as kelps. Nearly all of the 1500 known species inhabit temperate or cool coastal marine waters. Like golden algae, brown algae contain fucoxanthin, which gives them their characteristic colour. Their cell walls contain cellulose and a mucilaginous polysaccharide, alginic acid. This alginic acid, called **algin** when extracted, is an essentially tasteless substance used to thicken such diverse products as ice cream, salad dressing, jellybeans, cosmetics, and floor polish. Brown algae are also harvested as food crops and fertilizer.

Life cycles among the brown algae are typically complex and in many species consist of alternating haploid and diploid generations **(Figure 24.21)**. The large structures that we recognize as kelps and other brown seaweeds are diploid **sporophytes**, so called because they give rise to haploid spores by meiosis. The spores, which are flagellated swimming cells, germinate and divide by mitosis to form an independent, haploid **gametophyte** generation. The gametophytes give rise to haploid gametes, the egg and sperm cells. Most brown algal gametophytes are multicellular structures only a few centimetres in diameter. Cells in the gametophyte, produced by mitosis, differentiate to form nonmotile eggs or flagellated, swimming sperm cells. The sperm cells have the two different types of

worldswildlifewonders/Shutterstock.com

Figure 24.20
A sea otter (*Enhydra lutris*) wrapped in kelp.

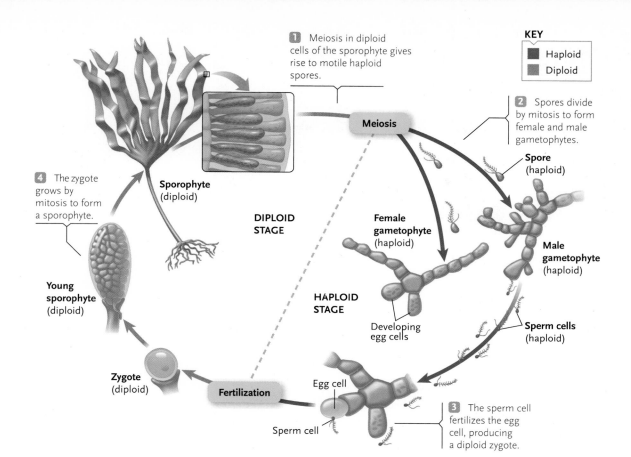

1 Meiosis in diploid cells of the sporophyte gives rise to motile haploid spores.

KEY

■	Haploid
■	Diploid

Figure 24.21
The life cycle of the brown alga *Laminaria*, which alternates between a diploid sporophyte stage and a haploid gametophyte stage.

Meiosis

2 Spores divide by mitosis to form female and male gametophytes.

Spore (haploid)

Sporophyte (diploid)

DIPLOID STAGE

Female gametophyte (haploid)

Male gametophyte (haploid)

4 The zygote grows by mitosis to form a sporophyte.

HAPLOID STAGE

Developing egg cells

Sperm cells (haploid)

Young sporophyte (diploid)

Zygote (diploid)

Fertilization

Egg cell

Sperm cell

3 The sperm cell fertilizes the egg cell, producing a diploid zygote.

flagella characteristic of the heterokont protists. Fusion of egg and sperm produces a diploid zygote that grows by mitotic divisions into the sporophyte generation. This complex life cycle is very similar to that of land plants (see Chapter 26).

24.4c Rhizara Are Eukaryotes with Filamentous Pseudopods

Amoeba (*amoibe* = change) is a descriptive term for a single-celled protist that moves by means of pseudopodia, as described earlier in this chapter (see Figure 24.6, p. 517). Several major groups of protists contain amoebas, which are similar in form but are not all closely related. Amoebas in the Rhizaria produce stiff, filamentous pseudopodia, and many produce hard outer shells, also called *tests*. We consider here two heterotrophic groups of amoebas, the Radiolaria and the Foraminifera, and a third, photosynthesizing group, the Chlorarachniophyta.

Radiolaria. Radiolarians (*radiolus* = small sunbeam) are marine organisms characterized by a glassy internal skeleton and **axopods,** slender, raylike strands of cytoplasm supported internally by long bundles of microtubules **(Figure 24.22a, b, p. 528).** This glassy skeleton is heavy—when radiolarians die, their skeletons sink to the ocean floor—so how do radiolarians

keep afloat? The axopods provide buoyancy, as do the numerous vacuoles and lipid droplets in the cytoplasm. Axopods are also involved in feeding: prey stick to the axopods and are then engulfed, brought into the cell, and digested in food vacuoles.

Radiolarian skeletons that accumulate on the ocean floor become part of the sediment, which, over time, hardens into sedimentary rock. The presence of radiolarians in such rocks is very useful to the oil industry as indicators of oil-bearing strata.

Foraminifera: Forams. These organisms take their name from the perforations in their shells (*foramen* = little hole), through which extend long, slender strands of cytoplasm supported internally by a network of needlelike spines. Their shells consist of organic matter reinforced by calcium carbonate (Figure 24.22c–e). Most foram shells are chambered, spiral structures that, although microscopic, resemble those of molluscs.

Like radiolarians, forams live in marine environments. Some species are planktonic, but they are most abundant on sandy bottoms and attached to rocks along the coasts. Forams feed in a manner similar to that of radiolarians: they engulf prey that adhere to the strands and conduct them through the holes in the shell into the central cytoplasm, where they are digested in food vacuoles. Some forams have algal symbionts that carry out

a. Radiolarian

Perennou Nuridsany/Science Source

b. Radiolarian skeleton

Eye of Science / Science Source

c. Living foram

Jerry McCormick-Ray/Science Source

d. Foram shells

Eric V. Grave/Science Source

├─┤
10 μm

├─┤
10 μm

e. Foram body plan

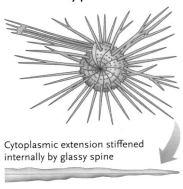

Cytoplasmic extension stiffened
internally by glassy spine

Redrawn from V. & J. Pearse and M. & R. Buchsbaum, *Living Invertebrates*, The Boxwood Press, 1987.

Figure 24.22

(a) A living radiolarian. **(b)** The internal skeletons of a radiolarian. Bundles of microtubules support the cytoplasmic extensions of the radiolarian. **(c)** A living foram, showing the cytoplasmic strands extending from its shelf. **(d)** Empty foram shells. **(e)** The body plan of a foram. Needlelike, glassy spines support the cytoplasmic extensions of the forams.

photosynthesis, allowing them to live as both heterotrophs and autotrophs.

Marine sediments are typically packed with the shells of dead forams. The sediments may be hundreds of feet thick: the White Cliffs of Dover in England are composed primarily of the shells of ancient forams. Most of the world's deposits of limestone and marble contain foram shells; the great pyramids of ancient Egypt are built from blocks cut from fossil foram deposits. Because distinct species lived during different geologic periods, they are widely used to establish the age of sedimentary rocks containing their shells. As they do with radiolarian species, oil prospectors use forams as indicators of hydrocarbon deposits because layers of forams often overlie oil.

Chlorarachniophyta. Chlorarachniophytes are amoebas that contain chloroplasts and thus are photosynthetic. However, they combine this mode of nutrition with heterotrophy, engulfing food with the many filamentous pseudopodia that extend from the cell surface.

24.4d The Unikont Supergroup Includes Slime Moulds and Most Amoebas

The Unikonts include most of the amoebas other than those in Rhizaria, as well as the cellular and plasmodial slime moulds. All members of this group use pseudopods for locomotion and feeding for all or part of their life cycles.

Amoebas. Amoebas of the Unikonts are single-celled organisms that are abundant in marine and freshwater environments and in the soil. All amoebas are microscopic, although some species can grow to 5 mm in size and so are visible with the naked eye. Some amoebas are parasitic, such as the 45 species that infect the human digestive tract. One of these parasites, *Entamoeba histolytica,* causes amoebic dysentery. Cysts of this amoeba contaminate water supplies and soil in regions with inadequate sewage treatment. When ingested, a cyst breaks open to release an amoeba that feeds and divides rapidly in the digestive tract. Enzymes released by the amoebas destroy cells lining the intestine, producing the ulcerations, painful cramps, and debilitating diarrhea characteristic of the disease. Amoebic dysentery afflicts millions of people worldwide; in less developed countries, it is a leading cause of death among infants and small children.

However, most amoebas are heterotrophs that feed on bacteria, other protists, and bits of organic matter. Unlike the stiff, supported pseudopodia of Rhizaria, pseudopods of amoebas extend and retract at any point on their body surface and are unsupported by any internal cellular organization—amoebas are thus "shape-shifters." How can an amoeba capture a

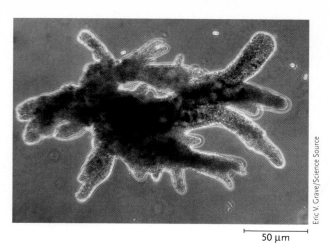

Figure 24.23
An amoeba capturing prey with pseudopods.

fast-moving organism? As an amoeba moves, its cytoplasm doesn't just move but also changes state, from a more liquid state to a more solid state and back again, allowing the amoeba to send out pseudopodia in different directions very quickly. These fast-moving pseudopods can capture even fast-swimming prey such as ciliates **(Figure 24.23).**

Amoebas reproduce only asexually, via binary fission. In unfavourable environmental conditions, some amoebas can form a cyst, essentially by rolling up and secreting a protective membrane. They survive as cysts until favourable conditions return.

Slime Moulds. After a very wet spring in 1973, residents of Dallas, Texas, were alarmed to see large, yellow blobs that resembled scrambled eggs *crawling* on their lawns. People thought it was an alien invasion. Luckily, a local biologist was able to prevent mass panic by identifying the blobs as slime moulds, unusual heterotrophic protists. Slime moulds exist for part of their lives as individuals that move by amoeboid motion but then come together in a coordinated mass—essentially, a large amoeba—that ultimately differentiates into a stalked structure called a **fruiting body,** in which spores are formed.

There are two major evolutionary lineages of slime moulds: the **cellular slime moulds** and the plasmodial slime moulds, which differ in cellular organization. Both types of slime moulds have been of great interest to scientists because of their ability to differentiate into fruiting bodies with stalks and spore-bearing structures. This differentiation is much simpler than the complex developmental pathways of other eukaryotes, providing a unique opportunity to study cell differentiation at its most fundamental level. Slime moulds also respond to stimuli in their environment, moving away from bright light and toward food. We have learned a great deal about eukaryotic signalling pathways, cell differentiation, and cell movement from studies of slime moulds.

Slime moulds live on moist, rotting plant material such as decaying leaves and bark. The cells engulf particles of dead organic matter, along with bacteria, yeasts, and other microorganisms, and digest them internally. They can be a range of colours: brown, yellow, green, red, and even violet or blue.

These organisms exist primarily as individual cells, either separately or as a coordinated mass. Among the 70 or so species of cellular slime moulds, *Dictyostelium discoideum* is best known. Its life cycle begins when a haploid spore lands in a suitably moist environment containing decaying organic matter **(Figure 24.24, p. 530).** The spore germinates into an amoeboid cell that grows and divides mitotically into separate haploid cells as long as the food source lasts. When the food supply dwindles, some of the cells release a **chemical signal** in pulses; in response, the amoebas move together and form a sausage-shaped mass that crawls in coordinated fashion like a slug. Some "slugs," although not much more than a millimetre in length, contain more than 100 thousand individual cells. At some point, the "slug" stops moving and differentiates into a stalked fruiting body, with some cells becoming spores, whereas others form the stalk. The cells that form the stalk die in the process, essentially sacrificing themselves so that a stalk can form. Why is formation of a stalk so crucial? Raising the spore-forming cells higher up in the air increases the likelihood that spores will be carried away by air currents and dispersed farther away from the parent. Because the cells forming the "slug" and fruiting body are all products of mitosis, this is asexual reproduction.

Cellular slime moulds also reproduce sexually: two haploid cells fuse to form a diploid zygote (also shown in Figure 24.24) that enters a dormant stage. Eventually, the zygote undergoes meiosis, producing four haploid cells that may multiply inside the spore by mitosis. When conditions are favourable, the spore wall breaks down, releasing the cells. These grow and divide into separate amoeboid cells.

Plasmodial Slime Moulds. Plasmodial slime moulds exist primarily as a multinucleate **plasmodium,** in which individual nuclei are suspended in a common cytoplasm surrounded by a single plasma membrane. (This is not to be confused with *Plasmodium,* the genus of apicomplexans that causes malaria.) There are about 500 known species of plasmodial slime moulds. The plasmodium **(Figure 24.25a, b, p. 530)** flows and feeds by phagocytosis like a single huge amoeba—a single cell that contains thousands to millions or even billions of diploid nuclei surrounded by a single plasma membrane. The plasmodium, which may range in size from a few centimetres to more than a metre in diameter, typically moves in thick, branching strands connected by thin sheets. The movements occur by cytoplasmic streaming, driven by actin microfilaments and myosin.

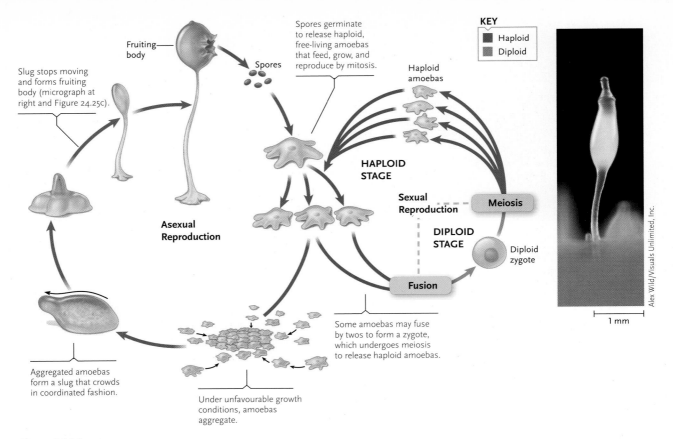

Fruiting body

Slug stops moving and forms fruiting body (micrograph at right and Figure 24.25c).

Spores germinate to release haploid, free-living amoebas that feed, grow, and reproduce by mitosis.

Spores

Haploid amoebas

KEY

| Haploid |
| Diploid |

HAPLOID STAGE

Sexual Reproduction

Meiosis

DIPLOID STAGE

Diploid zygote

Asexual Reproduction

Fusion

Some amoebas may fuse by twos to form a zygote, which undergoes meiosis to release haploid amoebas.

Aggregated amoebas form a slug that crowds in coordinated fashion.

Under unfavourable growth conditions, amoebas aggregate.

Alex Wild/Visuals Unlimited, Inc.

1 mm

Figure 24.24
Life cycle of the cellular slime mould *Dictyostelium discoideum*. The light micrograph shows a mature fruiting body.

a.

b.

c.

George Barron

George Barron

Greg Thorn

Figure 24.25
Slime moulds. **(a** and **b)** Plasmodia of slime moulds. **(c)** Fruiting bodies of slime mould.

Klaus-Peter Zauner, *University of Southampton*; Soichiro Tsuda, *University of Glasgow*; and Yukio-Pegio Gunjia, *Kobe University*

Robots controlled by slime moulds? Far from being a bizarre science fiction story, slime moulds may be the future of robotics, as demonstrated in research carried out by Klaus-Peter Zauner (University of Southampton, United Kingdom) and his collaborators at Kobe University in Japan, Soichiro Tsuda and Yukio-Pegio Gunjia. Zauner and colleagues grew *Physarum polycephalum* in a six-pointed star shape on an electrical circuit and connected the circuit to a six-legged robot with a computer interface. Each point of the *Physarum* plasmodium star corresponded to one leg of the

Klaus-Peter Zauner/New Scientist

Figure 1
A slime mould (image on screen) is able to direct the movement of a robot (in the foreground).

robot **(Figure 1).** When light was shone on certain parts of the robot, sensors mounted on the robot detected the light and illuminated the corresponding part of the plasmodium, which responded by moving away from the light, sending the robot scuttling into dark corners. One goal of Zauner's research in molecular computing is to incorporate this biological control right into the robot. Harnessing the capacity of living organisms to sense and respond to complex environments would give robots greater autonomy than is possible with control by computer programs.

These plasmodia are what the people in Dallas thought were aliens invading; after a period of heavy rain, plasmodia will sometimes crawl out of the woods to appear on lawns or the mulch of flowerbeds.

At some point, often in response to unfavourable environmental conditions, fruiting bodies form on the plasmodium. At the tips of the fruiting bodies, nuclei become enclosed in separate cells. These cells undergo meiosis, forming haploid, resistant spores that are released from the fruiting bodies and carried by water or wind. If they reach a favourable environment, the spores germinate to form gametes that fuse to form a diploid zygote. The zygote nucleus then divides repeatedly without an accompanying division of the cytoplasm, forming many diploid nuclei suspended in the common cytoplasm of a new plasmodium.

Plasmodial slime moulds are particularly useful in research because they become large enough to provide ample material for biochemical and molecular analyses. Actin and myosin extracted from *Physarum polycephalum,* for example, have been much used in studies of actin-based motility. A further advantage of plasmodial slime moulds is that the many nuclei of a plasmodium usually replicate and pass through mitosis in -synchrony, making them useful in research that tracks the changes that take place in the cell cycle. More recently, slime moulds have been used in robotics research, as outlined in "People behind Biology," 24.2.

Also included in the Unikonts are the choanoflagellates. Opisthokonta (*opistho* = posterior; *kontos* =

flagellum) are named for the single, posterior flagellum found at some stage in the life cycle of these organisms. This diverse group includes the choanoflagellates, protists thought to be the ancestors of fungi and animals.

Choanoflagellata (*choanos* = collar) are named for the collar surrounding the flagellum that the protist uses to feed and, in some species, to swim **(Figure 24.26).** The collar resembles an upside-down lampshade and is made up of small, fingerlike projections (microvilli) of the plasma membrane. As the flagellum moves water through the collar, these projections engulf bacteria and particles of organic matter in the water.

About 150 species of choanoflagellates live in either marine or freshwater habitats. Some species are mobile, with the flagellum pushing the cells along (in the same way that animal sperm are propelled by their flagella), but most choanoflagellates are *sessile* (attached by a stalk to a surface). A number of species are colonial

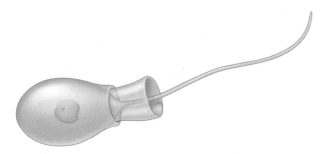

Figure 24.26
A choanoflagellate.

with a cluster of cells on a single stalk; these colonial species are of great interest to biologists studying the evolution of multicellularity in animals.

Why are choanoflagellates thought to be the ancestor of animals? Both molecular and morphological data indicate that a choanoflagellate type of protist gave rise to animals: For example, there are many morphological similarities between choanoflagellates and the collar cells (choanocytes) of sponges as well as the cells that act as excretory organisms in flatworms and rotifers (see Chapter 27). Comparisons of nucleic acid sequences done to date also support the hypothesis that choanoflagellates are the closest living relatives to animals. Molecular data also indicate that a choanoflagellate-like organism was also likely the ancestor of the fungi (see Chapter 25).

24.4e Plantae Include the Red and Green Algae and Land Plants

The Plantae supergroup consists of the red and green algae, which are protists, and the land plant. These three groups of photoautotrophs share a common evolutionary origin. Here we describe the two types of algae; we discuss land plants and how they evolved from green algae in Chapter 26.

Rhodophyta: The Red Algae. Nearly all of the 4000 known species of red algae, which are also known as the Rhodophyta (*rhodon* = rose), are small marine seaweeds **(Figure 24.27)**. Fewer than 200 species are found in freshwater lakes and streams or in soils. If you have had sushi, then you have eaten red algae: *Porphyra* is harvested for use as the *nori* wrapped around fish and rice.

Rhodophyte cell walls contain cellulose and mucilaginous pectins that give red algae a slippery texture. These pectins are widely used in industry and science. Extracted **agar** is used as a culture medium

in the laboratory and as a setting agent for jellies and desserts. **Carrageenan** is used to thicken and stabilize paints, dairy products such as ice cream, and many other emulsions.

Some species secrete calcium carbonate into their cell walls; these coralline algae are important in building coral reefs—in some places, they play a bigger role in reef building than do corals.

Red algae are typically multicellular organisms, with diverse morphologies, although many have plantlike bodies composed of stalks bearing leaflike blades. Although most are free-living autotrophs, some are parasites that attach to other algae or plants.

Although most red algae are reddish in colour, some are greenish purple or black. The colour differences are produced by accessory pigments, *phycobilins,* that mask the green colour of their chlorophylls. Phycobilins absorb the shorter wavelengths of light (green and blue-green light) that penetrate to the ocean depths, allowing red algae to grow at deeper levels than any other algae. Some red algae live at depths up to 260 m if the water is clear enough to transmit light to these levels.

Red algae have complex reproductive cycles involving alternation between diploid sporophytes and haploid gametophytes. No flagellated cells occur in the red algae; instead, gametes are released into the water to be brought together by random collisions in currents.

Chlorophyta: The Green Algae. The green algae or Chlorophyta (*chloros* = green) carry out photosynthesis using the same pigments as plants, whereas other photosynthetic protists contain pigment combinations that are very different from those of land plants. This shared pigment composition is one line of evidence that one lineage of green algae was the ancestor of land plants. With at least 16 thousand species, green algae show more diversity than any other algal group. They also have very diverse morphologies, including single-celled, colonial, and multicellular species **(Figure 24.28, p. 534; see also Figure 24.1d, p. 514)**. Multicellular forms have a range of morphologies, including filamentous, tubular, and leaflike forms. Most green algae are microscopic, but some range upward to the size of small seaweeds.

Most green algae live in freshwater aquatic habitats, but some are marine, whereas others live on rocks, soil surfaces, or tree bark, or even in snow. Other organisms rely on green algae to photosynthesize for them by forming symbiotic relationships. For example, lichens are symbioses between green algae and fungi (see Chapter 25), and many animals, such as the sea slugs described in "Life on the Edge" 24.3, contain green algal chloroplasts, or entire green algae, as symbionts in their cells.

Life cycles among the green algae are as diverse as their body forms. Many can reproduce either

Filamentous red alga

© Sabena Jane Blackbird/Alamy

Figure 24.27
Red algae. *Antithamnion plumula*, showing the filamentous and branched body form most common among red algae.

LIFE ON THE EDGE 24.3

Solar-Powered Animals

We associate photosynthesis with plants and some algae, but in a few exceptional cases, animals steal chloroplasts and use them to capture solar energy through photosynthesis. In a sense, these animals are doing the same as the ancestors of the eukaryotic cells that captured photosynthetic prokaryotes and thus acquired the ability to carry out oxygenic photosynthesis (see Chapter 7). There are many examples of sequestration of chloroplasts by species of the animal phyla Platyhelminthes, Porifera, Cnidaria, Mollusca, and Urochordata.

Some sea slugs, such as the *Elysia chlorotica* shown in **Figure 1**, use specialized teeth on their radulae (*radula* = tonguelike organ) to cut into algal cells so that they can suck out chloroplasts and other cell contents. In the slug's stomach, chloroplasts are engulfed by phagocytosis and then moved to areas below the epidermis. Slugs such as E. chlorotica can live for at least five months on the energy generated by these "solar panels"—the chloroplasts they have engulfed. This story is an example of how a chemoheterotrophic organism (the slug) acquires the ability to be photoautotrophic (engage in photosynthesis). How the slugs control the functions of the chloroplasts remains unknown. Have slugs also taken over some of the chloroplast or algal genome? At least one species, *Elysia crispata*, has genes from chloroplasts in its genomic DNA. For more about these slugs, see Chapter 27.

Dr. Mary Tyler & Dr. Mary Rumpho, University of Maine, (2008), "Horizontal gene transfer of the algal nuclear gene psbO to the photosynthetic sea slug Elysia chlorotica," PNAS, 105 (46), 17868, Copyright 2008 National Academy of Sciences, U.S.A.

Figure 1
The solar-powered sea slug, Elysia chlorotica, *an animal that extracts chloroplasts from algae and uses them to produce food.*

Such solar-powered symbioses also occur between algae and vertebrate animals: Ryan Kerney, a biologist at Dalhousie University in Halifax, Nova Scotia, recently discovered that green algae (*Oophila amblystomatis*) colonize embryos of spotted salamanders (*Ambystoma gracilis*), becoming part of the animals' bodies. The presence of the algae allows the salamanders to capture solar energy using the chloroplasts in the algal cells. The presence of algae in these salamanders' eggs was known for over 100 years, but Kerney was the first to show that the algae actually invade the developing embryos. Through photosynthesis, the algae provide nutrients and oxygen to the embryos, and, in turn, they feed on the salamanders' waste products. This intimate symbiosis doesn't persist in mature salamanders, which have opaque skin and live mostly underground. Many questions remain to be answered: How do the algae get into the eggs? Why doesn't the salamander's immune system perceive the algae as invaders and destroy them? And are there other solar-powered vertebrates?

---

sexually or asexually, and some alternate between haploid and diploid generations. Gametes in different species may be undifferentiated flagellated cells or differentiated as a flagellated sperm cell and a nonmotile egg cell. Most common is a life cycle with a multicellular haploid phase and a single-celled diploid phase **(Figure 24.29, p. 534).**

Among all the algae, the green algae are the most closely related to land plants, based on molecular, biochemical, and morphological data. Evidence of this close relationship includes not only the shared photosynthetic pigments, but also the use of starch as storage reserve and the same cell wall composition.

Which green alga might have been the ancestor of modern land plants? The evidence points to a group known as the **charophytes** as being most similar to the algal ancestors of land plants. This does not mean that modern-day charophytes are the ancestors of land plants but rather that the two groups have a common ancestor. Charophytes, including *Chara* **(Figure 24.30, p. 535),** *Spirogyra, Nitella,* and *Coleochaete,* live in freshwater ponds and lakes. Their ribosomal RNA and chloroplast DNA sequences are more closely related to plant sequences than those of any other green alga. We discuss the evolution of land plants from an algal ancestor more thoroughly in Chapter 26.

a. Single-celled green alga

Borut Furlan/WaterFrame/Getty Images

c. Multicellular green alga

Marevision/age fotostock/Getty Images

1 cm

b. Colonial green alga

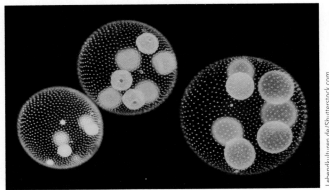

Lebendkulturen.de/Shutterstock.com

200 μm

Figure 24.28

Green algae. **(a)** A single-celled green alga, *Acetabularia*, which grows in marine environments. Each individual in the cluster is a large single cell with a rootlike base, stalk, and cap. **(b)** A colonial green alga, *Volvox*. Each green dot in the spherical wall of the colony is a potentially independent, flagellated cell. Daughter colonies can be seen within the parent colony. **(c)** A multicellular green alga, *Ulva*, common to shallow seas around the world.

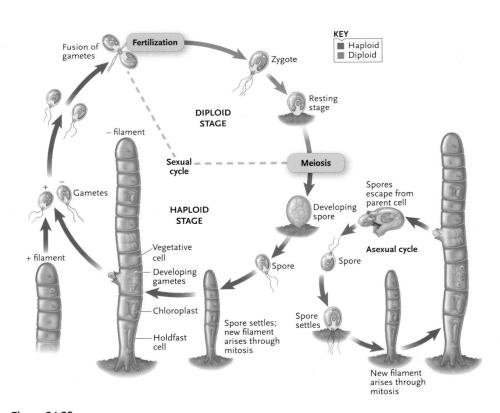

Figure 24.29

The life cycle of the green alga *Ulothrix*, in which the haploid stage is multicellular and the diploid stage is a single cell, the zygote. + and − are morphologically identical mating types ("sexes") of the alga.

Reproductive structures

Dr. John Clayton, National Institute of Water and Atmospheric Research, New Zealand

Figure 24.30

The charophyte *Chara*, representative of a group of green algae that may have given rise to the plant kingdom.

STUDY BREAK

1. For each of the protist groups listed below, indicate the cell structure that characterizes the group: apicomplexans, dinoflagellates, euglenoids, radiolarians.
2. Which eukaryotic supergroups contain amoeboid forms?
3. What is the major difference between cellular slime moulds and plasmodial slime moulds?

24.5 Some Protist Lineages Arose from Primary Endosymbiosis and Others from Secondary Endosymbiosis

We have encountered chloroplasts in a number of eukaryotic organisms in this chapter: red and green algae, euglenoids, dinoflagellates, stramenopiles, chlorarachniophytes, and land plants. How did these chloroplasts evolve? Unlike the endosymbiotic event that gave rise to mitochondria, endosymbiosis involving photoautotrophs happened more than once, resulting in the formation of a wide range of photosynthetic eukaryotes.

About 1 bya, the first chloroplasts evolved from free-living photosynthetic prokaryotic organisms (cyanobacteria) ingested by eukaryote cells that had already acquired mitochondria (see Chapter 3). In some cells, the cyanobacterium was not digested but instead formed a symbiotic relationship with the engulfing host cell—it became an endosymbiont, an independent organism living inside another organism. Over evolutionary time, the prokaryotic organism lost genes no longer required for independent existence and transferred most of its genes to the host's nuclear genome. In this process, the endosymbiont became an organelle. As explained in Chapter 3, moving genes from the endosymbiont to the nucleus would have given the host cell better control of cell functioning.

The chloroplasts of red algae, green algae, and land plants result from evolutionary divergence of the p-hotosynthetic eukaryotes formed from this primary endosymbiotic event (as shown in the top part of **Figure 24.31, p. 536**). Organisms that originated from this event have chloroplasts with two membranes, one from the plasma membrane of the engulfing eukaryote and the other from the plasma membrane of the cyanobacterium.

This **primary endosymbiosis** was followed by at least three **secondary endosymbiosis** events, each time involving different heterotrophic eukaryotes engulfing a photosynthetic eukaryote, producing new evolutionary lineages (see Figure 24.31). For example, s-econdary endosymbiosis involving red algae engulfed by a nonphotosynthetic eukaryote gave rise to the stramenopile algae and the alveolates.

Independent endosymbiotic events involving green algae and nonphotosynthetic eukaryotes produced euglenoids and chlorarachniophytes.

Organisms that formed via secondary endosymbiosis have chloroplasts surrounded by additional membranes acquired from the new host. For example, chlorarachniophytes have plastids with four membranes (see Figure 24.31). The new membranes correspond to the plasma membrane of the engulfed phototroph and the food vacuole membrane of the host.

In sum, the protists are a highly diverse and ecologically important group of organisms. Their complex evolutionary relationships, which have long been a subject of contention, are now being revised as new information is discovered, including more complete genome sequences. A deeper understanding of protists is also contributing to a better understanding of their recent descendants, the fungi, plants, and animals. We turn to these descendants in the next four chapters, beginning with the fungi.

STUDY BREAK

In primary endosymbiosis, a nonphotosynthetic eukaryotic cell engulfed a photosynthetic cyanobacterium. How many membranes surround the chloroplast that evolved?

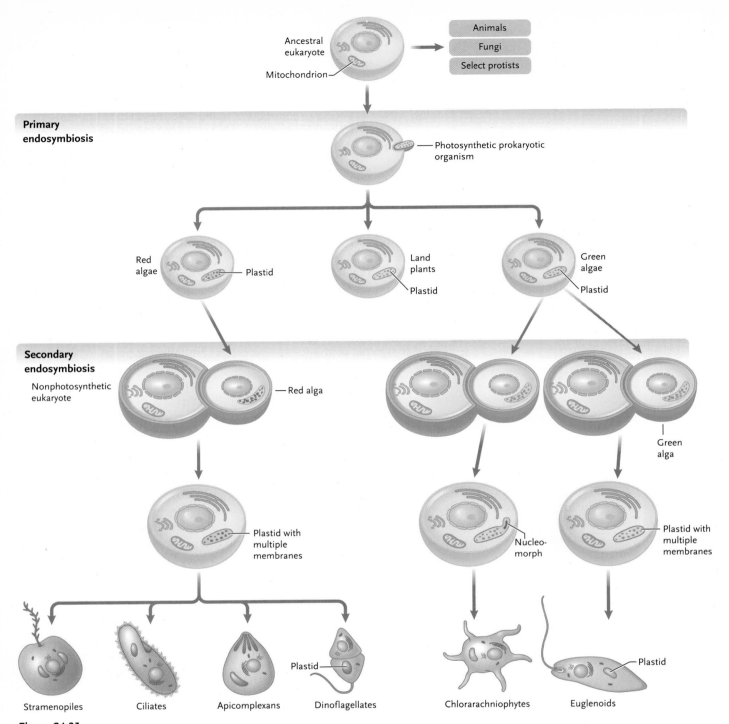

Figure 24.31

The origin and distribution of plastids among the eukaryotes by primary and secondary endosymbiosis.

Review

24.1 The Vast Majority of Eukaryotes Are Protists

- As eukaryotes, protists contain organelles including mitochondria and, sometimes, chloroplasts. Mitochondria evolved once via primary endosymbiosis, the engulfing of a free-living prokaryotic organism that became an organelle over evolutionary time. Some photosynthetic protists were formed via primary endosymbiosis involving a photosynthetic prokaryotic organism; others arose via more complex endosymbiotic events (see Section 24.5).

24.2 Characteristics of Protists

- Eukaryotes are divided into supergroups, a taxonomic level above kingdom. The vast majority of lineages within these supergroups are protists. Protists are eukaryotes that differ from fungi in having motile stages in their life cycles and cellulose cell walls. Unlike land plants, they lack roots, stems, and leaves and do not retain embryos in parental tissue. Unlike animals, protists are often unicellular; they lack collagen, nerve cells, and an internal digestive tract.

24.3 Protists' Diversity Is Reflected in Their Metabolism, Reproduction, Structure, and Habitat

- Most protists are aerobic organisms that live as autotrophs or as heterotrophs or by a combination of both nutritional modes. Some are symbionts living in or among the cells of other organisms.

- Protists live in aquatic or moist terrestrial habitats or as parasites within animals. They may be single-celled, colonial, or multicellular organisms, and they range in size from microscopic to some of Earth's largest organisms.

- Some protists are the most complex single cells known because of the wide variety of cytoplasmic structures they have; most are able to move by means of flagella, cilia, or pseudopodia.

- Reproduction may be asexual, by mitotic divisions, or sexual, involving meiosis and the union of gametes in fertilization.

24.4 The Eukaryotic Supergroups and Their Key Protist Lineages

- Excavates are almost all single-celled, heterotrophic protists that swim using flagella (Figures 24.7, 24.8). One lineage, the euglenoids, obtained plastids via secondary endosymbiosis. Other lineages, such as the diplomonads and parabasalids (Figure 24.9), lack "typical" mitochondria but often have organelles derived from mitochondria.

- Chromalveolates include ciliates (Figure 24.10), which swim using cilia and have complex cytoplasmic structures, including both micronuclei and macronuclei; apicomplexans, nonmotile parasites of animals; and dinoflagellates, which have two flagella that propel them in a "whirling" motion and are primarily marine organisms (Figure 24.11). Many alveolates are photosynthetic.

- Chromalveolates also include diatoms, photosynthetic single-celled organisms covered by a glassy silica shell (Figure 24.18); golden algae, photosynthetic, mostly colonial forms (Figure 24.3); brown algae, primarily multicellular marine forms that include large seaweeds (Figure 24.19); and the funguslike Oomycota, which often grow as masses of microscopic filaments and live as saprophytes or parasites, secreting enzymes that digest organic matter in their surroundings (Figure 24.17). Many stramenopiles (heterokonts) have flagella only on reproductive cells.

- Rhizaria are amoebas with filamentous pseudopods supported by internal cellular structures (Figure 24.22). Many produce hard outer shells. Radiolarians are primarily marine organisms that secrete a glassy internal skeleton. Foraminifera are marine, single-celled organisms that form chambered, spiral shells containing calcium. Both groups engulf prey that adhere to thin extensions of their cells. Chlorarachniophytes engulf food using their pseudopodia.

- Unikonts include most amoebas (Figure 24.23) and two types of slime moulds, cellular (which move as individual cells; Figure 24.24) and plasmodial (which move as large masses of nuclei sharing a common cytoplasm; Figure 24.25). Amoebas in this group are heterotrophs that are abundant in marine and freshwater environments and in the soil. They move by extending pseudopodia. Also included in the Unikont supergroup are opisthokonts, a broad group of eukaryotes including the choanoflagellates. Choanoflagellates have a single flagellum surrounded by a collar of fingerlike membrane projections. A choanoflagellate type of protist was likely the ancestor of animals.

- Plantae includes the red and green algae, as well as land plants. Red algae are typically multicellular, primarily photosynthetic organisms of marine environments with complex life cycles. Green algae are single-celled, colonial, multicellular species that live primarily in freshwater habitats (Figure 24.28) and carry out photosynthesis by mechanisms similar to those of plants.

- Opisthokonts are a broad group of eukaryotes that includes the choanoflagellates, which have a single flagellum surrounded by a collar of fingerlike membrane projections. A choanoflagellate type of protist was likely the ancestor of animals.

24.5 Some Protist Lineages Arose from Primary Endosymbiosis and Others from Secondary Endosymbiosis

- Several groups of protists, as well as land plants, contain chloroplasts, which arose via endosymbiosis events (Figure 24.31). In a primary endosymbiosis event, a eukaryotic cell engulfed a cyanobacterium, which became an organelle, the chloroplast. Evolutionary divergence from this ancestral phototrophic organism produced the red algae, green algae, and land plants. Other photosynthetic protists were produced by secondary endosymbiosis, in which a nonphotosynthetic eukaryote engulfed a photosynthetic eukaryote.

Questions

Self-Test Questions

1. Which group of protists move through viscous fluids using both freely beating flagella and a flagellum buried in a fold of cytoplasm, and cause a sexually transmitted disease in humans?
   a. Ciliates
   b. Parabasalids
   c. Euglenoids
   d. Diplomonads

2. Diplomonads are characterized by which of the following features?
   a. Cells with two functional nuclei and multiple flagella. *Giardia* is an example.
   b. A mouthlike gullet and a hairlike surface. *Paramecium* is an example.
   c. Nonmotility, parasitism, and sporelike infective stages. *Toxoplasma* is an example.
   d. Large protein deposits. Movement is by two flagella, which are part of an undulating membrane. *Trypanosoma* is an example.

3. Which of the following groups is the greatest contributor to protist fossil deposits?
   a. oomycetes
   b. brown algae
   c. golden algae
   d. diatoms

4. Which of the following groups has the distinguishing characteristic of gas-filled bladders and a cell wall composed of alginic acid?
   a. oomycetes
   b. brown algae
   c. golden algae
   d. diatoms

5. *Plasmodium* is transmitted to humans by the bite of a mosquito (*Anopheles*) and engages in a life cycle with infective spores, gametes, and cysts. To which group does this infective protist belong?
   a. oomycetes
   b. euglenoids
   c. dinoflagellates
   d. apicomplexans

6. The ancestor of land plants is thought to have belonged to which group of protists?
   a. red algae
   b. diatoms
   c. green algae
   d. golden algae

7. In oil exploration, the presence of shells is an indicator of oil-rich rock layers. To which group of protists would these shells belong?
   a. diatoms
   b. foraminiferans
   c. golden algae
   d. red algae

8. To which supergroup do the living representatives of the group of organisms thought to be ancestral to animals belong?
   a. Excavates
   b. Rhizaria
   c. Unikonts
   d. Chromalveolates

9. Which of the following statements about cellular slime moulds is correct?
   a. They are autotrophs.
   b. They move using cilia.
   c. They reproduce only asexually.
   d. They form a fruiting body that produces spores.

10. The latest stage for evolving the double membrane seen in modern-day algal chloroplasts is thought to involve the combining of two organisms. What are the two organisms?
    a. two ancestral photosynthetic prokaryotic organisms
    b. two ancestral nonphotosynthetic prokaryotic organisms
    c. a nonphotosynthetic eukaryote with a photosynthetic eukaryote
    d. a photosynthetic prokaryotic organism with a nonphotosynthetic eukaryote

Questions for Discussion

1. We have seen that, as a group, protists use three kinds of motility: flagella, cilia, and amoeboid movement. There are some cells in your body that also use each of these three forms of motility. Name an example of cells that use each form.

2. Photosynthetic protists (sometimes referred to as algae) are often thought of as single-celled plants. What features differentiate these protists from land plants? Would it be correct to consider the protists known as protozoa as single-celled animals? Explain why or why not.

3. Why is it harder to treat human diseases caused by protists, such as *Giardia*, than diseases caused by bacteria?

4. Many protists are able to produce cysts or other resting stages in their life cycle. What is the advantage of producing these resting structures?

5. You place a marine green algal cell and a *Paramecium* into fresh water, which is hypotonic to both cells. Water will flow into both cells. Which cell, if either, will burst? Explain your rationale.

25

The mushroom-forming fungus *Inocybe fastigiata*, a forest-dwelling species that commonly lives in close association with conifers and hardwood trees.

Fungi

WHY IT MATTERS

If you were asked what the first crop on Earth was and which organisms grew it, you would probably think of corn, wheat, or some other crop plant grown by humans. But you'd be wrong—the first domesticated crop was a fungus, and the first farmers were a certain group of ants over 50 million years ago (mya), whereas humans did not start farming until about 10 thousand years ago. Researchers have used molecular data combined with fossil evidence to determine when ants first domesticated their fungal crop. Today, these leaf-cutter ants (Tribe Attini) of Central and South America **(Figure 25.1a, p. 540)** still grow certain fungi in gardens. Just as humans do, the ants plant their crop, fertilize it, weed it, and then feed on it. The ants harvest leaves, flowers, and other plant parts and carry these back to their nests, where they are added to the fungal gardens in the nests (Figure 25.1b).

The ants plant small pieces of the plant material in the garden, placing bits of the fungus on each piece. They fertilize the garden with their excrement and graze on the fungal filaments. In fact, although the ants collect a wide range of plant matter, they never eat any of it directly—their sole food source is the fungus. When a queen ant leaves her birth nest to start a new nest, she carries a bit of fungus in her mouth and uses it to start a garden in the new nest. The ants' habitat contains ample supplies of other foods, so why have these ants developed this complex and rather bizarre lifestyle? What benefit do they gain by devoting their lives to looking after a fungus? The answer lies in the ability of the fungus to unlock the nutrients tied up in plant tissue. Cellulose is the most abundant organic molecule on Earth, but most organisms cannot get at the carbon it contains as they lack the enzymes needed to break apart the bonds in the molecule. Fungi are among the few organisms that can digest cellulose, so by forming a partnership with fungi, these ants gain access to a continuous source of carbon. In return, the fungus gains a secure habitat in which it doesn't have to compete with other organisms for a food source. Recent

Figure 25.1
(a) Leaf-cutter ants.
(b) Fungal garden of leaf-cutter ants.

a.

Tim Flach/Stone/Getty Images

b.

Alex Wild/Visuals Unlimited, Inc.

research has revealed that this ancient symbiosis is more complex than previously known (see "People behind Biology," 25.1, p. 541).

Although we often associate fungi with decay and decomposition, many fungi, such as those cultivated by leaf-cutter ants, instead live by forming symbiotic associations with other organisms. The vast majority of plants obtain soil mineral nutrients via a symbiotic relationship with soil fungi. Humans have also harnessed the metabolic activities of certain fungi to obtain substances ranging from flavourful cheeses and wine to bread and therapeutic drugs such as penicillin and the immunosuppressant cyclosporine. And, as you know from previous chapters, species such as the yeast *Saccharomyces cerevisiae* and the mould *Neurospora crassa* have long been pivotal model organisms in studies of DNA structure and function and in the development of genetic engineering methods. On the other hand, fungi collectively are the single greatest cause of plant diseases, and many species cause disease in humans and other animals. Some even produce carcinogenic toxins.

Evidence suggests that fungi were present on land at least 760 mya and possibly much earlier. Their presence on land was likely crucial for the successful colonization of land by plants, which relied on symbiotic associations with the fungi to obtain nutrients from the nutrient-poor soils of early land environments. In the course of the intervening millennia, evolution equipped fungi with a remarkable ability to break down a wide range of compounds, ranging from living and dead organisms and animal wastes to groceries, clothing, paper, and wood—even photographic film. Along with heterotrophic bacteria, they have become Earth's premier decomposers **(Figure 25.2).** Despite their profound impact on ecosystems and other life forms, most of us have only a passing acquaintance with the fungi—perhaps limited to the mushrooms on our pizza or the invisible but annoying types that cause skin infections, such as athlete's foot. This chapter provides you with an overview of fungal biology. We begin with the features that set fungi apart from all other organisms and discuss the diversity of fungi existing today before revisiting associations between fungi and other organisms.

25.1 General Characteristics of Fungi

We begin our survey of fungi by examining the features that distinguish fungi from other forms of life, how fungi obtain nutrients, and adaptations for reproduction and growth that enable fungi to spread far and wide through the environment.

Fungi are heterotrophic eukaryotes that obtain carbon by breaking down organic molecules synthesized by other organisms. Although all fungi are heterotrophs, fungi can be divided into two broad groups based on how they obtain carbon. If a fungus obtains carbon from nonliving material, it is a **saprotroph.** Fungi that decompose dead plant and animal tissues, for example, are saprotrophs. If a fungus obtains carbon from living organisms, it is a **symbiont.** Symbiosis is the living together of two (or sometimes more)

Nancy Kennedy/Shutterstock.com

Figure 25.2
Example of a wood decay fungus: sulfur shelf fungus (*Polyporus*).

PEOPLE BEHIND BIOLOGY 25.1

Cameron Currie, *University of Wisconsin-Madison*

Experimental Research Box

Discovery has been defined as "seeing what everyone else has seen and thinking what no one else has thought." Even though ant–fungal mutualism has been studied since 1874, in 1999 a graduate student at the University of Toronto, Cameron Currie, discovered a whole new dimension to this mutualism. For years, researchers studying this symbiosis had wondered how the ants kept their gardens free of competing fungi. The conditions created by the ants are ideal for many other fungi besides the garden fungus, and, as you know, fungal spores are everywhere—yet the ant gardens are pure monocultures of a single fungus. What prevents other fungi from invading the gardens? Biologists had thought that the ants kept other fungi out simply by weeding the gardens, removing all traces of invading fungi, and so keeping fungal competitors at bay. But Currie discovered that there is a third

symbiont at work, and this organism keeps out other fungi. Currie noticed that ants had a whitish substance on their abdomens; other researchers had previously noticed this crust as well but assumed it was just part of the ant's exoskeleton **(Figure 1a).** When Currie took a closer look, he discovered that the crust was actually a bacterium of the genus *Pseudonocardia;* these bacteria are actinomycetes, which are known to produce antibiotics (e.g., streptomycin). On further exploration, Currie found that this bacterium produced an antibiotic that specifically and completely inhibited growth of a parasitic fungus, *Escovopsis*, which is the greatest threat to the gardens. If *Escovopsis* isn't stopped, it will overgrow the desirable fungus and take over the garden. This groundbreaking research, demonstrating that mutualisms do not necessarily involve just two species, was done while Currie was still a student. Since

finishing his Ph.D., Currie has gone on to show that the ants' bodies have changed over evolutionary time to create and maintain a favourable environment for their bacteria. The bacteria live in specialized crypts on the ant's body that are associated with glands; secretions from these glands provide nutrients for the bacteria (Figure 1b, c). So not only did ants invent agriculture long before humans, but also they used microbes to produce antibiotics long before we thought of doing so. An interesting aspect of antibiotic use by the ants is that even though the parasitic fungus has been exposed to the antibiotic produced by the bacterium for a very long time, it has not become resistant to the antibiotic. Perhaps we can learn something from the ants about preventing antibiotic resistance, which has rendered many of the antibiotics we rely on ineffective (you read about antibiotic resistance in Chapter 22).

a.

b.

c.

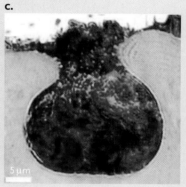

Figure 1

(a) *Leaf-cutter ant showing bacterial "crust."* **(b)** *Crypts (white spots) on the ant's body house bacteria.* **(c)** *A crypt on the body of a leaf-cutter ant.*

Photos a–c: From Cameron R. Currie, Michael Poulsen, John Mendenhall, Jacobus J. Boomsma, and Johan Billen, "Coevolved crypts and exocrine glands support mutualistic bacteria in fungus-growing ants," *Science*, Vol. 311, 6 January 2006, pp. 81–83. Reprinted with permission from AAAS.

organisms for extended periods; symbiotic relationships range along a continuum from **parasitism**, in which one organism benefits at the expense of the other, to **mutualism**, in which both organisms benefit. Although we often think of fungi as decomposers, fully half of all identified fungi live as symbionts with another organism.

Regardless of their nutrient source, fungi feed by **absorptive nutrition:** they secrete enzymes into their environment, breaking down large molecules into smaller soluble molecules that can then be absorbed into their cells. This mode of nutrition means that fungi cannot be stationary, as they would then deplete all of the food in their immediate environment.

a. Fungal hyphae

b. Mycelium and fruiting body of a mushroom-forming fungus

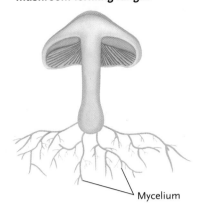

Mycelium

c. Mycelium on leaf litter

Figure 25.3

(a) Fungal hyphae forming a mycelium in culture in a Petri dish. **(b)** Sketch of the mycelium of a mushroom-forming fungus, which consists of branching septate hyphae. **(c)** Mycelium on leaf litter.

Instead, fungi have evolved the ability to proliferate quickly through their environment, digesting nutrients as they grow. How can fungi proliferate so quickly? Although some fungi grow as unicellular yeasts, which reproduce asexually by **budding** or binary fission (see Figure 25.14, p. 549), most are composed of **hyphae** ("web"; singular, *hypha*) **(Figure 25.3a)**, fine filaments that spread through whatever substrate the fungus is growing in—soil, decomposing wood, your skin—forming a network or **mycelium** (Figure 25.3b, c). Hyphae are essentially tubes of cytoplasm surrounded by cell walls made of chitin, a polysaccharide also found in the exoskeletons of insects and other arthropods.

Hyphae grow only at their tips, but because a single mycelium contains many, many tips, the entire mycelium grows outward very quickly. Together, this **apical growth** and absorptive nutrition account for much of the success of fungi. As the hyphal tips extend, they exert a mechanical force, allowing them to push through their substrate, releasing enzymes and absorbing nutrients as they go. Fungal species differ in the particular digestive enzymes they synthesize, so a substrate that is a suitable food source for one species may be unavailable to another. Although there are exceptions, fungi typically thrive only in moist environments, where they can directly absorb water, dissolved ions, simple sugars, amino acids, and other small molecules. When some of a mycelium's hyphal filaments contact a source of food, growth is channelled in the direction of the food source.

Nutrients are absorbed at the porous tips of hyphae; small atoms and molecules pass readily through these tips, and then transport mechanisms move them through the underlying plasma membrane. Some hyphae have regular cross-walls or **septa** ("fences" or "walls"; singular, *septum*) that separate a

hypha into compartments, whereas others lack septa and are effectively one large cell **(Figure 25.4)**. But even septate hyphae should be thought of as interconnected compartments rather than separate cells, as all septa have pores that allow cytoplasm and, in some fungi, even nuclei and other large organelles to flow through the mycelium. By a mechanism called *cytoplasmic streaming* (flowing of cytoplasm and organelles around a cell or, in this case, a mycelium), nutrients obtained by one part of a mycelium can be translocated to other nonabsorptive regions, such as reproductive structures.

When a fungus releases enzymes into its substrate, it faces competition from bacteria and other organisms for the nutrients that are now available. How can a fungus prevent these competitors from stealing the nutrients that it has just expended energy and resources to obtain? Many fungi produce antibacterial compounds and toxins that inhibit the growth of competing organisms. Many of these compounds are **secondary metabolites**, which are not required for day-to-day survival but are beneficial to the fungus. As we will see, many of these compounds not only are important in the life of a fungus but also benefit organisms

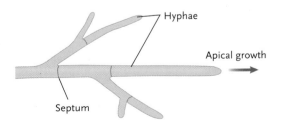

Hyphae

Apical growth

Septum

Figure 25.4

Septa. In some fungi, septa divide each hypha into separate compartments.

associated with the fungus. Many are also of commercial or medical importance to humans; for example, the antibiotic penicillin is a secondary metabolite produced by a species of *Penicillium*.

Fungi reproduce by spores, and this spore production can be amazingly prolific, with some species of fungi producing billions of spores per day **(Figure 25.5)**. These spores are microscopic, featherlight, and able to survive in the environment for extended periods after they are released. Reproducing via such spores allows fungi to be opportunists, germinating only when favourable conditions exist and quickly exploiting food sources that occur unpredictably in the environment. Releasing vast numbers of spores, as some fungi do, improves the odds that the spores will germinate and produce a new individual.

Spores can be produced asexually or sexually; some fungi produce both asexual and sexual spores at different stages of their lives. Sexual reproduction in fungi is quite complex. In all organisms, sexual reproduction involves three stages: the fusion of two haploid cells (**plasmogamy**), bringing together their two nuclei in one common cytoplasm; this cytoplasmic fusion is usually quickly followed by nuclear fusion (**karyogamy**) in most organisms; nuclear fusion is followed by meiosis to produce genetically distinct haploid cells. As we will see, fungi are unique in that these events can be separated in time for durations ranging from seconds to many years.

STUDY BREAK

1. What physical features distinguish fungi from other organisms?
2. How do fungi reduce competition for resources?
3. By what means do fungi reproduce? Why is this mode of reproduction advantageous?
4. What is the advantage of mycelial growth?

25.2 Evolution and Diversity of Fungi

25.2a Fungi Were Present on Earth by at Least 760 Million Years Ago

For many years, fungi were classified as plants because the earliest classification schemes had only two kingdoms, plants and animals. Fungi, like plants, have cell walls and do not move as animals did, so they were grouped with plants. As biologists learned more about the distinctive characteristics of fungi, however, it became clear that fungi should be treated as a separate kingdom.

CONCEPT FIX The idea that fungi are most closely related to plants has persisted, but this is a misconception! The discovery of chitin in fungal cells and recent comparisons of DNA and RNA sequences all indicate that fungi and animals are more closely related to each other than they are to other eukaryotes. The close biochemical relationship between fungi and animals may explain why fungal infections are typically so resistant to treatment and why it has proved rather difficult to develop drugs that kill fungi without damaging their human or other animal hosts. ◼

Analysis of the sequences of several genes suggests that the lineages leading to animals and fungi likely diverged between 760 mya and 1 billion years ago (bya). What were the first fungi like? We do not know for certain: phylogenetic studies indicate that fungi first arose from a single-celled, flagellated protist similar to choanoflagellates (see Chapter 24)—the sort of organism that does not fossilize well. Although traces of what may be fossil fungi exist in rock formations nearly 1 billion years old, the oldest fossils that we can confidently assign to the modern **kingdom Fungi** appear in rock strata laid down in the late Proterozoic (900–570 mya).

Figure 25.5
Spore production by fungal fruiting bodies. Some fruiting bodies can release billions of spores per day.

© Andrew Darrington / Alamy

Table 25.1

Summary of Fungal Phyla

Phylum	Body Type	Key Feature	
Chytridiomycota (chytrids)	One to several cells	Motile spores propelled by flagella; usually asexual	
Zygomycota (zygomycetes)	Hyphal	Sexual stage in which a resistant zygospore forms for later germination	
Glomeromycota (glomeromycetes)	Hyphal	Hyphae associated with plant roots, forming arbuscular mycorrhizas	
Ascomycota (ascomycetes)	Hyphal	Sexual spores produced in sacs called asci	
Basidiomycota (basidiomycetes)	Hyphal	Sexual spores formed on club-shaped cells called basidia of a prominent fruiting body (basidiocarp)	

25.2b Once They Appeared, Fungi Radiated into Several Major Lineages

Most likely, the first fungi were aquatic. When other kinds of organisms began to colonize land, they may well have brought fungi along with them. For example, researchers have discovered what appear to be mycorrhizas—symbiotic associations of a fungus and a plant—in fossils of some of the earliest known land plants. The final section of this chapter examines mycorrhizas more fully.

Over time, fungi diverged into the strikingly diverse lineages that we consider in the rest of this section **(Table 25.1).** Today, there are over 60 thousand described species of fungi, with at least 1.6 million more that have not yet been described.

As the lineages diversified, different adaptations associated with reproduction arose. For example, you'll notice that the structures in which sexual spores are formed and mechanisms by which spores are dispersed became larger and more elaborate over evolutionary time. Traditionally, therefore, biologists have classified fungi primarily by the distinctive structures produced in sexual reproduction. These features are

still useful indicators of the phylogenetic standing of a fungus, but the powerful tools of molecular analysis are bringing many revisions to our understanding of the evolutionary journey of fungi.

The evolutionary origins and lineages of fungi have been obscure ever since biologists began puzzling over the characteristics of this group. With the advent of molecular techniques for research, these topics have become extremely active and exciting areas of biological research that may shed light on fundamental events in the evolution of all eukaryotes. Currently, we recognize five phyla of fungi, known formally as the Chytridiomycota, Zygomycota, Glomeromycota, Ascomycota, and Basidiomycota **(Figure 25.6).** However, we know now that two of these phyla, the chytridiomycota and the zygomycota, are not monophyletic (i.e., they are taxa that do not contain only one ancestor and all of its descendants), so the classification scheme presented in Figure 25.6 will soon change to reflect this new information. Why do classifications of organisms change so often? Bear in mind that classification schemes such as those presented here are hypotheses that explain our best understanding of evolutionary relationships among organisms at any one time; like any other hypotheses, classification schemes are open

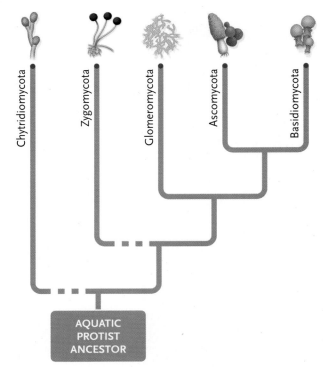

Figure 25.6

A phylogeny of fungi. This scheme represents a widely accepted view of the general relationships between major groups of fungi, but it may well be revised as new molecular findings provide more information. The dashed lines indicate that two groups, the chytrids and the zygomycetes, are probably paraphyletic—they include subgroups that are not all descended from a single ancestor.

to revision as we find out more about the organisms. Molecular data also suggest that some other eukaryotic organisms currently classified elsewhere may actually be fungi; we haven't included those organisms in this chapter but instead will focus on the groups of fungi that are best understood. Even though fungal classification will change greatly over the next few years, we summarize the major phyla recognized today as a way of illustrating the diversity of this group of organisms.

Phylum Chytridiomycota. The Chytridiomycota are likely the most ancient group of fungi, as they retain several traits characteristic of an aquatic lifestyle. For example, chytrids (as they are commonly called) are the only fungi that produce flagellated, motile spores **(Figure 25.7a);** these spores use chemotaxis (movement in response to a chemical gradient) to locate suitable substrates. Chytrids live in soil or freshwater habitats, wherever there is at least a film of water through which their motile spores can swim.

Most chytrids are saprotrophs, organisms that obtain nutrients by breaking down dead organic matter, although some are symbionts in the guts of cattle and other herbivores, where they break down cellulose to provide carbon for their hosts, and still others are parasites of animals, plants, algae, or other fungi. These tiny fungi also cause a disease, chytridiomycosis, that is one cause of the decline in amphibian species worldwide. Globally, at least 43% of all amphibian species are declining in population, and nearly 33% are threatened with extinction. Although many factors contribute to amphibian decline, including habitat loss, fragmentation, and increasing levels of environmental pollutants, chytridiomycosis has been linked to the decline of amphibian populations in Australia, New Zealand, central and South America, and parts of Europe. This disease has wiped out an estimated two-thirds of the

species of harlequin frogs (*Atelopus*) in the American tropics (Figure 25.7b). The epidemic has correlated with the rising average temperature in the frogs' habitats, an increase credited to global warming. Studies show that the warmer environment provides optimal growing temperatures for the chytrid pathogen. How does infection by a chytrid kill these animals? The fungus colonizes the skin of amphibians (Figure 25.7b), which interferes with the electrolyte balance and functioning of organs because amphibians take up water and exchange gases through their skins.

Most chytrids are unicellular, although some live as chains of cells and have rhizoids (branching filamentous extensions) that anchor the fungus to its substrate and that may also absorb nutrients from the substrate (Figure 25.7a). The vegetative stage of most chytrids is haploid; asexual reproduction involves the formation of a **sporangium**, in which motile spores are formed. A few chytrids reproduce sexually, via male and female gametes that fuse to form a diploid zygote. This cell may form a mycelium that gives rise to sporangia, or it may directly give rise to either asexual or sexual spores.

Zygomycota. This group of fungi includes the moulds on fruit and bread familiar to many of us and takes its name from the structure formed in sexual reproduction, the **zygospore (Figure 25.8, p. 546).** Many zygomycetes are saprotrophs that live in soil, feeding on organic matter. Their metabolic activities release mineral nutrients in forms that plant roots can take up. Some zygomycetes are parasites of insects (and even other zygomycetes), and some wreak havoc on human food supplies, spoiling stored grains, bread, fruits, and vegetables **(Figure 25.9, p. 546).** Others, however, have become major players in commercial enterprises, where they are used in manufacturing products that range from industrial pigments to pharmaceuticals

a. Chytridiomycosis in a frog

Skin surface

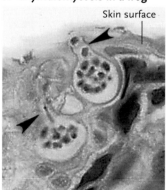

Centers for Disease Control and Prevention

b. Harlequin frog

Pedro Bernardo/Shutterstock.com

Figure 25.7

Chytrids. **(a)** Chytridiomycosis, a fungal infection, shown here in the skin of a frog. The two arrows point to flask-shaped spore-producing cells of the parasitic chytrid *Batrachochytrium dendrobatis*, which has devastated populations of harlequin frogs **(b)**.

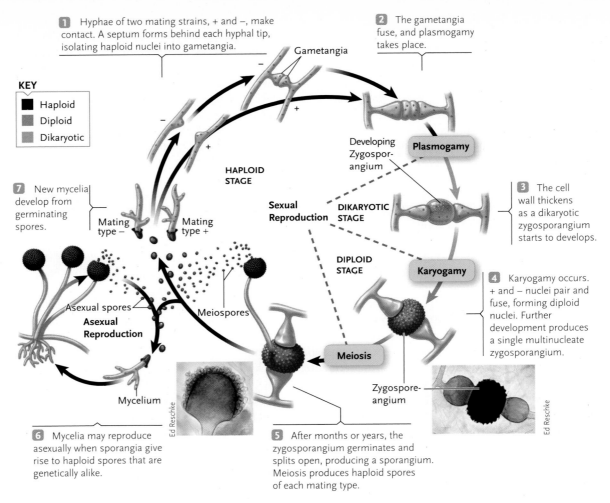

1 Hyphae of two mating strains, + and −, make contact. A septum forms behind each hyphal tip, isolating haploid nuclei into gametangia.

2 The gametangia fuse, and plasmogamy takes place.

KEY

- ■ Haploid
- ■ Diploid
- ■ Dikaryotic

Gametangia

HAPLOID STAGE

7 New mycelia develop from germinating spores.

Developing Zygospor-angium

Plasmogamy

Sexual Reproduction

DIKARYOTIC STAGE

3 The cell wall thickens as a dikaryotic zygosporangium starts to develops.

Mating type −

Mating type +

DIPLOID STAGE

Karyogamy

4 Karyogamy occurs. + and − nuclei pair and fuse, forming diploid nuclei. Further development produces a single multinucleate zygosporangium.

Asexual spores

Meiospores

Asexual Reproduction

Meiosis

Zygospore-angium

Mycelium

Ed Reschke

Ed Reschke

6 Mycelia may reproduce asexually when sporangia give rise to haploid spores that are genetically alike.

5 After months or years, the zygosporangium germinates and splits open, producing a sporangium. Meiosis produces haploid spores of each mating type.

Figure 25.8

Life cycle of the bread mould *Rhizopus stolonifer*, a zygomycete. Asexual reproduction is common, but different mating types (+ and −) also reproduce sexually. In both cases, haploid spores are formed and give rise to new mycelia.

Figure 25.9

Zygomycete fungus growing on strawberries.

humbak/Shutterstock.com

such as steroids (e.g., anti-inflammatory drugs). Zygo-mycetes are also used in the production of fermented foods such as tempeh.

Most zygomycetes consist of a haploid mycelium that lacks regular septa, although some groups have septa, and in others, septa form to wall off reproductive structures and aging regions of the mycelium. Sexual reproduction occurs when mycelia of different **mating types** (known as + and − types, rather than male and female) produce specialized hyphae that grow toward each other and form sex organs (**gametangia**) at their tips (see Figure 25.8, steps 1 and 2). How do the

gametangia find each other? Pheromones secreted by each mycelium stimulate the development of sexual structures in the complementary strain and cause gam-etangia to grow toward each other. The gametangia fuse, forming a thick-walled structure, a **zygosporangium** (see Figure 25.8, step 3), which can remain dormant for months or years, allowing the zygomycete to survive unfavourable environmental conditions. Eventually, meiosis occurs in the zygospo-rangium, forming a meiosporangium that will produce haploid spores by meiosis. (see Figure 25.8, step 5). Note that meiosis does not always produce gametes! We often tend to characterize meiosis as the formation of gametes, probably because we are so familiar with how sexual reproduction occurs in humans and other animals. But in many organisms, such as fungi and plants, meiosis results in the formation of haploid spores.

Like other fungi, however, zygomycetes also repro-duce asexually, as shown in steps 6 and 7 of Figure 25.8. When a haploid spore lands on a favourable substrate, it germinates and gives rise to a branching mycelium.

a. Sporangia of Rhizopus nigricans

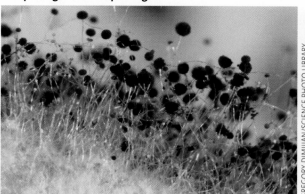

GREGORY DIMIJIAN/SCIENCE PHOTO LIBRARY

b. Sporangia (dark sacs) of *Pilobolus*

POWER AND SYRED/SCIENCE PHOTO LIBRARY

500 µm

Figure 25.10
Two of the numerous strategies for spore dispersal by zygomycetes. **(a)** The sporangia of *Rhizopus stolonifer,* shown here on a slice of bread, release powdery spores that are easily dispersed by air currents. **(b)** In *Pilobolus,* the spores are contained in a sporangium (the dark sac) at the end of a stalked structure. When incoming rays of sunlight strike a light-sensitive portion of the stalk, turgor pressure (pressure against a cell wall due to the movement of water into the cell) inside a vacuole in the swollen portion becomes so great that the entire sporangium may be ejected outward as far as 2 m—a remarkable feat given that the stalk is only 5 to 10 mm tall.

Some of the hyphae grow upward, and saclike sporangia form at the tips of these aerial hyphae. Inside the sporangia, the asexual cycle comes full circle as new haploid spores arise through mitosis and are released.

The black bread mould *Rhizopus stolonifer* may produce so many charcoal-coloured sporangia in asexual reproduction **(Figure 25.10a)** that mouldy bread looks black. The spores released are lightweight, dry, and readily wafted away by air currents. In fact, winds have dispersed *R. stolonifer* spores just about everywhere on Earth, including the Arctic. Another zygomycete, *Pilobolus* (Figure 25.10b), forcefully spews its sporangia away from the dung in which it grows. A grazing animal may eat a sporangium on a blade of grass; the spores then pass through the animal's gut unharmed and begin the life cycle again in a new dung pile.

Glomeromycota. Until recently, fungi in the phylum Glomeromycota were classified as zygomycetes based on morphological similarities such as the lack of regular septa. However, these fungi are quite dissimilar to zygomycetes in many ways—for example, sexual reproduction is unknown in this group of fungi, with spores usually forming asexually simply by walling off a section of a hypha **(Figure 25.11)**—causing many

a. Arbuscules (black) in leek root colonized by arbuscular mycorrhizal fungus

Science VU/R. Hussey/Visuals Unlimited, Inc.

b. Arbuscules inside root

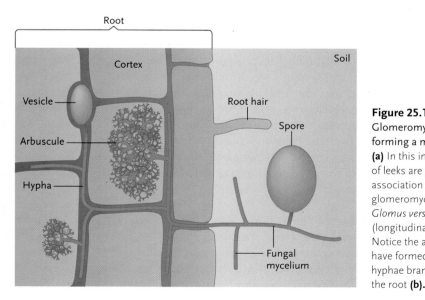

Root

Soil

Cortex

Vesicle

Root hair

Arbuscule

Spore

Hypha

Fungal mycelium

Figure 25.11
Glomeromycete fungus forming a mycorrhiza. **(a)** In this instance, the roots of leeks are growing in association with the glomeromycete fungus *Glomus versiforme* (longitudinal section). Notice the arbuscules that have formed as fungal hyphae branch after entering the root **(b)**.

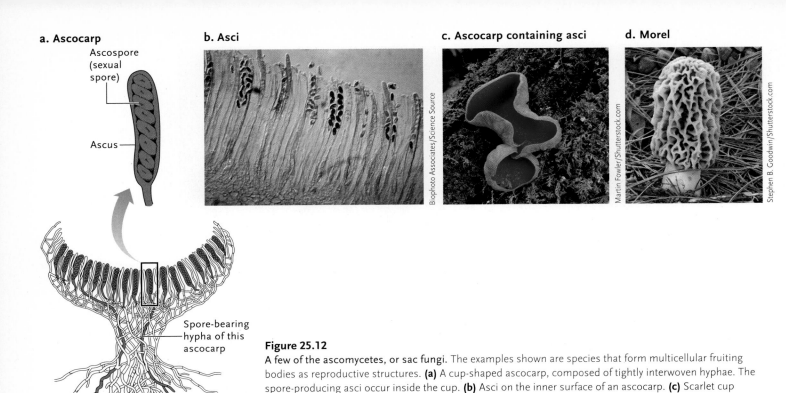

a. Ascocarp

Ascospore (sexual spore)

Ascus

Spore-bearing hypha of this ascocarp

b. Asci

Biophoto Associates/Science Source

c. Ascocarp containing asci

Martin Fowler/Shutterstock.com

d. Morel

Stephen B. Goodwin/Shutterstock.com

Figure 25.12

A few of the ascomycetes, or sac fungi. The examples shown are species that form multicellular fruiting bodies as reproductive structures. **(a)** A cup-shaped ascocarp, composed of tightly interwoven hyphae. The spore-producing asci occur inside the cup. **(b)** Asci on the inner surface of an ascocarp. **(c)** Scarlet cup fungus (*Sarcoscypha*). **(d)** A true morel (*Morchella esculenta*), a prized edible fungus.

researchers to question the inclusion of these fungi in the phylum Zygomycota. Recent evidence from molecular studies resulted in these fungi being placed in their own phylum.

The 160 known members of this phylum are all specialized to form **mycorrhizas**, or symbiotic associations with plant roots. This group of fungi has a tremendous ecological importance as they collectively make up roughly half of the fungi in soil and form mycorrhizas with many land plants, including most major crop species, such as wheat and maize. Mycelia of these fungi colonize the roots of host plants and also proliferate in the soil around the plants. Inside the roots, hyphae penetrate cell walls and branch repeatedly to form **arbuscules** ("little trees") (see Figure 25.11). The branches of each arbuscule are enfolded by the cell's plasma membrane, forming an interface with a large surface area through which nutrients are exchanged between the plant and the fungus. Some glomeromycetes also form vesicles inside roots, which store nutrients and can also act as spores. The fungus obtains sugars from the plant and in return provides the plant with a steady supply of dissolved minerals that it has obtained from the surrounding soil. We take a closer look at mycorrhizas in Section 25.3.

Ascomycota. The phylum Ascomycota takes its name from the saclike structures (**asci**; singular, *ascus*) in which spores are formed in sexual reproduction. These asci are often enclosed in a fruiting body (**ascocarp**) **(Figure 25.12a, b, c)**. However, some ascomycetes are

yeasts or filamentous fungi with a yeast stage, which reproduce asexually by budding or binary fission (see Figure 25.14). Ascomycetes are much more numerous than chytrids, zygomycetes, or glomeromycetes, with more than 30 thousand identified species.

Some ascomycetes are very useful to humans. One species, the orange bread mould *Neurospora crassa*, has been important in genetic research, including the elucidation of the one gene–one enzyme hypothesis (see Chapter 13). *Saccharomyces cerevisiae*, which produces the ethanol in alcoholic beverages and the carbon dioxide that leavens bread, is also a model organism used in genetic research. By one estimate, it has been the subject of more genetic experiments than any other eukaryotic microorganism. This multifaceted phylum also includes gourmet delicacies such as truffles (*Tuber melanosporum*) and the succulent morel *Morchella esculenta* (Figure 25.12d).

Many ascomycetes are saprotrophs, playing a key role in the breakdown of cellulose and other polymers. Ascomycetes are also common in symbiotic associations, forming mycorrhizas and lichens (see Section 25.3). A few ascomycetes prey on various agricultural insect pests—some are even carnivores that trap their prey in nooses **(Figure 25.13a)**—and thus have potential for use as biological pesticides.

However, other ascomycetes are devastating plant pathogens, including the blue-stain fungi that are associated with mountain pine beetles and contribute to the death of beetle-infested trees (Figure 25.13b). Several ascomycetes can be serious pathogens of humans.

a. A trapping ascomycete

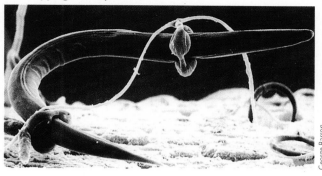

George Barron

b. Stump of pine tree infected with blue-stain fungus

U.S. Forest Service

Figure 25.13

(a) Nematode-trapping fungus. Hyphae of this ascomycete (*Arthobotrys*) form nooselike rings. When a prey organism enters the loop, rapid changes in ion concentration draw water into the loop by osmosis. The increased turgor pressure causes the noose to tighten, trapping its prey. Enzymes produced by the fungus then break down the nematode's tissues. **(b)** Stump of a pine tree infected with blue-stain fungus; the fungus grows into the tree's water-conducting tissue, blocking the flow of water.

Yeast cells

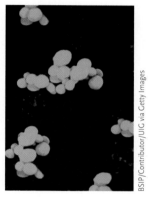

BSIP/Contributor/UIG via Getty Images

Figure 25.14

Candida albicans, the cause of yeast infections of the mouth and vagina.

The yeast *Candida albicans* **(Figure 25.14)** infects mucous membranes, especially of the vagina and mouth, causing a condition called thrush. Another yeast, *Pneumocystis jirovecii*, causes virulent pneumonia in AIDS patients and other immunocompromised people.

Claviceps purpurea, a parasite on rye and other grains, causes ergotism, a disease marked by vomiting; hallucinations; convulsions; and, in severe cases, gangrene and even death. It has even been suggested that this fungus was the cause of the Salem witch hunts of seventeenth century New England, as discussed in "Molecule behind Biology," 25.2, p. 550. Other ascomycetes cause nuisance infections, such as athlete's foot and ringworm.

Most ascomycetes grow as haploid mycelia with regular septa; large pores in the septa allow organelles, including nuclei, to move with cytoplasm through the mycelium. Sexual reproduction generally involves fusion of hyphae from mycelia of + and − mating types **(Figure 25.15, p. 551)**. The cytoplasms of the two

hyphae fuse, but fusion of the nuclei is delayed, resulting in the formation of **dikaryotic hyphae** that contain two separate nuclei and thus are referred to as $n + n$ rather than n or $2n$. Sacs (asci) form at the tips of these dikaryotic hyphae; inside the asci, the two nuclei fuse, forming a diploid zygote nucleus, which then undergoes meiosis to produce four haploid nuclei. Mitosis usually follows, resulting in the formation of eight haploid spores (**ascospores**).

Unlike zygomycetes, ascomycetes do not produce asexual spores in sporangia. Instead, modified hyphae produce numerous asexual spores called **conidia** ("dust"; singular, *conidium*), such as those seen when powdery mildew attacks grasses, roses, and other common garden plants **(Figure 25.16a, p. 551)**. The mode of conidial production varies from species to species, with some ascomycetes producing chains of conidia, whereas in others, the conidia are produced on a hypha in a series of "bubbles," rather like a string of detachable beads (Figure 25.16b). Either way, conidia are formed and released much more quickly than zygomycete spores.

As you can see from Figure 25.15, these asexual reproductive structures look very different from the sexual stages and are often not formed at the same time or under the same conditions as the sexual stage of the life cycle. These differences resulted in the asexual stages of many ascomycetes being classified as separate organisms from the sexual stages of the same species. Since fungal classification traditionally relied on features produced in sexual reproduction, these asexual stages could not be placed in any of the phyla; instead, researchers grouped them together in an artificial group called the Deuteromycota (also known as Fungi Imperfecti, or the "imperfect fungi"— imperfect meaning that a sexual stage is absent). Well-known examples of fungi once classified as deuteromycetes are *Penicillium* and *Aspergillus*. Certain species of *Penicillium* (Figure 25.16b) are the source

MOLECULE BEHIND BIOLOGY 25.2

Lysergic Acid

Was a fungus responsible for the Salem witch trials? In Salem, Massachusetts, in 1692–1693, several women were tried and found guilty of witchcraft. Their accusers were young women who had been experiencing bizarre symptoms: hallucinations, convulsions, a sensation of "prickling" of the skin, and even paralysis. Further evidence of witchcraft was the fact that cattle and other animals also suffered these symptoms. What was the real cause of these symptoms? Were they an example of mass hysteria? Or is there a biological explanation?

The symptoms reported by the "bewitched" girls match those of someone who has eaten flour made from wheat infected by the ascomycete fungus *Claviceps purpurea*. Ascospores of this fungus germinate when they land on the flower of a grass plant, such as wheat. The fungus grows quickly and, by the end of the growing season, forms a tough mass of hyphae known as a **sclerotium** in the seed head of the grass **(Figure 1)**. If the seed head isn't harvested, the sclerotia will fall to the ground, where they remain over winter. In the spring, the sclerotia will germinate, producing numerous fruiting bodies borne on stalks. However, if the fungus has infected a commercial grain crop, such as wheat, sclerotia are easily harvested along with the plants' seed heads and often end up being ground into flour along with the grain. In medieval times, if the weather favoured development of the fungus, up to 30% of some grain harvests were evidently not grain but sclerotia!

Figure 1
Sclerotium of Claviceps purpurea *in a grass seed head.*

Sclerotia produce many alkaloids, including lysergic acid **(Figure 2)**, which causes a range of symptoms, including hallucinations, convulsions, a sensation of ants crawling over the body, limb distortions, and dementia. These symptoms match those of the supposedly bewitched people of Salem in 1692. Further support for ergotism being the cause of the bewitching is the fact that most of the victims were adolescents, who are most suspectible to the effects of ergot alkaloids. Furthermore, the fact that cattle and other domestic animals would also have eaten infected grain and also presented the same symptoms as the "victims" suggests that ergot, not mass hysteria, was involved. Lysergic acid was purified in 1943 by a chemist (Albert Hofmann) to produce the

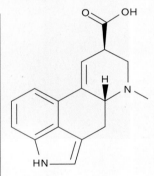

Figure 2
Structure of lysergic acid.

psychoactive drug LSD. Researchers hoped that this drug would be useful in psychotherapy, but its negative effects outweighed the benefits, and this line of research was dropped. However, other ergot alkaloids are used as treatment for migraine headaches.

of the penicillin family of antibiotics, whereas others produce the aroma and distinctive flavours of Camembert and Roquefort cheeses. Strains of *Aspergillus* grow in damp grain or peanuts. Their metabolic wastes, known as aflatoxins, can cause cancer in humans who eat the poisoned food over an extended period. With the development of molecular sequencing techniques, many fungi that were classified as deuteromycetes can now be reassigned to the appropriate phylum; most are ascomycetes, but some

are basidiomycetes, which also produce conidia in asexual reproduction.

Basidiomycota. The 24 thousand or so species of fungi in the phylum Basidiomycota include the mushroom-forming species, bracket fungi, stinkhorns, smuts, rusts, and puffballs **(Figure 25.17, p. 552)**. The common name for this group is club fungi, due to the club-shaped cells (**basidia**; singular, *basidium*) in which sexual spores are produced.

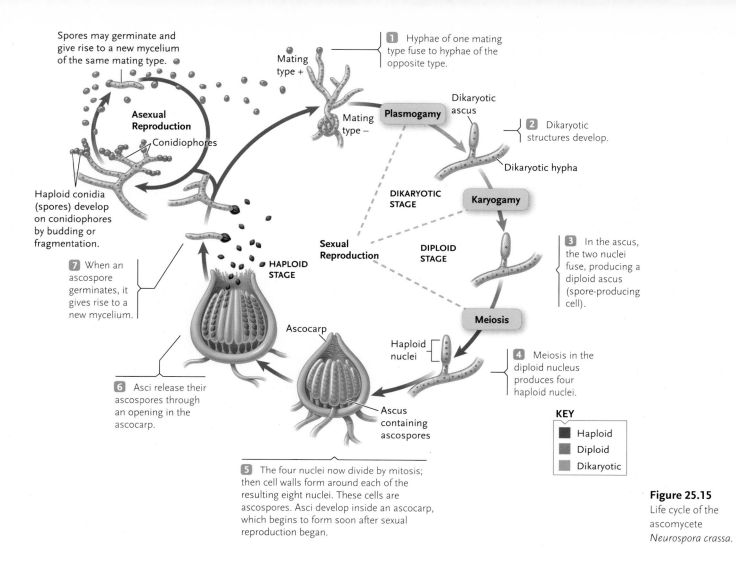

Spores may germinate and give rise to a new mycelium of the same mating type.

Asexual Reproduction

Conidiophores

1 Hyphae of one mating type fuse to hyphae of the opposite type.

Mating type +

Mating type −

Plasmogamy

Dikaryotic ascus

2 Dikaryotic structures develop.

Dikaryotic hypha

Haploid conidia (spores) develop on conidiophores by budding or fragmentation.

DIKARYOTIC STAGE

Sexual Reproduction

Karyogamy

HAPLOID STAGE

DIPLOID STAGE

3 In the ascus, the two nuclei fuse, producing a diploid ascus (spore-producing cell).

7 When an ascospore germinates, it gives rise to a new mycelium.

Meiosis

Ascocarp

Haploid nuclei

4 Meiosis in the diploid nucleus produces four haploid nuclei.

6 Asci release their ascospores through an opening in the ascocarp.

Ascus containing ascospores

KEY

■	Haploid
■	Diploid
■	Dikaryotic

5 The four nuclei now divide by mitosis; then cell walls form around each of the resulting eight nuclei. These cells are ascospores. Asci develop inside an ascocarp, which begins to form soon after sexual reproduction began.

Figure 25.15
Life cycle of the ascomycete *Neurospora crassa*.

Many basidiomycetes produce enzymes for digesting cellulose and lignin and are important decomposers of woody plant debris. Very few organisms can degrade lignin due to its very complex, irregular structure **(Figure 25.18, p. 552)**. The ability to degrade lignin also enables some basidiomycetes to break down complex organic compounds such as DDT, PCBs, and other persistent environmental pollutants that are structurally similar to lignin. Bioremediation of contaminated sites by these fungi is a very active research area.

A surprising number of basidiomycetes, including the prized edible oyster mushrooms (*Pleurotus ostreatus*), can also trap and consume small animals such as rotifers and nematodes by secreting paralyzing toxins or gluey substances that immobilize the prey, in a manner similar to that shown earlier for ascomycetes (Figure 25.13, p. 549). As is the case for insectivorous plants, such as the pitcher plants (*Saracenea purpurea*), discussed in Chapters 20 and 31, this adaptation gives the fungus access to a rich source of molecular nitrogen, an essential nutrient that is often scarce in terrestrial habitats. For example, the wood that is the substrate for many basidiomycetes is high in carbon but low in nitrogen; many wood-decay fungi have been found to be carnivorous, obtaining supplemental nitrogen from various invertebrates.

Some basidiomycetes form mycorrhizas with the roots of forest trees, as discussed later in this chapter. Recent research has shown that these mycorrhizas can be drawn into associations with achlorophyllous plants

a.

NIGEL CATTLIN/SCIENCE PHOTO LIBRARY

b.

Biophoto Associates/Science Source

Figure 25.16
(a) Powdery mildew on leaves. **(b)** Conidia of *Penicllium*. Note the rows of conidia (asexual spores) atop the elongate cells that produce them.

a. Coral fungus

b. Shelf fungus

c. White-egg bird's nest fungus

d. Fly agaric mushroom

e. Scarlet hood

Figure 25.17

Examples of basidiomycetes, or club fungi. **(a)** The light red coral fungus *Ramaria*. **(b)** The shelf fungus *Polyporus*. **(c)** The white-egg bird's nest fungus *Crucibulum laeve*. Each tiny "egg" contains spores. Raindrops splashing into the "nest" can cause "eggs" to be ejected, thereby spreading spores into the surrounding environment. **(d)** The fly agaric mushroom *Amanita muscaria*, which causes hallucinations. **(e)** The scarlet hood *Hygrophorus*.

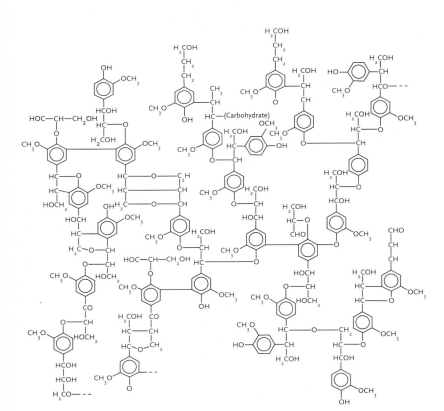

(plants that lack chlorophyll and so cannot carry out photosynthesis), which thus obtain nutrients from the trees via shared mycorrhizal fungi. Other basidiomycetes, the rusts and smuts, are parasites that cause serious diseases in wheat, rice, and other plants. Still others produce millions of dollars worth of the common edible button mushroom (*Agaricus bisporus*) sold in grocery stores. *Amanita muscaria* (Figure 25.17d) has been used in the religious rituals of ancient societies in Central America, Russia, and India. Other species of this genus, including the death cap mushroom *Amanita phalloides*, produce deadly toxins. The *A. phalloides* toxin, called α-amanitin, halts gene transcription, and hence protein synthesis, by inhibiting the activity of RNA polymerase. Within 8 to 24 hours of ingesting as little as 5 mg of the mushroom, vomiting and diarrhea begin. Later, kidney and liver cells start to degenerate; without intensive

Figure 25.18

A portion of a lignin molecule. Unlike most other biopolymers, lignin is not composed of regularly repeating monomers but instead is a complex polymer of various phenylpropane units, joined together by a range of diverse bonds, making it very difficult to degrade.

medical care, death can follow within a few days. You can read more about the effect of amanitin on gene expression in Chapter 13.

Most basidiomycetes are mycelial, although some grow as yeasts. The mycelium of many basidiomycetes contains two different, separate nuclei as a result of fusion between two different haploid mycelia and is termed a **dikaryon** ($n + n$) **(Figure 25.19)**. A dikaryotic mycelium is formed following fusion of the two haploid mycelia when both types of nuclei divide and migrate through the mycelium such that each hyphal compartment contains two dissimilar nuclei.

Basidiomycete fungi can grow for most of their lives as dikaryon mycelia—a major departure from an ascomycete's short-lived dikaryotic stage. After an extensive mycelium develops, favourable environmental conditions trigger the formation of fruiting bodies (**basidiocarps**), in which basidia develop. A basidiocarp consists of tight clusters of hyphae; the feeding mycelium is buried in the substrate. The shelf-like bracket fungi visible on trees are basidiocarps, as are the structures we call mushrooms and toadstools. Each mushroom is a short-lived reproductive body consisting of a stalk and a cap; basidia develop on "gills," the sheets of tissue on the underside of the cap. Inside each basidium, the two nuclei fuse; meiosis follows, resulting in the formation of four haploid **basidiospores** on the outside of the basidium (see Figure 25.19). Why does the fungus expend energy and resources on such elaborate spore-dispersal structures? A layer of still air

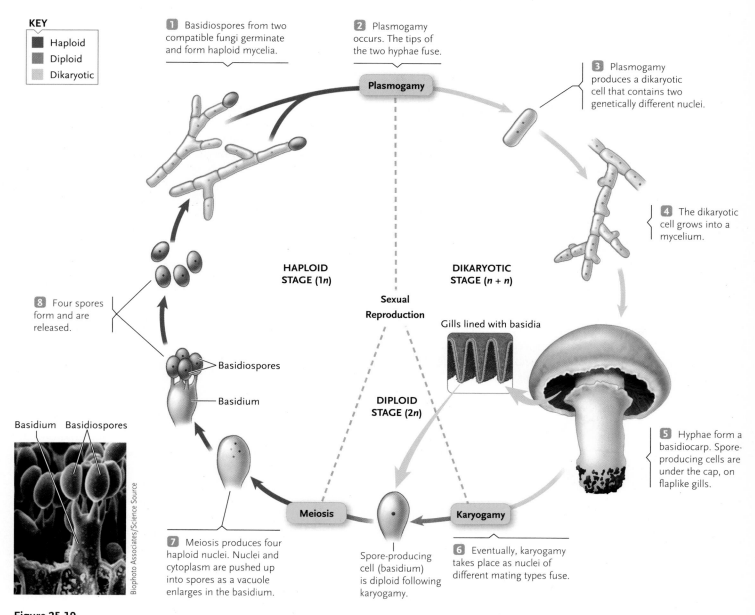

Figure 25.19
Generalized life cycle of the basidiomycete *Agaricus bisporus*, a species known commonly as the button mushroom. During the dikaryotic stage, cells contain two genetically different nuclei, shown here in different colours. Inset: Micrograph showing basidia and basidiospores.

occurs just above the ground (and any other surface); by elevating the basidia above this layer, the fungus increases the likelihood that its spores will be carried away by the wind.

CONCEPT FIX People often assume, when they see mushrooms sprouting from the ground, that each mushroom is an individual. But that's not true—mushrooms and other fungal fruiting bodies are produced by mycelia growing through their substrate. A mycelium, not a mushroom or fruiting body, is the "individual." ⬡

The prolonged dikaryon stage in basidiomycetes allows them many more opportunities for producing sexual spores than in ascomycetes, in which the dikaryon state is short lived. Basidia can produce huge numbers of spores—some species can produce 100 million spores *per hour* during reproductive periods, day after day! Basidiomycete mycelia can live for many years and spread over large areas. The largest organism on Earth could be the mycelium of a single individual of the basidiomycete *Armillaria ostoyae*, which spreads over 8.9 km^2 of land in eastern Oregon. This organism weighs at least 150 tonnes and is likely at least 2400 years old, making it not only the largest but also one of the heaviest and oldest organisms on Earth.

As for ascomycetes, asexual reproduction in basidiomycetes involves formation of conidia or budding in yeast forms such as *Cryptococcus gattii,* which causes cryptococcal disease in humans. A virulent strain of *C. gattii*, first reported from Vancouver Island in 1999, has since spread to the northwestern United States and California. Normally, only people with weakened immune systems, such as transplant recipients and cancer patients, are at risk from fungal pathogens, but *C. gattii* is different, causing disease in healthy people. The disease starts when spores of the fungus, which lives in trees and soil, are inhaled. In the lungs, the spores germinate to produce yeast cells that proliferate by budding in the warm, moist lung environment; the yeast cells then spread to the central nervous system via the bloodstream. The disease is characterized by a severe cough, fever, and, if the nervous system is affected, seizures and other neurological symptoms.

STUDY BREAK

1. What evidence is there that fungi are more closely related to animals than to plants?
2. Name the five phyla of the kingdom Fungi, and describe the reproductive adaptations that distinguish them.
3. What are the two main differences between asexual spores produced by zygomycetes and asexual spores produced by ascomycetes?
4. Fungi reproduce sexually or asexually, but for many species, the life cycle includes an unusual stage not seen in other organisms. What is this genetic condition, and what is its role in the life cycle?

25.3 Fungal Lifestyles

As mentioned earlier, fungi can be categorized as saprotrophs or symbionts, depending on whether they obtain nutrients from living organisms or from dead organic matter. It is important to remember that the categories of *saprotroph* and *symbiont* were created as separate categories to classify fungi, but fungi are very versatile organisms, and many fungi are capable of acting as both symbionts and saprotrophs at different times or under different conditions. Most people are more familiar with the role of fungi as saprotrophs (decomposers) rather than as symbionts, so in this section, we take a brief look at saprotrophy and then spend more time looking at fungal symbioses.

25.3a Some Fungi Are Saprotrophs

With their adaptations for efficient extracellular digestion, fungi are masters of the decay so vital to terrestrial ecosystems (see Figure 25.2, p. 540). For instance, in a single autumn, one elm tree can shed 200 kg of withered leaves! Without the metabolic activities of saprotrophic fungi and other decomposers such as bacteria and other organisms (e.g., earthworms), natural communities would rapidly become buried in their own detritus (dead organic matter). Even worse, without decomposers to break down this detritus, the soil would become depleted of nutrients, making further plant growth impossible. As fungi (and other decomposers) digest the dead tissues of other organisms, they also make a major contribution to the recycling of the chemical elements those tissues contain. For instance, over time, the degradation of organic compounds by saprotrophic fungi helps return key nutrients such as nitrogen and phosphorus to ecosystems. But the prime example of this recycling virtuosity involves carbon. The respiring cells of fungi and other decomposers give off carbon dioxide, liberating carbon that would otherwise remain locked in the tissues of dead organisms. Each year, this activity recycles a vast amount of carbon to plants, the primary producers of nearly all ecosystems on Earth.

However, there is a downside to the impressive enzymatic abilities of saprotrophic fungi; for example, when they decompose materials that are part of our houses, they can cause major economic and health problems. Fungi growing on wood and drywall following flooding or water damage to a building **(Figure 25.20a)** not only weaken the structural integrity of the building but also can be health hazards. The airborne spores of these fungi act as allergens, and some can also cause more serious health problems—for example, some fungi can colonize and grow in sinus cavities. Another example is dry rot, which causes millions of dollars in damage to buildings in Europe, Asia, and Australia (Figure 25.20b). Dry rot is notorious not only because it causes widespread and

Figure 25.20
(a) Mould growth following flooding. **(b)** Mycelium of dry rot (*Serpula lacrymans*) emerging through a wall.

costly damage but also because the responsible fungus, *Serpula lacrymans,* seems to have the mysterious ability to break down dry wood completely, which should not be possible—as described above, wood decay usually happens once wood becomes wet. Does this fungus really have the amazing ability to break down dry wood? In fact, this fungus is as dependent on water for growth as any other, but it can form specialized mycelial cords, which very efficiently transport water and nutrients over long distances through concrete, bricks, and other unfavourable substrates until the fungus at last finds wood. Then the mycelial cords release water into the substrate, allowing the fungus to spread through the wood and begin the process of decay.

25.3b Some Fungi Are Symbionts

Symbiotic associations range from mutualism, in which both partners benefit, to parasitism, in which one partner benefits at the expense of the other. Many fungal parasites are pathogens, parasites that cause disease symptoms in their hosts. We have discussed several examples of fungal diseases in humans and other animals earlier in this chapter. In this section, we will focus on fungi as mutualists. Many fungi are partners in mutually beneficial interactions with animals or photosynthetic organisms; some of these associations shaped the evolution of life on Earth and still play major roles in the functioning of ecosystems today. Chapter 30 discusses the general features of symbiotic associations more fully; here we are interested in some examples of the symbioses fungi form with other organisms.

Lichens Are Associations between a Fungus and One or More Photosynthetic Organisms. A **lichen** is a compound organism formed by an association between a fungus, an ascomycete or sometimes a basidiomycete, and a green alga and/or a cyanobacterium. Lichens may grow as crusts on rocks, bark, or soil; as flattened leaflike forms; or as radially symmetrical cups, treelike structures, or hairlike strands **(Figure 25.21, p. 556)**. Lichens have vital ecological roles and important human uses. Lichens secrete acids that eat away at rock, breaking it down and converting it to soil that can support plants. Many animals, such as caribou (*Rangifer tarandus*), rely on lichens for their winter forage. Some environmental chemists monitor air pollution by monitoring lichens, most of which cannot grow in heavily polluted air because they cannot discriminate between pollutants and mineral nutrients present in the atmosphere. Just as they do for mineral nutrients, lichens efficiently absorb airborne pollutants and concentrate them in their tissues. Humans use lichens as sources of dyes and perfumes, as well as medicines. Lichen chemicals are currently being explored as a source of natural pesticides.

The fungus (called the **mycobiont**) makes up most of the body (**thallus**) of the lichen, with the photosynthetic partner (**photobiont**) usually confined to a thin layer inside the lichen thallus (see Figure 25.21a). Some lichens have a green algal photobiont inside the thallus and a cyanobacterial photobiont contained in "pockets" on or in the thallus. Because lichens are composite organisms, it may seem odd to talk of lichen "species," but biologists do give lichens binomial names, based on the mycobiont. More than 13 500 different lichen species are recognized, each a unique combination of a particular species of fungus and one or more species of photobiont. As you might expect for a compound organism made up of two (or even three) organisms, reproduction can be complicated: it is not enough for each organism to reproduce itself, because formation of a new lichen requires that both partners be dispersed and end up together. Many lichens reproduce asexually, by specialized fragments such as the **soredia** (singular, *soredium*), shown in Figure 25.21b. Each soredium consists of photobiont cells wrapped in hyphae; the soredia can be dispersed by water, wind, or passing animals.

Inside the thallus, specialized hyphae wrap around and sometimes penetrate photobiont cells, which become the fungus's sole source of carbon. Often the

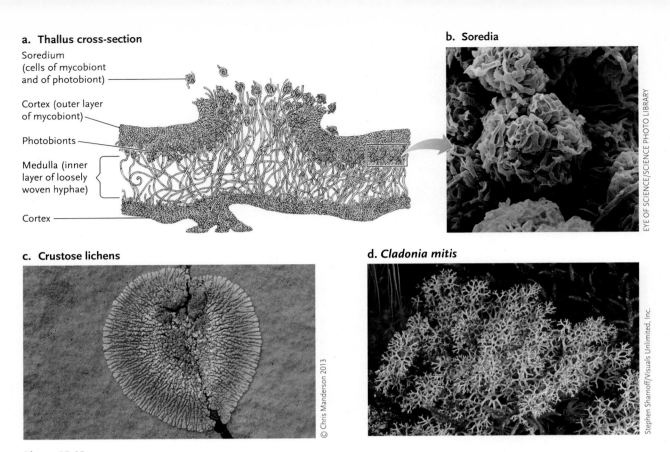

a. Thallus cross-section

Soredium
(cells of mycobiont
and of photobiont)

Cortex (outer layer
of mycobiont)

Photobionts

Medulla (inner
layer of loosely
woven hyphae)

Cortex

b. Soredia

EYE OF SCIENCE/SCIENCE PHOTO LIBRARY

c. Crustose lichens

© Chris Manderson 2013

d. *Cladonia mitis*

Stephen Sharnoff/Visuals Unlimited, Inc.

Figure 25.21

Lichens. **(a)** Diagram of a cross-section through the thallus of the foliose lichen *Lobaria verrucosa*. Soredia **(b),** which contain both hyphae and algal cells, are a type of dispersal fragment by which lichens reproduce asexually. **(c)** Crustose lichens. **(d)** *Cladonia mitis*, a branching, treelike lichen.

mycobiont absorbs up to 80% of the carbohydrates produced by the photobiont. Benefits for the photobiont are less clear cut, in part because the drain on nutrients hampers its growth and because the mycobiont often controls reproduction of the photobiont. In one view, many and possibly most lichens are parasitic symbioses, with the fungus enslaving the photobiont. On the other hand, although it is relatively rare to find a lichen photobiont species living independently in the same conditions under which the lichen survives, it may eke out an enduring existence as part of a lichen; some lichens have been dated as being more than 4000 years old! Studies have also revealed that at least some green algae clearly benefit from the relationship. Such algae are sensitive to desiccation and intense ultraviolet radiation. Sheltered by the lichen's fungal tissues, a green alga can thrive in locales where alone it would perish. Clearly, we still have much to learn about the physiological interactions between lichen partners.

Lichens often live in harsh, dry microenvironments, including on bare rock and wind-whipped tree trunks. Some lichens actually live *inside* rocks (see "Life on the Edge," Box 25.3, p. 558). Unlike plants, lichens do not control water loss from their tissues; instead, their water status reflects that of their environment,

and some lichens may dry out and re-wet several times a day. Lichens are very slow growing, even though the photobiont may have photosynthetic rates comparable to those of free-living species. What happens to all of the carbohydrates made in photosynthesis if they are not used to fuel growth? The mycobiont takes much of the carbohydrate made by the photobiont and uses it to synthesize secondary metabolites and other compounds that allow the lichen to survive the repeated wet–dry cycles and extreme temperatures common in their habitats. These compounds give lichens their vibrant colours and may also inhibit grazing on lichens by slugs and other invertebrates. The mycobiont uses other lichen chemicals to control the photobiont; some chemicals regulate photobiont reproduction, whereas others cause photobiont cells to "leak" carbohydrates to the mycobiont.

Mycorrhizas Are Symbiotic Associations between Fungi and Plant Roots. You might have learned in previous courses that plant roots are responsible for taking up soil nutrients. For most plants, however, this is not true: the roots of most plants are colonized by mycorrhizal fungi, which have mycelia that extend out into the soil far beyond the root zone of the plant and which take up most of the nutrients used by the plant **(Figure 25.22)**. Mycorrhizas, or "fungus roots," are mutualistic symbioses between certain soil-dwelling fungi and plant roots. Mycorrhizal plants greatly enhance the uptake of various nutrients, especially phosphorus and nitrogen, from soil (as discussed in Chapter 37) because the fungal mycelium has a tremendous surface area for absorbing mineral ions from a large volume of the surrounding soil.

As well, some mycorrhizal fungi can access sources of nutrients that are not available to plants; for example, certain basidiomycete fungi can penetrate directly into rocks and extract nutrients, which are then transported to their plant hosts. Other mycorrhizal associations involve the carnivorous basidiomycetes described above, which can obtain nitrogen by trapping and killing soil invertebrates and then transferring nitrogen from their prey to their host plants. By forming partnerships with these fungi, mycorrhizal plants gain access to nutrient sources that nonmycorrhizal plants do not. In exchange for soil nutrients, the plants provide the mycorrhizal fungi with sugars produced through photosynthesis. Mycorrhizas are generally mutualisms, representing a win–win situation for the partners. For plants that inhabit soils poor in mineral ions, such as in **tropical rain forests**, mycorrhizal associations are crucial for survival. Likewise, in temperate forests, species of spruce, oak, pine, and some other trees die unless mycorrhizal fungi are present **(Figure 25.23)**. There are at least seven different types of mycorrhizas, but the most common types are ectomycorrhizas and arbuscular mycorrhizas.

Arbuscular mycorrhizas are the oldest and most abundant type of mycorrhiza, formed by glomeromycete fungi and a wide range of plants, including nonseed plants and most flowering plants. In this type of mycorrhiza, fungal hyphae penetrate the cells of the root, forming arbuscules as described above (see Figure 25.11, p. 547). Fossils show that arbuscular mycorrhizas were common among ancient land plants, and some biologists have speculated that they might have been crucial for the colonization of land by plants by enhancing the transport of water and minerals to the plants.

Ectomycorrhizas evolved more recently and involve basidiomycetes and some ascomycetes. In these mycorrhizas, fungal hyphae form a sheath or mantle around a root (see Figure 25.22) and also grow between, but not inside, the root cells of their plant hosts. Ectomycorrhizal associations are very common with trees, such as the conifers of Canada's **boreal forest** and coastal rain forests. The extensive root system of a single mature pine may be studded with ectomycorrhizas involving dozens of fungal species.

a. Mycorrhizal symbiosis between Lodgepole pine and mycorrhizal fungus.

b. Mycorrhiza

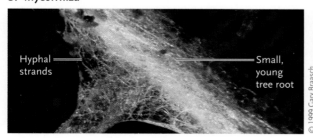

Hyphal strands

Small, young tree root

© 1999 Gary Braasch

Figure 25.22
Ectomycorrhizas. **(a)** Lodgepole pine, *Pinus contorta*, seedling grown in symbiosis with an ectomycorrhizal fungus. Notice the extent of the mycorrhizal fungal mycelium compared with the above-ground portion of the seedling, which is only about 4 cm tall. **(b)** Mycorrhiza of a hemlock tree.

Science VU/R.Roncadori/Visuals Unlimited, Inc.

Figure 25.23
Effect of mycorrhizal fungi on plant growth. The six-month-old juniper seedlings on the left were grown in sterilized low-phosphorus soil inoculated with a mycorrhizal fungus.

Cryptoendolithic Lichens

We tend to think of Antarctica as completely covered in ice, but some valleys of this continent are completely lacking in ice **(Figure 1a).** These dry valleys may look barren, but they are home to many endoliths—organisms that live in a narrow band under the surface of porous rocks. Predominant among these endoliths are cryptoendo-lithic lichens ("crypto" = hidden; "endo" = inside; "lith" = rock) (Figure 1b). These lichens lack the stratified layers typical of most other lichens; instead, hyphae and clusters of photobiont cells grow around and between the rock crystals, and the lichen that forms is embedded inside the rock. Enough light penetrates the translucent surface layer of rock to allow photosynthesis. Studying endolithic organisms not only helps us understand the diversity of life on Earth but also may be a model for life on other planets. If some organisms can live in such extreme conditions here on Earth, could similar kinds of organisms also exist elsewhere in the universe?

a.

© University of Canterbury—Christchurch, New Zealand

b.

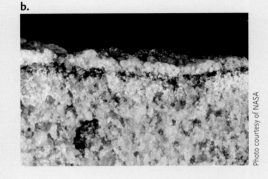

Photo courtesy of NASA

Figure 1
(a) *Antarctic dry valley.* **(b)** *Cryptoendolithic lichen.*

The musky-flavoured truffles (*Tuber melanosporum*) prized by gourmets are ascomycetes that form ectomycorrhizal associations with oak trees (*Quercus* spp.).

For plants, the benefits of being mycorrhizal extend beyond enhanced uptake of soil nutrients. In some cases, mycorrhizal fungi enhance a plant's defences against pathogens, and nutrients can be transferred among mycorrhizal plants via shared mycorrhizal fungal hyphae. Mycorrhizal fungi may, in fact, play a major role in shaping plant communities and ecosystems.

Endophytes Are Fungi Living in the Above-Ground Tissues of Plants. Just as the roots of many plants are colonized by fungi, so too are leaves and shoots **(Figure 25.24).** Although some of these fungi are pathogens, many others evidently peacefully coexist with their plant hosts.

Biologists have known about the presence of these leaf endophytes for some time, but recent discoveries have revealed a startling diversity of these fungi, sometimes within a single plant. Samples of plants from temperate regions have been revealed to have tens of different species of endophytes in a single plant, but tropical plants are truly impressive, with several reports of hundreds of different types of endophytes being isolated from a single plant. Most of these endophytes have not yet been identified to species as researchers have not yet observed sexual stages, so it is difficult to know how many species of endophytes are really living in these tropical plants. A bigger question is, what are these endophytes doing in these leaves? Are they mutualists, like mycorrhizal fungi? In many cases, we simply don't know enough about the interaction between the fungus and its host to answer these questions, but in some cases, the fungi do benefit their plant hosts by producing toxins that deter herbivores. Synthesis of toxins and other secondary metabolites has made these endophytes of great potential importance to humans. For example, the anticancer drug taxol (sold under the tradename Taxol) was originally isolated from the bark of the Pacific yew tree (*Taxus brevifolia*). Production of taxol from this source was limited since the tree is quite rare and makes only a small amount of taxol. However, researchers later discovered that a fungal endophyte living in the needles of the Pacific yew also makes taxol—as do other endophytes living in completely different tree species. Evidence indicates that taxol inhibits the growth of other fungi, so these endophytes may be producing it to protect themselves. Did the genes to produce taxol get transferred from the fungi to the plant? Such horizontal gene transfer is known to have occurred in the evolution of organelles such as mitochondria. The possibility that the genes necessary for biosynthesis of taxol were transferred from the endophyte to its host plant is intriguing, but, as of yet, there is no conclusive evidence to support this idea. Unlike the yew trees that

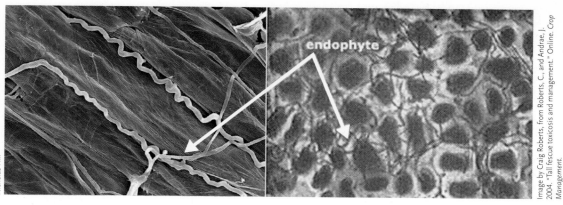

endophyte

Figure 25.24
Leaf endophytes growing inside plant leaves.

were the original source of taxol, these endophytic fungi can be grown very easily in the lab, so we may be able to produce large amounts of this promising anticancer drug very easily. What other sources of medicines are out there, hiding inside plants? The possibility of finding new antibiotics and medicinal compounds makes saving rain forests even more urgent as not only the trees are disappearing but the endophytes inside them as well.

Even though fungi are not closely related to plants in an evolutionary sense, you can see that relationships between fungi and plants play important roles in the lives of both types of organisms. Many saprotrophic and parasitic fungi depend on plants or their products as a source of carbon. Plants rely on fungi for nutrients either directly through mycorrhizal relationships or indirectly through the role of fungi as decomposers. The very first land plants likely relied on mycorrhizal associations to survive in the new harsh environments they faced. In the next chapter, we look at how land plants evolved and diversified.

STUDY BREAK

1. Describe the difference between a saprotroph and a symbiont.
2. What is a lichen? Explain how each partner contributes to the whole organism.
3. What benefit does a plant derive from being mycorrhizal?
4. What are the two most common types of mycorrhizas? How do they differ?
5. What is an endophyte? Why is its relationship with its plant hosts of interest to medical researchers?

Review

25.1 General Characteristics of Fungi

- Fungi can occur as single-celled yeasts or as multicellular filamentous organisms.

- A fungal mycelium consists of filamentous hyphae that grow throughout the substrate on which the fungus feeds (Figure 25.3). A cell wall of chitin surrounds the plasma membrane, and in most species, septa partition the hyphae into cell-like compartments. Pores in septa permit cytoplasm and sometimes organelles to move between hyphal cells.

- Fungi gain nutrients by extracellular digestion and absorption at hyphal tips. Saprotrophic species feed on nonliving organic matter and are key decomposers contributing to the recycling of carbon and other nutrients in ecosystems. Many fungi are symbionts, obtaining nutrients from organic matter of living hosts; these symbioses range from parasitism, in which the fungus benefits at the expense of its host, to mutualism, in which both the fungus and its host benefit.

- All fungi reproduce via spores generated either asexually or sexually (Figure 25.5). Some types also may reproduce asexually by budding or fragmentation of the parent body. Sexual reproduction usually has two stages. First, in plasmogamy, the cytoplasm of two haploid cells fuses, producing a cell that contains a haploid nucleus from each parent. In karyogamy, the nuclei fuse and form a diploid zygote; this stage is delayed in some phyla, resulting in a prolonged dikaryon $(n + n)$ condition. Meiosis then generates haploid spores.

25.2 Evolution and Diversity of Fungi

- Fungi have traditionally been classified mainly on the basis of the structures formed in sexual reproduction. When a sexual phase cannot be detected, or is absent from the life cycle, the specimen is assigned to an informal grouping, the Deuteromycete fungi. Currently, five main phyla of fungi are recognized (Figure 25.6):

- Chytridiomycetes are the only fungi that produce motile, flagellated spores. Many are parasites, including the species responsible for chytridiomycosis, a disease contributing to the worldwide decline in amphibian populations (Figure 25.7).

- Zygomycetes have aseptate hyphae. Asexual reproduction involves production of spores by sporangia. Sexual reproduction occurs by way of hyphae that occur in + and − mating types; haploid nuclei in the hyphae function as gametes. Further development produces the zygosporangium, which may remain dormant for a time. When the zygosporangium breaks dormancy, it produces a stalked sporangium containing haploid spores of each mating type, which are released (Figure 25.8).

- Glomeromycetes form arbuscular mycorrhizas, the most widespread type of mycorrhiza (Figure 25.11). They reproduce asexually, by way of spores formed from hyphae.

- Ascomycetes reproduce both asexually, via chains of haploid asexual spores called conidia, and sexually, via production of haploid ascospores in saclike cells called asci. In the most complex species, asci are produced in reproductive bodies called ascocarps (Figure 25.12).

- Most Basidiomycete species reproduce only sexually. Club-shaped basidia develop on a basidiocarp (the fruiting body or mushroom) and bear sexual spores on their surface. When dispersed, these basidiospores may germinate and give rise to a haploid mycelium (see Figure 25.19).

25.3 Fungal Lifestyles

- All fungi are heterotrophs but can obtain carbon by degrading dead organic matter (as saptrotrophs) or from living hosts (as symbionts). The two lifestyles are not mutually exclusive, with many fungi—such as the mycorrhizal fungi that also prey on invertebrates—combining these two modes of nutrition.

- Some basidiomycete fungi form a mutualistic symbiosis with leaf-cutter ants (see Figure 1, "People behind Biology," 25.1); the ants raise the fungi, which is the sole crop on which they feed. Recently, it was discovered that there is another partner in this ancient symbiosis, an actinomycete bacterium that lives on the ants' bodies and contributes to keeping parasitic fungi out of their fungal gardens.

- Many ascomycetes and a few basidiomycetes enter into symbioses with green algae and/or cyanobacteria to produce a compound organism known as a lichen. Fungal hyphae form the bulk of the lichen body (thallus); the hyphae entwine the algal cells that supply the lichen's carbohydrates, most of which are absorbed by the fungus (Figure 25.21).

- Fungi in the Glomeromycota, Ascomycota, and Basidiomycota form symbiotic associations known as mycorrhizas with plant roots. Hyphae of mycorrhizal fungi proliferate in the soil beyond plant roots and make mineral ions and, in some cases, organic forms of nutrients available to the plant. Some mycorrhizal associations also increase plant defences against pathogens. In turn, the fungus obtains carbohydrates and possibly other growth-enhancing substances from the plant (Figures 25.22 and 25.23).

- Endophytic fungi occur in the above-ground parts of many plants (Figure 25.24); this type of plant–fungus symbiosis is not as well understood as are mycorrhizas, but at least some endophytic fungi are known to produce toxins that deter herbivores.

Questions

Self-Test Questions

1. Which of the following traits is common to all fungi?
   a. parasitism
   b. septate hyphae
   c. reproduction via spores
   d. a prolonged dikaryotic phase

2. Which of the following is/are the chief characteristic(s) traditionally used to classify fungi into the major fungal phyla?
   a. cell wall features
   b. sexual reproductive structures
   c. adaptations for obtaining water
   d. nutritional dependence on nonliving organic matter

3. At lunch, you eat a mushroom, some truffles, a little Camembert cheese, and a bit of mouldy bread. Which group of fungi is NOT represented in this meal?
   a. Zygomycota
   b. Ascomycota
   c. Basidiomycota
   d. Glomeromycota

4. Which of the following fungal reproductive structures is diploid?
   a. ascospore
   b. zygospor-angium
   c. basidiocarp
   d. gametangium

5. Which of the following features characterizes a zygomycete?
   a. septate hyphae
   b. + and − mating strains
   c. mostly sexual reproduction
   d. a life cycle in which karyogamy does not occur

6. What is the reason that some fungi were placed in the Deuteromycetes or Fungi Imperfecti, rather than in a phylum?
   a. They form flagellated spores.
   b. They grow as single cells, rather than as hyphae.
   c. They lack a sexual reproductive stage in their life cycle.
   d. They lack an asexual reproductive stage in their life cycle.

7. Which of the following is the most accurate definition of a mushroom?
   a. a collection of saclike cells called asci
   b. the nutrient-absorbing region of an ascomycete
   c. the nutrient-absorbing region of a basidiomycete
   d. a reproductive structure formed only by basidiomycetes

8. What does it mean to classify a fungus as a saprotroph?
   a. The fungus has external digestion.
   b. The fungus forms extensive mycelia in the soil.
   c. The fungus obtains nutrients from organic matter.
   d. The fungus obtains nutrients from a living organism.

9. Which of the following best describes a lichen?
   a. an association between a green alga and a fungus
   b. an association between a basidiomycete and an ascomycete
   c. a fungus that breaks down rock to provide nutrients for an alga
   d. an organism that spends half of its life cycle as a photosymbiont and the other half as a mycobiont

10. What benefit do mycorrhizal fungi obtain from the plants with which they associate?
   a. increased nitrogen uptake
   b. a regular supply of water
   c. carbon in the form of sugars
   d. the ability to decompose organic material

Questions for Discussion

1. A mycologist wants to classify a specimen that appears to be a new species of fungus. To begin the classification process, what kinds of information on structures and/or functions must the researcher obtain to assign the fungus to one of the major fungal groups?

2. In a natural setting—a pile of horse manure in a field, for example—the sequence in which various fungi appear illustrates **ecological succession**, the replacement of one species by another in a community. The earliest fungi are the most efficient opportunists because they can form and disperse spores most rapidly. In what order would you expect representatives from each phylum of fungi to appear on the manure pile? Why?

3. As the text noted, conifers and some other types of plants cannot grow properly if their roots do not form associations with fungi. These associations provide the plant with minerals such as nitrogen and phosphate and in return fungi receive carbohydrates synthesized by the plant. In some instances, however, the plant receives proportionately more nutrients than the fungus does. Would you still classify such associations as mutualisms?

4. What evidence would you look for to determine whether the association between a plant and an endophyte was mutualistic?

5. Why is it more difficult to develop drugs against fungal infections of humans than bacterial infections?

26

Monotropa uniflora, a heterotrophic plant that lacks chlorophyll.

Plants

WHY IT MATTERS

You are out for a walk in a forest near your home; you are busy thinking about other things and so are not paying close attention to the plants that you're walking by—they are just a pleasing green background for your walk. Suddenly, a small white plant, like the one shown in the photo above, catches your eye. At least you think it's a plant. But aren't all plants green? How can there be a completely white plant?

What you have found is a plant known as ghost flower or Indian pipe (*Monotropa uniflora*), which does not produce chlorophyll and so cannot photosynthesize.

CONCEPT FIX We often assume that all plants are photoautotrophs, making their own organic carbon molecules from atmospheric CO_2 and sunlight. But some plants, such as *Monotropa*, are completely heterotrophic, living on organic carbon obtained from other plants. And other plants that do have chlorophyll supplement their carbon supply by being heterotrophic in low light levels or under other conditions that limit photosynthesis. How do heterotrophic plants get carbon? Some directly parasitize green plants, but others, like *Monotropa*, feed on neighbouring photosynthetic plants through shared root-colonizing fungi (mycorrhizal fungi; see Chapter 25). So, contrary to popular belief, not all plants are photosynthetic and green. ⬡

So if being green isn't an unifying feature of all plants, what is? What features could you look for to determine whether this *Monotropa* is a plant? What characteristics set plants apart from other organisms? And how did plants evolve? In this chapter, we investigate these questions and look at the adaptations to terrestrial life that have made plants so successful. And land plants *are* very successful: they can thrive in habitats where no animal can survive for long and some are able to grow much larger and live much longer than any animal. Together with photosynthetic bacteria and protists, plant tissues provide the nutritional foundation for nearly all ecosystems on Earth. Humans also use plants as sources of medicinal drugs, wood for building, fibres used in paper and clothing, and a wealth of other products. The partnership between humans and plants has a long evolutionary history: we first domesticated cereal plants 9000 years ago, but this was not the earliest relationship between humans and plants. Our early ancestors, like modern-day primates, would have relied heavily on plants in their diet.

Despite the long history between plants and humans, there is still much about plant biology that we don't understand and many questions that remain to be answered.

We start this chapter by considering the defining characteristics of plants and then look at the evolution of plants and their adaptations to life on land; we conclude by looking at the diversity of land plants.

26.1 Defining Characteristics of Land Plants

Land plants are eukaryotes; as we learned from the *Monotropa* example, not all are capable of photosynthesizing, but almost all plants are photoautotrophs (organisms that use light as their energy source and carbon dioxide as their carbon source; see Chapter 7). Like animals, all land plants are multicellular, but if you took a piece of tissue from *Monotropa* and looked at it under the microscope, you'd see that, unlike animal cells, plant cells have walls, which are made of cellulose. All plants are sessile or stationary (not able to move around); no terrestrial animals are sessile, although some aquatic ones are. Plants also differ from animals in having an **alternation of generations** life cycle.

In most animals, the diploid stage dominates the life cycle and produces gametes—sperm or eggs—by meiosis. Gametes are the only haploid stage, and it is short-lived: fusion of gametes produces a new diploid organism (some animals, for example, social insects such as bees and wasps, have a different life cycle). In other organisms, such as many green algae, the haploid stage dominates the life cycle; the haploid alga spends much of its life producing and releasing gametes into the surrounding water. The single-celled zygote is the only diploid stage and divides by meiosis to produce spores that give rise to the haploid stage again.

In contrast, land plants have two multicellular stages in their life cycles, one diploid and one haploid **(Figure 26.1)**. The diploid generation produces spores and is called a **sporophyte** (*phyte* = plant, hence "spore-producing plant"). The haploid generation produces gametes by mitosis and is called a **gametophyte** ("gamete-producing plant"). The haploid phase of the plant life cycle begins in specialized cells of the sporophyte, where haploid spores are produced by meiosis. So in plants meiosis produces spores, not gametes. Spores are single haploid cells with fairly thick cell walls. When a spore germinates, it divides by mitosis to produce a multicellular haploid gametophyte. A gametophyte's function is to nourish and protect the forthcoming sporophyte generation. Each generation

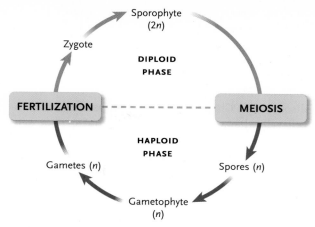

Figure 26.1

Overview of the alternation of generations, the basic pattern of the plant life cycle. The relative dominance of haploid and diploid phases is different for different plant groups.

gives rise to the other—hence the name *alternation of generations* for this life cycle.

The final defining feature of land plants is that the embryo (new sporophyte generation) is retained inside gametophyte tissue. The reasons for retention of embryos in parental tissue and for the rather complex life cycle will become clearer after we've looked at the evolution of plants and their transition onto land.

STUDY BREAK

1. What features of land plants differentiate them from other eukaryotes, for example, from fungi? From animals?
2. What is an alternation of generations life cycle? How does this differ from the life cycle of most animals?
3. What does meiosis produce in plants?
4. Differentiate between a gametophyte and a sporophyte in terms of ploidy and what is produced.

26.2 The Transition to Life on Land

Ages ago, along the shores of the ancient ocean, the only sound was the rhythmic muffled crash of waves breaking in the distance. There were no birds or other animals, no plants with leaves rustling in the breeze. In the preceding eons, cells that produce oxygen as a by-product of photosynthesis had evolved, radically changing Earth's atmosphere. Solar radiation had converted much of the oxygen into a dense ozone layer—a shield against lethal doses of ultraviolet radiation, which had kept early organisms below the water's surface. Now, they could populate the land.

Cyanobacteria were probably the first to adapt to **intertidal zones** and then to spread into shallow, coastal streams. Later, green algae and fungi made the same journey. Around 480 million years ago (mya), one group of green algae, living near the water's edge, or perhaps in a moist terrestrial environment, became the ancestors of modern plants. Several lines of evidence indicate that these algae were charophytes (a group discussed in Chapter 24): both groups have cellulose cell walls, they store energy captured during photosynthesis as starch, and their light-absorbing pigments include both chlorophyll *a* and chlorophyll *b*. Molecular data also support the relationship between the charophytes and the land plants. Like other green algae, the charophyte lineage that produced the ancestor of land plants arose in water and has aquatic descendants today **(Figure 26.2)**. Yet because terrestrial environments pose very different challenges than aquatic environments, evolution in land plants produced a range of adaptations crucial to survival on dry land.

The algal ancestors of plants probably invaded land about 450 mya. We say "probably" because the fossil record is sketchy in pinpointing when the first truly terrestrial plants appeared, and many important stages in evolution are not represented in the fossil record. Even in more recent deposits, the most commonly found plant fossils are just microscopic bits and pieces; easily identifiable parts such as leaves, stems, roots, and reproductive parts seldom occur together. Whole fossilized plants are extremely rare. Adding to the challenge, some chemical and structural adaptations to life on land arose independently in several plant lineages. Despite these problems, botanists have been able to gain insight into several innovations and overall trends in plant evolution.

While the ancestors of land plants were making the transition to a fully terrestrial life, some remarkable adaptive changes unfolded. For example, the earliest land plants were exposed to higher levels of harmful UV radiation than their aquatic ancestors had experienced. Gradual changes in existing metabolic pathways resulted in the ability to synthesize simple phenylpropanoids, molecules that absorb UV radiation, which enhanced the plants' ability to live on land. Where did these new metabolic pathways and associated enzyme functions come from? They did not simply appear because the plants needed them.

CONCEPT FIX The idea that evolution involves organisms "trying" to adapt or that natural selection gives organisms what they need to survive is one of the major misconceptions about evolution. Natural selection cannot sense what a species "needs," and organisms cannot try to adapt: if some individual organisms in the population have traits that allow them to survive and reproduce more in that environment than other individuals, then they will pass on these traits to more offspring, and the frequency of the traits in the population will increase. But the organism cannot "try" to get the right genes. Research shows that new enzyme functions usually follow duplication of genes, which can occur in various ways (e.g., an error during crossing-over of meiosis). Mutations in the second copy of a gene will not have negative effects on the host because the other copy retains its original function; thus over time the second copy tends to accumulate mutations. If the changes in this gene provide advantages to the host plant, then that gene is selected for. In this way, new enzyme functions and metabolic pathways evolve. ⬡

Eons of natural selection sorted out solutions to fundamental problems, among them avoiding desiccation, physically supporting the plant body in air, obtaining nutrients from soil, and reproducing sexually in environments where water would not be available for dispersal of eggs and sperm. With time, plants evolved features that not only addressed these problems but also provided access to a wide range of terrestrial environments. Those ecological opportunities opened the way for a dramatic radiation (rapid evolution and divergence; see Chapter 20) of varied plant species—and for the survival of plant-dependent organisms such as humans. Today the **kingdom Plantae** encompasses more than 300 thousand living species, organized in this textbook into 10 phyla. These modern plants range from mosses, horsetails, and ferns to conifers and flowering plants **(Figure 26.3)**.

26.2a Early Biochemical and Structural Adaptations Enhanced Plant Survival on Land

The greatest challenge plants had to overcome to survive on land was how to survive in the dry terrestrial conditions. Unlike most modern-day plants, the earliest land plants had neither a waterproof **cuticle** (a outer waxy layer that prevents water loss from plant tissues) nor tissues with sufficient mechanical strength to allow for upright growth. These limitations restricted

Figure 26.2

Chara, a stonewort. This representative of the charophyte lineage is known commonly as a stonewort due to the calcium carbonate that accumulates on its surface.

a. Mosses growing on rocks

© Chris Manderson 1995

b. A jack pine

Michael P. Gadomski/Science Source

c. An orchid

© iStockphoto.com/Don Enright

a. Cuticle on the surface of a leaf

Cuticle   Epidermal cell

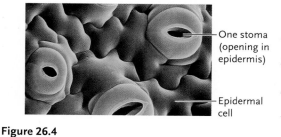

Epidermis

Jubal Harshaw/Shutterstock.com

b. Stomata

One stoma (opening in epidermis)

Epidermal cell

Jeremy Burgess/Science Source

Figure 26.3

Representatives of the kingdom Plantae. **(a)** Mosses growing on rocks. Mosses evolved relatively soon after plants made the transition to land. **(b)** A jack pine (*Pinus banksiana*). This species and other conifers belonging to the phylum Coniferophyta represent the gymnosperms. **(c)** An orchid, *Calypso bulbosa*, a showy example of a flowering plant.

Figure 26.4

Adaptations for limiting water loss. (a) A waxy cuticle, which covers the epidermis of land plants and helps reduce water loss. **(b)** Surface view of stomata in the epidermis (surface layer of cells) of a leaf. Stomata allow carbon dioxide to enter plant tissues and oxygen and water to leave.

these early plants to moist habitats and made it necessary for them to stay small and grow close to the ground. Like modern-day mosses, these plants were **poikilohydric** (*poikilo* = variable; *hydric* = relating to water), meaning that they have little control over their internal water content and do not restrict water loss. Instead, their water content fluctuates with moisture levels in their environment: as their habitat dries out, so do their tissues, and their metabolic activities virtually cease. When external moisture levels rise, they quickly rehydrate and become metabolically active. In other words, poikilohydric plants are drought tolerators that can survive drying out, while vascular plants, which regulate their internal water content and restrict water loss, are drought avoiders, with numerous adaptations to avoid drying out or with plant parts (e.g., underground stems) that can survive if the rest of the plant dries out. How are poikilohydric plants able to survive prolonged dehydration that would be lethal to most plants? This question is explored further in "Life on the Edge," Box 26.1.

Later-evolving plants were able to regulate water content and restrict water loss because they had cuticles covering their outer surfaces **(Figure 26.4a),** as well as **stomata** (singular, *stoma*; *stoma* = mouth), pores in the cuticle-covered surfaces (Figure 26.4b) that open and close to regulate water loss (and are the main route for carbon dioxide to enter leaves; see Chapter 35). These plants also had water-transport tissues that also provided support for upright growth, described further in Section 26.2c.

26.2b Symbiotic Associations with Fungi Were Likely Required for Evolution of Land Plants

The ancestor of land plants was not the first organism to colonize terrestrial habitats; certain bacteria, protists, and fungi had been present at least since the late Proterozoic (around 540 mya). Almost all modern-day plants form symbiotic associations, known as mycorrhizas, with certain soil fungi (see Chapter 25). In these associations, the fungus colonizes the plant's roots and grows prolifically in the soil beyond the root system, producing a very large network that takes up soil nutrients. **(Figure 26.5, p. 566).** Both partners generally benefit by a two-way exchange of nutrients: the plant provides the fungus with carbon, and the fungus increases the plant's supply of soil nutrients, which it is able to obtain much more efficiently than do the plant's own roots. Such mutually beneficial relationships may have been essential to the evolution of land plants and to their success in terrestrial habitats (see "People behind Biology," Box 26.2, p. 567), given that the first land plants lacked roots and that the soils of early Earth were nutrient poor.

26.2c Lignified Water-Conducting Cells Provided Strength and Support for Plants to Grow Upright

The earliest land plants remained small because they lacked the mechanical support necessary to grow taller.

Poikilohydric Plants

Most land plants, including our major crops, are killed if they dry out to the point of equilibrium with the water content of the air around them; this point is all too clearly illustrated by the terrible famines that result from drought in Africa and other regions of the world. But some plants are able to survive drying out to 10% absolute water content or less for months, and even years, in some cases **(Figure 1)**.

This ability is widespread among bryophytes but much less common in vascular plants: only about 50 species of seedless vascular plants have this ability in their sporophyte stage, along with about 300 species of angiosperms. Most of these desiccation-tolerant vascular plants, known as resurrection plants, live on rock outcrops in regions of southern Africa and Australia that receive only seasonal and sporadic rainfall. As far as we know, no gymnosperms have this ability.

How does dehydration kill a plant? Cellular water maintains membrane structure as well as the shapes of macromolecules such as enzymes and other proteins. Dehydration thus results in lethal changes to both membrane structure and

With kind permission from Springer Science + Business Media: *Planta*, "Molecular cloning of abscisic acid-modulated genes which are induced during desiccation of the resurrection plant," volume 181, April 1, 1990, pp. 27–34, Dorothea Bartels, figure: An illustration of the remarkable ability for extreme vegetative desiccation tolerance in an angiosperm species.

Figure 1
Desiccation-tolerant plant shown in a dehydrated state and following re-wetting.

macromolecular shape. A cell's metabolism also relies on water; as a cell dries out, metabolism first decreases and then ceases altogether. How do poikilohydric plants survive these changes that kill all other plants? We don't understand all of the

mechanisms at play, but we do know that part of the answer is accumulation of sugars (e.g., sucrose) in cells. These sugars and certain proteins replace the water in membranes and around macromolecules, preventing lethal changes in conformation. The high sugar content also converts the cytoplasm from its normal consistency to a thick, slow-moving liquid known as glass, immobilizing the cytoplasm. The cells are able to survive in a dehydrated state with metabolism slowed to a state of dormancy or "suspended animation." The cell walls of desiccation-tolerant plants are also more flexible, able to fold as the cell dries, allowing the entire cell to contract as it dries out.

These mechanisms come at a cost to the plant, limiting their growth and reproduction. We don't yet understand how tolerance restricts growth; once we have a better understanding of this relationship, we might be able to uncouple tolerance from slow growth and develop drought-tolerant plants with a higher productivity. This very active area of research clearly has practical applications in maintaining our food supply in the face of droughts and climate change.

Figure 26.5
Mycorrhizal fungus colonizing plant root and soil around the root.

Growing low to the ground helped them stay moist but was not very effective in capturing light: since all early land plants were low growing, there would have been intense competition for light. If any plant had been able to grow taller than its neighbours, it would have had a major advantage. But how could a plant support upright growth against the force of gravity? Plants require strengthening tissue to grow upright. And growing up and away from the ground surface also requires an internal water circulation system, since diffusion is not effective over longer distances. Some of the early land plants did have specialized water-conducting cells that transported water through the plant body, but these cells did not provide mechanical strength. Later land plants were able to synthesize lignin, a polymer of phenylpropanoids (the molecules mentioned earlier that absorb UV radiation). Why were these plants able to make lignin when earlier plants did not? Changes in Earth's atmosphere and climate altered certain biochemical pathways in plants, resulting in the excess formation of lignin; see

Kris Pirozynski and David Malloch, *Agriculture Canada*

If almost all land plants today depend to some extent on mutually beneficial relationships with mycorrhizal fungi, would the first land plants have been any different? The hypothesis that a mutualistic relationship with soil fungi was required for the evolution of land plants was first put forward in 1975 by two researchers at the Biosystematics Research Institute of Agriculture Canada.

In their 1975 paper outlining this hypothesis, Kris Pirozynski and David

Malloch pointed out that associations with fungi would have helped the earliest land plants avoid starvation: early soils would not have been as fertile as most modern-day soils, and nutrients would certainly not have been as abundant as in the aquatic environments in which the algal ancestor of land plants lived. Fungi are very adept at proliferating in their substrates and foraging for nutrients, which they take up via extracellular

enzymatic digestion (see Chapter 5). The earliest plants did not have roots, so forming a partnership with fungi would have greatly enhanced their uptake of nutrients. The fungi might also have protected the roots of its plant partner from root pathogens, as do modern-day mycorrhizal fungi.

Pirozynski and Malloch's hypothesis has since received strong support from both the fossil record and molecular data.

"Molecule behind Biology," Box 26.3. This lignin was deposited in cell walls, particularly in the water-conducting cells, providing support and rigidity to those tissues and allowing the plants to grow upright. These lignified water-conducting cells make up a tissue called xylem.

Xylem is one type of **vascular tissue** (*vas* = duct or vessel). Plants with this tissue (and the other type of vascular tissue, **phloem**, which conducts sugars through the plant body) are known as **vascular plants**. It is important to note that some plants, such as some mosses, that lack vascular tissues do have tissues that

conduct water and sugars through their bodies. These tissues are not the same as xylem and phloem—for example, their water-conducting cells do not have walls reinforced with lignin—and are likely not homologous with xylem and phloem, so they are not called vascular tissues. Thus, these plants are referred to as **nonvascular plants.** Chapter 35 explains how xylem and phloem perform these key internal transport functions.

Clearly, plants with lignified tissues had a clear benefit over plants lacking lignin and over time evolved to become the dominant plants in most habitats on

---

Coniferyl Alcohol, a Building Block of Lignin

Lignin is a polymer of several different monomers, including coniferyl alcohol **(Figure 1)**. These molecules are synthesized from the amino acid phenylalanine in a series of reactions in plant cell cytoplasm. The monomers are then transported through the cell membrane, where polymerization happens. We still do not fully understand how lignin is formed from monomers, but we do know that oxidative

enzymes are involved in polymerization; thus, oxygen is required for the process. Lignin is thought to have evolved due to the high oxygen levels in the atmosphere around 430 mya, which would have favoured the polymerization reaction.

Lignin is very difficult to degrade, with only a few fungi and bacteria able to break it down (see Chapter 25). Its accumulation in plant tissues would have meant that dead vascular plants, especially if large, would have decomposed more slowly than the earlier land plants, contributing to the formation of coal, one of today's fossil fuels. The forests of the Carboniferous period were dominated by large vascular seedless plants, which were

abundant in lignin. When these plants died and fell to the ground, they became buried in anaerobic sediments; even those that were not buried in such sediments would have been fairly slow to decompose due to their lignin content. Over geologic time, these buried remains became compressed and fossilized; today they form much of the world's coal reserves. This is why coal is called a "fossil fuel" and the Carboniferous period is called the Coal Age. Characterized by a moist climate over much of the planet and by the dominance of seedless vascular plants, the Carboniferous period continued for 150 million years, ending when climate patterns changed during the Paleozoic era.

Figure 1
Coniferyl alcohol, one of the monomers of lignin.

Earth. Ferns, conifers, and flowering plants—most of the plants you are familiar with—are vascular plants. Supported by lignin and with a well-developed vascular system, the body of a plant can grow very large. Extreme examples are the giant redwood trees of the northern California coast, some of which are more than 90 m tall. By contrast, non vascular plants lack lignin, although some do have simple internal transport systems, and are generally small (Table 26.1).

Vascular plants also have **apical meristems**, regions of constantly dividing cells near the tips of shoots and roots that produce all tissues of the plant body. Meristem tissue is the foundation for a vascular plant's extensively branching stem and root systems and is a central topic of Chapter 34.

26.2d Root and Shoot Systems Were Adaptations for Nutrition and Support

The body of a nonvascular plant is not differentiated into true roots and stems—structures that are fundamental adaptations for absorbing nutrients from soil and for support of an erect plant body. The evolution of sturdy stems—the basis of an aerial *shoot system*—went hand in hand with the capacity to synthesize lignin. To become large, land plants also require a means of anchoring aerial parts in the soil, as well as effective strategies for obtaining soil nutrients. **Roots**—anchoring structures that also absorb water and nutrients in association with mycorrhizal fungi—were the eventual solution to these problems. The earliest fossils showing clear evidence of roots are

from vascular plants, although the exact timing of this change is uncertain. Ultimately, vascular plants developed specialized **root systems**, which generally consist of underground, cylindrical absorptive structures with a large surface area that favours the rapid uptake of soil water and dissolved mineral ions. The root system has been called "the hidden half" of a plant: "half" refers to the fact that there is as much plant biomass below ground as there is above ground. And there are other similarities between above- and below-ground parts of plants: the fine roots of a root system go through regular cycles of growth and death, just as do the leaves of most plants. "Hidden" refers to the fact that the root system is hidden from our sight below ground, meaning that we cannot study it very easily. For this reason, we know less about root systems than about the above-ground parts of plants, although recent technological advances are changing this situation.

Above ground, the simple stems of early land plants also became more specialized, evolving into **shoot systems** in vascular plants. Shoot systems have stems and leaves that arise from apical meristems and that function in the absorption of light energy from the Sun and carbon dioxide from the air. Stems grew larger and branched extensively after the evolution of lignin. The mechanical strength of lignified tissues almost certainly provided plants with several adaptive advantages. For instance, a strong internal scaffold could support upright stems bearing leaves and other photosynthetic structures and so help increase the surface area for intercepting sunlight. Also, reproductive

Table 26.1	Trends in Plant Evolution Traits Derived from Algal Ancestor: Cell Walls with Cellulose, Energy Stored in Starch, Two Forms of Chlorophyll (*a* and *b*)				
Bryophytes	Ferns and Their Relatives	Gymnosperms	Angiosperms	Functions of This Trait in Land Plants	
Cuticle	———————————————————————————▶			Protection against water loss, pathogens	
Stomata	———————————————————————————▶			Regulation of water loss and gas exchange (CO_2 in, O_2 out)	
Nonvascular (although some have specialized water-conducting cells without lignin) ──▶	Vascular (have xylem and phloem) ————————————————▶			Internal tubes that transport water, nutrients	
	Lignin ————————————————————▶			Mechanical support for vertical growth	
	Apical meristem ——————————————▶			Branching shoot system	
	Roots, stems, leaves ————————————▶			Enhanced uptake, transport of nutrients, and enhanced photosynthesis	
Haploid phase dominant ——————▶	Diploid phase dominant ————————————▶			Genetic diversity	
One spore type (homospory) ——————▶	Homospory in most but heterospory (two spore types) in some ──▶	Heterospory ————————————▶		Promotion of genetic diversity	
Motile sperm	——————————————▶	Nonmotile sperm ——————————▶		Protection of gametes within parent body	
Seedless	——————————————▶	Seeds ——————————————▶		Protection of embryo	

a. Development of microphylls as an offshoot of the main vertical axis

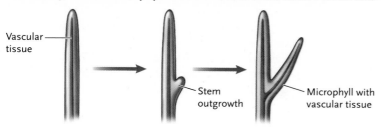

Vascular tissue

Stem outgrowth

Microphyll with vascular tissue

b. Development of megaphylls in a branching pattern

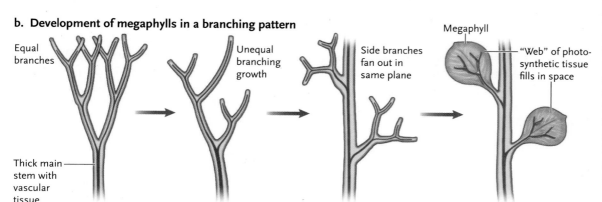

Equal branches

Thick main stem with vascular tissue

Unequal branching growth

Side branches fan out in same plane

Megaphyll

"Web" of photosynthetic tissue fills in space

Figure 26.6

Evolution of leaves. **(a)** One type of early leaflike structure may have evolved as off-shoots of the plant's main vertical axis; there was only one vein (transport vessel) in each leaf. Today, the seedless vascular plants known as lycophytes (club mosses) have this type of leaf. **(b)** In other groups of seedless vascular plants, leaves arose in a series of steps that began when the main stem evolved a branching growth pattern. Small side branches then fanned out and photosynthetic tissue filled the space between them, becoming the leaf blade. With time, the small branches modified into veins.

structures borne on aerial stems might serve as platforms for more efficient launching of spores from the parent plant.

Structures we think of as "leaves" arose several times during plant evolution. In general, leaves represent modifications of stems and can be divided into two types. Microphylls are narrow leaves with only one vein or strand of vascular tissue, while megaphylls are broader leaves with multiple veins. **Figure 26.6** illustrates the basic steps of possible evolutionary pathways by which these two types of leaves evolved. In some early plants, microphylls may have evolved as flaplike extensions of the main stem. In contrast, megaphylls likely evolved from modified branches when photosynthetic tissue filled in the gaps between neighbouring branches.

Other land plant adaptations were related to the demands of reproduction in a dry environment. As described in more detail shortly, these adaptations included multicellular chambers that protect developing gametes and a multicellular embryo that is sheltered inside the tissues of a parent plant.

26.2e In the Plant Life Cycle, the Diploid Phase Became Dominant

As early plants moved into drier habitats, their life cycles were also modified considerably. The haploid gametophyte phase became physically smaller and less complex and had a shorter life span, whereas the opposite occurred with the diploid sporophyte phase. In mosses and other nonvascular plants, the sporophyte is a little larger and longer lived than in green algae, and in vascular plants, the sporophyte is clearly larger and more complex and lives much longer than the gametophyte **(Figure 26.7)**. When you look at a pine tree, for example, you see a large,

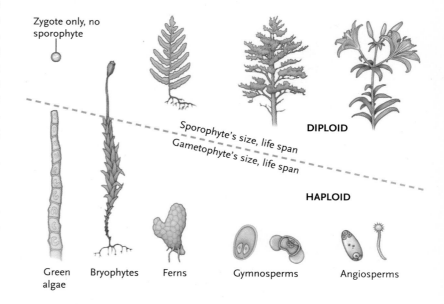

Zygote only, no sporophyte

Sporophyte's size, life span
Gametophyte's size, life span

DIPLOID

HAPLOID

Green algae   Bryophytes   Ferns   Gymnosperms   Angiosperms

Figure 26.7

Evolutionary trend from dominance of the gametophyte (haploid) generation to dominance of the sporophyte (diploid) generation, represented here by existing species ranging from a green alga (*Ulothrix*) to a flowering plant. This trend developed as early plants colonized habitats on land. In general, the sporophytes of vascular plants are larger and more complex than those of bryophytes, and their gametophytes are smaller and less complex. In this diagram, the fern represents seedless vascular plants.

long-lived sporophyte. The sporophyte generation begins after fertilization, when the zygote divides by mitosis to produce a multicellular diploid organism. Its body will eventually develop capsules called **sporangia** (*angium* = vessel or chamber, hence, "spore-producing chambers"; singular, *sporangium*), which produce spores by meiosis.

Why did the diploid phase become dominant over evolutionary time? Many botanists hypothesize that the trend toward "diploid dominance" reflects the advantage of being diploid in land environments; if there is only one copy of DNA, as in a haploid plant, and if a deleterious mutation occurs or if the DNA is damaged (e.g., by UV radiation, which is a greater problem on land than in aquatic habitats), the consequences could be fatal. In contrast, the sporophyte phase of that plant is diploid and so has a "backup" copy of the DNA that can continue to function normally even if one strand is damaged. However, it is important to remember that the land plants that do have a dominant haploid stage, such as mosses, are very successful plants in certain habitats. The lack of a dominant diploid stage has certainly not caused them to become extinct.

26.2f Some Vascular Plants Evolved Separate Male and Female Gametophytes

When a plant makes only one type of spore, it is said to be **homosporous** ("same spore") **(Figure 26.8a).**

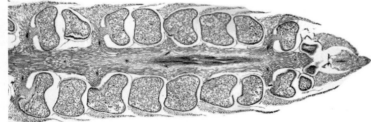

a. *Lycopodium*

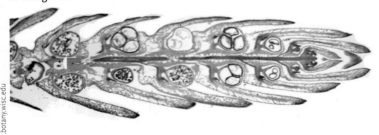

b. *Selaginella*

Photographer: Michael Clayton. University of Wisconsin Plant Teaching Collection, http://botit.botany.wisc.edu

Figure 26.8
Longitudinal sections through strobili of two lycophytes, **(a)** *Lycopodium* and **(b)** *Selaginella*. *Lycopodium* is a homosporous plant that produces spores of only one type, as can be seen in (a). Note that the sporangia of *Lycopodium* are all the same. The *Selaginella* strobilus shown here is from a heterosporous plant, which produces megasporangia (containing a few large megaspores) and microsporangia (containing numerous small microspores) in the same strobilus.

Usually, a gametophyte that develops from such a spore is bisexual—it can produce both sperm and eggs. However, some homosporous plants have ways to produce male and female sex organs on different gametophytes or to otherwise prevent self-fertilization, as described below in ferns. The sperm have flagella and are motile because they must swim through liquid water to encounter eggs.

Other vascular plants, including gymnosperms and angiosperms, are **heterosporous** (Figure 26.8b). They produce two types of spores—one type is smaller than the other—in two different types of sporangia. The smaller spores are **microspores**, which develop into male gametophytes, and the larger **megaspores** will develop into female gametophytes. Heterospory and the development of gametophytes inside spore walls are important steps in the evolution of the seed, as we will see further on.

As you will read in a later section, the evolution of seeds and related innovations, such as pollen grains and pollination, helped spark the rapid diversification of plants in the Devonian period, 408 to 360 mya. In fact, so many new fossils appear in Devonian rocks that paleobotanists—scientists who specialize in the study of fossil plants—have thus far been unable to determine which fossil lineages gave rise to the modern plant phyla. Clearly, however, as each major lineage came into being, its characteristic adaptations included major modifications of existing structures and functions **(Figure 26.9).** The next sections fill out this general picture, beginning with the plants that are the living representatives of the earliest land plants.

STUDY BREAK

1. What features do land plants share with their closest living relatives, the charophyte algae? What features differentiate the two groups?
2. How did mycorrhizal fungi fulfill the role we associate with roots in early land plants?
3. What is the main difference between the specialized water-conducting cells present in some nonvascular plants and those of vascular plants? How did this difference influence the evolution of vascular plants?
4. How did plant adaptations such as a root system, a shoot system, and a vascular system collectively influence the evolution of land plants?
5. Describe the difference between homospory and heterospory, and explain how heterospory paved the way for other reproductive adaptations in land plants.

570 | UNIT SIX DIVERSITY OF LIFE

NEL

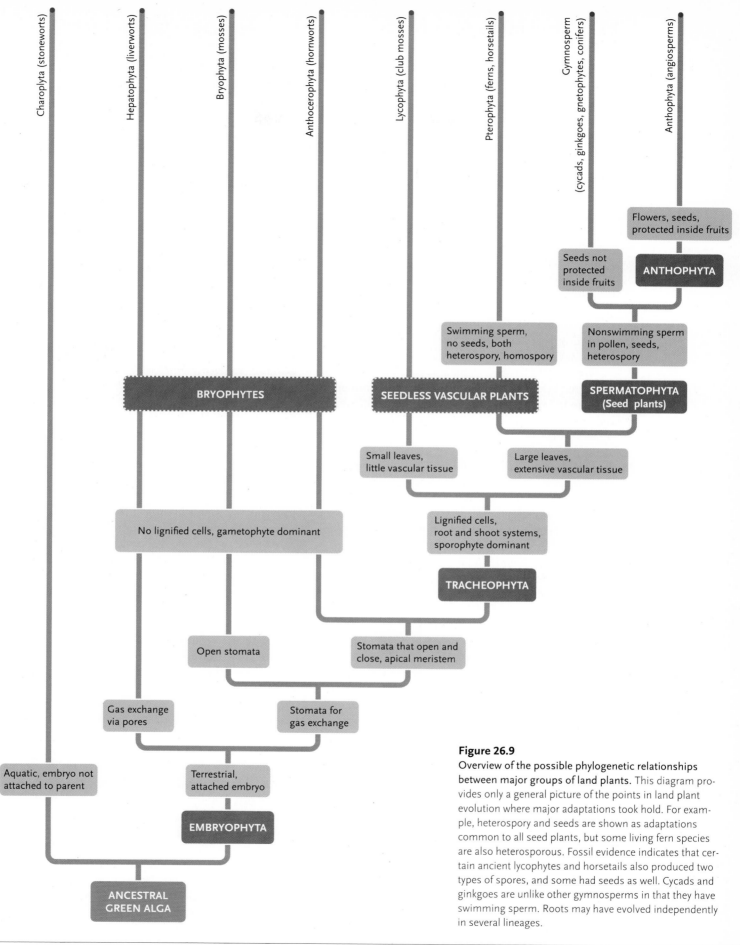

Figure 26.9

Overview of the possible phylogenetic relationships between major groups of land plants. This diagram provides only a general picture of the points in land plant evolution where major adaptations took hold. For example, heterospory and seeds are shown as adaptations common to all seed plants, but some living fern species are also heterosporous. Fossil evidence indicates that certain ancient lycophytes and horsetails also produced two types of spores, and some had seeds as well. Cycads and ginkgoes are unlike other gymnosperms in that they have swimming sperm. Roots may have evolved independently in several lineages.

26.3 Bryophytes: Nonvascular Land Plants

The **bryophytes** (*bryon* = moss)—liverworts, hornworts, and mosses—are important both ecologically and economically. As colonizers of bare land, their small bodies trap particles of organic and inorganic matter, helping to build soil on bare rock and stabilizing soil surfaces with a biological crust in harsh places such as coastal dunes, inland deserts, and embankments created by road construction. In boreal forests and arctic tundras, bryophytes constitute as much as half of the biomass, and they are crucial components of the **food web** that supports animals in these ecosystems. People have long used *Sphagnum* and other absorbent "peat" mosses (which typically grow in bogs and fens) for everything from primitive diapers and filtering whiskey to increasing the water-holding capacity of garden soil. Peat moss has also found use as a fuel; each day, the Rhode generating station in Ireland, one of several that use peat in that nation, burns 2000 tonnes of peat to produce electricity.

Bryophytes have a combination of traits that allow them to bridge aquatic and land environments. Because bryophytes lack cells strengthened by lignin and are poikilohydric, it is not surprising that they are small and commonly grow on wet sites along creek banks (see Figure 26.3a, p. 565); in bogs, swamps, or the dense shade of damp forests; and on moist tree trunks or rooftops. However, some mosses live in very dry environments, such as **alpine tundra** and **arctic tundra (Figure 26.10)**. Being poikilohydric enables them to live in such seemingly inhospitable habitats (see "Life on the Edge," Box 26.1, p. 566).

Bryophytes retain many of the features of their algal ancestors: they produce flagellated sperm that must swim through water to reach eggs, which is another reason they are small: the sperm must be able to swim between plants in a film of water (e.g., from rain or dew), which is only possible if the plants are relatively close to the ground. They also lack xylem and phloem (although some do have specialized conductive tissues). Bryophytes have parts that are rootlike, stemlike, and leaflike. However, the "roots" are **rhizoids** that serve only to anchor the plant to its substrate and do not take up any water or nutrients from the substrate. Bryophyte "stems" and "leaves" are not considered to be true stems and leaves like those of vascular plants because they lack vascular tissue and because they did not evolve from the same structures as vascular plant stems and leaves did. (Said another way, stems and leaves are not homologous in bryophytes and vascular plants.)

In other ways, bryophytes are clearly adapted to land. The sporophytes (but not the longer-lived gametophytes) of some species have a water-conserving cuticle and stomata. And, as is true of all plants, the

a.

b.

Figure 26.10
Bryophytes of arid habitats: **(a)** moss growing on exposed rock; **(b)** mosses and other plants in alpine tundra.

bryophyte life cycle has both multicellular gametophyte and sporophyte phases, but the sporophyte is permanently associated with the gametophyte (it never becomes independent of the gametophyte) and lives for a shorter time than the gametophyte. **Figure 26.11** shows the green, leafy gametophyte of a moss plant, with diploid sporophytes attached to it by slender stalks. Bryophyte gametophytes produce gametes inside a protective organ called a **gametangium** (plural, *gametangia*). The gametangia in which bryophyte eggs form are flask-shaped structures called **archegonia** (*archi* = first; *gonos* = seed). Flagellated sperm form in rounded gametangia called **antheridia** (*antheros* = flowerlike; singular, *antheridium*). The sperm swim through a film of water to the archegonia to fertilize eggs. Each fertilized egg gives rise to a diploid embryo sporophyte, which stays attached to the gametophyte and produces spores—and the cycle repeats.

Despite these similarities to more complex plants, bryophytes are unique in several ways. Unlike vascular plants, the gametophyte is much longer lived than the sporophyte and is photosynthetic, whereas the sporophyte remains attached to the gametophyte and depends on the gametophyte for much of its nutrition.

Bryophytes are not a monophyletic group (i.e., they did not all evolve from a common ancestor); instead, the various bryophytes evolved as separate lineages, in parallel with vascular plants.

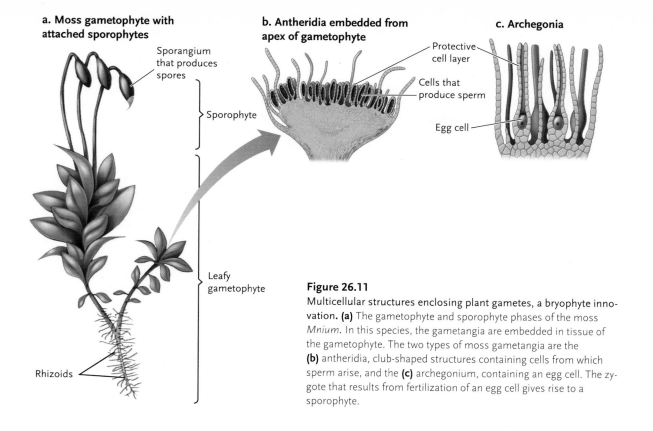

a. Moss gametophyte with attached sporophytes

Sporangium that produces spores

Sporophyte

Leafy gametophyte

Rhizoids

b. Antheridia embedded from apex of gametophyte

Cells that produce sperm

c. Archegonia

Protective cell layer

Egg cell

Figure 26.11

Multicellular structures enclosing plant gametes, a bryophyte innovation. **(a)** The gametophyte and sporophyte phases of the moss *Mnium*. In this species, the gametangia are embedded in tissue of the gametophyte. The two types of moss gametangia are the **(b)** antheridia, club-shaped structures containing cells from which sperm arise, and the **(c)** archegonium, containing an egg cell. The zygote that results from fertilization of an egg cell gives rise to a sporophyte.

26.3a Liverworts Resemble the First Land Plants

Liverworts make up the phylum **Hepatophyta**, so called because early herbalists thought that these small plants were shaped like the lobes of the human liver (*hepat* = liver; *wort* = herb). The resemblance might be a little vague to modern eyes: while some of the 6000 species of liverworts consist of a flat, branching, ribbonlike plate of tissue closely pressed against damp soil, other liverworts are leafy and superficially resemble mosses, although the arrangement of leaves is different **(Figure 26.12)**. This simple body, called a **thallus** (plural, *thalli*), is the gametophyte generation. Threadlike rhizoids anchor the gametophytes to their substrate. None have true stomata, the openings that regulate gas exchange in most other land plants, although some species do have pores. They lack some features present in the other two groups of bryophytes; this evidence, together with molecular data, suggests that the first land plants likely resembled modern-day liverworts.

We will look at one genus, *Marchantia* (see Figure 26.12), as an example of liverwort reproduction.

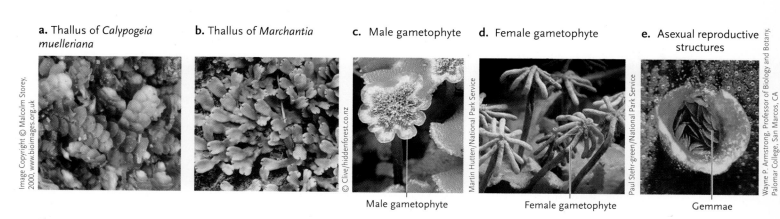

a. Thallus of *Calypogeia muelleriana*

b. Thallus of *Marchantia*

c. Male gametophyte

Male gametophyte

d. Female gametophyte

Female gametophyte

e. Asexual reproductive structures

Gemmae

Figure 26.12

Examples of liverworts. **(a)** Thallus of a leafy liverwort, *Calypogeia muelleriana*. **(b)** Thallus of the thalloid liverwort *Marchantia*, the only liverwort to produce **(c)** male and **(d)** female gametophytes on separate plants. *Marchantia* and some other liverworts also reproduce asexually by way of **(e)** gemmae, multicellular vegetative bodies that develop in tiny cups on the plant body. Gemmae can grow into new plants when splashing raindrops transport them to suitable sites.

Separate male and female gametophytes produce sexual organs (antheridia and archegonia) on tall stalks (Figure 26.12c, d). The motile sperm released from antheridia swim through surface water to reach the eggs inside archegonia. After fertilization, a small, diploid sporophyte develops inside the archegonium, matures there, and produces haploid spores by meiosis. During meiosis, sex chromosomes segregate, so some spores have the male genotype and others the female genotype. As in other liverworts, the spores develop inside jacketed sporangia that split open to release the spores. A spore that is carried by air currents to a suitable location germinates and gives rise to a haploid gametophyte, which is either male or female. *Marchantia* and some other liverworts can also reproduce asexually by way of **gemmae** (*gem* = bud; singular, *gemma*), small cell masses that form in cuplike growths on a thallus (Figure 26.12e). Gemmae can grow into new thalli when rainwater splashes them out of the cups and onto an appropriately moist substrate.

26.3b Many Mosses Have Specialized Cells for Water and Nutrient Transport

Chances are that you have seen, touched, or sat on at least some of the approximately 10 thousand species of mosses, and the use of the name **Bryophyta** for this phylum underscores the fact that mosses are the best-known bryophytes, forming tufts or carpets of vegetation on the surface of rocks, soil, or bark.

The moss life cycle, diagrammed in **Figure 26.13,** begins when a haploid (*n*) spore lands on a wet soil surface. After the spore germinates, it elongates and branches into a filamentous web of tissue called a **protonema** ("first thread"), which can become dense enough to colour the surface of soil, rocks, or bark

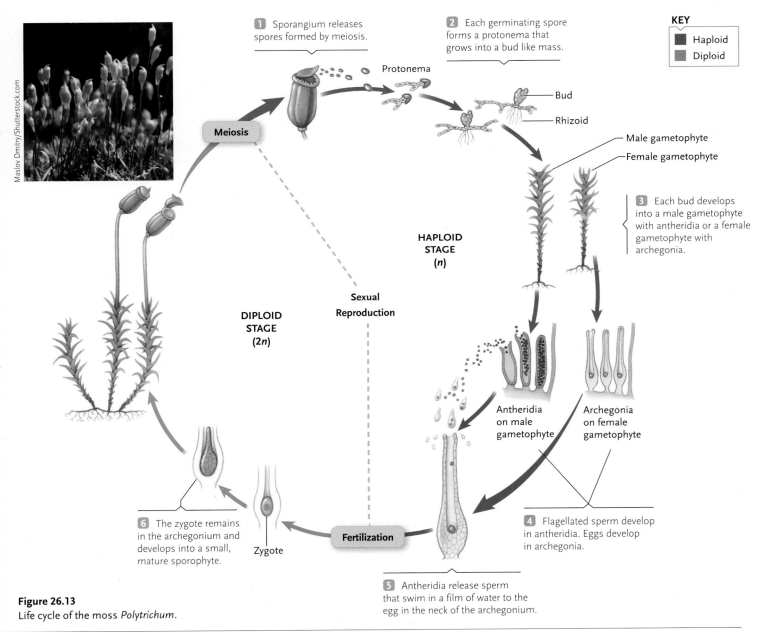

KEY

■ Haploid
■ Diploid

1 Sporangium releases spores formed by meiosis.

2 Each germinating spore forms a protonema that grows into a bud like mass.

Protonema

Meiosis

Bud

Rhizoid

Male gametophyte

Female gametophyte

3 Each bud develops into a male gametophyte with antheridia or a female gametophyte with archegonia.

HAPLOID STAGE (*n*)

DIPLOID STAGE (2*n*)

Sexual Reproduction

Antheridia on male gametophyte

Archegonia on female gametophyte

4 Flagellated sperm develop in antheridia. Eggs develop in archegonia.

Fertilization

Zygote

6 The zygote remains in the archegonium and develops into a small, mature sporophyte.

5 Antheridia release sperm that swim in a film of water to the egg in the neck of the archegonium.

Maslov Dmitry/Shutterstock.com

Figure 26.13
Life cycle of the moss *Polytrichum*.

visibly green. After several weeks of growth, the bud-like cell masses on a protonema develop into leafy, green gametophytes anchored by rhizoids. A single protonema can be extremely prolific, producing bud after bud, thus giving rise to a dense clone of genetically identical gametophytes. Leafy mosses may also reproduce asexually by gemmae produced at the surface of rhizoids and on above-ground parts.

Antheridia and archegonia are produced at the tips of male and female gametophytes, respectively. Propelled by flagella, sperm released from antheridia swim through a film of dew or rainwater and down a channel in the neck of the archegonium, attracted by a chemical gradient secreted by each egg. Fertilization produces the new sporophyte generation inside the archegonium, in the form of diploid zygotes that develop into small, mature sporophytes, each consisting of a sporangium on a stalk. Moss sporophytes may eventually develop chloroplasts and nourish themselves photosynthetically, but initially they depend on the gametophytes for food. Even after a moss sporophyte begins photosynthesis, it still must obtain water, carbohydrates, and some other nutrients from the gametophyte.

Certain moss gametophytes are structurally complex, with features similar to those of higher plants. For example, some species have a central strand of conducting tissue. One kind of tissue is made up of elongated, thin-walled, dead and empty cells that conduct water. In a few mosses, the water-conducting cells are surrounded by sugar-conducting tissue resembling the phloem of vascular plants. These tissues did not give rise to the xylem and phloem of vascular plants, however.

26.3c Hornworts Share a More Recent Ancestor with Vascular Plants

Roughly 100 species of hornworts make up the phylum **Anthocerophyta**. Like some liverworts, a hornwort gametophyte has a flat thallus, but the sporangium of the sporophyte phase is long and pointed, like a horn **(Figure 26.14)**, and splits into two or three ribbonlike sections when it releases spores. Sexual reproduction occurs in basically the same way as in liverworts, and hornworts also reproduce asexually by fragmentation as pieces of a thallus break off and develop into new individuals. While the gametophyte is the dominant stage of the hornwort life cycle, hornworts differ from other nonvascular plants in that their sporophytes can become free-living plants that are independent of the gametophyte! Recent genetic research into evolutionary relationships among the major groups of land plants indicates that hornworts are the group of bryophytes that have a more recent common ancestor with vascular plants.

In the next section, we turn to the vascular plants, which have lignified water-conducting tissue. Without

Figure 26.14
The hornwort *Anthoceros*. The base of each long, slender sporophyte is embedded in the flattened, leafy gametophyte.

the strength and support provided by this tissue, as well as its capacity to move water and minerals efficiently throughout the plant body, large sporophytes could not have survived on land. Unlike bryophytes, modern vascular plants are monophyletic—all groups are descended from a common ancestor.

STUDY BREAK

1. Give some examples of bryophyte features that bridge aquatic and terrestrial environments.
2. Summarize the main similarities and differences among liverworts, hornworts, and mosses.
3. How do specific aspects of a moss plant's anatomy resemble those of vascular plants?

26.4 Seedless Vascular Plants

The first vascular plants did not produce seeds and were the dominant plants on Earth for almost 200-million years, until seed plants became abundant. The fossil record shows that seedless vascular plants were well established by the late Silurian, about 428 mya, and they flourished until the end of the Carboniferous, about 250 mya. Some living seedless vascular plants have certain bryophyte-like traits, whereas others have some characteristics of seed plants. On the one hand, like bryophytes, seedless vascular plants disperse themselves by releasing spores, and they have swimming sperm that require free water to reach eggs. On the other hand, as in seed plants, the sporophyte of a seedless vascular plant becomes independent of the gametophyte at a certain point in its development and has well-developed vascular tissues (xylem and phloem). Also, the sporophyte is the larger, longer-lived stage of the life cycle and the gametophytes are very small, with some even lacking chlorophyll.

Table 26.2 summarizes these characteristics and gives an overview of seedless vascular plant features within the larger context of modern plant phyla.

In the late Paleozoic era, seedless vascular plants were Earth's dominant vegetation. Some lineages have endured to the present, but, collectively, these survivors total fewer than 14 thousand species. The taxonomic relationships between various lines are still under active investigation, and comparisons of gene sequences from the genomes in chloroplasts, nuclei, and mitochondria are revealing previously unsuspected links between some of them. In this book, we assign seedless vascular plants to two phyla, the Lycophyta (club mosses and their close relatives; the common name "club moss" for lycophytes is misleading, as they are vascular plants, not mosses) and the Pterophyta (ferns, whisk ferns, and horsetails).

26.4a Early Seedless Vascular Plants Flourished in Moist Environments

What did the first vascular plant look like? There are no living relatives of the earliest vascular plants, so we rely on fossil data to answer this question. The extinct

Table 26.2	Plant Phyla and Major Characteristics		
Phylum	Common Name	Number of Species	Common General Characteristics
Bryophytes: nonvascular plants. Gametophyte dominant, free water required for fertilization, cuticle and stomata present in some.			
Hepatophyta	Liverworts	6000	Leafy or simple flattened thallus, rhizoids; spores in capsules. Moist, humid habitats.
Bryophyta	Mosses	10 000	Simple flattened thallus, rhizoids; hornlike sporangia. Moist, humid habitats.
Anthocerophyta	Hornworts	100	Feathery or cushiony thallus; some have hydroids; spores in capsules. Moist, humid habitats; colonizes bare rock, soil, or bark.
Seedless vascular plants: sporophyte dominant, free water required for fertilization, cuticle and stomata present.			
Lycophyta	Club mosses	1000	Small simple leaves, true roots; most species have sporangia on sporophylls. Mostly wet or shady habitats.
Pterophyta	Ferns, whisk ferns, horsetails	13 000	*Ferns:* Finely divided large leaves, sporangia often in sori. Habitats from wet to arid. *Whisk ferns:* Branching stem from rhizomes; sporangia on stem scales. Tropical to subtropical habitats. *Horsetails:* Hollow photosynthetic stem, scalelike leaves, sporangia in strobili. Swamps, disturbed habitats.
Gymnosperms: vascular plants with "naked" seeds. Sporophyte dominant, fertilization by pollination, cuticle and stomata present.			
Cycadophyta	Cycads	185	Shrubby or treelike with palmlike leaves, pithy stems; male and female strobili on separate plants. Widespread distribution.
Ginkgophyta	Ginkgo	1	Woody-stemmed tree, deciduous fan-shaped leaves. Male, female structures on separate plants. Temperate areas of China.
Gnetophyta	Gnetophytes	70	Shrubs or woody vines; one has strappy leaves. Male and female strobili on separate plants. Limited to deserts, tropics.
Coniferophyta	Conifers	550	Mostly evergreen, woody trees and shrubs with needlelike or scalelike leaves; male and female cones usually on same plant.
Angiosperms: plants with flowers and seeds protected inside fruits. Sporophyte dominant, fertilization by pollination, cuticle and stomata present. Major groups: monocots, eudicots.			
Anthophyta	Flowering plants	268 500+ (including monocots and dicots, as well as magnoliids, other basal angiosperms)	Wood and herbaceous plants. Nearly all land habitats, some aquatic.
Monocots	Grasses, palms, lilies, orchids, and others	(60 000)	One cotyledon; parallel-veined leaves common; bundles of vascular tissue scattered in stem; flower parts in multiples of three.
Eudicots	Most fruit trees, roses, beans, potatoes, and others	(200 000)	Most species have two cotyledons; net-veined leaves common; central core of vascular tissue in stem; flower parts in multiples of four or five.

genus *Rhynia* was one of the earliest ancestors of modern seedless vascular plants. Based on fossil evidence, the sporophytes of the first vascular plants, such as *Rhynia* and related genera **(Figure 26.15)**, lacked leaves and roots. Above-ground photosynthetic stems produced sporangia at the tips of branches. Below ground, the plant body was supported by **rhizomes**, horizontal modified stems that can penetrate a substrate and anchor the plant. *Rhynia*'s simple stems had a central core of xylem, an arrangement seen in many existing vascular plants. Mudflats and swamps of the damp Devonian period were dominated by *Rhynia* and related plants (Figure 26.15). Although these and other now-extinct phyla came and went, ancestral forms of both modern phyla of seedless vascular plants appeared.

Carboniferous forests were swampy places dominated by members of the phylum **Lycophyta**, and fascinating fossil specimens of this group have been unearthed in North America and Europe. One example is *Lepidodendron,* which had broad, straplike leaves and sporangia near the ends of the branches **(Figure 26.16a, p. 578)**. It also had xylem and other tissues typical of all modern vascular plants. Also abundant at the time were representatives of the phylum **Pterophyta**, including ferns and giants such as *Calamites*—huge horsetails that could have a trunk diameter of 30 cm. Some early seed plants were also present, including now-extinct fern like plants, called seed ferns, that bore seeds at the tips of their leaves (Figure 26.16b).

Characterized by a moist climate over much of the planet and by the dominance of seedless vascular plants, the Carboniferous period continued for 150 million years, ending when climate patterns changed during the Paleozoic era. Most modern seedless vascular plants are confined largely to wet or humid environments because they require external water for reproduction. However, some are poikilohydric and can survive in a dehydrated state for long periods of time (see "Life on the Edge," Box 26.1, p. 566).

26.4b Modern Lycophytes Are Small and Have Simple Vascular Tissues

Lycophytes were highly diverse 350 mya, when some tree-sized forms inhabited lush swamp forests. Today, however, such giants are no more. The most familiar of the 1000 or so living species of lycophytes are club mosses (e.g., species of *Lycopodium* and *Selaginella*), which grow on forest floors, in alpine meadows, and in some prairie habitats. **(Figure 26.17, p. 578)**. For example, *Selaginella densa* (Figure 26.17b) is a dominant plant in shortgrass prairies of western North America. Club moss sporophytes have upright or horizontal stems that contain xylem and bear small green leaves and roots. Sporangia are clustered at the bases of specialized leaves, called **sporophylls** (*phyll* = leaf; thus, sporophyll = "spore-bearing leaf"). Sporophylls are clustered into a **cone** or **strobilus** (plural, *strobili*) at the tips of stems. Most lycophytes are homosporous, but some are heterosporous, producing two types of spores that will in turn produce separate male and female gametophytes.

26.4c Ferns, Whisk Ferns, Horsetails, and Their Relatives Make Up the Diverse Phylum Pterophyta

Second in size only to the flowering plants, the phylum Pterophyta (*pteron* = wing) contains a large and diverse group of vascular plants—the 13 thousand or so species of ferns, whisk ferns, and horsetails. Most ferns, including some that are popular houseplants, are native to tropical and temperate regions. Some floating species are less than 1 cm across, whereas some tropical tree ferns grow to 25 m tall. Other species are adapted to life in arctic and alpine tundras, salty mangrove swamps, and semi-arid deserts.

Features of Ferns. The familiar plant body of a fern is the sporophyte phase **(Figure 26.18, p. 579)**, which produces an above-ground clump of leaves. Young leaves are tightly coiled, and as they emerge above the soil,

a. *Rhynia*

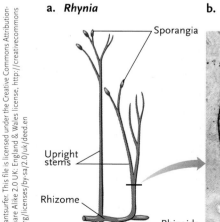

b. Cross-section of *Rhynia gwynne-vaughnii*

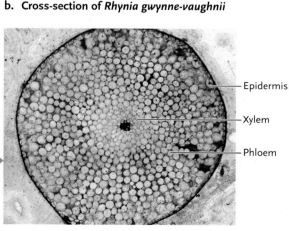

Figure 26.15

Rhynia, an early seedless vascular plant. **(a)** Fossil-based reconstruction of the entire plant, about 30 cm tall. **(b)** Cross-section of the stem, approximately 3 mm in diameter. This fossil was embedded in chert approximately 400 mya. Still visible in it are traces of the transport tissues xylem and phloem, along with other specialized tissues.

a. The lycophyte tree (*Lepidodendron*)   **b. Artist's depiction of a Coal Age forest**

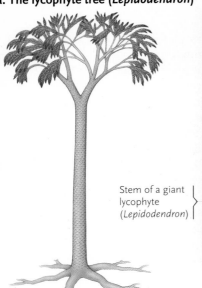

Stem of a giant lycophyte (*Lepidodendron*)

John Weinstein/Field Museum Library/Getty Images

Seed fern (*Medullosa*); probably related to the progymnosperms, which may have been among the earliest seed-bearing plants

Stem of a giant horsetail (*Calamites*)

Figure 26.16
Reconstruction of a lycophyte tree (*Lepidodendron*) and its environment. **(a)** Fossil evidence suggests that *Lepidodendron* grew to be about 35 m tall with a trunk 1 m in diameter. **(b)** Artist's depiction of a Coal Age forest.

Figure 26.17
Lycophytes. **(a)** *Lycopodium* sporophyte, showing the conelike strobili in which spores are produced. **(b)** *Selaginella densa* sporophytes.

a. *Lycopodium* sporophyte

b. *Selaginella densa* sporophytes

Strobilus

Ed Reschke

Dave Powell, USDA Forest Service, Bugwood.org

these fiddleheads (so named because they resemble the scrolled pegheads of violins) unroll and expand. The fiddleheads of some species are edible when cooked, tasting similar to fresh asparagus, but be sure you have collected the right type of fiddlehead—some species contain a carcinogen.

Sporangia are produced on the lower surface or margins of leaves. Often several sporangia are clustered into a rust-coloured **sorus** ("heap"; plural, *sori*) (see Figure 26.17). Spores released from sporangia develop into gametophytes, which are typically small, heart-shaped plants anchored to the soil by rhizoids. Antheridia and archegonia develop on the underside of gametophytes, where moisture is trapped. Inside an

antheridium is a globular packet of haploid cells, each of which develops into a helical sperm with many flagella. When water is present, the antheridium bursts, releasing the sperm. If mature archegonia are nearby, the sperm swim toward them, drawn by a chemical attractant that diffuses from the neck of the archegonium, which is open when free water is present.

In some ferns, antheridia and archegonia are produced on a single bisexual gametophyte. In other ferns, the first spores to germinate develop into bisexual gametophytes, which produce a chemical (antheridiogen) that diffuses through the substrate and causes all later-germinating spores to develop into male gametophytes. What is the advantage of producing a

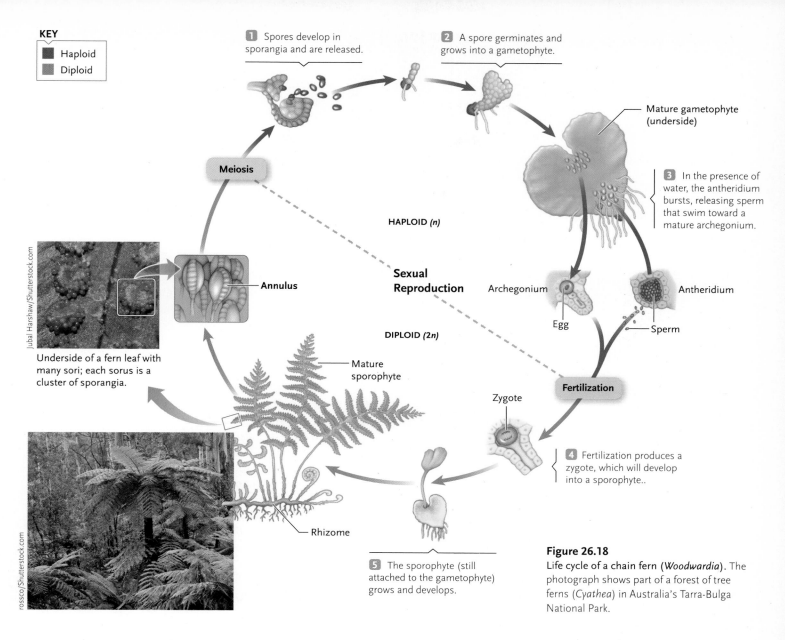

KEY
- ■ Haploid
- ■ Diploid

1 Spores develop in sporangia and are released.

2 A spore germinates and grows into a gametophyte.

Meiosis

HAPLOID (*n*)

Mature gametophyte (underside)

3 In the presence of water, the antheridium bursts, releasing sperm that swim toward a mature archegonium.

Annulus

Sexual Reproduction

Archegonium

Antheridium

Egg

Sperm

DIPLOID (*2n*)

Underside of a fern leaf with many sori; each sorus is a cluster of sporangia.

Jubal Harshaw/Shutterstock.com

Mature sporophyte

Fertilization

Zygote

4 Fertilization produces a zygote, which will develop into a sporophyte..

Rhizome

rossco/Shutterstock.com

5 The sporophyte (still attached to the gametophyte) grows and develops.

Figure 26.18
Life cycle of a chain fern (*Woodwardia*). The photograph shows part of a forest of tree ferns (*Cyathea*) in Australia's Tarra-Bulga National Park.

few bisexual gametophytes followed by many male gametophytes? If a bisexual gametophyte is surrounded by several male gametophytes that developed from other spores, it is more likely that eggs will be fertilized by sperm from one of the male gametophytes rather than by its own sperm, thus increasing the genetic diversity of the resulting zygote.

An embryo is retained on and nourished by the gametophyte for the first part of its life but soon develops into a young sporophyte larger than the gametophyte, with its own green leaf and root system. Once the sporophyte is nutritionally independent, the parent gametophyte degenerates and dies.

Features of Whisk Ferns. The whisk ferns and their relatives are represented by only 2 genera, with about 10 species in total; we look at just one genus, *Psilotum* **(Figure 26.19)**. Whisk ferns grow in tropical and subtropical regions, often as epiphytes.

Figure 26.19
Sporophytes of a whisk fern (*Psilotum*), a seedless vascular plant. Three-lobed sporangia occur at the ends of stubby branchlets; inside the sporangia, meiosis gives rise to haploid spores.

William Ormerod/Visuals Unlimited, Inc.

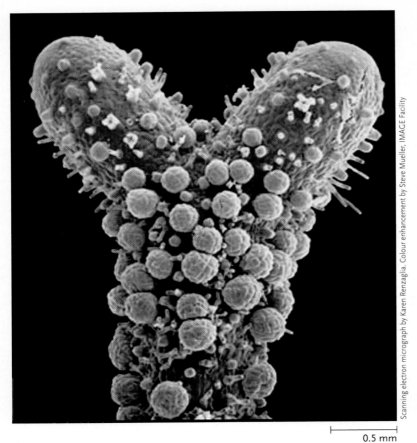

0.5 mm

Scanning electron micrograph by Karen Renzaglia. Colour enhancement by Steve Mueller, IMAGE Facility

Figure 26.20
Scanning electron micrograph image of the subterranean gametophyte of *Psilotum*. Antheridia have been coloured blue, and the smaller archegonia have been coloured pink.

The sporophytes of *Psilotum* resemble the extinct vascular plants in that they lack true roots and leaves. Instead, small, leaflike scales adorn an upright, green, branching stem, which arises from a horizontal rhizome system anchored by rhizoids. Symbiotic fungi colonize the rhizoids, increasing the plant's uptake of soil nutrients (read more about these mycorrhizal fungi in Chapter 25). The stem is photosynthetic and bears sporangia above the small scales. Gametophytes of *Psilotum* are nonphotosynthetic and live underground **(Figure 26.20)**; like the sporophyte, they obtain nutrients via symbioses with mycorrhizal fungi.

Features of Horsetails. The ancient relatives of modern-day horsetails included treelike forms taller than a two-storey building. Only 15 species in a single genus, *Equisetum*, have survived to the present **(Figure 26.21)**. Horsetails grow in moist soil along streams and in disturbed habitats, such as roadsides and beds of railway tracks. Their sporophytes typically have underground rhizomes and roots that anchor the rhizome to the soil. Small, scalelike leaves are arranged in whorls about a photosynthetic stem that is stiff and gritty because horsetails accumulate silica in their tissues. Pioneers used them to scrub out pots and pans—hence their other common name, "scouring rushes."

As in lycophytes, *Equisetum* sporangia are borne in strobili. Haploid spores germinate within a few days to produce gametophytes, which are free-living plants about the size of a small pea.

a. Sporophyte stem   **b.** Sporangia

Strobilus, an aggregation of sporangia and sporophylls at the tip of the horsetail sporophyte

© William Ferguson

W. H. Hodges

Copyright by Curtis Clark

c. This longitudinal section through a horsetail's strobilus shows sporangia containing spores formed by meiosis.

Figure 26.21
A species of *Equisetum*, the horsetails. **(a)** Vegetative stem. **(b)** Strobili, which bear sporangia. **(c)** Close-up of sporangium and associated structures on a strobilus.

26.4d Some Seedless Vascular Plants Are Heterosporous

Most seedless vascular plants are homosporous, but some (e.g., some lycophytes and some ferns) are heterosporous, producing microspores and megaspores in separate sporangia (see Figure 26.8, p. 570). Both types of spores are usually shed from sporangia and germinate on the ground some distance from the parent plant. In many heterosporous plants, the gametophytes produced by the spores develop inside the spore wall; this **endosporous** development provides increased protection for the gametes and, later, for the developing embryo. The microspore gives rise to a male gametophyte, which produces motile sperm. At maturity, the microspore wall will rupture, releasing the sperm, which swim to the female gametophyte; water is thus still required for fertilization in these plants. The megaspore produces a female gametophyte inside the spore wall; archegonia of this gametophyte produce eggs, as in other seedless plants.

1. Compare the lycophyte and bryophyte life cycles with respect to the sizes and longevity of gametophyte and sporophyte phases.
2. Summarize the main similarities and differences in the life cycles of lycophytes, horsetails, whisk ferns, and ferns.
3. Define *sorus* and *strobilus*. How are these two structures similar?

them contributed to the radiation of gymnosperms into land environments.

As a prelude to our survey of modern gymnosperms, we begin by considering some of these innovations.

26.5 Gymnosperms: The First Seed Plants

Gymnosperms are the conifers and their relatives. The earliest fossils identified as gymnosperms are found in Devonian rocks. By the Carboniferous, when nonvascular plants were dominant, many lines of gymnosperms, including conifers, had also evolved. These radiated during the Permian period; the Mesozoic era that followed, 65 to 248 mya, was the age not only of the dinosaurs but of the gymnosperms as well.

The evolution of gymnosperms marked sweeping changes in plant structures related to reproduction. The evolution of gymnosperms included important reproductive adaptations—pollen and pollination, the ovule, and the seed. The fossil record has not revealed the sequence in which these changes arose, but all of

26.5a Major Reproductive Adaptations Occurred as Gymnosperms Evolved

The word *gymnosperm* is derived from the Greek *gymnos,* meaning naked, and *sperma,* meaning seed. As this name indicates, gymnosperms produce seeds that are exposed, not enclosed in fruit, as are the seeds of other seed plants.

Ovules: Increased Protection for Female Gametophyte and Egg. How did seeds first arise? Think about the heterosporous plants described in the previous section and picture two steps that would lead us toward the development of a seed. In the first step, spores are not shed from the plant but instead are retained inside sporangia on the sporophyte. In the second step, the number of megaspores is reduced to just one per sporangium (i.e., four megaspores are produced by meiosis, but only one survives). These two steps result in retention of a single megaspore inside a megasporangium on a plant **(Figure 26.22).** As in all land plants, the megaspore will give rise to a female gametophyte; because this is a heterosporous plant, the gametophyte will develop inside the megaspore wall and inside the megasporangium. Physically connected to the sporophyte and surrounded

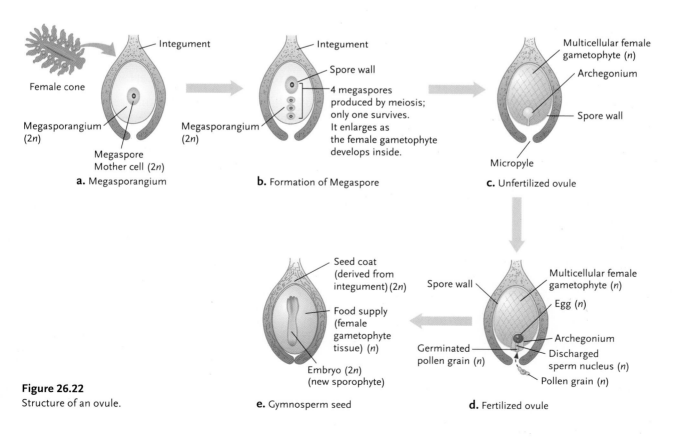

Figure 26.22
Structure of an ovule.

a. Megasporangium
b. Formation of Megaspore
c. Unfertilized ovule
d. Fertilized ovule
e. Gymnosperm seed

by protective layers, a female gametophyte no longer faces the same risks of predation or environmental assault that can threaten a free-living gametophyte.

This new structure, of an egg developing inside a gametophyte that is retained not only inside the spore wall but also inside megasporangial tissue, is an **ovule.** When fertilized, an ovule becomes a **seed:** the fertilized egg will produce an embryo surrounded by nutritive tissue, all encased in sporangial tissue that has become a seed coat.

When you look at Figure 26.22, p. 581, you can see that the megasporangium is surrounded by extra layers of sporophyte tissue, which would add additional protection for gametes and embryos, but this tissue, along with that of the megasporangium, has also created a problem: how can sperm get to the egg now that the gametophyte is enclosed inside these layers of tissue? The solution is similar to that of internal fertilization in animals: there needs to be a male structure that can penetrate sporophyte tissue and release sperm inside the female gametophyte. In the next section, we look at the male gametophyte in seed plants.

Pollen: Eliminating the Need for Water in Reproduction.

As for megaspores, the microspores of gymnosperms (and other seed plants) are not dispersed. Instead, they are retained inside microsporangia and are enveloped in additional layers of sporophyte tissue. As in other heterosporous plants, each microspore produces a male gametophyte, which develops inside the microspore wall. This male gametophyte is very small relative to those of nonseed plants—it is made of only a few cells—and is called a **pollen grain.** Pollen grains are transferred to female reproductive parts via air currents or on the bodies of animal pollinators; this transfer is known as **pollination.** When the pollen grain lands on female tissue, the pollen grain germinates to produce a **pollen tube (Figure 26.23),** a cell that grows through female gametophyte tissue by invasive growth and carries the nonmotile sperm to the egg.

Pollen and pollination were enormously important adaptations for gymnosperms because the shift to nonswimming sperm, along with a means for delivering them to female gametes, meant that reproduction no longer required liquid water. The only gymnosperms that have retained swimming sperm are the cycads and ginkgoes described below, which have relatively few living species and are restricted to just a few native habitats.

Seeds: Protecting and Nourishing Plant Embryos.

As described above, a seed is the structure that forms when an ovule matures, after a pollen grain reaches it and a sperm fertilizes the egg. Seeds consist of three basic parts: (1) the embryo sporophyte; (2) the tissues surrounding the embryo containing nutrients that nourish it until it becomes established as a seedling with leaves and roots; and (3) a tough, protective outer seed coat **(Figure 26.24).** This complex structure makes seeds ideal packages for sheltering an embryo from drought, cold, or other adverse conditions. As a result, seed plants enjoy a tremendous survival advantage over species that simply release spores to the environment. Encased in a seed, the embryo can also be transported far from its parent, as when ocean currents carry coconut seeds ("coconuts" protected in large, buoyant fruits) hundreds of kilometres across the sea. As discussed in Chapter 36, some plant embryos housed in seeds can remain dormant for months or years before environmental conditions finally prompt them to germinate and grow.

26.5b Modern Gymnosperms Include Conifers and a Few Other Groups Represented by Relatively Few Species That Tend to Be Restricted to Certain Climates

Today there are about 800 gymnosperm species. The sporophytes of nearly all are large trees or shrubs, although a few are woody vines.

Economically, gymnosperms, particularly conifers, are vital to human societies. They are sources of lumber, paper pulp, turpentine, and resins, among other products. They also have huge ecological importance. Their habitats range from **tropical forests** to deserts, but gymnosperms are most dominant in the cool-temperate

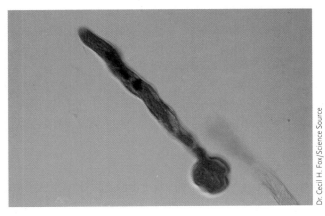

Figure 26.23
Pollen tube extending from germinating pollen grain at top right.

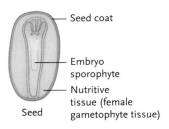

Figure 26.24
Generalized view of the seed of a pine, a gymnosperm.

zones of the Northern and Southern Hemispheres. They flourish in poor soils, where flowering plants don't compete as well. In Canada, for example, gymnosperms make up most of the boreal forests that cover about one-third of the country's landmass. Our survey of gymnosperms begins with the conifers, and then we will look at the cycads, ginkgoes, and gnetophytes—the latter two groups are remnants of lineages that have all but vanished from the modern scene.

Conifers Are the Most Common Gymnosperms.
About 80% of all living gymnosperm species are members of one phylum, the **Coniferophyta**, or conifers ("cone-bearers"). Examples are pines, spruces, and firs. Coniferous trees and shrubs are longer lived, and anatomically and morphologically more complex, than any sporophyte phase we have discussed so far. Characteristically, they form woody cones, and most have needlelike leaves that are adapted to dry environments. For instance, needles have a thick cuticle, sunken stomata, and a fibrous epidermis, all traits that reduce the loss of water vapour.

Pines and many other gymnosperms produce resins, a mix of organic compounds that are by-products of metabolism. Resin accumulates and flows in long resin ducts through the wood, inhibiting the activity of wood-boring insects and certain microbes. Pine resin extracts are the raw material of turpentine and (minus the volatile terpenes) the sticky rosin used to treat violin bows. Fossil resin is known as amber and is commonly used in jewellery; amber often contains fossilized insects or even small animals.

We know a great deal about the pine life cycle **(Figure 26.25, p. 584)**, so it is a convenient model for gymnosperms. Male cones are relatively small and delicate (about 1 cm long) and are borne on the lower branches. Each cone consists of many sporophylls with two microsporangia on their undersides. Inside the microsporangia, **microspores** are produced by meiosis. Each microspore then undergoes mitosis to develop into a winged pollen grain—an immature male gametophyte. At this stage, the pollen grain consists of four cells, two that will degenerate and two that will function later in reproduction.

Young female cones develop higher in the tree, at the tips of upper branches. Ovules are produced on modified sporophylls. Inside each ovule, four megaspores are produced by meiosis, but only one survives to develop into a megagametophyte. This female gametophyte develops slowly, becoming mature only when pollination is under way; in a pine, this process takes well over a year. The mature female gametophyte is a small oval mass of cells with several archegonia at one end, each containing an egg.

Each spring, air currents release vast numbers of pollen grains from male cones—by some estimates, billions may be released from a single pine tree. The extravagant numbers ensure that at least some pollen grains will land on female cones. The process is not as random as it might seem: studies have shown that the contours of female cones create air currents that can favour the "delivery" of pollen grains near the cone scales. After pollination, the two remaining cells of the pollen grain divide, one producing sperm by mitosis, the other producing the pollen tube that grows toward the developing gametophyte. When a pollen tube reaches an egg, the stage is set for fertilization, the formation of a zygote, and early development of the plant embryo. Often fertilization occurs months to a year after pollination. Once an embryo forms, a pine seed—which, remember, includes the embryo, female gametophyte tissue, and seed coat—is eventually shed from the cone. The seed coat protects the embryo from drying out, and the female gametophyte tissue serves as its food reserve. This tissue makes up the bulk of a "pine nut."

Cycads Are Restricted to Warmer Climates.
During the Mesozoic era, the **Cycadophyta** (*kykas* = palm), or cycads, flourished along with the dinosaurs. About 185 species have survived to the present, but they are confined to the tropics and subtropics.

At first glance, you might mistake a cycad for a small palm tree **(Figure 26.26, p. 585)**. Some cycads have massive cones that bear either pollen or ovules. Air currents or crawling insects transfer pollen from male plants to the developing gametophyte on female plants. Poisonous alkaloids that may help deter insect predators occur in various cycad tissues. In tropical Asia, some people consume cycad seeds and flour made from cycad trunks, but only after rinsing away the toxic compounds. Much in demand from fanciers of unusual plants, cycads in some countries are uprooted and sold in what amounts to a black-market trade, greatly diminishing their numbers in the wild.

Ginkgoes Are Limited to a Single Living Species.
The phylum **Ginkgophyta** has only one living species, the ginkgo (or maidenhair) tree (*Ginkgo biloba*), which grows wild today only in warm-temperate forests of central China. Ginkgo trees are large, diffusely branching trees with characteristic fan-shaped leaves **(Figure 26.27, p. 585)** that turn a brilliant yellow in autumn. Nursery-propagated male trees are often planted in cities because they are resistant to insects, disease, and air pollutants. The female trees are equally pollution resistant, but gardeners avoid them because their seeds produce a foul odour that only a ginkgo could love. The leaves and seeds have been used in traditional Chinese medicine for centuries. The extract of the leaves is one of the most intensely investigated herbal medicines; although studies have not found any conclusive evidence for claims that the extract improves memory, there is some evidence that it does assist in blood flow and so may be effective in the treatment of circulatory disorders.

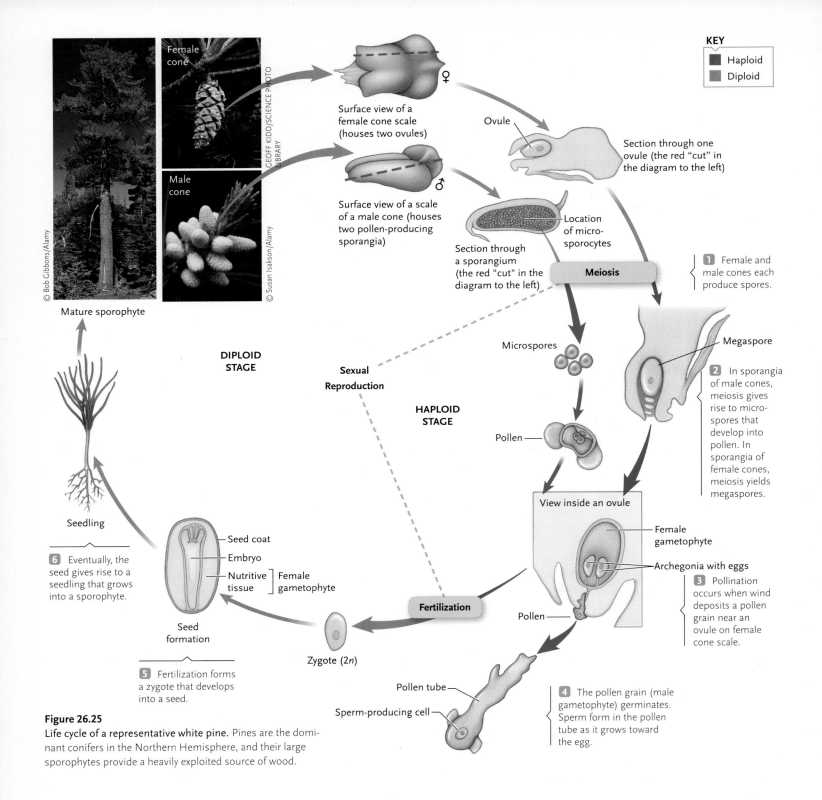

KEY
- ■ Haploid
- ■ Diploid

Female cone

Male cone

© Bob Gibbons/Alamy

© Susan Isakson/Alamy

GEOFF KIDD/SCIENCE PHOTO LIBRARY

Mature sporophyte

Surface view of a female cone scale (houses two ovules) ♀

Surface view of a scale of a male cone (houses two pollen-producing sporangia) ♂

Ovule

Section through one ovule (the red "cut" in the diagram to the left)

Location of micro-sporocytes

Section through a sporangium (the red "cut" in the diagram to the left)

Meiosis

DIPLOID STAGE

Sexual Reproduction

HAPLOID STAGE

Microspores

Megaspore

Pollen

1 Female and male cones each produce spores.

2 In sporangia of male cones, meiosis gives rise to micro-spores that develop into pollen. In sporangia of female cones, meiosis yields megaspores.

Seedling

6 Eventually, the seed gives rise to a seedling that grows into a sporophyte.

Seed coat
Embryo
Nutritive tissue } Female gametophyte

Seed formation

5 Fertilization forms a zygote that develops into a seed.

Zygote (2n)

Fertilization

View inside an ovule

Female gametophyte

Archegonia with eggs

Pollen

3 Pollination occurs when wind deposits a pollen grain near an ovule on female cone scale.

Pollen tube

Sperm-producing cell

4 The pollen grain (male gametophyte) germinates. Sperm form in the pollen tube as it grows toward the egg.

Figure 26.25
Life cycle of a representative white pine. Pines are the domi-nant conifers in the Northern Hemisphere, and their large sporophytes provide a heavily exploited source of wood.

Gnetophytes Include Simple Seed Plants with Intriguing Features. The phylum Gnetophyta contains three genera—*Gnetum, Ephedra,* and *Welwitschia*—that together include about 70 species. Moist, tropical regions are home to about 30 species of *Gnetum,* which includes both trees and leathery-leafed vines (lianas). About 35 species of *Ephedra* grow in desert regions of the world **(Figure 26.28a–c).**

Of all the gymnosperms, *Welwitschia* is the most bizarre. This seed-producing plant grows in the hot deserts of southwest Africa. The bulk of the plant is a deep-reaching taproot. The only exposed part is a woody, disk-shaped stem that bears cone-shaped strobili and leaves. The plant never produces more than two strap-shaped leaves, which split lengthwise repeatedly as the plant grows older, producing a rather scraggly pile (Figure 26.28d).

1. What are the four major reproductive adaptations that evolved in gymnosperms?
2. What are the basic parts of a seed, and how is each one adaptive?
3. Summarize the main similarities and differences among ginkgoes, cycads, gnetophytes, and conifers.
4. Describe some features that make conifers structurally more complex than other gymnosperms.

a. *Ephedra* plant

b. *Ephedra* male cone

c. *Ephedra* female cone

d. *Welwitschia* plant with female cones

Figure 26.26
The cycad *Zamia* showing a large, terminal female cone and fernlike leaves.

Figure 26.28
Gnetophytes. **(a)** Sporophyte of *Ephedra,* with close-ups of its **(b)** pollen-bearing cones and **(c)** seed-bearing cone, which develop on separate plants. **(d)** Sporophyte of *Welwitschia mirabilis,* with seed-bearing cones.

Figure 26.27
Ginkgo biloba. **(a)** A ginkgo tree. **(b)** A fossilized ginkgo leaf compared with a leaf from a living tree. The fossil formed at the Cretaceous–Tertiary boundary. Even though 65 million years have passed, the leaf structure has not changed much. **(c)** Pollen-bearing cones and **(d)** fleshy-coated seeds of the *Ginkgo.*

a. Ginkgo tree

b. Fossil and modern ginkgo leaves

c. Male cone

d. Ginkgo seeds

26.6 Angiosperms: Flowering Plants

Of all plant phyla, the flowering plants, or **angiosperms**, are the most successful today. At least 260 thousand species are known **(Figure 26.29,** shows a few examples), and botanists regularly discover new ones in previously unexplored regions of the tropics. The word *angiosperm* is derived from the Greek *angeion* ("vessel") and *sperma* ("seed"). The "vessel" refers to the modified sporophyll, called a *carpel,* which surrounds and protects the ovules. Carpels are located in the centre of **flowers**, reproductive structures that are a defining feature of angiosperms. Another defining feature is the **fruit**—botanically speaking, a structure that helps protect and disperse seeds.

In addition to having flowers and fruits, angiosperms are the most ecologically diverse plants on Earth, growing on dry land and in **wetlands**, fresh water, and the seas. Angiosperms range in size from tiny duckweeds that are about 1 mm long to towering *Eucalyptus* trees more than 60 m tall.

26.6a The Fossil Record Provides Little Information about the Origin of Flowering Plants

The evolutionary origin of angiosperms has confounded plant biologists for well over a hundred years. Charles Darwin called it the "abominable mystery" because flowering plants appear suddenly in the fossil record, without a fossil sequence that links them to any other plant groups. As with gymnosperms, attempts to reconstruct the earliest flowering plant lineages have produced several conflicting classifications and family trees. Some paleobotanists hypothesize that flowering plants arose in the Jurassic period; others propose that

a. Flowering plants in a desert

b. Alpine angiosperms

c. Triticale, a grass

d. The carnivorous plant Venus flytrap

Figure 26.29

Flowering plants. Diverse photosynthetic species are adapted to nearly all environments, ranging from **(a)** deserts to **(b)** snowlines of high mountains. **(c)** Triticale, a hybrid grain derived from parental stocks of wheat (*Triticum*) and rye (*Secale*), is one example of the various grasses used by humans. **(d)** The carnivorous plant Venus flytrap (*Dionaea muscipula*) grows in nitrogen-poor soils and traps insects as an additional source of nitrogen.

they evolved in the Triassic from now-extinct gymnosperms or from seed ferns. However, progress in this area does not rely solely on fossil evidence; molecular data can be used to test hypotheses, and the combination of molecular, morphological, and fossil evidence offers great promise in solving this mystery.

The fossil record has yet to reveal obvious transitional organisms between flowering plants and either gymnosperms or seedless vascular plants. As the Mesozoic era ended and the modern Cenozoic era began, great extinctions occurred among both plant and animal kingdoms. Gymnosperms declined, and dinosaurs disappeared. Flowering plants, mammals, and social insects flourished, radiating into new environments. Today we live in what has been called "the age of flowering plants."

26.6b Angiosperms Are Subdivided into Several Groups, Including Monocots and Eudicots

Angiosperms are assigned to the phylum **Anthophyta**, a name that derives from the Greek *anthos,* meaning flower. The great majority of angiosperms are classified as either monocots or eudicots, which are differentiated on the basis of morphological features such as the number of flower parts and the pattern of vascular tissue in stems and leaves. The two groups also differ in terms of the morphology of their embryos: **monocot** embryos have a single leaf like structure called a cotyledon, whereas **eudicot** ("true dicots") embryos generally have two cotyledons (see Table 26.1, p. 568).

Botanists currently recognize several other groups of plants in addition to eudicots and monocots, but figuring out the appropriate classification for and relationships among these other groups is an ongoing challenge and an extremely active area of plant research. In this chapter, we focus only on monocots and eudicots.

There are at least 60 thousand species of monocots, including 10 thousand grasses and 20 thousand orchids. **Figure 26.30a** gives some idea of the variety of living monocots, which include grasses, palms, lilies, and orchids. The world's major crop plants (wheat, corn, rice, rye, sugar cane, and barley) are all monocots and are all domesticated grasses. Eudicots are even more diverse, with nearly 200 thousand species (Figure 26.30b). They include flowering shrubs and trees, most

a. Representative monocots

Wheat (*Triticum*)

Trillium (*Trillium*)

Western wood lily
(*Lilium philadelphicum*)

b. Representative eudicots

Wild rose (*Rosa acicularis*)

Twinflower (*Linnaea borealis*)

Claret cup cactus
(*Echinocereus triglochidiatus*)

Figure 26.30

Examples of monocots and eudicots. **(a)** Representative monocots: wheat (*Triticum*), trillium (*Trillium*), and Western wood lily (*Lilium philadelphicum*). **(b)** Representative eudicots: wild rose (*Rosa acicularis*), twinflower (*Linnaea borealis*), and cactus (*Echinocereus triglochidiatus*).

nonwoody (herbaceous) plants, and cacti. We will take a closer look at angiosperms in Chapter 36, which focuses on the structure and function of flowering plants.

26.6c Many Factors Contributed to the Adaptive Success of Angiosperms

Flowering plants likely originated about 140 mya. It took only about 40 million years—a short span in geologic time—for angiosperms to eclipse gymnosperms as the prevailing form of plant life on land. Several factors fuelled this adaptive success. As with other seed plants, the large, diploid sporophyte phase dominates a flowering plant's life cycle, and the sporophyte retains and nourishes the much smaller gametophytes. But flowering plants also show some evolutionary innovations not seen in gymnosperms.

More Efficient Transport of Water and Nutrients. Where gymnosperms have only one type of water-conducting cell in their xylem, angiosperms have an additional, more specialized type of cell that is larger and open ended and thus moves water more rapidly from roots to shoots (see Chapter 34). Also, modifications in angiosperm phloem tissue allow it to more efficiently transport sugars produced in photosynthesis through the plant body.

Enhanced Nutrition and Physical Protection for Embryos. Other changes in angiosperms increased the likelihood of successful reproduction and dispersal of offspring. For example, a two-step double-fertilization process in the ovules of flowering plants produces both an embryo and a unique nutritive tissue (called endosperm) that nourishes the embryonic sporophyte **(Figure 26.31)**. The ovule containing a female gametophyte is enclosed within an ovary, part of the carpel, which shelters the ovule against desiccation and against attack by herbivores or pathogens. After fertilization, an ovary develops into a fruit that not only protects seeds but also helps disperse them—for instance, when an animal eats a fruit, seeds may pass through the animal's gut none the worse for the journey and be released in a new location in the animal's feces. Above all, angiosperms have flowers, the unique reproductive organs that you will read much more about in Chapter 36.

26.6d Angiosperms Coevolved with Animal Pollinators

The evolutionary success of angiosperms is due not only to the adaptations just described but also to the efficient mechanisms of transferring pollen to female reproductive parts. Whereas a conifer depends on air currents to disperse its pollen, as do such angiosperms as grasses, many angiosperms coevolved with pollinators—insects, bats, birds, and other animals that transfer pollen from male floral structures to female reproductive parts, often while obtaining nectar. Nectar is a sugar-rich liquid secreted by flowers to attract pollinators. Pollen itself is a reward for some pollinators, such as bees, that use it as a food resource. So, while plants benefit from their animal pollinators, there is also a cost to the plant in providing a reward to the pollinator. **Coevolution** occurs when two or more species interact closely in the same ecological setting. A heritable change in one species affects selection pressure operating between them, so that the other species evolves as well. Over time, plants have coevolved with their pollinating animals.

In general, a flower's reproductive parts are positioned so that visiting pollinators will brush against them. In addition, many floral features correlate with the morphology and behaviour of specific pollinators. For example, reproductive parts may be located above nectar-filled floral tubes that are the same length as the feeding structure of a preferred pollinator. Nectar-sipping bats **(Figure 26.32a, p. 590)** and moths forage by night. They pollinate intensely sweet-smelling flowers with white or pale petals that are more visible than coloured petals in the dark. The long, thin mouthparts of moths and butterflies reach nectar in narrow floral tubes or floral spurs. The Madagascar hawkmoth uncoils a mouthpart the same length—an astonishing 22 cm—as the narrow flower of the orchid it pollinates, *Angraecum sesquipedale* (Figure 26.32b). Red and yellow flowers attract birds (Figure 26.32c), which have good daytime vision but a poor sense of smell. Hence, bird-pollinated plants do not squander metabolic resources to make fragrances. By contrast, flowers of species that are pollinated by beetles or flies may smell like rotten meat, dung, or decaying matter. This trickery by the plants is known as signal mimicry: the plant uses visual and olfactory signals to trick a pollinator into visiting it; some of these plants provide no nutritional reward for their pollinators at all. Daisies and other fragrant flowers with distinctive patterns, shapes, and red or orange components attract butterflies, which forage by day.

Bees see ultraviolet light and visit flowers with sweet odours and parts that appear to humans as yellow, blue, or purple (Figure 26.32d). Produced by pigments that absorb ultraviolet light, the colours form patterns called "nectar guides" that attract bees—which may pick up or drop off pollen during the visit. Here, as in our other examples, flowers contribute to the reproductive success of plants that bear them.

In this chapter, we have introduced some of the strategies that plants use to meet the challenges of life on Earth; they face the same challenges as animals and other terrestrial organisms (attract a mate,- reproduce, disperse offspring, and survive unfavourable conditions) but have had to find ways to do all of these without being able to move around (they are sessile).

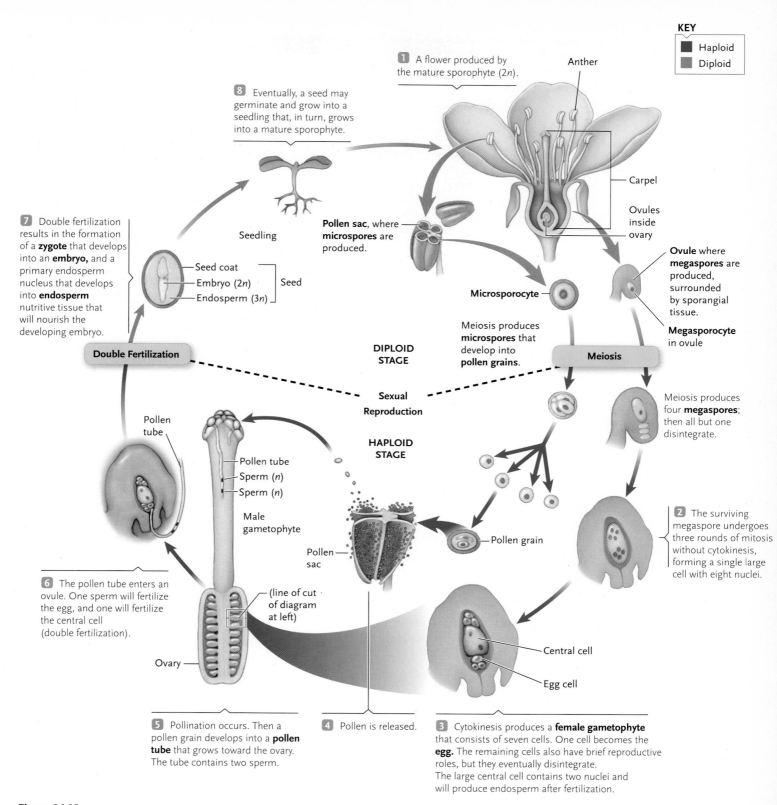

KEY
- Haploid
- Diploid

1 A flower produced by the mature sporophyte (2n).

Anther

Carpel

Ovules inside ovary

8 Eventually, a seed may germinate and grow into a seedling that, in turn, grows into a mature sporophyte.

Pollen sac, where **microspores** are produced.

Ovule where **megaspores** are produced, surrounded by sporangial tissue.

Seedling

7 Double fertilization results in the formation of a **zygote** that develops into an **embryo,** and a primary endosperm nucleus that develops into **endosperm** nutritive tissue that will nourish the developing embryo.

Seed coat
Embryo (2n)   Seed
Endosperm (3n)

Microsporocyte

Megasporocyte in ovule

Double Fertilization

DIPLOID STAGE

Meiosis produces **microspores** that develop into **pollen grains.**

Meiosis

Meiosis produces four **megaspores;** then all but one disintegrate.

Sexual Reproduction

HAPLOID STAGE

Pollen tube

Pollen tube
Sperm (n)
Sperm (n)

Male gametophyte

Pollen grain

2 The surviving megaspore undergoes three rounds of mitosis without cytokinesis, forming a single large cell with eight nuclei.

Pollen sac

6 The pollen tube enters an ovule. One sperm will fertilize the egg, and one will fertilize the central cell (double fertilization).

(line of cut of diagram at left)

Central cell

Egg cell

Ovary

5 Pollination occurs. Then a pollen grain develops into a **pollen tube** that grows toward the ovary. The tube contains two sperm.

4 Pollen is released.

3 Cytokinesis produces a **female gametophyte** that consists of seven cells. One cell becomes the **egg.** The remaining cells also have brief reproductive roles, but they eventually disintegrate. The large central cell contains two nuclei and will produce endosperm after fertilization.

Figure 26.31
Life cycle of a typical flowering plant. Double fertilization is a notable feature of the cycle. The male gametophyte delivers two sperm to an ovule. One sperm fertilizes the egg, forming the embryo, and the other fertilizes the endosperm-producing cell, which nourishes the embryo.

a. Bat pollinating a giant saguaro

b. Hawkmoth pollinating an orchid

c. Hummingbird visiting a hibiscus flower

d. Bee-attracting pattern of a marsh marigold

Visible light   UV light

Figure 26.32

Coevolution of flowering plants and animal pollinators. The colours and configurations of some flowers, and the production of nectar or odours, have coevolved with specific animal pollinators. **(a)** At night, nectar-feeding bats sip nectar from flowers of the giant saguaro cactus (*Carnegia gigantea*), transferring pollen from flower to flower in the process. **(b)** The hawkmoth (*Xanthopan morganii praedicta*) has a proboscis long enough to reach nectar at the base of the equally long floral spur of the orchid *Angraecum sesquipedale*. **(c)** A ruby-throated hummingbird (*Archilochus colubris*) sipping nectar from a hibiscus blossom (*Hibiscus*). The long, narrow bill of hummingbirds coevolved with long, narrow floral tubes. **(d)** Under ultraviolet light, the bee-attracting pattern of a gold-petalled marsh marigold becomes visible to human eyes.

Many of these topics are followed up in more detail in the chapters dealing with plant biology (Chapters 34 to 38).

The next two chapters introduce animals. As you read these chapters, look for similarities and differences in how they have addressed the challenges of life compared to plants.

STUDY BREAK

1. What are the advantages and costs to plants of using animals to disperse their pollen?
2. List at least three adaptations that have contributed to the evolutionary success of angiosperms as a group.

Review

aplia

To access course materials such as Aplia and other companion resources, please visit www.NELSONbrain.com.

26.1 Defining Characteristics of Land Plants

- Land plants are multicellular eukaryotes with cellulose cell walls. Most, but not all, are photoautotrophs. All have an alternation of generations life cycle (Figure 26.1), although which generation is dominant varies among groups of plants, and all retain embryos inside parental tissue.

26.2 The Transition to Life on Land

- Plants are thought to have evolved from charophyte green algae between 425 and 490 mya.
- Adaptations to terrestrial life in the earliest land plants include poikilohydry, multicellular chambers that protect developing gametes, and an embryo sheltered inside a parent plant.
- Other key evolutionary trends among land plants included symbiotic associations with fungi (Figure 26.5); the development of vascular tissues (including lignified water-conducting tissue), root systems, and shoot systems; lignified stems and leaves equipped

with stomata (Figure 26.4b); increasing dominance by the diploid sporophyte generation; and a shift from homospory to heterospory.

- Gametophytes became reduced in size (Figure 26.7), male gametophytes (pollen) became specialized for dispersal without liquid water, and female gametophytes became increasingly protected inside sporophyte tissues.

26.3 Bryophytes: Nonvascular Land Plants

- Existing nonvascular land plants, or bryophytes, include the liverworts, hornworts, and mosses Figures 26.12–26.14).
- Bryophytes produce flagellated sperm that swim through free water to reach eggs. They lack xylem and phloem (although some have specialized conductive tissues), lignified tissues, roots, stems, and leaves. The gametophyte phase is dominant.

26.4 Seedless Vascular Plants

- Existing seedless vascular land plants include the lycophytes (club mosses), whisk ferns, horsetails, and ferns (Figures 26.16–26.21). Like bryophytes, they release spores and have swimming sperm. Unlike bryophytes, they have well-developed vascular tissues. The sporophyte generation is dominant and independent of the gametophyte.

- Most seedless vascular plants are homosporous, but some are heterosporous (Figure 26.8).

26.5 Gymnosperms: The First Seed Plants

- Gymnosperms (conifers and their relatives; Figures 26.22–26.28), together with angiosperms (flowering plants), are the seed-bearing vascular plants. Reproductive innovations include pollination, the ovule, and the seed. Liquid water is not required for reproduction.
- During the Mesozoic, gymnosperms were the dominant land plants. Today conifers are the primary vegetation of forests at higher latitudes and elevations and have important economic uses as sources of lumber and other products.

26.6 Angiosperms: Flowering Plants

- Angiosperms (Anthophyta) have dominated the land for more than 100 million years and are currently the most diverse plant group (Figures 26.29 and 26.30).
- The angiosperm vascular system moves water and sugars through the plant body more efficiently than that of gymnosperms. Reproductive adaptations include a protective ovary around the ovule, endosperm, flowers that attract pollinators, and fruits that protect and disperse seeds.

Questions

Self-Test Questions

1. Which of the following correctly describes an evolutionary trend that occurred as land plants evolved?
   a. becoming seedless
   b. producing only one type of spore
   c. producing nonmotile gametes
   d. haploid generation becoming dominant

2. Which of the following occurs in the life cycle of both mosses and angiosperms?
   a. The sporophyte is the dominant generation.
   b. The gametophyte is the dominant generation.
   c. Spores develop into sporophytes.
   d. The sporophyte produces spores.

3. The evolution of which of the following features freed land plants from requiring water for reproduction?
   a. lignified stems
   b. fruits and roots
   c. seeds and pollen
   d. flowers and leaves

4. Which of the following statements about archegonia is correct?
   a. They are found in all land plants.
   b. They are found in all land plants except seed plants.
   c. They are found in all land plants except angiosperms.
   d. They are found in nonvascular land plants but not in vascular plants.

5. Which of the following statements about antheridia is correct?
   a. They are found in all land plants.
   b. They are found in all land plants except seed plants.
   c. They are found in all land plants except angiosperms.
   d. They are found in nonvascular land plants but not in vascular plants.

6. Which of the options correctly pairs a plant group with its phylum?
   a. Hepatophyta: cycads
   b. Pterophyta: horsetails
   c. Bryophyta: gnetophytes
   d. Coniferophyta: angiosperms

7. A homeowner notices moss growing between bricks on his patio. Closer examination reveals tiny brown stalks with cuplike tops emerging from green leaflets. Which of the following are these brown structures?
   a. antheridia
   b. archegonia
   c. the gametophyte generation
   d. the sporophyte generation

8. Horsetails are most closely related to which of the following plant groups?
   a. club mosses and ferns
   b. mosses and whisk ferns
   c. liverworts and hornworts
   d. gnetophytes and gymnosperms

9. In which of the following groups is the evolution of true roots first seen?
   a. mosses
   b. conifers
   c. liverworts
   d. seedless vascular plants

10. Arrange the following adaptations to terrestrial life in the order in which they first appeared during the evolution of land plants:

1. seeds	a. 1, 2, 3, 4
2. vascular tissue	b. 2, 3, 4, 1
3. gametangia	c. 2, 3, 1, 4
4. flowers	d. 3, 2, 1, 4

Questions for Discussion

1. Working in the field, you discover a fossil of a previously undescribed plant species. The specimen is small and may not be complete; the parts you have do not include any floral organs. What evidence would you need to classify the fossil as a seedless vascular plant with reasonable accuracy? What evidence would you need to distinguish between a fossil lycopod and a fern?

2. Compare the size, anatomical complexity, and degree of independence of a moss gametophyte, a fern gametophyte, a Douglas fir (conifer) female gametophyte, and a dogwood (angiosperm) female gametophyte. Which one is the most protected from the external environment? Which trends in plant evolution does your work on this question bring to mind?

3. How has the relative lack of fossil early angiosperms affected our understanding of this group?

4. One of the major challenges plants faced in living on land was coping with the reduced availability of water. Contrast the water-use strategy that the earliest land plants (and mosses living today) use with that used by vascular plants. Suggest some advantages and disadvantages of each strategy.

5. Some vascular plants have motile sperm and some do not. How do male and female gametes meet in (a) plants that have swimming sperm and (b) those that do not?

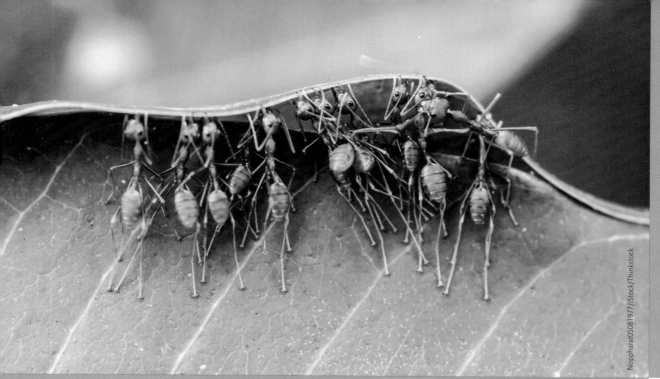

Workers of the weaver ant, *Oecophylla*, engaged in the construction of their nest, which is formed from living leaves, curled or folded to form an envelope held together by silk secreted by the larvae.

Noppharat0508197/iStock/Thinkstock

Diversity of Animals 1: Sponges, Radiata, Platyhelminthes, and Protostomes

WHY IT MATTERS

Beginning about 540 million years ago (mya) in the Early Cambrian, conditions were ripe for rapid development of the marine fauna, and an explosion of new forms of animals appeared, particularly in warm and shallow seas bordering continents. About 505 mya, a series of mud slides carried the animals that lived at the edge of a submarine cliff, the Cathedral Escarpment, over its edge and buried them in fine silt. That mud and its contained fossils formed shale, a sedimentary rock with layers that are easily split apart. During continent and mountain building, the shale beds, now known as the Burgess Shale, came to lie in the Canadian Rocky Mountains of eastern British Columbia in what is now Yoho National Park.

In 1909, Charles Walcott, an American paleontologist working in the Burgess Shale area, located a rich bed of very strange fossils that were exquisitely preserved, including the soft parts. Reconstructions have revealed not only familiar animals such as trilobites and sponges but also many truly bizarre animals **(Figure 27.1, p. 594).** For example, *Opabinia* was about as long as a tube of lipstick and had five eyes on its head and a single anterior grasping organ, which was probably used to catch prey. The smaller *Hallucigenia* had seven pairs of hard spines on its back and what appear to have been seven pairs of softer ventral protuberances that probably functioned for locomotion. Some of the other organisms look like early chordates, but many do not resemble any living animals and may represent phyla never previously described. Furthermore, some of the animals from the Burgess Shale have moved between phyla as more and more details about them became available.

a. *Opabinia*

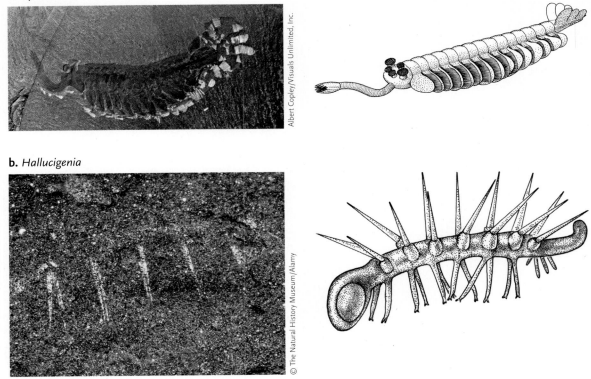

b. *Hallucigenia*

Figure 27.1

Animals of the Burgess Shale. **(a)** *Opabinia* had five eyes and a grasping organ on its head. **(b)** *Hallucigenia* had seven pairs of spines and soft protuberances.

The Burgess Shale and other similar sites elsewhere provide us with a snapshot of some of the animals that inhabited the coastal waters at the time of the Cambrian Explosion. Most of the bizarre forms did not survive the extinctions that were to come. Without these extinctions, some of the forms might have survived to found lineages completely different from those living today.

In this chapter, we introduce the general characteristics of animals and a phylogenetic hypothesis about their evolutionary history and classification. We also survey the major invertebrate phyla belonging to one lineage, the Protostomia. In Chapter 19, we defined the various levels used in the Linnaean system of classifying animals. In Chapter 28, we examine the other major animal lineage, the Deuterostomia, which includes the phylum Chordata and their nearest invertebrate relatives.

In this chapter we begin our consideration of animal diversity with protostomes, metazoans in which the blastopore develops into the mouth and the anus appears as a second opening. Protostomes include molluscs, annelids, and arthropods, as well as other smaller phyla not usually familiar to biology students. While some protostomes have simple body plans with little evidence of organ systems, others have fully developed organ systems. Some protostomes have shells and an extensive fossil record, while others are soft-bodied and not well known as fossils. Protostomes include species with lifestyles across the trophic spectrum from herbivores to carnivores and **detritivores.** Some protostomes are infamous as parasites, usually because they pose important health risks to humans. The other animals we include in this chapter are sponges (poriferans), radiatia, the jellyfish (cnidarians), and flatworms (platyhelminths).

27.1 What Is an Animal?

Most biologists recognize the **kingdom Animalia** as a monophyletic group that is easily distinguished from the other kingdoms.

27.1a All Animals Share Certain Structural and Behavioural Characteristics

Animals are eukaryotic, multicellular organisms. The cell membranes of adjacent animal cells are in direct contact with one another. This is different from plants and fungi, which have cell walls around the cells. Animal cells may be organized into different morphological types, reflecting their role in the functioning of the animal as a single unit.

All animals are **heterotrophs:** they depend on other life forms for their food, either by eating them directly or by living in a parasitic association with them. They use oxygen to metabolize their food through aerobic respiration, and most store excess energy as glycogen, oil, or fat.

All animals are **motile** (able to move from place to place) at some time in their lives. Most familiar animals are motile as adults. However, in some species, such as mussels and barnacles, only the young are motile; they eventually settle down as **sessile** (unable to move from one place to another) adults. All animals are able to perceive and respond to information about the environment in which they live.

Animals reproduce either asexually or sexually; in many groups, they switch from one mode to the other. Sexually reproducing species produce haploid **gametes** (eggs and sperm) that fuse to form diploid **zygotes** (fertilized eggs). For many invertebrates, development to the adult involves one or more **larval forms**. This pattern of development, in which a species can exist during development in two or more distinct forms, is referred to as **polymorphic development**. This developmental strategy is important for the success of many of the invertebrate groups, particularly those with a parasitic lifestyle.

27.1b The Animal Lineage Probably Arose from a Colonial Choanoflagellate Ancestor

Most biologists agree that the common ancestor of all animals was probably a colonial, flagellated protist that lived at least 700 Ma ago, during the Precambrian. It may have resembled the minute, sessile choanoflagellates (see Chapter 24) that live in both freshwater and marine habitats today. In 1874, the German embryologist Ernst Haeckel proposed a colonial, flagellated ancestor, suggesting that it was a hollow, ball-shaped organism with unspecialized cells. Its cells became specialized for particular functions, and a developmental reorganization produced a double-layered, sac-within-a-sac body plan **(Figure 27.2)**. The embryology of many living animals roughly parallels this hypothetical evolutionary transformation. He included this hypothetical organism among what he called the Metazoa (*meta* = more developed; *zoon* = animal) to distinguish them from the Protozoa.

STUDY BREAK

1. What characteristics distinguish animals from plants?
2. What early steps may have led to the first metazoans?

27.2 Key Innovations in Animal Evolution

Once established, the animal lineage diversified quickly into an amazing array of body plans. Biologists have used several key morphological innovations to unravel the evolutionary relationships of the major animal groups.

27.2a Tissues and Tissue Layers Appeared Early in Animal Evolution

In most Metazoans, the process of development gives rise to two or three layers of cells that eventually form **tissues**, groups of similar differentiated cells specialized for particular functions.

In most metazoans, embryonic tissues form as either two or three concentric **germ layers** (see Chapter 42). The innermost layer, the **endoderm**, eventually develops into the lining of the gut (digestive system) and, in some animals, respiratory organs. The outermost layer, the **ectoderm**, forms the external covering and nervous system. Between the two, the **mesoderm** forms the muscles of the body wall and most other structures between the gut and the external covering. Some animals have a **diploblastic** body plan based on two embryonic layers, endoderm and ectoderm, but most are **triploblastic**, having all three germ layers.

27.2b Most Animals Exhibit Either Radial or Bilateral Symmetry

The most obvious feature of an animal's body plan is its shape **(Figure 27.3, p. 596)**. Most animals are **symmetrical**; in other words, their bodies can be divided by

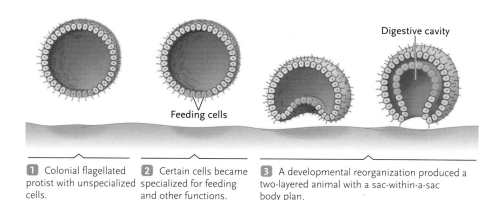

Figure 27.2

Animal origins. Many biologists believe that animals arose from a colonial, flagellated protist in which cells became specialized for specific functions and a developmental reorganization produced two cell layers. The cell movements illustrated here are similar to those that occur during the development of many animals, as described in Chapter 42.

Digestive cavity

Feeding cells

1 Colonial flagellated protist with unspecialized cells.

2 Certain cells became specialized for feeding and other functions.

3 A developmental reorganization produced a two-layered animal with a sac-within-a-sac body plan.

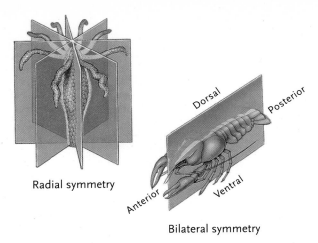

Figure 27.3
Patterns of body symmetry. Most animals have either radial or bilateral symmetry.

Radial symmetry

Bilateral symmetry

a plane into mirror-image halves. By contrast, most sponges have irregular shapes and are therefore **asymmetrical.**

All other phyla exhibit one of two body symmetry patterns (see Figure 27.3). The Radiata includes two phyla, Cnidaria (hydras, jellyfishes, and sea anemones) and Ctenophora (comb jellies), that are radially symmetrical. Their body parts are arranged regularly around a central axis, like spokes on a wheel. Thus, any cut down the long axis of a radially symmetrical animal divides it into matching halves.

All other metazoan phyla fall within the Bilateria, animals that have **bilateral symmetry.** In other words, on either side of the body's midline they have left and right sides that are mirror images of each other. Bilaterally symmetrical animals also have front (**anterior**) and back (**posterior**) ends, as well as upper (**dorsal**) and lower (**ventral**) surfaces. As they move through the environment, the anterior end encounters food, shelter, or enemies first. In bilaterally symmetrical animals, natural selection favoured **cephalization**, the development of an anterior head where sensory organs and nerve tissue are concentrated.

27.2c Many Animals Have Body Cavities That Surround Their Internal Organs

The body plans of many bilaterally symmetrical animals include a body cavity that separates the gut from the muscles of the body wall. **Acoelomate** animals (*a* = without; *koiloma* = cavity), such as flatworms (phylum Platyhelminthes), do not have such a cavity, rather a mass of cells derived largely from mesoderm that packs the region between gut and body wall **(Figure 27.4a).**

Pseudocoelomate animals (*pseudo* = false), including the roundworms (phylum Nematoda) and wheel animals (phylum Rotifera), have a **pseudocoelom**, a fluid-filled space between gut and muscles of the body wall that has no mesodermal lining around the endoderm

(Figure 27.4b). The muscles of the body wall, derived from mesoderm, form the outer lining of the pseudocoelom, and its inner lining is the gut, which lacks muscles. Internal organs lie within the pseudocoelom and are bathed by its fluid.

Coelomate animals have a **coelom**, a fluid-filled body cavity completely lined by mesoderm. In vertebrates, this lining takes the form of the **peritoneum**, a thin tissue derived from mesoderm (Figure 27.4c). The inner and outer layers of the peritoneum connect, forming **mesenteries**, membranes that surround the internal organs and suspend them within the coelom. In some arthropods and molluscs, the coelom has been displaced by the development of a **hemocoel**, resulting from an open circulatory system. This can be envisaged as consisting of a single large blood vessel that has expanded to fill the coelom. In these animals, the coelom persists around the gonads and, in some cases, the heart.

Biologists describe the body plan of pseudocoelomate and coelomate animals as a "tube-within-a-tube." The digestive system forms the inner tube, and the body wall forms the outer tube. The body cavity may serve a number of functions, such as the transport of nutrients and the products of metabolism, provision of an environment in which eggs and sperm can develop, a **hydrostatic skeleton** that provides a basis for locomotion (see Chapter 46), and an appropriate environment for the functioning of internal organs.

27.2d Developmental Patterns Mark a Major Divergence in Animal Ancestry

Embryological evidence suggests that bilaterally symmetrical animals are divided into two lineages, the **protostomes** and the **deuterostomes**, that differ in several developmental characteristics **(Figure 27.5, p. 598).**

Shortly after fertilization, an egg undergoes a series of cell divisions called **cleavage.** The first two cell divisions divide a zygote as you might slice an apple, cutting it into four wedges from top to bottom. In some animals, subsequent cell divisions occur at oblique angles to the vertical axis of the embryo, ultimately producing a mass in which each cell at the top of the embryo lies in the groove between the pair of cells below it (see the left side of Figure 27.5a). This pattern is called **spiral cleavage.** It is characteristic of most protostomes, although cleavage patterns in arthropods and some other groups are highly specialized. In deuterostomes, by contrast, the third cell division is perpendicular to the vertical axis of the embryo, cutting each of the four cells near its midsection. The fourth cell division is vertical, producing a mass of cells that are stacked directly above and below one another (see the right side of Figure 27.5a). This pattern is called **radial cleavage.**

Protostomes and deuterostomes often differ in the timing of important developmental events. During

a. Acoelomate animals

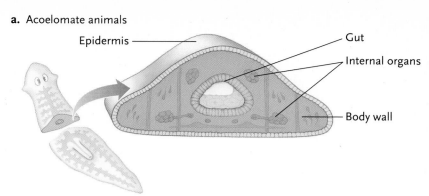

b. Pseudocoelomate animals

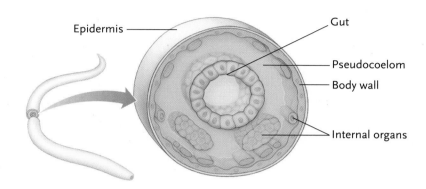

c. Coelomate animals

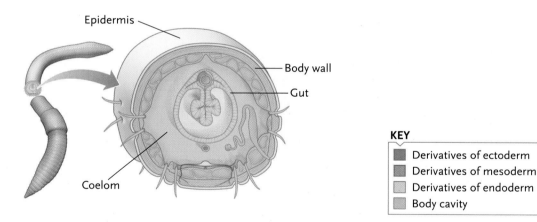

KEY

■	Derivatives of ectoderm
■	Derivatives of mesoderm
▨	Derivatives of endoderm
▨	Body cavity

Figure 27.4
Body plans of triploblastic animals.
(a) In acoelomate animals, no body cavity separates the gut and body wall.
(b) In pseudocoelomate animals, the pseudocoelom forms between the gut (a derivative of endoderm) and the body wall (a derivative of mesoderm).
(c) In coelomate animals, the coelom is completely lined by peritoneum (a derivative of mesoderm).

cleavage, certain genes are activated at specific times, determining a cell's developmental path and ultimate fate. Many protostomes undergo **determinate cleavage:** each cell's developmental path is determined as the cell is produced. Thus, one cell isolated from a two- or four-cell protostome embryo cannot develop into a functional embryo or larva. By contrast, many deuterostomes have **indeterminate cleavage:** the developmental fates of cells are determined later. A cell isolated from a four-cell deuterostome embryo will develop into a functional embryo. In humans, the two cells produced by the first cleavage division sometimes separate and develop into identical twins.

As development proceeds, the **blastopore,** an opening on the surface of the embryo, eventually connects the **archenteron** (developing gut) to the outside environment (see Figure 27.5b). Later in development, a second opening at the opposite end of the embryo transforms the pouchlike gut into a digestive tube (see Figure 27.5c). The traditional view of the difference between protostomes (*proto* = first; *stoma* = mouth) and deuterostomes (*deuteros* = second) is that in protostomes the blastopore develops into the mouth and the second opening forms the anus. In deuterostomes, the blastopore develops into the anus and the second opening becomes the mouth. But an alternative view has emerged from recent genetic studies. Specifically, the genes *brachyury* and *goosecoid* are expressed as the mouth develops in the acoel *Convolutriloba longifissura,* and the mouths of other bilaterians develop the same way. This

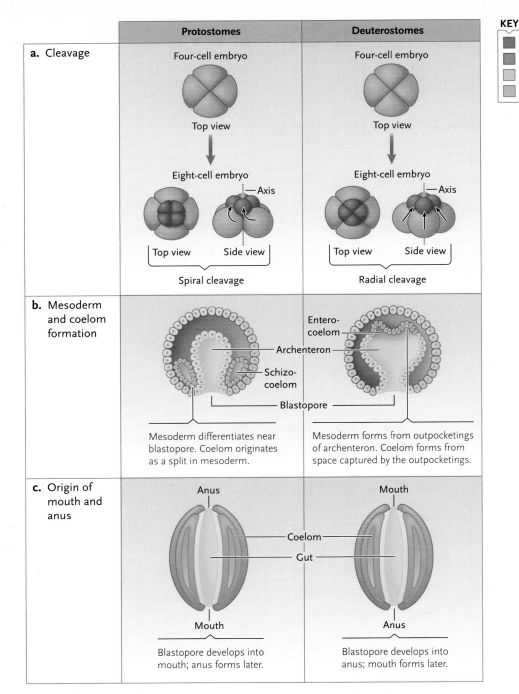

	Protostomes	Deuterostomes
a. Cleavage	Four-cell embryo Top view Eight-cell embryo Top view / Side view — Axis Spiral cleavage	Four-cell embryo Top view Eight-cell embryo Top view / Side view — Axis Radial cleavage
b. Mesoderm and coelom formation	Archenteron Schizo-coelom Blastopore Mesoderm differentiates near blastopore. Coelom originates as a split in mesoderm.	Entero-coelom Archenteron Blastopore Mesoderm forms from outpocketings of archenteron. Coelom forms from space captured by the outpocketings.
c. Origin of mouth and anus	Anus Coelom Gut Mouth Blastopore develops into mouth; anus forms later.	Mouth Coelom Gut Anus Blastopore develops into anus; mouth forms later.

KEY
- Derivatives of ectoderm
- Derivatives of mesoderm
- Derivatives of endoderm
- Body cavity

Figure 27.5

Protostomes and deuterostomes. The two lineages of coelomate animals differ in **(a)** cleavage patterns, **(b)** the origin of mesoderm and the coelom, and **(c)** the polarity of the digestive system.

makes the mouths in acoels and other bilaterians homologous. Furthermore, other genes, such as *caudal, orthopedia,* and *brachyury,* are expressed in a small area of the hind gut. This finding raises the possibility that the development of the anus (and a through gut) may have evolved independently in different lineages of bilaterians.

Protostomes and deuterostomes differ in the origin of mesoderm and coelom (see Figure 27.5b). In most protostomes, mesoderm originates from a few specific cells near the blastopore. As the mesoderm grows and develops, it splits into inner and outer layers. The space between the layers forms a **schizocoelom** (*schizo* = split). In deuterostomes, mesoderm forms from outpocketings of the archenteron. The space pinched off by the outpocketings forms an **enterocoelom** (*entero* = intestine).

Several other characteristics differ in protostomes and deuterostomes. For example, the nervous system of protostomes is positioned on the ventral side of the body, and their brain surrounds the opening of the digestive tract. By contrast, the nervous system and brain of deuterostomes lie on the dorsal side of the body.

27.2e Segmentation Divides the Bodies of Some Animals into Repeating Units

Some phyla in both protostome and deuterostome lineages exhibit varying degrees of **segmentation**, the production of body parts as repeating units. During development, segmentation first arises in the mesoderm, the middle tissue layer that produces most of the body's bulk. In vertebrates, segmentation is obvious in the embryo, and in the adult there is evidence of segmentation in the vertebral column (backbone), ribs, and associated muscles, as well as the nervous system. Among invertebrates, segmentation is pronounced in annelids (earthworms and their relatives), where each segment, visible externally as a ring, has its own set of muscles, ganglion (collection of nerve cells), and excretory structures. Arthropods (insects and their relatives) are also segmented, although some segments may be specialized, bearing, for example, wings or reproductive structures.

The advantages of segmentation lie principally in movement, but to different degrees. In vertebrates, with their articulated backbone and with each segment having its own muscles, segmentation permits the S-shaped side-to-side motion—think of fish or snakes. Annelids are capable of similar motion, but many of them live in burrows or tubes. The ability to expand segments by contracting muscles of adjacent segments assists this lifestyle. The articulated stiffened cuticle of arthropods serves as a point of attachment for muscles, providing significant leverage and strength (see Chapter 46). Arthropods have taken advantage of the existence of segmental appendages to assign special functions, such as locomotion, reproduction, or gas exchange, to particular appendages.

CONCEPT FIX Many people believe that invertebrate animals are protostomes. The truth is that the deuterostomes include many species (some whole phyla) that lack backbones (they are invertebrates). No members of the phylum Echinodermata have a backbone, and not all of the Chordata, the phylum including the vertebrates, have backbones. ⬡

STUDY BREAK

1. What is a tissue, and what three primary tissue layers are present in the embryos of most animals? Explain the function of each layer.
2. What kind of symmetry does an earthworm have?
3. What is the function of the coelom, and what is the importance of the fluid?

27.3 An Overview of Animal Phylogeny and Classification

For many years, biologists used the morphological innovations and embryological patterns described above, together with evidence from the fossil record, to trace the phylogenetic history of animals (see Chapter 20). That evidence led to the construction of phylogenetic trees, which were broadly accepted as reasonable hypotheses about the relatedness of various phyla. Thus, phyla with similar developmental and morphological patterns were regarded as sharing common ancestries. For example, annelids and arthropods, both schizocoelous, segmented coelomates, were seen as sharing a common ancestor, a view supported by the fossil record and by the existence of the Phylum Onychophora, which has some of the characteristics of both phyla. Increasingly, however, biologists are using molecular sequence data to reanalyze animal relationships.

27.3a Molecular Analyses Have Refined Our Understanding of Animal Phylogeny

Molecular analyses of animal relationships are often based on nucleotide sequences in small subunit ribosomal RNA and mitochondrial DNA and, more recently, the sequences of specific genes. These analyses are used to construct molecular cladograms (see Chapter 20). **Figure 27.6, p. 600,** is a phylogenetic tree developed from a number of cladograms based on molecular sequences. It represents, as do all such trees, a working hypothesis that explains the information that is now available.

The phylogenetic tree based on molecular characters includes the major lineages that biologists had defined using the morphological innovations and embryological characters described above. For example, molecular data confirm the distinctions between the Radiata and the Bilateria. They also confirm the separation of deuterostome phyla from all others within the Bilateria.

The sponges are to some degree a special case. The way that tissues form during embryology in the most primitive sponges differs from the other Metazoa and therefore sponges had, in the past, been regarded as lacking "true" tissues. Sponges were therefore placed in the Parazoa, a separate subkingdom, distinct from the Eumetazoa, the remaining metazoans. But the most recent molecular evidence includes sponges with the other Metazoa in a single monophyletic lineage. Sponges are distinct from other metazoans because they do not form distinct nervous tissue and are asymmetrical. Nonetheless, molecular data confirm that sponges and other Metazoa have a common ancestor.

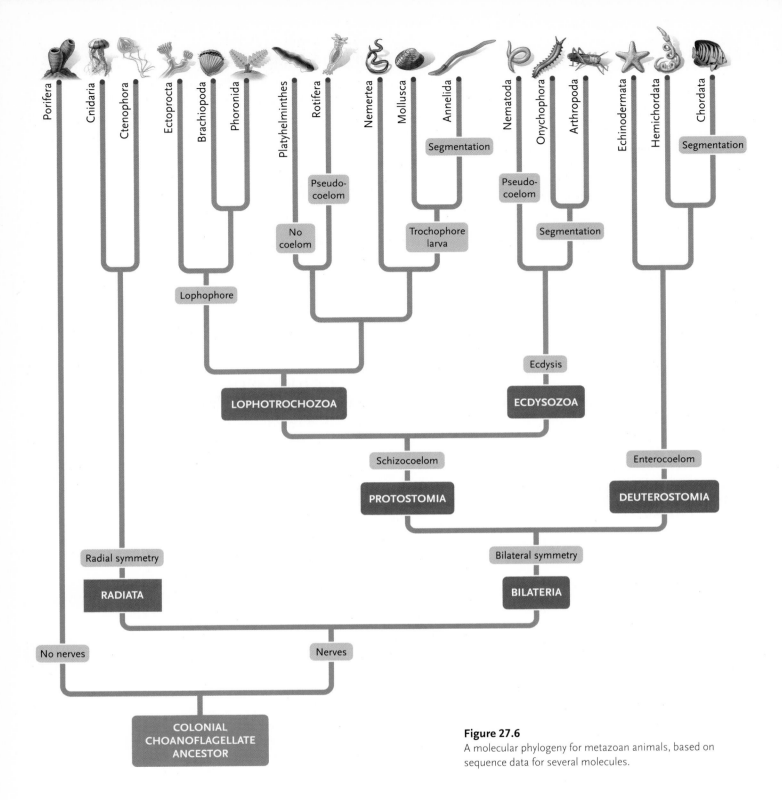

Figure 27.6

A molecular phylogeny for metazoan animals, based on sequence data for several molecules.

Molecular phylogeny confirms the Protostomia and Deuterostomia as separate lineages within the Metazoa. Protostomia is, in turn, subdivided into two major lineages, Lophotrochozoa and Ecdysozoa, groupings not previously recognized. The name Lophotrochozoa (*lophos* = crest; *troch* = wheel; *zoa* = animals, plural of *zoon*) refers to both the "lophophore," a feeding structure found in three phyla (illustrated in Figure 27.15, p. 607), and the "trochophore," a type of larva found in annelids and molluscs (illustrated in Figure 27.23, p. 614). The name Ecdysozoa (*ekdero* = strip off the skin) refers to the cuticle that these species secrete and periodically replace; the shedding of the cuticle is called **ecdysis**.

27.3b Molecular Phylogeny Reveals Surprising Patterns in the Evolution of Key Morphological Innovations

Molecular phylogeny has forced biologists to reevaluate the evolution of several important morphological innovations. Traditional phylogenies based on morphology and embryology implied that the absence of a body cavity, the acoelomate condition, was ancestral and that the presence of a body cavity, the pseudocoelomate or coelomate condition, was derived. But the molecular tree provides a very different view. It suggests that the schizocoelomate condition is ancestral, having evolved in the common ancestor of the lineage. If that hypothesis is correct, then the acoelomate condition of flatworms may represent the evolutionary *loss* of the schizocoelom, *not* an ancestral condition. Similarly, the molecular tree hypothesizes that the pseudocoelom evolved independently in rotifers (Lophotrochozoa, phylum Rotifera) and in roundworms (Ecdysozoa, phylum Nematoda) as modifications of the ancestral schizocoelom.

Traditional phylogenies also suggested that the segmented body plan of several protostome phyla was inherited from a segmented common ancestor and that segmentation arose independently in the chordates by convergent evolution. The molecular tree, by contrast, suggests that segmentation evolved independently in *three* lineages: segmented worms (Lophotrochozoa, phylum Annelida), arthropods and velvet worms (Ecdysozoa, phyla Arthropoda and Onychophora), and chordates (Deuterostomia, phylum Chordata).

The hypothesis based on molecular studies and represented in Figure 27.6, is the framework we use for our consideration of the major invertebrate phyla. It is important to recognize, however, that phylogenetic trees are always provisional. In the future, new data may lead to revisions of the phylogeny.

STUDY BREAK

1. How is molecular analysis used in creating phylogenetic trees?
2. Describe the way molecular phylogeny has changed how biologists view the absence of the coelom.

27.4 Phylum Porifera

Sponges **(Figure 27.7)** are mostly marine, with a small number of species living in fresh water. They have no particular symmetry, are completely sessile as adults, and obtain their food by filtering it from the water. Sponges have been abundant since the Cambrian, and about 8000 living species are known. They range in size from 1 cm to 2 m.

Figure 27.7

Asymmetry in sponges. The shapes of sponges vary with their habitats. Those that occupy calm waters, such as this stinker vase sponge (*Ircinia campana*), may be lobed, tubular, cuplike, or vaselike.

© Jeff Rotman/Alamy

Their body plan **(Figure 27.8, p. 602)** is simple: sponges can be regarded as sacs, with a cavity, the **spongocoel**, opening to the environment via an **osculum** (osteopore). There are two layers of organized cells. The **pinacoderm** (epithelium) consists of the cells on the outside of the sponge, pinacocytes. The inner layer of cells, lining the cavity, are **choanocytes**, each with a **flagellum** surrounded by microvilli. The two layers are separated by a gelatinous matrix, the **mesohyl**, with **amoeboid** cells called **archaeocytes** that move throughout the mesohyl by typical amoeboid movement. The wall of the bag is perforated by a number of pores lined by porocytes, specialized derivatives of the pinacocytes.

Almost all sponges are **suspension (or filter) feeders.** The action of choanocytes sets up a unidirectional current by which water enters the spongocoel through the porocytes and leaves via the osculum. Flow rates can be adjusted by the porocytes, which are capable of contraction, suggesting communication among the cells in spite of the absence of nerves. Particles of food are captured by choanocytes and passed to the mesohyl, where they are ingested and digested within archaeocytes, which may also store reserves.

Some archaeocytes may become specialized to form spicules, extracellular rigid supporting structures that give shape to the sponge. Spicules are microscopic rigid structures of various shapes (depending on the species) composed of a calcareous or siliceous (silicon) material. Collagen is also found in the mesohyl, as is spongin, a collagen-like protein. Collagens and spongin are produced by archaeocytes, which are **totipotent** (like stem cells), with the capacity to differentiate into any of the cell types found in sponges, including eggs and sperm.

Most sponges are **monoecious:** individuals produce both sperm and eggs. Sperm are released into the spongocoel and then out into the environment; eggs **(oocytes)** remain in the mesohyl, where sperm from other sponges, drawn in with water, are captured by choanocytes and carried to oocytes. Early development occurs within the sponge and produces a ciliated larva, the **dispersal** stage. Sponges have various types of

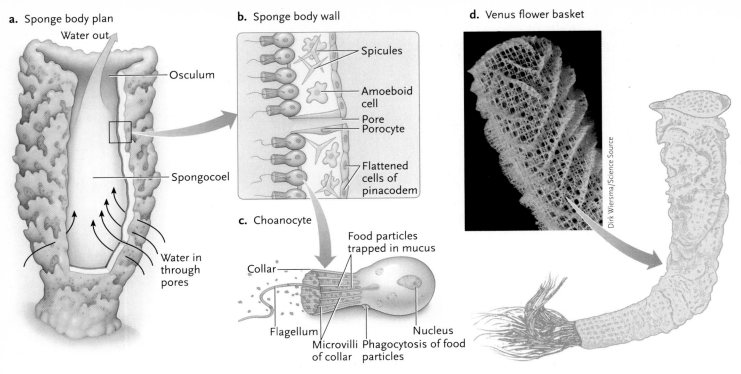

a. Sponge body plan
Water out
Osculum
Spongocoel
Water in through pores

b. Sponge body wall
Spicules
Amoeboid cell
Pore
Porocyte
Flattened cells of pinacoderm

c. Choanocyte
Collar
Food particles trapped in mucus
Flagellum
Microvilli of collar
Phagocytosis of food particles
Nucleus

d. Venus flower basket
Dirk Wiersma/Science Source

Figure 27.8

The body plan of sponges. Most sponges have **(a)** simple body plans and **(b)** relatively few cell types. **(c)** Beating flagella on the choanocytes create a flow of water through incurrent pores, into the spongocoel, and out through the osculum. **(d)** Venus flower basket (Euplectella species), a marine sponge, has spicules of silica fused into a rigid framework.

larvae. While some are free swimming, others use their cilia to crawl over the substrate. There is evidence that some larvae avoid light to select a location to settle, where they undergo **metamorphosis** (a reorganization of form) into sessile adults. Some sponges also reproduce asexually; small fragments break off an adult and grow into new sponges. Many species, particularly those in fresh water, also produce **gemmules**, clusters of cells with a resistant covering that allows them to survive unfavourable conditions. Gemmules germinate into new sponges when conditions improve.

Even with a very simple basic body plan, sponges have achieved remarkable diversity. Sponges formed very large reefs during the Mesozoic, and a modern reef of sponges, originating at the end of the last ice age, has been found off the west coast of Canada. It is being studied by Verena Tunnicliffe's lab at the University of Victoria.

Many sponges serve as refuges for other species. Bacteria and cyanobacteria can be found in the mesohyl and, in some species, within archaeocytes. A curious relationship with another species occurs in the Venus flower basket, *Euplectella aspergillum* (see Figure 27.8d). Male and female shrimp (*Spongicola* species) may enter the spongocoel when small, feed on material brought in by the sponge, and grow large enough that they are unable to leave. The pair of shrimp spend their entire lives in the prison formed by the elaborate basket of spicules.

One species, *Asbestopluma hypogea,* catches small arthropods that become entangled in hook-shaped spicules on the surface. The prey are then encased in filamentous structures and digested. Choanocytes are absent in this sponge.

STUDY BREAK

1. Do sponges exhibit symmetry? If so, what type?
2. How does a sponge gather food from its environment?

27.5 Metazoans with Radial Symmetry

Unlike sponges, the remaining metazoans have some form of symmetry and well-differentiated tissues, including nerves, that develop from distinct layers in the embryo. In this section, we describe metazoans with **radial symmetry**, a body plan that permits the detection of stimuli from all directions. This is an effective adaptation for life in open water, freshwater or marine.

Two phyla of soft-bodied organisms, Cnidaria and Ctenophora, have radial symmetry and nerves. Both phyla possess a **gastrovascular cavity** with a single opening, the mouth. Gas exchange and excretion can occur by diffusion because no cell is far from a body surface.

The radiate phyla have a diploblastic body plan with only inner and outer tissue layers, the **gastrodermis** (an endoderm derivative) and the **epidermis** (an ectoderm derivative), respectively. Most species also possess a gelatinous **mesoglea** (*meso* = middle; *glea* = glue) between the two layers. The mesoglea contains widely dispersed fibrous and amoeboid cells, recalling the organization of the mesohyl in sponges.

27.5a Phylum Cnidaria

Nearly all of the 8900 species in the phylum Cnidaria (*cnid* = stinging nettle, a plant with irritating hairs) live in the sea. Their body plan is organized around a saclike gastrovascular cavity, and the mouth is ringed with tentacles, which push food into it. Cnidarians may be vase-shaped, upward-pointing **polyps** or bell-shaped, downward-pointing **medusae (Figure 27.9).** Most polyps attach to a substrate at the *aboral* (opposite the mouth) end, while medusae are unattached and float.

Cnidarians are the simplest animals that exhibit a division of labour among irreversibly specialized tissues (see Figure 27.9c) and that have nerve cells. The gastrodermis includes sensory receptor cells, gland cells, and phagocytic nutritive cells. Gland cells secrete enzymes for the **extracellular digestion** of food, which is then engulfed by nutritive cells and exposed to **intracellular digestion**. The epidermis includes sensory cells, contractile cells, and cells specialized for prey capture.

Cnidarians prey on crustaceans, fishes, and other animals. The epidermis includes unique cells, **cnidocytes**, each armed with a stinging **nematocyst (Figure 27.10, p. 604).** The nematocyst contains an encapsulated, coiled thread that is fired at prey or predators, sometimes releasing a toxin through its tip. Discharge of nematocysts may be triggered by touch, vibrations, or chemical stimuli. The toxin can paralyze small prey by disrupting nerve cell membranes. The painful stings of some jellyfishes and certain corals result from the discharge of nematocysts.

Cnidarians engage in directed movements by contracting specialized ectodermal cells with fibres that resemble those in muscles. In medusae, the mesogleal jelly serves as a deformable skeleton against which contractile cells act. Rapid contractions narrow the bell, forcing out jets of water that propel the animal. Polyps use their water-filled gastrovascular cavity as a hydrostatic skeleton. When some cells contract, fluid within the chamber is shunted about, changing the body's shape and moving it in a particular direction.

The **nerve net**, which threads through both tissue layers, is a simple nervous system that coordinates responses to stimuli (see Chapter 44). Although there is no recognizable "brain," there are control and coordination centres, particularly in a ring of nerves encircling the mouth. In spite of its structural simplicity, the nerve net permits directed swimming movements so the animal can escape predators.

Many cnidarians exist in only the polyp or the medusa form, but some have a life cycle that alternates between them **(Figure 27.11, p. 605).** In the alternating type, the polyp often produces new individuals

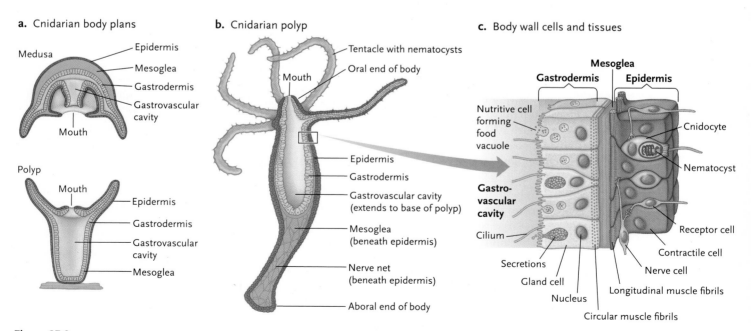

a. Cnidarian body plans

Medusa
- Epidermis
- Mesoglea
- Gastrodermis
- Gastrovascular cavity
- Mouth

Polyp
- Mouth
- Epidermis
- Gastrodermis
- Gastrovascular cavity
- Mesoglea

b. Cnidarian polyp

- Tentacle with nematocysts
- Oral end of body
- Mouth
- Epidermis
- Gastrodermis
- Gastrovascular cavity (extends to base of polyp)
- Mesoglea (beneath epidermis)
- Nerve net (beneath epidermis)
- Aboral end of body

c. Body wall cells and tissues

Mesoglea
Gastrodermis   Epidermis
- Nutritive cell forming food vacuole
- Cnidocyte
- Nematocyst
- Gastrovascular cavity
- Cilium
- Receptor cell
- Secretions
- Contractile cell
- Gland cell
- Nerve cell
- Nucleus
- Longitudinal muscle fibrils
- Circular muscle fibrils

Figure 27.9

The cnidarian body plan. **(a)** Cnidarians exist as either polyps or medusae. **(b)** The body of both forms is organized around a gastrovascular cavity, which extends all the way to the aboral end of the animal. **(c)** The two tissue layers in the body wall, the gastrodermis and the epidermis, include a variety of cell types.

a. *Hydra* consuming a crustacean

b. Cnidocytes

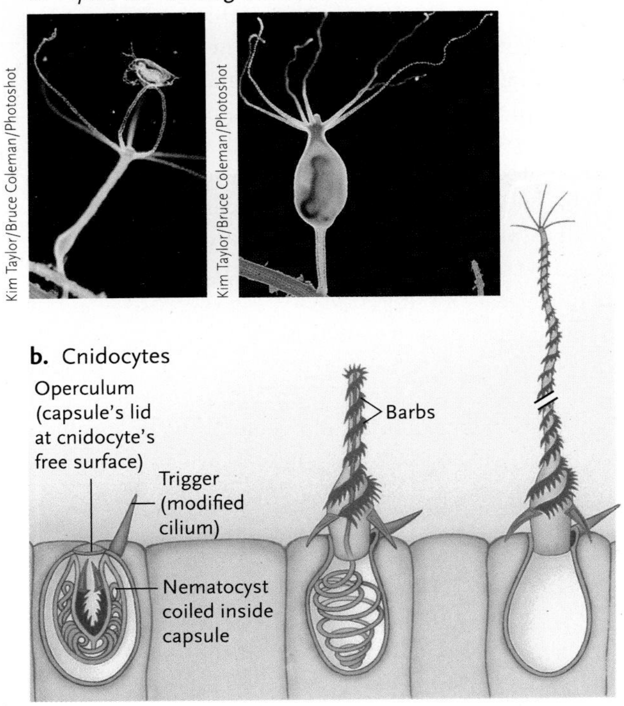

Operculum
(capsule's lid
at cnidocyte's
free surface)

Trigger
(modified
cilium)

Nematocyst
coiled inside
capsule

Barbs

Figure 27.10

Predation by cnidarians. (a) A polyp of a freshwater *Hydra* captures a small crustacean with its tentacles and swallows it whole. **(b)** Cnidocytes, special cells on the tentacles, encapsulate nematocysts, which are discharged at prey.

asexually from buds that break free of the parent (see Chapter 47). The medusa is often the sexual stage, producing sperm and eggs, which are released into the water. Sexual reproduction results in a ciliated, nonfeeding larval stage, the planula, that eventually settles and undergoes metamorphosis into the polyp form. The four classes of Cnidaria differ in the form that predominates in the life cycle.

Class Hydrozoa. Most of the 2700 species in the class Hydrozoa have both polyp and medusa stages in their life cycles (see Figure 27.11). The polyps form sessile colonies that develop asexually from one individual. A colony can include thousands of polyps, which may be specialized for feeding, defence, or reproduction. They share food through their connected gastrovascular cavities. A few warm-water species secrete a calcareous skeleton and form large colonies. These hydrocorals are different from the anthozoans that form **coral reefs** (see Class Anthozoa, below).

Some pelagic hydrozoans have both polyp and medusoid forms present in the same colony, which functions as an individual organism. The majestic Portuguese man-of-war jellyfish, for example, has the medusoid bell modified to form a gas-filled sail (see "Nematocycsts," p. 720). The hydroid form is represented by feeding and reproductive polyps dangling from the sail (see Chapter 30).

Unlike most Hydrozoa, freshwater species of *Hydra* (see Figure 27.10a) live as solitary polyps that attach temporarily to rocks, twigs, and leaves. Under favourable conditions, hydras reproduce by budding. Under adverse conditions, they produce eggs and sperm. Zygotes, formed by fertilization, are encapsulated in a protective coating but develop and grow when conditions improve. There is no larval stage; the eggs hatch into small *Hydra*.

Class Scyphozoa. The medusa stage predominates in the 200 species of the class Scyphozoa or jellyfish **(Figure 27.12a)**. They range from 2 cm to more than 2 m in diameter. Nerve cells near the margin of the bell control their tentacles and coordinate the rhythmic activity of contractile cells, which move the animal. Specialized sensory cells are clustered at the edge of the bell: statocysts (see Chapter 45) sense gravity, and ocelli are sensitive to light. Scyphozoan medusae are either male or female, releasing gametes into the water, where fertilization takes place.

Class Cubozoa. Most of the 20 known species of box jellyfish, the Cubozoa (Figure 27.12b), exist as cube-shaped medusae only a few centimetres tall; the largest species grows to 25 cm in height. Nematocyst-rich tentacles grow in clusters from the four corners of the box-like medusa, and groups of light receptors and image-forming eyes occur on the four sides of the bell. The eyes have lenses and retinas. Unlike the scyphozoan jellyfish, cubozoans are active swimmers. They eat small fish and invertebrates, immobilizing their prey with one of the deadliest toxins produced by animals. Cubozoans live in tropical and subtropical coastal waters, where they sometimes pose a serious threat to swimmers: the nematocysts of some species can inflict considerable pain to, and may kill, humans.

Class Anthozoa. The Anthozoa includes 6000 species of corals and sea anemones **(Figure 27.13, p. 606)**. Anthozoans exist only as polyps, which have a more complex structure than the Hydrozoa. A muscular pharynx leads into the gastrovascular cavity, and the body often consists of compartments partially separated by vertical membranes called septa. They reproduce by budding or fission. Most also reproduce sexually, producing eggs that develop into ciliated larvae. Corals (see Figure 27.13a) are always sessile and colonial. Their ciliated larvae settle and metamorphose into polyps that produce colonies by budding. Most species of corals build calcium carbonate skeletons that sometimes accumulate into gigantic underwater reefs. A coral reef usually contains more than one species of anthozoan. The energy needs of corals are partly fulfilled by the photosynthetic activity of symbiotic protists that live within the anthozoans. For this reason, corals are restricted to shallow water, where sunlight can penetrate.

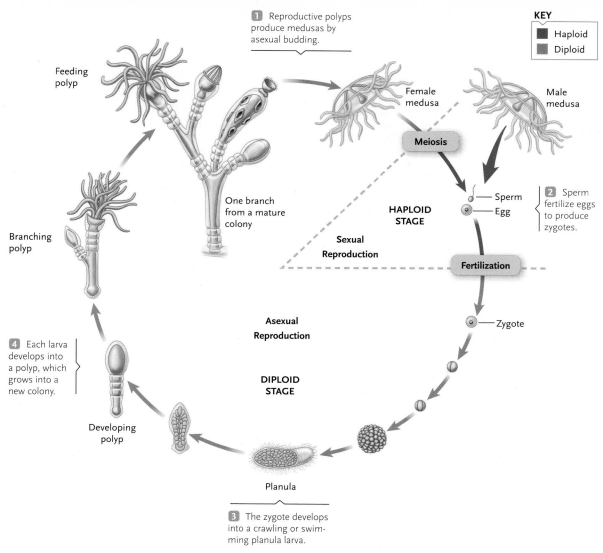

1 Reproductive polyps produce medusas by asexual budding.

KEY
■ Haploid
■ Diploid

Feeding polyp

Female medusa

Male medusa

One branch from a mature colony

Meiosis

HAPLOID STAGE

Sperm

Egg

2 Sperm fertilize eggs to produce zygotes.

Sexual Reproduction

Fertilization

Branching polyp

Zygote

Asexual Reproduction

4 Each larva develops into a polyp, which grows into a new colony.

DIPLOID STAGE

Developing polyp

Planula

3 The zygote develops into a crawling or swimming planula larva.

Figure 27.11
Life cycle of *Obelia*. The life cycle of *Obelia*, a colonial hydrozoan, includes both polyp and medusa stages.

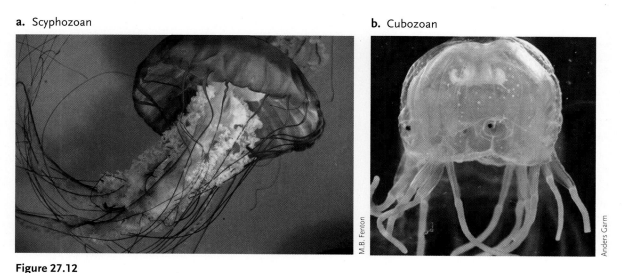

a. Scyphozoan

b. Cubozoan

M.B. Fenton

Anders Garm

Figure 27.12
Scyphozoans and cubozoans. **(a)** Most scyphozoans, like the sea nettle (*Chrysaora* species), live as floating medusae. Their tentacles trap prey, and the long oral arms transfer it to the mouth on the underside of the bell. **(b)** Cubozoans, unlike most jellyfish, are active swimmers and can change direction abruptly. They have several light-sensitive organs, but only four of them, two of which are clearly visible here, form images.

Coral

Tentacle of one polyp

scubaluna/iStock/Thinkstock

Interconnected
skeletons of polyps
of a colonial coral

Figure 27.13

Anthozoans. (a) Many corals are colonial, and their polyps build a hard skeleton of calcium carbonate. The skeletons accumulate to form coral reefs in shallow tropical waters. **(b)** A sea anemone detaches from its substrate to escape from a predatory sea star.

Sea anemones (see Figure 27.13b), by contrast, are soft-bodied, solitary polyps, ranging from 1 to 10 cm in diameter. They occupy shallow coastal waters. Most species are sessile, but some move by crawling slowly or by using the gastrovascular cavity as a hydrostatic skeleton.

27.5b Phylum Ctenophora

The 100 species of comb jellies in the marine phylum Ctenophora (*ctenos* = comb; *phor* = to carry) also have radial symmetry, mesoglea, and feeding tentacles. However, they differ from Cnidaria in significant ways. They lack nematocysts, they expel some waste through anal pores located at the opposite end to the mouth, and certain of their tissues appear to be of mesodermal origin. These transparent and often luminescent (light-producing) animals range in size from a few millimetres to 30 cm in diameter, with tentacles up to 1 m or more in length **(Figure 27.14)**.

Ctenophores move by beating cilia arranged on eight longitudinal plates that resemble combs. They are the largest animals to use

cilia for locomotion, but they are feeble swimmers. Nerve cells coordinate the animals' movements, and a gravity-sensing statocyst helps them maintain an upright position. Most species have two tentacles with specialized cells that discharge sticky filaments to entrap small animals floating in the sea, particularly small crustaceans. The food-laden tentacles are drawn across the mouth. Others lack tentacles and take large prey by a single gulp of the mouth. Some species that attack Cnidaria incorporate the nematocysts from the prey and use them in feeding (see Chapter 30). Ctenophores are hermaphroditic, producing gametes in cells that line the gastrovascular cavity. Eggs and sperm are expelled through the mouth or from special pores, and fertilization occurs in the open water.

STUDY BREAK

1. How do cnidarians capture, consume, and digest their prey?
2. Describe the differences between a polyp and a medusa.
3. Which group of cnidarians has only a polyp stage in its life cycle?
4. What do ctenophores eat, and how do they collect their food?

27.6 Lophotrochozoan Protostomes

The remaining organisms described in this chapter are in the group Bilateria because of their bilateral symmetry. They have a greater variety of tissues, some of

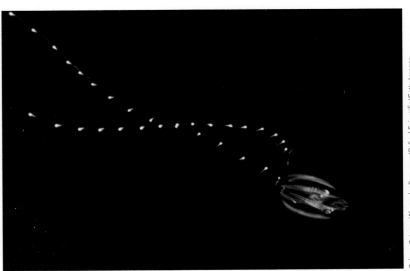

Cultura Science/Alexander Semenov/Oxford Scientific/Getty Images

Figure 27.14

Ctenophores. The comb jelly *Pleurobrachia* collects microscopic prey on its two long sticky tentacles and then wipes the food-laden tentacles across its mouth.

which are developed into organ systems. Most of the phyla have a coelom or pseudocoelom. With bilateral symmetry and sensory organs that are concentrated at the anterior end of the body, most bilaterians can make directed movements in pursuit of food or mates or to escape danger. Organ systems can operate more efficiently than simple tissues. For example, animals that have a tubular digestive system surrounded by a space (the coelom) use muscular contractions of the digestive system to move ingested food past specialized epithelial cells that break it down and absorb the breakdown products.

Molecular analyses group eight of the Bilateria phyla into the Lophotrochozoa, one of the two main protostome lineages (see Figure 27.6, p. 600).

27.6a Three Lophophorate Phyla Share a Distinctive Feeding Structure

Three small groups of mostly marine (a few also occur in fresh water) coelomate animals, the phyla Brachiopoda, Ectoprocta, and Phoronida, have a **lophophore**, a circular or U-shaped fold with one or two rows of hollow, ciliated tentacles surrounding the mouth **(Figure 27.15)**. Molecular sequence data and the lophophore suggest that these phyla have a common ancestry.

The coelomic cavity extends into the lophophore, which looks like a crown of tentacles at the anterior end of the animal. The lophophore is involved in the capture of food and serves as a site for gas exchange. Most lophophorates are sessile suspension feeders (see Chapter 48) as adults. Movement of cilia on the tentacles brings food-laden water toward the lophophore, where the tentacles capture small organisms and debris, and the cilia transport them to the mouth. The lophophorates have a complete digestive system, which is U-shaped in most species, with the anus lying outside the ring of tentacles.

Phylum Ectoprocta. The Ectoprocta (sometimes called Bryozoa or Polyzoa) are tiny colonial animals that occupy mainly marine habitats (see Figure 27.15a). They secrete a hard covering over their soft bodies. The lophophore is normally retracted into a chamber at the anterior end of the animal and extended when the animal feeds. Each colony, which may include more than a million individuals, is produced asexually by a single animal. Ectoproct colonies are permanently attached to solid substrates, where they form encrusting mats, bushy upright growths, or jellylike blobs. Sexual reproduction involves the production of eggs and sperm in the coelom. The sperm are shed through special pores. Fertilization may be internal or external, and the zygote gives rise to a ciliated larva that eventually settles and undergoes metamorphosis. Nearly 5000

living species are known, and about 50 of them live in fresh water.

Phylum Brachiopoda. The brachiopods, or lampshells, have two calcified shells that are secreted on the animal's dorsal and ventral sides (see Figure 27.15b). Most species attach to substrates with a stalk that protrudes through one of the shells. The lophophore is held within the two shells, and the animal feeds by opening its shell and drawing water over its tentacles. The animal has well-developed organs, such as a heart that propels blood through a number of interconnected sinuses and specialized excretory organs. Eggs and sperm are produced in different individuals (dioecious), and fertilization is external. The zygote gives rise to a ciliated larva.

Phylum Phoronida. The 18 or so species of phoronid worms vary in length from a few millimetres to 25 cm (see Figure 27.15c). They usually build tubes of chitin, a polymer of N-acetylglucosamine (see "Molecule behind Biology," Box 46.1, p. 1145), in soft ocean sediments or on hard substrates and feed by protruding the lophophore from the top of the tube. Phoronids reproduce both sexually and by budding. The animals are monoecious (both eggs and sperm produced by one individual). A ciliated feeding larva is produced that settles, undergoes metamorphosis, secretes a tube, and develops into an adult.

a. Ectoprocta (*Plumatella repens*)

b. Brachiopoda (*Terebraulina septentrionalis*)

c. Phoronida (*Phoronis*)

Andrew J. Martinez/Science Source

© blickwinkel/Hecker/Alamy

Andrew J. Martinez/Science Source

Figure 27.15

Lophophorate animals. Although the lophophorate animals differ markedly in appearance, they all use a lophophore to acquire food.

27.6b Phylum Platyhelminthes

The 13 thousand flatworm species in the phylum Platyhelminthes (*plat* = flat; *helminth* = worm) live in aquatic (freshwater and marine) and moist terrestrial habitats. Some are parasitic. Like cnidarians, flatworms can swim or float in water, but they are also able to crawl over surfaces. They range from less than 1 mm to more than 20 m in length, and most are just a few millimetres thick. Free-living species eat live prey or decomposing carcasses, whereas parasitic species derive their nutrition from the tissues of living hosts.

Like the radiate phyla, flatworms are acoelomate, but they have a complex structural organization that reflects their triploblastic construction **(Figure 27.16)**. In those with a gut (some parasitic forms lack this organ), endoderm lines the digestive cavity with cells specialized for the chemical breakdown and absorption of ingested food. A single opening serves as both mouth and anus. Mesoderm, the middle tissue layer, produces muscles and reproductive organs. Ectoderm produces a ciliated epidermis, the nervous system, and the **flame cell** system, a simple excretory system (see Chapter 50). Flatworms lack circulatory or respiratory systems, but because all cells of their dorsoventrally (top-to-bottom) flattened bodies are near an interior or exterior surface, diffusion supplies them with nutrients and oxygen.

The flatworm nervous system includes two or more longitudinal ventral nerve cords interconnected by numerous smaller nerve fibres, like rungs on a ladder. An anterior **ganglion**, a concentration of nervous system tissue that serves as a primitive "brain," integrates their behaviour (see Chapter 44). Most free-living species have **ocelli** or "eye spots" that distinguish light from dark and chemoreceptor organs that sense chemical cues.

The phylum Platyhelminthes includes four classes, defined largely by their anatomical adaptations to free-living or parasitic habits. One class, Turbellaria, is free living, whereas the remaining three classes are parasitic, obtaining their nutrition from the tissues of another animal, the host.

Class Turbellaria. Most free-living flatworms (class Turbellaria) live in the sea **(Figure 27.17)**, where they may be brightly coloured. The familiar planarians and a few others live in fresh water or on land and are drab. Turbellarians swim by undulating the body wall musculature or crawl across surfaces by using muscles and cilia to glide on mucus trails produced by the ventral epidermis. Some terrestrial turbellarians are relatively large and prey on other invertebrates. For example, *Microplana termitophaga* waits at the entrance to termite colonies in Africa and entangles the prey in the slime it produces. Other species may gang up on large snails or other animals (see Figure 32.9, Chapter 32).

The gastrovascular cavity in free-living flatworms is similar to that in cnidarians. Food is ingested and wastes are eliminated through a single opening, the

Digestive system

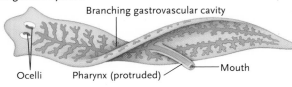

Nervous system

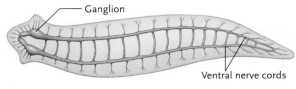

Reproductive system

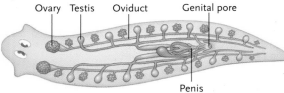

Excretory system

Figure 27.16

Flatworms. The phylum Platyhelminthes, exemplified by a freshwater planarian, have well-developed digestive, excretory, nervous, and reproductive systems. Because flatworms are acoelomate, their organ systems are embedded in a solid mass of tissue between the gut and the epidermis.

Figure 27.17

Turbellaria. A few turbellarians, such as *Pseudoceros dimidiatus*, are colourful marine worms.

mouth, located on the ventral surface. Most turbellarians acquire food with a muscular **pharynx** that connects the mouth to the digestive cavity (see Figure 27.16, top). Chemicals secreted into the saclike cavity digest ingested items, after which cells throughout the gastrovascular surface engulf food particles and subject them to intracellular digestion. In some species, the digestive cavity is highly branched, increasing the surface area for digestion and absorption.

Nearly all turbellarians are hermaphroditic, with complex reproductive systems (see Figure 27.16 second from bottom). When they mate, each partner functions simultaneously as a male and a female. The eggs of most species hatch directly into small worms, but ciliated larvae occur in a few marine turbellarians. Many free-living species also reproduce asexually by simply separating the anterior half of the animal from the posterior half. Both halves subsequently regenerate the missing parts.

Class Monogenea. Flukes (classes Trematoda and Monogenea) are parasites that obtain nutrients from host tissues. Monogenea flukes are **ectoparasites** that attach to the gills or skin of aquatic vertebrates. They have an anterior sucker surrounding the mouth and a more posterior sucker. The suckers may be equipped with hooks.

Reproduction occurs by internal fertilization. The eggs are released into water and hatch as ciliated larvae. The larvae attach to a new host and undergo metamorphosis.

Class Trematoda. Adult trematodes **(Figure 27.18)** are all internal parasites of vertebrates, but their development involves two or more host species in their life cycle. They are sometimes called digenean (two hosts) flukes. The host species in which sexual reproduction occurs is the primary host, and other hosts, usually invertebrates, are secondary hosts. Thus, the same individual will encounter very different environments during its life and may have two or more very different larval stages during development (see "Polymorphic

Development," p. 612). Like the monogeneans, trematodes normally have two suckers, one of which is around the mouth. The unciliated epidermis is a syncytium (the cells are interconnected without separating membranes). Trematodes can be found in many vertebrates, including humans, where they may cause some serious diseases. Infestations of trematodes may alter the behaviour of afflicted animals, making them more vulnerable to predators, often the host for the adult stage of the trematode.

Class Cestoda. Tapeworms **(Figure 27.19, p. 610)** are parasitic in the intestines of vertebrates, their primary host. They lack a mouth or digestive system and absorb nutrients from the host's intestinal contents across the syncytial epithelium. The anterior end is modified as a **scolex**, consisting of hooks and/or suckers that allow it to attach to the wall of the intestine. The remainder of the worm consists of a series of identical units, **proglottids**, each with its own reproductive system. Proglottids are generated just posterior to the scolex and become progressively more mature near the tail. The posterior, fully mature units break off or burst and are passed out with the feces. Worms may consist of only a few proglottids, but many species have 2000 to 3000, and such worms may be 10 m in length, occupying the entire length of the human small intestine.

Each proglottid contains a complete set of reproductive organs producing both sperm and eggs. Fertilization is internal and may involve a neighbouring worm, or the worm may be self-fertilizing. Each proglottid may contain as many as 50 thousand eggs. Further development varies with the species, but, typically, the egg must be eaten by an appropriate intermediate host, usually an arthropod, in which it undergoes development into a series of larval stages. The life cycle is completed when an appropriate primary host eats an infected intermediate host. For example, the adult tapeworm *Hymenolepis diminuta* lives in rat intestines. The eggs in rat feces are eaten by flour beetles, where larvae develop and form cysts. Rats become infected when they eat the beetles. Humans can also become infected by unwittingly consuming infected beetles that may live in dry breakfast cereals. The tapeworms that infest our domestic animals, notably dogs and cats, can pose a serious health threat to humans.

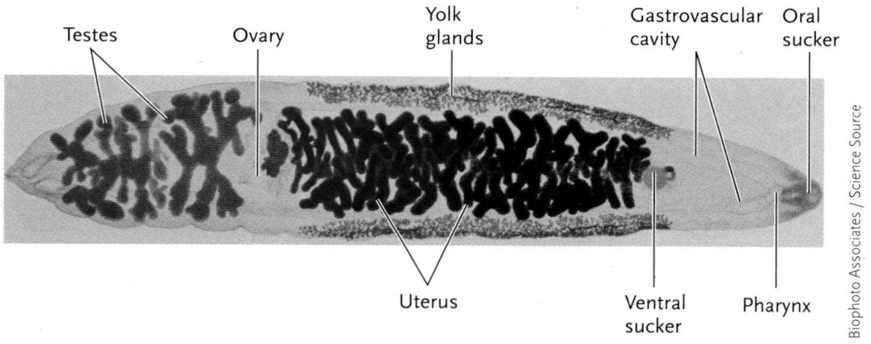

27.6c Phylum Rotifera

Most of the 1800 species in the pseudocoelomate phylum Rotifera (*rota* = wheel; *fera* = to bear) live in fresh water, and a few are marine **(Figure 27.20, p. 610)**. Most are less than 0.5 mm long, but a few range up

Figure 27.18

Trematoda. The hermaphroditic Chinese liver fluke (*Opisthorchis sinensis*) uses a well-developed reproductive system to produce thousands of eggs.

Labels: Testes, Ovary, Yolk glands, Gastrovascular cavity, Oral sucker, Uterus, Ventral sucker, Pharynx

Biophoto Associates / Science Source

Figure 27.19

Cestoda. **(a)** Tapeworms have long bodies composed of a series of proglottids, each of which produces thousands of fertilized eggs. **(b)** The anterior end is a scolex with hooks and suckers that attach to the host's intestinal wall.

a. Tapeworm

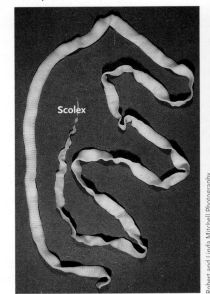

Scolex

Robert and Linda Mitchell Photography

b. Scolex

SPL/Science Source

Figure 27.20

Phylum Rotifera.

(a) Despite their small size, rotifers, such as *Philodina roseola*, have complex body plans and organ systems. **(b)** This rotifer, another *Philodina* species, is laying eggs.

a. Rotifer body plan

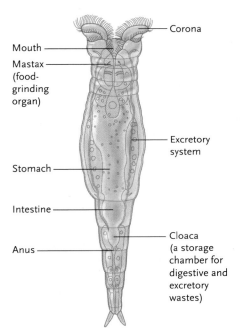

Corona
Mouth
Mastax (food-grinding organ)
Excretory system
Stomach
Intestine
Cloaca (a storage chamber for digestive and excretory wastes)
Anus

b. Rotifer laying eggs

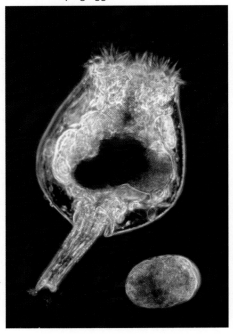

M. I. Walker/Science Source

to 3 mm. They exhibit *eutely*, a mode of development in which cell division ceases early and subsequent growth is by cell enlargement. In spite of their size, they have well-developed digestive, reproductive, excretory, and nervous systems. In some habitats, rotifers make up a large part of the zooplankton (tiny animals that float in open water). Some species, however, are attached to the substrate and move only a little. Others may form colonies, and some live in pitcher plants (see Chapter 31).

Rotifers use coordinated movements of cilia, arranged in a wheel-like **corona** around the head, to propel themselves in the environment. Cilia also bring

food-laden water to their mouths. Ingested microorganisms are conveyed to the **mastax**, a toothed grinding organ, and then passed to the stomach and intestine. Rotifers have a **complete digestive system**: food enters through the mouth, and undigested waste is voided through a separate anus.

The life history patterns of some rotifers are adapted to the ever-changing environments in small bodies of water. During most months, rotifer populations of these species include only females that reproduce by **parthenogenesis** (the development of unfertilized eggs; see Chapter 42). In this particular form of parthenogenesis, females produce diploid eggs by mitosis that develop into females. When environmental conditions deteriorate, females produce eggs by meiosis. If these eggs remain unfertilized, they develop into haploid males that produce sperm. If the haploid eggs are fertilized, they produce diploid female zygotes. The fertilized eggs have durable shells and food reserves to survive drying or freezing.

27.6d Phylum Nemertea

The 650 species of ribbon worms or proboscis worms vary from less than 1 cm to 30 m in length **(Figure 27.21)**. Most species are marine, but a few occupy moist terrestrial habitats. The often brightly coloured ribbon worms have no obvious coelom and use a ciliated epidermis to glide over a film of secreted mucus. Ribbon worms have a complete digestive tract with a mouth and an anus. They have a circulatory system in which fluid flows through **circulatory vessels** that carry nutrients and oxygen to tissues and remove wastes. They have a muscular, mucus-covered proboscis, a tube that can be everted (turned inside out) through a separate pore to capture prey. The proboscis is housed within a chamber, the **rhynchocoel**, which is unique to this phylum (see Figure 27.21b).

Nemerteans are aggressive predators. The proboscis may have a barb that is used to impale the prey, or the proboscis may wrap around the prey in a form of stranglehold. Many nemerteans are burrowing animals, living in tubes that protect them from predators. The life cycle includes a microscopic ciliated larva.

27.6e Phylum Mollusca

Most of the 100 thousand species of fleshy molluscs in the coelomate phylum Mollusca (*moll* = soft), including clams, snails, octopuses, and their relatives, are marine. However, many clams and snails occupy fresh water habitats, and some snails live on land. Molluscs vary in length from clams less than 1 mm across to the giant squids that can exceed 18 m in length.

The mollusc body is divided into three regions: the visceral mass, head-foot, and mantle **(Figure 27.22, p. 613)**. The **visceral mass** contains the digestive, excretory, and reproductive systems and the heart. The muscular **head-foot** often provides the major means of locomotion. In the more active groups, the head area of the head-foot region is well defined and carries sensory organs and a brain. The mouth often includes a toothed **radula**, which scrapes food into small particles or drills through the shells of prey.

Many molluscs are covered by a protective shell of calcium carbonate secreted by the **mantle**, a folding of the body wall that may enclose the visceral mass. The mantle also defines a space, the **mantle cavity**, housing the **gills**, delicate respiratory structures with an enormous surface area (see Chapter 49). In most molluscs, cilia on the mantle and gills generate a steady flow of water into the mantle cavity.

Most molluscs have an **open circulatory system** in which **hemolymph**, a bloodlike fluid, leaves the circulatory vessels and bathes tissues directly. Hemolymph pools in spaces called **sinuses** and then drains into vessels that carry it back to the heart (see Figure 40.3, Chapter 40).

The sexes are usually separate, although many snails are hermaphroditic. Fertilization may be internal or external. In some snails, eggs and sperm are produced simultaneously in the same organ, an ovotestis. In others, the hermaphroditism is serial, with younger snails producing sperm and older individuals switching to egg production. Fertilization is often internal in these organisms, and in simultaneous hermaphrodites, there is a mutual exchange of sperm during copulation. Sperm may be stored for long periods before being used. In some terrestrial snails, a calcium "love dart" may be fired into one of the partners preceding a mutual exchange of sperm. Dr. Ron Chase of McGill University has shown that mucus coating of the dart makes it more likely that the shooter's sperm will be used to fertilize the eggs.

a. Ribbon worm

b. Ribbon worm anatomy

Figure 27.21
(a) The flattened, elongated bodies of ribbon worms, such as genus *Lineus*, are often brightly coloured. **(b)** Ribbon worms have a complete digestive system and a specialized cavity, the rhynchocoel, that houses a protrusible proboscis.

Polymorphic Development

Most of the protostomes have a capacity for developmental polymorphism. During development from egg to adult, the organism may assume different morphologies. Most commonly, the immature form is referred to as a larva. In insects, for example, the caterpillar that hatches from the egg and and grows through a number of moults is a feeding stage very different from the adult butterfly, the distributive and reproductive stage. The transformation from larva to adult stage is accomplished by metamorphosis.

In other cases, particularly in sessile marine animals, the larval stage functions in distribution. Often this is an inconspicuous ciliated stage, such as the trochophore larva of molluscs and marine annelids (see **Figure 27.23, p. 614),** that drifts with ocean currents. It settles to the ocean floor in response to some signal and metamorphoses

into the form that will eventually become the adult. Some Cnidaria may have three distinct forms (see Figure 27.11, p. 605). This capacity for assuming different forms during development is particularly important for two lifestyles, parasitism and social insects.

Populations of animals that live in other organisms are faced with a particular challenge. The environment in which they live, the host, is discontinuous in space: one host is not connected to another. It is also discontinuous in time: the host eventually dies. It is thus essential for a parasite to move from one part of this discontinuous environment to another if the parasite population is to survive. Moving to another host may involve a period as a free-living form or further development in an alternative host. During its life cycle, the parasite is thus obliged to experience two or

more different environments. These different environments have favoured different developmental stages, often a series of morphologically distinct larvae. In the Chinese liver fluke (*Opisthorchis sinensis*), for example **(Figure 1),** the egg is eaten by a snail and hatches into a ciliated larva, the miracidium.

Almost immediately, metamorphosis occurs, involving extensive reorganization of the larva. This produces the sporocyst. Groups of embryonic cells, called "germ balls" **(Figure 2),** within the body cavity of the larva develop to produce additional larval stages, which are morphologically distinct from the sporocyst. In each stage, groups of embryonic germ balls are reserved to give rise to increased numbers of the next stage. The snail eventually releases enormous numbers of another larval form, the free-swimming cercaria, that

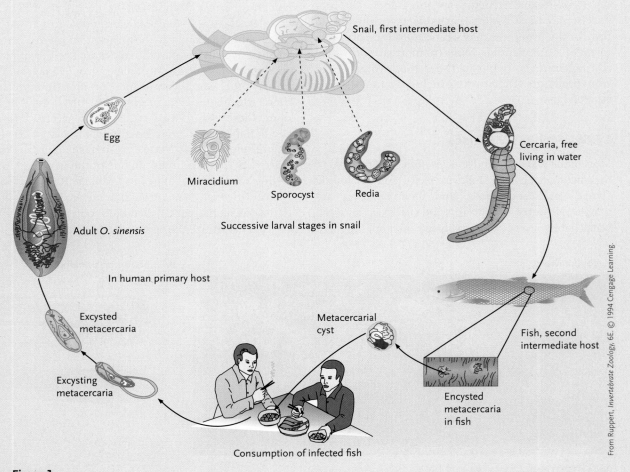

Figure 1
The life cycle of the liver fluke Opisthorchis sinensis. *Humans become infected by eating raw fish, and the adult fluke lives in the liver and bile duct. Estimates of the number of persons infected range up to 30 million.*

From Ruppert, Invertebrate Zoology, 6E. © 1994 Cengage Learning.

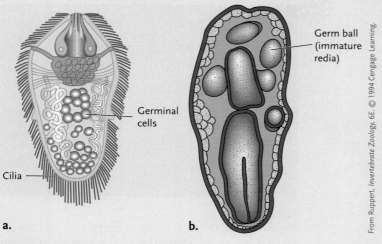

Germinal cells

Germ ball (immature redia)

Cilia

a. **b.**

From Ruppert, *Invertebrate Zoology*, 6E. © 1994 Cengage Learning.

Figure 2

The transformation of a miracidium (a) *into a sporocyst* (b), *involving the development of totipotent germ cells (stem cells) into germ balls that will form several copies of the next larval stage.*

characterized by different castes (individuals that perform particular tasks on behalf of the entire colony), which usually differ in morphology. The difference among the castes is particularly pronounced in termites **(Figure 3).**

Entomological Society of America

Figure 3

Termite castes. The queen termite in the centre of the photograph is surrounded by workers, called pseudergates. One soldier, with an enlarged head and mandibles, is also visible. At each moult, depending on the conditions in the colony, a pseudergate may develop into another pseudergate, embark on development to a soldier, or develop into a winged supplementary reproductive caste that will leave the colony to found a new one.

enters another intermediate host, a fish, where it forms a cyst in the muscle. This cyst will develop into the adult fluke if it is consumed by a human. The existence of populations of totipotent stem cells in flatworms has made possible the developmental polymorphism on which parasitism depends. The development of the cells is directed into different pathways appropriate for each parasitic stage.

Developmental polymorphism is also a feature of social insects. Social insects live in colonies and are

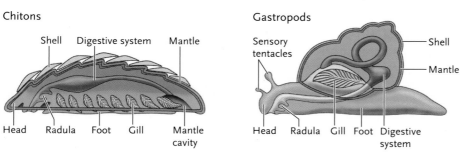

Chitons

Shell   Digestive system   Mantle

Head   Radula   Foot   Gill   Mantle cavity

Gastropods

Sensory tentacles

Shell

Mantle

Head   Radula   Gill   Foot   Digestive system

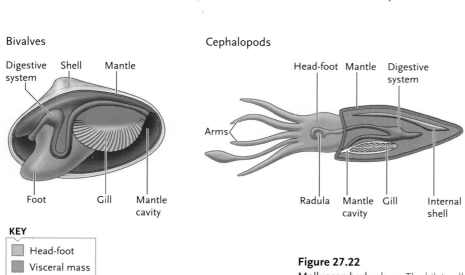

Bivalves

Digestive system   Shell   Mantle

Foot   Gill   Mantle cavity

Cephalopods

Head-foot   Mantle   Digestive system

Arms

Radula   Mantle cavity   Gill   Internal shell

KEY

☐ Head-foot
☐ Visceral mass
■ Mantle

Figure 27.22

Molluscan body plans. The bilaterally symmetrical body plans of molluscs include a muscular head-foot, a visceral mass, and a mantle.

Figure 27.23

Trochophore larva. At the conclusion of their embryological development, both molluscs and annelids typically pass through a trochophore stage. The top-shaped trochophore larva has a band of cilia just anterior to its mouth.

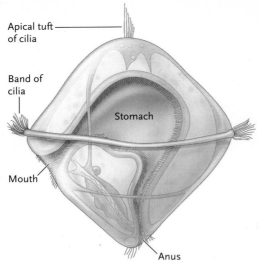

Apical tuft of cilia

Band of cilia

Stomach

Mouth

Anus

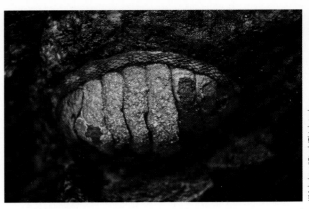

Figure 27.24

Polyplacophora. Chitons live on rocky shores, where they use their foot and mantle to grip rocks and other hard substrates. This chiton (*Mopalia ciliata*) lives in Monterey Bay, California.

The zygotes of marine species often develop into free-swimming, ciliated **trochophore** larvae (Figure 27.23), typical of both this phylum and the phylum Annelida, which we describe in Section 27.6f. In some molluscs, the trochophore develops into a second larval stage, called a **veliger**, before metamorphosing into an adult. In some snails, the larval stage may occur only within the egg. Squids and octopuses have no larval stage, and eggs hatch into miniature replicas of the adult. Although members of the phylum have common characteristics, they have evolved an extraordinary diversity in form and lifestyle, ranging from sessile clams to the agile octopus capable of learned behaviour. The phylum includes seven classes. We examine the four most commonly encountered classes below.

Class Polyplacophora. The 600 species of chitons (Polyplacophora; *poly* = many; *plak* = plate) are sedentary molluscs that graze on algae along rocky marine coasts. The oval, bilaterally symmetrical body has a dorsal shell divided into eight plates that allow it to

conform to irregularly shaped surfaces **(Figure 27.24)** and to roll into a ball when disturbed or theatened. When exposed to strong wave action, a chiton uses the muscles of its broad foot to maintain a tenacious grip, and the mantle's edge functions like a suction cup to hold fast to the substrate.

Class Gastropoda. Snails and slugs (Gastropoda; *gaster* = belly; *pod* = foot) are the largest molluscan group, numbering 40 thousand species **(Figure 27.25)**. The class exhibits a wide range of morphologies and lifestyles. Aquatic and marine species use gills to acquire oxygen, but in terrestrial species, a modified mantle cavity functions as an air-breathing lung. Some snails have the opening into the mantle cavity extended as a tubular siphon. Gastropods feed on algae, vascular plants, or animal prey. Some are scavengers, and a few are parasites.

The visceral mass of most snails is housed in a coiled or cone-shaped shell that is balanced above the rest of the body, much as you balance a backpack full of books (see Figure 27.25a, b). Most shelled species

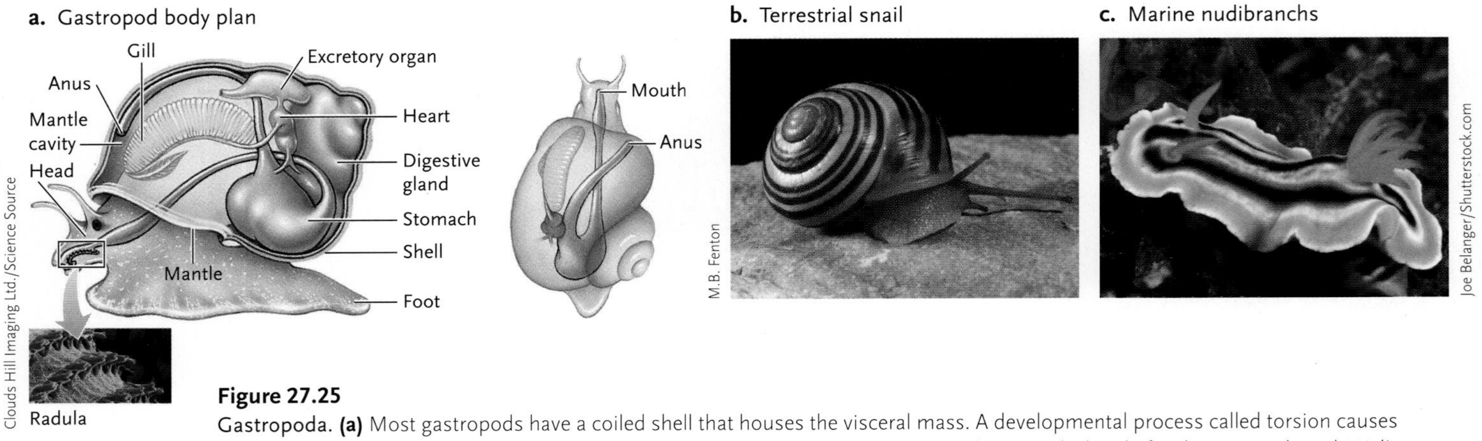

a. Gastropod body plan

Gill
Anus
Mantle cavity
Head
Excretory organ
Heart
Digestive gland
Stomach
Shell
Mantle
Foot

Radula

Mouth
Anus

b. Terrestrial snail

c. Marine nudibranchs

Figure 27.25

Gastropoda. **(a)** Most gastropods have a coiled shell that houses the visceral mass. A developmental process called torsion causes the digestive and excretory systems to eliminate wastes into the mantle cavity, near the animal's head. **(b)** The terrestial snail (*Helix pomatia*) is a typical terrestrial gastropod. **(c)** Nudibranchs, like this pair of Spanish shawl nudibranchs (*Flabellina iodinea*), are shell-less marine snails.

undergo **torsion** during development. Differential growth rates and muscle contractions twist the developing visceral mass and mantle a full 180° relative to the head and foot. These events begin in the larva before the shell is established and thus are not dictated by the coiling of the shell. Indeed, a snail that has undergone torsion to the right may exist in a left-handed shell. Among the many results of this developmental manoeuvre is the relocation of the mantle cavity to the anterior, allowing the head and foot to be withdrawn into the shell. In some snails, the **operculum** is an ovoid disk of protein fortified with calcium that can be used to close the entrance to the shell. This permits the snail to survive unfavourable conditions.

Some gastropods, including terrestrial slugs and colourful nudibranchs (sea slugs), are shell-less, a condition that leaves them somewhat vulnerable to predators (see Figure 27.25c). Some nudibranchs consume cnidarians and then transfer undischarged nematocysts to projections on their dorsal surface, where these "borrowed" stinging capsules provide protection (see Chapter 30).

Because many of its neurons are large and easily accessed and identifiable, the nudibranch *Aplysia* has been widely used to explore fundamental questions in neurobiology. For example, Dr. Wayne Sossin's lab at the Montreal Neurological Institute at McGill University is examining the biochemical and molecular basis of memory and learning in *Aplysia*.

The nervous and sensory systems of gastropods are well developed. Tentacles on the head include chemical and touch receptors; the eyes detect changes in light intensity but do not form images. The importance and relative sophistication of the nervous system is well illustrated by some limpets. *Patella vulgaris* is a gastropod with a conical shell **(Figure 27.26)** that lives on rocks in the intertidal zone. During low tide, it is exposed to the air and its foot and mucus secretions combine to fasten it closely to the rock. During development, the edges of its shell grow to conform to irregularities in the rock, increasing the protection against drying. As the rising tide covers the limpet, it moves about, foraging and feeding on algae. About an hour before the falling tide would once again expose it to desiccation, it returns to its precise location so that it can seal itself against exposure to air. This involves not only precise navigation but also a sense of time attuned to the tides.

Class Bivalvia. The 8000 species of clams, scallops, oysters, and mussels (Bivalvia; *bi* = two; *valv* = folding door) are restricted to aquatic habitats. They are enclosed within a pair of shells, hinged together dorsally by an elastic ligament. Contraction of the **adductor muscles** closes the shell and stretches the ligament. When the muscles relax, the stretched ligament opens the shell **(Figure 27.27, p. 616)**. Although some bivalves are tiny, the giant clams of the South Pacific can be more than 1 m across and weigh 225 kg.

Adult mussels and oysters are sessile and permanently attached to hard substrates. However, many clams are mobile and use their muscular foot to burrow in sand or mud. Some bivalves, such as young scallops, swim by rhythmically clapping their valves together, forcing a current of water out of the mantle cavity (see Figure 27.27b). The "scallops" that we eat are their well-developed adductor muscles.

Bivalves have a reduced head and lack a radula. Part of the mantle forms two tubes called *siphons* (see Figure 27.27a). Beating of cilia on the gills and mantle carries water into the mantle cavity through the **incurrent siphon** and out through the excurrent siphon. Incurrent water carries dissolved oxygen and particulate food to the gills, where oxygen is absorbed. Mucus strands on the gills trap food, which is then transported by cilia to *palps*, where final sorting takes place; acceptable bits are carried to the mouth. The excurrent water carries away metabolic wastes and feces.

Despite their sedentary existence, bivalves have moderately well-developed nervous systems: sensory organs that detect chemicals, touch, and light and statocysts to sense their orientation. When they encounter pollutants, many bivalves stop pumping water and close their shells. When confronted by a predator, some burrow into sediments or swim away.

Class Cephalopoda. The 600 living species of octopuses, squids, and nautiluses constituting the class Cephalopoda (*cephal* = head; *pod* = foot) are active marine predators and include the fastest and most intelligent invertebrates **(Figure 27.28, p. 616)**. They vary in length from a few centimetres to 18 m. Ammonites are well-known fossil cephalopods.

M.B. Fenton

Figure 27.26
The common limpet, *Patella vulgata*.

a. Bivalve body plan

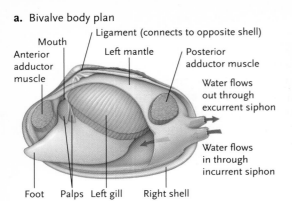

Mouth
Anterior adductor muscle
Ligament (connects to opposite shell)
Left mantle
Posterior adductor muscle
Water flows out through excurrent siphon
Water flows in through incurrent siphon
Foot
Palps
Left gill
Right shell

b. Bivalve locomotion

© age fotostock Spain, S.L. /Alamy

c. Geoduck

Tom McHugh/Science Source

Figure 27.27

Bivalvia. **(a)** Bivalves are enclosed in a hinged two-part shell. Part of the mantle forms a pair of water-transporting siphons. **(b)** When threatened by a predator (in this case, a sea star), some scallops clap their shells together rapidly, propelling the animal away from danger. **(c)** The geoduck (*Panope generosa*) is a clam with enormous muscular siphons.

a. Squid   **b.** Octopus   **c.** Chambered nautilus

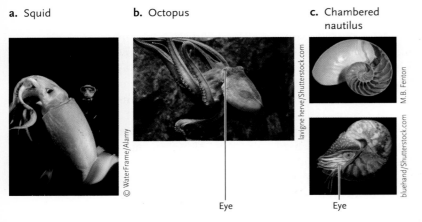

© WaterFrame/Alamy
lavigne herve/Shutterstock.com
M.B. Fenton
bluehand/Shutterstock.com

Eye   Eye

d. Internal anatomy of a squid

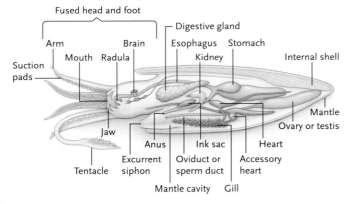

Fused head and foot
Arm
Mouth   Radula
Brain
Digestive gland
Esophagus   Stomach
Kidney
Internal shell
Suction pads
Jaw
Anus
Excurrent siphon
Oviduct or sperm duct
Ink sac
Mantle cavity
Gill
Heart
Accessory heart
Ovary or testis
Mantle
Tentacle

Figure 27.28

Cephalopoda. **(a)** Squids, such as *Dosidicus gigas*, and **(b)** octopuses, such as *Octopus vulgaris*, are the most familiar cephalopods. **(c)** The chambered nautilus (*Nautilus macromphalus*) and its relatives retain an external shell. **(d)** Like other cephalopods, the squid body includes a fused head and foot; most organ systems are enclosed by the mantle.

The cephalopod body has a fused head and foot. The head comprises the mouth and eyes. The ancestral "foot" forms a set of arms, which are equipped with suction pads, adhesive structures, or hooks. Cephalopods use their arms to capture prey and a pair of beaklike jaws to bite or crush it. Venomous secretions often speed the captive's death. Some species use their radula to drill through the shells of other molluscs.

Cephalopods have a highly modified shell. Octopuses have no remnant of a shell at all. In squids and cuttlefishes, the shell is reduced to a stiff internal support. Only the chambered nautilus (see Figure 27.28c) and its relatives retain an external shell; spaces (chambers) in the shell regulate the animal's buoyancy. Species in the genus *Nautilus* are clearly cephalopods because the foot is modified in a way that is characteristic of that class. But they have retained an elegant, chambered shell, a body plan that is very successful, since essentially identical animals can be found among the Cambrian fossils.

Squids (see Figure 27.28a, d) move by a kind of jet propulsion. When muscles in the mantle relax, water enters the mantle cavity. When they contract, a jet of water is squeezed out through a funnel. By manipulating the position of the mantle and funnel, the animal can control the rate and direction of its locomotion. While escaping, many species simultaneously release a dark fluid ("ink") that obscures their direction of movement. Octopuses and squids are able to change colour rapidly by the migration of various pigments in special pigment cells called chromatophores. Many squids have light-emitting cells called photophores.

Cephalopods are the only molluscs to have a **closed circulatory system.** The heart and accessory hearts speed the flow of hemolymph through blood vessels and gills, enhancing the uptake of oxygen and release of carbon dioxide.

Cephalopods have larger brains than other molluscs, and their brains are more complex than any other invertebrate. Giant nerve fibres connect the brain with the muscles of the mantle, enabling quick responses to food or danger (see Chapter 44).

The image-forming eyes of cephalopods, complete with lens and retina, are similar to those of vertebrates (see Chapter 45). The same basic plan for an eye has arisen independently in the cubozoan Cnidaria, the cephalopods, and the vertebrates and represents an example of convergent evolution. Cephalopods are also highly intelligent. Octopuses, for example, learn to recognize objects with distinctive shapes or colours and can be trained to approach or avoid them.

Cephalopods have separate sexes and elaborate courtship rituals. Males store sperm within the mantle cavity and use a specialized tentacle to transfer packets of sperm into the female's mantle cavity, where fertilization occurs. The young hatch with an adult body form.

27.6f Phylum Annelida

The 15 thousand species of segmented worms in the phylum Annelida (*annelis* = ring) occupy marine, freshwater, and moist terrestrial habitats. They range from a few millimetres to as much as 3 m in length. Terrestrial annelids eat organic debris, whereas aquatic species consume algae, microscopic organisms, detritus, or other animals. They have a complete digestive system, with the mouth at the anterior end and the anus at the rear.

The annelid body is highly segmented: the body wall muscles and some organs, including respiratory surfaces; parts of the nervous, circulatory, and excretory systems; and the coelom are divided into similar repeating units **(Figure 27.29)**. Body segments are separated by transverse partitions called **septa** (singular, *septum*). The digestive system and major blood vessels are not segmented and run the length of the animal.

The body wall muscles of annelids have both circular and longitudinal layers. Alternate contractions of these muscle groups allow annelids to make directed movements, using the coelom as a hydrostatic skeleton. The outer covering of annelids is a flexible cuticle that grows with the animal; it is not moulted. All annelids except leeches also have chitin-reinforced bristles, called **setae** (sometimes written *chaetae*; singular, *seta*), which protrude outward from the body wall. Setae anchor the worm against the substrate, providing traction.

Annelids have a closed circulatory system. The blood of most annelids contains hemoglobin or another oxygen-binding pigment. Oxygen diffusing across the cuticle may be picked up by capillaries in the skin to be transported to the tissues.

The excretory system is composed of paired **metanephridia** (singular, *metanephridium*) (see Figure 27.29d and Chapter 50), which usually occur in all body segments posterior to the head. The nervous system is well developed, with local control centres (ganglia) in every segment; a simple brain in the head; and sensory organs that detect chemicals, moisture, light, and touch.

Most freshwater and terrestrial annelids are hermaphroditic, and worms exchange sperm when they mate. Newly hatched worms have an adult morphology.

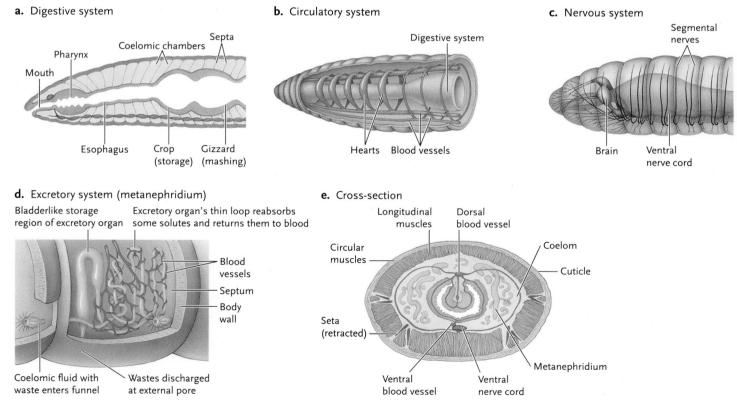

Figure 27.29

Segmentation in the phylum Annelida. Although the digestive system **(a)**, the longitudinal blood vessels **(b)**, and the ventral nerve cord **(c)** form continuous structures, the coelom **(a)**, blood vessels **(b)**, nerves **(c)**, and excretory organs **(d)** appear as repeating structures in most segments. The body musculature **(e)** includes both circular and longitudinal layers that allow these animals to use the coelomic chambers as a hydrostatic skeleton.

Some terrestrial annelids also reproduce asexually by fragmenting and regenerating missing parts. Marine annelids usually have separate sexes and release gametes into the sea for fertilization. The zygotes develop into trochophore larvae that add segments, gradually assuming an adult form.

Annelids are divided into three classes.

Class Polychaeta. The 10 000 species of bristle worms (Polychaeta; *poly* = many; *chaeta* = bristles) are primarily marine **(Figure 27.30)**. Many live under rocks or in tubes constructed from mucus, calcium carbonate secretions, grains of sand, and small shell fragments. Their setae project from well-developed **parapodia**

a. Feather duster worm

orlandin/Shutterstock.com

b. Polychaete feeding structures

c. Polychaete setae

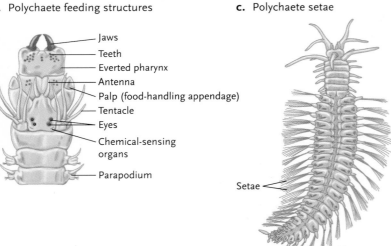

- Jaws
- Teeth
- Everted pharynx
- Antenna
- Palp (food-handling appendage)
- Tentacle
- Eyes
- Chemical-sensing organs
- Parapodium

Setae

Figure 27.30
Polychaeta. **(a)** The tube-dwelling feather duster worm (*Sabella melanostigma*) has mucus-covered tentacles that trap small food particles. **(b)** Some polychaetes, such as *Nereis*, actively seek food; when they encounter a suitable tidbit, they evert their pharynx, exposing sharp jaws that grab the prey and pull it into the digestive system. **(c)** Many marine polychaetes (such as *Proceraea cornuta*, shown here) have numerous setae, which they use for locomotion.

(singular, *parapodium* = closely resembling a foot), fleshy lateral extensions of the body wall used for locomotion and gas exchange. Sense organs are concentrated on a well-developed head.

Many crawling or swimming polychaetes are predatory, using sharp jaws in a protrusible muscular pharynx to grab small invertebrate prey. Other species graze on algae or scavenge organic matter. A few tube dwellers draw food-laden water into the tube by beating their parapodia; most others collect food by extending feathery, ciliated, mucus-coated tentacles.

Class Oligochaeta. Most of the 3500 species of oligochaete worms (*oligo* = few) are terrestrial **(Figure 27.31)**, but they are restricted to moist habitats because they quickly dehydrate in dry air or soil. They range in length from a few millimetres to more than 3 m. Terrestrial oligochaetes, the earthworms, are nocturnal, spending their days in burrows that they excavate. They are important scavengers, assisting in mixing and aerating soil and converting plant and animal debris to nutrients useful to plants. Aquatic species live in mud or detritus at the bottom of lakes and rivers. Earthworms have complex organ systems (see Figure 27.29, p. 617), and they sense light and touch at both ends of the body. In addition, they have moisture receptors, an important adaptation in organisms that must stay wet to allow gas exchange across the skin.

Class Hirudinea. Most of the 500 species of leeches (*hirudo* = leech) live in fresh water and suck the blood of vertebrates. These blood feeders have dorsoventrally flattened, tapered bodies with a sucker at each end. Although the body wall is segmented, the coelom is reduced and not partitioned. About a quarter of the known species are not blood feeders but prey on other invertebrates. Almost all of the leeches live in fresh water, but a few marine species are known. Some leeches are terrestrial, living in the moist tropics and feeding on warm-blooded vertebrates.

M.B. Fenton

Figure 27.31
Oligochaeta. Earthworms (genus *Lumbricus*) generally move across the ground surface at night.

LIFE ON THE EDGE 27.1

Hydrothermal Vents

Hydrothermal vents on the sea bed, discovered in 1970, are equivalent to miniature undersea volcanoes. Superheated water emerges from them at temperatures up to 400°C, laden with sulfides **(Figure 1)**. The Endeavour Hot Vent Area, over 2000 m deep, lies 250 km off the coast of Vancouver Island. It has been explored by Verena Tunnicliffe at the University of Victoria and was declared a Marine Protected Area by the Canadian government in 2003. A range of invertebrates, particularly molluscs and annelids, flourish near the vents. Many of these species are new to science, and they are larger and more numerous than the fauna nearby. They experience at least brief exposure to temperatures as high as 50°C. More important, however, is the absence of sunlight, which deprives them of a source of food from plants, and the presence of sulfides. Sulfide is normally toxic, but associated with the vents are mats of bacteria that utilize sulfide for energy production and growth. The metazoans feed on these, and some of the invertebrates have symbiotic bacteria that rely on the sulfides.

Figure 1

A hot vent smoker, surrounded by organisms specific to the environment. Inset: Tube worms, originally placed in a separate Phylum Vestimentifera. Molecular analysis places them among the annelids.

Canadian Scientific Submersible Facility

Blood-feeding leeches attach to the host with the posterior sucker and use their sharp jaws on the anterior sucker to make a small, often painless, triangular incision. A sucking apparatus draws blood from the prey, and a special secretion prevents the host's blood from coagulating. Leeches have a highly branched gut that allows them to consume huge blood meals **(Figure 27.32)**. For centuries, doctors used medicinal leeches (*Hirudo medicinalis*) to "bleed" patients; today, surgeons still use them to drain excess fluid from tissues after reconstructive surgery, reducing swelling until the patient's blood vessels regenerate and resume this function.

Leech before feeding

M.B. Fenton

Leech after feeding

Oxford Scientific/Getty Images

Figure 27.32

Hirudinea. Parasitic leeches consume huge blood meals, as shown by these before and after photos of a medicinal leech (*Hirudo medicinalis*). Because suitable hosts are often hard to locate, gorging allows a leech to take advantage of any host it finds.

STUDY BREAK

1. What characteristic reveals the close evolutionary relationship of ectoprocts, brachiopods, and phoronid worms?
2. Describe the three regions of the mollusc body.
3. Which organ systems exhibit segmentation in most annelid worms?

27.7 Ecdysozoan Protostomes

The three phyla in the protostome group Ecdysozoa all have an external cuticle secreted by epidermal cells. The cuticle serves as protection from harsh environmental conditions and helps parasitic species resist host defences. It also permits these animals to change the nature of the covering, which is important if the life stages live in different environments. Although many of these animals live in aquatic or moist terrestrial habitats, a tough exoskeleton allows many, particularly the insects, to thrive on dry land.

27.7a Phylum Nematoda

Members of the phylum Nematoda (*nemata* = thread) are round worms, often tapered at each end. The numerous species of free-living worms are

Figure 27.33
Phylum Nematoda. Many roundworms are animal parasites, like these *Anguillicola crassus*, shown here inside the swim bladder of an eel.

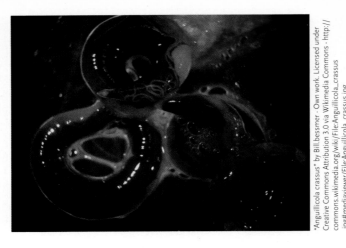

microscopic, reaching a size of at most a few millimetres. Some parasitic species, however, are larger **(Figure 27.33)**, and some are very large. The record is held by *Placentonema gigantissima*, a parasite of the placenta of the sperm whale: it may reach 9 m in length! Although superficially very similar in morphology, nematodes have achieved remarkable diversity. One species lives only in vinegar, and another is found in beer vats. In marine sediments, concentrations of a million or more worms per square metre have been reported. A hectare of farm soil may contain a billion or more nematodes. A single rotting fruit on the ground will contain tens of thousands of worms. Many species are parasitic in plants and animals, and some cause serious diseases in humans. Although fewer than 20 thousand species have been described, it is generally agreed that the number of living species is at least 100 thousand.

The nematode cuticle, often complex in structure, is composed of collagen-like proteins secreted by an epidermis that is often syncytial **(Figure 27.34)**. The cuticle is replaced four times during the life of the nematode, and in some cases, the characteristics of each of the cuticles differ. Moulting of the cuticle is not necessary for the worms to increase in size, and growth usually occurs between moults. *Ascaris lumbricoides*, a common intestinal parasite in humans, represents an extreme case, growing in length from about 6 mm after the final moult to an adult of more than 20 cm. Nematodes, like rotifers, exhibit eutely, having few or no cell divisions in somatic cells after hatching.

This characteristic, together with a transparent cuticle, has made *Caenorhabditis elegans*, which has fewer than 1000 cells, a very useful model for studying development. It eats bacteria such as *E. coli* and can be reared easily in the lab. Because the number of cells is so small, the developmental fate of each cell in the embryo has been documented using techniques such as microinjection and laser microsurgery. The genome is small and has been completely sequenced, yielding about 17 thousand genes. *C. elegans* is widely used as a model to explore general questions in developmental biology. For example, Dr. David Baillie at Simon Fraser University in Vancouver uses *C. elegans* to ask questions such as how many of the genes are essential to normal development. He has determined which mutations are lethal and estimates that about 4000 to 5000 genes of the 17 thousand are essential for development in *C. elegans*.

Growth in nematodes occurs by an increase in cell size, and in a large nematode such as *Ascaris*, a muscle cell may be more than a centimetre in length. There are no cilia or flagella in nematodes; the spermatozoa move by amoeboid motion. A single layer of muscles forms part of the body wall. These muscles, like those of flatworms, do not receive nerves but make contact with the ventral, dorsal, or lateral nerve cords by long extensions of the muscle (see Figure 27.34).

The nervous system consists of dorsal, lateral, and particularly prominent ventral nerve cords, with a nerve ring surrounding the pharynx at the anterior. Nematodes respond to various chemicals, likely through sense organs located in pits at the anterior end. Some are sensitive to light, possibly through a general sensitivity of their nerves, but others have pigmented eye spots.

The gut of nematodes is a simple tube consisting of a single layer of epithelial cells. There are no muscles surrounding the gut, and food is propelled through the digestive system by a muscular pharynx. The pumping action of the pharynx is also responsible for maintaining pressure in the pseudocoelom. The resulting stiffness produces a hydrostatic skeleton. The cuticle has some elasticity, so that a contraction of muscles in the dorsal part of the worm results in an expansion on the ventral side. Alternate contraction and relaxation produces the dorsoventral wave that characterizes the movement of most nematodes.

The sexes of most nematodes are separate, and fertilization is internal. A few nematodes are hermaphroditic and may be self-fertilizing. Others may

Figure 27.34
Cross-section of a female *Ascaris*, a typical nematode.

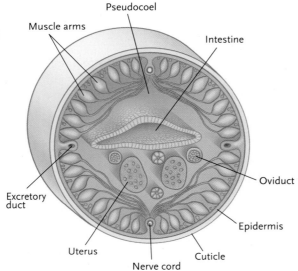

Pseudocoel
Muscle arms
Intestine
Excretory duct
Oviduct
Epidermis
Uterus
Cuticle
Nerve cord

be parthenogenetic. Fertilized females, particularly those of parasitic species, produce huge numbers of eggs. The intestinal parasite *A. lumbricoides* may produce 200 thousand eggs per day for about 10 months.

Nematodes are particularly successful as parasites and have invaded representatives of most other phyla. *A. lumbricoides* infects about a quarter to a third of the entire human population. Most wild vertebrates harbour one or more species of nematode parasites. The replaceable cuticle contributes to this success for it allows the worm to produce a different cuticle in each different environment that it encounters as it moves from a free-living form to a parasitic form or from one host to another.

27.7b Phylum Onychophora

The 65 living species of velvet worms (Onychophora; *onux* = claw; *phor* = to bear) live under stones, logs, and forest litter in moist temperate and tropical habitats in the southern hemisphere. They range in size from 15 mm to 15 cm and feed on small invertebrates and plants. Living onychophorans are all terrestrial, but fossils are known from marine environments.

Onychophorans have a flexible cuticle, superficially segmented bodies, and numerous pairs of unjointed legs **(Figure 27.35)**. Like annelids, they have pairs of excretory organs in most segments. But unlike annelids, no internal septa separate the segments; they have an open circulatory system, a specialized respiratory system similar to that of insects, and relatively large brains, jaws, and tiny claws on their feet. Many produce live young, which, in some species, are nourished within a uterus (see Chapter 39). The sexes are separate, and fertilization is internal.

Fossil onychophorans are known from the Cambrian (they are represented in the Burgess Shale), and the body plan has not changed much since then, suggesting that this highly specialized group of animals represented one of the successes in the experiments of the Cambrian speciation.

27.7c Phylum Arthropoda

If the Mesozoic was the age of the dinosaurs, we are living in the age of the arthropods (*arthros* = joint; *poda* = feet). About three-quarters of all living species of animals are arthropods, a phylum that includes insects, spiders, scorpions, crustaceans, centipedes, millipedes, and the extinct trilobites.

Arthropods have a segmented body encased in a rigid **exoskeleton**. This external covering is a complex of chitin (see "Molecule behind Biology," Box 46.1, p. 1145) and glycoproteins. In some marine and freshwater groups, such as crabs and lobsters, the cuticle is hardened with calcium carbonate. In terrestrial forms, such as insects, a surface layer of wax provides protection from dehydration. The exoskeleton is thin and flexible at the joints between body segments and at the joints of appendages. Contractions of muscles attached to the exoskeleton move individual body parts like levers, allowing highly coordinated movements and patterns of locomotion.

Although the exoskeleton has obvious advantages, it is nonexpandable and could limit the growth of the animal. Arthropods periodically develop a new cuticle beneath the old one, which they shed in the complex process of ecdysis **(Figure 27.36)**. The new cuticle is soft

M.B. Fenton

Figure 27.36
Ecdysis in insects. Like all other arthropods, this cicada (*Graptopsalatsia nigrofusca*) sheds its old exoskeleton as it grows.

© FLPA/Alamy

Figure 27.35
Phylum Onychophora. Members of the small phylum Onychophora, such as species in the genus *Dnycophor*, have segmented bodies and unjointed appendages.

and usually pleated, allowing for expansion after ecdysis. After shedding the old cuticle, arthropods swell with water or air before the new one hardens. They are especially vulnerable to predators at these times.

Primitively, each segment had a pair of lateral appendages, often specialized for locomotion, gas exchange, eating, or reproduction. As arthropods evolved, however, body segments became grouped in various ways. Each region, along with its highly modified paired appendages, is specialized, and the structure and function of the regions vary greatly among groups.

The coelom of arthropods is greatly reduced, and another cavity, the hemocoel, is filled with bloodlike hemolymph. The heart pumps the hemolymph through an open circulatory system, bathing tissues directly.

Because the hardened cuticle does not permit the easy passage of O_2 and CO_2, arthropods have specialized mechanisms for gas exchange. Marine and freshwater species, such as crabs and lobsters, rely on diffusion across gills that are specialized appendages, usually assisted by currents established by the appendages. The terrestrial groups have developed unique respiratory systems (see Chapter 49).

Many arthropods are equipped with a highly organized central nervous system, touch receptors, chemical sensors, image-forming **compound eyes**, and, in some, hearing organs. These are described in Chapters 44 and 45.

The phylogeny of this huge and diverse phylum has been a difficult and disputed subject for many years. Hexapods, and specifically insects, have been regarded by some as most closely related to myriapods, based largely on shared anatomical characters such as Malpighian tubules for excretion and one pair of antennae. Others, however, relate insects more closely to Crustacea, based on other morphological characteristics such as similarities in mouthparts and walking appendages. Molecular studies, including analysis of mitochondrial DNA and *Hox* genes, support the view that hexapods and crustacea are paraphyletic (see Chapter 20, for a definition of paraphyletic) and that myriapods and chelicerates form a separate paraphyletic grouping. This is an active area in research, and other hypotheses may be developed.

We follow the traditional definition of five *subphyla*, partly because this classification adequately reflects arthropod diversity and partly because no alternative hypothesis has been widely adopted by experts.

Subphylum Trilobita. The trilobites (*tri* = three; *lob* = lobed), now extinct, were among the most numerous animals in the shallow Paleozoic seas. Most were ovoid, dorsoventrally flattened, and heavily armoured, with two deep longitudinal grooves that divided the body into one median and two lateral lobes **(Figure 27.37)**. The head included a pair of sensory

M.B. Fenton

Figure 27.37
Subphylum Trilobita. Trilobites, such as the *Olenellus gilberti*, bore many pairs of relatively undifferentiated appendages.

antennae and compound eyes, and the segmented thorax and abdomen had pairs of identical appendages, each with two branches. The inner branch was used for locomotion, and the outer, consisting of a number of fine filaments, was used as a gill or in filter feeding.

The position of trilobites in the fossil record indicates that they were among the earliest arthropods. Thus, biologists are confident that their three body regions and unspecialized appendages represent ancestral traits in the phylum. Although there were numerous species, indicating a high degree of success, trilobites disappeared in the Permian mass extinction.

Subphylum Chelicerata. In spiders, ticks, mites, scorpions, and horseshoe crabs (subphylum Chelicerata; *cheol* = claw; *cera* = horn), the first pair of appendages, the **chelicerae**, are fanglike structures used for biting prey. The second pair of appendages, the **pedipalps**, serve as grasping organs, sensory organs, or walking legs. All chelicerates have two major body regions, the **cephalothorax** (a fused head–thorax) and the **abdomen.** The group originated in shallow Paleozoic seas, but most living species are terrestrial. They vary in size from less than 1 mm to 20 cm; all are predators or parasites.

The 60 thousand species of spiders, scorpions, mites, and ticks (class Arachnida) represent the vast majority of chelicerates **(Figure 27.38)**. Arachnids have four pairs of walking legs on the cephalothorax and highly modified chelicerae and pedipalps. In some spiders, males use their pedipalps to transfer packets of sperm to females. Scorpions use them (the "claws") to shred food and to grasp one another during courtship. Many predatory arachnids have excellent vision, provided by up to four pairs of simple eyes on the cephalothorax. Scorpions and some spiders also have unique pocket like respiratory organs called **book lungs** (see Chapter 49), derived from abdominal appendages.

a. Wolf spider

b. Spider anatomy

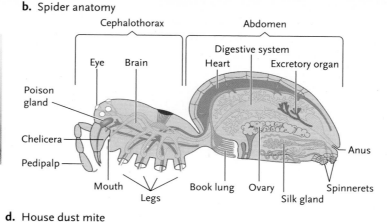

Cephalothorax

Abdomen

Eye Brain

Digestive system

Heart

Excretory organ

Poison
gland

Chelicera

Pedipalp

Mouth

Legs

Book lung Ovary

Silk gland

Anus

Spinnerets

c. Scorpion

d. House dust mite

Chelicerae

Figure 27.38

Subphylum Chelicerata, Class Arachnida. **(a)** The wolf spider (*Lycosa* species) is harmless to humans. **(b)** The arachnid body plan includes a cephalothorax and an abdomen. **(c)** Scorpions have a stinger at the tip of the segmented abdomen. Many, such as *Centruroides sculpuratus*, protect their eggs and young. **(d)** House dust mites (*Dermatophagoides pteronyssinus*), shown in a scanning electron micrograph, feed on microscopic debris.

Like most other arachnids, spiders subsist on a liquid diet. They use their chelicerae to inject paralyzing poisons and digestive enzymes into prey and then suck up the partly digested tissues. Many spiders are economically important predators, helping to control insect pests. Only a few are a threat to humans. The toxin of a black widow (*Latrodectus mactans*) causes paralysis, and the toxin of the brown recluse (*Loxosceles reclusa*) destroys tissues around the site of the bite.

Although many spiders hunt actively, others capture prey on silken threads secreted by **spinnerets**, which are modified abdominal appendages. Some species weave the threads into complex, netlike webs. The silk is secreted as a liquid protein but quickly polymerizes. Spiders also use silk to make nests, to protect their egg masses, as a safety line when moving through the environment, and to wrap prey for later consumption. Spider silk is extremely tough, and the material from some spiders exceeds the tensile strength of steel. It is also highly elastic. These properties have led to proposals for its use in fabrics. A Canadian company has developed transgenic goats that produce spider silk in their milk.

Most mites are tiny, but they have a big impact. Some are serious agricultural pests that feed on plant sap. Others cause mange (patchy hair loss) or painful and itchy welts on animals. House dust mites, which feed on the dried skin cast off by humans, cause allergic reactions in many people. Ticks, which are generally larger than mites, are blood feeders that often transmit pathogens, such as those causing Rocky Mountain spotted fever and Lyme disease.

The subphylum Chelicerata also includes five species of horseshoe crabs (class Merostomata), an ancient lineage that has not changed much over its 350-Ma history **(Figure 27.39)**. Horseshoe crabs are carnivorous bottom feeders in shallow coastal waters. Beneath their characteristic shell, they have one pair of chelicerae; a pair of pedipalps; four pairs of walking legs; and a set of paperlike gills, derived from ancestral walking legs. A component of horseshoe crab blood is important to the pharmaceutical industry, where it is used to test for the presence of endotoxins resulting from bacterial contamination during manufacture.

Subphylum Crustacea. The 35 thousand species of shrimps, lobsters, crabs, and their relatives in the subphylum Crustacea (*crusta* = shell) represent a lineage that emerged more than 500 Ma ago **(Figure 27.40, p. 624)**. They are abundant in marine and freshwater habitats. A few species, such as sowbugs and pillbugs, live in moist, sheltered terrestrial environments. In many crustaceans, two and, in some cases, all three of

Figure 27.39

Marine chelicerates.

Horseshoe crabs, such as *Limulus polyphemus*, are included in the Merostomata.

a. Crab

b. Lobster

Ivan Kuzmin/Shutterstock.com

c. Lobster anatomy

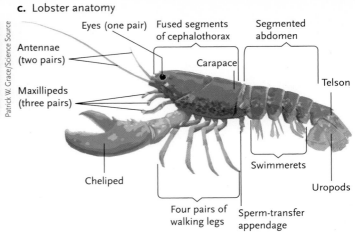

Patrick W. Grace/Science Source

Figure 27.40

Decapod crustaceans. **(a)** Crabs, such as this ghost crab in the genus *Ocypode*, and **(b)** lobsters (*Homarus americanus*) are typical decapod crustaceans. The abdomen of a crab is shortened and wrapped under the cephalothorax, producing a compressed body. **(c)** Lobsters bear 19 pairs of distinctive appendages; one pair of mandibles and two pairs of maxillae are not illustrated in this lateral view.

the arthropod body regions (head, **thorax**, and abdomen) may be fused. Fusion of the head and thorax into a cephalothorax is a common pattern. In some, the exoskeleton forms a **carapace**, a protective covering that extends backward from the head. Crustaceans vary in size from water fleas less than 1 mm long to lobsters that can grow to 60 cm in length and weigh as much as 20 kg.

Crustaceans generally have five characteristic pairs of appendages on the head (see Figure 27.40c). Most have two pairs of sensory antennae and three pairs of mouthparts. The latter include one pair of **mandibles**, which move laterally to bite and chew, and two pairs of **maxillae** (singular, *maxilla*), which hold and manipulate food. Numerous paired appendages posterior to the mouthparts vary among groups. Ancestrally, crustacean appendages were divided into two branches at the base, but many living species have unbranched appendages.

Most crustaceans are active animals that exhibit complex movements during locomotion and in the performance of other behaviours. These activities are coordinated by elaborate sensory and nervous systems, including chemical and touch receptors in the antennae, compound eyes, statocysts on the head, and sensory hairs embedded in the exoskeleton throughout the body. The nervous system is similar to that in annelids, but the ganglia, particularly those forming the brain, are larger and more complex. Larger species have complex, feathery gills derived from appendages tucked beneath the carapace. Metabolic wastes such as ammonia are excreted by diffusion across the gills or, in larger species, by **antennal glands**, located in the head.

The sexes are typically separate, and courtship rituals are often complex. Eggs are usually brooded on the surface of the female's body or beneath the carapace. Many have free-swimming larvae that, after undergoing a series of moults, gradually assume an adult form.

The subphylum includes so many different body plans that it is usually divided into six classes with numerous subclasses and orders. The crabs, lobsters, and shrimps (class Malacostraca, order Decapoda; *deka* = 10; *poda* = foot) number more than 10 thousand species. The vast majority of decapods are marine, but a few shrimps, crabs, and crayfishes occupy freshwater habitats. Some crabs also live in moist terrestrial habitats, where they scavenge dead vegetation, clearing the forest floor of debris.

All decapods exhibit extreme specialization of their appendages. In the American lobster, for example, each of the 19 pairs of appendages is different (see Figure 27.40c). Behind the antennae, mandibles, and maxillae, the thoracic segments have three pairs of maxillipeds, which shred food and pass it up to the mouth; a pair of large chelipeds (pinching claws); and four pairs of walking legs. The abdominal appendages include a pair specialized for sperm transfer (in males only); swimmerets for locomotion and for brooding eggs; and uropods, which, in combination with the telson (the tip of the abdomen), make a fan-shaped tail.

Representatives of several crustacean classes—fairy shrimps, amphipods, water fleas, ostracods, and copepods **(Figure 27.41)**—live as plankton in the upper waters of oceans and lakes. Most are only a few millimetres long but are present in huge numbers. They feed on microscopic algae or detritus and are themselves food for larger invertebrates, fishes, and some suspension-feeding marine mammals such as the baleen whales. Planktonic crustaceans are among the most abundant animals on Earth. The total biomass of a single species, *Euphausia superba,* is estimated at 500 million tonnes, more than the total mass of humans.

Adult barnacles (class Maxillopoda, subclass Cirripedia; *cirrus* = curl of hair; *poda* = foot) are sessile marine crustaceans that live within a strong, calcified, cup-shaped shell **(Figure 27.42).** Their free-swimming larvae attach permanently to substrates—rocks, wooden pilings, the hulls of ships, the shells of molluscs, and

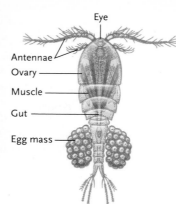

Figure 27.41

Copepods. Tiny crustaceans, such as this cope-pod (*Calanus* species on the left, *Cyclops* species on the right), occur by the billions in freshwater and marine plankton.

Eye
Antennae
Ovary
Muscle
Gut
Egg mass

Lebendkulturen.de/Shutterstock.com

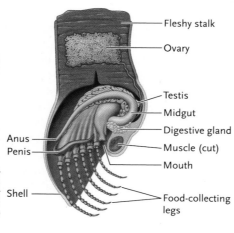

Figure 27.42

Barnacles. Gooseneck barnacles (*Lepas anatifera*) attach to the underside of float-ing debris. Like other barnacles, they open their shells and extend their feathery legs to collect particulate food from seawater.

Fleshy stalk
Ovary
Testis
Midgut
Digestive gland
Anus
Penis
Muscle (cut)
Mouth
Shell
Food-collecting legs

© FLPA/Alamy

even the skin of whales—and secrete the shell, which is a modified exoskeleton. To feed, barnacles open the shell and extend six pairs of feathery legs. The beating legs capture microscopic plankton and transfer it to the mouth. Unlike most crustaceans, barnacles are hermaphroditic.

Subphylum Myriapoda. The 3000 species of centi-pedes (class Chilopoda) and 10 thousand species of millipedes (class Diplopoda) are classified together in the subphylum Myriapoda (*murias* = 10 thousand *poda* = foot). Myriapods have two body regions: a head and a segmented trunk **(Figure 27.43).** The head bears one pair of antennae, and the trunk bears one (centipedes) or two (millipedes) pairs of walking legs on most of its many segments. Myriapods are terrestrial, and many species live under rocks or dead leaves. Centipedes are fast and voracious predators; they generally feed on invertebrates, but some eat small vertebrates. Although most species are less than 10 cm long, some grow to 25 cm. The millipedes are slow but powerful herbi-vores or scavengers. The largest species attain a length of nearly 30 cm.

Subphylum Hexapoda. The subphylum Hexapoda (*hex* = six) includes the class Insecta, as well as some other smaller classes. In terms of sheer numbers and

a.

Jason Poston/Shutterstock.com

b.

Tom McHugh/Science Source

Figure 27.43

Millipedes and centipedes. (a) Millipedes, such as *Spirobolus* species, feed on living and decaying vegetation. They have two pairs of walking legs on most segments. **(b)** Like all centi-pedes, this one, shown feeding on a mouse, is a vora-cious predator. Cen-tipedes have one pair of walking legs per segment.

diversity, the approximately 1 million species of insects are the most successful animals on Earth, occupying virtually every terrestrial and aquatic habitat. They were among the first animals to colonize terrestrial habitats, where most species still live. The oldest Hexapod fossils date from the Devonian, about 400 Ma ago, and the first insect fossils appeared shortly after. Insects are generally small, ranging from 0.1 mm to 30 cm in length. The class is divided into about 30 orders **(Figure 27.44)**.

The insect body plan always includes a head, a thorax, and an abdomen **(Figure 27.45)**. The head is equipped with multiple mouthparts, a pair of compound eyes, and one pair of sensory antennae. The thorax has three pairs of walking legs and often one or two pairs of wings. Adult insects are the only invertebrates capable of flight. The origin of wings is uncertain. The traditional view holds that they are new structures arising as outgrowths of the body wall. However, on the basis of both fossil and molecular evidence, one of the foremost researchers in the field, Jarmila Kukalova-Peck of Carleton University in Ottawa, maintains that wings are derived from branches of a proximal (near the body) segment of the leg.

Insects exchange gases through a specialized **tracheal system** (see Chapter 49), a branching network of tubes that carries oxygen from small openings in the exoskeleton to individual cells throughout the body.

Insects excrete nitrogenous wastes through specialized **Malpighian tubules** (see Chapter 50) that transport wastes to the digestive system for disposal with the feces. These two organ systems also appear in some of the terrestrial chelicerates. Since these are not paraphyletic with hexapods, this is another example of convergent evolution.

Insect sensory systems are diverse and complex. Besides a pair of image-forming compound eyes, many insects have light-sensing ocelli on their heads. Many also have hairs, sensitive to touch, on their antennae, legs, and other regions of the body. Chemical receptors are particularly common on the legs and feet, allowing the identification of food. Many groups of insects have sound receptors to detect predators and potential mates. The familiar chirping of crickets, for example, is a mating call emitted by males that may repel other males and attract females. The beetles of the family Lampyridae emit light signals from their abdomens to attract mates (see Chapter 44).

As a group, insects use an enormous variety of materials as food, and their mouthparts may be modified to reflect the nature of the food source **(Figure 27.46)**. The basic plan is reflected in a plant feeder such as a locust or a generalized feeder such as a cockroach. The *labrum* is an anterior flaplike extension of the front of the head that covers the mouthparts and has sensory structures. The mouthparts themselves are modified

a. Silverfish (Thysanura, *Ctenolepisma longicaudata*) are primitive wingless insects.

b. Dragonflies, like the flame skimmer (Odonata, *Libellula saturata*), have aquatic larvae that are active predators; adults capture other insects in mid-air.

c. Male praying mantids (Mantodea, *Mantis religiosa*) are often eaten by the larger females during or immediately after mating.

d. This rhinoceros beetle (Coleoptera, *Dynastes granti*) is one of more than 250 thousand beetle species that have been described.

e. Fleas (Siphonoptera, *Hystrichopsylla dippiei*) have strong legs with an elastic ligament that allows these parasites to jump on and off their animal hosts.

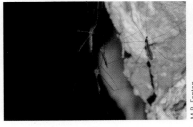

f. Crane flies (Diptera, *Tipula* species) look like giant mosquitoes, but their mouthparts are not useful for biting other animals; the adults of most species live only a few days and do not feed at all.

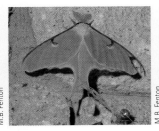

g. The luna moth (Lepidoptera, *Actias luna*), like other butterflies and moths, has wings that are covered with colourful microscopic scales.

h. Like many other ant species, fire ants (Hymenoptera, *Solenopsis invicta*) live in large cooperative colonies. Fire ants—named for their painful sting—were introduced into southeastern North America, where they are now serious pests.

Figure 27.44

Insect diversity. Insects are grouped into about 30 orders, 8 of which are illustrated here.

External anatomy of a grasshopper

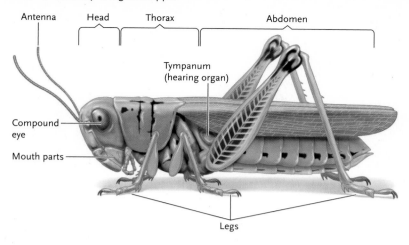

Internal anatomy of a female grasshopper

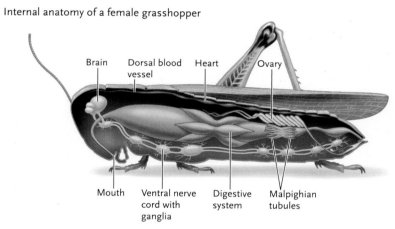

Figure 27.45

The insect body plan. Insects have a distinct head, thorax, and abdomen. Of all the internal organ systems, only the dorsal blood vessel, ventral nerve cord, and some muscles are strongly segmented.

pair of appendages, which is well supplied with sensory structures and palps. This ancestral mandibulate pattern, with mouth parts representing three of the six segments that form the insect head, is modified in various ways to accommodate different modes of feeding. In some biting flies, such as mosquitoes, the mouthparts are piercing structures, with a narrow channel to suck up blood. In butterflies and moths, the mouthparts include a long proboscis to drink nectar. In houseflies, the mouthparts are adapted for sopping up food that has been moistened by its saliva.

Life on land requires internal fertilization (see Chapter 41). In insects, males may produce packets of sperm enclosed in spermatophores and insert them into the female ducts, or sperm transfer may be direct via a penis. Sperm are stored in the female until used to fertilize eggs at the time of egg laying. The eggs of most insects are covered with a waterproof shell before they are fertilized and have one or more minute pores to permit the entry of sperm.

Parthenogenesis occurs in a number of species. In aphids, not only are the females parthenogenetic at times when food plants are abundant, they also produce live young, and development is so telescoped that embryos within the mother already have embryos within their ovaries. This results in an enormously rapid increase in population when conditions are favourable. Under less favourable conditions, normal sexual reproduction occurs. In a few species, parthenogenesis is the only mode of reproduction and males are unknown.

appendages. The paired mandibles are chewing organs, and behind those are paired maxillae abundantly supplied with sense organs, particularly on its *palps* (jointed projections), which act to scoop the food. The most posterior is the *labium,* representing a fused

a. Grasshopper

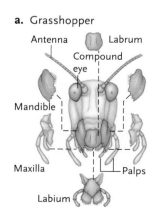

b. Housefly

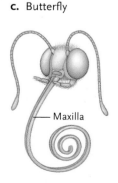

c. Butterfly

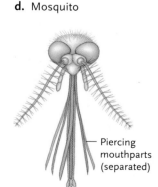

d. Mosquito

Figure 27.46

Specialized insect mouthparts. The **(a)** ancestral chewing mouthparts have been modified during evolution, allowing different insects to **(b)** sponge up food, **(c)** drink nectar, and **(d)** pierce skin to drink blood.

PEOPLE BEHIND BIOLOGY 27.2
V.B. Wigglesworth, *Cambridge University*

V.B. Wigglesworth was a British researcher who founded the subject of insect physiology. He had an active research career that spanned seven decades, from 1928 to 1991. He discovered the utility of the blood-sucking bug *Rhodnius prolixus* **(Figure 1)** for experimental work. Unfed, the insect remains in a state of suspended development. However, when it takes a blood meal, development to the next stage begins. Wigglesworth used this signal, together with clever surgical approaches **(Figure 2)**, to establish the basic facts of the hormonal control of development in insects. Although he is perhaps best known for this work, he also established the basic facts of insect digestion, insect excretion, the operation of the Malpighian tubules, the operation of the tracheal system, and the properties of the cuticle. He used a keen sense of observation to identify appropriate experimental questions and devised and carried out clever experimental approaches to answer the questions he investigated.

Plate II a opposite p. 45 of *The Physiology of Insect Metamorphosis* Issue 1 by V. B. Wigglesworth. Cambridge Univeresity Press, 1954. Reprinted with the permission of Cambridge University Press.

Figure 2
One of Wigglesworth's surgical procedures. The nymph on the right was decapitated within a day of feeding, before the hormones from the head governing moulting were secreted; therefore, it will not develop. The insect on the left was decapitated after the hormones were released and will develop normally. The two are joined so that their hemocoels are connected (a procedure called parabiosis). They will both initiate the formation of a new cuticle driven by hormones from the insect on the left.

Ken Davey

Figure 1
The adult female of Rhodnius prolixus.

---

a. Incomplete metamorphosis

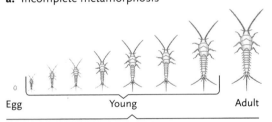

Egg   Young   Adult

Some wingless insects, like silverfish (order Thysanura), do not undergo a dramatic change in form as they grow.

b. Metamorphosis without a pupa

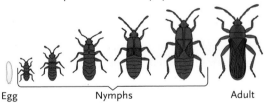

Egg   Nymphs   Adult

Some insects, such as the Order Hemiptera, undergo a metamorphosis that involves no major reorganization in form apart from the development of wings.

Figure 27.47
Patterns of postembryonic development in insects.

c. Metamorphosis with a pupa

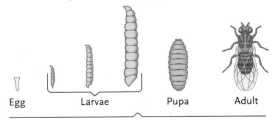

Egg   Larvae   Pupa   Adult

Fruit flies (order Diptera) and many other insects undergo a total reorganization of their internal and external anatomy when they pass through the pupal stage of the life cycle.

After it hatches from an egg, an insect passes through a series of developmental stages called **instars.** Several hormones control development and ecdysis, which marks the passage from one instar to the next. Insects exhibit one of three basic patterns of postembryonic development **(Figure 27.47).** Primitive, wingless species (order Thysanura) simply grow and shed their exoskeleton, undergoing only minimal changes in morphology. Early instars lack scales on their cuticle, and the appearance of scales corresponds

Insect Juvenile Hormone

The juvenile hormone of insects is a family of hormones, each differing only slightly in structure **(Figure 1)**. Its existence was first demonstrated by the English researcher V.B. Wigglesworth (see "People behind Biology," Box 27.2) in 1948. Using surgical procedures, he showed that the corpus allatum, an endocrine organ just behind the brain, was the source of a factor governing metamorphosis. In the presence of the factor, the insect remained larval, and in its absence, metamorphosis occurred, leading to the adult insect. In the adult, the hormone governs egg production and other elements of reproduction. The precise structure of the first member of the family of molecules was not elucidated until 1967. Because juvenile hormone is an oil, it passes easily through the cuticle of insects **(Figure 2)**. This raised the possibility that mimics of the hormone might be useful as insecticides. Since the hormone has no obvious counterpart in other animals, it was argued that insecticides based on the hormone

Figure 1
JH III, the most common of the family of molecules used as juvenile hormone in insects.

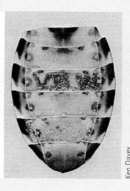

Ken Davey

Figure 2
Juvenile hormone passes easily through the cuticle. In this experiment, Wigglesworth applied the appropriate concentration of the hormone to a localized area on the dorsal surface of a last-stage larva of Rhodnius prolixus *during the process of forming the adult cuticle. On the left, he applied the hormone to a single segment, and on the right, he applied it in the form of his initials. When the insects moulted to the adult, the new cuticle in the treated portions retained the characteristics of the larval cuticle, whereas the rest of the insect exhibited normal adult cuticle. This photo also shows that in some insects with "incomplete metamorphosis," the changes in morphology may be very great.*

should be safe. Some compounds have emerged as useful pesticides and have been particularly useful in controlling mosquito larvae in water bodies and fleas on pets.

to reproductive maturity. Moulting cycles may continue after reproductive maturity. Other species undergo what is often called **incomplete metamorphosis**. They hatch from the egg as a nymph, which lacks functional wings. In many species, such as grasshoppers (order Orthoptera), the nymphs resemble the adults. In other insects, such as dragonflies (order Odonata), the aquatic nymphs are morphologically very different from the adults. Even in insects following this developmental pattern, the adult form differs by more than the abrupt development of the wings. The nature or colour of the cuticle may differ (see "Molecule behind Biology," Box 27.3). The terminal segments are reorganized to produce the external genitalia. In general, however, the descendants of the cells present in the first instar produce these changes.

Most insects undergo **complete metamorphosis**: the larva that hatches from the egg differs greatly from the adult. Larvae and adults often occupy different habitats and consume different food. The larvae (caterpillars, grubs, or maggots) are often worm shaped, with chewing mouthparts. They grow and moult several times, retaining their larval morphology. Before

they transform into sexually mature adults, they spend a period of time as a sessile **pupa**. During this stage, most of the larval tissues are destroyed and replaced by groups of embryonic cells, called *disks*, that have been in place since hatching. Although these cells are not obviously differentiated, their developmental fate is determined. Thus, there are antennal disks, eye disks, wing disks, and so on. The process is fundamentally different from that in insects with incomplete metamorphosis. In the latter, existing cells are reprogrammed at the last moult to produce the adult form, whereas in insects with complete metamorphosis, entirely new cells, programmed during embryogenesis to produce adult tissues, are involved.

Moths, butterflies, beetles, wasps, and flies are examples of insects with complete metamorphosis. Their larval stages specialize in feeding and growth, whereas the adults are adapted for dispersal and reproduction. In some species, the adults never feed, relying on the energy stores accumulated during the larval stage. This mode of development has been highly successful. The four principal orders with a pupa—Lepidoptera, Coleoptera, Hymenoptera, and

Diptera—account for about two-thirds of all known species of animals.

The evolution of insects has been characterized by innovations in morphology, life cycle patterns, locomotion, feeding, and habitat use. Insects' well-developed nervous systems govern exceptionally complex patterns of behaviour, including parental care, a habit that reaches its zenith in the colonial social insects, the termites, ants, bees, and wasps (see "Polymorphic Development," p. 612). The factors that contribute to the insects' success also make them our most aggressive competitors. They destroy agricultural crops, stored food, wool, paper, and timber. They feed on blood from humans and domesticated animals, sometimes transmitting disease-causing pathogens such as malaria as they do so. Nevertheless, insects are essential members of terrestrial ecological communities. Many species pollinate flowering plants, including important crops. Many others attack or parasitize species that are harmful to human activities. Most insects are a primary source of food for other animals. Some make useful products, such as honey, shellac, beeswax, and silk, and many human cultures use them for food.

STUDY BREAK

1. What are the advantages of moulting in nematodes?
2. If an arthropod's rigid exoskeleton cannot be expanded, how does the animal grow?
3. How does the number of body regions differ among the four subphyla of living arthropods?
4. How do the life stages differ between insects that have incomplete metamorphosis and those that have complete metamorphosis?

Review

To access course materials such as Aplia and other companion resources, please visit www.NELSONbrain.com.

27.1 What Is an Animal?

- Animals are eukaryotic, multicellular organisms that are differentiated from plants by heterotrophy, motility, and direct contact between adjacent cells.
- Animals probably arose in the Precambrian from a hollow sphere of colonial flagellates that reorganized as a double-layered sac-within-a-sac.

27.2 Key Innovations in Animal Evolution

- Tissues, groupings of identical cells specialized to perform specific functions, are organized into two or three tissue layers: ectoderm; endoderm; and, in those animals with three layers, mesoderm. In some sponges, the specialized cells may be capable of dedifferentiation.
- Some animals exhibit radial symmetry; most exhibit bilateral symmetry. Bilaterally symmetrical animals have left and right sides, dorsal and ventral sides, and anterior and posterior ends.
- Acoelomate animals have no body cavity. Pseudocoelomate animals have a body cavity between the derivatives endoderm and mesoderm. Coelomate animals have a body cavity that is entirely lined by derivatives of mesoderm. The cavities are filled with fluid that separates and protects the organs and in some cases functions as a hydrostatic skeleton.
- Two lineages of animals differ in developmental patterns. Most protostomes exhibit spiral, determinate cleavage; the coelom (when present) is a schizocoelom; and the blastopore develops into the mouth. Deuterostomes have radial symmetry, indeterminate cleavage, and an enterocoelom, and their blastopore becomes the anus.
- The development of many protostomes includes a larval stage. This polymorphic development allows sessile animals to be distributed, permits parasitic forms to exist in widely different environments, and avoids competition between the young and the adults.
- Four animal phyla exhibit segmentation.

27.3 An Overview of Animal Phylogeny and Classification

- Sequence analyses of highly conserved structures such as rRNA, mitochondrial DNA, and DNA coding for specific proteins can be compared in various species. The closer the similarity, the more closely related the species are assumed to be. Phylogenetic trees based on such data have confirmed some relationships based on developmental and morphological data and challenged others.
- The Radiata includes animals with two tissue layers and radial symmetry, and the Bilateria includes animals with three tissue layers and bilateral symmetry.
- Bilateria is further subdivided into Protostomia and Deuterostomia. The phylogeny based on molecular evidence divides the Protostomia into the Lophotrochozoa and the Ecdysozoa.
- Molecular phylogeny suggests that ancestral protostomes had a coelom and that acoelomate and pseudocoelomate conditions were derived from the coelomate.
- Segmentation arose independently in three lineages: the annelids, the Onychophora/Arthropoda, and the Chordata.

27.4 Phylum Porifera

- Sponges (phylum Porifera) are asymmetrical animals, many with limited integration of cells in their bodies.
- The body of many sponges is a water-filtering system with incurrent pores, a spongocoel, and an osculum through which water exits the body. Flagellated choanocytes draw water into the body and capture particulate food.

27.5 Metazoans with Radial Symmetry

- The two major radiate phyla have two well-developed tissue layers with a gelatinous mesoglea between them. They lack organ systems but have well-developed nerve nets. All are aquatic or marine.
- The hydrozoans, jellyfishes, sea anemones, and corals (phylum Cnidaria) are predators that capture prey with tentacles and stinging nematocysts.
- The life cycles of cnidarians may include polyps, medusae, or both. Anthozoans lack a medusa stage, whereas in jellyfish (Scyphozoa and Cubozoa), medusae are prominent and hydroids may be absent; both are present in Hydrozoa.
- The small, translucent comb jellies (phylum Ctenophora) use long, sticky tentacles to capture particulate food. They are weak swimmers that use rows of cilia for locomotion.

27.6 Lophotrochozoan Protostomes

- The taxon Lophotrochozoa includes eight phyla that share either a characteristic type of larva or a specialized feeding structure.
- Flatworms (phylum Platyhelminthes) are either free living or parasitic. Free-living species have well-developed digestive, excretory, reproductive, and nervous systems. Parasitic flukes and tapeworms live within or upon animal hosts. They attach to hosts with suckers or hooks, and they produce numerous eggs. Some organ systems may be greatly reduced in parasitic species.
- The wheel animals (phylum Rotifera) are tiny and abundant inhabitants of freshwater and marine ecosystems. Movements of cilia in the corona control their locomotion and bring food to their mouths. Many are parthenogenetic.
- Three small phyla (Ectoprocta, Brachiopoda, and Phoronida) all use a lophophore to feed on particulate matter. Brachiopods live within a two-part shell; ectoprocts form flattened or branching colonies; and phoronids are small, usually tube-dwelling worms.
- The ribbon worms (phylum Nemertea) are elongate and often colourful animals with a proboscis housed in a unique structure, the rhynchocoel.
- Chitons, snails, clams, octopuses, and their relatives (phylum Mollusca) have fleshy bodies that are often enclosed in a hard shell. The molluscan body plan includes a head-foot, a visceral mass, and a mantle.
- Segmented worms (phylum Annelida) generally exhibit segmentation of the coelom and of the muscular, circulatory, excretory, respiratory, and nervous systems. Polychaetes have segmental appendages used in locomotion and gas exchange. Leeches have reduced segmentation.

27.7 Ecdysozoan Protostomes

- The taxon Ecdysozoa includes three phyla that periodically shed their cuticle.
- Roundworms (phylum Nematoda) feed on decaying organic matter or parasitize plants or animals. Locomotion depends on muscles contracting against a hydrostatic skeleton provided by a fluid-filled pseudocoel. Moulting is not essential for growth, but it permits changing the nature of the cuticle to accommodate different environments.
- The velvet worms (phylum Onychophora) have segmented bodies and unjointed legs. Some species bear live young, which develop in a uterus.
- The arthropods (phylum Arthropoda) are the most diverse animals on Earth. Their segmented bodies are often differentiated into distinct regions, and their jointed appendages are specialized for feeding, locomotion, or reproduction. They shed their firm, water-resistant exoskeleton to accommodate growth or to begin a new stage of the life cycle. Arthropods have an open circulatory system; numerous sense organs that provide input to a complex nervous system; and, in some groups, highly specialized respiratory and excretory systems.
- Arthropods are divided into five subphyla. The extinct trilobites (subphylum Trilobita), with three-lobed bodies and relatively undifferentiated appendages, were abundant in Paleozoic seas.
- Spiders, ticks, mites, scorpions, and horseshoe crabs (subphylum Chelicerata) have a cephalothorax and an abdomen; two pairs of appendages on the head serve in feeding.
- Lobsters, crabs, and their relatives (subphylum Crustacea) have a carapace that covers the cephalothorax, as well as highly modified appendages, including five pairs on the head.
- The centipedes and millipedes (subphylum Myriapoda) are largely terrestrial. They have a head and an elongate, segmented trunk.
- Insects and their relatives (subphylum Hexapoda) are also largely terrestrial. Insects have three body regions, three pairs of walking legs on the thorax, and a pair of antennae and three pairs of feeding appendages on the head.
- Most insects undergo metamorphosis. In incomplete metamorphosis, the cells of the nymph are reprogrammed to produce adult structures. In complete metamorphosis, an additional stage, the pupa, permits entirely new adult structures to replace larval cells.

Questions

Self-Test Questions

1. Which of the following characteristics is NOT typical of most animals?
   a. heterotrophic
   b. sessile
   c. radially symmetrical
   d. multicellular

2. Which term refers to a body cavity that separates the digestive system from the body wall but is NOT completely lined with mesoderm?
   a. schizocoelom
   b. mesentery
   c. peritoneum
   d. pseudocoelom

3. Which part of a mollusc secretes the shell?
   a. the visceral mass
   b. the trochophore
   c. the head-foot
   d. the mantle

4. Which of the following is NOT a result of polymorphic development?
   a. castes in social insects
   b. the alternation of generations in Cnidaria
   c. the pupal stage in insects
   d. the adult octopus

5. Ecdysis refers to a process in which
   a. bivalves use siphons to pass water across their gills
   b. arthropods and nematodes shed their cuticles
   c. cnidarians build skeletons of calcium carbonate
   d. squids escape from predators in a cloud of ink

6. The Burgess Shale contains fossils from the
   a. Precambrian   c. Devonian
   b. Pleistocene   d. Cambrian

7. Which of the following are NOT part of the cnidarian body plan?
   a. notochord   c. hydrozoa
   b. mantle   d. metazoa

8. Protostomes and deuterostomes differ markedly in which of the following?
   a. the pattern of embryological development
   b. the origin of the anus and the mouth
   c. the pattern of fertilization
   d. the presence of a mantle cavity

9. Which of the following does NOT belong to the Mollusca?
   a. tapeworms   c. limpets
   b. leeches   d. lobsters

10. In which of the following does the ventral solid nerve chord occur?
    a. Chordata   c. Platyhelminthes
    b. Cnidaria   d. Arthropoda

Questions for Discussion

1. Many invertebrate species are hermaphroditic. What selective advantages might this characteristic offer? In what kinds of environments might it be most useful?

2. In terms of numbers of species, insects are the dominant life form on Earth, but the individuals are also smaller in size than many other groups. What has contributed to their success, and why are they not larger?

3. The egg of the human parasite *A. lumbricoides* hatches in the small intestine. What experiments would you do to test whether this is a result of the egg shell being digested by the intestinal enzymes?

4. What is a parasite? In which groups do we find parasitic animals?

5. What role does the coelom play in the development of protostomes?

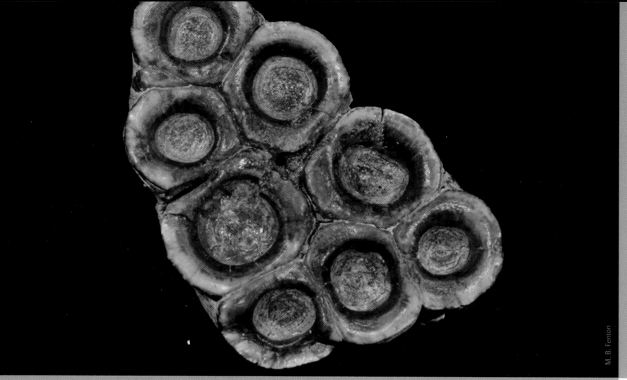

28

The molar tooth of an extinct mammal (Desmostylus) was used to crush and grind marine plants. Paleontologists discovered that Desmostylus was more walruslike than seacow (manatee)-like only when they found almost-complete skeletons.

Diversity of Animals 2: Deuterostomes: Vertebrates and Their Closest Relatives

WHY IT MATTERS

Based on molecular evidence, Xenoturbellida was identified as a new phylum in November 2006. This new phylum is closely related to the Chordata (traditional phylogeny in **Figure 28.1a, p. 634**), which includes vertebrates, among them *Homo sapiens*. Specifically, a phylogenetic analysis of 170 nuclear proteins and 13 mitochondrial proteins was used to derive the phylogeny that placed Xenoturbellida among the Deuterostomes (Figure 28.1b, p. 635). Look at some of the organisms **(Figure 28.2, p. 636)** arranged in the phylogeny.

Xenoturbellida (such as *Xenoturbella bocki;* see Figure 28.2f) was originally described in 1949. They are delicate, ciliated marine worms with simple body plans. They lack a gut with two openings, organized gonads, excretory structures, and a coelom. The nervous system is a diffuse net with no brain. Until 2006, there were more questions than answers about their phylogenetic position, and even to what phylum they should belong. At first, *X. bocki* was thought to be a turbellarian flatworm (Platyhelminthes), but it was also identified as a possible hemichordate or echinoderm (based on similarities in the nerve net). The details of *Xenoturbella*'s cilia are like those of hemichordates, which could indicate that it is really an acoelomorph flatworm.

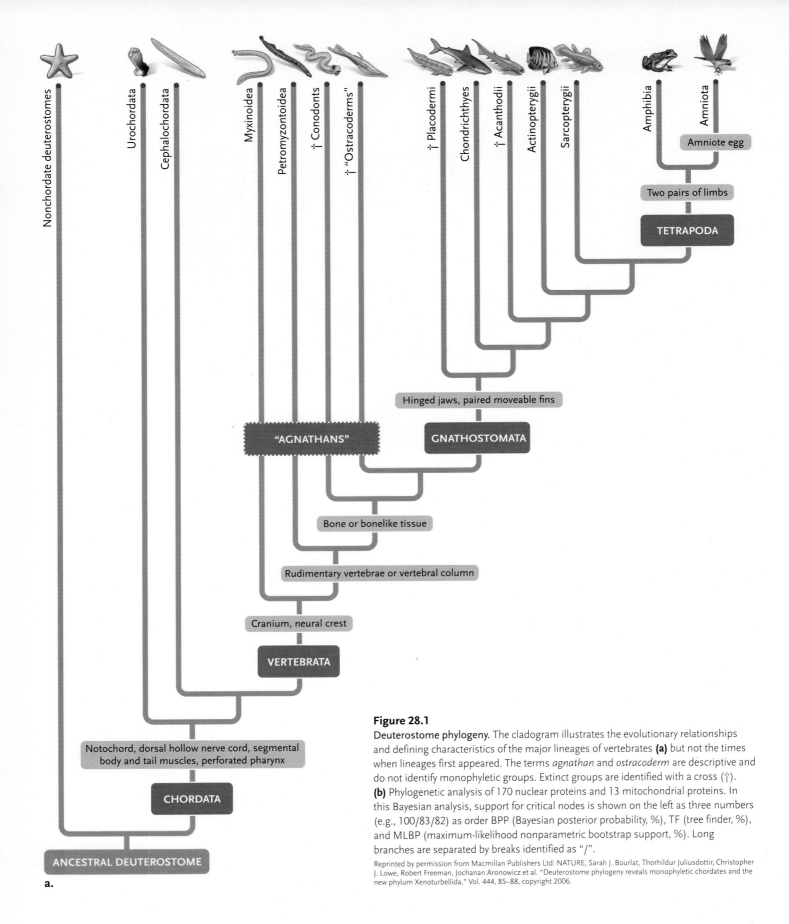

Figure 28.1

Deuterostome phylogeny. The cladogram illustrates the evolutionary relationships and defining characteristics of the major lineages of vertebrates **(a)** but not the times when lineages first appeared. The terms *agnathan* and *ostracoderm* are descriptive and do not identify monophyletic groups. Extinct groups are identified with a cross (†). **(b)** Phylogenetic analysis of 170 nuclear proteins and 13 mitochondrial proteins. In this Bayesian analysis, support for critical nodes is shown on the left as three numbers (e.g., 100/83/82) as order BPP (Bayesian posterior probability, %), TF (tree finder, %), and MLBP (maximum-likelihood nonparametric bootstrap support, %). Long branches are separated by breaks identified as "/".

Reprinted by permission from Macmillan Publishers Ltd: NATURE, Sarah J. Bourlat, Thorhildur Juliusdottir, Christopher J. Lowe, Robert Freeman, Jochanan Aronowicz et al. "Deuterostome phylogeny reveals monophyletic chordates and the new phylum Xenoturbellida," Vol. 444, 85–88, copyright 2006.

Labels in figure:

Nonchordate deuterostomes
Urochordata
Cephalochordata
Myxinoidea
Petromyzontoidea
† Conodonts
† "Ostracoderms"
† Placodermi
Chondrichthyes
† Acanthodii
Actinopterygii
Sarcopterygii
Amphibia
Amniota

Amniote egg
Two pairs of limbs
TETRAPODA
Hinged jaws, paired moveable fins
"AGNATHANS"
GNATHOSTOMATA
Bone or bonelike tissue
Rudimentary vertebrae or vertebral column
Cranium, neural crest
VERTEBRATA
Notochord, dorsal hollow nerve cord, segmental body and tail muscles, perforated pharynx
CHORDATA
ANCESTRAL DEUTEROSTOME

a.

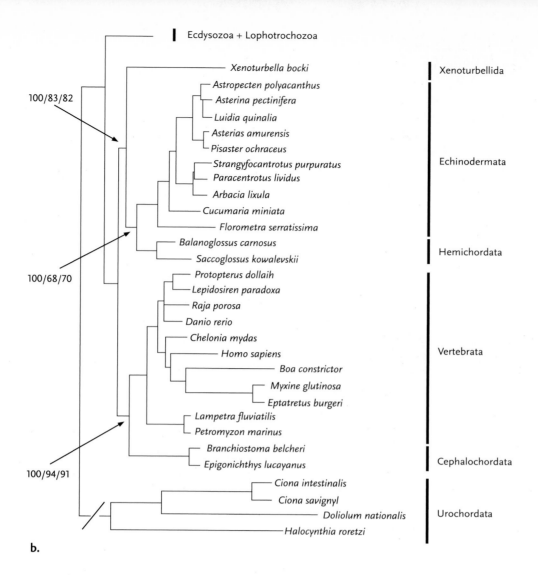

b.

In 1997, analysis of molecular phylogenetic data had been used to place *Xenoturbella* in the Mollusca, specifically among the bivalves. This arrangement was supported by the discovery of bivalvelike eggs and larvae within specimens of *Xenoturbella*. In 1997, it was easy to believe the molecular argument.

How can molecular data be challenged? In 1998, an alternative explanation was offered: mollusc genetic information appeared inside *Xenoturbella* because it eats molluscs. When the molluscan genetic information is ignored, *Xenoturbella* is clearly a deuterostome, most closely related to Echinodermata (see Figure 28.1b).

Thus, making correct choices about classification (see Chapter 20) means looking beyond appearance and may also require careful consideration of molecular data.

Although vertebrates are the best-known deuterostomes, echinoderms, hemichordates, and tunicates are also deuterostomes. In deuterostomes, the anus develops from the blastopore and the mouth arises as a separate opening, making them fundamentally different from protostomes. Deuterostome species show the same trophic diversity as protostomes, but with few species specialized as parasites. Vertebrates are classified as chordates, first known as small, filter-feeding animals. Vertebrates have an internal skeleton usually made of bone, partly accounting for their rich fossil record. Vertebrates include the largest animals known on Earth, the enormous plankton-feeding ichthyosaurs and whales.

28.1 Deuterostomes

Membership in the Deuterostomia (Greek, *deutero* = second; *stomia* = opening) is restricted to animals in which the anus develops from the blastopore and the mouth from a second opening. At first glance, deuterostome animals—such as echinoderms, chordates, and hemichordates, let alone *Xenoturbella*—are not obviously similar, reflecting modifications of their bodies that mask underlying developmental and genetic features.

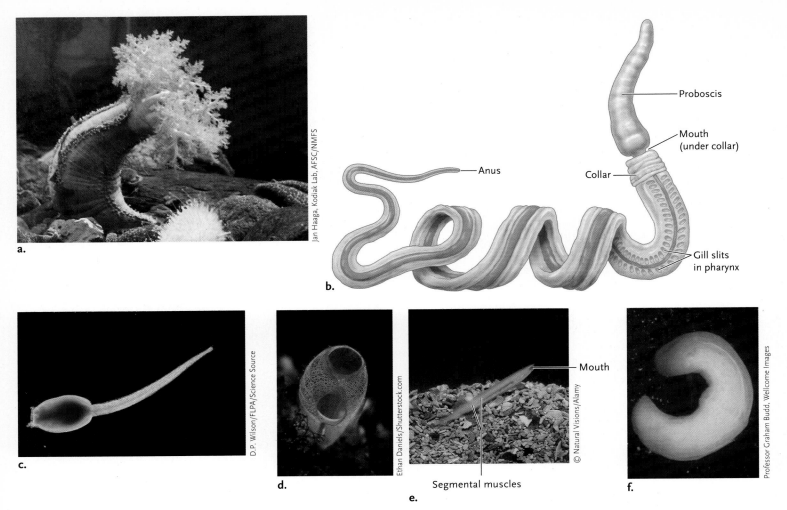

Figure 28.2

Deuterostomes. (a) Holothuroidea. A sea cucumber (*Cucumaraia miniata*) extends its tentacles, which are modified tube feet. **(b)** Phylum Hemichordata. Acorn worms draw food- and oxygen-laden water in through the mouth and expel it through gill slits in the anterior region of the trunk. **(c)** Urochordates. A tadpole-like tunicate larva will metamorphose into a sessile adult. **(d** and **e)** Cephalochordates. The unpigmented skin of an adult lancelet (*Brachiostoma* species) reveals its segmented body wall muscles. **(f)** A *Xenoturbella bocki* does not look very similar to any of the other animals illustrated here.

STUDY BREAK

Give one difference between protostomes and deuterostomes.

28.2 Phylum Echinodermata

The phylum Echinodermata (*echino* = spiny; *derm* = skin) includes 6500 species of sea stars, sea urchins, sea cucumbers, brittle stars, and sea lilies. These slow-moving or sessile bottom-dwelling animals are important herbivores and predators living in oceans from the shallow coastal waters to the depths. The phylum was diverse in the Paleozoic, but only a remnant of that fauna remains. Echinoderms vary in size from less than 1 cm in diameter to more than 50 cm long. Adult echinoderms develop from bilaterally symmetrical, free-swimming larvae. As the larvae develop, they assume a secondary radial symmetry, often organized around five rays or "arms" **(Figure 28.3)**. Many echinoderms have an oral surface, with the mouth facing the substrate, and an aboral surface facing in the opposite direction. Virtually all echinoderms have an internal skeleton made of calcium-stiffened ossicles that develop from mesoderm. In some groups, fused ossicles form a rigid container called a *test*. In most species with these features, spines or bumps project from the ossicles.

The internal anatomy of echinoderms is unique among animals (see Figure 28.3). They have a well-defined coelom and a complete digestive system (see Figure 28.3e) but no excretory or respiratory systems, and most have only a minimal circulatory system. In many, gases are exchanged and metabolic wastes eliminated through projections of the epidermis and peritoneum near the base of the spines. Given their radial symmetry, there is no head or central brain; the nervous system is organized around nerve cords that encircle the mouth and branch into the radii. Sensory cells are abundant in the skin.

a. Asteroidea: This sea star (*Fromia milleporella*) lives in the intertidal zone.

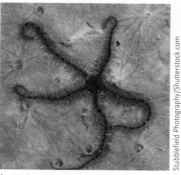

b. Ophiuroidea: A brittle star (*Ophiothrix swensonii*) perches on a coral branch.

c. Echinoidea: A sea urchin (*Strongylocentrotus purpuratus*) grazes on algae.

d. Crinoidea: A feather star (*Himerometra robustipinna*) feeds by catching small particles with its numerous tentacles.

Figure 28.3

Echinoderm diversity. (a)–(d) Echinoderms exhibit secondary radial symmetry, usually organized as five rays around an oral–aboral axis. The coelom **(e)** is well developed in echinoderms, as illustrated by this cutaway diagram of a sea star. The water vascular system **(f)**, unique in the animal kingdom, operates the tube feet. Tube feet **(g)** are responsible for locomotion. Note the pedicillariae on the upper surface of the star's arm.

e. Internal anatomy

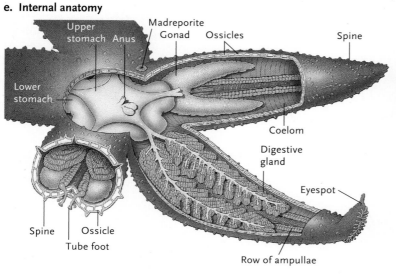

f. Water vascular system

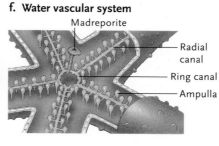

g. Tube feet

Echinoderms move using a system of fluid-filled canals, the *water vascular system* (see Figure 28.3f). In a sea star, for example, water enters the system through the madreporite, a sievelike plate on the aboral surface. A short tube connects it to the *ring canal,* which surrounds the **esophagus**. The ring canal branches into five *radial canals* that extend into the arms. Each radial canal is connected to numerous *tube feet* that protrude through holes in the plates. Each tube foot has a mucus-covered, sucker like tip and a small muscular bulb, the *ampulla,* that lies inside the body. When an ampulla contracts, fluid is forced into the tube foot, causing it to lengthen and attach to the substrate (see Figure 28.3g). When the tube foot contracts, it pulls the animal along. As the tube foot shortens, water is forced back into the ampulla, and the tube foot releases its grip on the substrate. The tube foot can then take another step forward, reattaching to the substrate. Although each tube foot has limited strength, the coordinated action of hundreds or even thousands of them is so strong that they can hold an echinoderm to a substrate even against strong wave action.

Echinoderms have separate sexes, and most reproduce by releasing gametes into the water. Radial cleavage is so clearly apparent in the transparent eggs of some sea urchins that they are commonly used to demonstrate cleavage in introductory biology laboratories. A few echinoderms reproduce asexually by splitting in half and regenerating the missing parts. Other echinoderms regenerate body parts lost to predators. Four-day-old sand dollars (*Dendraster excentricus*) asexually clone themselves in response to the odour of fish (in mucus), apparently a defensive response.

Echinoderms are divided into six groups, the most recently described (1986) being the sea daisies (Concentricycloidea). These small, medusa-shaped animals occupy sunken, waterlogged wood in the deep sea. Sunken ships are often important habitats for these and other marine organisms. The five other groups, described below, are more diverse and better known.

28.2a Asteroidea: Starfish or Sea Stars

Sea stars live on rocky shorelines to depths of 10 000 m. Many are brightly coloured. The body consists of a central disk surrounded by 5–20 radiating "arms" (see Figure 28.3a), with the mouth centred on the oral surface. The ossicles of the endoskeleton are not fused, permitting flexibility of the arms and disk. **Pedicellariae** are small

pincers at the base of short spines. They are used to remove debris that falls onto the animal's aboral surface (see Figure 28.3g). Many sea stars eat invertebrates and small fishes. Species that consume bivalve molluscs grasp the two valves with tube feet and slip their everted stomachs between the bivalve's shells **(Figure 28.4)**. The stomach secretes digestive enzymes that dissolve the mollusc's tissues. Some sea stars are destructive predators of corals, endangering many reefs.

28.2b Ophiuroidea: Brittle Stars

The 2000 species of brittle stars and basket stars occupy roughly the same range of habitats as sea stars. Their bodies have a well-defined central disk and slender, elongated arms that are sometimes branched (see Figure 28.3b). Ophiuroids can crawl fairly swiftly across substrates by moving their arms in a coordinated fashion. As their common name implies, the arms are delicate and easily broken, an adaptation allowing them to escape from predators with only minor damage. Brittle stars feed on small prey, suspended plankton, or detritus that they extract from muddy deposits.

28.2c Echinoidea: Sea Urchins and Sand Dollars

The 950 species of sea urchins and sand dollars lack arms (see Figure 28.3c). Their ossicles are fused into solid tests that provide excellent protection but

Figure 28.4
Sea star feeding on a mussel. Even when the tide is out in Haida Gwaii, sea stars hunt mussels.

restrict flexibility. The test is spherical in sea urchins and flattened in sand dollars. These animals use tube feet in locomotion. Five rows of tube feet emerge through pores in the test. Most echinoids have movable spines, some with poison glands. A jab from some tropical species can cause a careless swimmer severe pain and inflammation. Echinoids graze on algae and other organisms that cling to surfaces. In the centre of an urchin's oral surface is a five-part nipping jaw that is controlled by powerful muscles. Some species damage kelp beds, disrupting the habitat of young lobsters and other crustaceans. Echinoid ovaries are a gourmet delicacy in many countries, making these animals a prized natural resource.

28.2d Holothuroidea: Sea Cucumbers

Sea cucumbers are elongated animals that lie on their sides on the ocean bottom (see Figure 28.2a, p. 636); they number about 1500 species. Although they have five rows of tube feet, their endoskeleton is reduced to widely separated microscopic plates. The body, which is elongated along the oral–aboral axis, is soft and fleshy, with a tough, leathery covering. Modified tube feet form a ring of tentacles around the mouth. The central disk and mouth point upward rather than toward the substrate. Some species secrete a mucus net that traps plankton or other food particles. The net and tentacles are inserted into the mouth, where the net and trapped food are ingested. Other species extract food from bottom sediments. Many sea cucumbers exchange gases through an extensively branched respiratory tree arising from the rectum, the part of the digestive system just inside the anus at the aboral end of the animal. A well-developed circulatory system distributes oxygen and nutrients to tissues throughout the body.

Sea cucumbers are actually home for a specialized symbiotic fish. *Carapus bermudensis,* the pearl fish, enters sea cucumbers' cloacal opening tail first. The cloaca is the chamber receiving urine, feces, and reproductive products. Pearl fish are members of a group that usually live in the tubes of other animals, including the cavities of bivalves. These fishes have elongated, thin bodies. They have lost pelvic fins and scales, and the anal opening has moved forward to a position under the head. This adaptation ensures that the fish defecates outside the body of the sea cucumber. These fishes use olfactory cues to find the "correct" host.

28.2e Crinoidea: Sea Lilies and Feather Stars

The 600 living species of sea lilies and feather stars are the surviving remnants of a diverse and abundant fauna 500 million years ago (mya) (see Figure 28.3d, p. 637). Most species occupy marine waters of medium depth. Between five and several hundred branched arms surround the disk that contains the mouth. New arms are added as a crinoid grows larger.

Figure 28.5
Fossil crinoid stems. Ossicles making up the stems of crinoids are commonly fossilized. The individual ossicles are from the Devonian of Ontario. The section of complete stem is *Encrinus liliiformis* from the Triassic of Germany. Scale is in millimetres.

The branches of the arms are covered with tiny, mucus-coated tube feet that trap suspended microscopic organisms. Sessile sea lilies have the central disk attached to a flexible stalk that can reach 1 m in length. By contrast, adult feather stars can swim or crawl weakly, attaching temporarily to substrates. The disks making up sea lily stalks, called ossicles, are common fossils in many deposits **(Figure 28.5)**.

STUDY BREAK

1. What are echinoderms? How do adult echinoderms develop?
2. Use a table to compare an echinoderm and a human by system: (a) digestive, (b) excretory, (c) respiratory, (d) circulatory, and (e) nervous.
3. Using a sea star as an example, describe how echinoderms move.

28.3 Phylum Hemichordata

The 80 species of **acorn worms** making up this phylum take their name from *hemi,* meaning half, and *chord,* referring to the phylum Chordata. Hemichordates have a stomochord that superficially resembles the notochord of chordates. Acorn worms are sedentary marine animals living in U-shaped tubes or burrows in coastal sand or mud. Their soft bodies range in length from

2 cm to 2 m and are organized into an anterior proboscis, a tentacled collar, and an elongated trunk (see Figure 28.2b, p. 636). They use the muscular, mucus-coated proboscis to construct burrows and trap food particles. Acorn worms also have pairs of gill slits in the pharynx, the part of the digestive system just posterior to the mouth. Beating cilia create a flow of water, which enters the pharynx through the mouth and exits through the gill slits. As water passes through, suspended food particles are trapped and shunted into the digestive system, and gases are exchanged across the partitions between gill slits. The dorsal nerve cord, coupled with feeding and respiration, reflects a close evolutionary relationship between hemichordates and chordates. Pterobranchia, or sea angels, are the other class of animals in this phylum. These uncommon marine animals are colonial and live in tubes. They superficially resemble some cnidarians.

STUDY BREAK

How do hemichordates feed?

28.4 Phylum Chordata

This phylum includes evolutionary lines of invertebrates, the Urochordata and Cephalochordata, as well as the more diverse line, the Vertebrata. A notochord, a dorsal hollow nerve cord, and gill slits (a perforated pharynx) are key morphological features distinguishing chordates from all other Deuterostomes. These features occur during at least some time in a chordate's life cycle. Chordates also have segmental muscles in the body wall and tail **(Figure 28.6)**. Collectively, these structures enable higher levels of activity and unique modes of aquatic locomotion, as well as more efficient feeding and oxygen acquisition.

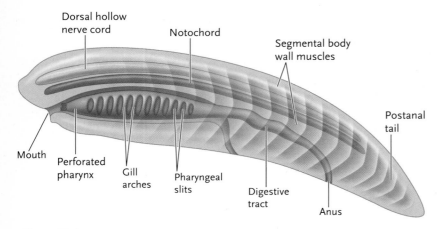

Figure 28.6
Diagnostic chordate characteristics. Chordates have a notochord, a dorsal hollow nerve cord, pharangeal (gill) slits (a perforated pharynx), and a muscular postanal tail with segmental body wall and tail muscles. Other basic features are shown as well, but they are not unique to chordates.

Early in chordate embryonic development, the **notochord** (*noto* = back; *chord* = string), a flexible rod, develops from mesoderm dorsal to the developing digestive system. The notochord is constructed of fluid-filled cells surrounded by tough connective tissue. It supports the embryo from head to tail. The notochord is the skeleton of invertebrate chordates, serving as an anchor for body wall muscles. When these muscles contract, the notochord bends but does not shorten. Waves of contractions pass down one side of the animal and then up the other, sweeping the body and tail back and forth in a smooth and continuous movement. Thus, the chordate body swings left and right during locomotion, propelling the animal forward. The chordate tail, which is posterior to the anus, provides most of the propulsion in some aquatic species. Segmentation allows each muscle block to contract independently. Unlike the bodies of annelids and other nonchordate invertebrates, the chordate body does not shorten when the animal is moving. Remnants of the notochord persist as gelatinous disks between the vertebrae of some adult vertebrates.

The central nervous system of chordates is a hollow nerve cord on the dorsal side of the embryo (see Chapter 42). Most nonchordate invertebrates have ventral, solid nerve cords. In vertebrates, an anterior enlargement of the nerve cord forms the brain. In invertebrates, an anterior concentration of nervous system tissue is a **ganglion** and may be referred to as a "brain."

Gill (pharyngeal) slits mean that the chordate pharynx is perforated. The pharynx is the part of the digestive system just behind the mouth. **Gill slits** are paired openings originating as exit holes for water that carried particulate food into the mouth, allowing chordates to gather food by filtration. Invertebrate chordates also collect oxygen and release carbon dioxide across the walls of the pharynx. In fishes, gill arches have evolved as supporting structures between the slits in the pharynx. Invertebrate chordates and fishes retain a perforated pharynx throughout their lives. In most air-breathing vertebrates, the slits are present only during embryonic development and in some larvae.

28.4a Subphylum Urochordata: Sea Squirts and Tunicates

The 2500 species of urochordates (*uro* = tail) float in surface waters or attach to substrates in shallow marine habitats. Sessile adults of many species secrete a gelatinous or leathery "tunic" around their bodies and squirt water through a siphon when disturbed. Adults can attain lengths of several centimetres (**Figure 28.7**; see also Figure 28.2c, p. 636). In the most common group of sea squirts (Ascidiacea), swimming larvae have notochords, dorsal hollow nerve cords, and gill slits, features lacking in the sessile adults. Larvae eventually attach to substrates and transform into sessile adults.

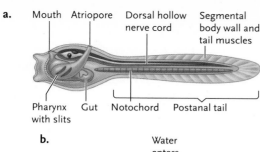

a. Mouth Atriopore Dorsal hollow Segmental
                      nerve cord   body wall and
                                     tail muscles

Pharynx Gut Notochord Postanal tail
with slits

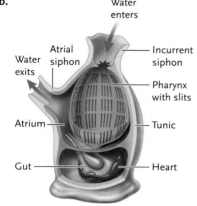

b.
                        Water
                        enters

        Atrial   Incurrent
Water   siphon   siphon
exits

                                  Pharynx
                                  with slits

Atrium   Tunic

Gut   Heart

Figure 28.7
Diagrams of urochordates. The tadpolelike tunicate larva
(a) metamorphoses into an adult, a sessile filter-feeder.
(b) In the adult, the atriopore becomes the atrial siphon.

During metamorphosis, larvae lose most traces of the notochord, dorsal nerve cord, and tail, and their basketlike pharynx enlarges. In adults, beating cilia pull water into the pharynx through an incurrent siphon. A mucus net traps particulate food, which is carried with the mucus to the gut. Water passes through the gill slits, enters a chamber called the **atrium**, and is expelled through the **atrial siphon** along with digestive wastes and carbon dioxide. Oxygen is absorbed across the walls of the pharynx. In some urochordates, the larvae are neotenous, acquiring the ability to reproduce and remaining active throughout their life cycles.

28.4b Subphylum Cephalochordata: Lancelets

All 28 species of cephalochordates (*cephalo* = head) live in warm, shallow marine habitats, where they lie mostly buried in sand (see Figure 28.2d, e, p. 636).

Although generally sedentary, they have well-developed body wall muscles and a prominent notochord. Most species are included in the genus *Branchiostoma* (formerly *Amphioxus*). Lancelet bodies, which are 5 to 10 cm long, are pointed at both ends like the double-edged surgical tools for which they are named **(Figure 28.8)**. Adults have light receptors on the head as well as chemical sense organs on tentacles that grow from the **oral hood**. Lancelets use cilia to draw food-laden water through hundreds of pharyngeal slits; water flows into the atrium and is expelled through the **atriopore**. Most gas exchange occurs across the skin.

28.4c Subphylum Vertebrata: Vertebrates

Species in this subphylum have a distinct head making them craniate, and most have a **backbone (spine)** made up of individual bony vertebrae (see Chapters 20 and 42). This internal skeletal feature provides structural support for muscles and protects the nervous system and other organs. In addition, the internal skeleton and attached muscles allow most vertebrates to move rapidly. Vertebrates are the only animals that have bone, a connective tissue in which cells secrete the mineralized matrix that surrounds them (see Chapter 46). One vertebrate lineage, cartilaginous fishes (class Chondrichthyes), may have lost its bone over evolutionary time. These animals, mostly sharks and rays, have skeletons of cartilage, a dense, flexible connective tissue that can be a developmental precursor of bone (see Chapters 42 and 46).

At the anterior end of the vertebral column, the head is usually protected by a bony **cranium** or skull. The backbone surrounds and protects the dorsal nerve cord, and the bony cranium surrounds the brain. The cranium, vertebral column, ribs, and sternum (breastbone) make up the **axial skeleton.** Most vertebrates also have a **pectoral girdle** anteriorly and a **pelvic girdle** posteriorly that attach bones in the fins or limbs to the axial skeleton. The bones of the two girdles and the appendages constitute the appendicular skeleton.

Vertebrates have neural crest cells (see Chapter 42), a unique cell type distinct from endoderm, mesoderm, and ectoderm. Neural crest cells arise next to the developing nervous system but migrate throughout the body. Neural crest cells ultimately contribute to

Lancelet anatomy

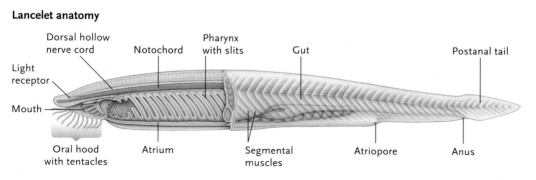

Dorsal hollow   Pharynx
nerve cord   Notochord   with slits   Gut   Postanal tail

Light
receptor

Mouth

Oral hood   Atrium   Segmental   Atriopore   Anus
with tentacles   muscles

Figure 28.8
A drawing of the internal anatomy of an adult lancelet (*Branchiostoma*).

uniquely vertebrate structures such as parts of the cranium, teeth, sensory organs, **cranial nerves**, and the medulla (the interior part) of the adrenal glands.

The brains of vertebrates are larger and more complex than those of invertebrate chordates. Moreover, the vertebrate brain is divided into three regions, the forebrain, midbrain, and hindbrain, each governing distinct nervous system functions (see Chapter 44).

STUDY BREAK

1. List four morphological features distinguishing chordates from other deuterostomes.
2. Explain the purpose and structure of gill slits.
3. What are tunicates and lancelets? To which subphyla do they belong? What characteristics of swimming tunicate larvae are missing from the sessile adults?

28.5 The Origin and Diversification of Vertebrates

Biologists have used embryological, molecular, and fossil evidences to trace the origin of vertebrates and to chronicle the evolutionary diversification of the group to which humans belong. We suspect that vertebrates arose from a cephalochordate-like ancestor through duplication of genes that regulate development.

Vertebrates appear to be more closely related to cephalochordates than to urochordates (see Figure 28.1, p. 634). The change from cephalochordate-like creatures to vertebrates was marked by the emergence of neural crest cells, bone, and other vertebrate traits. Biologists hypothesize that an increase in the number of genes that control the expression of other genes (homeotic) may have facilitated the development of more complex anatomy. (For more about homeotic genes, see Chapter 42). When it comes to organization, there is no compelling reason to believe that "more complex" is superior to "simple."

Hox genes are homeotic genes that influence the three-dimensional shape of the animal and the locations of important structures such as eyes, wings, and legs, particularly along the head-to-tail axis of the body. *Hox* genes are arranged on chromosomes in a particular order, forming the *Hox* gene complex. Each gene in the complex governs the development of particular structures. Animal groups with the simplest structure, such as cnidarians, have two *Hox* genes. Those with more complex anatomy, such as insects, have 10. Chordates typically have up to 13 or 14. Lineages with many *Hox* genes generally have more complex anatomy than those with fewer *Hox* genes.

Molecular analyses reveal that the entire *Hox* gene complex was duplicated several times in the evolution of vertebrates, producing multiple copies of all the genes in the *Hox* complex **(Figure 28.9)**. The cephalochordate *Branchiostoma* has one *Hox* gene

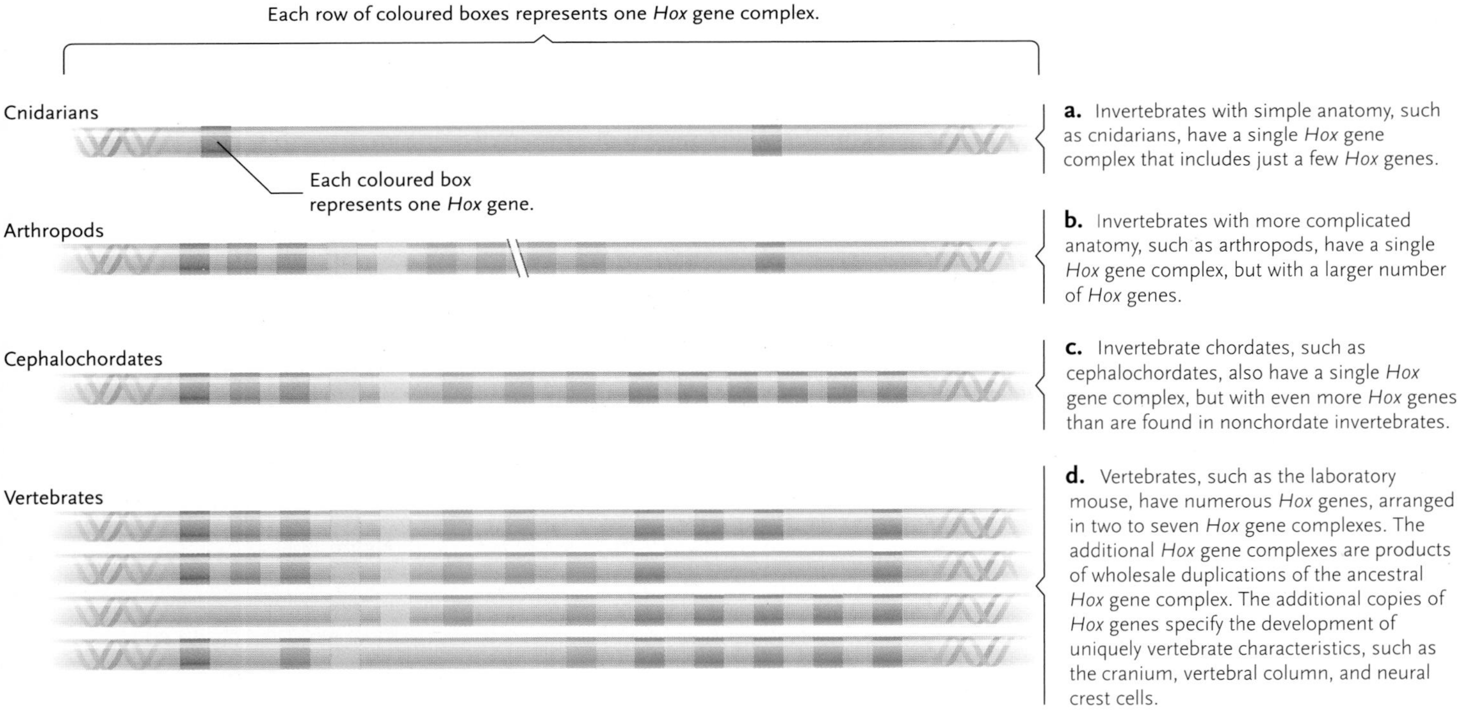

Each row of coloured boxes represents one *Hox* gene complex.

Each coloured box represents one *Hox* gene.

a. Invertebrates with simple anatomy, such as cnidarians, have a single *Hox* gene complex that includes just a few *Hox* genes.

b. Invertebrates with more complicated anatomy, such as arthropods, have a single *Hox* gene complex, but with a larger number of *Hox* genes.

c. Invertebrate chordates, such as cephalochordates, also have a single *Hox* gene complex, but with even more *Hox* genes than are found in nonchordate invertebrates.

d. Vertebrates, such as the laboratory mouse, have numerous *Hox* genes, arranged in two to seven *Hox* gene complexes. The additional *Hox* gene complexes are products of wholesale duplications of the ancestral *Hox* gene complex. The additional copies of *Hox* genes specify the development of uniquely vertebrate characteristics, such as the cranium, vertebral column, and neural crest cells.

Figure 28.9
Hox genes and the evolution of vertebrates. The *Hox* genes in different animals appear to be homologous, indicated here by their colour and position in the complex. Vertebrates have many more individual *Hox* genes than invertebrates, and the entire *Hox* gene complex was duplicated in the vertebrate lineage.

complex, whereas hagfish, the most ancestral living vertebrate, has two. All vertebrates with jaws have at least four sets of *Hox* genes, and some fishes have seven. Evolutionary biologists who study development hypothesize that the duplication of *Hox* genes and other tool-kit genes allowed the evolution of new structures. Although original copies of these genes maintained their ancestral functions, duplicate copies were available to assume *new* functions, leading to the development of novel structures such as the vertebral column and jaws. These changes coincided with the adaptive radiation of vertebrates.

The oldest known vertebrate fossils are from the early Cambrian (about 550 mya) in China. Both *Myllokunmingia* and *Haikouichthys* were fish-shaped animals about 3 cm long **(Figures 28.10** and **28.11)**. In both species, the brain was surrounded by a cranium of fibrous connective tissue or cartilage. They also had segmental body wall muscles and fairly well-developed fins, but neither shows any evidence of bone.

The early vertebrates gave rise to numerous descendants (see Figure 28.1a, p. 634), which varied

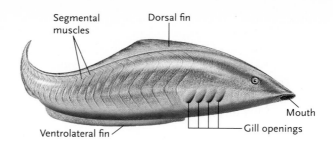

Figure 28.10

Cambrian agnathan, *Haikouichthys*, was more like a hagfish than a lamprey but generally similar to an ammocoetes larva of lampreys, living agnathans.

greatly in anatomy, physiology, and ecology. New feeding mechanisms and locomotor structures were correlated with their success. Today, vertebrates occupy nearly every habitat on Earth and eat virtually all other organisms. Biologists tend to identify vertebrates with four key morphological innovations: cranium, vertebrae, bone, and neural crest cells. We must remember that these structures did not evolve spontaneously.

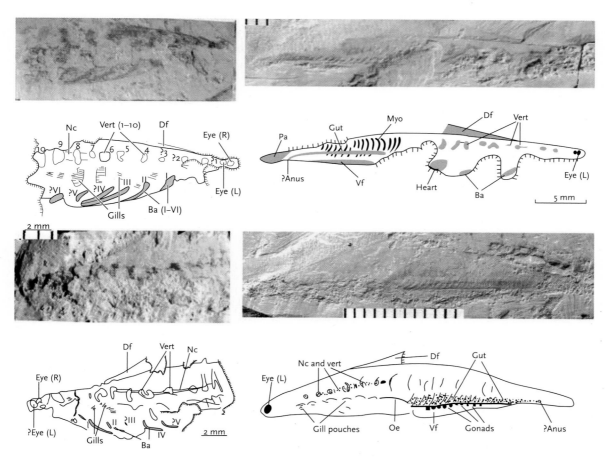

Figure 28.11

A diagram showing an early vertebrate. *Myllokunmingia* is one of the earliest vertebrates yet discovered. This species had no bones and was about 30 cm long. The labels identify features, some with abbreviations, including the following: Ap—anterior plates; Ba—branchial arches; Df—dorsal fin; Myo—myosepta; Nc—notochord; Nc and Vert—notochord with vertebral elements; Nos—nostril; Ns—nasal sacs; Oc—otic capsule; Oe—esophagus; Pa—postanal tail; Vert—vertebral elements; Vf—ventral fin fold; L—left; R—right.

Reprinted by permission from Macmillan Publishers Ltd: NATURE, "Head and backbone of the early Cambrian vertebrate Haikouichthys", Shu, D.-G., S. C. Morris, J. Han, Z-F. Zhang, K. Yasui, P. Janvaier, L. Chen, X-L. Zhang, J-N. Liu, and H-Q. Liu, vol. 421: 526–529, copyright 2003.

Important biological changes during the evolution of vertebrates included improved access to energy (food), which involved mobility and jaws, combined with effective aerobic metabolism (access to oxygen).

The earliest vertebrates lacked jaws (Agnatha, *a* = not; *gnath* = jawed), but Agnatha is not a monophyletic group. Although most became extinct by the end of the Paleozoic, two ancestral lineages, Myxinoidea (hagfishes) and Petromyzontoidea (lampreys), survive today. All other vertebrates have movable jaws and form the monophyletic lineage **Gnathostomata** (*gnath* = jawed; *stoma* = mouth). The first jawed fishes, the Acanthodii and Placodermi, are now extinct, but several other lineages of jawed fishes are still abundant. Included are Chondrichthyes, fishes with cartilaginous skeletons (sharks, skates, chimaeras), and Teleostei (actinopterygians and sarcopterygians), with bony endoskeletons. Although all jawless vertebrates and most jawed fishes are restricted to aquatic habitats, mudskippers (*Periophthalmus* species) and climbing perch (*Anabas* species) regularly venture onto land. Many fish have developed lunglike structures for breathing atmospheric oxygen, but most use gills to extract dissolved oxygen from water. Lungs may be an ancestral trait in vertebrates.

Gnathostomata also includes the monophyletic lineage **Tetrapoda** (*tetra* = four; *pod* = foot), most of which use four limbs for locomotion. Many tetrapods are amphibious, semiterrestrial, or terrestrial, although some, such as sea turtles and porpoises, have secondarily returned to aquatic habitats. Adult tetrapods generally use lungs to breathe atmospheric oxygen. Within the Tetrapoda, one lineage, the Amphibia (such as frogs and salamanders), typically needs standing water to complete its life cycle. Another lineage, the Amniota, comprises animals with specialized eggs that can develop on land. Shortly after their appearance, amniotes diversified into three lineages, one ancestral to living mammals; another to living turtles; and a third to lizards, snakes, alligators, and birds.

STUDY BREAK

1. What is the function of a backbone?
2. What marked the change from a cephalochordate-like creature to a vertebrate?
3. What is a *Hox* gene, and how does it influence the diversity of vertebrates?

28.6 Agnathans: Hagfishes and Lampreys, Conodonts, and Ostracoderms

Lacking jaws, the earliest vertebrates used a muscular pharynx to suck water containing food particles into the mouth, and used gills both to acquire dissolved oxygen and to filter food from the water. The agnathans that flourished in the Paleozoic varied greatly in size and shape and possessed different combinations of vertebrate characters.

Lampreys and hagfishes, the two living groups of agnathans, have skeletons composed entirely of cartilage. Although as yet no fossilized lampreys or hagfishes have been found before the Devonian, the absence of bone in their living descendants suggests that they arose early in vertebrate history, before the evolution of bone. The first fossil lamprey is known from the Devonian of South Africa **(Figure 28.12)**.

a. Living jawless fishes

Hagfish

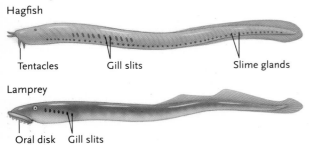

Tentacles   Gill slits   Slime glands

Lamprey

Oral disk   Gill slits

b. Mouth of a lamprey

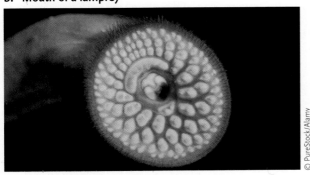

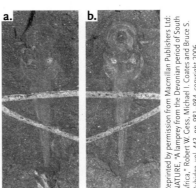

Reprinted by permission from Macmillan Publishers Ltd: NATURE, "A lamprey from the Devonian period of South Africa," Robert W. Gess, Michael I. Coates and Bruce S. Rubidge, vol. 443, pp. 981–984, copyright 2006.

10 mm

Figure 28.12

Living agnathans. Two groups of jawless fishes, the hagfishes and the lampreys **(a)**, are shown as diagrams with a photograph of a lamprey **(b)**. Also shown is the fossil and diagram of a Devonian lamprey from South Africa.

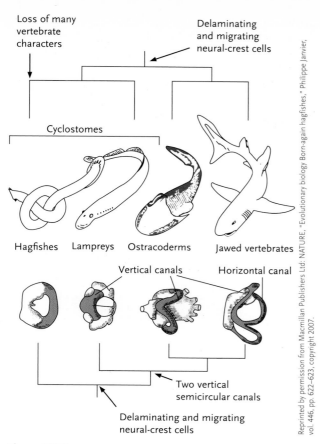

Figure 28.13

Two alternatives for hagfishes. The top tree implies that they are vertebrates that have lost features and the bottom one that they are a sister group of all other vertebrates. One striking difference is the presence of one semicircular canal in hagfishes and at least two in all other vertebrates. The truth remains elusive.

Hagfishes and lampreys have a well-developed notochord but no true vertebrae or paired fins. Their skin lacks scales. Individuals grow to a maximum length of about 1 m (see Figure 28.12). Two possible phylogenies for hagfishes and other vertebrates are presented **(Figure 28.13),** but at this time, there are too few data to decide which is most likely to be correct.

The axial skeletons of the 60 living species of hagfishes include only a cranium and a notochord. No specialized structures surround the dorsal nerve cord. Hagfishes are marine scavengers that burrow in sediments on continental shelves. They feed on invertebrate prey and on dead or dying fishes. In response to

predators, they secrete an immense quantity of sticky, noxious slime. When no longer threatened, a hagfish ties itself into a knot and wipes the slime from its body. The life cycle of a hagfish lacks a larval stage.

The 38 living species of lamprey have a more specialized axial skeleton than hagfishes. Their notochord is surrounded by dorsally pointing cartilage that partially covers the nerve cord, perhaps representing an early stage in the evolution of the vertebral column. About half of the living lamprey species are parasitic as adults and use the sucking disk around their mouths to attach to the bodies of fish (or other prey), rasp a hole in the host's body, and ingest body fluids. In most species, sexually mature adults migrate from the ocean or a lake to the headwaters of a stream, where they reproduce and then die. The filter-feeding **ammocoetes** larvae of lampreys resemble adult cephalochordates. They burrow into mud and develop for as long as seven years before metamorphosing and migrating to the sea or lake to live as adults.

Conodonts and ostracoderms were early jawless vertebrates with bony structures. Conodonts are mysterious bonelike fossils, mostly less than 1 mm long, occurring in oceanic rocks from the early Paleozoic through the early Mesozoic. Called **conodont** elements, these abundant fossils were originally described as supporting structures of marine algae or feeding structures of ancient invertebrates. Recent analyses of their mineral composition reveal that they were made of dentine, a bonelike component of vertebrate teeth. In the 1980s and 1990s, many questions about conodonts were answered by the discovery of fossils of intact conodont animals with these elements.

We now know that conodonts were elongate, soft bodied animals, 3–10 cm long. They had a notochord, a cranium, segmental body wall muscles, and large, movable eyes **(Figure 28.14a).** The conodont elements at the front of the mouth were forward-pointing, hook-shaped structures (the original fossils) apparently used in the collection of food. Conodont elements in the pharynx were stouter, making them suitable for crushing food. Paleontologists now classify conodonts as vertebrates, the earliest ones with bonelike structures.

Ostracoderms (*ostrac* = shell; *derm* = skin) include an assortment of jawless fishes representing several evolutionary lines that lived from the

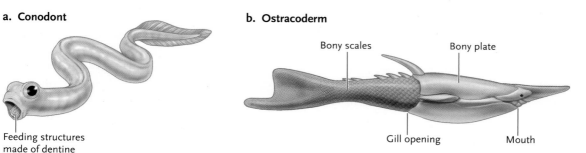

a. Conodont

Feeding structures made of dentine

b. Ostracoderm

Bony scales

Bony plate

Gill opening

Mouth

Figure 28.14

Extinct agnathans. (a) Conodonts were elaborate, soft-bodied animals with bonelike feeding structures in the mouth and pharynx.

(b) *Pteropsis*, an ostracoderm, had large bony plates on its head and small body scales on the rest of its body. It was about 6 cm long.

Figure 28.15

The evolution of jaws. In two early lineages of jawed fishes (Acanthodii and Placodermi), the upper jaw (**maxillae, premaxillae**) was firmly attached to the cranium, while the lower jaw moved up and down. This meant an inflexible mouth that simply snapped open and shut. Acanthodians and placoderms had bony internal skeletons.

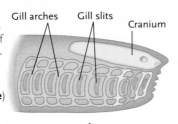

a. Jaws evolved from gill arches in the pharynx of jawless fishes.

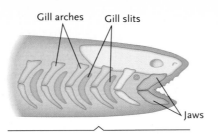

b. In early jawed fishes, the upper jaw was firmly attached to the cranium.

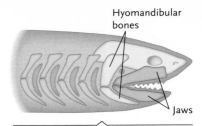

c. In later jawed fishes, the jaws were supported by the hyomandibular bones, which were derived from a second pair of gill arches.

Ordovician through the Devonian (Figure 28.14b, p. 645). Like their invertebrate chordate ancestors, ostracoderms probably used the pharynx to draw water with food particles into the mouth and used gills to filter food from water. The muscular pharynx was more efficient than that of agnathans, using currents generated by cilia. Greater flow rates allowed ostracoderms to collect food more rapidly and achieve larger body sizes. Although most ostracoderms were much smaller, some were 2 m long.

The skin of ostracoderms was heavily armoured with bony plates and scales. Although some had paired lateral extensions of their bony armour, they could not move them in the way living fishes move paired fins. Ostracoderms lacked a true vertebral column, but they had rudimentary support structures surrounding the nerve cord. Ostracoderms had other distinctly vertebrate-like characteristics. Their head shields indicate that their brains had the three regions (forebrain, midbrain, and hindbrain) typical of all later vertebrates (see Chapter 44).

STUDY BREAK

1. How did the earliest vertebrates feed without jaws?
2. Compare the hagfish and the lamprey based on body structure, feeding habits, and life cycles.

28.7 Jawed Fishes: Jaws Expanded the Feeding Opportunities for Vertebrates

The first gnathostomes were jawed fishes. Jaws meant that they could eat more than just filtered food particles and take larger food items with higher energy content. The renowned anatomist and paleontologist A.S. Romer (see "People behind Biology," Box 18.1, p. 422) described the evolution of jaws as "perhaps the greatest of all advances in vertebrate history." Hinged jaws allow vertebrates to grasp, kill, shred, and crush large

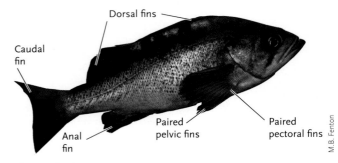

Figure 28.16

Fish fins. Most fishes have both paired and unpaired fins.

food items. Some species also use their jaws for defence, for grooming, to construct nests, and to transport young. Jaws may serve more than one purpose.

Embryological evidence suggests that jaws evolved from paired gill arches in the pharynx of a jawless ancestor (**Figure 28.15**). One pair of ancestral **gill arches** formed bones in the upper and lower jaws, whereas a second pair was transformed into the **hyomandibular bones** that braced the jaws against the cranium. Nerves and muscles of the ancestral suspension-feeding pharynx control the movement and actions of jaws. Jawed fishes also had fins, first appearing as folds of skin and movable spines that stabilized locomotion and deterred predators. Movable fins appeared independently in several lineages, and by the Devonian, most jawed fishes had unpaired (dorsal, anal, and caudal) and paired (pectoral and pelvic) fins (**Figure 28.16**).

28.7a Class Acanthodii

The spiny "sharks" (*acanth* = spine) persisted from the late Ordovician through the Permian. Most of these sharklike fishes were less than 20 cm long, with small, light scales, streamlined bodies, well-developed eyes, large jaws, and numerous teeth (**Figure 28.17a**). Although acanthodians were not true sharks, they were probably fast swimmers and efficient predators. Many of them lived in fresh water. Most had a row of ventral spines and fins with internal skeletal support on each side of the body. The anatomy of acanthodians suggests a close relationship to bony fishes of today.

a. Spiny shark

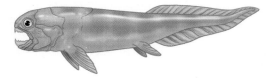

b. Placoderm

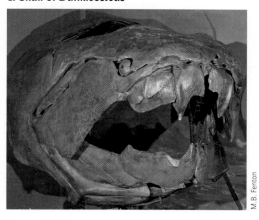

c. Skull of *Dunkleosteus*

M.B. Fenton

Figure 28.17
Early gnathostomes. *Climatius*, an acanthodian (spiny shark) **(a)**, was small, about 8 cm long. The placoderm **(b)** *Dunkleosteus* was gigantic, growing to 10 m in length. Although some acanthodians had teeth, placoderms had only sharp cutting edges. The 3-m-long skull of a *Dunkleosteus* **(c)** demonstrates how impressive placoderms could be.

28.7b Class Placodermi

The placoderms (*plac* = plate; *derm* = skin) appeared in the Silurian and diversified in the Devonian and Carboniferous but left no direct descendants. Some, such as *Dunkleosteus* species (Figure 28.17b, c), reached lengths of 10 m. The bodies of placoderms were covered with large, heavy plates of bone anteriorly and smaller scales posteriorly. Their jaws had sharp cutting edges but no separate teeth, and their paired fins had internal skeletons and powerful muscles.

28.7c Class Chondrichthyes

The cartilaginous fishes (*chondr* = cartilage; *ichthy* = fish) are represented today by about 850 living species of sharks, skates and rays, and chimeras. As the name implies, their skeletons are entirely cartilaginous. However, the absence of bone is a derived trait because all earlier fishes had bony armour or bony endoskeletons. Most living chondrichthyans are grouped into two subclasses, the **Elasmobranchii** (skates, rays, and sharks; **Figure 28.18**) and the **Holocephali** (chimeras). Most are marine predators. With about 40 living species, holocephalians are the only cartilaginous fishes with an operculum (gill cover).

Skates and rays are dorsoventrally flattened (see Figure 28.18a) and swim by undulating their enlarged pectoral fins. Most are bottom dwellers that often lie partly buried in sand. They eat hard-shelled invertebrates (such as molluscs), which they crush with rows of flattened teeth (see Figure 28.18). The largest species, the manta ray (*Manta birostris*), measures 6 m across and eats plankton in the open ocean. Some rays

a. Manta ray

Masa Ushioda/age fotostock/Getty Images

b. Galapagos shark

c. Swell shark egg case

Photos.com

© BRUCE COLEMAN INC./Alamy

Figure 28.18
Chondricthyes.
(a) Skates and rays, such as the manta ray (*Manta birostris*), as well as **(b)** sharks, such as the Galapagos shark (*Carcharhinus galapagensis*), are grouped in the Elasmobranchii. The eggs of many sharks **(c)** include a large yolk that nourishes the developing embryo.

have electric organs that stun prey with shocks of as much as 200 volts. There are species of freshwater skates and rays in some rivers in the tropics; for example, in the Mekong River basin, some *Himantura chaophraya* are 2 m across.

Sharks (see Figure 28.18b, p. 647) are among the oceans' dominant predators. Flexible fins, lightweight skeletons, streamlined bodies, and the absence of heavy body armour allow most sharks to rapidly pursue prey. Their livers often contain **squalene**, an oil that is lighter than water, which increases their buoyancy. The great white shark (*Carcharodon carcharias*), the largest living predatory species of shark, can be 10 m long. At 18 m, the whale shark (*Rhincodon typus*) is the world's largest fish, and it eats only plankton. Sharks' teeth are designed for cutting. *Isisius plutodus*, the cookie-cutter shark, uses piercing teeth in its upper jaw to attach to its prey, biting with the lower jaw and its cutting teeth while rotating its body. The feeding process removes a disk of flesh from the prey. The combination of serrated teeth and flexible extensible jaws makes the effects of shark bites astonishing and frightening.

Elasmobranchs have remarkable adaptations for acquiring and processing food. Their teeth develop in whorls under the fleshy parts of the mouth. New teeth migrate forward as old, worn teeth break free **(Figure 28.19).** In many sharks, the upper jaw is loosely attached to the cranium, and it swings down during feeding. As the jaws open, the mouth spreads wide, sucking in large, hard-to-digest chunks of prey, which are swallowed intact, allowing hurried eating. Although the elasmobranch digestive system is short, it includes a corkscrew-shaped **spiral valve**, which slows the passage of material and increases the surface area available for digestion and absorption.

Elasmobranchs also have well-developed sensory systems. In addition to vision and olfaction, they use **electroreceptors** to detect weak electric currents produced by other animals. Their **lateral line system**, a row of tiny sensors in canals along both sides of the body, detects vibrations in water (see Figure 45.5, p. 1102). They use urea as an osmolyte that makes their body fluids more concentrated than sea water. Freshwater skates have much lower concentrations of urea in their blood than their saltwater relatives do (for more about osmoregulation, see Chapter 50).

Chondrichthyans have evolved numerous reproductive specializations. Males have a pair of organs, the **claspers**, on the pelvic fins, which help transfer sperm into the female's reproductive tract. Fertilization occurs internally. In many species, females produce yolky eggs with tough leathery shells (see Figure 28.18c). Others retain the eggs within the oviduct until the young hatch. A few species nourish young in utero (see Chapters 41 and 42).

28.7d The Bony Fishes

In terms of diversity (numbers of species) and sheer numbers of individuals, fishes with bony endoskeletons (cranium, vertebral column with ribs, and bones supporting their movable fins) are the most successful of all vertebrates. The endoskeleton provides lightweight support compared with the bony armour of ostracoderms and placoderms, enhancing their locomotor efficiency. Some bony fishes have cartilaginous skeletons, but they are not chondrichthyans.

Bony fishes have numerous adaptations that increase swimming efficiency. The scales of most bony fishes are small, smooth, and lightweight, and their bodies are covered with a protective coat of mucus that retards bacterial growth and minimizes drag as water flows past the body.

Bony fishes first appeared in the Silurian and rapidly diversified into two lineages, Actinopterygii and Sarcopterygii. The ray-finned fishes (Actinopterygii; *acti* = ray; *ptery* = fin) have fins supported by thin and flexible bony rays, whereas the fleshy-finned fishes (Sarcopterygii; *sarco* = flesh) have fins supported by muscles and an internal bony skeleton. Ray-finned

Figure 28.19

Elasmobranch teeth. Barndoor skates (*Dipturus laevis;* see also Chapter 32) **(a)** are specialized for crushing hard prey, such as bivalve molluscs. Cookie-cutter sharks (*Isistius plutodus*) **(b)** have cutting teeth. In (a) and (b), the replacement pattern of the teeth (from back to front of the jaws) is obvious.

a.
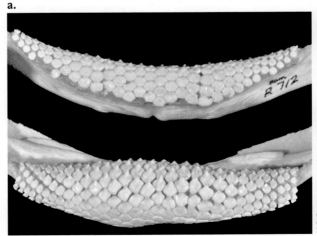

b.
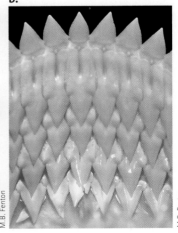

a. Lake sturgeon

© Mark Conlin/Alamy

b. Long-nosed gar

Gary Meszaros/Science Source

Figure 28.20

Ancestral actinopterygians (ray-finned bony fishes). Lake sturgeon (*Acipenser fulvescens*) **(a)**, and a long-nosed gar (*Lepidosteus sasteus*) **(b)**, are living representatives of early ray-finned fishes.

fishes are more diverse as measured by numbers of species and today vastly outnumber fleshy-finned fishes. The ~30 thousand living species of bony fishes occupy nearly every aquatic habitat and represent more than 95% of living fish species. Adults range from 1 cm to more than 6 m in length. In the Yangtze River basin, *Pseuphurus glodius*, the Chinese paddlefish, can weigh up to 500 kg.

Class Actinopterygii. Sturgeons **(Figure 28.20a)** and paddlefishes, the most ancestral members of this group, are characterized by mostly cartilaginous skeletons. These large fishes live in rivers and lakes of the Northern Hemisphere. Sturgeons eat detritus and invertebrates, whereas paddlefish eat plankton. Gars (Figure 28.20b) and bowfins are remnants of a more recent radiation. They occur in the eastern half of North America, where they eat fish and other prey. Gars are protected from predators by a heavy coat of bony scales.

The subclass Teleosteii represents the latest radiation of Actinopterygii, one that produced a wide range of body forms **(Figure 28.21, p. 650).** Teleosts have an internal skeleton made almost entirely of bone. On either side of the head, the **operculum**, a flap of the body wall, covers a chamber that houses the gills. Sensory systems (see Chapter 45) generally include large eyes, a lateral line system, sound receptors, chemoreceptive nostrils, and taste buds.

Variations in jaw structure allow different teleosts to consume plankton, macroalgae, invertebrates, or other vertebrates. Teleosts exhibit remarkable adaptations for feeding and locomotion. When some teleosts open their mouths, bones at the front of the jaws swing forward to create a circular opening. Folds of skin extend backward, forming a tube through which they suck food (see Figure 28.21f). Like Chondrichthyes, Actinopterygii exhibit great variation in tooth structure **(Figure 28.22, p. 651).** Species such as piranhas (*Sarrasalamus*) are notorious for their bites. Other species have teeth specialized for crushing hard prey, such as bivalve molluscs. Whereas the piranha's

teeth are on the premaxilla, maxilla, and mandible (as they are in mammals and many other vertebrates), the crushing teeth of ray-finned fishes often occur on the bones of the pharynx.

In many modern ray-finned fishes, a gas-filled **swim bladder** serves as a hydrostatic organ that increases buoyancy (see Figure 28.21a). The swim bladder is derived from an ancestral air-breathing lung that allowed early actinopterygians to gulp air, supplementing gill respiration in aquatic habitats, where dissolved oxygen concentration is low.

Many have symmetrical tail fins posterior to the vertebral column that provide power for locomotion. Their pectoral fins often lie high on the sides of the body, providing fine control over swimming. Some species use pectoral fins for acquiring food, for courtship, and for care of eggs and young. Some teleosts use pectoral fins for crawling on land (e.g., mudskippers, *Periophthalmus* species, and climbing perch, *Anabas* species) or gliding in the air (flying fish, family Exocoetidae).

Most marine species produce small eggs that hatch into larvae that live among the plankton. Eggs of freshwater teleosts are generally larger and hatch into tiny versions of the adults. Parents often care for their eggs and young, fanning oxygen-rich water over them, removing fungal growths, and protecting them from predators. Some freshwater species, such as guppies, give birth to live young (see "On the Road to Vivipary," p. 998).

Class Sarcopterygii. The two groups of fleshy-finned fishes—lobe-finned fishes and lungfishes—are represented by only eight living species **(Figure 28.23, p. 651).** Although lobe-finned fishes were once thought to have been extinct for 65 million years, a living coelacanth (*Latimeria chalumnae*) was discovered in 1938 near the Comoros Islands, off the southeastern coast of Africa. A population of these metre-long fishes live at depths of 70–600 m, feeding on other fishes and squid. Remarkably, a second population of coelacanths was discovered in 1998, 10 000 km east of the Comoros,

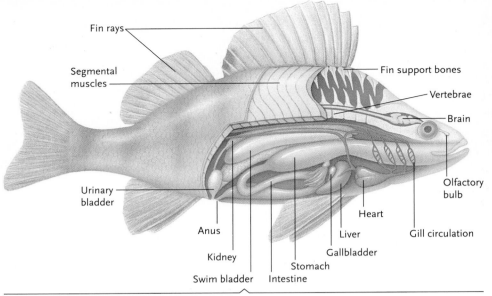

Fin rays

Segmental muscles

Fin support bones

Fin support bones

Vertebrae

Brain

Olfactory bulb

Urinary bladder

Gill circulation

Anus

Liver

Gallbladder

Kidney

Stomach

Heart

Swim bladder

Intestine

a. Teleost internal anatomy

b. Sea horses, like the northern sea horse (*Hippocampus hudsonius*), use a prehensile tail to hold on to substrates; they are weak swimmers.

c. The long, flexible body of a spotted moray eel (*Gymnothorax moringa*) can wiggle through the nooks and crannies of a reef.

d. Flatfishes, like this European flounder (*Platichthys flesus*), lie on one side and leap at passing prey.

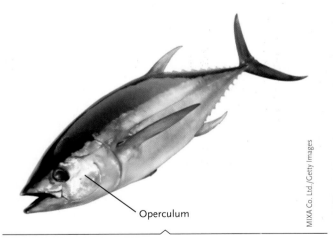

Operculum

e. Open ocean predators, like the yellowfin tuna (*Thunnus albacares*), have strong, torpedo-shaped bodies and powerful caudal fins.

f. Kissing Gouramis (*Helostoma temmincki*) extend their jaws into a tube that sucks food into the mouth.

Figure 28.21
Teleost diversity. Although all teleosts (bony fish) share similar internal features, their diverse shapes adapt them to different diets and types of swimming.

a.
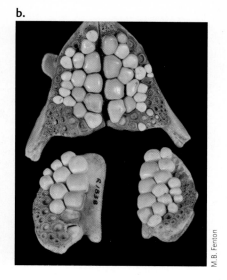
b.

Figure 28.22
Teleost teeth. Like Chrondricthyes, bony fishes have also developed cutting **(a)** and crushing **(b)** teeth. The cutting teeth are those of a piranha (*Sarrasalamus* species); the crushing teeth are from a black drum (*Pogones cromis*).

M.B. Fenton

M.B. Fenton

a. Coelacanth

AlessandroZocc/Shutterstock.com

b. Australian lungfish

Tom McHugh/Science Source

Figure 28.23
Sarcopterygians. The coelocanth (*Latimeria chalumnae*) **(a)** is now one of two living species of lobe-finned fishes. The Australian lungfish (*Neoceratodus forsteri*) **(b)** is one of six living lungfish species.

when a specimen was found in an Indonesian fish market. Analyses of the DNA of the Indonesian specimen indicated that it is a distinct species (*Latimeria menadoensis*).

Lungfishes have changed relatively little over the last 200 million years. Six living species are distributed on southern continents. Australian lungfishes live in rivers and pools, using their lungs to supplement gill respiration when dissolved oxygen concentration is low. South American and African species live in swamps and use their lungs for breathing during the annual dry season, which they spend encased in a mucus-lined burrow in the dry mud. When the rains begin, water fills the burrow and the fishes awaken from dormancy. During their periods of dormancy, these fishes excrete urea.

STUDY BREAK

1. What did the evolution of jaws mean for fish?
2. What anatomical and physiological characteristics make sharks dominant ocean predators?
3. What is the lateral line system? What does it do?

28.8 Early Tetrapods and Modern Amphibians

The fossil record suggests that tetrapods evolved in the late Devonian from a group of fleshy-finned fishes, the Osteolepiformes. Osteolepiformes and early tetrapods shared several derived characteristics, including dental and cranial features. Specifically, both had infoldings of tooth surfaces that probably increased the functional area of the tooth. They also shared shapes and positions of bones on the dorsal side of their crania and in their appendages.

Some problems of moving onto land were identified earlier. During dry periods in swampy, late Devonian habitats, drying pools may have forced osteolepiform ancestors to move overland to adjacent pools that still had water. During these excursions, the fish may have found that land plants, worms, and arthropods provided abundant food, and oxygen was more readily available in air than in water. Furthermore, there may well have been fewer terrestrial predators at that time, but this interpretation is open to question.

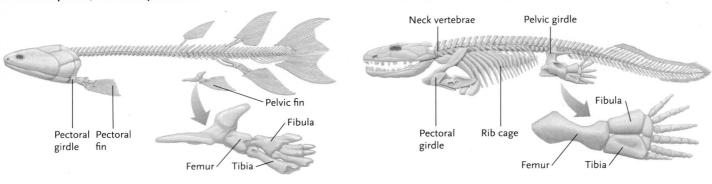

a. *Eusthenopteron,* an osteolepiform fish

b. *Ichthyostega,* an early tetrapod

Neck vertebrae · Pelvic girdle

Pelvic fin

Fibula

Pectoral girdle · Pectoral fin

Femur · Tibia

Fibula

Pectoral girdle · Rib cage

Femur · Tibia

Figure 28.24

Evolution of tetrapod limbs. The limb skeleton of osteolepiform fishes such as **(a)** *Eusthenopteron* is homologous to that of early tetrapods, such as **(b)** *Ichthyostega*. Although *Ichthyostega* retained many fishlike characteristics, its pectoral girdle was completely freed from the cranium, and it had a heavy ribcage. Fossils of its forefoot have not yet been discovered.

Osteolepiformes **(Figure 28.24a)** usually had strong, stout fins that allowed them to crawl on mud. Of particular importance were crescent-shaped bones in their vertebral columns that provided strong intervertebral connections. Their nostrils led to sensory pits housing olfactory (odour) receptors (see Chapter 45). They almost certainly had lungs, allowing them to breathe atmospheric oxygen. Like living lungfishes, they could also have excreted urea or uric acid rather than ammonium, which is toxic.

The earliest tetrapod with nearly complete skeletal data is the semiterrestrial, metre-long *Ichthyostega* (Figure 28.24b). Compared with its fleshy-finned ancestors, *Ichthyostega* had a more robust vertebral column, sturdier limb girdles and appendages, a ribcage that protected its internal organs (including lungs), and a neck. Fishes lack necks because the pectoral girdle is fused to the cranium. In *Ichthyostega,* several vertebrae separated the pectoral girdle and the cranium, allowing the animal to move its head to scan the environment and capture food. *Ichthyostega* retained a fishlike lateral line system, caudal fin, and scaly body covering.

Life on land also required changes in sensory systems. In fishes, the body wall picks up sound vibrations and transfers them directly to sensory receptors. Sound waves are harder to detect in air. The appearance of a **tympanum** (ear drum) in early tetrapods apparently allowed them to detect vibrations in air associated with airborne sounds. The tympana are specialized membranes on either side of the head. The tympanum connects to the **stapes**, a bone homologous to the hyomandibula, which had supported the jaws of fishes (see Figure 20.14, Chapter 20). The stapes, in turn, transfers vibrations to the sensory cells of an inner ear.

28.8a Class Amphibia: Frogs and Toads, Salamanders, and Caecilians

Most of the 6000+ living species of amphibians (*amphi* = both; *bios* = life) are small, and their skeletons contain fewer bones than those of Paleozoic tetrapods such

as *Ichthyostega.* All living amphibians are carnivorous as adults, but the aquatic larvae of some are herbivores. Fossil amphibians, such as *Eryops* **(Figure 28.25),** were quite large and predatory.

The thin, scaleless skin of most living amphibians is well supplied with blood vessels and can be a major site of gas exchange. To operate in oxygen uptake, the skin must be moist and thin enough to bring blood into close contact with air. Having moist skin limits amphibians to moist habitats. Many species of living amphibians keep their skin surfaces moist, and some are lungless, but most use lungs in gaseous exchange. The evolution of lungs was accompanied by modifications of the heart and circulatory system that increase the efficiency with which oxygen is delivered to body tissues (see Chapter 42). Some adult anurans have a waxy coating on their skin, making them as waterproof as lizards **(Figure 28.26).**

The life cycles of many amphibians include larval and adult stages. In frogs, larvae (tadpoles) hatch from fertilized eggs and eventually metamorphose into adults (see Chapter 42). The larvae of most frog species are aquatic, but adults may live their lives in water (be aquatic), move between land and water (be amphibious), and live entirely on land (be terrestrial). Some salamanders are pedomorphic (see Chapter 17), which means that the larval stage attains sexual maturity without changing its form or moving to land. Some frogs and salamanders reproduce on land, omitting the larval stage altogether. In these species, tiny adults emerge directly from fully developed eggs. However, the eggs of terrestrial breeders dry out quickly unless they are laid in moist places.

Modern amphibians are represented by three lineages **(Figure 28.27, p. 654),** but the evolutionary origin of frogs, salamanders, and caecilians has remained unresolved. The 2008 description of a small fossil from the Lower Permian of Texas suggests that frogs and salamanders have a relatively close common ancestor, whereas caecilians are distantly related to them.

Figure 28.25
A fossil amphibian. This amphibian, *Eryops*, from the Texas Permian, was about 1.8 m long and was strikingly different from living amphibians.

M.B. Fenton

a.

Figure 28.26
Waterproof frogs. **(a)** *Chiromantis xerampelina* from southern Africa and **(b)** *Phyllomedusa sauvagii* from South America make their skin waterproof with a waxy secretion. These frogs are as waterproof as chameleons. They also excrete uric acid to further conserve water.

M.B. Fenton

b.

M.B. Fenton

Populations of practically all amphibians have declined rapidly in recent years. These declines are probably due to exposure to acid rain, high levels of ultraviolet B radiation, and fungal and parasitic infections. Another major factor in the decline of amphibians may be habitat splitting, the human-induced disconnection of habitats essential to the survival of amphibians. This aspect of **habitat fragmentation** (see "People behind Biology," Box 31.2, p. 770) can cause adult amphibians to move across inhospitable habitat (roads, power line rights-of-way) to reach breeding habitats.

Anura. The 3700 species of frogs and toads (*an* = not; *ura* = tail) have short, compact bodies, and the adults lack tails. Their elongated hind legs and webbed feet allow them to hop on land or to swim. A few species are adapted to dry habitats, encasing themselves in mucus cocoons to withstand periods of drought.

Urodela. The 400 species of salamanders (*uro* = tail; *del* = visible) have an elongated, tailed body and four legs. They walk by alternately contracting muscles on either side of the body, much the way fishes swim. Species in the most diverse group, the lungless salamanders, are fully terrestrial throughout their lives, using their skin and the lining of the throat for gas exchange.

a. A frog

b. A salamander

c. A caecelian

Figure 28.27

Living amphibians. Anurans **(a),** such as the northern leopard frog (*Rana pipiens*), have compact bodies and long hind legs. Urodeles **(b),** such as the red-spotted newt (*Notophthalmus viridescens*), have an elongate body and four legs. Caecilians **(c),** such as *Caecelia nigricans* from Colombia, are legless burrowers.

Gymnophonia. The 200 species of caecelians (*gymno* = naked; *ophioneos* = snakelike) are legless, burrowing animals with wormlike bodies. They occupy tropical habitats throughout the world. Unlike other extant amphibians, caecilians have small bony scales embedded in their skin. Fertilization is internal, and females give birth to live young. In some species, the mother's skin produces a milklike substance for the young, which use specialized teeth to collect it from the mother's body (see Chapter 42).

STUDY BREAK

1. Present four lines of evidence suggesting that tetrapods arose from Osteolepiformes.
2. Why was the development of the tympanum important to life on land?
3. What characteristics allow amphibians to use their skin as a major site of gas exchange?

28.9 The Origin and Mesozoic Radiations of Amniotes

The amniote lineage arose during the Carboniferous, a time when seed plants and insects began to invade terrestrial habitats, providing additional food and cover for early terrestrial vertebrates. Amniotes take their name from the **amnion**, a fluid-filled sac that surrounds the embryo during development (see Chapter 42). Although the fossil record includes many skeletal remains of early amniotes, it provides little direct information about soft body parts and physiology. Three key features of living amniotes allow life on dry land and liberate them from reliance on standing water. The changes involve being waterproof and producing waterproof eggs.

- First, skin is waterproof: keratin and lipids in the cells make skin relatively impermeable to water.

- Second, **amniote (amniotic) eggs** can survive and develop on dry land because they have four specialized membranes and a hard or leathery shell perforated by microscopic pores **(Figure 28.28).** Amniote eggs are resistant to desiccation. The membranes protect the developing embryo and facilitate gas exchange and excretion. The shell mediates the exchange of air and water between the egg and its environment. Developing amniote embryos can excrete uric acid, which is stored in the allantois of the embryo, which will later become the bladder. Generous supplies of **yolk** in the egg are the developing embryo's main energy source, whereas **albumin** supplies nutrients and water. There is no larval stage, and hatchling amniotes are miniature versions of the adult. Amniote eggs are the ancestral condition, but they

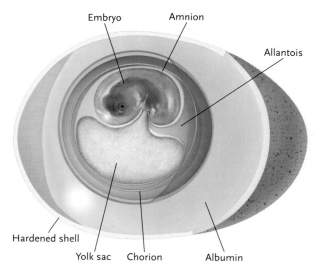

Embryo   Amnion

Allantois

Hardened shell

Yolk sac   Chorion   Albumin

Figure 28.28

The amniote egg. A water-retaining egg with four specialized membranes surrounded by a hard or leathery shell allowed amniotes and their descendants to reproduce in dry environments. The chorion surrounds the amnion which in turn surrounds the amniotic fluid.

are circumvented in most mammals (and some reptiles; see Chapter 18).

- Third, some amniotes produce urea and/or uric acid as a waste product of nitrogen metabolism (see Chapter 44). Although ammonia (NH_3^+) is less expensive (metabolically) to produce, it is toxic and must be flushed away with water. Urea is much less toxic than NH_3^+ and therefore easier to store and to void. Uric acid is even less toxic and, because it is insoluble, it can be stored or voided without risk while conserving water.

The abundance and diversity of fossils of amniotes indicate that they were extremely successful, quickly replacing many nonamniote species in terrestrial habitats. During the Carboniferous and Permian, amniotes produced three major radiations: synapsids, anapsids, and diapsids **(Figure 28.29, p. 656)**, distinguishable by the numbers of bony arches in the temporal region of the skull (in addition to the openings for the eyes (see **Figure 28.30, p. 657**). The bony arches delimit fenestrae, openings in the skull that allow space for contraction (and expansion) of large and powerful jaw muscles.

Synapsids (Figure 28.30a), a group of small predators, were the first offshoot from ancestral amniotes. Synapsids (*syn* = with; *apsid* = connection) had one temporal arch on each side of the head. They emerged late in the Permian, and mammals are their living descendants.

Anapsida (Figure 28.30b), the second lineage (*an* = not), had no temporal arches and no spaces on the sides of the skull. Turtles are living representatives of this group.

Diapsida (Figure 28.30c, d; *di* = two) are the third lineage and included most Mesozoic amniotes. Diap-

sids had two temporal arches, and their descendants include the dinosaurs, as well as extant lizards and snakes, crocodilians, and birds. Arguably, birds are other examples of living diapsids.

28.9a Extinct Diapsids

Early diapsids differentiated into two lineages, **Archosauromorpha** (*archo* = ruler; *sauro* = lizard; *morph* = form) and **Lepidosauromorpha** (*lepi* = scale), which differed in many skeletal characteristics. Archosaurs (archosauromorphs), or "ruling reptiles," include crocodilians, pterosaurs, and dinosaurs. Crocodilians first appeared during the Triassic. They have bony armour and a laterally flattened tail, which is used to propel them through water. Pterosaurs, now extinct, were flying predators of the Jurassic and Cretaceous **(Figure 28.31, p. 657)**. The smallest were sparrow sized; the largest had wing spans of 11 m. Some evidence indicates that pterosaur wings attached to the side of their bodies at about the hips.

Two lineages of dinosaurs, "lizard-hipped" saurischians and "bird-hipped" ornithischians, proliferated in the Triassic and Jurassic **(Figure 28.32, p. 657)**. Saurischians included bipedal carnivores and quadrupedal herbivores. Some carnivorous saurischians **(Figure 28.33a, p. 658)** were swift runners, and some had short forelimbs (e.g., *Tyrannosaurus rex*, which was 12 m long and stood 6 m high; Figure 28.33b). One group of small carnivorous saurischians, the deinonychsaurs, is ancestral to birds (see Figure 28.33a).

By the Cretaceous, some herbivorous saurischians were gigantic, and many had long, flexible necks. *Apatosaurus* (previously known as *Brontosaurus*) was 25 m long and may have weighed 50 000 kg **(Figures 28.34 and 28.35, p. 658)**. The largely herbivorous ornithischian dinosaurs had large, chunky bodies. This lineage included armoured or plated dinosaurs (*Ankylosaurus* and *Stegosaurus*), duck-billed dinosaurs (*Hadrosaurus*), horned dinosaurs (*Styracosaurus*), and some with remarkably thick skulls (*Pachycephalosaurus*). Ornithischians were most abundant in the Jurassic and Cretaceous.

Lepidosaurs (Lepidosauromorpha) are the second major lineage of diapsids. This diverse group included both marine and terrestrial animals. Fossil lepidosaurs include champosaurs (see Figure 28.30d, p. 657), which were freshwater fish eaters, and the marine, fish-eating plesiosaurs, with long, paddlelike limbs they used like oars **(Figure 28.36, p. 658)**. Fossil lepidosaurs also included ichthyosaurs (see Figure 20.20, p. 464), porpoise like animals with laterally flattened tails. Like today's whales, ichthyosaurs were highly specialized for marine life and did not return to land to lay eggs. Indeed, it appears that ichthyosaurs, like today's whales, gave birth to live young. Squamates, the living lizards and snakes, are the third important group within this lineage. *Sphenodon*, the tuatara, is the last living genus of a once diverse group of lizard like squamates.

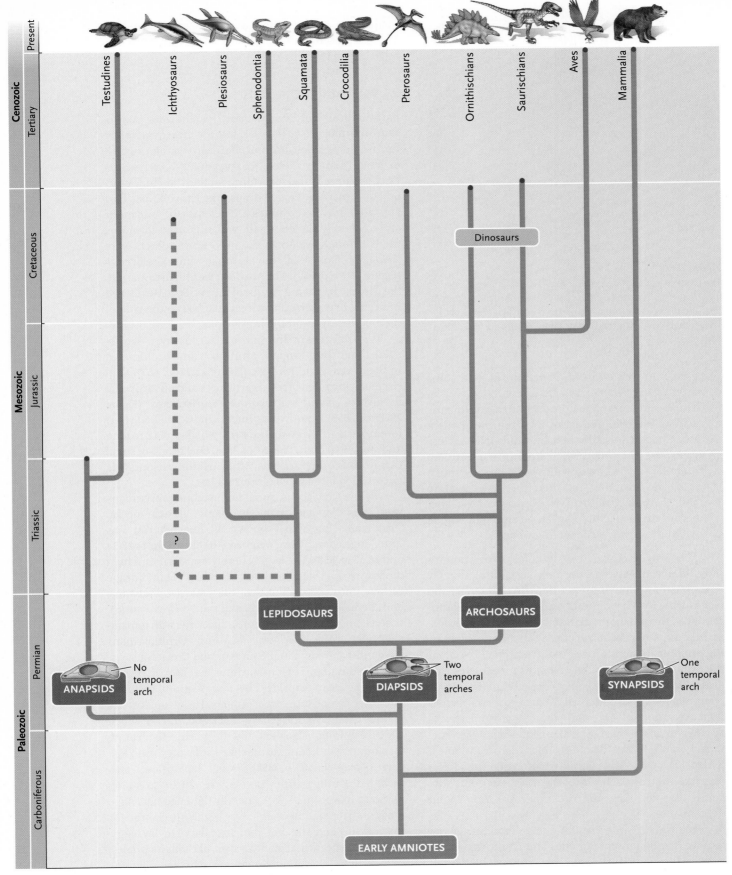

Figure 28.29

Amniote ancestry. The early amniotes gave rise to three lineages (anapsids, synapsids, and diapsids) and numerous descendants. The lineages are distinguished by the number of bony arches in the temporal region of the skull (indicated on the small icons).

a.

b.

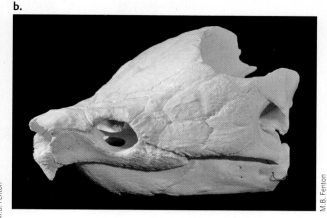

Figure 28.30
Skulls of reptiles. The synapsid condition **(a),** shown by *Dimetrodon;* the anapsid condition **(b),** shown by a snapping turtle; and the diapsid conditions shown by *Camarosaurus* **(c)** and *Champsosaurus* **(d).**

c.

d.

a.

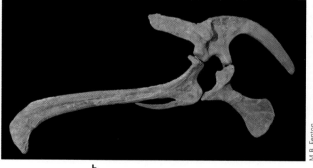

b.

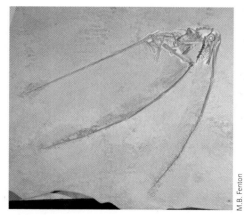

Figure 28.31
Rhamphorynchus meunsteri, a pterosaur with a wing span of about 1.7 m. Note the impressions of the wing membranes, the teeth, and the long tail. This species is known from the Upper Jurassic of Germany.

Figure 28.32
Ornithischian **(a)** and saurischian **(b)** dinosaurs differed in their pelvic structures. The ornithischian is a hadrasaur (duck-billed dinosaur), the saurischian an Albertosaurus. In each case, the **acetabulum,** the socket receiving the head of the femur, is the large elliptical area in the middle.

Figure 28.33
Saurischian dinosaurs. Whereas *Ornitholestes hermanii* **(a)** stood less than 1 m at the shoulder, the fearsome *Tyrannosaurus rex* **(b)** was about 12 m long. Both were carnivores.

a.

b.

a. *Lambeosaurus*

b. *Stegoceras*

a.

b.

Figure 28.34
Ornithischian dinosaurs. These herbivores ranged in size from the 15-m-long **(a)** *Lambeosaurus lambei*, a smaller, thick-skulled **(b)** *Stegoceras*, to the 10-m-long **(c)** *Triceratops horridus*.

c. *Triceratops*

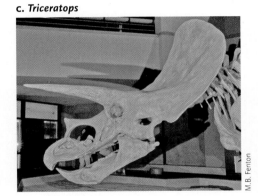

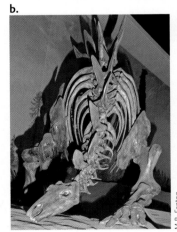

b.

Figure 28.35
Large, lumbering herbivores. Other ornithischian dinosaurs included the 18-m-long *Camarasaurus supremus* **(a)** and the 9-m-long *Stegosaurus armatus* **(b).** The latter had distinctive plates along its back.

Figure 28.36
Paddles. Paddlelike forelimbs developed in sea turtles **(a)** and plesiosaurs **(b)**, *Trinacromerum bonneri*).

a.

b.

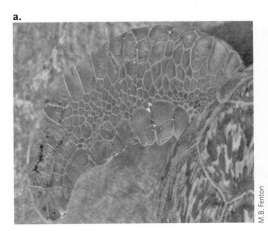

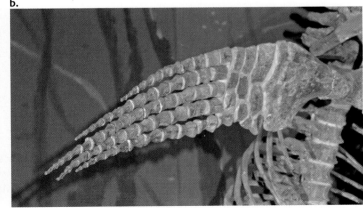

a.

b.

c.

d.

e.

f.

M.B. Fenton

Figure 28.37
Reptile teeth. As usual, teeth reflect the dietary habits of vertebrates. Herbivorous dinosaurs, **(a)** *Diplodocus longus* and **(b)** hadrosaur, had teeth adapted for gathering plant material (a) and grinding it (b). They differ from those of a carnivorous dinosaur (*Daspletosaurus*) **(c)** or a fish-eating reptile such as a champhosaur **(d).** A tooth of a *Tyrannosaurus rex* changed distinctly over its length. The biting part of the tooth **(e)** had enamel and serrated edges. There was no enamel on the part of the tooth located within the socket of the skull **(e, f).**

The teeth of reptiles provide important clues about their diets **(Figure 28.37)** and show interesting parallels with the teeth of other vertebrates.

STUDY BREAK

1. Where do amniotes get their name? Why are amniote eggs resistant to desiccation?
2. What three key features liberate living amniotes from reliance on standing water?
3. Name and describe three major radiations of amniotes during the Carboniferous and Permian.

28.10 Subclass Testudinata: Turtles and Tortoises

The turtle body plan, largely defined by a bony, boxlike shell, has changed little since the group first appeared during the Triassic **(Figure 28.38)**. A turtle's ribs are fused to the inside of the shell, and in contrast to other tetrapods, the pectoral and pelvic girdles lie within the ribcage. The shell is formed from large keratinized scales covering the bony plates.

The 250 living species occupy terrestrial, fresh water, and marine habitats. They range from 8 cm to 2 m in length. Turtles use a keratinized beak in feeding, whether they eat animal or plant material. When threatened, most species retract into their shells. Many species are now endangered because adults are hunted for meat

a. **The turtle skeleton**

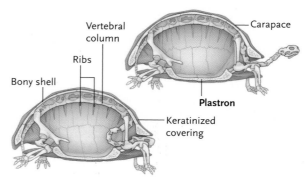

Vertebral column

Ribs

Carapace

Bony shell

Plastron

Keratinized covering

b. **An aquatic turtle**

Paul J. Fusco/Science Source

Figure 28.38
Testudines. Most turtles **(a)** can withdraw their heads and legs into a bony shell. Aquatic turtles **(b),** such as the eastern painted turtle (*Chrysmys picta*), often bask in the Sun to warm up. The sunlight may also help to eliminate parasites that cling to the turtle's skin.

and their eggs are eaten by humans and other predators. Young are often collected for the pet trade, and the beaches favoured as nesting sites by marine species are too often used as tourist attractions (see Chapter 32).

STUDY BREAK

Describe the body plan of a turtle.

28.11 Living Diapsids: Sphenodontids, Squamates, and Crocodilians

28.11a Infraclass Lepidosaura, Order Rhynchocephalia: The Tuatara

Sphenodon punctatus is one of two living species of sphenodontids (*sphen* = wedge; *dont* = tooth) or tuataras, a lineage that was diverse in the Mesozoic

(**Figure 28.39a**). These lizardlike animals are best known as tetrapods with a "third" or pineal eye, a reflection of earlier vertebrates such as lampreys with pineal eyes (see also photoreceptors in Chapter 1). They survive on a few islands off the coast of New Zealand. Adults are about 60 cm long. They live in dense colonies, where males and females defend small territories. They often share underground burrows with seabirds and eat invertebrates and small vertebrates. They are primarily nocturnal and maintain low body temperatures during periods of activity. Their survival is threatened by two introduced predators, cats and rats.

28.11b Infraclass Lepidosaura, Order Squamata: Lizards and Snakes

Lizards and snakes (Figure 28.39b) are covered by overlapping, keratinized scales (*squam* = scale) that protect against dehydration. Squamates periodically shed their skin while growing, much the way

C. Allan Morgan/Photolibrary/Getty Images

a. Sphenodontia includes the tuatara (*Sphenodon punctatus*) and one other species.

Stephen Dalton / Science Source

b. Basilisk lizards (*Basiliscus basiliscus*) escape from predators by running across the surface of streams.

Stephen Dalton/Science Source

c. A western diamondback rattlesnake (*Crotalus atrox*) of the American southwest bares its fangs with which it injects a powerful toxin into prey.

ChrisKrugerSA/iStock/Thinkstock

d. Crocodilia includes semiaquatic predators, like this resting African Nile crocodile (*Crocodylus niloticus*), that frequently bask in the Sun.

Figure 28.39
Living nonfeathered diapsids.

arthropods shed their exoskeletons (see Chapter 27). Most squamates regulate their body temperature behaviourally (see Chapter 50), so they are active only when weather conditions are favourable. They shuttle between sunny and shady places to warm up or cool down as needed.

Most of the 3700 lizard species are less than 15 cm long, but Komodo dragons (*Varanus komodoensis*) grow to nearly 3 m in length (see Figure 45.11, Chapter 45). Lizards occupy a wide range of habitats and are especially common in deserts and the tropics. One species (*Lacerta vivipara*) occurs within the Arctic Circle. Most lizards eat insects, although some consume leaves or meat.

The 2300 species of snakes evolved from a lineage of lizards that lost their legs over evolutionary time. Streamlined bodies make snakes efficient burrowers or climbers (Figure 28.39c). Many subterranean species are 10 or 15 cm long, whereas the giant constrictors may grow to 10 m. Unlike lizards, all snakes are predators that swallow prey whole. Compared with their lizard ancestors, snake skull bones are reduced in size and connected to each other by elastic ligaments. This gives snakes a remarkable capacity to stretch their mouths. Some snakes can swallow food items that are larger than their heads (see Chapter 48). Snakes also have well-developed sensory systems for detecting prey. The flicking tongue carries airborne molecules to sensory receptors in the roof of the mouth (see "Forked Tongues," p. 1118). Most snakes can detect vibrations on the ground, and some, like rattlesnakes, have heat-sensing organs (see Figure 45.26, Chapter 45). Many snakes kill by constriction, which suffocates prey, whereas other species produce venoms, toxins that immobilize, kill, and partially digest prey (see "Molecule behind Biology," Box 30.1, p. 718).

28.11c Infraclass Archosauria, Order Crocodylia: Crocodiles, Alligators, and Gavials

The 21 species of alligators and crocodiles, along with the birds, are the living remnants of the archosaurs (Figure 28.39d). Australian saltwater crocodiles (*Crocodylus porosus*) are the largest, growing to 7 m in length. Crocodilians are aquatic predators that eat other vertebrates. Striking anatomical adaptations distinguish them from living lepidosaurs, including a four-chambered heart that is homologous to the heart in birds, analogous to this structure in mammals. In some crocodilians, muscles that originate on the pubis insert on the liver and pericardium. When these muscles contract, the liver moves toward the tail, creating negative pressure in the chest cavity and drawing air in. This situation is analogous to the role of the diaphragm in mammals.

American alligators (*Alligator mississippiensis*) exhibit strong maternal behaviour, perhaps reflecting their relationship to birds. Females guard their nests ferociously and, after the young hatch, free their offspring from the nest. The young stay close to the mother for about a year, feeding on scraps that fall from her mouth and living under her watchful protection.

Many species of alligators and crocodiles are endangered because their habitats have been disrupted by human activities. They have been hunted for meat and leather and because larger individuals are predators of humans. There is hope, however, as some populations of *A. mississippiensis* have recovered in the wake of efforts to protect them. In Africa and Australia, crocodiles are farmed for their meat and skin.

In the past, crocodilians were more diverse in body form than they are today. *Dakosaurus andiniensis,* a Jurassic–Cretaceous marine crocodilian from western South America, differed dramatically from typical crocodilians **(Figure 28.40, p. 662).**

STUDY BREAK

1. How do snakes kill their prey?
2. What features of crocodilians are homologous to those of birds? Which ones are analogous to those of mammals?

28.12 Aves: Birds

Birds (Aves; *avis* = bird) appeared in the Jurassic as descendants of carnivorous, bipedal dinosaurs (see Figure 20.18, Chapter 20). Birds belong to the archosaur lineage, and their evolutionary relationship to dinosaurs is evident in their skeletal anatomy and in the scales on their legs and feet. Powered flight gave birds access to new **adaptive zones**, likely contributing to their astounding evolutionary success **(Figure 28.41, p. 663).** Some species of birds are flightless, and some of these are bipedal runners. Other birds are weak fliers.

Three skeletal features associated with flight in birds are the **keeled sternum** (breastbone), the **furculum** (wishbone), and the uncinate processes on the ribs **(Figure 28.42, p. 663).** The keel on the sternum anchors the flight muscles (see Figure 28.41c); the furculum acts like a spring; and the uncinate processes, which effect overlap of adjoining ribs, give the ribcage strength and anchor intercostal muscles. In flightless species, the sternum often lacks a keel (see Figure 28.42), an exception being penguins that "fly" through the water. However, flightless species often have uncinate processes.

Birds' skeletons are light and strong (see Figure 28.41c). The skeleton of a 1.5 kg frigate bird (*Fregata magnificens*) weighs just 100 g, far less than the mass

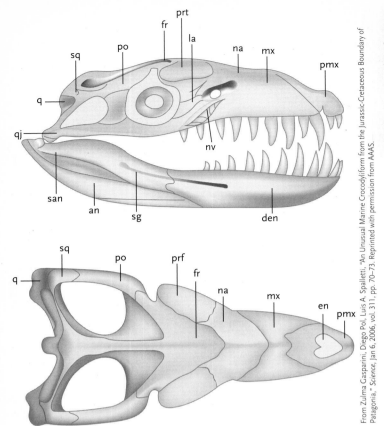

From Zulma Gasparini, Diego Pol, Luis A. Spalletti, "An Unusual Marine Crocodyliform from the Jurassic-Cretaceous Boundary of Patagonia," *Science*, Jan 6, 2006, vol. 311, pp. 70–73. Reprinted with permission from AAAS.

Figure 28.40

Crocodilians. *Dakosaurus andinensis*, a crocodile from the Jurassic–Cretaceous boundary in Patagonia **(a)**, has a more robust skull and jaw than a more typical member of the group **(b)**, *Alligator mississipiensis*. The skull of *Dakosaurus* is more rounded than the wedge-shaped skull of *Alligator*. Bones are abbreviated: an—angular; den—dentary; en—external nares; eoc—exoccipital; fr—frontal; ic—internal carotid formane; la—lacrimal; na—nasal; nv—-neurovascular formina; pmx—premaxilla; po—postorbital; prf—prefrontal; pt—pterygoid; q—quadrate; qj—quadratojugal; san—surangular; sg—surangular groove; soc—supraoccipital; sq—squamosal.

of its feathers. Although the skeleton of a 20 g mammal weighs the same as that of a 20 g bird, the bird's bones are larger and lighter. Most birds have hollow limb bones with small supporting struts that criss-cross the internal cavities. Birds have reduced numbers of separate bony elements in the wings, skull, and vertebral column (especially the tail), so the skeleton is rigid. The bones associated with flight are generally large, and the wingbones are long (see Figure 28.41).

All extant birds **(Figure 28.43, p. 664)** use a keratinized bill for feeding rather than teeth, which are dense and heavy. Many species have a long, flexible neck that allows them to use their bills for feeding, grooming, nest building, and social interactions. Birds' soft internal organs are modified to reduce mass. Most birds lack a urinary bladder, so uric acid paste is eliminated with digestive wastes. Females have only one ovary and never carry more than one mature egg at a time. Eggs are laid as soon as they are shelled. Egg sizes give an indication of the range of size in birds **(Figure 28.44, p. 664)**. Birds range in size from a bee hummingbird (*Mellisuga helenae*) at 2 g to ostriches (*Struthio camelus*) at about 150 kg. The size spectrum is illustrated by a comparison of breast bones **(Figure 28.45, p. 664)**.

All birds have **feathers** (see Figure 28.41d), sturdy, lightweight structures derived from scales in the skin of their reptilian ancestors. Each feather has numerous barbs and barbules with tiny hooks and grooves that maintain the feathers' structures, even during vigorous activity. Flight feathers on the wings provide lift, whereas contour feathers streamline the surface of the body. Down feathers form an insulating cover close to the skin. Moulting replaces feathers once or twice each year. But not all animals with feathers are birds. Several extinct archosaurs had feathers, but these animals had none of the adaptations for flight.

Other adaptations for flight allow birds to harness the energy needed to power their flight muscles. Their metabolic rates are 8 to 10 times as high as those of comparably sized reptiles, allowing them to process energy-rich food rapidly. A complex and efficient respiratory system (see Chapter 49) and a four-chambered heart (see Chapter 40) enable them to consume and distribute oxygen efficiently. As a consequence of high rates of metabolic heat production, most birds maintain a high and constant body temperature (see Chapter 50).

Flying birds were abundant by the Cretaceous. Even in the Jurassic, *Archaeopteryx* had a furculum and was capable of at least limited flight. Until 2008, two main theories purported to explain the evolution of flight in birds. Proponents of the *top-down* theory argued that ancestral birds lived in trees and glided down from them in pursuit of insect prey. Gliding and access to prey are key elements of this theory. Proponents of the *bottom-up* theory proposed that a

a. Wing movements of an owl during flight

© A & J Visage / Alamy

b. Skeletal system of birds

Skull

Radius

Ulna

Pectoral girdle

Humerus

Scapula

Furculum (wishbone)

Pelvic girdle

Coracoid

Keeled sternum

c. Pectoral girdle and flight muscles of bird in frontal view

Humerus

Tendon

Humerus

Scapula

Furculum (wishbone)

Sternum

Pectoralis major (lowers wings)

Supracoracoideus (raises wings)

Keel of sternum

Internal structure of bird limb bones

d. Feather structure

Barbule

Barb

Shaft

Figure 28.41

Adaptations for flight in birds. The flapping movements **(a)** of a bird's wing provide thrust for forward momentum and lift to counteract gravity. The bird skeleton **(b)** includes a boxlike trunk, short tail, long neck, lightweight skull and beak, and well-developed limbs. In large birds, limb bones are hollow. Two sets of flight muscles **(c)** originate on the keeled sternum: one set raises the wings, whereas the other lowers them. Flexible feathers **(d)** form an airfoil on the wing surface.

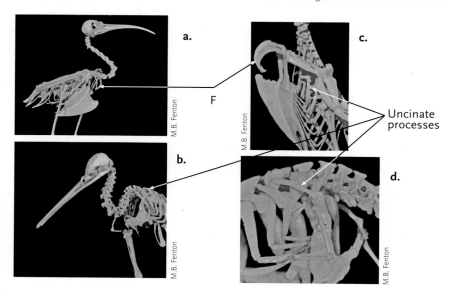

a.

b.

c.

d.

F

M.B. Fenton

Uncinate processes

Figure 28.42

Bird skeletons and flight. Compared are the thoracic skeletons of a Hudsonian Curlew (*Numenius phaeopus*) **(a)**, a kiwi (*Apteryx australis*) **(b, d)**, and a penguin **(c)**. Note the wishbones (furcula F—plural of furculum), as well as keels on the sterna of the curlew and the penguin but not on the kiwi. Neither the penguin nor the kiwi can fly, but the penguin "flies" in water. The wings of the kiwi are drastically reduced **(d)**, and there is no furculum (wishbone), which is obvious in the penguin and the curlew. All three species have distinct uncinate processes on the ribs **(b, c, d)**.

Figure 28.43
Bird diversity.

a. The Laysan albatross (Procellariiformes, *Phoebastria immutabilis*) has the long thin wings typical of birds that fly great distances.

b. The roseate spoonbill (Ciconiformes, *Ajaia ajaja*) uses its bill to strain food particles from water.

c. The bald eagle (Falconiformes, *Haliaeetus leucocephalus*) uses its sharp bill and talons to capture and tear apart prey.

d. A European nightjar (Caprimulgiformes, *Caprimulgus europaeus*) uses its wide mouth to capture flying insects.

e. A ruby-throated hummingbird. (Apodiformes, *Archilochus colubris*) hovers before a hibiscus blossom to drink nectar from the base of the flower.

f. The chestnut-backed chickadee (Passeriformes, *Parus rufescens*) uses its thin bill to probe for insects in dense vegetation.

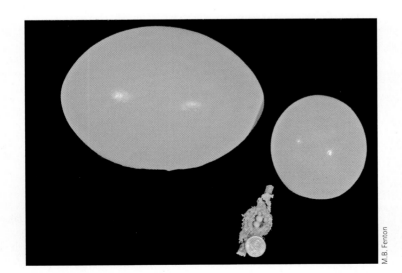

Figure 28.44
Bird eggs. Bird eggs range in size from those of elephant birds (left, *Aepyornis* of Madagascar) to ostriches (*Struthio camelus*, right) and a hummingbird (bottom). The scale, a Canadian $2 coin, is 2.8 cm in diameter.

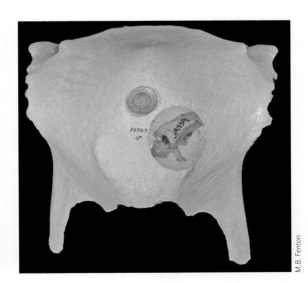

Figure 28.45
Breastbones of birds. Compared are the unkeeled breastbone of an ostrich (*Struthio camelus*) and the keeled breastbone of a hummingbird (*Trochilus polytmus*). A Canadian $2 coin (2.8 cm in diameter) is shown for scale.

protobird was a runner (cursorial) and ran in pursuit of prey and jumped up to catch it.

In 2008, Kenneth P. Dial and two colleagues proposed the *ontogenic–transitional wing* (OTW) hypothesis to explain the evolution of flight in birds. They asserted that the transitional stages leading to the development of flight in modern birds corresponded to its evolutionary development. Key to the OTW theory is the observation that in developing from flightless hatchlings to flight-capable juveniles, individual birds move their protowings in the same ways as adults move fully developed wings. Dial and his colleagues noted that flap-running allows as yet flightless birds to move over obstacles. The OTW theory provides another look at the evolution of flight, and its predictions can be tested with fledglings of extant species. The combination of wings and bipedalism is central to the OTW hypothesis. Birds are bipedal, and pterosaurs may have been. Bats, however, are not bipedal, so the OTW hypothesis does not explain the evolution of flight in that group.

The first known radiation of birds produced the enantiornithines ("opposite" birds), the dominant birds of the Jurassic and Cretaceous. Ornithurines are modern birds **(Figure 28.46).** Like dinosaurs, many mammals, and other organisms, the enantiornithines did not survive the extinctions that marked the end of the Cretaceous (see Chapter 32). Many enantiornithines flew, reflected by keeled sterna, furcula, and other "modern" skeletal features. Others, such as *Hesperornis*, were swimmers that used their feet for propulsion and, unlike penguins, had unkeeled sterna **(Figure 28.47, p. 666).** Ornithurines include modern groups of wading birds and seabirds, first known from late Cretaceous rocks. Woodpeckers, perching birds, birds of prey, pigeons, swifts, the flightless ratites, penguins, and some other groups were all present by the end of the Oligocene. Birds continued to diversify through the Miocene.

The ~ 9000 living bird species show extraordinary ecological specializations built on the same body plan. Living birds are traditionally classified into nearly 30 orders. A bird's bill usually reflects its diet. Seed and nut eaters, such as finches and parrots, have deep, stout bills that crack hard shells. Carnivorous hawks and carrion-eating vultures have sharp beaks to rip flesh. Nectar-feeding hummingbirds and sunbirds have long slender bills to reach into flowers, although

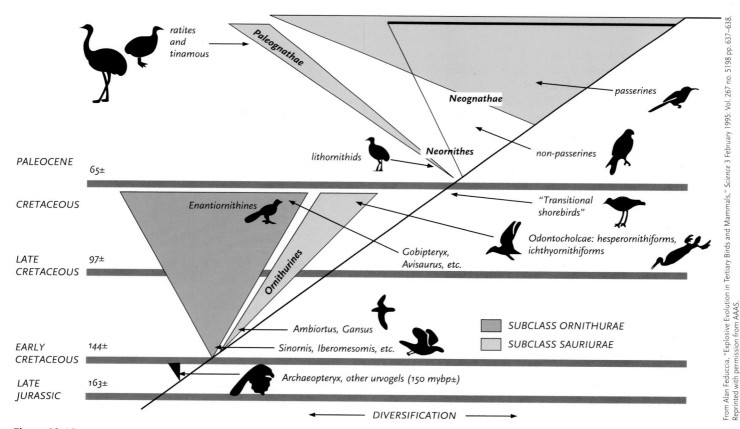

From Alan Feduccia, "Explosive Evolution in Tertiary Birds and Mammals," *Science* 3 February 1995: Vol. 267 no. 5198 pp. 637–638. Reprinted with permission from AAAS.

Figure 28.46

Evolution of birds. Enantiornithines, or opposite birds, were dominant in the Mesozoic but coexisted with ornithurine (more modern) birds in the early Cretaceous. The enantiornithines did not survive the extinctions at the end of the Cretaceous. By the Miocene, passerine birds became the dominant landbirds. Names of genera for some fossil birds make it easier to find out more about these animals.

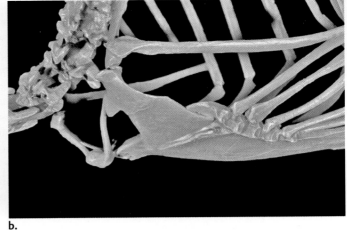

Figure 28.47

Skull **(a)** and sternum **(b)** of Hesperornis, a Cretaceous enantiornithine bird. Note the teeth, along with an unkeeled sternum and a furculum. This diving bird swam with its feet rather than its wings. The skull is 25 cm long.

many perching birds also have slender bills to feed on insects. The bills of ducks are modified to extract particulate matter from water.

Birds also differ in the structure of their feet and wings. Predators have large, strong talons (claws), whereas ducks and other swimming birds have webbed feet that serve as paddles. Long-distance fliers such as albatrosses have narrow wings, whereas species that hover at flowers have short, broad wings. The wings of penguins and similar species are so specialized for swimming that they are incapable of aerial flight.

All birds have well-developed sensory and nervous systems, and their brains are proportionately larger than those of comparably sized diapsids. Large eyes provide sharp vision, and most species also have good hearing, which nocturnal hunters such as owls use to locate prey. Vultures and some other species have a good sense of smell, which they use to find food. Migrating birds use polarized light, changes in air pressure, and Earth's magnetic field for orientation (see Chapter 45).

Many birds exhibit complex social behaviour, including courtship, territoriality, and parental care. Many species use vocalizations and visual displays to challenge other individuals or attract mates. Most raise their young in nests, using body heat to incubate eggs. The nest may be a simple depression on a gravel beach, a cup woven from twigs and grasses, or a feather-lined hole in a tree.

Many bird species make semiannual long-distance migrations (see Chapter 47). Golden plovers (*Pluvialis dominica*) and the godwit (*Limosa lipponica*) migrate over 20 000 km a year going to and from their summer and winter ranges. Migrations are a response to seasonal changes in climate. Birds travel toward the tropics as winter approaches. In spring, they return to high latitudes to breed and to use seasonally abundant food sources.

CONCEPT FIX Some people think that birds can fly because of air spaces between their cells. In reality, many of the bones of birds are laminated structures with hollows that reduce the density of their skeletons, but this is true even of flightless birds such as ostriches. Birds, bats, pterosaurs, and insects fly because they have wings and muscles to flap them in addition to other morphological and physiological specializations.

STUDY BREAK

1. What three skeletal features are associated with bird flight? Which ones are missing in flightless birds?
2. What adaptations make flight possible in birds and pterosaurs?
3. What characteristics maintain the structure of feathers and make them important to flight in birds?

28.13 Mammalia: Monotremes, Marsupials, and Placentals

Mammals are part of the synapsid lineage, the first of the amniotes to diversify. During the late Paleozoic, medium- to large-sized synapsids were the most abundant vertebrate predators in terrestrial habitats. Therapsids were one successful and persistent branch of synapsids. Therapsids were relatively mammal-like in their legs, skulls, jaws, and teeth and represented an early radiation of synapsids. By the end of the Triassic, the earliest mammals (most of them no bigger than a rat) had appeared. Several lineages of early mammals, such as multituberculates (see Chapter 32) and the lineage that includes the Mesozoic beaver (see Figures 18.6 and 18.7, Chapter 18), persisted and even flourished through much of the Mesozoic. These mammals coexisted with dinosaurs and other diapsids, as well as with the enantiornithine birds.

Paleontologists hypothesize that most Mesozoic mammals were nocturnal, perhaps to avoid diurnal predators and/or overheating. There are two living mammalian lineages **(Figure 28.48)**: the egg-laying Prototheria (or Monotremata) and the live-bearing Theria (marsupials and placentals).

Several features distinguish mammals from other vertebrates, but mammalian diversity makes it difficult to generalize absolutely about definitive characteristics. Living mammals are relatively easy to recognize. They are usually furry and have a diaphragm (a sheet of muscle separating the chest cavity from the viscera); most are **endothermic** (warm-blooded) and bear live young. In mammals, most blood leaves the heart through the **left aortic arch** (the main blood vessel leaving the heart; see Chapter 40). Mammals have two occipital condyles where the skull attaches to the neck, as well as a secondary palate (the plate of bones forming the roof of the mouth). They are **heterodont** and **diphyodont (Figure 28.49, p. 668)**. Heterodont means that different teeth are specialized for different jobs; diphyodont means that there are two generations of teeth (milk or deciduous teeth and adult teeth). But some mammals have no teeth, and others lay eggs. The secondary palate allows mammals to breathe while sucking, without releasing hold on the nipple—an essential part of nursing.

Endothermy means that mammals typically maintain an elevated and stable body temperature so that they can be active under different environmental conditions. They can do this because of their metabolic rates and insulation. Heterodont teeth make mammals more efficient at mechanically dealing with their food (chewing), reducing the lag between the time food is consumed and when the energy in it is available to the consumer. Heterodont teeth are correlated with improved jaw articulation, in mammals between the dentary (lower jaw) and squamosal (bone on the skull). The diaphragm means that mammals are reasonably efficient at breathing, and the circulatory system with a four-chambered heart makes them efficient at internal circulation of resources or collection of wastes. Milk is a rich food source, and by feeding it to their young, female mammals provide the best opportunity for growth and development. The **cortex** of the brain is central to information processing and learning. Mammals' brains are another key to their evolutionary success.

28.13a The Mammalian Radiation: Variations on Mammals

The egg-laying Prototheria (*proto* = first; *theri* = wild beast), also called Monotremata, and the live-bearing Theria are the two groups of living mammals. Among the Theria, the Metatheria (*meta* = between), also

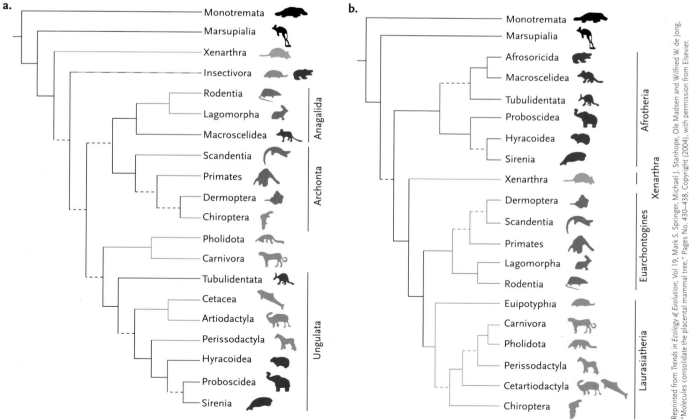

Figure 28.48

Modern mammals. Prevailing phylogenies of mammals derived from **(a)** morphological and **(b)** molecular data.

Reprinted from *Trends in Ecology & Evolution*, Vol 19, Mark S. Springer, Michael J. Stanhope, Ole Madsen and Wilfried W. de Jong, "Molecules consolidate the placental mammal tree," Pages No. 430–438, Copyright (2004), with permission from Elsevier.

Figure 28.49

Mammal teeth. In most mammals, the teeth are diphyodont, meaning that milk (deciduous) teeth are replaced by permanent teeth. The skull of a vampire bat (*Desmodus rotundus*) clearly shows four deciduous teeth (arrows), as well as permanent teeth **(a).** The teeth of mammals are also heterodont **(b),** meaning that different teeth are specialized to do different jobs. In this bear (*Ursus americana*), incisors (i), a canine (c), **premolars** (p), and molars (m) are obvious.

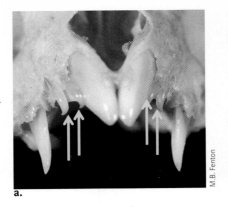

a.

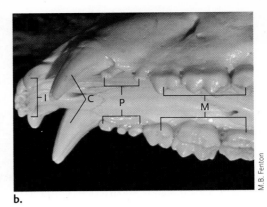

b.

called marsupials, and the Eutheria (*eu* = good), or placentals, differ in their reproductive adaptations.

Monotremata. The **monotremes** (*mono* = one; *trema* = perforation) are represented by three living species that occur only in the Australian region **(Figure 28.50).** Females lay leathery shelled eggs, and newly hatched young lap up milk secreted by modified sweat glands (mammary glands) on the mother's belly. The duck-billed platypus (*Ornithorhynchus anatinus*) lives in burrows along riverbanks and feeds on aquatic invertebrates. The two species of echidnas or spiny anteaters (*Tachyglossus aculeatus* and *Zaglossus bruijnii*) feed on ants or termites.

Marsupialia. Represented by 240 species, marsupials (*marsupion* = purse) (Metatheria) are characterized by short **gestation** periods. The young are briefly (as few as 8 to 10 days in some species and up to 30 days in others) nourished in the uterus via a placenta and are then born at an early stage of development. Newborns use their forelimbs to drag themselves from the vagina and across the mother's belly fur to her abdominal pouch, the marsupium, where they complete their development attached to a teat. Marsupials are prevalent among the native mammals of Australia and are also diverse in South America **(Figure 28.51).** One species, the opossum (*Didelphis virginiana*), occurs as far north as Canada. South America once had a diverse marsupial fauna, which declined after the Isthmus of Panama bridged the seaway between North and South America (see Chapter 19), allowing placental mammals to move southward.

Placental mammals (Eutheria) are represented by 4000 living species. They complete embryonic development in the mother's uterus, nourished through a **placenta** until they reach an advanced stage of development **(viviparous).** Some species, such as humans, are helpless at birth **(altricial),** but others, such as horses, are born with fur and are quickly mobile **(precocial).** Biologists divide the eutherians into about 18 orders, of which only 8 have more than 50 living species **(Figure 28.52).** Rodents (Rodentia) make up about 45%

a. Short-nosed echidna

b. Duck-billed platypus

Figure 28.50

Monotremes. The short-nosed echidna (*Tachyglossus aculeatus*) **(a),** is terrestrial. The duck-billed platypus (*Ornithorhynchus anatinus*) **(b),** raises its young in a streamside burrow.

a.

b.

c.

Figure 28.51

Marsupials. **(a)** A kangaroo (*Macropus giganteus*) carries her "joey" in her pouch; **(b)** a male koala (*Phascolarctos cinereus*) naps; and **(c)** an opossum from Guyana (*Didelphis* species) emerges from its den after dark to feed.

of eutherian species, and bats (Chiroptera) make up another 22%. We belong to the primates, along with 169 other species, representing about 5% of the current mammalian diversity.

Some eutherians are obviously specialized for locomotion. Although whales and dolphins (order Cetacea) and manatees and dugongs (order Sirenia) are descended from terrestrial ancestors, they are aquatic (mainly marine) and can no longer function on land. By contrast, seals and walruses (order Carnivora) feed under water but rest and breed on land. Bats (order Chiroptera) use wings for powered flight.

a. The capybara (Rodentia, *Hydrochoerus hydrochaeris*), the largest rodent, feeds on vegetation in South American wetlands.

b. Most bats, like the Eastern small-footed bat (*Myotis leibii*), are nocturnal predators on insects.

c. Walruses (Carnivora, *Obodenus rosmarus*) feed primarily on marine invertebrates in frigid arctic waters.

d. The black rhinoceros (Perissodactyla, *Diceros bicornis*) feeds on grass in sub-Saharan Africa.

e. Arabian camels (Artiodactyla, *Camelus dromedarius*) use enlarged foot pads to cross hot desert sands.

Figure 28.52
Eutherian diversity.

Although early mammals appear to have been insectivorous, the diets of modern eutherians are diverse. Odd-toed ungulates (*ungula* = hoof) such as horses and rhinoceroses (order Perissodactyla), even-toed ungulates such as cows and camels (order Artiodactyla), and rabbits and hares (order Lagomorpha) all eat vegetation. Some of the vegetarians use fermentation to digest cellulose (see Chapter 48). **Carnivores** (order Carnivora) usually consume other animals, but some, such as the giant panda (*Ailuropoda melanoluca*), are vegetarians. Most bats eat insects, but some feed on flowers, fruit, or nectar, and some, the vampires, consume blood. Many whales and dolphins prey on fishes and other animals, but some eat plankton. Some groups, including rodents and primates, feed opportunistically on both plant and animal matter. Ants and termites are the preferred food of a variety of mammals, both prototherian and therian.

STUDY BREAK

1. How do monotremes differ from marsupials and placentals?
2. How do marsupials and placentals differ from one another?
3. What are four distinctive features of mammals?

28.14 Evolutionary Convergence and Mammalian Diversity: Tails to Teeth

In the discussion on the Mesozoic beaver (see Figures 18.6 and 18.7, Chapter 18), we learned that a dorsoventrally flattened tail occurred in a Mesozoic mammal and today occurs in a monotreme (duck-billed platypus) and in beavers (*Castor* species). Evolutionary convergences in design features such as these are common in mammals. Another good example is the development of protective spines (quills) from hairs. These occur in spiny anteaters (monotremes), porcupines (rodents), and hedgehogs and tenrecs (insectivores).

Another striking example of convergence among mammals is provided by the teeth and lumbar vertebrae of the Mesozoic *Fruitafossor windscheffeli* **(Figure 28.53)** and some living xenarthrans (armadillos and sloths). Like sloths and armadillos, *Fruitafossor* had round molars with open roots **(Figure 28.54)**. Also like sloths and armadillos, *Fruitafossor* had processes in its lower back (**lumbar vertebrae**) known only from living xenarthrans. Although the teeth and vertebral structures converge between *Fruitafossor* and xenarthrans, *Fruitafossor* is not closely related to any living mammals. We do not know if *Fruitafossor* had other features of mammals such as mammary glands, a diaphragm, and vivipary.

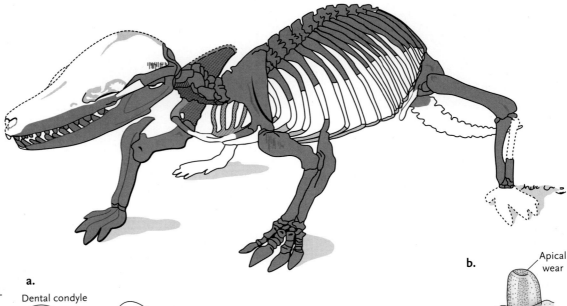

Figure 28.53
Fruitafossor, a mammal of the mid-Jurassic. Dark outlined bones are known from fossils; other bones are presumed. The *Fruitafossor* had round teeth with open roots **(a, b)**.

a.
Dental condyle

3 mm

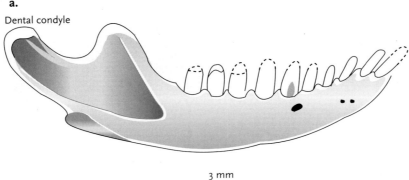

b.
Apical wear

1 mm

Open root-end

mg

From Zhe-Xi Luo, John R. Wible, "A Late Jurassic Digging Mammal and Early Mammalian Diversification," *Science*, Apr 1, 2005, vol. 308, pp. 103–107. Reprinted with permission from AAAS.

Figure 28.54
Convergences in mammals. Like the *Fruitafossor* (Figure 28.53), round teeth with open roots are well known from edentate mammals like armadillos and tubulidentates, the aardvark (shown here).

But *Fruitafossor's* bones, particularly its jaw joints and occipital condyles, make it a mammal.

28.14a Mammalian Teeth: Diversity of Form and Function

As in other vertebrates, mammals' teeth provide a good indication of diet. Some molars (cheek teeth) with W-shaped cusps cut and crush food **(Figure 28.55a, b)**, whereas others mainly crush (Figure 28.55c, d). Grinding teeth have appeared in a wide range of forms

in mammals **(Figure 28.56)** and show considerable variation in the details of their design.

Animals such as the walrus **(Figure 28.57a, p. 672)** have tusks for digging and small flat molars for crushing the shells of bivalves (compare with Figures 28.19a, p. 648, and 28.22b, p. 651). The molars of *Desmostylus* have an artistic circular pattern (see photo on opening page of this chapter). These Miocene mammals were thought to have resembled Sirenia, the dugongs and manatees. Later discoveries revealed that they had massive limbs and, in body form, looked more

a. **b.** **c.** **d.**

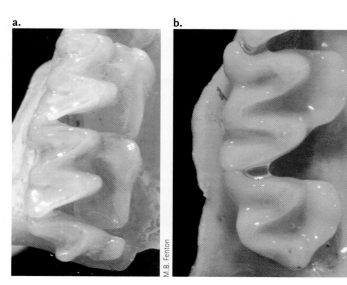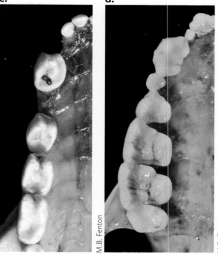

Figure 28.55
Molars that cut and crush **(a, b)**, and those that crush **(c, d)**. The molars (cheek teeth) of a mole **(a)**, an insectivorous bat **(b)**, an Old World fruit bat **(c)**, and a New World fruit bat **(d)**. The mole is *Condylura*, the insectivorous bat a *Taphozous*, the Old World fruit bat a *Pteropus*, the New World fruit bat a *Brachyphylla*.

a. **b.** **c.** **d.**

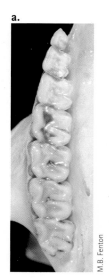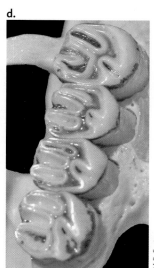

Figure 28.56
Molar teeth for crushing plant material (mostly leaves and stems). Look carefully at the details of the molars of a hyrax (*Heterohyrax brucei* from Africa) **(a)**, a three-toed sloth (*Bradypus tridactylus* from South America) **(b)**, a black rhino (*Diceros bicornis* from Africa) **(c)**, and a porcupine (*Erethizon dorsatum* from North America) **(d)**. The hyrax belongs to the order Hyracoidea, the sloth to the order Xenarthra, the rhino to the order Persissodactyla, and the porcupine to the order Rodentia.

a.

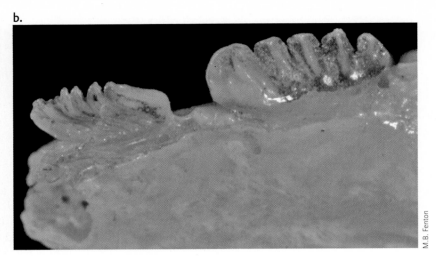

b.

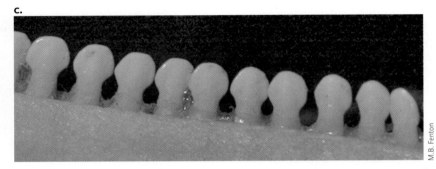

c.

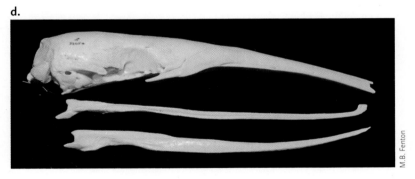

d.

Figure 28.57

Teeth (or no teeth) for different jobs. Whereas walruses (*Odobenus rosmarus*) **(a)** use tusks (which are canine teeth) for digging clams, which they crush with their molars, flying lemurs (*Cynocephalus*) **(b)** use comblike lower incisor teeth to comb their fur, and dolphins (*Tursiops*) **(c)** have rows of similar teeth (homodont) for grasping fish. The giant anteater (*Myrmecophaga*) **(d)** eats ants and termites and lacks teeth.

like modern hippos. They presumably used their teeth for crushing aquatic vegetation.

Mammals that eat mainly ants and termites (see Figure 28.57d) often lack teeth entirely. This way of life has been exploited by mammals across the tropics. The ant and termite eaters of the New World (armadillos—*Dasypus* spp—anteaters—*Myrmecophaga*) are different evolutionary lineages (see Figure 28.48, p. 667) than the African aardvark (*Orycteropus*, order Tubulidantata) or scaly anteater (*Manis*, order Pholidota). Australia has both monotreme and marsupial anteaters. The most astonishing anteater is the aardwolf, a variation on a hyena **(Figure 28.58)**; both of these species belong to the order Carnivora.

Whereas reptiles, amphibians, fish, and sharks can replace teeth many times (e.g., Figure 28.19, p. 648), mammals replace them only once. Teeth wear with age **(Figure 28.59).** When the teeth are worn out, the animal can no longer feed itself properly and dies.

Elephants deal with this problem by having only four active molars in the jaw at any one time. The new molar grows in from the back (Figure 28.59), replacing the worn one. In rodents and some other mammals (and also in hydrasaur dinosaurs), molar (and for rodents and lagomorphs, incisor) teeth grow continuously. Here the teeth are curved so that pressure during biting is not directed at the points of growth (see Figures 18.7b, Chapter 18, and 47.4, Chapter 47).

STUDY BREAK

1. What features are found in most mammals and distinguish them from other vertebrates?
2. How is heterodont different from diphyodont?
3. Distinguish among monotremes, marsupials, and placentals.

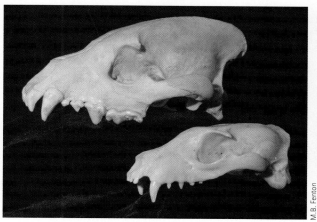

Figure 28.58
Divergence! Spotted hyena (top, *Crocuta crocuta*) and aardwolf (bottom, *Proteles capensis*) are in the same family (Hyaenidae). The spotted hyena is a carnivorous scavenger with massive teeth capable of cutting tendons and crushing bones. The aardwolf eats mainly ants and termites and has reduced teeth (and a differently shaped skull). The *Crocuta* skull is about 30 cm long. Both belong to the order Carnivora.

Figure 28.59
Tooth wear and replacement. Elephants (*Loxodonta africana*) have four functional molars in the mouth at any one time (one in each jaw quadrant). New molars push into the tooth row from the back.

MOLECULE BEHIND BIOLOGY 28.1

Geckel

Geckos **(Figure 1)** can cling to vertical and even inverted smooth surfaces such as window glass. They do this using specialized keratinous setae (fine hairs) on their feet. There are spatulate extensions at the end of each seta. The other element in the sticking ability of geckos is an adhesive. But the remarkable feature of geckos is that the system allows rapid detachment of the foot. As many people have learned using Crazy Glue and its equivalents, getting stuck to something is easy; getting unstuck is a different story.

As remarkable as geckos, some mussels (see Chapter 27) secrete a specialized adhesive with a high concentration of catecholic amino acid 3,4-dihydroxyl-phenlalanine (DOPA). DOPA allows the mussels to cling firmly to wet surfaces. In contrast, the adhesive ability of geckos is diminished by full immersion in water.

Geckel is a new hybrid adhesive **(Figure 2)** combining the adhesive features of those used by geckos and mussels. Geckel is a thin layer of a synthetic polymer that retains its adhesive properties in dry and wet environments for more than 1000 contact cycles.

Work with geckos (probably *Rhoptropus biporosus*) from the Namib desert sheds light on the evolutionary background of their extraordinary clinging power. These small geckos weigh about 2 g and show great mobility on the variety of substrates they encounter—rough, undulant, and unpredictable, often providing few points of adhesion. The adhesive pads under the geckos' toes allow them to cling to the full spectrum of surfaces, and their ability to stick to glass is coincidental.

Figure 1
Geckos, such as this one, can walk on (stick to) glass.

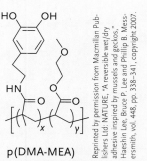

p(DMA-MEA)

Figure 2 *Geckel, a fabricated adhesive that mimics those produced and used by geckos and mussels.*

Reprinted by permission from Macmillan Publishers Ltd: NATURE, "A reversible wet/dry adhesive inspired by mussels and geckos," Haeshin Lee, Bruce P. Lee and Phillip B. Messersmith, vol. 448, pp. 338–341, copyright 2007.

Norman John (Jack) Berrill, *McGill University*, 1903–1996

Growing up on the edge of a city gave Jack Berrill ready access to the fauna and flora of the surrounding countryside: the woods, the streams, the ponds. As a first-year premedical student at the University of Bristol, he was offered the chance to join a zoological expedition to the Indian Ocean, a life-changing experience that drew him into zoology. In 1929, he completed his Ph.D. (University College, London, England) and became an assistant professor of zoology at McGill University in Montreal, Quebec, where he remained until his retirement.

Berrill is distinguished for his research on development, morphogenesis, and regeneration in the Tunicata (phylum Urochordata). He chose these animals for study because they were transparent, allowing him to see what was going on inside them with minimum disturbance. His contribution to our knowledge of tunicates was profound and continues to strongly influence the field.

In addition to his research, Berrill was also an innovative teacher. He used 16 mm film (silent and in black and white) to illustrate some of his lectures. He provided commentary as he showed the film, stopping here and there, backing up, and then continuing. One can only imagine what he could have done with videos and PowerPoint.

But Berrill's contributions did not stop there. He also wrote 10 books that popularized zoology, such as *The Living Tide* (1951), *Inherit the Earth* (1966), and *Animals in Action* (1972).

Berrill influenced the lives of others through his research. His published findings opened new avenues of thought. He also influenced people through his teaching and nontechnical publications.

Review

aplia

To access course materials such as Aplia and other companion resources, please visit www.NELSONbrain.com.

28.2 Phylum Echinodermata

- Echinodermata are slow-moving or sessile bottom-dwelling animals. They are important herbivores and predators that occur from shallow coastal waters to the oceans' depths. Adult echinoderms develop from bilaterally symmetrical, free-swimming larvae. Developing larvae assume a secondary radial symmetry, often organized around five rays or arms. The table below compares echinoderms and humans.

System	Echinoderms	Humans
digestive	complete system	complete system
excretory	not present	complete system
respiratory	not present	complete system
circulatory	minimal system	closed, complete system
nervous system	no head or central	cephalized system

- Water enters the fluid-filled canals of the water vascular system through the madreporite (a sievelike plate) on the aboral surface. A tube connects the madreporite to the ring canal, which surrounds the esophagus. The ring canal branches into radial canals that extend into each arm and is connected to numerous mucus-covered tube feet. When ampullae, small muscular bulbs in each tube foot, contract, they force water into the tube foot, causing it to lengthen and attach to a substrate. The tube foot contracts, pulling the animal along and pushing water back into the ampulla; this causes the tube foot to release the substrate. Echinoderms reproduce either sexually or asexually. Sexual reproduction is usually achieved by releasing gametes into the water. Asexual reproduction involves clonal budding and may be stimulated by the odour of a predator.

- Asteroidea (sea stars) consist of a central disk surrounded by 5 to 20 radiating arms. Small pincers at the base of pedicellariae (short spines) are used to remove debris that falls onto the animal's surface. The ossicles of their endoskeleton are not fused, permitting flexibility of the arms and disk. Most sea stars eat invertebrates and small fish.

- Ophiuroidea (brittle stars and basket stars) have a well-defined central disk and slender, elongated arms that are sometimes branched. They crawl swiftly across substrates by moving their arms in a coordinated fashion. They feed on small prey, suspended plankton, or detritus extracted from muddy deposits.

- Echinoidea (sea urchins and sand dollars) lack arms. Their ossicles are fused into tests that provide excellent protection but restrict flexibility. Echinoids use tube feet in locomotion. They graze on algae and other organisms that cling to marine surfaces.

- Holothuroidea (sea cucumbers) have a reduced endoskeleton consisting of widely separated microscopic plates. They have five rows of tube feet and a soft body elongated along the oral–aboral axis. Modified tube feet form a ring of tentacles around the mouth. Some species secrete a mucus net that traps plankton or other food particles. The net and tentacles are inserted into the mouth, where the food is ingested. Other species extract food from bottom sediments.

- Crinoidea (sea lilies and feather stars) have five to several hundred branched arms surrounding the disk containing the mouth. Branches of the arms are covered with mucus-coated tube feet that trap suspended microscopic organisms. Sessile sea lilies have a central disk attached to a flexible stalk that can reach a metre in length, whereas adult feather stars swim or crawl weakly.

28.3 Phylum Hemichordata

- Hemichordates (acorn worms) use a muscular, mucus-coated proboscis to construct burrows and trap food particles. Drawn in by beating cilia, water enters the pharynx and exits through the pharyngeal gill slits. As the water passes, suspended food is trapped and directed to the digestive system, while gases are exchanged across the partitions between gill slits.

28.4 Phylum Chordata

- A notochord, a dorsal hollow nerve cord, and gill slits distinguish chordates from all other deuterostomes. Gill slits are sets of paired openings in the pharynx. Water is drawn into the mouth, food is filtered out, and the water passes through the pharynx, where gas exchange takes place, and then out through the gill slits.

- Tunicates (sea squirts) belong to the subphylum Urochordata. They float in surface waters or attach to substrates in shallow waters. Many are sessile as adults and secrete a gelatinous or leathery "tunic" around their bodies. They squirt water through a siphon when disturbed. Some urochordates have larvae that resemble cephalochordates (or lancelets), and in a few species, these larvae are neotenous. Adult lancelets are mainly sedentary, lying partly buried in sand of shallow marine waters. They have well-developed body wall muscles and a prominent notochord. Adults have light receptors on the head and chemical sense organs on tentacles.

28.5 The Origin and Diversification of Vertebrates

- The internal skeleton of vertebrates provides structural support for muscles and protects the nervous system and other internal organs. The backbone surrounds and protects the dorsal nerve cord, and a bony cranium provides protection for the brain. The backbone acts as a place for muscle attachments, which allows quick movement.

- Homeotic (Hox) genes influence the three-dimensional shape of an animal and the locations of structures such as eyes, wings, and legs. Hox genes are arranged on chromosomes in a specific order to form the Hox gene complex. Each gene in the complex governs the development of particular structures. Species with simple anatomy have fewer Hox genes than more complex species, which have duplicated copies. These duplicate copies assumed new functions, directing the development of novel structures, such as the vertebral column and jaws.

- Vertebrates have four characteristic morphological innovations: cranium, vertebrae, bone, and neural crest cells.

28.6 Agnathans: Hagfishes and Lampreys, Conodonts, and Ostracoderms

- Agnatha are primitive vertebrates that use a muscular pharynx to suck water containing food particles into their mouths and gills to filter the food and perform gas exchange. Hagfishes and lampreys have a well-developed notochord but lack true vertebrae and paired fins. The hagfish skeleton is a cranium and a notochord. These marine scavengers feed on invertebrate prey and dead fish. They lack a larval stage. Lampreys have a more derived axial skeleton than hagfishes. Their notochord is surrounded by cartilage that partially covers the nerve cord, whereas hagfish

have no specialized structures surrounding the nerve cord. Some species of lampreys are parasitic as adults, attaching to a host. Ammocoetes, the larval stage of lampreys, resemble cephalochordates and may develop for up to seven years before metamorphosing into adults.

28.7 Jawed Fishes: Jaws Expanded the Feeding Opportunities for Vertebrates

- Jaws meant that fishes could feed on larger items of food with higher energy content. Jaws also function to defend against predators; groom; transport young; and grasp, kill, and shred food items.

- Flexible fins, lightweight skeletons, streamlined bodies, and an absence of heavy body armour allow sharks to pursue prey rapidly. Sharks and their relatives have squalene, an oily substance contained within the liver. Squalene is lighter than water and increases the animals' buoyancy.

- The lateral line system of elasmobranchs and other fishes consists of a row of tiny sensors in canals along both sides of the body. This system allows detection of vibrations in water, which can be used when hunting.

- In many bony fishes, a gas-filled swim bladder serves as a hydrostatic organ to increase buoyancy. Bony fish also have small, smooth, lightweight scales and bodies that are covered with a protective coat of mucus that retards bacterial growth and smoothes the flow of water past the body.

28.8 Early Tetrapods and Modern Amphibians

- Osteolepiformes (fleshy-finned fishes) and tetrapods had infoldings of their tooth surfaces. The shapes and positions of bones on the dorsum and side of their crania and in their appendages were similar. Osteolepiformes had strong fins enabling them to crawl on mud (making their move onto land) and possessed vertebral columns with crescent-shaped bones for support. Osteolepiformes had lungs allowing them to breathe atmospheric oxygen. They could excrete urea or uric acid rather than ammonium.

- The body wall of fish picks up sound vibrations and directly transfers them to sensory receptors. Sound waves are harder to detect in air. The development of a tympanum, or eardrum, allowed tetrapods to detect airborne vibrations and transfer them to the sensory cells of their inner ear.

- Amphibians have thin, scaleless skin, well supplied with blood vessels. Since some oxygen and carbon dioxide enters the body across a thin layer of water, most amphibians need moist skin, restricting them to aquatic or wet terrestrial habitats. Many amphibians need access to free-standing water to reproduce.

28.9 The Origin and Mesozoic Radiations of Amniotes

- Amniotes get their name from the amnion, a fluid-filled sac surrounding the embryo during development. Amniote eggs are resistant to desiccation because the developing embryos excrete uric acid that is stored in the allantois.

- Amniote eggs have four specialized membranes that protect the embryo and facilitate gas exchange and excretion. They also have a hard or leathery shell perforated by microscopic pores that mediates the exchange of air and water between the egg and its environment. Keratin and lipids are partly responsible for making the skin waterproof.

28.10 Subclass Testudinata: Turtles and Tortoises

- There have been three major radiations of amniotes: anapsids, synapsids, and diapsids, distinguishable by the numbers of bony arches in the temporal region of the skull. The bony arches allow space for contraction (and expansion) of large and powerful jaw muscles. Anapsids lacked temporal arches, synapsids have one pair of temporal arches, and diapsids have two pairs of arches.

- Surviving anapsids are turtles and tortoises. A turtle's body is defined by a bony, boxlike shell, which includes a dorsal carapace and a ventral plastron. Its ribs are fused to the inside of the carapace, and the pectoral and pelvic girdles lie within the ribcage. Large keratinized scales cover the bony plates that form the shell.

28.11 Living Diapsids: Sphenodontids, Squamates, and Crocodilians

- Diapsids evolved in two lines, lepidosaurs and archosaurs. Lepidosaurs include snakes and lizards and many extinct forms. Snakes and lizards use olfactory and vibrational cues to detect prey. Some even have thermal perception.

- Living archosaurs include crocodilians, animals with a four-chambered heart that is homologous to the heart in birds. Some crocodilians have muscles that originate on the pubis and insert on the liver. When these muscles contract, the liver moves toward the tail, creating negative pressure in the chest cavity. This situation is analogous to the role of the diaphragm in mammals.

28.12 Aves: Birds

- Birds' ability to fly reflects a keeled sternum (breastbone), a furculum (wishbone), and uncinate processes on the ribs. These main adaptations, coupled with lightweight, strong bones and feathers, contributed to the success of birds. Flightless birds often lack a keeled sternum. Most birds have hollow limb bones with small supporting struts that criss-cross the internal cavities. Birds have fewer separate bony

elements in the wings, skull, and vertebral column, so the skeleton is light and rigid. All modern birds have replaced dense and heavy teeth with a lightweight keratinized bill. Birds have much higher metabolic rates than comparably sized reptiles do, and they depend on energy-rich food. A complex and efficient respiratory system and a four-chambered heart enable them to consume and distribute oxygen efficiently. Other adaptations include modification of internal organs to reduce weight, elimination of a urinary bladder so that uric paste is eliminated with digestive wastes, and laying eggs as soon as they are shelled.

- Each feather has numerous barbs and barbules with tiny hooks and grooves that maintain the feathers' structure, even during vigorous activity. Flight feathers on the wings provide lift, whereas contour feathers streamline the surface of the body. Down feathers form an insulating cover close to the skin.

- The top-down theory suggests that ancestral birds lived in trees and glided down from those trees in pursuit of insect prey. The bottom-up theory proposes a cursorial ancestor that ran along in pursuit of prey and jumped up to catch it. The ontogenic–transitional wing (OTW) hypothesis is a third effort to explain the evolution of flight in birds. Its proponents suggest that flapping protowings gave the ancestors of birds greater mobility.

28.13 Mammalia: Monotremes, Marsupials, and Placentals

- Most living mammals are furry and endothermic (warm-blooded). Mammals usually bear live young and have a diaphragm, a left aortic arch leaving the heart, and two occipital condyles. Mammals also have a secondary palate and are heterodont (teeth specialized for different jobs) and diphyodont (two generations of teeth, milk or deciduous teeth and adult teeth). Heterodont teeth make mammals more efficient at mechanically dealing with their food (chewing), reducing the lag time between consumption of food and availability of the food's energy.

- Monotremes lay leathery shelled eggs. When newborns hatch, they lap up milk secreted by the mammary glands located on the mother's belly. Marsupials have short gestation periods of as few as 8 to 10 days. Young are born at an early stage of development and complete their development attached to a teat in the abdominal pouch (the marsupium) of their mother. Placental mammals complete embryonic development in the mother's uterus, nourished through a placenta until they reach a fairly advanced stage of development.

Questions

Self-Test Questions

1. Which phylum includes animals with a water vascular system?
   a. Echinodermata
   b. Hemichordata
   c. Chordata
   d. Arthropoda

2. Which of the following is NOT a characteristic of all chordates?
   a. a notochord
   b. a segmented nervous system
   c. a dorsal hollow nerve cord
   d. a perforated pharynx

3. Which group of vertebrates has adaptations allowing reproduction on land?
   a. agnathans
   b. gnathostomes
   c. amniotes
   d. ichthyosaurs

4. Which group of fishes has the most living species today?
   a. actinopterygians
   b. chondrichthyans
   c. acanthodians
   d. ostracoderms

5. Which is true about modern amphibians?
   a. They closely resemble their Paleozoic ancestors.
   b. They always occupy terrestrial habitats as adults.
   c. They never occupy terrestrial habitats as adults.
   d. They are generally larger than their Paleozoic ancestors.

6. Which one of the following key adaptations allows amniotes to occupy terrestrial habitats?
   a. the production of carbon dioxide as a metabolic waste product
   b. an unshelled egg protected by jellylike material
   c. a dry skin largely impermeable to water
   d. a lightweight skeleton with hollow bones

7. Which of the following characteristics are central to powered flight in birds?
   a. webbed feet, long legs, feathers
   b. efficient respiratory and excretory systems, flight muscles
   c. elongated forelimbs, keeled breast bone, flight muscles
   d. feathers, furculum, eyes

8. Which of the following characteristics did NOT contribute to the evolutionary success of mammals?
   a. extended parental care of young
   b. an erect posture and flexible hip and shoulder joints
   c. specializations of the teeth and jaws
   d. high metabolic rate and homeothermy

9. Why are Echinodermata and Chordata deuterostomes?
   a. because each has a notochord
   b. because neither has spiral cleavage
   c. because both have radial cleavage
   d. because both have a mouth

10. Why are lamprey eels and hagfish chordates?
    a. because both have a notochord, gill slits, and a dorsal hollow nerve chord
    b. because both have mouths
    c. because neither have paired fins
    d. because neither have jaws

Questions for Discussion

1. Most sharks and rays are predatory, but the largest species feed on plankton. Construct a hypothesis to explain this observation. How would you test your hypothesis?

2. What selection pressures did tetrapods face when they first ventured onto land? What characteristics allowed them to meet these pressures?

3. Use binoculars to observe several species of birds in different environments, such as lakes and forests. How are their beaks and feet adapted to their habitats and food habits?

4. Imagine that you unearthed the complete fossilized remains of a mammal. How would you determine its diet?

5. What evidence suggests that birds are really only specialized reptiles and should not be considered a distinct class (see also Chapter 19).

29

The large number of gulls (*Larus* species) is obvious at a landfill site near Thunder Bay, Ontario. The population of gulls reflects the local population of humans.

Population Ecology

WHY IT MATTERS

Controlling rabies in wildlife involves understanding many aspects of biology, from populations to epidemiology and behaviour. Rabies, from the Latin *rabere* (to rage or rave), affects the nervous system of terrestrial mammals. Caused by a Lyssavirus, rabies is usually spread by bites because the virus accumulates in the saliva of infected animals. Before 1885, when Louis Pasteur in France developed a vaccine for it, rabies was common in Europe, and many people died from it every year. In 2007, the World Health Organization estimated that worldwide, more than 50 000 people die annually from rabies, usually people in the developing world. Between 1980 and the end of 2000, 43 people in the United States and Canada died of rabies.

Animals with *furious* rabies become berserk, attacking anything and everything in their path, a behaviour that spreads the virus and helps ensure its survival. Animals with paralytic rabies (*dumb rabies*) suffer from increasing paralysis that progresses forward from the hindlimbs. Animals with either manifestation of rabies can spread the disease by biting when there is virus in their saliva. Paralysis of the throat muscles means that rabid animals cannot swallow the saliva they produce, so they appear to foam at the mouth.

Rabies is almost invariably fatal once an animal or a human shows clinical symptoms of the disease, so immunization of someone exposed to the disease should start as soon as possible after exposure. Since 1980, human diploid vaccines have been commonly available, raising the level of protection against rabies.

From the 1960s to the 1990s, a visit to almost any rural hospital in southern Ontario would have revealed at least one farmer receiving postexposure rabies shots. During that time, red foxes (*Vulpes vulpes*) were the main vector for rabies in Ontario, and cows (*Bos taurus*) exposed to rabies through fox bites in turn exposed farmers to the virus. Many farmers are accustomed to treating choking cows by reaching into the cow's gullet to clear an obstruction. A farmer dealing with a rabid cow could have been scratched and exposed to the virus, and then, after the cow died of rabies, the farmer would have received postexposure rabies shots. Rabies transmitted to cows from foxes posed a threat to human lives and was a significant drain on the economy through compensation paid to farmers whose cattle succumbed to the disease.

In 1967, 4-year-old Donna Featherstone of Richmond Hill, Ontario, died of rabies after being bitten by a stray cat. The resulting public outcry set the stage for a rabies eradication program in Ontario. Controlling fox rabies in southern Ontario was achieved by a combination of innovation and knowledge of basic biology. There were three phases: (a) developing an oral vaccine, (b) developing a means of vaccinating foxes, and (c) monitoring the impact of the program on the fox population.

First, two main baits for the oral vaccine were developed, and one, Evelyn, Rocketniki, Abelseth (ERA), was a modified live virus replicated in tissues of the mouth and throat. ERA successfully stimulated seroconversion in red foxes and vaccinated them against rabies. Second, foxes were vaccinated by eating ERA-containing baits scented with chicken. The baits were small, the size of restaurant packets of jam, and easy to distribute widely from low-flying aircraft, allowing vaccination of foxes across large areas of southern Ontario. Third, each bait contained tetracycline, a biomarker absorbed into the system of any mammal that ate the bait. Once in the body, some tetracycline penetrated the dentine of the animals' teeth, especially in younger individuals. Biologists sectioned and stained teeth from foxes taken by trappers. In the sections, bands of tetracyline in tooth rings **(Figure 29.1)** identified foxes that had taken baits, and biologists established that over 70% of red foxes had been vaccinated by this method.

Before the bait vaccination program, on average 211 cattle annually died of rabies in southern Ontario. The baiting program started in 1989, and by 1996, rabies in cattle dropped to an average of 11 cases a year and the levels of rabies in foxes in Ontario were dramatically reduced. The example demonstrates how problems in biology are solved by combined approaches, from population biology, behaviour, immunology, and epidemiology.

The purpose of this chapter is to introduce you to ecology in general and population ecology in particular. For more details about the incidence of a disease in a population, see Section 29.8.

Ecology is the study of the relationships among species and between species and the environments in which they occur. Studies of populations are fundamental to ecology and involve everything from the numbers of individuals to their age structure and patterns of reproduction. This connects with work on the life histories of species, often with a focus on

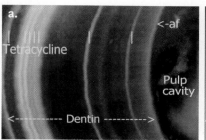 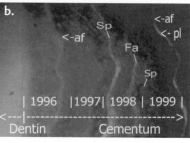

Reprinted from *Rabies*, Alan C. Jackson, "Rabies Control in Wildlife" by David H. Johnston, Rowland R. Tinline, Pages No. 445–471, Copyright 2003, with permission from Elsevier.

Figure 29.1

Tetracycline rings in carnivore teeth. Yellow fluorescent lines from ingestion of rabies baits with tetracycline as a biomarker. The sections are undecalcified, ultraviolet fluorescent × 100. **(a)** Coyote (*Canis latrans*) tooth with seven daily tetracycline lines from vaccine baits. **(b)** Canine tooth of a four-year-old raccoon (*Procyon lotor*) with yearly tetracycline lines in cementum. af = autofluorescent collagen; Fa = fall baits, 1998; pl = periodontal ligament; Sp = spring baits, 1998 and 1999.

reproduction and survival. Changes in the numbers of individuals of a species may lead to an explosion in numbers or to extinction, and both biotic and abiotic factors can influence changes in populations. The basics of population biology and ecology apply as much to our own species as they do to others.

29.1 The Science of Ecology

Ecology encompasses two related disciplines. In basic ecology, major research questions relate to the distribution and abundance of species and how they interact with each other and the physical environment. Using these data as a baseline, workers in **applied ecology** develop conservation plans and amelioration programs to limit, repair, and mitigate ecological damage caused by human activities (see also Chapter 32). Ecology has its roots in descriptive natural history dating back to the ancient Greeks. Modern ecology was born in 1870 when the German biologist Ernst Haeckel coined the term (from *oikos* = house). Contemporary researchers still gather descriptive information about ecological relationships, often as the starting point for other studies. Although ecological research is dominated by hypothetico-deductive approaches, initial inductive approaches allow biologists to generate appropriate hypotheses about how systems function. Research in ecology is often linked to work in genetics, physiology, anatomy, behaviour, paleontology, evolution, geology, geography, and environmental science. Many ecological phenomena, such as climate change, occur over huge areas and long time spans, so ecologists must devise ways to determine how environments influence organisms and how organisms change the environments in which they live. The responses of biological systems to climate change illustrate how the ecology of an organism (or group of organisms) reflects the impact(s) of a range of extrinsic and intrinsic factors on individuals. The range of points of impact includes physiological, reproductive, and energetic factors.

Ecology can be divided into four increasingly complex and inclusive levels of organization. First, in **organismal ecology**, researchers study organisms to determine the genetic, biochemical, physiological, morphological, and behavioural adaptations to the abiotic environment (see *The Purple Pages*). Second, in **population ecology**, researchers focus on groups of individuals of the same species that live together. Population ecologists study how the size and other characteristics of populations change in space and time. Third, in **community ecology**, biologists examine populations of different species that occur together in one area (are **sympatric**). Community ecologists study interactions between species, analyzing how predation, competition, and environmental disturbances influence a community's development, organization, and structure (see Chapter 31). Fourth, those studying **ecosystem ecology** explore how nutrients cycle and energy flows between the biotic components of an **ecological community** and the abiotic environment (see Chapter 31).

Ecologists can create hypotheses about ecological relationships and how they change through time or differ from place to place. Some formalize these ideas in mathematical models that express clearly defined, but hypothetical, relationships among important variables in a system. Manipulation of a model, usually with the help of a computer, can allow researchers to ask what would happen if some of the variables or their relationships changed. Thus, researchers can simulate natural events and large-scale experiments before investing time, energy, and money in fieldwork and laboratory work. Bear in mind that mathematical models are no better than the ideas and assumptions they embody, and useful models are constructed only after basic observations have defined the relevant variables.

Ecologists use field or laboratory studies to test predictions of their hypotheses about relationships among variables in systems. In controlled experiments, researchers compare data from an experimental treatment (involving manipulation of one or more variables) with data from a control (in which nothing is changed). In some cases, *natural experiments* can be conducted because of the patterns of distribution and/or behaviour of species. This has the advantage of allowing ecologists to test predictions about how systems are operating without manipulating variables. Two species of fish, cutthroat trout (*Oncorhynchus clarki*) and Dolly Varden char (*Salvelinus malma*), live in coastal lakes of British Columbia. Some lakes have either trout or char, but others contain both species. The natural distributions of these fishes allowed researchers to measure the effect of each species on the other. In lakes in which both species live, each restricts its activities to fewer areas and eats a smaller variety of prey than it does in lakes in which it occurs alone.

29.2 Population Characteristics

CONCEPT FIX Many people believe that the sizes of populations of animals and plants will increase until net demands for food (and other resources) exceed the supply. This crisis of carrying capacity leads to crashes in populations and even to extinction. Under natural conditions, however, interactions among individuals (of the same or different species) usually cause populations to stop growing well before they reach carrying capacity. In many populations, there are natural cycles of numbers. ⬡

Seven characteristics can be described for any population.

29.2a Geographic Range Is Determined by the Boundaries of Distribution

Populations have characteristics that transcend those of the individuals making up the populations. Every population has a **geographic range**, the overall spatial boundaries within which it lives. Geographic ranges vary enormously. A population of snails might inhabit a small tide pool, whereas a population of marine phytoplankton might occupy an area that is orders of magnitude as large. Every population also occupies a **habitat**, the specific environment in which it lives, as characterized by its biotic and abiotic features. Ecologists also measure other population characteristics, such as size, distribution in space, and age structure.

29.2b Population Density Is Based on the Number of Individuals per Unit Area

Population size is the number of individuals making up the population at a specified time (N_t). **Population density** is the number of individuals per unit area or per unit volume of habitat. Species with a large body size generally have lower population densities than those with a small body size **(Figure 29.2)**. Although population size and density are related measures, knowing a population's density provides more information about its relationship to the resources it uses. If a population of 200 oak trees occupies 1 hectare (ha; 10 000 m²), the population density is 200 × 10 000 m⁻² or 1 tree per 50 m². But if 200 oaks are spread over 5 ha, the density is 1 tree per 250 m². Clearly, the second

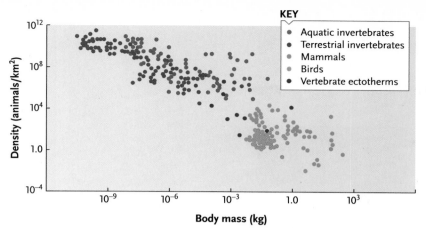

KEY
- Aquatic invertebrates
- Terrestrial invertebrates
- Mammals
- Birds
- Vertebrate ectotherms

Figure 29.2
Population density and body size. Population density generally declines with increasing body size among animal species. There are similar trends for other organisms.

population is less dense than the first, and its members will have greater access to sunlight, water, and other resources.

Ecologists measure population size and density to monitor and manage populations of endangered species, economically important species, and agricultural pests. For large-bodied species, a simple head count could provide accurate information. For example, ecologists survey the size and density of African elephant populations by flying over herds and counting individuals **(Figure 29.3)**. Researchers use a variation on that technique to estimate population size in tiny organisms that live at high population densities. To estimate the density of aquatic phytoplankton, for example, you might collect water samples of known volume from representative areas in a lake and count them by looking through a microscope. These data allow you to estimate population size and density based on the estimated volume of the entire lake. In other cases, researchers use the mark–release–recapture sampling technique (see "Capture–Recapture," p. 682). One ongoing challenge is measuring population

size in organisms that are clones, for example, stands of poplar trees (*Populus* spp.).

29.2c Population Dispersion Is the Distribution of Individuals in Space

Populations can vary in their **dispersion**, the spatial distribution of individuals within the geographic range. Ecologists define three theoretical patterns of dispersion: *clumped, uniform,* and *random* **(Figure 29.4, p. 682).**

Clumped dispersion (see Figure 29.4a) is common and occurs in three situations. First, suitable conditions are often patchily distributed. Certain pasture plants, for instance, may be clumped in small, scattered areas where cowpats had fallen for months, locally enriching the soil. Second, populations of some social animals (see Chapter 47) are clumped because mates are easy to locate within groups, and individuals may cooperate in rearing offspring, feeding, or defending themselves from predators. Third, populations can be clumped when species reproduce by asexual clones that remain attached to the parents.

Figure 29.3
Counting elephants. It is easy to think that large animals such as African elephants (*Loxodonta africana*) would be easy to count from the air **(a).** This may or may not be true, depending on vegetation. But it can be easy to overlook animals, particularly young ones **(b),** in the shade.

Capture–Recapture

Research Method Box

Ecologists use the mark–release–recapture technique to estimate the population size of mobile animals that live within a restricted geographic range. To do this, a sample of organisms (n_1) is captured, marked, and released. Ideally, the marks (or tags) are permanent and do not harm the tagged animal. Insects and reptiles are often marked with ink or paint, birds with rings (bands) on their legs, and mammals with ear tags or collars.

Later, a second sample (n_2) of the population is captured. In the second sample, the proportion of marked (n_2m) to unmarked individuals is used to estimate the total population (x) of the study area by solving the following equation for x:

$$n_1/x = n_{2m}/n_2$$

Assume that you capture a sample of 120 butterflies **(Figure 1)**, mark each one, and release them. A week later, you capture a sample of 150 butter-

M.B. Fenton

Figure 1
This butterfly has been captured and marked before release in a capture–recapture experiment.

flies, 30 that you marked. Thus, you had marked 30 of 150, or 1 of every 5 butterflies, on your first field trip. Because you captured 120 individuals on that first excursion, you would estimate that the total population size is 120 × (150/30) = 600 butterflies.

The capture–recapture technique is based on several assumptions that are critical to its accuracy: (1) being marked has no effect on survival, (2) marked and unmarked animals mix randomly in the population, (3) no migration into or out of the population takes place during the estimating period, and (4) marked individuals are just as likely to be captured as unmarked individuals. (Sometimes animals become "trap shy" or "trap happy," a violation of the fourth assumption.)

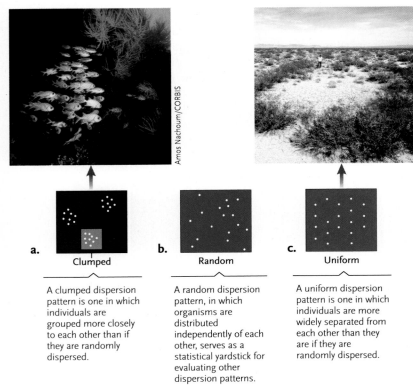

Amos Nachoum/CORBIS

Dan Suzio/Science Source

a.
Clumped

A clumped dispersion pattern is one in which individuals are grouped more closely to each other than if they are randomly dispersed.

b.
Random

A random dispersion pattern, in which organisms are distributed independently of each other, serves as a statistical yardstick for evaluating other dispersion patterns.

c.
Uniform

A uniform dispersion pattern is one in which individuals are more widely separated from each other than they are if they are randomly dispersed.

Figure 29.4
Dispersion patterns. A clumped pattern **(a)** is evident in fish that live in social groups. A random pattern **(b)** of dispersion appears to be rare in nature, where it occurs in organisms that are neither attracted to nor repelled by conspecifics. Nearly uniform patterns **(c)** are demonstrated by creosote bushes (*Larrea tridentata*) near Death Valley, California.

Aspen trees and sea anemones reproduce this way and often occur in large aggregations (see Chapter 19). Clumping may also occur in species in which seeds, eggs, or larvae lack dispersal mechanisms and offspring grow and settle near their parents.

Uniform distributions can occur when individuals repel one another because resources are in short supply. Creosote bushes are uniformly distributed in the dry scrub deserts of the U.S. Southwest (see Figure 29.4c). Mature bushes deplete the surrounding soil of water and secrete toxic chemicals, making it impossible for seedlings to grow. This chemical warfare is called *allelopathy*. Moreover, seed-eating ants and rodents living at the bases of mature bushes eat any seeds that fall nearby. In these situations, the distributions of species of plants and animals can be uniform and interrelated. Territorial behaviour, the defence of an area and its resources, can also produce **uniform dispersion** in some species of animals, such as nests in colonies of colonial birds (see Chapter 47).

Random dispersion (see Figure 29.4b) occurs when environmental conditions do not vary much within a habitat, and individuals are neither attracted to nor repelled by others of their species (conspecifics). Ecologists use formal statistical definitions of *random* to establish a theoretical baseline for assessing the pattern of distribution. In cases of random dispersion, individuals are distributed unpredictably. Some spiders, burrowing clams, and rainforest trees exhibit random dispersion.

Black-Footed Ferret, *Mustela nigripes*

Black-footed ferrets **(Figure 1)** are crepuscular and nocturnal hunters of the prairie. Weighing 0.6 to 1.1 kg, these weasel relatives (family Mustelidae, order Carnivora) were once abundant in western North America, from Texas in the United States to Saskatchewan and Alberta in Canada. Like other mustelids, males are larger than females. In the wild, these predators probably fed mainly on prairie dogs (*Cynomys* species) and lived around prairie dog towns. Litters range in size from one to five. Females bear a single litter a year, and males and females are sexually mature at age one year.

By 1987, *M. nigripes* was probably extinct in the wild. The last known wild population was discovered near Meeteetse, Wyoming, in 1981. Seven animals from this population were captured and brought into captivity and served as the genetic founders for a captive breeding program. Over 4800 juvenile black-footed ferrets were produced by this program, and wildlife officials began to release captive-bred animals into suitable habitats.

At Shirley Basin, Wyoming, 228 captive-born black-footed ferrets were received between 1991 and 1994. By 1996, only 25 were observed in the wild and by 1997, only 5. This decline reflected the impact of diseases, specifically canine distemper and plague. In 1996, it seemed that the reintroductions would fail, and *M. nigripes* would again be extinct in the wild.

In 2003, however, 52 black-footed ferrets were observed in the field at Shirley Basin, and, since then, the population has increased significantly **(Figure 2)**. The increase reflects an intrinsic rate of increase (*r*) of 0.47, which reflects success in the first year of life and derives from a combination of survival and fertility.

There appears to be hope for the future of *M. nigripes*. It remains to be determined if the genetic bottleneck (see Chapter 18) that the species has endured will prove to be an important handicap to long-term survival.

Figure 1

Mustela nigripes, *the black-footed ferret. This critically endangered carnivore from the North American prairie shows evidence of a comeback.*

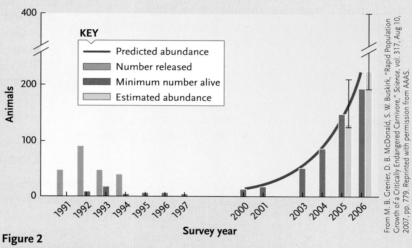

Figure 2

Population growth. Black-footed ferrets in Shirley Basin, Wyoming, have shown rapid population growth. The 95% confidence limits suggest a population of 192 to 401 in 2006.

From M. B. Grenier, D. B. McDonald, S. W. Buskirk, "Rapid Population Growth of a Critically Endangered Carnivore," *Science*, vol. 317, Aug 10, 2007, pp. 779. Reprinted with permission from AAAS.

Whether the spatial distribution of a population appears to be clumped, uniform, or random depends partly on the size of the organisms and of the study area. Oak seedlings may be randomly dispersed on a spatial scale of a few square metres, but over an entire mixed hardwood forest, they are clumped under the parent trees. Therefore, dispersion of a population depends partly on the researcher's scale of observation.

In addition, the dispersion of animal populations often varies through time in response to natural environmental rhythms. Few habitats provide a constant supply of resources throughout the year, and many animals move from one habitat to another on a seasonal cycle, reflecting the distribution of resources such as food. Tropical birds and mammals are often widely dispersed in deciduous forests during the wet season, when food is widely available. During the dry season, these species crowd into narrow *gallery forests* along watercourses, where evergreen trees provide food and shelter.

29.2d Age Structure Is the Numbers of Individuals of Different Ages

All populations have an **age structure**, a statistical description of the relative numbers of individuals in each age class (discussed further in Chapter 30). Individuals can be categorized roughly as prereproductive (younger than the age of sexual maturity), reproductive, or postreproductive (older than the maximum age of reproduction). A population's age structure reflects its recent growth history and predicts its future growth potential. Populations composed of many prereproductive individuals obviously grew rapidly in the recent past. These populations will continue to grow as young individuals mature and reproduce.

29.2e Generation Time Is the Average Time between Birth and Death

Another characteristic that influences a population's growth is its **generation time**, the average time between the birth of an organism and the birth of its offspring. Generation time is usually short in species that reach sexual maturity at a small body size **(Figure 29.5).** Their populations often grow rapidly because of the speedy accumulation of reproductive individuals.

29.2f Sex Ratio: Females : Males

Populations of sexually reproducing organisms also vary in their **sex ratio**, the relative proportions of males and females. In general, the number of females in a population has a bigger impact on population growth than the number of males because only females actually produce offspring. Moreover, in many species, one male can mate with several females, and the number of males may have little effect on the population's reproductive output. In northern elephant seals (see Chapter 18), mature bulls fight for dominance on the beaches where the seals mate. Only a few males may ultimately inseminate a hundred or more females. Thus, the presence of other males in the group may have little effect on the size of future generations. In animals that form lifelong pair bonds, such as geese and swans, the numbers of males and females influence reproduction in the population.

29.2g The Proportion Reproducing Is the Incidence of Reproducing Individuals in a Population

Population ecologists try to determine the proportion of individuals in a population that are reproducing. This issue is particularly relevant to the conservation of any species in which individuals are rare or widely dispersed in the habitat (see Chapter 32).

STUDY BREAK

1. What is the difference between geographic range and habitat?
2. What are the three types of dispersion? What is the most common pattern found in nature? Why?
3. What is the common pattern of generation time among bacteria, protists, plants, and animals?

29.3 Demography

Populations grow larger through the birth of individuals and the **immigration** (movement into the population) of organisms from neighbouring populations. Conversely, death and **emigration** (movement out of the population) reduce population size. **Demography** is the statistical study of the processes that change a population's size and density through time.

Ecologists use demographic analysis to predict a population's growth. For human populations, these data help governments anticipate the need for social services such as schools, hospitals, and chronic care facilities. Demographic data allow conservation ecologists to develop plans to protect endangered species. Demographic data on northern spotted owls (*Strix occidentalis caurina*) helped convince the courts to restrict logging in the owl's primary habitat, the old-growth forests of the Pacific Northwest. Life tables and survivorship curves are among the tools ecologists use to analyze demographic data.

29.3a Life Tables Show the Number of Individuals in Each Age Group

Although every species has a characteristic life span, few individuals survive to the maximum age possible. Mortality results from starvation, disease, accidents,

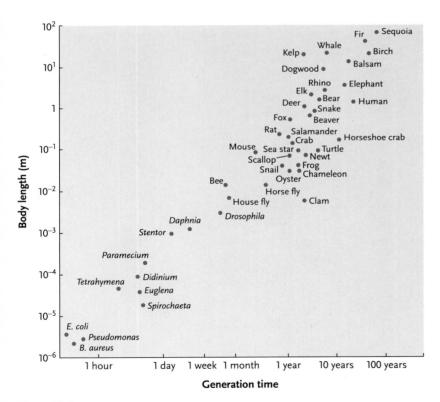

Figure 29.5
Generation time and body size. Generation time increases with body size among bacteria, protists, plants, and animals. The logarithmic scale on both axes compresses the data into a straight line.

predation, or inability to find a suitable habitat. Life insurance companies first developed techniques for measuring mortality rates (known as actuarial science), and ecologists adapted these approaches to the study of nonhuman populations.

A **life table** summarizes the demographic characteristics of a population **(Table 29.1)**. To collect life table data for short-lived organisms, demographers typically mark a **cohort**, a group of individuals of similar age, at birth and monitor their survival until all members of the cohort die. For organisms that live more than a few years, a researcher might sample the population for one or two years, recording the ages at which individuals die and then extrapolating these results over the species' life span. The approach to the timing of collection of data about reproduction and longevity will depend on the details of the species under study.

In any life table, life spans of organisms are divided into age intervals of appropriate length. For short-lived species, days, weeks, or months are useful, whereas for longer-lived species, years or groups of years will be better. Mortality can be expressed in two complementary ways. **Age-specific mortality** is the proportion of individuals alive at the start of an age interval that died during that age interval. Its more cheerful reflection, **age-specific survivorship**, is the proportion of individuals alive at the start of an age interval that survived until the start of the next age interval. Thus, for the data shown in Table 29.1, the age-specific mortality rate during the 3- to 6-month age interval is $195/722 = 0.270$, and the age-specific survivorship rate is $527/722 = 0.730$. For any age interval, the sum of age-specific mortality and age-specific survivorship must equal 1. Life tables also summarize the proportion of the cohort that survived to a particular age, a statistic identifying the probability that any randomly selected newborn will still be alive at that age. For the 3- to 6-month age interval in Table 29.1, this probability is $722/843 = 0.856$.

Life tables also include data on **age-specific fecundity**, the average number of offspring produced by surviving females during each age interval. Table 29.1 shows that plants in the 3- to 6-month age interval produced an average of 300 seeds each. In some species, including humans, fecundity is highest in individuals of intermediate age. Younger individuals have not yet reached sexual maturity, and older individuals are past their reproductive prime. However, fecundity increases steadily with age in some plants and animals.

29.3b Survivorship Curves Graph the Timing of Deaths of Individuals in a Population

Survivorship data are depicted graphically in a **survivorship curve**, which displays the rate of survival for individuals over the species' average life span. Ecologists have identified three generalized survivorship curves (blue lines in **Figure 29.6, p. 686**), although most organisms exhibit survivorship patterns falling between these idealized patterns.

Type I curves reflect high survivorship until late in life (see Figure 29.6a, p. 686). They are typical of large animals that produce few young and provide them with extended care, which reduces juvenile mortality. Large mammals, such as Dall mountain sheep, produce only one or two offspring at a time and nurture them through their vulnerable first year. At that time, the young are better able to fend for themselves and are at

Table 29.1	Life Table for a Cohort of 843 Individuals of the Grass *Poa annua* (Annual Bluegrass)					
Age Interval (in months)	Number Alive at Start of Age Interval	Number Dying during Age Interval	Age-Specific Mortality Rate	Age-Specific Survivorship Rate	Proportion of Original Cohort Alive at Start of Age Interval	Age-Specific Fecundity (Seed Production)
0–3	843	121	0.144	0.856	1.000	0
3–6	722	195	0.270	0.730	0.856	300
6–9	527	211	0.400	0.600	0.625	620
9–12	316	172	0.544	0.456	0.375	430
12–15	144	90	0.625	0.375	0.171	210
15–18	54	39	0.722	0.278	0.064	60
18–21	15	12	0.800	0.200	0.018	30
21–24	3	3	1.000	0.000	0.004	10
24–	0	—	—	—	—	—

Source: *Population Ecology*, Begon, M., and M. Mortimer. Copyright © 1981 John Wiley and Sons. Reproduced with permission of Blackwell Publishing Ltd.

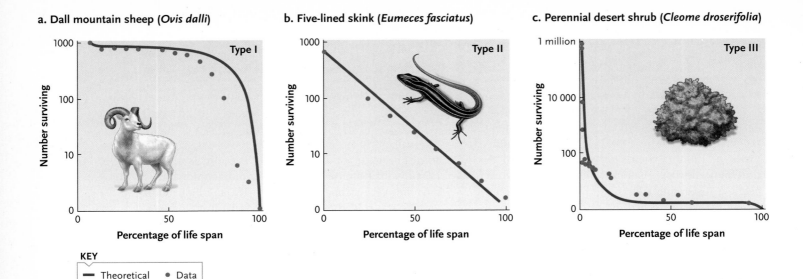

a. Dall mountain sheep (*Ovis dalli*)

Type I

Number surviving

1000

100

10

0

0   50   100

Percentage of life span

b. Five-lined skink (*Eumeces fasciatus*)

Type II

Number surviving

1000

100

10

0

0   50   100

Percentage of life span

c. Perennial desert shrub (*Cleome droserifolia*)

Type III

Number surviving

1 million

10 000

100

0

0   50   100

Percentage of life span

KEY

— Theoretical   • Data

Figure 29.6

Survivorship curves. The survivorship curves of many organisms (pink) roughly match one of three idealized patterns (blue).

lower risk for mortality (compared with younger animals). The picture of survivorship in mammals could change if one starts with the time of conception, as opposed to birth. The change would reflect problems of pregnancy (see Chapter 41) and health of mothers.

Type II curves reflect a relatively constant rate of mortality in all age classes, a pattern that produces steadily declining survivorship (see Figure 29.6b). Many lizards, such as the five-lined skink, as well as songbirds and small mammals, face a constant probability of mortality from predation, disease, and starvation and show a type II pattern.

Type III curves reflect high juvenile mortality, followed by a period of low mortality once offspring reach a critical age and size (see Figure 29.6c, in which the vertical scale is logarithmic). *Cleome droserifolia*, a desert shrub from the Middle East, experiences extraordinarily high mortality in its seed and seedling stages. Researchers estimate that for every 1 million seeds produced, fewer than 1000 germinate, and only about 40 individuals survive their first year. Once a plant becomes established, however, its likelihood of future survival is higher, and the survivorship curve flattens out. Many plants, insects, marine invertebrates, and fishes exhibit type III survivorship.

STUDY BREAK

1. What is the relationship between age-specific mortality and age-specific survivorship? If the age-specific mortality is 0.384, what is the age-specific survivorship?
2. What is age-specific fecundity?
3. Describe three survivorship curves. Which curve describes humans? Songbirds? Insects?

29.4 The Evolution of Life Histories

Analysis of life tables reveals how natural selection affects an organism's **life history**, which includes the lifetime patterns of growth, maturation, and reproduction. Ecologists study life histories to understand trade-offs in the allocation of resources to these three activities. The results of their research suggest that natural selection adjusts the allocation of resources to maximize an individual's number of surviving offspring.

Every organism is constrained by a finite **energy budget**, the total amount of energy it can accumulate and use to fuel its activities. An organism's energy budget is like a savings account. When the individual accumulates more energy than it needs, it makes deposits to this account, storing energy as starch, glycogen, or fat. When the individual expends more energy than it harvests, it makes withdrawals from its energy stores. But unlike a bank account, an organism's energy budget cannot be overdrawn, and no loans against future "earnings" are possible.

Just as humans find clever ways to finance their schemes, many organisms use different ways to mortgage their operations. Organisms that enter states of inactivity or dormancy can maximize the time over which they use stored energy. An extreme example is animals and plants that can survive freezing, an obvious strategy for conserving energy. Hibernation and estivation in animals are other examples (see Chapter 50). Hibernating animals use periods of reduced body temperature to survive prolonged periods of cold weather. Estivation is inactivity during prolonged periods of high temperatures. Specialized spores can be resistant to heat and desiccation. Migrating birds on long flights get energy by metabolizing fat as well as other body structures, such as muscle or digestive tissue. Organisms use the energy they harvest for three broadly defined functions: maintenance (the preservation of good physiological condition), growth, and reproduction. When an organism devotes energy to any one of these functions, the balance in its energy budget is reduced, leaving less energy for other functions.

A fish, a deciduous tree, and a mammal illustrate the dramatic variations existing in life history patterns. Larval coho salmon (*Oncorhynchus kisutch*) hatch in the headwaters of a stream, where they feed and grow for about a year before assuming their adult body form and swimming to the ocean. They remain at sea for a year or two, feeding voraciously and growing rapidly. Eventually, using a Sun compass and geomagnetic and chemical cues, salmon return to the rivers and streams where they hatched. The fishes swim upstream. Males prepare nests and try to attract females. Each female lays hundreds or thousands of relatively small eggs. After breeding, the body condition of males and females deteriorates, and they die.

Most deciduous trees in the temperate zone, such as oaks (genus *Quercus*), begin their lives as seeds (acorns) in late summer. The seeds remain metabolically inactive until the following spring or a later year. After germinating, seedling trees collect nutrients and energy and continue to grow throughout their lives. Once they achieve a critical size, they may produce thousands of acorns annually for many years. Thus, growth and reproduction occur simultaneously through much of the trees' life.

European red deer (*Cervus elaphus*) are born in spring, and the young remain with their mothers for an extended period, nursing and growing rapidly. After weaning, the young feed on their own. Female red deer begin to breed after reaching adult size in their third year, producing one or two offspring annually until they are about 16 years old, when they reach their maximum life span and die.

How can we summarize the similarities and differences in the life histories of these organisms? All three species harvest energy throughout their lives. Salmon and deciduous trees continue to grow until old age, whereas deer reach adult size fairly early in life. Salmon produce many offspring in a single reproductive episode, whereas deciduous trees and deer reproduce repeatedly. However, most trees produce thousands of seeds annually, whereas deer produce only one or two young each spring.

What factors have produced these variations in life history patterns? Life history traits, like all population characteristics, are modified by natural selection. Thus, organisms exhibit evolutionary adaptations that increase the fitness of individuals. Each species' life history is, in fact, a highly integrated "strategy" or suite of selection-driven adaptations.

In analyzing life histories, ecologists compare the number of offspring with the amount of care provided to each by the parents. They also determine the number of reproductive episodes in the organism's lifetime and the timing of first reproduction. Because these characteristics evolve together, a change in one trait is likely to influence others.

29.4a Fecundity versus Parental Care: Cutting Your Losses

If a female has a fixed amount of energy for reproduction, she can package that energy in various ways. A female duck with 1000 units of energy for reproduction might lay 10 eggs with 100 units of energy per egg. A salmon, which has higher fecundity, might lay 1000 eggs with 1 unit of energy in each. The amount of energy invested in each offspring before it is born is **passive parental care** provided by the female. Passive parental care is provided through yolk in an egg; endosperm in a seed; or, in mammals, nutrients that cross the placenta.

Many animals also provide **active parental care** to offspring after their birth. In general, species producing many offspring in a reproductive episode (e.g., the coho salmon) provide relatively little active parental care to each offspring. In fact, female coho salmon, each producing 2400 to 4500 eggs, die before their eggs even hatch. Conversely, species producing few offspring at a time (e.g., European red deer) provide much more care to each one. A red deer doe nurses its single fawn for up to eight months before weaning it.

29.4b How Often to Breed: Once or Repeatedly?

The number of reproductive episodes in an organism's life span is a second life history characteristic adjusted by natural selection. Some organisms, such as coho salmon, devote all of their stored energy to a single reproductive event. Any adult that survives the upstream migration is likely to leave some surviving offspring. Other species, such as deciduous trees and red deer, reproduce more than once. In contrast to salmon, individuals of these species devote only some of their energy budget to reproduction at any time, with the balance allocated to maintenance and growth. Moreover, in some plants, invertebrates, fishes, and reptiles, larger individuals produce more offspring than smaller ones. Thus, one advantage of using only part of the energy budget for reproduction is that continued growth may result in greater fecundity at a later age. However, if an organism does not survive until the next breeding season, the potential advantage of putting energy into maintenance and growth is lost.

29.4c Age at First Reproduction: When to Start Reproducing

Individuals that first reproduce at the earliest possible age may stand a good chance of leaving some surviving offspring. But the energy they use in reproduction is not available for maintenance and growth. Thus, early reproducers may be smaller and less healthy than individuals that delay reproduction in favour of other functions. Conversely, an individual that delays reproduction may increase its chance of survival and its future fecundity by becoming larger or more experienced. But there is always some chance that it will die before the next breeding season, leaving no offspring at all. Therefore, a finite energy budget and the risk of mortality establish a tradeoff in the timing of first reproduction. Mathematical models suggest that delayed reproduction will be favoured by natural selection if a sexually mature individual has a good chance of surviving to an older age, if organisms grow larger as they age, and

if larger organisms have higher fecundity. Early reproduction will be favoured if adult survival rates are low, if animals do not grow larger as they age, or if larger size does not increase fecundity. These characteristics apply more readily to some animals and plants than they do to others. Among animals, the features discussed above apply more readily to vertebrate than to invertebrate animals. Parasitic organisms may have quite different patterns of life history.

Life history characteristics vary from one species to another, and they can vary among populations of a single species. Predation differentially influences life history characteristics in natural populations of guppies (*Poecilia reticulata*) in Trinidad (see "Life Histories of Guppies").

Life Histories of Guppies

Some years ago, drenched with sweat and with fishnets in hand, two ecologists were engaged in fieldwork on the Caribbean island of Trinidad. They were after guppies (*Poecilia reticulata*), small fish most of us see in pet shops. In their native habitats, guppies bear live young in shallow mountain streams **(Figure 1),** and John Endler and David Reznick were studying the environmental variables influencing the evolution of their life history patterns.

Male guppies are easy to distinguish from females. Males stop growing at sexual maturity. They are smaller, and their scales have bright colours that serve as visual signals in intricate courtship displays. Females are drably coloured and continue to grow larger throughout their lives. In

the mountains of Trinidad, guppies live in different streams, even in different parts of the same stream. Two other species of fish eat guppies **(Figure 2)**. In some streams, a small killifish (*Rivulus hartii*) preys on immature guppies but does not have much success with the larger adults. In other streams, a large pike–cichlid (*Crenicichla alta*) prefers mature guppies and rarely hunts small, immature ones.

Reznick and Endler found that the life history patterns of guppies vary among streams with different predators. In streams with pike–cichlids, male and female guppies mature faster and begin to reproduce at a smaller size and younger age than their counterparts in streams where killifish live. Female guppies from pike–cichlid streams reproduce more often, producing smaller and more numerous young. These differences allow guppies to avoid some predation. Those in pike–cichlid streams

begin to reproduce when they are smaller than the size preferred by that predator. Those from killifish streams grow quickly to a size that is too large to be consumed by killifish **(Figure 3)**.

Although these life history differences were correlated with the

Male guppy (right) that shared a stream with pike–cichlids (below)

Male guppy (right) that shared a stream with killifish (below)

David Reznick/University of California, Riverside; Mark Smith/Science Source

David Reznick/University of California, Riverside; DEA/C. DANI/Contributor/ Getty Images

Figure 2
Male guppies from streams where pike–cichlids live (top) are smaller and more streamlined and have duller colours than those from streams where killifish live (bottom). The pike–cichlid prefers to eat large guppies, and the killifish feeds on small guppies. Guppies are shown approximately life sized; adult pike–cichlids grow to 16 cm in length, and adult killifish grow to 10 cm.

David Reznick/University of California, Riverside

Figure 1
David Reznick surveys a shallow stream in the mountains of Trinidad.

distributions of the two predatory fishes, they might result from some other, unknown differences between the streams. Endler and Reznick investigated this possibility with controlled laboratory experiments. They shipped groups of live guppies to California, where they bred guppies from each kind of stream for two generations. Both types of experimental populations were raised under identical conditions in the absence of predators. Even in the absence of predators, the two types of experimental populations retained their life history differences. These results provided evidence of a genetic (heritable) basis for the observed life history differences.

Endler and Reznick also examined the role of predators in the *evolution* of the size differences **(Figure 4)**. They raised guppies for many generations in the laboratory under three experimental conditions: some alone, some with killifish, and some with pike–cichlids. As predicted, the guppy lineage subjected to predation by killifish became larger at maturity. Individuals that were small at maturity were frequently eaten, and their reproduction was limited. The lineage raised with pike–cichlids showed a trend toward earlier maturity. Individuals that matured at a larger size faced a greater likelihood of being eaten before they had reproduced.

When they first visited Trinidad, Endler and Reznick had introduced guppies from a pike–cichlid stream to another stream that contained killifish but no pike–cichlids or guppies. There, 11 years later, guppy populations had changed. As the researchers predicted, the guppies became larger and reproduced more slowly, characteristics typical of natural guppy populations that live and die with killifish.

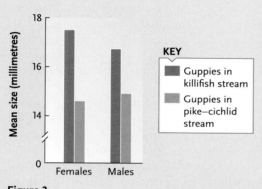

Figure 3

Guppies in streams occupied by pike–cichlids are smaller than those in streams occupied by killifish.

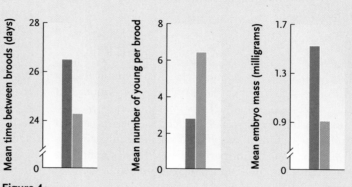

Figure 4

Female guppies from streams occupied by pike–cichlids reproduce more often (shorter time between broods) and produce more young per brood and smaller young (lower embryo mass) than females living in streams occupied by killifish.

STUDY BREAK

1. Organisms use energy for what three main operations?
2. Explain passive and active parental care in humans.
3. When would early reproduction be favoured?

29.5 Models of Population Growth

We now examine two mathematical models of population growth, exponential and logistic. **Exponential** models apply when populations experience unlimited growth. **Logistic** models apply when population growth is limited, often because available resources are finite. These simple models are tools that help ecologists refine their hypotheses, but neither provides entirely accurate predictions of population growth in nature. In the simplest versions of these models, ecologists define births as the production of offspring by any form of reproduction and ignore the effects of immigration and emigration.

29.5a Exponential Models: Populations Taking Off

Populations sometimes increase in size for a period of time with no apparent limits on their growth. In models of exponential growth, population size increases steadily by a constant ratio. Populations of bacteria and prokaryotes provide the most obvious examples, but multicellular organisms also sometimes exhibit exponential population growth.

Bacteria reproduce by binary fission. A parent cell divides in half, producing two daughter cells, and each can divide to produce two granddaughter cells. Generation time in a bacterial population is simply the time between successive cell divisions. If no bacteria in the population die, the population doubles in size each generation.

Bacterial populations grow quickly under ideal temperatures and with unlimited space and food. Consider a population of the human intestinal bacterium *Escherichia coli,* for which the generation time can be as short as 20 minutes. If we start with a population of one bacterium, the population doubles to two cells after one generation, to four cells after two generations, and to eight cells after three generations **(Figure 29.7)**. After only 8 hours (24 generations), the population will number almost 17 million. And after a single day (72 generations), the population will number nearly 5×10^{21} cells. Although other bacteria grow more slowly than *E. coli,* it is no wonder that pathogenic bacteria, such as those causing cholera or plague, can quickly overtake the defences of an infected animal.

When populations of multicellular organisms are large, they can grow exponentially, as we shall see below for our own species. In any event, over a given time period,

change in population size = number of births − number of deaths.

We express this relationship mathematically by defining N as the population size; ΔN (pronounced "delta N") as the change in population size; Δt as the time period during which the change occurs; and B and D as the numbers of births and deaths, respectively, during that time period. Thus, $\Delta N/\Delta t$ symbolizes the change in population size over time, and

$$\Delta N/\Delta t = B - D.$$

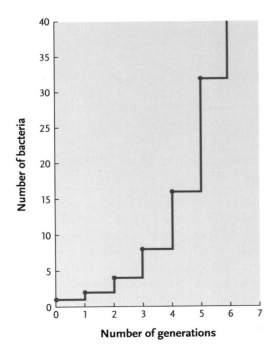

Figure 29.7

Bacterial population growth. If all members of a bacterial population divide simultaneously, a plot of population size over time forms a stair-stepped curve in which the steps get larger as the number of dividing cells increases.

The above equation applies to any population for which we know the exact numbers of births and deaths. Ecologists usually express births and deaths as per capita (per individual) rates, allowing them to apply the model to a population of any size. The per capita birth rate (b) is the number of births in the population during the specified time period divided by the population size: $b = (B/N)$. Similarly, the per capita death rate, d, is the number of deaths divided by the population size: $d = (D/N)$.

If in a population of 2000 field mice, 1000 mice are born and 200 mice die during 1 month, $b = 1000/2000 = 0.5$ births per individual per month, and $d = 200/2000 = 0.1$ deaths per individual per month. Of course, no mouse can give birth to half an offspring, and no individual can die one-tenth of a death. But these rates tell us the per capita birth and death rates *averaged over all mice in the population.* Per capita birth and death rates are always expressed over a specified time period. For long-lived organisms, such as humans, time is measured in years. For short-lived organisms, such as fruit flies, time is measured in days. We can calculate per capita birth and death rates from data in a life table.

Now we can revise the population growth equation to use per capita birth and death rates instead of the actual numbers of births and deaths. The change in a population's size during a given time period ($\Delta N/\Delta t$) depends on the per capita birth and death rates, as well as on the number of individuals in the population. Mathematically, we can write

$$\Delta N/\Delta t = B - D = bN - dN = (b - d)N$$

or, in the notation of calculus,

$$dN/dt = (b - d)N.$$

This equation describes the **exponential model of population growth.** (Note that in calculus, dN/dt is the notation for the population growth rate. The d in dN/dt is *not* the same d we use to symbolize the per capita death rate.)

The difference between the per capita birth rate and the per capita death rate, $b - d$, is the **per capita growth rate** of the population, symbolized by r. Like b and d, r is always expressed per individual per unit time. Using the per capita growth rate, r, in place of $b - d$, the exponential growth equation is written

$$dN/dt = rN.$$

If the birth rate exceeds the death rate, r has a positive value ($r > 0$), and the population is growing. In our example with field mice, r is $0.5 - 0.1 = 0.4$ mice per mouse per month. If, on the other hand, the birth rate is lower than the death rate, r has a negative value ($r < 0$), and the population is shrinking. In populations in which the birth rate equals the death rate, r is zero, and the population's size is not changing-a situation known as **zero population growth,** or ZPG. Even under ZPG, births and deaths still occur, but the numbers of births and deaths cancel each other out.

Populations will grow as long as the per capita growth rate is positive ($r > 0$). In our hypothetical population of field mice, we started with $N = 2000$ mice and calculated a per capita growth rate of 0.4 mice per individual per month. In the first month, the population grows by $0.4 \times 2000 = 800$ mice **(Figure 29.8).** At the start of the second month, $N = 2800$ and r is still 0.4. Thus, in the second month, the population grows by $0.4 \times 2800 = 1120$ mice. Notice that even though r remains constant, the *increase* in population size grows each month because more individuals are reproducing. In less than two years, the mouse population will increase to more than one million! A graph of exponential population growth has a characteristic J shape, getting steeper through time. The population grows at an ever-increasing pace because the change in a population's size depends on the number of individuals in the population and its per capita growth rate.

Imagine a hypothetical population living in an ideal environment with unlimited food and shelter; no predators, parasites, or disease; and a comfortable abiotic environment. Under such circumstances (admittedly unrealistic), the per capita birth rate is very high; the per capita death rate is very low; and the per capita growth rate, r, is as high as it can be. This maximum per capita growth rate, symbolized r_{max}, is the population's **intrinsic rate of increase.** Under these ideal conditions, our exponential growth equation is

$$dN/dt = r_{max}N.$$

When populations grow at their intrinsic rate of increase, population size increases very rapidly. Across a wide variety of protists and animals, r_{max} varies inversely with generation time: species with a short generation time have higher intrinsic rates of increase than those with a long generation time **(Figure 29.9, p. 692).**

The exponential model predicts unlimited population growth. But we know from even casual observations that population sizes of most species are somehow limited. We are not knee-deep in bacteria, rosebushes, or garter snakes. What factors limit the growth of populations? As a population gets larger, it uses more vital resources, perhaps leading to a shortage of resources. In this situation, individuals may have less energy available for maintenance and reproduction, causing decreases in per capita birth rates and increases in per capita death rates. Energy in food is not always equally available, and when an animal spends time handling food to eat it, the ratio of cost (handling) to benefit (energy in the food) diminishes, affecting return on investment. Such rate changes can affect a population's per capita growth rate, causing population growth to slow or stop.

29.5b Logistic Models: Populations and Carrying Capacity (K)

Environments provide enough resources to sustain only a finite population of any species. The maximum number of individuals that an environment can support indefinitely is termed its **carrying capacity,** symbolized as K. K is defined for each population. It is a property of the environment that can vary from one habitat to another and in a single habitat over time. The spring and summer flush of insects in temperate habitats supports large populations of insectivorous birds. But fewer insects are available in autumn and winter, causing a seasonal decline in K for birds, so autumnal migrations occur in birds seeking more food and less inclement weather. Other cycles are annual, such as variation in water levels in wetlands from year to year.

Month	Old Population Size		Net Monthly Increase		New Population Size
1	2 000	+	800	=	2 800
2	2 800	+	1 120	=	3 920
3	3 920	+	1 568	=	5 488
4	5 488	+	2 195	=	7 683
5	7 683	+	3 073	=	10 756
6	10 756	+	4 302	=	15 058
7	15 058	+	6 023	=	21 081
8	21 081	+	8 432	=	29 513
9	29 513	+	11 805	=	41 318
10	41 318	+	16 527	=	57 845
11	57 845	+	23 138	=	80 983
12	80 983	+	32 393	=	113 376
13	113 376	+	45 350	=	158 726
14	158 726	+	63 490	=	222 216
15	222 216	+	88 887	=	311 102
16	311 102	+	124 441	=	435 543
17	435 543	+	174 217	=	609 760
18	609 760	+	243 904	=	853 664
19	853 674	+	341 466	=	1 195 1340

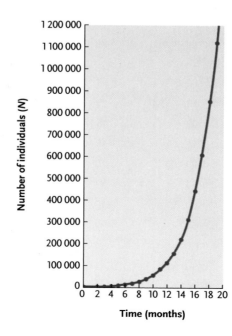

Figure 29.8
Exponential population growth. Exponential population growth produces a J-shaped curve when population size is plotted against time. Although the per capita growth rate (r) remains constant, the increase in population size gets larger every month because more individuals are reproducing.

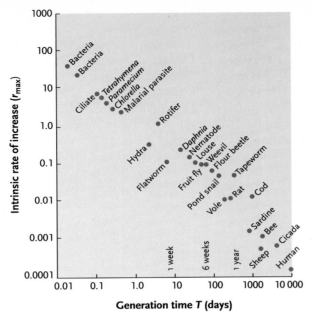

Figure 29.9
Generation time and r_{max}. The intrinsic rate of increase (r_{max}) is high for protists and animals with short generation times and low for those with long generation times.

a. The predicted effect of N on r

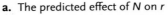
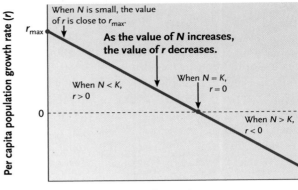

b. Population size through time

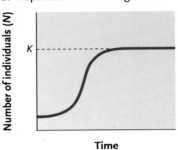

Figure 29.10
The logistic model of population growth. The logistic model **(a)** assumes that the per capita population growth rate (r) decreases linearly as population size (N) increases. The logistic model also predicts that population size **(b)** increases quickly at first but then slowly approaches carrying capacity (K).

The **logistic model of population growth** assumes that a population's per capita growth rate, r, decreases as the population gets larger **(Figure 29.10)**. In other words, population growth slows as the population size approaches K. The mathematical expression $K - N$ tells us how many individuals can be added to a population before it reaches K. The expression $(K - N)/K$ indicates what percentage of the carrying capacity is still available.

To create the logistic model, we factor the impact of K into the exponential model by multiplying r_{max} by $(K - N)/K$ to reduce the per capita growth rate (r) from its maximum value (r_{max}) as N increases:

$$dN/dt = r_{max}N(K - N)/K.$$

The calculation of how r varies with population size is straightforward **(Table 29.2)**. In a very small population (N much smaller than K), plenty of resources are available; the value of $(K - N)/K$ is close to 1. Here the per capita growth rate (r) approaches the maximum possible (r_{max}). Under these conditions, population growth is close to exponential. If a population is large (N close to K), few additional resources are available. Now the value of $(K - N)/K$ is small, and the per capita growth rate (r) is very low. When the size of the population exactly equals K, $(K - N)/K$ becomes 0, as does the population growth rate, the situation defined as ZPG.

The logistic model of population growth predicts an S-shaped graph of population size over time, with the population slowly approaching K and remaining at that level **(Figure 29.11)**. According to this model, the population grows slowly when the population size is small because few individuals are reproducing. It also

Table 29.2	The Effect of *N* on *r* and Δ*N** in a Hypothetical Population Exhibiting Logistic Growth in which *K* equals 2000 and r_{max} is 0.04 per capita per year		
N (population size)	$(K - N)/K$ (% of *K* available)	$r = r_{max}(K - N/K)$ (per capita growth rate)	$\Delta N = rN$ (change in *N*)
50	0.975	0.0390	2
100	0.950	0.0380	4
250	0.875	0.0350	9
500	0.750	0.0300	15
750	0.625	0.0250	19
1000	0.500	0.0200	20
1250	0.375	0.0150	19
1500	0.250	0.0100	15
1750	0.125	0.0050	9
1900	0.050	0.0020	4
1950	0.025	0.0010	2
2000	0.000	0.0000	0

*Δ*N* rounded to the nearest whole number.

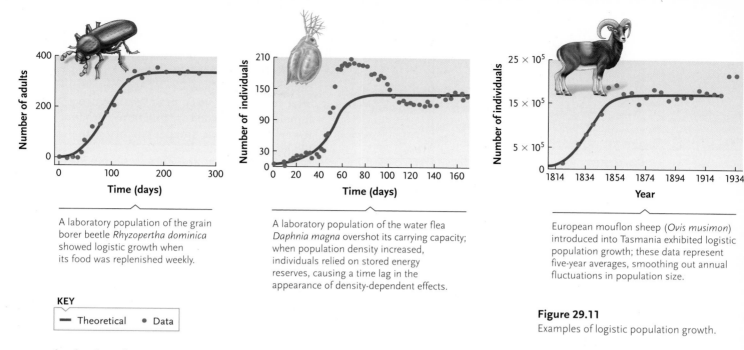

A laboratory population of the grain borer beetle *Rhyzopertha dominica* showed logistic growth when its food was replenished weekly.

A laboratory population of the water flea *Daphnia magna* overshot its carrying capacity; when population density increased, individuals relied on stored energy reserves, causing a time lag in the appearance of density-dependent effects.

European mouflon sheep (*Ovis musimon*) introduced into Tasmania exhibited logistic population growth; these data represent five-year averages, smoothing out annual fluctuations in population size.

KEY

— Theoretical   • Data

Figure 29.11
Examples of logistic population growth.

grows slowly when the population size is large because the per capita population growth rate is low. The population grows quickly (dN/dt is highest) at intermediate population sizes, when a sizable number of individuals are breeding and the per capita population growth rate (r) is still fairly high (see Table 29.2).

The logistic model assumes that vital resources become increasingly limited as a population grows. Thus, the model is a mathematical portrait of **intraspecific** (within species) **competition**, the dependence of two or more individuals in a population on the same limiting resource. For mobile animals, limiting resources could be food, water, nesting sites, and refuges from predators. For sessile species, space can be a limiting resource. For plants, sunlight, water, inorganic nutrients, and growing space can be limiting. The pattern of uniform dispersion described earlier often reflects intraspecific competition for limited resources.

In some very dense populations, accumulation of poisonous waste products may reduce survivorship and reproduction. Most natural populations live in open systems where wastes are consumed by other organisms or flushed away. But the build-up of toxic wastes is common in laboratory cultures of microorganisms. For example, yeast cells ferment sugar and produce ethanol as a waste product. Thus, the alcohol content of wine usually does not exceed 13% by volume, the ethanol concentration that poisons yeasts that are vital to the wine-making process.

How well do species conform to the predictions of the logistic model? In simple laboratory cultures, relatively small organisms, such as *Paramecium,* some crustaceans, and flour beetles, often show an S-shaped pattern of population growth (Figure 29.11 left,

middle). Moreover, large animals introduced into new environments sometimes exhibit a pattern of population growth that matches the predictions of the logistic model (Figure 29.11 right).

Nevertheless, some assumptions of the logistic model are unrealistic. For example, the model predicts that survivorship and fecundity respond immediately to changes in a population's density. Many organisms exhibit a delayed response (a **time lag**) because fecundity has been determined by resource availability at some time in the past. This may reflect conditions that prevailed when individuals were adding yolk to eggs or endosperm to seeds. Moreover, when food resources become scarce, individuals may use stored energy reserves to survive and reproduce. This delays the impact of crowding until stored reserves are depleted and means that population size may overshoot K (see Figure 29.11 middle). Deaths may then outnumber births, causing the population size to drop below K, at least temporarily. Time lags often cause a population to oscillate around K.

The assumption that the addition of new individuals to a population always decreases survivorship and fecundity is unrealistic. In small populations, modest population growth may not have much impact on survivorship and fecundity. In fact, most organisms probably require a minimum population density to survive and reproduce. Some plants flourish in small clumps that buffer them from physical stresses, whereas a single individual living in the open would suffer adverse effects. In some animal populations, a minimum population density is necessary for individuals to find mates. Determining the minimum viable population for a species is an important issue in conservation biology (see Chapter 32).

29.6 Population Regulation

What environmental factors influence population growth rates and control fluctuations in population size? Some factors affecting population size are **density dependent** because their influence increases or decreases with the density of the population. Intraspecific competition and predation are examples of density-dependent environmental factors. The logistic model includes the effects of density dependence in its assumption that per capita birth and death rates change with population density.

Numerous laboratory and field studies have shown that crowding (high population density) decreases individual growth rate, adult size, and survival of plants and animals **(Figure 29.12)**. Organisms living in extremely dense populations are unable to harvest enough resources. They grow slowly and tend to be small, weak, and less likely to survive.

Gardeners understand this relationship and thin their plants to achieve a density that maximizes the number of vigorous individuals that survive to be harvested.

Crowding has a negative effect on reproduction **(Figure 29.13)**. When resources are in short supply, each individual has less energy for reproduction after meeting its basic maintenance needs. Hence, females in crowded populations produce either fewer offspring or smaller offspring that are less likely to survive.

In some species, crowding stimulates developmental and behavioural changes that may influence the density of a population. Migratory locusts can develop into either solitary or migratory forms in the same population. Migratory individuals have longer wings and more body fat, characteristics that allow long-distance dispersal. High population density increases the frequency of the migratory form. Thus, many locusts move away from the area of high density **(Figure 29.14)**, reducing the size and thus the density of the original population.

Although these data about locusts confirm the assumptions of the logistic equation, they do not prove that natural populations are regulated by density-dependent factors. Experimental evidence is necessary to provide a convincing demonstration that an increase in population density causes population size to decrease, whereas a decrease in density causes it to increase.

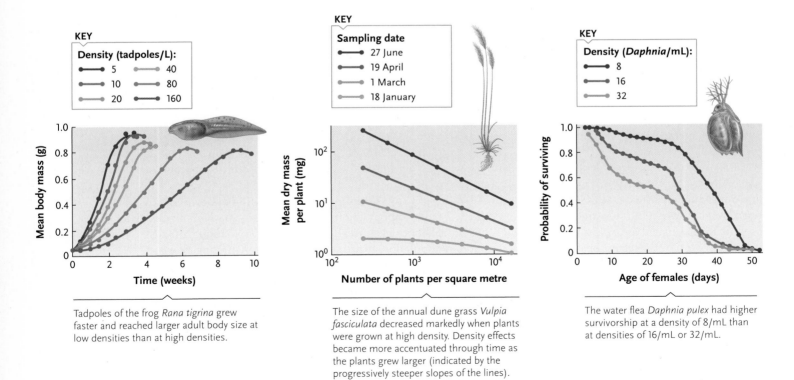

Tadpoles of the frog *Rana tigrina* grew faster and reached larger adult body size at low densities than at high densities.

The size of the annual dune grass *Vulpia fasciculata* decreased markedly when plants were grown at high density. Density effects became more accentuated through time as the plants grew larger (indicated by the progressively steeper slopes of the lines).

The water flea *Daphnia pulex* had higher survivorship at a density of 8/mL than at densities of 16/mL or 32/mL.

Figure 29.12
Effects of crowding on individual growth, size, and survival.

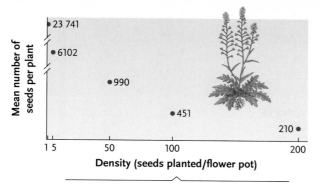

The number of seeds produced by shepherd's purse (*Capsella bursa-pastoris*) decreased dramatically with increasing density in experimental plots.

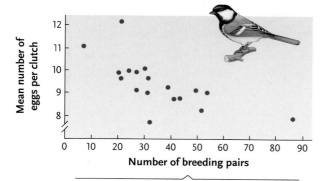

The mean number of eggs produced by the Great Tit (*Parus major*), a woodland bird, declined as the number of breeding pairs in Marley Wood increased.

Figure 29.13
Effects of crowding on fecundity.

Figure 29.14
A swarm of locusts. Migratory locusts (*Locusta migratoria*) moving across an African landscape can devour their own weight in plant material every day.

In the 1960s, Robert Eisenberg experimentally increased the numbers of aquatic snails (*Lymnaea elodes*) in some ponds, decreased them in others, and maintained natural densities in control ponds. Adult survivorship did not differ between experimental and control treatments. But there was a gradient in egg production from few eggs (snails in high-density ponds), to more (control density), to most (low density). Furthermore, survival rates of young snails declined as density increased. After four months, densities in the two experimental groups converged on those in the control, providing strong evidence of density-dependent population regulation.

At this stage, intraspecific competition appears to be the primary density-dependent factor regulating population size. Competition between populations of different species can also exert density-dependent effects on population growth (see Chapter 30). The Allee effect occurs when *r* begins to decline after *N* falls below some threshold. This is another example of a density-dependent regulator.

But predation can also cause density-dependent population regulation. As a particular prey species becomes more numerous, predators may consume more of it because it is easier to find and catch. Once a prey species exceeds some threshold density, predators may consume a larger percentage of its population, a density-dependent effect. On rocky shores in California, sea stars feed mainly on the most abundant invertebrate there. When one prey species becomes common, predators feed on it disproportionately, drastically reducing its numbers. Then they switch to now more abundant alternative prey.

Sometimes several density-dependent factors influence a population at the same time. On small islands in the West Indies, spiders are rare wherever lizards (*Ameiva festiva, Anolis carolinensis,* and *Anolis sagrei*) are abundant but common where the lizards are rare or absent. To test whether the presence of lizards limits the abundance of spiders, David Spiller and Tom Schoener built fences around plots on islands where these species occur. They eliminated lizards from experimental plots but left them in control plots. After two years, spider populations in some experimental plots were five times as dense as those in control plots, suggesting a strong impact of lizard populations on spider populations **(Figure 29.15, p. 696).** In this situation, lizards had two density-dependent effects on spider populations. First, lizards ate spiders, and, second, they competed with them for food. Experimental evidence made it possible for biologists to better understand the situation.

Predation, parasitism, and disease can cause density-dependent regulation of plant and animal populations. Infectious microorganisms (e.g., rabies) spread quickly in a crowded population. In addition, if crowded individuals are weak or malnourished, they are more susceptible to infection and may die from diseases that healthy organisms would survive. Effects on survival can be direct or indirect.

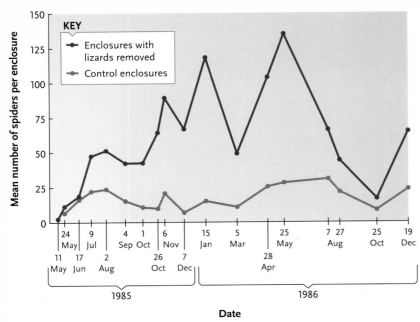

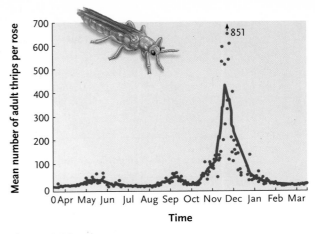

Figure 29.16

Booms and busts in a thrips population. Populations of the Australian insect *Thrips imaginis* grow exponentially when conditions are favourable during spring (which begins in September in the southern hemisphere). But the populations crash in summer when hot and dry conditions cause high mortality.

Experimental Research Figure 29.15 Populations of spiders (*Metepeira daytona*) on a small island in the Bahamas are influenced by the presence of lizards. Note how much higher the population densities of spiders are in the absence than in the presence of lizards.

few individuals survive in remaining flowers, and they are the stock from which the population grows exponentially the following spring.

29.6a Density-Independent Factors: Reducing Population in Spite of Density

Some populations are affected by **density-independent** factors that reduce population size regardless of its density. If an insect population is not physiologically adapted to high temperature, a sudden hot spell may kill 80% of them whether they number 100 or 100 000. Fires, earthquakes, storms, and other natural disturbances can contribute directly or indirectly to density-independent mortality. Because such factors do not cause a population to fluctuate around its *K*, these density-independent factors can reduce but do not regulate population size.

Density-independent factors have a particularly strong effect on populations of small-bodied species that cannot buffer themselves against environmental change. Their populations grow exponentially for a time, but shifts in climate or random events cause high mortality before populations reach a size at which density-dependent factors would regulate their numbers. When conditions improve, populations grow exponentially, at least until another density-independent factor causes them to crash again. A small Australian insect, a thrip (*Thrips imaginis*), eats the pollen and flowers of plants in the rose family. These thrips can be abundant enough to damage blooms. Populations of thrips grow exponentially in spring, when many flowers are available and the weather is warm and moist **(Figure 29.16).** But their populations crash predictably during summer because thrips do not tolerate hot and dry conditions. After the crash, a

29.6b Interactions between Density-Dependent and Density-Independent Factors: Sometimes Population Density Affects Mortality

Density-dependent factors can interact with density-independent factors and limit population growth. Food shortage caused by high population density (a density-dependent factor) may lead to malnourishment. Malnourished individuals may be more likely to succumb to the stress of extreme weather (a density-independent factor).

Populations can be affected by density-independent factors in a density-dependent manner. Some animals retreat into shelters to escape environmental stresses, such as floods or severe heat. If a population is small, most individuals can be accommodated in available refuges. But if a population is large (exceeds the capacity of shelters), only a proportion will find suitable shelter. The larger the population, the greater the percentage of individuals exposed to the stress(es). Thus, although the density-independent effects of weather limit populations of thrips, the availability of flowers in summer (a density-dependent factor) regulates the size of the starting populations of thrips the following spring. Hence, both density-dependent and density-independent factors influence the size of populations of thrips.

Other explanations focus on extrinsic control, such as the relationship between a cycling species and its food or predators. A dense population may exhaust its food supply, increasing mortality and decreasing reproduction. The die-off of large numbers of African elephants in Tsavo National Park in Kenya is an example of the impact of overpopulation. There elephants overgrazed vegetation in most of the park habitat. In 1970, the combination of overgrazing and a drought caused high mortality of elephants. The picture is not always clear, because experimental food supplementation does not always prevent decline in mammal populations. This suggests some level of intrinsic control.

29.6c Life History Characteristics: Evolution of Strategies for Population Growth

Even casual observation reveals tremendous variation in how rapidly population sizes change in different species. New weeds often appear in a vegetable garden overnight, whereas the number of oak trees in a forest may remain relatively stable for years. Why do only some species have the potential for explosive population growth? The answer lies in how natural selection has moulded life history strategies adapted to different ecological conditions. Some ecologists recognize two quite different life history patterns, **r-selected** species and **K-selected** species **(Table 29.3; Figure 29.17)**.

On the face of it, r-selected species are adapted to rapidly changing environments, and many have at least some of the features outlined in Table 29.3. The success of an r-selected life history depends on flooding the environment with a *large quantity* of young because only some may be successful. Small body size means that compared with larger-bodied species, r-selected species lack physiological mechanisms to buffer them from environmental variation. Populations of r-selected species can be so reduced by changes in abiotic environmental factors (e.g., temperature or moisture) that they never grow large enough to reach K and face a shortage of limiting resources. In these cases, K cannot be estimated by researchers, and changes in population size are not accurately described by the logistic model of population growth. Although r-selected species appear to have poor tolerance of environmental change, they are said to be adapted to rapidly changing environments.

At the same time, K-selected species have at least some of the features outlined for them in Table 29.3. These organisms survive the early stages of life (type I or type II survivorship), and a low r_{max} means that their populations grow slowly. The success of a K-selected life history is linked to the production of a relatively small number of high-quality offspring that join an already well-established population. Generalizations about r-selected and K-selected species are misleading. We can recognize this by comparing two species of mammals.

Peromyscus maniculatus, deer mice, occur widely in North America. In southern Ontario, adults weigh 12 to 31 g, females produce average litters of four (range two to eight), and each can have four or five

Table 29.3	Characteristics of *r*-Selected and *K*-Selected Species	
Characteristic	*r*-Selected Species	*K*-Selected Species
Maturation time	Short	Long
Life span	Short	Long
Mortality rate	Usually high	Usually low
Reproductive episodes	Usually one	Usually several
Time of first reproduction	Early	Late
Clutch or brood size	Usually large	Usually small
Size of offspring	Small	Large
Active parental care	Little or none	Often extensive
Population size	Fluctuating	Relatively stable
Tolerance of environmental change	Generally poor	Generally good

litters a year. Females become sexually mature at age two months and breed in their first year. Occasionally, deer mice live to age three years in the wild. Throughout their extensive range in North America, *Myotis lucifugus,* little brown bats, weigh 7 to 12 g; females bear a single young per litter and have one litter per year. Females may breed a year after they are born, but many wait until they are two years old. In the wild, little brown bats can live over 30 years. Using these data, one small mammal (deer mouse) is an r-strategist, whereas another (little brown bat) is a K-strategist. To complicate matters, deer mice living in Kananaskis in the mountains near Calgary mature at one year and may have two litters per year, typically five young per litter. Compared to little brown bats, Kananaskis deer mice are r-strategists. Compared to Ontario deer mice, they are more like K-strategists.

a. An r-selected species　　**b. A K-selected species**

Figure 29.17
Life history differences. An r-selected species, **(a)** *Chenopodium quinoa*, matures in one growing season and produces many tiny seeds. Quinoa was a traditional food staple for indigenous people of North and South America. A K-selected species, **(b)** *Cocos nucifera*, a coconut palm, grows slowly and produces a few large seeds repeatedly during its long life.

Biologists may find the idea of *r*-strategists and *K*-strategists useful, but too often the idea means imposing some human view of the world on a natural system. *K*-strategists and *r*-strategists may be more like beauty, defined by the eye of the beholder. Elephants (*Loxodonta africana*, *Loxodonta cyclotis*, *Elephas maximus*) are big and meet all *K*-strategist criteria. Many insects are small but in all other respects meet the criteria considered typical of *K*-strategists because of their patterns of reproduction. Codfish (*Gadus morhua*) are big (compared to insects or bats) but meet most of the criteria used to identify *r*-strategists, such as their patterns of reproduction.

29.6d Population Cycles: Ups and Downs in Numbers of Individuals

Population densities of many insects, birds, and mammals in the northern hemisphere fluctuate between species-specific lows and highs in a multiyear cycle. Arctic populations of small rodents (*Lemmus lemmus*) vary in size over a 4-year cycle, whereas snowshoe hares (*Lepus americanus*), ruffed grouse (*Bonasa umbellus*), and lynx have 10-year cycles. Ecologists

documented these cyclic fluctuations more than a century ago, but none of the general hypotheses proposed to date explain cycles in all species. Availability and quality of food, abundance of predators, prevalence of disease-causing microorganisms, and variations in weather can influence population growth and declines. Furthermore, food supply and predators for a cycling population are themselves influenced by a population's size.

Theories of intrinsic control suggest that as an animal population grows, individuals undergo hormonal changes that increase aggressiveness, reduce reproduction, and foster dispersal. The dispersal phase of the cycle may be dramatic. When populations of Norway lemming (*Lemmus lemmus*), a rodent that lives in the Scandinavian Arctic, reach their peak density, aggressive interactions drive younger and weaker individuals to disperse. The dispersal of many thousands of lemmings during periods of population growth has sometimes been incorrectly portrayed in nature films as a suicidal mass migration.

Cycles in populations of predators could be induced by time lags between populations of predators and prey and vice versa **(Figure 29.18)**. The 10-year

Ed Cesar/Science Source

Figure 29.18

The predator–prey model. Predator–prey interactions may contribute to density-dependent regulation of both populations. A mathematical model **(a)** predicts cycles in the numbers of predators and prey because of time lags in each species' responses to changes in the density of the other. (Predator population size is exaggerated in this graph.) **(b)** Canada lynx (*Lynx canadensis*) and snowshoe hare (*Lepus americanus*) were often described as a typical cyclic predator–prey interaction. The abundances of lynx (red line) and snowshoe hare (blue line) are based on counts of pelts trappers sold to the Hudson's Bay Company over a 90-year period. Recent research shows that population cycles in snowshoe hares are caused by complex interactions between the snowshoe hares, its food plants, and its predators.

b. Lynx and hare population sizes through time

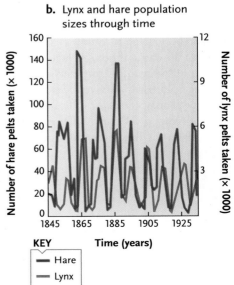

a. Predictions of a predator–prey model

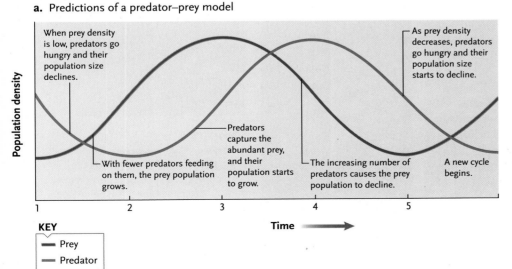

cycles of snowshoe hares and their feline predators, Canada lynx, were often cited as a classic example of such an interaction. But snowshoe hare populations can exhibit a 10-year fluctuation even on islands where lynx are absent. Thus, lynx are not solely responsible for population cycles in snowshoe hares. To further complicate matters, the database demonstrating fluctuations was often the numbers of pelts purchased by the Hudson's Bay Company. Here, fur price influenced the trapping effort and the numbers of animals harvested. This economic reality brought into question the relationship between the numbers of pelts and actual population densities of lynx and snowshoe hares.

Charles Krebs and his colleagues studied hare and lynx interactions with a large-scale, multiyear experiment in Kluane in the southern Yukon. Using fenced experimental areas, they could add food for snowshoe hares, exclude mammalian predators, or apply both experimental treatments while monitoring unmanipulated control plots. When mammalian predators were excluded, densities of snowshoe hares approximately doubled relative to controls. Where food was added, densities of snowshoe hares tripled relative to controls. In plots where food was added and predators were excluded, densities of snowshoe hares increased 11-fold compared with controls. Krebs and his colleagues concluded that neither food availability nor predation is solely responsible for population cycles in snowshoe hares. They postulated that complex interactions between snowshoe hares, their food plants, and their predators generate cyclic fluctuations in populations of snowshoe hares.

STUDY BREAK

1. What are density-dependent factors? Why do dense populations tend to decrease in size?
2. Define density-independent factors, and give some examples.
3. Describe two key differences between r-selected species and K-selected species.

29.7 Human Population Growth

How do human populations compare with those of other species? The worldwide human population was over 7 billion in 2014. Like many other species, humans live in somewhat isolated populations that vary in their demographic traits and access to resources. Although many of us live comfortably, at least a billion people are

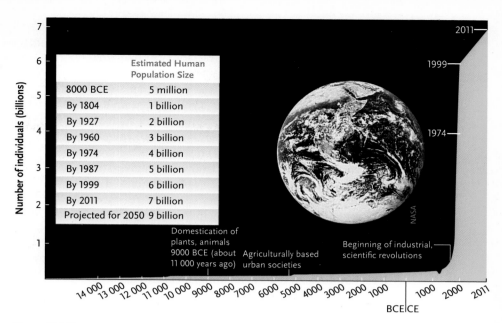

Estimated Human Population Size	
8000 BCE	5 million
By 1804	1 billion
By 1927	2 billion
By 1960	3 billion
By 1974	4 billion
By 1987	5 billion
By 1999	6 billion
By 2011	7 billion
Projected for 2050	9 billion

Figure 29.19

Human population growth. The worldwide human population grew slowly until 200 years ago, when it began to increase explosively. The dip in the mid-fourteenth century represents the death of 60 million Asians and Europeans from the bubonic plague. The table shows how long it took for the human population to add each billion people.

malnourished or starving, lack access to clean drinking water, and live without adequate shelter or health care.

For most of human history, our population grew slowly, reflecting the impact of a range of restraints. Over the past two centuries, the worldwide human population has grown exponentially **(Figure 29.19)**. Demographers identified three ways in which we have avoided the effects of density-dependent regulating factors.

First, humans have expanded their geographic range into virtually every terrestrial habitat, alleviating competition for space. Our early ancestors lived in tropical and subtropical grasslands, but by 40 thousand years ago, they had dispersed through much of the world. Their success resulted from their ability to solve ecological problems by building fires, assembling shelters, making clothing and tools, planning community hunts, and sharing information. Vital survival skills spread from generation to generation and from one population to another because language allowed communication of complex ideas and knowledge.

Second, we have increased K in habitats we occupy, isolating us, as a species, from restrictions associated with access to resources. This change began to occur about 11 thousand years ago, when populations in different parts of the world began to shift from hunting and gathering to agriculture (see Chapter 33). At that time, our ancestors cultivated wild grasses and other plants, diverted water to irrigate crops, and used domesticated animals for food and labour. Innovations such as these increased the availability of food, raising both K and rates of population growth. In the mid-eighteenth century, people harnessed the energy in fossil fuels, and industrialization began in western Europe and North America. Food supplies and *K* increased again, at least in industrialized countries, largely through the use of synthetic fertilizers, pesticides, and efficient methods of transportation and food distribution.

Third, advances in public health reduced the effects of critical population-limiting factors such as malnutrition, contagious diseases, and poor hygiene. Over the past 300 years, modern plumbing and sewage treatment, removal of garbage, and improvements in food handling and processing, as well as medical discoveries, have reduced death rates sharply. Births now greatly exceed deaths, especially in less industrialized countries, resulting in rapid population growth. Note, however, that problems of hygiene and access to fresh water and food had been solved in some societies at least hundreds of years ago. Rome, for example, had a population of about 1 million people by 2 CE, and this was supported by an excellent infrastructure for importing and distributing food, providing fresh water, and dealing with human wastes.

29.7a Age Structure and Economic Growth: Phases of Development

Where have our migrations and technological developments taken us? It took about 2.5 million years for the human population to reach 1 billion, 80 years to reach the second billion, and only 12 years to jump from 5 billion to 6 billion and another 12 years to reach 7 billion (see the inset table in Figure 29.19, p. 699). Rapid population growth now appears to be an inevitable consequence of our demographic structure and economic development.

29.7b Population Growth and Age Structure: Not All Populations Are the Same

In 2011, the worldwide annual growth rate for the human population averaged about 1.15% (*r* = 0.0115 new individuals per individual per year). Population experts expect that rate to decline, but even so, the human population will probably exceed 9 billion before 2050.

In 2000, population growth rates of individual nations varied widely, ranging from much less than 1% to more than 3% **(Figure 29.20a).** Industrialized countries of Western Europe have achieved nearly ZPG, but other countries, particularly those in Africa, Latin America, and Asia, will experience huge increases over the next 20 or 25 years (Figure 29.20b).

For all long-lived species, differences in age structure are a major determinant of differences in population growth rates **(Figure 29.21).** There are three basic patterns in the graphs in Figure 29.21. In the first, in countries with ZPG, there are approximately equal numbers of people of reproductive and prereproductive ages. The ZPG situation is exacerbated when reproductives have very few offspring, meaning that prereproductives may not even replace themselves in the population. Second, in countries with negative growth (without immigration), postreproductives outnumber reproductives, and these populations will not experience a growth spurt when today's children reach reproductive age. Third are countries with rapid growth, where reproductives vastly outnumber postreproductives.

Countries with rapid growth have a broad-based age structure (pattern three, above), with many youngsters

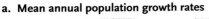

a. Mean annual population growth rates

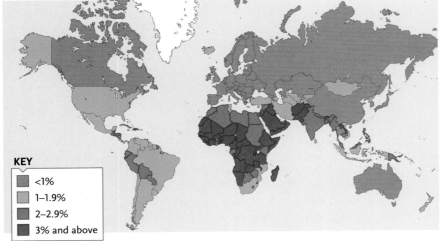

KEY
- <1%
- 1–1.9%
- 2–2.9%
- 3% and above

b. Projected population sizes for 2025

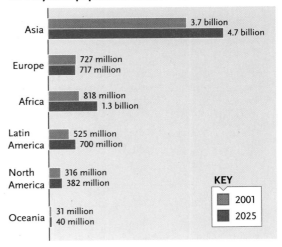

Asia — 3.7 billion / 4.7 billion
Europe — 727 million / 717 million
Africa — 818 million / 1.3 billion
Latin America — 525 million / 700 million
North America — 316 million / 382 million
Oceania — 31 million / 40 million

KEY
- 2001
- 2025

Figure 29.20
Local variation in human population growth rates. In 2001, **(a)** average annual population growth rates varied among countries and continents. In some regions **(b)**, the population is projected to increase greatly by 2025 (red) compared with the population size in 2001 (orange). The population of Europe is likely to decline.

a. Hypothetical age distributions for populations with different growth rates

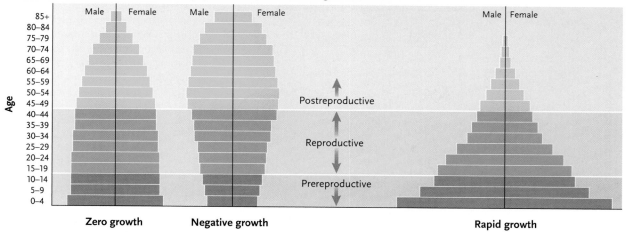

b. Age pyramids for the United States and Mexico in 2000

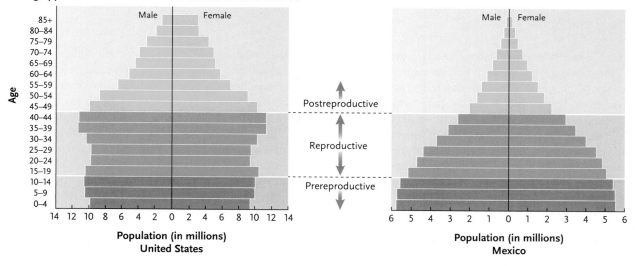

Figure 29.21

Age structure diagrams. Age structure diagrams **(a)** differ for countries with zero, negative, and rapid population growth rates. The width of each bar represents the proportion of the population in each age class. Age structure diagrams for the United States and Mexico **(b)** in 2000 (measured in millions of people) suggest that these countries will experience different growth rates.

born during the previous 15 years. Worldwide, more than one-third of the human population falls within this prereproductive base. This age class will soon reach sexual maturity. Even if each woman produces only two offspring, populations will continue to grow rapidly because so many individuals are reproducing. This situation can be described as a *population bomb*.

The age structures of the United States and Mexico differ, which has consequences for population growth in the two jurisdictions. Remember the potential importance of immigration and emigration when considering the longer-term impact of the population bomb.

29.7c Population Growth and Economic Development: Interconnections

The relationship between a country's population growth and its economic development can be depicted by the **demographic transition model (Figure 29.22,**

p. 702). This model describes historical changes in demographic patterns in the industrialized countries of western Europe. Today, we do not know if it accurately predicts the future for developing nations.

According to this model, during a country's preindustrial stage, birth and death rates are high, and the population grows slowly. Industrialization begins a *transitional* stage, when food production rises, and health care and sanitation improve. Death rates decline, resulting in increased rates of population growth. Later, as living conditions improve, birth rates decline, causing a drop in rates of population growth. When the industrial stage is in full swing, population growth slows dramatically. Now people move from countryside to cities, and urban couples often choose to accumulate material goods instead of having large families. ZPG is reached in the *postindustrial* stage. Eventually, the birth rate falls below the death rate, *r* falls below zero, and population size begins to decrease.

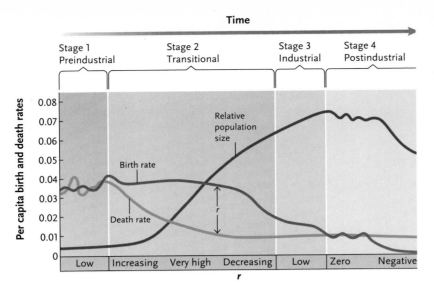

Figure 29.22

The demographic transition. The demographic transition model describes changes in the birth and death rates and relative population size as a country passes through four stages of economic development. The bottom bar describes the net population growth rate, *r*.

Today, the United States, Canada, Australia, Japan, Russia, and most of western Europe are in the industrial stage. Their growth rates are slowly decreasing. In Germany, Bulgaria, and Hungary (and other European countries), birth rates are lower than death rates, and populations are shrinking, indicating entry into the postindustrial stage. Kenya and other less industrialized countries are in the transitional stage, but they may not have enough skilled workers or enough capital to make the transition to an industrialized economy. For these reasons, many poorer nations may be stuck in the transitional stage. Developing countries experience rapid population increase because they experience declines in death rates associated with the transitional stage without the decreases in birth rates typical of industrial and postindustrial stages.

29.7d Controlling Reproductive Output: Planned Reproduction

Most governments realize that increased population size is now the major factor causing resource depletion, excessive pollution, and an overall decline in quality of life. The principles of population ecology demonstrate that slowing the rate of population growth and effecting an actual decline in population size can be achieved only by decreasing the birth rate or increasing the death rate. Increasing mortality is neither a rational nor a humane means of population control. Some governments use **family planning programs** in an attempt to lower birth rates. In other countries, any form of family planning is unlawful. This topic is discussed further in Chapter 32, where we will see that education of women is a vital undertaking.

To achieve ZPG, the average replacement rate should be just slightly higher than two children per couple. This is necessary because some female children die before reaching reproductive age. Today's replacement rate averages about 2.5 children in less industrialized countries with higher mortality rates in prereproductive cohorts and 2.1 in more industrialized countries. However, even if each couple on Earth produced only 2 children, the human population would continue to grow for at least another 60 years (the impact of the population bomb). Continued population growth is inevitable because today's children, who outnumber adults, will soon mature and reproduce. The worldwide population will stabilize only when the age distributions of all countries resemble that for countries with ZPG.

Family planning efforts encourage women to delay their first reproduction. Doing so reduces the average family size and slows population growth by increasing generation time (see Figure 29.9, p. 692). Imagine two populations in which each woman produces two offspring. In the first population, women begin reproducing at age 32 years, and in the second, they begin reproducing at age 16 years. We can begin with a cohort of newborn baby girls in each population. After 32 years, women in the first population will be giving birth to their first offspring, but women in the second population will be new grandmothers. After 64 years, women in the first population will be new grandmothers, but women in the second population will witness the birth of their first great-great grandchildren (if their daughters also bear their first children at age 16 years). Obviously, the first population will grow much more slowly than the second.

29.7e The Future: Where Are We Going?

Homo sapiens has arrived at a turning point in our cultural evolution and in our ecological relationship with Earth. Hard decisions await us, and we must make them soon. All species face limits to their population growth, and it is naive to assume that our unique abilities

John (Jack) S. Millar, *Western University*

Jack Millar, a professor of biology at Western University in London, and his students study the life histories of small mammals such as mice, voles, and wood rats. They do most of this work in the field, mainly at sites in the Kananaskis Valley in southwestern Alberta. The work involves trapping the small mammals and marking them so that they can recognize them later. Recaptures of known individuals allow the researchers to track the performances of individuals. Millar and his students have been following populations of deer mice for over 20 years, and their records have allowed them to ask basic questions about life history.

Most deer mice born in any year in the Kananaskis Valley are dead by the end of September of that year. Although a few females are able to bear two litters in a year, most do not. Compared with deer mice living in southwestern Ontario (see Section 29.6c), the mice in Kananaskis are barely hanging on. But Millar and his students were interested to learn what factors limit age at first reproduction and the ability of a female to breed

more than once a year. Female mice given protein-rich diets (cat food) were sometimes able to breed in their first summer, suggesting a strong influence of food quality on life history traits.

With data on deer mice covering a span of more than 20 years, Millar was able to explore the possible effects of climate change on deer mice living in the Kananaskis Valley. Specifically, between 1985 and 2003, female deer mice typically conceived their first litters on May 2, and the first births occurred on May 26. There were no statistically significant changes in the timing of first births, although the average temperatures in early May had declined by about 2°C during this period. Spring breeding of the deer mice was not related to temperatures or snowfall. The decline in temperature had no effect on the mice's reproductive success. Changes in photoperiod appear to be responsible for initiating reproductive activity in the mice. Their access to protein did affect their reproductive output.

Discoveries about diseases associated with wildlife are a side

benefit of Millar's endeavours. The work that Millar and his students have done provided a different view of the role of beavers in the spread of giardiasis, also known as "beaver fever." Giardiasis is caused by infections of a protozoan species in the genus *Giardia*. Humans can be infected if they drink water containing spores or trophozoites of *Giardia* species. Humans with giardiasis suffer from intestinal distress. As the name beaver fever implies, these aquatic rodents have been presumed to be the source of human infections. Using specimens, some provided by Millar and his students, P.M. Wallis and colleagues determined that 20 of 21 red-backed voles (*Clethrionomys gapperi*) were infected by *Giardia*—a much higher rate of infection than any other small mammals and also higher than beavers (2 of 50 infected).

Millar's work has demonstrated how long-term experimental research involving both observations and experiments can shed light on life history strategies.

---

exempt us from the laws of population growth. We have postponed the action of most factors that limit population growth, but no amount of invention and intervention can expand the ultimate limits set by resource depletion and a damaged environment. We now face two options for limiting human population growth: we can make a global effort to limit our population growth, or we can wait until the environment does it for us.

Return to Figure 29.19, p. 699 and observe that only the bubonic plague (also known as the Black Death) caused any deflection from the trajectory of the curve tracking growth in the human population. The plague appears to have been spread into Europe by the Mongols. The plague, long established in China, arrived in the Mongol summer capital of Shangdu in 1332. By 1351, the population of China had been reduced by 50 to 66%. By 1345, the plague had reached Feodosija in Ukraine (then Kaffa). Between 1340 and 1400, it is estimated that the population of Africa declined from 80 million to 68 million and the world population from 450 million to between 350 million and 375 million. These data do not include the Americas because the plague did not reach there until about 1600.

To put these percentages in context, consider human deaths associated with World War II. In this conflict, Great Britain lost less than 1% of its population, France about 1.5%, and Germany 9.1%. In Poland and Ukraine, where there was a postwar famine, 19% of the human populations there are said to have died.

These sobering figures remind us that we are animals, vulnerable to many of the factors that affect other species on Earth. Now, look back at the chapter-opening image and note the large numbers of gulls at a landfill site. The gulls and the landfill illustrate a fundamental point in population biology, namely, the ability of populations to reach large numbers and have large environmental impacts whether the species are gulls or people.

STUDY BREAK

1. In what three ways have humans avoided the effects of density-dependent regulation factors?
2. What is a population bomb?
3. What does family planning encourage women to do?

Progesterone

The advent of birth control pills **(Figure 1)** had a great impact on the behaviour of people. Women using birth control pills had more control over their fertility than others. Central to the development of an effective oral contraceptive was a change in the molecular structure of progesterone **(Figure 2a)**. Specifically, the addition of a CH_3 group (Figure 2b) meant that the new molecule, megestrol, had the same effect on a woman's reproductive system, but it was not quickly metabolized and remained in the system long enough to have the desired effect (suppressing ovulation). Similarly, slight modifications to the estradiol molecule turned it into ethinylestradiol **(Figure 3)**. Megestrol is an analogue of progesterone, and ethinylestradiol is an analogue of estradiol.

Today, biologists working in zoos use a variety of birth control methods to control the fertility of animals in their collections. For critically endangered species such as black-footed ferrets (see "Black-Footed Ferret, *Mustela nigripes,*" p. 683), this means using information about cycles of fertility to maximize reproductive output.

For animals whose populations are growing at a rapid pace, birth control gives keepers the chance to control growth of the populations. The same principles apply to working with organisms in the wild, but getting African elephants to take their birth control pills has not proven to be easy.

Hormones and their -analogues are common in untreated municipal wastewaters. In some cases, male fish exposed to these wastewaters are becoming feminized. Specifically, some male fish produce vitellogenin mRNA and protein, substances normally associated with the maturation of oocytes in females. Males thus exposed produce early-stage eggs in their testes. This feminization occurs in the presence of estrogenic substances, including natural estrogen (17b-estradiol) and the synthetic estrogen 17a-ethinylestradiol.

Do a few feminized male fish in the population matter? Karen A. Kidd and six colleagues conducted a seven-year whole-lake experiment in northwestern Ontario (the Experimental Lakes Area). Male fathead minnows (*Pimephales promelas*) **(Figure 4)** chronically exposed to low levels (5–6 ng·L^{-1}) of estrogenic substances showed feminizing effects and the development of intersex males, whereas females had altered oogenesis. The situation led to the near-extinction of fathead minnows in the experimental lake.

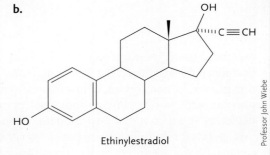

Figure 1
Birth control pills, a selection of products.

M.B. Fenton

a.

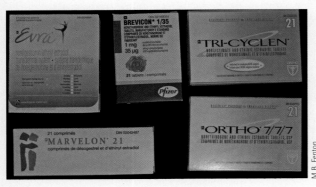

Progesterone

b.

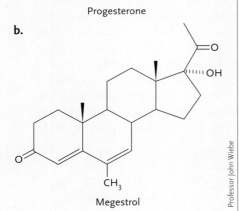

CH_3

Megestrol

Professor John Wiebe

Figure 2
Progesterone and the synthetic megestrol.

a.

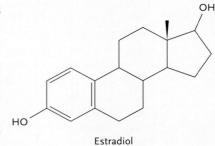

OH

HO

Estradiol

b.

OH

$C\equiv CH$

HO

Ethinylestradiol

Professor John Wiebe

Figure 3
Estradiol and the synthetic ethinylestradiol.

Konrad P. Schmidt

Figure 4
Pimphales promelas, *the fathead minnow.*

29.8 Sisyphean Problems in Population Biology

According to the myth, Sisyphus revealed a secret of Zeus, the supreme Greek god, so in Hades, Sisyphus faced a life of eternal frustration because the boulder he had to push up a hill always rolled back down. The equivalent of this situation is a recurring theme in population biology. For example, on many oceanic islands, eradication of rats and cats is a Sisyphean problem because when either introduced pest is at levels of high populations, it can be easy to kill them with a relatively small investment per cat or per rat. Control or, better still, eradication of cats and rats may reduce the threat to native species (see also Chapter 32). When their populations diminish, the effort and cost per kill increases. In difficult financial times, it may be easy to relent on control measures and use the funds elsewhere. But as soon as the pressure is removed from the populations of rats and cats, the numbers increase, starting the cycle anew.

Sisyphean problems also occur in disease control. The prevalence rates of malaria in children (ages 2 to 10 years) in Zanzibar illustrates this problem **(Figure 29.23)**. Zanzibar, an archipelago consisting of several islands off the east coast of Africa, has a population of 1.3 million people, compared to 1.1 billion in Africa. A recent study showed that the annual antimalarial efforts in Zanzibar prevent about 600 000 cases

of and about 3300 deaths from malaria. The cost is about US$1183 per death averted and US$34.50 per impact of reducing the incidence of the disease.

The data on the prevalence of the malarial parasite in children demonstrate the importance of maintaining antimalarial programs even in the face of success measured as fewer cases of malaria and fewer deaths from it. Arrival of people already infected with the malarial parasite in Zanzibar is part of the problem. The same situation applies to the control of measles, which, unlike malaria, can be controlled by vaccination. This problem is Sisyphean because as soon as efforts to control malaria are cut back, the incidence of the disease increases.

Infestations of bedbugs are yet another Sisyphean problem. The population biology roots of such problems lie in the patterns of population growth: the impact of r (intrinsic rate of increase), N (population size), and K (carrying capacity). Perhaps Sisyphus's challenge was modest compared to that posed by some problems in population biology!

STUDY BREAK

1. Why is it appropriate to consider malaria control as a Sisyphean problem?
2. Why is the malaria situation different between Zanzibar and continental Africa?

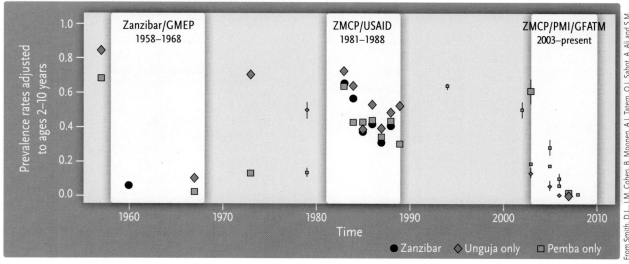

From Smith, D.L., J.M. Cohen, B. Moonen, A.J. Tatem, O.J. Sabot, A. Ali and S.M. Mugheiry, 2011. "Solving the Sisyphean problem of malaria in Zanzibar." *Science*, vol. 332: 1384–1385. Reprinted with permission from AAAS.

Figure 29.23

The incidence of malaria parasites in children between 1958 and 2010. During periods of concerted efforts to reduce or eliminate malaria, the incidence of parasites in children drops, only to rebound when control measures are stopped. Large symbols represent regional and national surveys; smaller symbols, smaller-scale surveys. GMEP was part of the Global Malaria Eradication Programme, ZMCP was the Zanzibar Malaria Control Programme assisted by USAID, and ZMCP/PMI/GFATM is the Zanzibar Malaria Control Programme assisted by the U.S. President's Malaria Initiative and the Global Fund to Fight AIDS, Tuberculosis, and Malaria. Here Zanzibar refers to all of the islands; Pemba and Unguja are specific islands included in the archipelago.

Review

To access course materials such as Aplia and other companion resources, please visit www.NELSONbrain.com.

29.1 The Science of Ecology

- Organismal ecology is the study of organisms to determine adaptations to the abiotic environment, including morphological, physiological, biochemical, behavioural, and genetic adaptations. Population ecologists document changes in size and other characteristics of populations of species over space and time. Community ecologists study sympatric populations, the interactions among them, and how these interactions affect the community's growth. Interactions may include predation and competition. Ecosystem ecologists study nutrient cycling and energy flow through the biotic and the abiotic environment.

- Mathematical models express hypotheses about ecological relationships and different variables, allowing researchers to manipulate the model and document resulting changes. In this way, researchers can simulate natural events before investing in lab work.

- Experimental and control treatments are necessary because they allow ecologists to separate cause and effect.

29.2 Population Characteristics

- Geographic range is the overall spatial boundary around a population. Individuals in the population often live in a specific habitat within the range.

- A lower population density means that individuals have greater access to resources such as sunlight and water. The capture−mark−recapture technique assumes that (1) a mark has no effect on an individual's survival, (2) marked and unmarked individuals mix randomly, (3) there is no migration throughout the estimation period, and (4) marked and unmarked individuals are equally likely to be caught.

- Three types of dispersion are clumped, uniform, and random. Clumped is most common in nature because suitable conditions are usually patchily distributed and animals often live in social groups. Asexual reproduction patterns can also lead to clumped aggregations.

- Generation time increases with body size.

- The number of males in a population of mammals has little impact on population growth because females bear the costs of reproduction (pregnancy and lactation), thus limiting population growth. Sea horses are different because males get pregnant.

29.3 Demography

- Age-specific mortality and age-specific survivorship deal with age intervals. In any one interval, age-specific mortality is the proportion of individuals that died during that time. Age-specific survivorship is the number surviving during the interval. The two values must sum to 1. For example, if age-specific survivorship is 0.616, then age-specific mortality is $1 - 0.616$ or 0.384.

- Age-specific fecundity is the average number of offspring produced by surviving females during each age interval.

- In a type I curve, high survivorship at a young age decreases rapidly later in life. Type I curves are common for large animals, including humans. In a type II curve, the relationship is linear because there is a constant rate of mortality across the life span. Songbirds fit in this category. A type III curve shows high mortality at a young age that stabilizes as individuals grow older and larger. Insects fall into this category.

29.4 The Evolution of Life Histories

- Maintenance, growth, and reproduction are the three main energy-consuming processes.

- Passive care occurs in animals that simply lay eggs and leave them or, in mammals, as nutrients cross the placenta from the mother to the developing baby. Active care involves nursing and other care provided after birth.

- Salmon have a short life span and devote a great deal of energy to reproduction. Deciduous trees may reproduce more than once and use only some energy in any reproductive event, balancing reproduction and growth.

- Early reproduction is favoured if adult survival rates are low or if, when animals age, they do not increase in size. In this case, fecundity does not increase with size.

29.5 Models of Population Growth

- An exponential model is used when a population has unlimited growth.

- dN/dt = change in a population's size during a given time period; b = per capita birth rate; d = per capita death rate; N = number of individuals in the population; $b - d$ = per capita growth rate = r.

- When $r > 0$, the birth rate exceeds the death rate, and the population is growing. When $r < 0$, the birth rate is less than the death rate, and the population is shrinking. When $r = 0$, the birth and death rates are equal, and the population is neither growing nor shrinking. The intrinsic rate of increase (r_{max}) is the maximum per capita growth rate. This value usually varies inversely with generation time, so a shorter generation time means a higher r_{max}.

- Intraspecific competition occurs when two or more individuals of the same species depend on the same limiting resource. For deer, this could include food, water, or refuge from predators.

- A logistic model has the following pattern: when the population growth is low, the population is small. At intermediate population sizes, growth is more rapid because more individuals breed and r is high. When population growth approaches K (carrying capacity), competition increases, r decreases, and the growth of the population is reduced.

29.6 Population Regulation

- Density-dependent factors include intraspecific competition and predation. At high density, fewer resources are available for individuals, which, in turn, use more energy in maintenance needs and less in reproduction. Offspring produced at higher population densities are often smaller in number or size and less likely to survive. At high population levels, adults may be smaller and weaker.

- Density-independent factors, such as fire, earthquakes, storms, floods, and other natural disturbances, reduce a population size regardless of density.

- r-selected species often have large numbers of small young, whereas K-selected species usually have small numbers of larger young. Other possibilities may include characteristics from Table 29.3.

- Extrinsic control includes interactions between individuals in a population and their food and predators. Once a food supply is exhausted, reproduction will decrease and mortality will increase. Intrinsic control can be hormonal changes within a population that cause increased aggressiveness, faster dispersal, and reduced reproduction. Aggression can cause weaker individuals to be forced to disperse to reduce the population density.

29.7 Human Population Growth

- Humans have avoided the effects of density-dependent regulation factors by expanding their geographic range into virtually every habitat, increasing K through agriculture, and reducing population-limiting factors resulting from poor hygiene, malnutrition, and contagious diseases.

- A population bomb is when many offspring are born in one time period, first forming the prereproductive base. At sexual maturity, populations can grow rapidly because of the large number of individuals in this cohort.

- The preindustrial stage is characterized by slow population growth, as birth and death rates are high. The transitional stage has better health care and sanitation, as well as increased food production. In the transitional stage, there is a decline in death rates, allowing population growth, but birth rates eventually decline as living conditions improve. In the industrial stage, there is slow population growth as family size decreases because couples choose to have fewer children and accumulate more material goods. In the postindustrial stage, the population size decreases as the birth rate falls below the death rate.

- Family planning encourages families to delay first reproduction, decreasing the size of the average family and, in turn, reducing the population size as generation time increases. Decisions about reproduction should involve couples.

29.8 Sisyphean Problems in Population Biology

- Sisyphean problems in biology range from the challenge of eradicating pests (such as introduced rats and cats) to the problem of controlling diseases such as malaria. In either case, when the incidence of the pests or the disease drops below some level and control operations stop, the populations of pests and the incidence of disease rebound.

Questions

Self-Test Questions

1. Ecologists sometimes use mathematical models to do which of the following tasks?
   a. simulate natural events before conducting detailed field studies
   b. make basic observations about ecological relationships in nature
   c. collect survivorship and fecundity data to construct life tables
   d. determine the geographic ranges of populations

2. Which term can be used to describe the number of individuals per unit area or volume of habitat?
   a. dispersion pattern
   b. density
   c. size
   d. age structure

3. Suppose that one day you caught and marked 90 butterflies in a population. A week later, you returned to the population and caught 80 butterflies, including 16 that had been marked previously. What is the size of the butterfly population?
   a. 170
   b. 450
   c. 154
   d. 186

4. What does a uniform dispersion pattern imply about the members of a population?
   a. They work together to escape from predators.
   b. They use resources that are patchily distributed.
   c. They may experience intraspecific competition for vital resources.
   d. They have no ecological interactions with each other.

5. What does the model of exponential population growth predict about the per capita population growth rate (r)?
   a. r does not change as a population gets larger.
   b. r gets larger as a population gets larger.
   c. r gets smaller as a population gets larger.
   d. r is always at its maximum level (r_{max}).

6. If a population of 1000 individuals experiences 452 births and 380 deaths in 1 year, what is the value of r for this population?
   a. 0.072/individual/year
   b. 0.452/individual/year
   c. 0.380/individual/year
   d. 0.820/individual/year

7. According to the logistic model of population growth, what happens to the absolute number of individuals by which a population grows during a given time period?
   a. It gets steadily larger as the population size increases.
   b. It gets steadily smaller as the population size increases.
   c. It remains constant as the population size increases.
   d. It is highest when the population is at an intermediate size.

8. Which example might reflect density-dependent regulation of population size?
   a. An exterminator uses a pesticide to eliminate carpenter ants from a home.
   b. Mosquitoes disappear from an area after the first frost.
   c. Northeast storms blow over and kill all willow trees along a lake.
   d. A clam population declines in numbers in a bay as the number of predatory herring gulls increases.

9. Which pattern is a K-selected species likely to exhibit?
   a. a type I survivorship curve and a short generation time
   b. a type II survivorship curve and a short generation time
   c. a type III survivorship curve and a short generation time
   d. a type I survivorship curve and a long generation time

10. Which one of the following is a reason that human populations have sidestepped factors that usually control population growth?
    a. Agriculture and industrialization have increased the carrying capacity for our species.
    b. The population growth rate (r) for the human population has always been small.
    c. The age structure of human populations has no impact on its population growth.
    d. Plagues have killed off large numbers of humans at certain times in the past.

Questions for Discussion

1. Do you expect to see a genetic bottleneck effect in *Mustela nigripes* populations in the wild in the future? If so, how long will they take to appear?

2. Design an income tax policy and social services plan that would encourage people to have either larger or smaller families.

3. Many city-dwellers have noted that the density of cockroaches in apartment kitchens appears to vary with the habits of the occupants. People who wrap food carefully and clean their kitchen frequently tend to have fewer arthropod roommates than those who leave food on kitchen counters and clean less often. Interpret these observations from the viewpoint of a population ecologist.

4. Why have bedbugs reemerged as all-too-common pests in many dwellings? How can they be controlled?

5. How could the use of composting by urban gardeners affect local populations of raccoons? Is there a "green" solution to the perceived problem posed by urban raccoons?

M.B. Fenton

A hovering rubythroated hummingbird (*Archilochus colubris*) approaches a flower.

Population Interactions and Community Ecology

WHY IT MATTERS

The late Robert Whittaker, a well-known ecologist, referred to birds such as hummingbirds as "ornaments" of evolution. He made this observation based on the reality that although they were spectacular and distinctive, many of these plant-pollinating and seed-dispersing birds did not appear to contribute much to the overall productivity of ecosystems. This is an example of one important challenge facing biologists working to understand the nature and details of interactions among the species that coexist in and constitute communities. How can we assess detailed interactions among sympatric species?

If you watched hummingbirds (opening photograph) visiting flowers, it would be easy to believe that they subsisted on the nectar and perhaps pollen that are available from flowers. But if you watch the same birds bringing food to their nestlings, it would seem that they eat mainly insects. So which data set is more indicative of the role these "ornaments" play in the ecosystems in which they occur?

By examining and describing morphological specializations, biologists who study bats identified tropical and subtropical species apparently specialized for nectar feeding. Long snouts, small teeth **(Figure 30.1, p. 710)**, and extensible tongues **(Figure 30.2, p. 710)** are features of most of the bats that visit flowers and pollinate plants. Flowers pollinated mainly by bats are chiropterphilous, quite different from those pollinated by birds (ornithophilous). At flowers, the bats pump blood into their tongues, thus extending them into the flowers they visit. Papillae at the tips of their tongues soak up nectar. Flowers pollinated by insects and birds often have nectar guides, visual patterns that guide the pollinator to the nectar. The positioning of nectaries, the areas of nectar production and storage, ensures that the pollinator

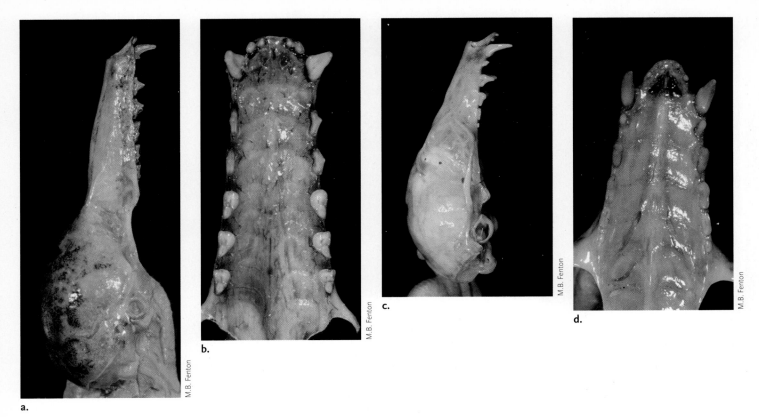

Figure 30.1
The skulls of nectar-feeding bats have long muzzles and relatively small teeth. This is true of a Mexican long-tongued bat (*Choeronycteris Mexicana*) **(a, b)** and a Woerman's bat (*Megaloglossus woermanii*) **(c, d).** (b) and (d) show a closer view of the teeth. The Mexican long-tongued bat is a New World leaf-nosed bat from the Neotropics. Woermna's bat is an Old World fruit bat from Africa. Flower-visiting has arisen independently in both lineages of bats.

Figure 30.2
Two exposures of a Pallas' long-tongued bat with tongue extended into a banana flower. Bananas are a chiropterophilous species.

gets coated with pollen in the process of feeding. Some chiropterophilous flowers have acoustic nectar guides that allow the hovering bat to position itself to achieve best access to the nectar.

In earlier studies, biologists had collected and analyzed the contents of the droppings of bats

(Figure 30.3). This approach was at best tedious and provided only a general view of what the bats had eaten. DNA barcoding ("Barcode of Life," p. 730) changed the scene, perhaps indicating that the bats were not just ornaments in ecosystems.

The work of a team led by Canadian Beth (E. L.) Clare (currently in the Department of Biology and Chemistry at Queen Mary College in London, United Kingdom) overturned the conventional view of nectar-feeding bats. Their analysis revealed that Pallas' long-tongued bats frequently ate insects and fruit as well as nectar and pollen. The proof of this came from barcoding insect remains from the droppings of Pallas' long-tongued bat, observing the animals hunting insects, and monitoring their echolocation calls (see Chapter 47) and behaviour. The data set demonstrated that these bats actively hunted insects, and their echolocation calls made them relatively undetectable by many insects equipped with bat-detecting ears. The bats filled at least three trophic roles, eating insects, fruit, and nectar and pollen.

Earlier work with DNA barcode analysis of insect fragments in the droppings of insectivorous bats had revealed that six sympatric species in the same ecosystem in Jamaica showed relatively little overlap in diet, raising questions about the role of competition in

a.

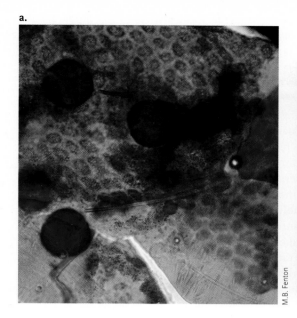

b.

M.B. Fenton

M.B. Fenton

Figure 30.3
A microscopic view of the contents of a dissected bat stool showing a mixture of pollen (four grains) and pieces of insect **(a)**. The bat that contributed the sample is Leach's single leaf bat (*Monophyllus redmani*) from Cuba **(b)**.

structuring this community. These results, combined with those about the diets of Pallas' long-tongued bats, give biologists a different view of the trophic structure of bat communities. Comparable analyses of the diets of other species, for example, fish and leeches, alerts us to the dynamic nature of communities and the diversity of interactions among the species that the communities comprise.

30.1 Interspecific Interactions

Interactions between species typically benefit or harm the organisms involved, although they may be neutral **(Table 30.1)**. Furthermore, where interactions with other species affect individuals' survival and reproduction, many of the relationships we witness today are the products of long-term evolutionary modification. Good examples range from predator–prey interactions to those associated with pollination or dispersal of seeds.

Interactions between species can change constantly, but remember that the interactions occur at the individual level. Some individuals of a species may be better adapted when another species exerts selection pressure on that species. This adaptation can, in turn, help these individuals exert selection pressure on the other species, which can exert selection pressure on the first species in the chain. The situation, known as **coevolution**, is defined as genetically based reciprocal adaptation in two or more interacting species. A good example is provided by the arms race between echolocating bats and insects with bat-detecting ears (see "Echolocation: Communication," Chapter 47).

Some coevolutionary relationships are straightforward. Ecologists describe the coevolutionary interactions between some predators and their prey as a race

Table 30.1	Population Interactions and Their Effects	
Interaction	Effects on Interacting Populations	
Predation	+/−	Predators gain nutrients and energy; prey are killed or injured.
Parasitism	+/−	Parasites gain nutrients and energy; hosts are injured or killed.
Herbivory	+/−	Herbivores gain nutrients and energy; plants are killed or injured.
Competition	−/−	Both competing populations lose access to some resources.
Commensalism	−/0	One population benefits; the other population is unaffected.
Mutualism	+/+	Both populations benefit.

in which each species evolves adaptations that temporarily allow it to outpace the other. When antelope populations suffer predation by cheetahs, natural selection fosters the evolution of faster antelopes. Faster cheetahs may be the result of this situation, and if their offspring are also fast, then antelopes will also become more fleet of foot. Other coevolved interactions provide benefits to both partners. Flower structures of different monkey-flower species have evolved characteristics that allow them to be visited by either bees or hummingbirds (see Figure 19.4, Chapter 19).

One can hypothesize a coevolutionary relationship between any two interacting species, but documenting the evolution of reciprocal adaptations is difficult. Coevolutionary interactions often involve more than two species, and most organisms experience complex interactions with numerous other species in their communities. Cheetahs take several prey species. Antelopes are prey for many species of predators, from cheetahs to lions, leopards, and hyenas, as well as some larger birds of prey. Not all predators use the same hunting strategy. Therefore, the simple portrayal of coevolution as taking place between two species rarely does justice to the complexity of these relationships.

STUDY BREAK

What is coevolution? Is it usually restricted to two species?

30.2 Getting Food

Because animals typically acquire nutrients and energy by consuming other organisms, **predation** (the interaction between predatory animals and the animal prey they consume) and **herbivory** (the interaction between herbivorous animals and the plants they eat) can be the most conspicuous relationships in ecological communities.

Both predators and **herbivores** have evolved characteristics allowing them to feed effectively. Carnivores use sensory systems to locate animal prey and specialized behaviours and anatomical structures to capture and consume it. Herbivores use sensory systems to identify preferred food or to avoid food that is toxic.

Rattlesnakes, such as species in the genus *Crotalus,* use heat sensors on pits in their faces (see Figure 45.26, Chapter 45) to detect warm-blooded prey. The snakes deliver venom through fangs (hollow teeth) by open-mouthed strikes on prey. After striking, the snakes wait for the venom to take effect and then use chemical sensors also on the roofs of their mouths to follow the scent trail left by the dying prey. The venom is produced in the snakes' salivary glands. It contains neurotoxins that paralyze prey and protease enzymes that begin to digest it. Elastic ligaments connecting the bones of the snakes' jaws (mandibles) to one another and the mandibles to the skull allow snakes to open their mouths very wide to swallow prey larger than their heads (see Figure 48.21, Chapter 48).

Herbivores have comparable adaptations for locating and processing their food plants. Insects use chemical sensors on their legs and mouthparts to identify edible plants and sharp mandibles or sucking mouthparts to consume plant tissues or sap. Herbivorous mammals have specialized teeth to harvest and grind tough vegetation (see Figure 28.57, Chapter 28). Herbivores, such as farmer ants (see Chapter 33), ruminants, and termites (see "Digesting Cellulose: Fermentation," Chapter 48), may also coopt other species to gain access to nutrients locked up in plant materials.

All animals select food from a variety of potential items. Some species, described as *specialists,* feed on one or just a few types of food. Among birds, Everglades

Kites (*Rostrhamus sociabilis*) eat only apple snails (*Pomacea paludosa*). Koalas (see Figure 28.51, Chapter 28) eat the leaves of only a few of the many available species of *Eucalyptus*. Other species, described as generalists, have broader tastes. Crows (genus *Corvus*) take food ranging from grain to insects to carrion. Bears (genus *Ursus*) and pigs (genus *Sus*) are as omnivorous as humans.

How does an animal select its food? Why pizza rather than salad? Mathematical models, collectively described as **optimal foraging theory**, predict that an animal's diet is a compromise between the costs and benefits associated with different types of food. Assuming that animals try to maximize their energy intake at any meal, their diets should be determined by the ratio of costs to benefits: the costs of obtaining the food versus the benefits of consuming it. Costs are the time and energy it takes to pursue, capture, and consume a particular kind of food. Benefits are the energy provided by that food. A cougar (*Felis concolor*) will invest more time and energy hunting a mountain goat (*Oreamnos americanus*) than a jackrabbit (*Lepus townsendii*), but the payoff for the cat is a bigger meal. One important element in food choice is the relative abundance of prey. *Encounter rate* is usually influenced by population density and can influence a predator's diet. For the cougar, encounter rate determines the time between jackrabbits, and when they are abundant, they can be a more economical target than larger, scarcer prey.

Food abundance affects food choice. When prey are scarce, animals often take what they can get, settling for food that has a higher cost-to-benefit ratio. When food is abundant, they may specialize, selecting types that provide the largest energetic return. Bluegill sunfishes eat *Daphnia* and other small crustaceans. When crustacean density is high, these fishes hunt mostly large *Daphnia,* which provide more energy for their effort. When prey density is low, bluegills eat *Daphnia* of all sizes **(Figure 30.4)**.

Think of yourself at a buffet. The array of food can be impressive, if not overwhelming. But your state of hunger, the foods you like, the ones you do not like, and any to which you are allergic all influence your selection. You may also be influenced by choices made by others. In your feeding behaviour, you betray your animal heritage.

STUDY BREAK

1. How do predators differ from herbivores? How are they similar?
2. Is a koala a generalist or a specialist? What is the difference?
3. What does optimal foraging theory predict? Describe the costs and benefits central to this theory.

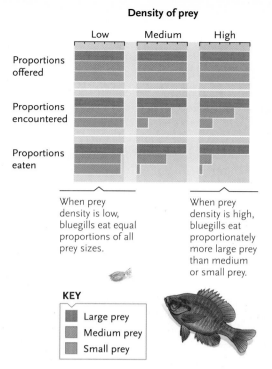

Density of prey

Low   Medium   High

Proportions offered

Proportions encountered

Proportions eaten

When prey density is low, bluegills eat equal proportions of all prey sizes.

When prey density is high, bluegills eat proportionately more large prey than medium or small prey.

KEY

Large prey
Medium prey
Small prey

Figure 30.4

An experiment demonstrating that prey density affects predator food choice. Bluegill sunfishes (*Lepomis macrochirus*) were offered equal numbers of small, medium, and large prey (*Daphnia magna*) at three different total densities of prey. Because large prey are easy to find, the fishes encountered them more often, especially at the highest prey densities, than either medium-sized or small prey. The fishes' choice of prey varied with prey density, but they always chose the largest prey available.

30.3 Defence

CONCEPT FIX Some people believe that "natural" products (chemicals) are beneficial to us, while artificial ones are potentially harmful. In reality, many plants produce chemicals (natural products) that are dangerous and even deadly—to humans and to other animals and to some plants. Contact with the leaves (or stems, roots, flowers, or berries) of poison ivy may be enough to convince you that not all plant products are beneficial. If not, you can read about conine (see Figure 33.31, Chapter 33), an active ingredient in poison hemlock, the poison that killed Socrates. ⬡

Predation and herbivory negatively affect the species being eaten, so it is no surprise that animals and plants have evolved mechanisms to avoid being caught and eaten. Some plants use spines, thorns, and irritating hairs to protect themselves from herbivores. Plant tissues often contain poisonous chemicals that deter herbivores from feeding. When damaged, milkweed plants (family Asclepiadaceae) exude a milky,

irritating sap **(Figure 30.5)** that contains poisons that affect the heart (cardiac glycosides). Even small amounts of cardiac glycosides are toxic to the heart muscles of some vertebrates. Other plants have compounds that mimic the structure of insect hormones, disrupting the development of insects that eat them. Most of these poisonous compounds are volatile, giving plants their typical aromas. Some herbivores have developed the ability to recognize these odours and avoid toxic plants. Some plants increase their production of toxic compounds in response to herbivore feeding. Potato and tomato plants damaged by herbivores have higher levels of protease-inhibiting chemicals. These compounds prevent herbivores from digesting the proteins they have eaten, reducing the food value of these plants.

30.3a Be Too Big to Tackle

Size can be a defence. At one end of the spectrum, this means being too small to be considered food. At the other end, it means being so big that few, if any, predators can succeed in attacking and killing the prey. Today, elephants and some other large herbivores (megaherbivores) are species with few predators (other than humans). But 50 thousand years ago, there were larger predators (see Figure 17.1, Chapter 17), including one species of "lion" that was one-third larger than an African lion.

30.3b Eternal Vigilance: Always Be Alert

A first line of defence of many animals is avoiding detection. This often means not moving, but it also means keeping a sharp lookout for approaching predators and the danger they represent **(Figure 30.6, p. 714)**. Animals that live in groups benefit from the multitude of eyes and ears that can detect approaching danger, so the risk of predation can influence group size and social interactions.

Figure 30.5

Protective latex sap. Milky sap laced with cardiac glycosides oozes from a cut milkweed (*Asclepias* species) leaf. Milky sap does not always mean dangerous chemicals; for example, the sap of dandelions is benign.

Figure 30.6
Eternally vigilant. The sentry of a group of meerkats (*Suricata suricatta*).

30.3c Avoid Detection: Freeze—Movement Invites Discovery

Many animals are cryptic, camouflaged so that a predator does not distinguish them from the background. Patterns such as the stripes of a zebra (*Equus burchellii*) make the animal conspicuous at close range, but at a distance, patterns break up the outline, rendering the animals almost invisible. Many other animals look like something that is not edible. Some caterpillars look like bird droppings, whereas other insects look like thorns or sticks. Neither bird droppings nor thorns are usually eaten by insectivores.

30.3d Thwarting Attacks: Take Evasive Action

Animals resort to other defensive tactics once they have been discovered and recognized. Running away is a typical next line of defence. Taking refuge in a shelter and getting out of a predator's reach are an alternative. African pancake tortoises (*Malacochersus tornieri*) are flat, as the name implies. When threatened, they retreat into rocky crevices and puff themselves up with air, becoming so tightly wedged that predators cannot extract them.

If cornered by a predator, offence becomes the next line of defence. This can involve displays intended to startle or intimidate by making the prey appear large and/or ferocious. Such a display might dissuade a predator or confuse it long enough to allow the potential victim to escape. Many animals use direct attack in these situations, engaging whatever weapons they have (biting, scratching, stinging, etc.). Direct attacks are not usually a good primary defence because they involve getting very close to the predator, something prey usually avoid doing.

30.3e Spines and Armour: Be Dangerous or Impossible to Attack

Other organisms use active defence in the form of spines or thorns **(Figure 30.7)**. North American porcupines (genus *Erethizon*) release hairs modified into sharp, barbed quills that when stuck into a predator, cause severe pain and swelling. The spines detach easily from the porcupine, and the nose, lips, and tongue of an attacker are particularly vulnerable. There are records of leopards (*Panthera pardus*) being killed by porcupine spines. In these instances, the damage to the leopards' mouths, combined with infection, was probably the immediate cause of death. Many other mammals, from monotremes (spiny anteaters) to tenrecs (insectivores from Madagascar, *Tenrec* species and *Hemicentetes* species), hedgehogs (*Erinaceus* species), and porcupines in the Old World, use the same defence. So do some fishes and many plants.

Other organisms are armoured **(Figure 30.8)**. Examples include bivalve and gastropod molluscs, chambered nautiluses, arthropods such as horseshoe crabs (*Limulus* species), trilobites (see Chapter 27), fishes such as catfish (*Siluriformes*), reptiles (turtles; see Figure 28.38, Chapter 28), and mammals (armadillos, scaly anteaters). We know a great deal about extinct species that were armoured (see Chapter 20) because they often made good fossils.

30.3f Chemical Defence Ranges from Bad Taste to Deadly

Like plants that produce chemicals to repel herbivores, many animals make themselves chemically unattractive. At one level, this can be as simple as smelling or tasting bad. Have you ever had a dog or a cat that was sprayed by a skunk (*Mephitis mephitis*)? Many animals vomit and defecate on their attackers. Skunks and bombardier beetles escalate this strategy by producing and spraying a noxious chemical. Other animals go beyond spraying. Many species of cnidarians, annelids, arthropods, and chordates produce dangerous toxins and deliver them directly into their attackers. These toxins may be synthesized by the user (e.g., snake venom; see "Molecule behind Biology," Box 30.1, p. 718) or sequestered from other sources, often plants

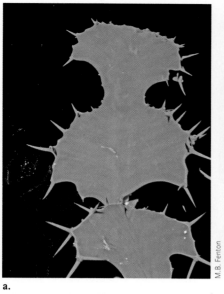

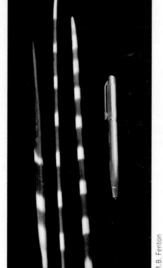

Figure 30.7
Defensive spines. Plants such as **(a)** the cowhorn euphorb (*Euphorbia grandicornis*) and **(b)** crown of thorns (*Euphorbia milli*) and animals such as **(c)** spiny anteaters (*Tachyglossus* species) and **(d)** porcupines (*Hystrix* species) use thorns or spines in defence. Pen shown for scale with quills in (d).

or other animals (see "Nematocysts," p. 720). Caterpillars of monarch butterflies are immune to the cardiac glycosides in the milkweed leaves they eat. They extract, concentrate, and store these chemicals, making the caterpillars themselves (and the adult butterflies) poisonous to potential predators. The concentrations of defensive chemicals may be higher in the animal than they were in its food. Cardiac glycosides persist through metamorphosis, making adult monarchs poisonous to vertebrate predators.

30.3g Warnings Are Danger Signals

Many animals that are noxious or dangerous are **aposematic**: they advertise their unpalatability with an appropriate display (**Figure 30.9**; see Chapter 18). Aposematic displays are designed to "teach" predators to avoid the signaller, reducing the chances of harm to would-be predators and prey. Predators that attack a brightly coloured bee or wasp and are stung learn to

Figure 30.8
Armour. Turtles and their allies (see Chapter 26) live inside shells. This leopard tortoise (*Geochelone pardalis*) is inspecting the remains of a conspecific. Armour does not guarantee survival.

Figure 30.9
Warning colours. This arrowhead frog gets its name from toxins in its skin that were used to poison arrowheads.

Figure 30.10

Mimicry. **(a)** Batesian mimics are harmless animals that mimic a dangerous one. The harmless drone fly (*Eristalis tenax*) is a Batesian mimic of the stinging honeybee (*Apis mellifera*). **(b)** Müllerian mimics are poisonous species that share a similar appearance. Two distantly related species of butterfly, *Heliconius erata* and *Heliconius melpomene*, have nearly indistinguishable patterns on their wings.

a. Batesian mimicry

Drone fly (*Eristalis tenax*), the mimic   Honeybee (*Apis mellifera*), the model

b. Müllerian mimicry

Heliconius erato   *Heliconius melpomene*

associate the aposematic pattern with the sting. Many predators quickly learn to avoid black-and-white skunks, yellow-banded wasps, or orange monarch butterflies because they associate the warning display with pain, illness, or severe indigestion.

But for every ploy there is a counterploy, and some predators eat mainly dangerous prey. Bee-eaters (family Meropidae) are birds that eat hymenopterans (bees and wasps). Some individual African lions specialize in porcupines, and animals such as hedgehogs (genus *Erinaceus*) seem able to eat almost anything and show no ill effects. Indeed, some hedgehogs first lick toads and then their own spines, anointing them with toad venom. Hedgehog spines treated with toad venom are more irritating (at least to people) than untreated ones, enhancing their defensive impact.

30.3h Mimicry Is Advertising, Whether True or False

If predators learn to recognize warning signals, it is no surprise that many harmless animals' defences are based on imitating (mimicking) species that are dangerous or distasteful. **Mimicry** occurs when one species evolves to resemble another **(Figure 30.10)**. **Batesian mimicry**, named for English naturalist Henry W. Bates, occurs when a palatable or harmless species (the **mimic**) resembles an unpalatable or poisonous one (the **model**). Any predator that eats the poisonous model and suffers accordingly will subsequently avoid other organisms that resemble it. However, the predator must survive the encounter. **Müllerian mimicry**, named for German zoologist Fritz Müller, involves two or more unpalatable

species looking the same, presumably to reinforce lessons learned by a predator that attacks any species in the mimicry complex.

For mimicry to work, the predator must learn (see Chapter 47) to recognize and then avoid the prey. The more deadly the toxin, the less likely an individual predator is to learn by its experience. In many cases, predators learn by watching the discomfort of a conspecific that has eaten or attacked an aposematic prey.

Plants often use toxins to protect themselves against herbivores. Is this also true of toxins in mushrooms (see Chapter 19)?

30.3i There Is No Perfect Defence

Helmets protect soldiers, skiers, motorcyclists, and cyclists, but not completely because no defence provides perfect protection. Some predators learn to circumvent defences. Many predators learn to deal with a diversity of prey species and a variety of defensive tactics. Orb web spiders confronting a captive in a web adjust their behaviour according to the prey. They treat moths differently from beetles, and they treat bees in yet another way. When threatened by a predator, headstand beetles raise their rear ends and spray a noxious chemical from a gland at the tip of the abdomen. This behaviour deters many would-be predators. But experienced grasshopper mice from western North America circumvent this defence. An experienced mouse grabs the beetle, averts its face (to avoid the spray), turns the beetle upside down so that the gland discharges into the ground, and eats the beetle from the head down **(Figure 30.11)**.

a. *Eleodes* **bettle**   **b. Grasshopper mouse**

Figure 30.11

Defence and learning. When confronted by a predator, **(a)** the headstand beetle (*Eleodes longicollis*) raises its abdomen and sprays a noxious chemical from its hind end. Experienced grasshopper mice (*Onychomys leucogaster*) **(b)** thwart the beetle's defence by grabbing it, turning it upside down, and eating it headfirst.

30.4 Competition

Different species using the same limiting resources experience **interspecific competition** (competition between species). Competing individuals may experience increased mortality and decreased reproduction, responses similar to the effects of intraspecific competition. Interspecific competition can reduce the size and population growth rate of one or more of the competing populations.

Community ecologists identify two main forms of interspecific competition. In **interference competition**, individuals of one species harm individuals of another species directly. Here animals may fight for access to resources, as when lions chase smaller predators, such as hyenas, jackals, and vultures, from their kills. Many plant species, including creosote bushes (see Figure 29.4, Chapter 29), release toxic chemicals into the soil, preventing other plants from growing nearby.

In **exploitative competition**, two or more populations use (*exploit*) the same limiting resource, and the presence of one species reduces resource availability for others. Exploitative competition need not involve snout-to-snout or root-to-root confrontations. In the deserts of the U.S. Southwest, many bird and ant species eat mainly seeds, and each seed-eating species may deplete the food supply available to others without necessarily encountering each other.

30.4a Competition and Niches: When Resources Are Limited

In the 1920s, the Russian mathematician Alfred J. Lotka and the Italian biologist Vito Volterra independently proposed a model of interspecific competition, modifying the logistic equation (see Chapter 29) to describe the effects of competition between two species. In their model, an increase in the size of one population reduces the population growth rate of the other.

In the 1930s, a Russian biologist, G. F. Gause, tested the model experimentally. He grew cultures of two *Paramecium* species (ciliate protozoans) under constant laboratory conditions, regularly renewing food and removing wastes. Both species feed on bacteria suspended in the culture medium. When grown alone, each species exhibited logistic growth. When grown together in the same dish, *Paramecium aurelia* persisted at high density, but *Paramecium caudatum* was almost eliminated **(Figure 30.12)**. These results inspired Gause to define the **competitive exclusion principle**. Populations of two or more species cannot coexist indefinitely if they rely on the same limiting resources and exploit them in the same way. One species inevitably harvests resources more efficiently; produces more offspring than the other; and, by its actions, negatively affects the other species.

Ecologists developed the concept of the **ecological niche** to visualize resource use and the potential for interspecific competition in nature. They define a population's niche by the resources it uses and the environmental conditions it requires over its lifetime. In this context, niche includes food, shelter, and nutrients, as well as nondepletable abiotic conditions such as light intensity and temperature. In theory, an almost infinite variety of conditions and resources could contribute to a population's niche. In practice, ecologists usually identify the critical resources for

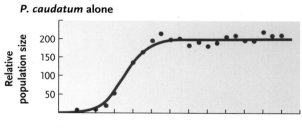

P. caudatum alone

P. aurelia alone

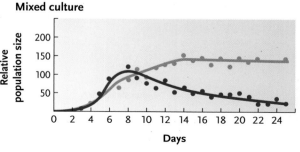

Mixed culture

Days

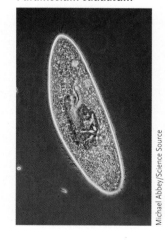

Paramecium caudatum

Michael Abbey/Science Source

Paramecium aurelia

Eric V. Crave/Science Source

Experimental Research Figure 30.12 Gause's experiments on interspecific competition in *Paramecium.*

MOLECULE BEHIND BIOLOGY 30.1

Taipoxin: Snake Presynaptic Phospholipase A$_2$ Neurotoxins

Snake venoms are typically a concoction of ingredients designed to immobilize and digest prey. Like those of the venom of nematocysts (see "Nematocysts," p. 720), the effects of snake venom can include symptoms associated with neurotoxins cardiotoxins, hemolytic actions, and digestion (necrosis) of tissues. Not all snakes (or other venomous animals) have the same venom.

Snake presynaptic phospholipase A$_2$ neurotoxins, or SPANs, have neurotoxic effects and work by blocking neuromuscular junctions. Phospholipase A$_2$ activity varies greatly among SPANs. Using mouse neuromuscular junction hemidiaphragm preparations and neurons in culture, M. Rigoni and seven colleagues explored the way in which SPANs work. They used SPANs from single-chain notexin (from *Notechis scutatus*, the eastern tiger snake); a two-subunit B-bungarotoxin (from *Bungarus multicinctus*, the many-banded krait); the three-subunit taipoxin (from *Oxyuranus scutellatus*, the taipan; **Figure 1**); and the five-unit textilotoxin (from *Pseudonaja textilis*, the eastern brown snake).

The results showed that administration of SPANs to neuromuscular junctions causes enlargement of the junctions and reduction in the contents of synaptic vesicles. SPANs also induce exocytosis of neurotransmitters. In other words, SPANs bind nerve terminals via receptors **(Figure 2)**; the

Figure 1

The taipan (Oxyuranus spp.) is an extremely venomous snake from northern Australia and southern New Guinea.

BMCL/Shutterstock.com

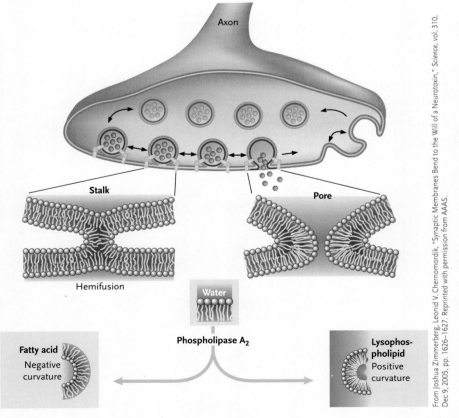

Figure 2

Taipoxin paralyzes the neuromuscular junction by causing membrane fusion between the plasma membranes (tan colour) of presynaptic ganglia and synaptic vesicles, forming a pore that allows mixing of lipids from inner (purple) and outer (green) leaflets. Although this is probably restricted by proteins (yellow ribbons) around the pore, phospholipase A$_2$ changes the curvature at the junction, resulting in the paralysis noted above.

From Joshua Zimmerberg, Leonid V. Chernomordik, "Synaptic Membranes Bend to the Will of a Neurotoxin," *Science*, vol. 310, Dec 9, 2005, pp. 1626–1627. Reprinted with permission from AAAS.

Axon

Stalk

Pore

Hemifusion

Water

Phospholipase A$_2$

Fatty acid
Negative curvature

Lysophospholipid
Positive curvature

results indicate that venoms can be used to further our understanding of what happens at neuromuscular junctions.

Among extant lepidosauran reptiles (snakes and lizards), two lineages have venom delivery systems, advanced snakes and helodermatid (gila monster) lizards. The view that the evolution of venom systems is fundamental to the radiation of snakes has been supported by the prevalence of venom among snakes and its restriction to just two species of lizards. Using tools of molecular genetics, B. G. Fry and 13 colleagues explored the early evolution of venom systems in lizards and snakes. The ancestral condition, in venomous lizards, is lobed, venom-secreting glands on upper and lower jaws. Advanced snakes and two lizards have more derived venom systems. They have one pair of venom glands, either upper or lower glands. Analysis of venoms indicates that snakes, iguanians (monitor lizards), and anguimorphs form a single clade **(Figure 3),** suggesting that venom is an ancestral trait in this evolutionary line of reptiles.

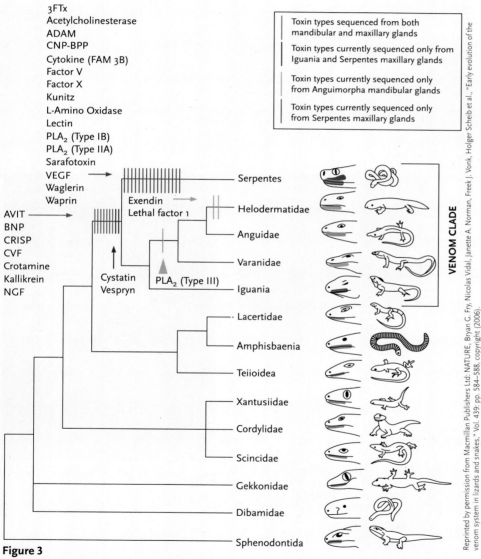

Reprinted by permission from Macmillan Publishers Ltd: NATURE, Bryan G. Fry, Nicolas Vidal, Janette A. Norman, Freek J. Vonk, Holger Scheib et al., "Early evolution of the venom system in lizards and snakes," Vol. 439: pp. 584–588, copyright (2006).

Figure 3

Snake and lizard phylogeny based on the appearance of venom. Shown here are the relative glandular development and appearance of toxin recruitment in squamate reptile phylogeny. Glands secreting mucus are blue, ancestral venom glands are red, and derived venom is orange. Elements in venom include the following: three-finger toxins (3FTx); a disintegrin and metalloproteinase (ADAM); C-type natriuretic peptide–bradykinin–potentiating peptide (CNP-BPP); cobra venom factor (CVF); nerve growth factor (NGF); and vascular endothelial growth factor (VEGF).

Nematocysts

Swimmers at ocean beaches in warmer parts of the world can be exposed to stings from the Portuguese man-o'-war. In the United States, at least three human deaths have been caused by exposure to the venom of its nematocysts **(Figure 1).** First aid for someone who has been stung includes (a) using sea water to flush away any tentacles still clinging to the victim (or picking them off if necessary), (b) applying ice or cold packs to the area of the sting(s) and leaving them in place for 5 to 15 minutes, (c) using an inhaled analgesic to reduce pain, and (d) seeking additional medical aid.

Beaches where Portuguese man-o'-war and other jellyfish may occur must be supervised by lifeguards who understand the danger. The inflatable bladders (sails; see **Figure 2**) make Portuguese man-o'-war easy to see in the water, and swimming is not permitted when they are near.

In 1968, lifeguards and others at a beach at Port Stephens, New South Wales, in Australia were surprised and concerned when they realized that some people had been stung by Portuguese man-o'-war when none of these animals had been spotted near the beach. The stinging animals turned out to be sea slugs, *Glaucus* species.

Since 1903, it had been known that sea slugs use nematocysts as a defence.

Glaucus atlanticus **(Figure 3)** feed on the cnidosacs that contain the nematocysts, preferentially selecting and storing those of Portuguese man-o'-war, which have two sizes of nematocysts. The sea slugs take the larger nematocysts that, when discharged, have the longest penetrants. It is likely that the same digestive processes other sea slugs use to extract chloroplasts can be used to extract nematocysts.

This situation demonstrates the versatility of defensive systems in animals.

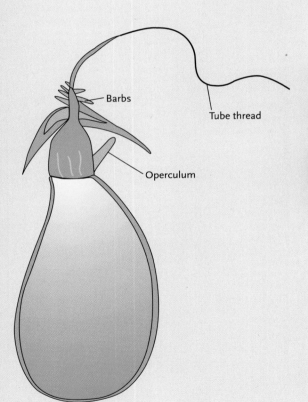

Figure 1
Nematocyst. Nematocysts are stinging cells occurring in animals in the phylum Cnidaria (see Chapter 27). Nematocysts of Physalia physalis *(Portuguese man-o'-war; Figure 2) contain toxic proteins and at least six or seven enzymes that can be injurious. Unpurified nematocyst venoms have several effects, some of which can be lethal. The venoms can be neurotoxic, cardiotoxic, or myotoxic or cause lysis of red blood cells or mitochondria. Like other venoms, nematocyst venom from the Portuguese man-o'-war can interfere with the transport of* Na^+ *and* Ca^{2+} *ions.*

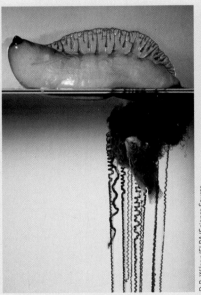

D.P. Wilson/FLPA/Science Source

Figure 2
Physalia physalis, *Portuguese man-o'-war.*

Dr. T.E. Thompson/Science Source

Figure 3
Glaucus atlanticus, *a sea slug that ingests nematocysts from* Physalia physalis.

which populations might compete. Sunlight, soil moisture, and inorganic nutrients are important resources for plants, so differences in leaf height and root depth, for example, can affect plants' access to these resources. Food type, food size, and nesting sites are important for animals. When several species coexist, they often use food and nest resources in different ways.

Ecologists distinguish the **fundamental niche** of a species, the range of conditions and resources it could tolerate and use, from its **realized niche**, the range of conditions and resources it actually uses in nature. Realized niches are smaller than fundamental niches, partly because all tolerable conditions are not always present in a habitat and partly because some resources are used by other species. We can visualize competition between two populations by plotting their fundamental and realized niches with respect to one or more resources **(Figure 30.13)**. If the fundamental niches of two populations overlap, they *might* compete in nature.

Observing that several species use the same resource does not demonstrate that competition occurs (or does not occur). All terrestrial animals consume oxygen but do not compete for oxygen because it is usually plentiful. Nevertheless, two general observations provide *indirect* evidence that interspecific competition may have important effects.

Resource partitioning occurs when several sympatric (living in the same place) species use different resources or the same resources in different ways. Although plants might compete for water and dissolved nutrients, they may avoid competition by partitioning these resources, collecting them from different depths in the soil **(Figure 30.14)**. This allows coexistence of different species.

Character displacement can be evident when comparing species that are sometimes sympatric and sometimes allopatric (living in different places). Allopatric populations of some animal species are morphologically similar and use similar resources, whereas sympatric populations are morphologically different and use different resources. Differences between sympatric species allow them to coexist without competing. Allen Keast studied honey-eaters (family Meliphagidae), a group of birds from Australia, to illustrate this situation. In mainland Australia, up to six species in the genus *Melithreptus* occur in some habitats. Just off the coast of Kangaroo Island, there are two species. When two species are sympatric, each feeds in a wider range of situations than when six species live in the same area, reflecting the use of broader niches. Behavioural and morphological differences are evident when species are compared between the different situations. Although well known for his work on birds, Keast also studied communities of fish. He spent most of his academic career at Queen's University in Kingston, Ontario.

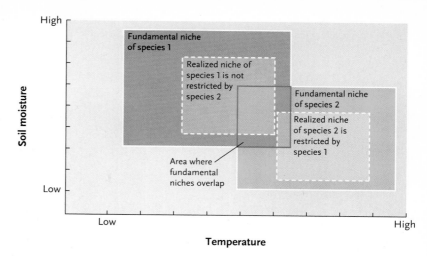

Figure 30.13

Fundamental versus realized niches. In this hypothetical example, both species 1 and species 2 can survive intermediate temperature conditions, as indicated by the shading where their fundamental niches overlap. Because species 1 actually occupies most of this overlap zone, its realized niche is not much affected by the presence of species 2. In contrast, the realized niche of species 2 is restricted by the presence of species 1, and species 2 occupies warmer and drier parts of the habitat.

Data on resource partitioning and character displacement suggest, but do not prove, that interspecific competition is an important selective force in nature. To demonstrate *conclusively* that interspecific competition limits natural populations, one must show that the presence of one population reduces the population size or density of its presumed competitor. In a classic field experiment, Joseph Connell examined

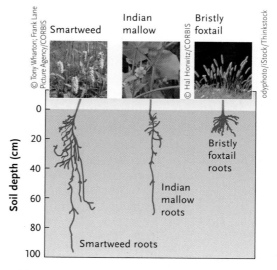

Figure 30.14

Resource partitioning. The root systems of three plant species that grow in abandoned fields partition water and nutrient resources in soil. Bristly foxtail grass (*Setaria faberi*) has a shallow root system, Indian mallow (*Abutilon theophrasti*) has a moderately deep taproot, and smartweed (*Polygonum pennsylvanicum*) has a deep taproot that branches at many depths.

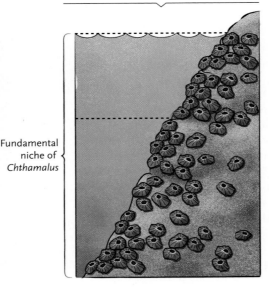

Realized niches before experimental treatments.

High tide

Chthamalus

Low tide

Balanus

Treatment 1: Remove *Balanus*
In the absence of *Balanus*, *Chthamalus* occupies both shallow water and deep water.

Fundamental niche of *Chthamalus*

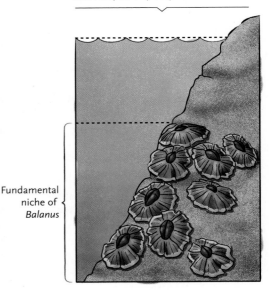

Treatment 2: Remove *Chthamalus*
In the absence of *Chthamalus*, *Balanus* still occupies only deep water.

Fundamental niche of *Balanus*

Experimental Research Figure 30.15 Demonstration of competition between two species of barnacles.

competition between two barnacle species **(Figure 30.15).** Connell first observed the distributions of both species of barnacles in undisturbed habitats to establish a reference baseline. *Chthamalus stellatus* is generally found in shallow water on rocky coasts, where it is periodically exposed to air. *Balanus balanoides* typically lives in deeper water, where it is usually submerged.

In the absence of *Balanus* on rocks in deep water, larval *Chthamalus* colonized the area and produced a flourishing population of adults. *Balanus* physically displaced *Chthamalus* from these rocks. Thus, interference competition from *Balanus* prevents *Chthamalus* from occupying areas where it would otherwise live. Removal of *Chthamalus* from rocks in shallow water did not result in colonization by *Balanus*. *Balanus* apparently cannot live in habitats that are frequently exposed to air. Connell concluded that there was competition between the two species. But competition was asymmetrical because *Chthamalus* did not affect the distribution of *Balanus*, whereas *Balanus* had a substantial effect on *Chthamalus*.

30.4b Symbiosis: Close Associations

Symbiosis occurs when one species has a physically close ecological association with another (*sym* = together; *bio* = life; *sis* = process). Biologists define three types of symbiotic interactions: commensalism, mutualism, and parasitism (see Table 30.1, p. 711).

In **commensalism,** one species benefits from and the other is unaffected by the interactions. Commensalism appears to be rare in nature because few species are unaffected by interactions with another. One possible example is the relationship between Cattle Egrets (*Bubulcus ibis,* birds in the heron family) and the large grazing mammals with which they associate **(Figure 30.16).** Cattle Egrets eat insects and other small animals that their commensal partners flush from grass. Feeding rates of Cattle Egrets are higher when they associate with large grazers than when they do not. The birds clearly benefit from this interaction, but the presence of birds has no apparent positive or negative impact on the mammals.

M.B. Fenton

Figure 30.16
Commensalism. Cattle Egrets (*Bubulcus ibis*) feed on insects and other small animals flushed by the movements of large grazing animals such as African elephants (*Loxodonta africana*).

a. Flowering yucca plant

SeanPavonePhoto/Shutterstock.com

b. Female yucca moth

Dan Suzio/Science Source

A female yucca moth uses highly modified mouthparts to gather the sticky pollen and roll it into a ball. She carries the pollen to another flower, and after piercing its ovary wall, she lays her eggs. She then places the pollen ball into the opening of the stigma.

c. Yucca moth larva

Dan Suzio/Science Source

When moth larvae hatch from the eggs, they eat some of the yucca seeds and gnaw their way out of the ovary to complete their life cycle. Enough seeds remain undamaged to produce a new generation of yuccas.

Figure 30.17
Mutualism between plants and animals. Several species of yucca plants (*Yucca* species) are each pollinated exclusively by one species of moth (*Tegeticula* species). The adult moth appears at the time of year when the plants are flowering. These species are so mutually interdependent that the larvae of each moth species can feed on only one species of yucca, and each yucca plant can be pollinated by only one species of moth. Extinction of the pollinator will usually lead to extinction of the plant. Most plant–animal mutualisms are less specific.

In **mutualism**, both partners benefit. Mutualism appears to be common and includes coevolved relationships between flowering plants and animal pollinators. Animals that feed on a plants' nectar or pollen carry the plants' gametes from one flower to another **(Figure 30.17)**. Similarly, animals that eat fruits disperse the seeds and "plant" them in piles of nutrient-rich feces. Mutualistic relationships between plants and animals do not require active cooperation, as each species simply exploits the other for its own benefit. Some associations between bacteria and plants are mutualistic. Perhaps the most important of these associations is between *Rhizobium* and leguminous plants such as peas, beans, and clover (see Chapter 37).

Mutualistic relationships between animal species are common. Cleaner fishes, small marine species, feed on parasites attached to the mouths, gills, and bodies of larger fishes **(Figure 30.18)**. Parasitized fishes hover motionless while cleaners remove their ectoparasites. The relationship is mutualistic because cleaner fishes get a meal, and larger fishes are relieved of parasites.

The relationship between the bull's horn acacia tree (*Acacia cornigera*) of Central America and a species of small ants (*Pseudomyrmex ferruginea*) is a highly coevolved mutualism **(Figure 30.19, p. 724)**. Each acacia is inhabited by an ant colony that lives in the tree's swollen thorns. Ants swarm out of the thorns to sting, and sometimes kill, herbivores that touch the tree. Ants also clip any vegetation that grows nearby. Acacia trees colonized by ants grow in a space free of herbivores and competitors, and occupied trees grow faster and produce more seeds than unoccupied trees. In return, the plants produce sugar-rich nectar consumed by adult ants and protein-rich structures that the ants feed to their larvae. Ecologists describe the coevolved mutualism between these species as *obligatory*, at least for the ants, because they cannot subsist on any other food sources.

Many animals eat honey and sometimes also the bees that produce it. In Africa, Greater Honey-Guides (*Indicator indicator*) are birds that use a special guiding display to lead humans to beehives. In one tribe of Kenyans, the honey-gathering Borans use a special whistle to call *I. indicator*. Boran honey-gatherers that work with Greater Honey-Guides are much more efficient at finding beehives than those working alone. When the honey-gatherer goes to the hive and raids it to obtain honey, Greater Honey-Guides help themselves to bee larvae, left-over honey, and wax. Although *I. indicator* are said also to guide ratels (honey badgers, *Mellivora capensis*) to beehives, there are no firm data supporting this assumption.

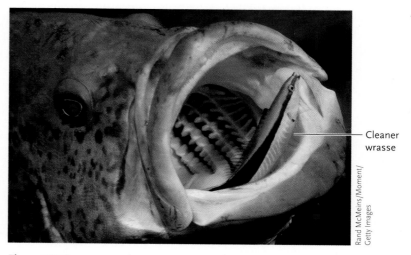

Cleaner wrasse

Rand McMeins/Moment/Getty Images

Figure 30.18
Mutualism between animal species. A large potato cod (*Epinephelus tukula*) from the Great Barrier Reef in Australia remains nearly motionless in the water while a striped cleaner wrasse (*Labroides dimidiatus*) carefully removes and eats ectoparasites attached to its lip. The potato cod is a predator; the striped cleaner wrasse is a potential prey. Here the mutualistic relationship supersedes the possible predator–prey interaction.

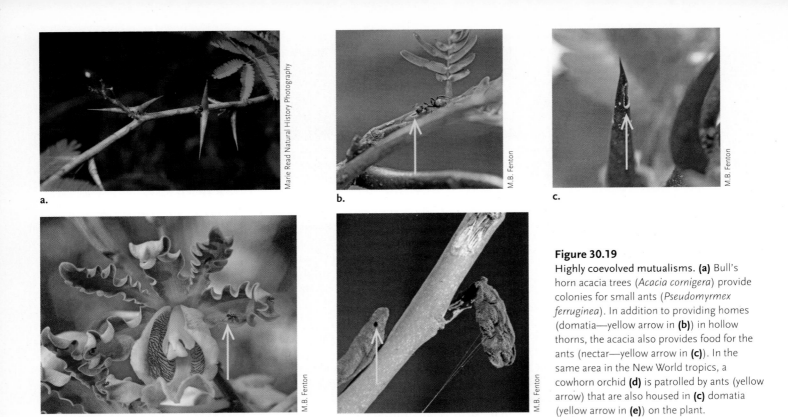

Figure 30.19
Highly coevolved mutualisms. **(a)** Bull's horn acacia trees (*Acacia cornigera*) provide colonies for small ants (*Pseudomyrmex ferruginea*). In addition to providing homes (domatia—yellow arrow in **(b)**) in hollow thorns, the acacia also provides food for the ants (nectar—yellow arrow in **(c)**). In the same area in the New World tropics, a cowhorn orchid **(d)** is patrolled by ants (yellow arrow) that are also housed in **(c)** domatia (yellow arrow in **(e)**) on the plant.

In **parasitism**, one species—the parasite—uses another—the host—in a way that is harmful to the host. Parasite–host relationships are often considered to be specialized predator–prey relationships because one population of organisms feeds on another. But parasites rarely kill their hosts because a dead host is not a continuing source of nourishment.

Endoparasites, such as tapeworms, flukes, and roundworms, live *within* a host. Many endoparasites acquire their hosts passively when a host accidentally ingests the parasites' eggs or larvae. Endoparasites generally complete their life cycle in one or two host individuals. Ectoparasites, such as leeches, aphids, and mosquitoes, feed on the exterior of the host. Most animal ectoparasites have elaborate sensory and behavioural mechanisms, allowing them to locate specific hosts, and they feed on numerous host individuals during their lifetimes. Plants such as mistletoes (genus *Phoradendron*) live as ectoparasites on the trunks and branches of trees; their roots penetrate the host's xylem and extract water and nutrients. These differ from epiphytes, such as bromeliads or Spanish moss, that use the host only as a base. Other plants are root parasites, for example, *Conopholis americana*.

Not all parasites feed directly on a host's tissues. Some bird species are brood parasites, laying their eggs in the host's nest. It is quite common for female birds such as Canvasback Ducks, Brown-headed Cowbirds, and Kirtland's Warblers (*Aythya valisineria*) to lay their eggs in the nests of conspecifics (members of the same species). Some species of songbirds often lay some eggs in the nests of others, a variation on hedging of genetic bets and on extra-pair copulations (see Chapters 41 and 47, respectively). Brood parasitism is the next level of escalation in this spectrum of parasitism. Brown-headed Cowbirds (*Molothrus ater*), like other brood parasites, always lay their eggs in the nest of other species, leaving it to the host parents to raise their young. This behaviour can have drastic repercussions for host species. Brown-headed Cowbirds, for instance, have played a large role in the near-extinction of Kirtland's Warblers (*Dendroica kirtlandii*).

The feeding habits of insects called parasitoids fall somewhere between true parasitism and predation. A female parasitoid lays her eggs in a larva or pupa of another insect species, and her young consume the tissues of the living host. But the parasitoid spends part of its life cycle as free living. It is the larval stage that usually kills the host. Because the hosts chosen by most parasitoids are highly specific, agricultural ecologists often try to use parasitoids to control populations of insect pests.

One of the most striking and perhaps startling example of symbioses is the rich biota of prokaryotes and Protozoa that inhabit our digestive tracts. This biota significantly expands our capacity for extracting nutrients and other important factors from the food we ingest. The producers of "probiotic" foods depend upon our being impressed by the importance of our symbionts.

STUDY BREAK

1. What is interspecific competition? What two types of interspecific competition have been identified by community ecologists?
2. Describe the competitive exclusion principle.
3. What is a species' ecological niche? How does a fundamental niche differ from a realized niche?

30.5 The Nature of Ecological Communities

The interactions that occur among species in an ecological community can be broadly categorized as antagonistic or mutually beneficial **(Figure 30.20)**. Trophic interactions are those associated with consumption (antagonistic)—one species eating another—the usual situation portrayed in food webs. Mutually beneficial interactions include, for example, those between flowering plants and their insect pollinators. Understanding community dynamics requires knowledge of the structure of community networks (e.g., food webs; see Chapter 31) as well as information about how structure influences the extinction or persistence of species. An overview of the dynamics of an ecological community

is obtained through a combination of fieldwork and attendant statistical analysis of data to document ecosystem architecture. The second element, knowledge of the influence of architecture on species persistence, emerges from mathematical modelling.

To explore the nature of ecosystems, Elisa Thebault and Colin Fontaine examined pollination (mutualistic) and plant–herbivore (trophic) systems **(Figure 30.21, p. 726)**. They found that the structure of the network favouring ecosystem **stability** differs between trophic (herbivore) and mutually beneficial (pollination) networks. In pollination networks, the elements are highly connected and nested, promoting stability of communities. In herbivore networks, stability is greater in structures that are compartmentalized and weakly connected. The work identifies features that affect the stability of ecosystems, potentially informing those working to effect conservation at the system level.

Ecotones, the borders between communities, are sometimes wide transition zones. Ecotones are generally species rich because they include plants and animals from both neighbouring communities, as well as some species that thrive only under transitional conditions. Although ecotones are usually relatively broad, places where there is a discontinuity in a critical resource or important abiotic factor may have a sharp community boundary. Chemical differences between soils derived from serpentine rock and sandstone establish sharp boundaries between communities of native California wildflowers and introduced European grasses.

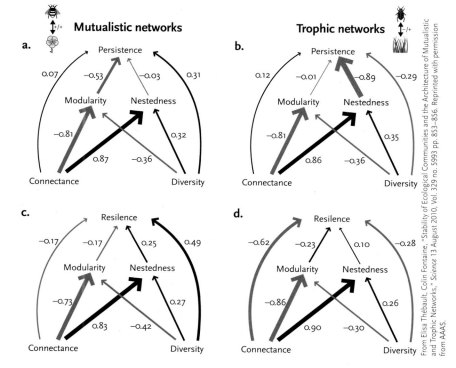

From Elisa Thébault, Colin Fontaine, "Stability of Ecological Communities and the Architecture of Mutualistic and Trophic Networks," *Science* 13 August 2010, Vol. 329 no. 5993 pp. 853–856. Reprinted with permission from AAAS.

Figure 30.20

The persistence **(a, b)** and resilience **(c, d)** of pollinating and herbivore ecosystem patterns are summarized, revealing important differences. The thickness of arrows, scaled to standardized coefficients, illustrates the relative strength of the effects. Red identifies negative effects; black, positive ones. There is a further comparison of the effects of connectance and diversity, comparing direct and indirect effects, considering modularity and nestedness. The numbers in each diagram indicate the coefficients along the path.

a. Interactive hypothesis

The interactive hypothesis predicts that species within communities exhibit similar distributions along environmental gradients (indicated by the close alignment of several curves over each section of the gradient) and that boundaries between communities (indicated by arrows) are sharp.

b. Individualistic hypothesis

The individualistic hypothesis predicts that species distributions along the gradient are independent (indicated by the lack of alignment of the curves) and that sharp boundaries do not separate communities.

c. Siskiyou Mountains

Most gradient analyses support the individualistic hypothesis, as illustrated by distributions of tree species along moisture gradients in Oregon's Siskiyou Mountains and Arizona's Santa Catalina Mountains.

d. Santa Catalina Mountains

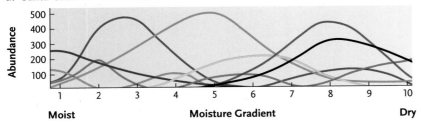

Figure 30.21
Two views of ecological communities.
Each graph line indicates a different species.

30.6 Community Characteristics

Growth forms (sizes and shapes) of plants vary markedly in different environments, so the appearances of plants can often be used to characterize communities. Warm, moist environments support complex vegetation with multiple vertical layers. Tropical forests include a canopy formed by the tallest trees, an understorey of shorter trees and shrubs, and a herb layer under openings in the canopy. Vinelike lianas and epiphytes grow on the trunks and branches of trees **(Figure 30.22).** In contrast, physically harsh environments are occupied by low vegetation with simple structure. Trees on mountainsides buffeted by cold winds are short, and the plants below them cling to rocks and soil.

Other environments support growth forms between these extremes.

Communities differ greatly in **species richness,** the number of species that live within them. The harsh environment on a low desert island may support just a few species of microorganisms, fungi, algae, plants, and arthropods. In contrast, tropical forests that grow under milder physical conditions include many thousands of species. Ecologists have studied global patterns of species richness (see Chapter 32) for decades. Today, as human disturbance of natural communities has reached a crisis point, conservation biologists try to understand global patterns of species richness to determine which regions of Earth are most in need of preservation.

The relative abundances of species vary across communities. Some communities have one or two abundant species and a number of rare species. In others, the species are represented by more equal numbers of individuals. In a **temperate deciduous forest** in southern Quebec, red oak trees (*Quercus rubra*) and sugar maples (*Acer saccharum*) might together account for nearly 85% of the trees. A tropical forest in Costa Rica may have more than 200 tree species, each making up a small percentage of the total.

Some Perils of Mutualism

Living organisms offer many examples of mutualistic interactions in which one species (or group of species) shows varying levels of dependence on another or others. Mutualistic situations can place species on the edge of survival. Where one species depends entirely on another, the extinction of one must lead to change or the extinction of both (e.g., Dodos—see Chapter 32—and yucca plants and their moths). There are many other examples of close relationships, including a desert melon (*Cucumis humifructus*) that depends perhaps entirely on aardvarks (*Orycteropus afer*) for dispersal of its seeds. Aardvarks sniff out the underground melons, dig them up, and eat them to obtain water. When aardvarks bury

their dung, they plant the melon's seeds and fertilize them. The survival of the melon depends on the aardvark but not vice versa.

Mutualistic interactions between species can be even more complex. In the African **savannah**, ants often live in mutualistic relationships with trees. In east Africa, whistling thorn acacia trees (*Acacia drepanolobium*) are host to four species of ants (see Figure 30.19, p. 724). One species of ant (*Crematogaster mimosae*) in particular depends on room (hollows in swollen thorns, called domatia) and board (carbohydrates secreted from extrafloral glands and the bases of leaves) provided by the trees. Another species of ant (*Crematogaster sjostedti*) also lives on the trees but usually nests in holes made by cerambycid beetles

that burrow into and harm the trees.

The ants, particularly *C. mimosae*, attack animals that attempt to browse on the foliage or branches of *A. drepanolobium*. They deter many herbivores, from large mammals to wood-boring beetles (such as cerambycids). If large, browsing mammals are excluded from the area, *A. drepanolobium* produce fewer domatia and fewer carbohydrates for *C. mimosae*. The decline in this species of ant leads to higher damage by cerambycid beetles and increases in populations of *C. sjostedti*.

Many other plants also use ants as mercenaries (see Figure 30.19), and it is becoming clear that survival of these systems depends on the continued presence of participating species.

Figure 30.22
Layered forests. Tropical forests, such as one near the Mazaruni River in Guyana (South America), include a canopy of tall trees and an understorey of short trees and shrubs. Huge vines (lianas) climb through the trees, eventually reaching sunlight in the canopy. Epiphytic plants grow on trunks and branches, increasing the structural complexity of the habitat.

The factors underlying diversity and community structure can be expected to vary among groups of organisms, and the interactions between very different groups of organisms can have positive effects on both. Using an experimental mycorrhizal plant system (see Chapter 25), H. Maherali and J. N. Klironomos found that after one year, the species richness of mycorrhizal fungi correlated with higher plant productivity. In turn, the diversity and species richness of mycorrhizal fungi were highest when their starting community had more distinct evolutionary lineages. This example illustrates the importance of diversity and interactions.

30.6a Measuring Species Diversity and Evenness: Calculating Indices

The number of species is the simplest measure of diversity, so a forest with four tree species has higher **species diversity** than one with two tree species. But there can

Figure 30.23

Species diversity. In this hypothetical example, each of three samples of forest communities (A, B, and C) contains 50 trees. Indices allow biologists to express the diversity of species and evenness of numbers (see Table 30.2).

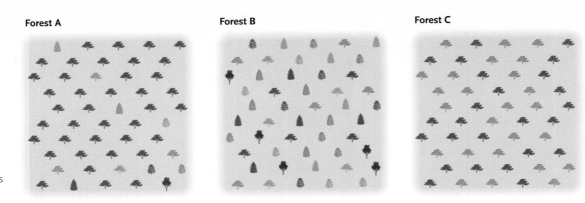

Forest A   Forest B   Forest C

be more to measuring diversity than just counting species. Biologists use indices of diversity to facilitate comparison of data sets documenting the numbers of species and of individuals. Shannon's index of diversity (H'), one commonly used measure, is calculated using the formula

$$H' = -\sum_{i=1}^{S} p_i \ln p_i$$

where S is the total number of species in the community (richness), p_i is the proportion of S made up by species I, and ln is the natural logarithm.

Another index, Shannon's evenness index (E_H), is calculated using the formula

$$E_H = \frac{H'}{\ln S}$$

where ln S is the natural logarithm of the number of species. Evenness is an indication of the mixture of species. Indices of diversity and evenness allow population biologists to objectively portray and compare the diversity of communities.

Use the two indices to compare the 3 forests of 50 trees each **(Figure 30.23)**. The number of species and number of individuals of each species in each forest are shown in **Table 30.2**. In Table 30.2, the values of H' and E_H indicate the diversity of the three hypothetical forests and the evenness of species representations. Lower values of H' and E_H suggest communities with few species (low H' values) or uneven distribution (low E_H values). Higher values of H' and E_H suggest a richer array of species with evenly distributed individuals.

Measures of diversity can be used to advantage. Ecologists refer to α diversity to represent the numbers of sympatric species in one community and β diversity to depict the numbers in a collection of communities. The number of herbivorous Lepidoptera species in one national park is α diversity, whereas β diversity is the number of species in the country in which the park is located. The trend to establish parks that cross international boundaries is a step toward recognizing the reality that political and biological boundaries can be quite different. Measures of diversity can be used directly in some conservation plans (see Chapter 32).

Table 30.2 | **Shannon's Indices for Measuring Diversity and Evenness**

Numbers of Individuals Per Species

	Forest A*	Forest B*	Forest C*
Species 1	39	5	25
Species 2	2	5	25
Species 3	2	5	0
Species 4	1	5	0
Species 5	1	5	0
Species 6	1	5	0
Species 7	1	5	0
Species 8	1	5	0
Species 9	1	5	0
Species 10	1	5	0
Shannon Indices			
H' diversity	0.6	2.3	0.7
E_h evenness	0.26	1.0	1.0

*Forests from Figure 30.23.

30.6b Trophic Interactions: Between Nourishment Levels

Every ecological community has trophic structure (*troph* = nourishment), comprising all plant–herbivore, predator–prey, host–parasite, and potential competitive interactions **(Figure 30.24)**. We can visualize the trophic structure of a community as a hierarchy of **trophic levels**, defined by the feeding relationships among its species (see Figure 30.24a). Photosynthetic organisms are primary producers, the first trophic level. Primary producers are photoautotrophs (*auto* = self) because they capture sunlight and convert it into chemical energy that is used to make larger organic molecules that plants can use directly. Plants are the main primary producers in terrestrial communities. Multicellular algae and plants are the major primary producers in shallow freshwater and marine environments, whereas photosynthetic protists and cyanobacteria play that role in deep, open water.

All **consumers** in a community (animals, fungi, and diverse microorganisms) are heterotrophs

a. Trophic levels **b. Marine food web**

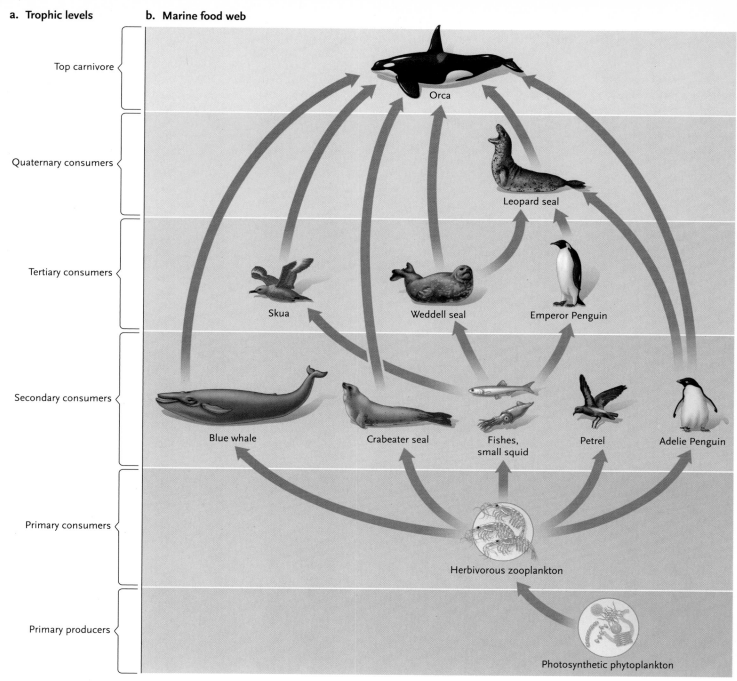

Figure 30.24
The marine food web off the coast of Antarctica.

(*hetero* = other) because they acquire energy and nutrients by eating other organisms or their remains. Animals are consumers. Herbivores (primary consumers) feed directly on plants and form the second trophic level. Secondary consumers (mesopredators) eat herbivores and form the third trophic level. Animals that eat secondary consumers make up the fourth trophic level, the **tertiary consumers.** At one meal, animals that are omnivores (e.g., humans, pigs, and bears) can act as primary, secondary, and tertiary consumers.

Detritivores (scavengers) form a separate and distinct trophic level. These organisms extract energy from organic detritus produced at other trophic levels. Detritivores include fungi, bacteria, and animals such as earthworms and vultures that ingest dead organisms, digestive wastes, and cast-off body parts such as leaves and exoskeletons. Decomposers, a type of detritivore, are small organisms, such as bacteria and fungi, that feed on dead or dying organic material. Detritivores and decomposers serve a critical ecological function because their activity reduces organic material to small inorganic molecules that producers can assimilate (see Chapter 22).

Although omnivores obviously do not fit exclusively into one trophic level, this can also be true of other organisms. Sea slugs that use chloroplasts or carnivorous plants are examples of species that do not fit readily into trophic categories.

30.6c Food Chains and Webs: Connections in Ecosystems

Ecologists use food chains and webs to illustrate the trophic structure of a community. Each link in a food chain is represented by an arrow pointing from food to consumer (see Figure 30.24b). Simple, straight-line food chains are rare in nature because most consumers feed on more than one type of food and because most organisms are eaten by more than one type of consumer. Complex relationships are portrayed as food webs—sets of interconnected food chains with multiple links.

In the food web for the waters off the coast of Antarctica (see Figure 30.24), primary producers and primary consumers are small organisms occurring in vast numbers. Microscopic diatoms (phytoplankton) are responsible for most photosynthesis, and small shrimplike krill (zooplankton) are the major primary consumers. These tiny organisms, in turn, are eaten by larger species such as fish and seabirds, as well as by suspension-feeding baleen whales. Some secondary consumers are eaten by birds and mammals at higher trophic levels. The top carnivore in this ecosystem, the orca, feeds on carnivorous birds and mammals.

Ideally, depictions of food webs would include all species in a community, from microorganisms to top consumer. But most ecologists simply cannot collect data on every species, particularly those that are rare or very small. Instead, they study links between the most important species and simplify analysis by grouping trophically similar species. Figure 30.24 categorizes the many different species of primary producers and primary consumers as phytoplankton and zooplankton, respectively.

Many biological *hot spots* (areas with many species) exist, from thermal vents on the floor of some oceans to deposits of bat guano in some caves. A more recently described example is icebergs drifting north from Antarctica. The icebergs can be hot spots of enrichment because of the nutrients and other materials they shed into surrounding waters. The water around two free-drifting icebergs (0.1 km^2 and 30.8 km^2 in area) was sampled in the Weddell Sea. High concentrations of chlorophyll, krill, and seabirds extended about 3.7 km around each iceberg. These data, reported by K. L. Smith Jr. and seven colleagues, demonstrate that icebergs can have substantial effects on pelagic ecosystems.

In the late 1950s, Robert MacArthur analyzed food webs to determine how the many links between trophic levels may contribute to a community's stability. The stability of a community is defined as its ability to maintain **species composition** and relative abundances when environmental disturbances eliminate some

Barcode of Life

To calculate the diversity of organisms using an index such as the Shannon–Weaver one described in the text, you need to know how many different species are in your sample. The diversity of species can be overwhelming, so it is difficult to provide a confident estimate of how many species remain undescribed. For many groups of organisms, there may be very few authorities able to identify species and provide descriptions of "new" species, the ones not yet described and therefore nameless. The Barcode of Life Data Systems, based in Guelph, Ontario, offers one alternative to the challenge of knowing how many species are in the sample you have just acquired, or the origin of a mysterious mouse found in a shipment of frozen chickens from Thailand.

The Barcode of Life project depends upon variation in the mitochondrial cytochrome *c oxidase 1* (CO1) gene consisting of about 650 nucleotides. This genetic barcode is embedded in almost every cell and offers biologists a chance to identify a species even if they have only a small sample of feathers or fur, a leaf, a seed, or a caterpillar. Since identification of some species depends upon having a whole adult specimen, being able to make an identification from an egg, a larva, or a hair offers enormous potential. Identification of organisms with different life stages can be particularly challenging. Using morphology, it can be easy to identify a butterfly or a frog, but much more difficult to identify its caterpillar or its tadpole.

The Barcode of Life project is based on polymerase chain reaction (PCR) technology, which allowed biologists to process 100 samples every three hours. Subsequent advances in genomic technology have increased our capacity for efficient sequencing of DNA. The combination of this potential, an army of researchers collecting specimens, and global positioning satellite (GPS) technology to document locations means that the Barcode of Life project can deliver accurate (to 97.5%) identifications of specimens in a short time. Further developments could see biologists and naturalists armed with appropriately programmed handheld devices to obtain in-field identifications.

One important consequence of this project is that biologists will have a fighting chance to document more fully the diversity of life on Earth. On one hand, this means realizing that one species of the butterfly *Astraptes fulgerator* is actually ten species, or that what people had thought were several species is, in fact, one. Protecting species through CITES, the Convention on International Trade in Endangered Species, means being able to name them so that they can be placed on a protected list. The Barcode of Life project should allow a merchant to be sure that the ivory being sold in her shop is from an extinct mammoth, rather than a living species said to be endangered (Chapter 32). The same applies to food species in a market—is that fish really what the label says?

Everyone has experience with barcode operations because they are used in many retail outlets and therefore we all know that barcodes and readers do not always work. These limitations, as well as biological ones associated with genetics of different species, make some organisms more appropriate for Barcode of Life approaches than others.

International Barcode of Life

M.B. Fenton

*Shown here is **(a)** the bar code of **(b)** a flying fringe-lipped bat* (Trachops cirrhosus).

species from the community. MacArthur hypothesized that in species-rich communities, where animals feed on many food sources, the absence of one or two species would have only minor effects on the structure and stability of the community as a whole. He proposed a connection between species diversity, food web complexity, and community stability.

Subsequent research has confirmed MacArthur's reasoning. The average number of links per species generally increases with increasing species richness. Comparative food web analysis reveals that the relative proportions of species at the highest, middle, and lowest trophic levels are reasonably constant across communities. In 92 communities, MacArthur found two or three prey species per predator species, regardless of species richness.

Interactions among species in most food webs can be complex, indirect, and hard to unravel. In contrast, rodents and ants living in desert communities of the U.S. Southwest potentially compete for seeds, their main food source. Plants that produce the seeds compete for water, nutrients, and space. Rodents generally prefer to eat large seeds, whereas ants prefer small seeds. Thus, feeding by rodents reduces the potential population sizes of plants that produce large seeds. As a result, the population sizes of plants that produce small seeds may increase, ultimately providing more food for ants (see Chapter 48). Compared with the Antarctic system described above (see Figure 30.24), this community is not particularly complex.

STUDY BREAK

1. Why are indices important for population biologists? What do Shannon's indices measure?
2. Differentiate between α and β diversity.
3. Are herbivores primary or secondary consumers? Which trophic level do they form? Where do omnivores belong?

30.7 Effects of Population Interactions on Community Structure

Observations of resource partitioning and character displacement suggested that some process had fostered differences in resource use among coexisting species, and competition provided the most straightforward explanation of these patterns.

Interspecific competition can cause local extinction of species or prevent new species from becoming established in a community, reducing its species richness. During the 1960s and early 1970s, ecologists emphasized competition as the primary factor structuring communities.

30.7a Competition: More Than One Species Competing for a Resource

To further explore the role of competition, ecologists undertook field experiments on competition in natural populations. The experiment on barnacles (see Figure 30.15, p. 722) is typical of this approach—the impact on one species' potential competitors of adding or removing another species changed patterns of distribution or population size. The picture that emerges from the results of these experiments is not clear, even to ecologists. In the early 1980s, Joseph Connell surveyed 527 published experiments on 215 species. He found that competition was demonstrated in roughly 40% of the experiments and more than 50% of species. At the same time, Thomas W. Schoener used different criteria to evaluate 164 experiments on approximately 400 species. He found that competition affected more than 75% of species.

It is not surprising that there is no single answer to the question about how competition works in and influences communities. Plant and vertebrate ecologists working with *K*-selected species generally believe that competition has a profound effect on species distributions and resource use. Insect and marine ecologists working with *r*-selected species argue that competition is not the major force governing community structure, pointing instead to predation or parasitism and physical disturbance. We know that even categorizing a species as *r*- or *K*-selected is open to discussion (see Chapter 29).

30.7b Feeding

Predators can influence the species richness and structure of communities by reducing the sizes of prey populations. On the rocky coast of British Columbia, different species that fill different trophic roles compete for attachment sites on rocks, a requirement for life on a wave-swept shore. Mussels are the strongest competitors for space, eliminating other species from the community (see "Effect of a Predator on the Species Richness of Its Prey"). At some sites, predatory sea stars preferentially eat mussels, reducing their numbers and creating space for other species to grow. Because the interaction between *Pisaster* and *Mytilus* affects other species as well, it qualifies as a strong interaction.

In the 1960s, Robert Paine used removal experiments to evaluate the effects of predation by *Pisaster* (see "Effect of a Predator on the Species Richness of Its Prey"). In predator-free experimental plots, mussels outcompeted barnacles, chitons, limpets, and other invertebrate herbivores, reducing species richness from 15 species to 8. In control plots containing predators, all 15 species persisted. Ecologists describe predators such as *Pisaster* as **keystone species**, defined as species with a greater effect on community structure than their numbers might suggest. Snowshoe hares (Chapter 29) are candidates to be keystone species in boreal forest ecosystems because they are prey for a range of predators. Pallas' long-tongued bats may emerge as keystone species because, as we have seen, they eat insects and fruit as well as nectar and pollen.

Herbivores also exert complex effects on communities. In the 1970s, Jane Lubchenco studied herbivory in a periwinkle snail, believed to be a keystone species on rocky shores in Massachusetts (see "The Complex Effects of a Herbivorous Snail on Algal Species Richness," p. 734). The features of plants and algae and the food preferences of animals that eat them can influence community structure.

STUDY BREAK

1. How does the importance of competition vary between *K*-selected and *r*-selected species?
2. Does predation or herbivory increase or decrease species richness? Explain.
3. What is a keystone species?

30.8 Effects of Disturbance on Community Characteristics

Recent research tends to support the individualistic view that many communities are not in equilibrium and that species composition changes frequently. Environmental disturbances such as storms, landslides, fires, floods, avalanches, and cold spells often eliminate some species and provide opportunities for others to become established. Frequent disturbances keep some ecological communities in a constant state of flux.

Physical disturbances are common in some environments. Lightning-induced fires commonly sweep through grasslands, powerful hurricanes often demolish patches of forest and coastal habitats, and waves wash over communities at the edge of the sea and sweep away organisms as well as landforms and other structures.

Joseph Connell and his colleagues conducted an ambitious long-term study of the effects of disturbance on coral reefs, shallow tropical marine habitats that are among the most species-rich communities on Earth. In some parts of the world, reefs are routinely battered by violent storms that wash corals off the substrate, creating bare patches in the reef. The scouring action of storms creates opportunities for coral larvae to settle on bare substrates and start new colonies.

From 1963 to 1992, Connell and his colleagues tracked the fate of the Heron Island Reef at the south end of Australia's Great Barrier Reef **(Figure 30.25)**. The inner flat and protected crests of the reef are sheltered from severe wave action during storms, whereas some pools and crests are routinely exposed to physical disturbance. Because corals live in colonies of variable size, the researchers monitored coral abundance by measuring the percentage of the substrate (i.e., the sea floor) that colonies covered. They revisited marked study plots at intervals, photographing and identifying individual coral colonies.

Effect of a Predator on the Species Richness of Its Prey

Biologists used a predatory sea star (*Pisaster ochraceus*) to assess the influence a predator can have on species richness and relative abundance of prey **(Figure 1)**. *P. ochraceus* preferentially eats mussels (*Mytilus californicus*), one of the strongest competitors for space in rocky intertidal pools. Robert Paine removed *Pisaster* from caged experimental study plots, leaving control study plots undisturbed, and then monitored the species richness of *Pisaster*'s invertebrate prey over many years.

Paine documented an increase in mussel populations in the experimental plots as well as complex changes in the feeding relationships among species in the intertidal food web **(Figure 2)**. When he removed *P. ochraceus*, the top predator in this food web, he observed a rapid decrease in the species richness of invertebrates and algae. Species richness on control plots did not change over the course of the experiment.

Predation by *P. ochraceus* prevents mussels from outcompeting other invertebrates on rocky shores.

Figure 1

A predatory sea star (Pisaster ochraceus) *feeding on a mussel* (Mytilus californicus).

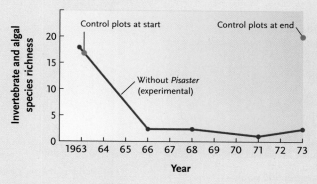

Figure 2

Changes in the species richness of invertebrates and algal species according to changes in populations of sea stars.

a. Exposed areas

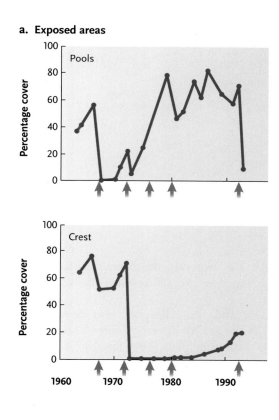

b. Sheltered areas

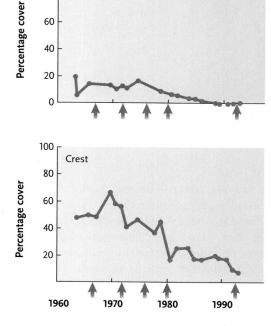

Figure 30.25

Major hydrodynamic disturbances to coral reefs, such as tsunamis and severe storms, **have important impacts on coral reefs.** Using oceanographic and engineering models, it is possible to predict the degree of dislodgement of benthic reef corals and, in this way, predict how coral shape and size indicate vulnerability to major disturbances. The use of these models is particularly important during times of climate change. The graphs show the effects of storms on corals. Five tropical cyclones (marked by grey arrows) damaged corals on the Heron Island Reef during a 30-year period. Storms reduced the percentage cover of corals in **(a)** exposed parts of the reef much more than in **(b)** sheltered parts of it. These data show that the 1970 event had the largest impact on some exposed and sheltered areas.

The Complex Effects of a Herbivorous Snail on Algal Species Richness

Experimental Research Box

Jane Lubchenco made enclosures that prevented periwinkle snails (*Littorina littorea*) from entering or leaving study plots in tide pools and on exposed rocks in rocky intertidal habitat **(Figure 1).** She then monitored the algal species composition in the plots, comparing them to the density of the periwinkles. In this way, she examined the influence of the periwinkles on the species richness of algae in intertidal communities.

The results varied dramatically between the study plots in tide pools and on exposed rocks. In tide pools, periwinkle snails preferentially ate *Enteromorpha*, the competitively dominant alga. At intermediate densities of *Enteromorpha*, the periwinkles remove some of these algae, allowing weakly competitive species to grow. The snails' grazing increases species richness. But grazing by periwinkles when *Enteromorpha* is at low or high densities reduces the species richness of algae in tide pools. On exposed rocks, where periwinkle snails rarely eat the competitively dominant alga *Chondrus*, feeding by snails reduces algal species richness **(Figure 2).**

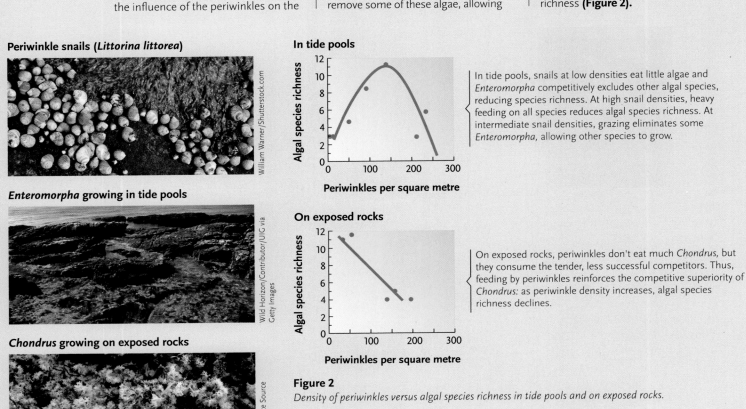

Periwinkle snails (*Littorina littorea*)

William Warner/Shutterstock.com

***Enteromorpha* growing in tide pools**

Wild Horizon/Contributor/UIG via Getty Images

***Chondrus* growing on exposed rocks**

Ted Kinsman/Science Source

Figure 1
The distribution of periwinkle snails and two kinds of algae.

In tide pools

In tide pools, snails at low densities eat little algae and *Enteromorpha* competitively excludes other algal species, reducing species richness. At high snail densities, heavy feeding on all species reduces algal species richness. At intermediate snail densities, grazing eliminates some *Enteromorpha*, allowing other species to grow.

On exposed rocks

On exposed rocks, periwinkles don't eat much *Chondrus*, but they consume the tender, less successful competitors. Thus, feeding by periwinkles reinforces the competitive superiority of *Chondrus:* as periwinkle density increases, algal species richness declines.

Figure 2
Density of periwinkles versus algal species richness in tide pools and on exposed rocks.

Five major cyclones crossed the reef during the 30-year study period. Coral communities in exposed areas of the reef were in a nearly continual state of flux. In exposed pools, four of the five cyclones reduced the percentage of cover, often drastically. On exposed crests, the cyclone of 1972 eliminated virtually all corals, and subsequent storms slowed the recovery of these areas for more than 20 years. In contrast, corals in sheltered areas suffered much less storm damage. Nevertheless, their coverage also declined steadily during the study as a natural consequence of the corals' growth. As colonies grew taller and closer to the ocean's surface, their increased exposure to air resulted in substantial mortality.

Connell and his colleagues also documented *recruitment,* the growth of new colonies from settling larvae, in their study plots. They discovered that the rate at which new colonies developed was almost always higher in sheltered than in exposed areas. Recruitment rates were extremely variable, depending in part on the amount of space that storms or coral growth had made available.

This long-term study of coral reefs illustrates that frequent disturbances prevent some communities from reaching an equilibrium determined by interspecific interactions. Changes in the coral reef community at Heron Island result from the effects of external disturbances that remove coral colonies from the reef, as well as internal processes (growth and recruitment) that either eliminate colonies or establish new ones. In this community, growth and recruitment are slow processes and disturbances are frequent. Thus, the community never attains equilibrium, and moderate levels of disturbance can foster high species richness.

The **intermediate disturbance hypothesis**, proposed by Connell in 1978, suggests that species richness is greatest in communities experiencing fairly frequent disturbances of moderate intensity. Moderate disturbances create openings for *r*-selected species to arrive and join the community while allowing *K*-selected species to survive. Thus, communities that experience intermediate levels of disturbance contain a rich mixture of species. Where disturbances are severe and frequent, communities include only *r*-selected species that complete their life cycles between catastrophes. Where disturbances are mild and rare, communities are dominated by long-lived *K*-selected species that competitively exclude other species from the community.

Several studies in diverse habitats have confirmed the predictions of the intermediate disturbance hypothesis. Colin R. Townsend and his colleagues studied the effects of disturbance at 54 stream sites in the Taieri River system in New Zealand. Disturbance occurs in these communities when water flow from heavy rains moves rocks, soil, and sand in the streambed, disrupting animal habitats. Townsend and his colleagues measured how much the substrate moved in different streambeds to develop an index of the intensity of disturbance. Their results indicate that species richness is highest in areas that experience intermediate levels of disturbance **(Figure 30.26)**.

Some ecologists have suggested that species-rich communities recover from disturbances more readily than less diverse communities. In the United States, David Tilman and his colleagues conducted large-scale

experiments in midwestern grasslands. They examined relationships between species number and the ability of communities to recover from disturbance. Grassland plots with high species richness recover from drought faster than plots with fewer species.

STUDY BREAK

1. What did Connell's 30-year study of coral reefs illustrate about the ability of communities to reach a state of equilibrium?
2. What is the intermediate disturbance hypothesis? Describe one study that supports this hypothesis.
3. How does species richness affect the rate of recovery following a disturbance?

30.9 Succession

Ecosystems change over time in a process called **succession**, the change from one community type to another.

30.9a Primary Succession: The First Steps

Primary succession begins when organisms first colonize habitats without soil, such as those created by erupting volcanoes and retreating glaciers **(Figure 30.27, p. 736)**. Lichens are often among the very first colonists (see Chapter 25), deriving nutrients from rain and bare rock. They secrete mild acids that erode rock surfaces, initiating the slow development of soil, which is enriched by the organic material lichens produce. After lichens modify a site, mosses (see Chapter 26) colonize patches of soil and grow quickly.

As soil accumulates, hardy, opportunistic plants (grasses, ferns, and broad-leaved herbs) colonize the site from surrounding areas. Their roots break up rock, and when they die, their decaying remains enrich the soil. Detritivores and decomposers facilitate these processes. As the soil becomes deeper and richer, increased moisture and nutrients support bushes and, eventually, trees. Late successional stages are often dominated by *K*-selected species with woody trunks and branches that position leaves in sunlight and large root systems that acquire water and nutrients from soil.

In the classical view of ecological succession, long-lived species, which replace themselves over time, eventually dominate a community, and new species join it only rarely. This relatively stable, late successional stage is called a **climax community** because the dominant vegetation replaces itself and persists until an environmental disturbance eliminates it and allows other species to invade. Local climate and soil conditions, the surrounding communities where colonizing species originate, and chance events determine the species composition of climax communities. We now know that even climax communities change slowly in response to environmental fluctuations.

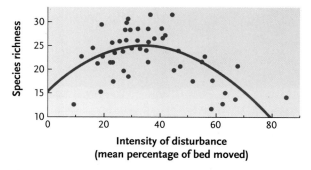

Figure 30.26

An observational study that supports the intermediate disturbance hypothesis. In the Taieri River system in New Zealand, species richness was highest in stream communities that experienced an intermediate level of disturbance.

1 The glacier has retreated about 8 m per year since 1794.

2 This site was covered with ice less than 10 years before this photo was taken. When a glacier retreats, a constant flow of melt water leaches minerals, especially nitrogen, from the newly exposed substrate.

3 Once lichens and mosses have established themselves, mountain avens (genus *Dryas*) grows on the nutrient-poor soil. This pioneer species benefits from the activity of mutualistic nitrogen-fixing bacteria, spreading rapidly over glacial till.

4 Within 20 years, shrubby willows (genus *Salix*), cottonwoods (genus *Populus*), and alders (genus *Alnus*) take hold in drainage channels. These species are also symbiotic with nitrogen-fixing microorganisms.

5 In time, young conifers, mostly hemlocks (genus *Tsuga*) and spruce (genus *Picea*), join the community.

6 As the years progress the smaller trees and shrubs are gradually replaced by larger trees.

Figure 30.27
Primary succession following glacial retreat. The retreat of glaciers at Glacier Bay, Alaska, has allowed ecologists to document primary succession on newly exposed rocks and soil.

30.9b Secondary Succession: Changes after Destruction

Secondary succession occurs after existing vegetation is destroyed or disrupted by an environmental disturbance, such as a fire, a storm, or human activity. The presence of soil makes disturbed sites ripe for colonization and may contain numerous seeds that germinate after disturbance. Early stages of secondary succession proceed rapidly, but later stages parallel those of primary succession.

30.9c Climax Communities: The Ultimate Ecosystems until Something Changes

Similar climax communities can arise from several different successional sequences. Hardwood forests can also develop in sites that were once ponds. During **aquatic succession**, debris from rivers and runoff accumulates in a pond, filling it to its margins. Ponds are first transformed into swamps, inhabited by plants adapted to a semisolid substrate. As larger plants get established, their high transpiration rates dry the soil, allowing other plant species to colonize. Given enough time, the site may become a meadow or forest in which an area of moist, low-lying ground is the only remnant of the original pond.

Because several characteristics of communities can change during succession, ecologists try to document how patterns change. First, because *r*-selected species are short lived and *K*-selected species are long lived, species composition changes rapidly in the early stages and more slowly in later stages of succession. Second, species richness increases rapidly during early stages because new species join the community faster than resident species become extinct. In later stages, species richness stabilizes or may even decline. Third, in terrestrial communities receiving sufficient rainfall, the maximum height and total mass of the vegetation increase steadily as large species replace small ones, creating the complex structure of the climax community.

Because plants influence the physical environment below them, the community itself increasingly moderates its **microclimate.** The shade cast by a forest canopy helps retain soil moisture and reduce temperature fluctuations. The trunks and canopy also reduce wind speed. In contrast, the short vegetation in an early successional stage does not effectively shelter the space below it.

Although ecologists usually describe succession in terms of vegetation, animals can show similar patterns. As the vegetation shifts, new resources become available, and animal species replace each other over time. Herbivorous insects, often with strict food preferences, undergo succession along with their food plants. And as herbivores change, so do their predators, parasites, and parasitoids. In old-field succession in eastern North America, different vegetation stages harbour a changing assortment of bird species **(Figure 30.28).**

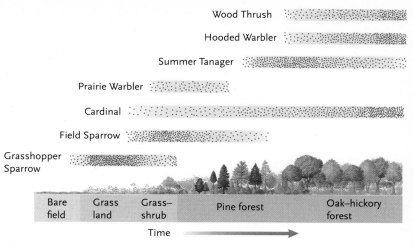

Figure 30.28

Succession in animals. Successional changes in bird species composition in an abandoned agricultural field in eastern North America parallel the changes in plant species composition. The residence times of several representative species are illustrated. The density of stippling inside each bar illustrates the density of each species through time.

Differences in dispersal abilities (see "Dispersal," p. 738), maturation rates, and life spans among species are partly responsible for ecological succession. Early successional stages harbour many *r*-selected species because they produce numerous small seeds that colonize open habitats and grow quickly. Mature successional stages are dominated by *K*-selected species because they are long lived. Nevertheless, coexisting populations inevitably affect one another. Although the role of population interactions in succession is generally acknowledged, ecologists debate the relative importance of processes that either facilitate or inhibit the turnover of species in a community.

30.9d Facilitation Hypothesis: One Species Makes Changes That Help Others

The **facilitation hypothesis** suggests that species modify the local environment in ways that make it less suitable for themselves but more suitable for colonization by species typical of the next successional stage. When lichens first colonize bare rock, they produce a small quantity of soil that is required by mosses and grasses that grow there later. According to this hypothesis, changes in species composition are both orderly and predictable because the presence of each stage facilitates the success of the next one. Facilitation is important in primary succession, but it may not be the best model of interactions that influence secondary succession.

30.9e Inhibition Hypothesis: One Species Negatively Affects Others

The **inhibition hypothesis** suggests that new species are prevented from occupying a community by species that are already present. According to this

Dispersal

Organisms often show astonishing dispersal abilities. In some cases, long-distance dispersal by plants in the Arctic is effected by the combination of strong winds and extensive expanses of ice and snow. The Svalbard Archipelago **(Figure 1)** is an interesting location for the study of plant dispersal. The islands were glaciated 20 thousand years ago, and it is likely that plants did not survive this condition. The fossil record indicates that plants have been present on Svalbard for fewer than 10 thousand years, although between 4000 and 9500 years ago, the climate was warmer there (by 1 to 2°C) than it is now.

Using DNA fingerprinting, I. G. Alsos and eight colleagues demonstrated that plant colonization of the Svalbard Archipelago has involved the arrival of plants from all possible adjacent regions **(Figure 2).** In eight of nine species, genetic evidence indicates multiple colonization events.

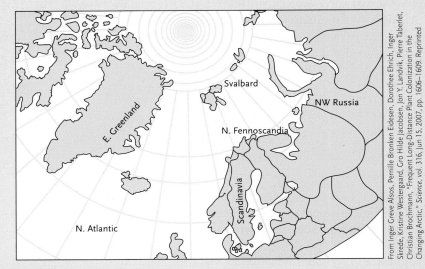

Figure 1
The location of the Svalbard Archipelago.

From Inger Greve Alsos, Pernille Bronken Eidesen, Dorothee Ehrich, Inger Skrede, Kristine Westergaard, Gro Hilde Jacobsen, Jon Y. Landvik, Pierre Taberlet, Christian Brochmann, "Frequent Long-Distance Plant Colonization in the Changing Arctic," Science, vol. 316, Jun 15, 2007, pp. 1606–1609. Reprinted with permission from AAAS.

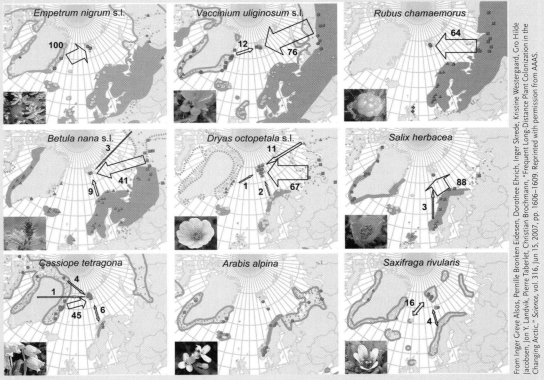

From Inger Greve Alsos, Pernille Bronken Eidesen, Dorothee Ehrich, Inger Skrede, Kristine Westergaard, Gro Hilde Jacobsen, Jon Y. Landvik, Pierre Taberlet, Christian Brochmann, "Frequent Long-Distance Plant Colonization in the Changing Arctic," Science, vol. 316, Jun 15, 2007, pp. 1606–1609. Reprinted with permission from AAAS.

Figure 2
Source regions for Svalbard plants. Shading shows the geographic distribution of nine species of plants, and dotted lines show the distributions of related species. The main genetic groups are represented by colours, although some populations (⁎) could not be assigned to a genetic group. Arrows identify **source populations,** and the numbers indicate the percentage allocation by source region.

Plants can obviously disperse without assistance from animals.

In other situations, plants disperse with the assistance of animals through pollination and seeds. Using *Prunus mahaleb*, the mahaleb cherry **(Figure 3)**, and genetic techniques, P. Jordano and two colleagues examined the role of birds and mammals in pollination and dispersing seeds. Small Passerine Birds dispersed seeds short distances (most less than 50 m) from the parent tree, whereas medium-sized birds (*Corvus corone* and *Turdus viscivorus*) usually dispersed seeds over longer distances (more than 110 m). Mammals (usually *Martes foina* and *Vulpes vulpes* but sometimes *Meles meles*) dispersed seeds about 500 m. The genetic work also indicated the extent of gene flow during pollination.

It is obvious that plants capable of self-fertilization or vegetative reproduction can be more effective colonists than those depending on outcrossing, especially with the help of animal pollinators.

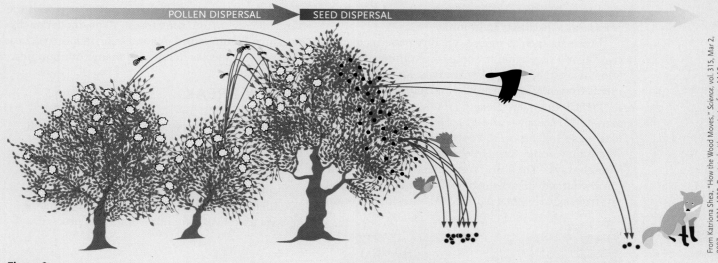

POLLEN DISPERSAL   SEED DISPERSAL

From Katriona Shea, "How the Wood Moves," *Science*, vol. 315, Mar 2, 2007, pp. 1231–1232. Reprinted with permission from AAAS.

Figure 3

The movement of pollen and seeds from mahaleb cherry trees. Gene flow occurs through pollination and seed dispersal (see Chapter 18).

hypothesis, succession is neither orderly nor predictable because each stage is dominated by the species that happened to have colonized the site first. Species replacements occur only when individuals of dominant species die of old age or when an environmental disturbance reduces their numbers. Eventually, long-lived species replace short-lived species, but the precise species composition of a mature community is open to question. Inhibition appears to play a role in some secondary successions. The interactions among early successional species in an old field are highly competitive. Horseweed inhibits the growth of asters that follow them in succession by shading aster seedlings and releasing toxic substances from their roots. Experimental removal of horseweed enhances the growth of asters, confirming the inhibitory effect.

30.9f   Tolerance Hypothesis: Species Tolerate One Another

The **tolerance hypothesis** asserts that succession proceeds because competitively superior species replace competitively inferior ones. According to this model, early-stage species neither facilitate nor inhibit the growth of later-stage species. Instead, as more species arrive at a site and resources become limiting, competition eliminates species that cannot harvest scarce resources successfully. In the Piedmont region of North America, young hardwood trees are more tolerant of shade than are young pine trees, and hardwoods gradually replace pines during succession. Thus, the climax community includes only strong competitors. Tolerance may explain the species composition of many transitional and mature communities.

At most sites, succession probably results from some combination of facilitation, inhibition, and tolerance, coupled with interspecific differences in dispersal, growth, and maturation rates. Moreover, within a community, the patchiness of abiotic factors strongly influences plant distributions and species composition. In deciduous forests of eastern North America, maples (*Acer* species) predominate on wet, low-lying ground, but oaks (*Quercus* species) are more abundant at higher and drier sites. Thus, a mature deciduous forest is often a mosaic of species and not a uniform stand of trees.

Disturbance and density-independent factors play important roles, in some cases speeding successional change. Moose (*Alces alces*) prefer to feed on deciduous shrubs in northern forests. This disturbance accelerates the rate at which conifers replace deciduous shrubs. On Isle Royale in Lake Superior, however, grazing by moose strongly affects balsam fir (*Abies balsamea*), their preferred food there. The net effect is a severe reduction in conifers and an increase in deciduous shrubs. Disturbance can also inhibit successional change, establishing a **disturbance climax** or **disclimax community**. In many grassland communities, periodic fires and grazing by large mammals kill seedlings of trees that would otherwise become established. Thus, disturbance prevents the succession from grassland to forest, and grassland persists as a disclimax community.

Animals such as moose can alter patterns of succession and vegetation in some communities, but the effect also extends to small mammals. Removal experiments involving kangaroo rats and plots of shrubland in the Chihuahuan Desert (southeastern Arizona) allowed J. H. Brown and E. J. Heske to demonstrate that these rodents were keystones in some systems where they occur. Kangaroo rats affect the plants in several ways. They are seed predators, and their burrowing activities disturb soils. Excluding kangaroo rats from experimental plots led to a threefold increase in the density of tall perennials and annual grasses **(Figure 30.29),** suggesting that by predation on seeds and burrowing, these rodents affected the vegetation in the experimental areas.

On a local scale, disturbances often destroy small patches of vegetation, returning them to an earlier successional stage. A hurricane, tornado, or avalanche may topple trees in a forest, creating small, sunny patches of open ground. Locally occurring *r*-selected species take advantage of newly available resources and quickly colonize the openings. These local patches then undergo succession that is out of step with the immediately surrounding forest. Thus, moderate disturbance, accompanied by succession in local patches, can increase species richness in many communities.

STUDY BREAK

1. What are the two types of succession? How do they differ?
2. What is a climax community? What determines the species composition of a climax community?
3. Identify and briefly describe the three hypotheses used to explain how succession proceeds.

30.10 Variations in Species Richness among Communities

Species richness often varies among communities according to a recognizable pattern. Two large-scale patterns of species richness—latitudinal trends and island patterns—have captured the attention of ecologists for more than a century.

30.10a Latitudinal Effects: From South to North

Ever since Darwin and Wallace travelled the globe (see Chapter 17), ecologists have recognized broad latitudinal trends in species richness. For many but not all plant and animal groups, species richness follows a latitudinal gradient, with the most species in the tropics and a steady decline in numbers toward the poles **(Figure 30.30).** Several general hypotheses may explain these striking patterns.

Some hypotheses propose historical explanations for the *origin* of high species richness in the tropics. The benign climate in tropical regions allows some tropical organisms to have more generations per year than their temperate counterparts. Small seasonal changes in temperature mean that tropical species may be less likely than temperate species to migrate from one habitat to another, reducing gene flow between geographically isolated populations (see Chapter 18). These factors may have fostered higher speciation rates

Figure 30.29
Predation and succession. Kangaroo rats (*Dipodomys*) were removed from the left side of the fence, which excluded them from the plot on the left. The top photograph was taken 5 years after the removal and the bottom one 13 years after. A large-seeded annual (after 5 years) and tall grasses are present in the *Dipodomys*-free plots.

From James H. Brown, Edward J. Heske, "Control of a Desert-Grassland Transition by a Keystone Rodent Guild," *Science*, vol. 250, Dec 21, 1990, pp. 1705–1707. Reprinted with permission from AAAS.

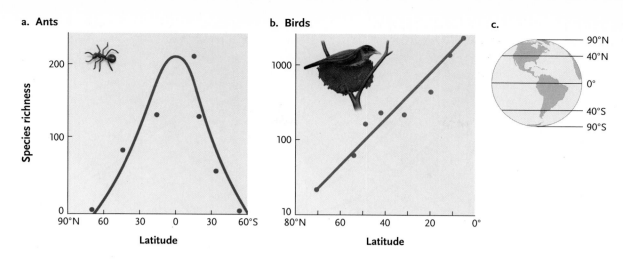

a. Ants

Species richness

Latitude

b. Birds

Latitude

c.

90°N
40°N
0°
40°S
90°S

Figure 30.30
Latitudinal trends in species richness. The species richness of many animals and plants varies with latitude **(c)** as illustrated here for **(a)** ants in North, Central, and South America and **(b)** birds in North and Central America. The species richness data used in (b) are based on records of where these birds breed.

in the tropics, accelerating the accumulation of species. Tropical communities may also have experienced severe disturbance less often than communities at higher latitudes, where periodic glaciations have caused repeated extinctions. Thus, new species may have accumulated in the tropics over longer periods of time.

Other hypotheses focus on ecological explanations for the *maintenance* of high species richness in the tropics. Some resources are more abundant, predictable, and diverse in tropical communities. Tropical regions experience more intense sunlight, warmer temperatures in most months, and higher annual rainfall than temperate and polar regions (see *The Purple Pages*). These factors provide a long and predictable growing season for the lush tropical vegetation, which supports a rich assemblage of herbivores, and through them many carnivores and parasites. Furthermore, the abundance, predictability, and year-round availability of resources allow some tropical animals to have specialized diets. Tropical forests support many species of fruit-eating bats and birds that could not survive in temperate forests where fruits are not available year-round.

Species richness may be a self-reinforcing phenomenon in tropical communities. Complex webs of population interactions and interdependency have coevolved in relatively stable and predictable tropical climates. Predator–prey, competitive, and symbiotic interactions may prevent individual species from dominating communities and reducing species richness.

30.10b Equilibrium Theory of Island Biogeography

In 1883, a volcanic eruption virtually obliterated the island of Krakatoa. Within 50 years, what was left of Krakatoa had been recolonized by plants and animals, providing biologists with a clear demonstration of the dispersal powers of many living species. The colonization of islands and the establishment of biological communities there have provided many natural experiments that have advanced our knowledge of ecology and populations. Islands are attractive sites for experiments

because although the species richness of communities may be stable over time, the species composition is often in flux as new species join a community and others drop out. In the 1960s, Robert MacArthur and Edward O. Wilson used islands as model systems to address the question of why communities vary in species richness. Islands provide natural laboratories for studying ecological phenomena, just as they do for evolution (see Chapter 17). Island communities can be small, with well-defined boundaries, and are isolated from surrounding communities.

MacArthur and Wilson developed the **equilibrium theory of island biogeography** to explain variations in species richness on islands of different size and different levels of isolation from other landmasses. They hypothesized that the number of species on any island was governed by give and take between two processes: the immigration of new species to an island and the extinction of species already there **(Figure 30.31, p. 742)**.

According to their model, the mainland harbours a *species pool* from which species immigrate to offshore islands. Seeds and small arthropods are carried by wind or floating debris. Animals such as birds arrive under their own power. When only a few species are on an island, the rate at which new species immigrate to the island is high. But as more species inhabit the island over time, the immigration rate declines because fewer species in the mainland pool can still arrive on the island as *new* colonizers (see Chapter 17). Once some species arrive on an island, their populations grow and persist for variable lengths of time. Other immigrants die without reproducing. As the number of species on an island increases, the rate of species extinction also rises. Extinction rates increase over time partly because more species can go extinct there. In addition, as the number of species on the island increases, competition and predator–prey interactions can reduce the population sizes of some species and drive them to extinction.

According to MacArthur and Wilson's theory, an equilibrium between immigration and extinction determines the number of species that ultimately occupy an island (see Figure 30.30a). Once that equilibrium has

a. Immigration and extinction rates

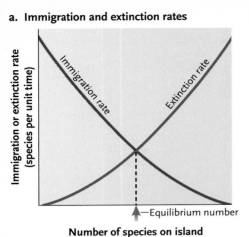

Number of species on island

The number of species on an island at equilibrium (indicated by the arrow) is determined by the rate at which new species immigrate and the rate at which species already on the island go extinct.

b. Effect of island size

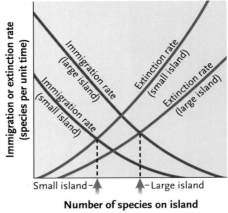

Number of species on island

Immigration rates are higher and extinction rates lower on large islands than on small islands. Thus, at equilibrium, large islands have more species.

c. Effect of distance from mainland

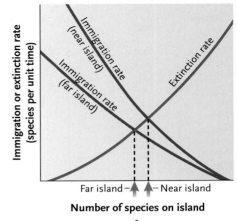

Number of species on island

Organisms leaving the mainland locate nearby islands more easily than distant islands, causing higher immigration rates on near islands. Thus, near islands support more species than far ones.

Figure 30.31
Predictions of the theory of island biogeography. The horizontal axes of the graphs are time.

been reached, the number of species remains relatively constant because one species already on the island becomes extinct in about the same time it takes a new one to arrive. The model does not specify which species immigrate or which ones already on the island become extinct. It simply predicts that the number of species on the island is in equilibrium, although species composition is not. The ongoing processes of immigration and extinction establish a constant turnover in the roster of species that live on any island.

The MacArthur–Wilson model also explains why some islands harbour more species than others. Large islands have higher immigration rates than small islands because they are larger targets for dispersing organisms. Moreover, large islands have lower extinction rates because they can support larger populations and provide a greater range of habitats and resources. At equilibrium, large islands have more species than small islands do (see Figure 30.31b). Islands near the mainland have higher immigration rates than distant islands because dispersing organisms are more likely to arrive at islands close to their point of departure. Distance does not affect extinction rates, so, at equilibrium, nearby islands have more species than distant islands (see Figure 30.31c).

The equilibrium theory's predictions about the effects of area and distance are generally supported by data on plants and animals **(Figure 30.32)**. Experimental work has verified some of the theory's basic assumptions. Amy Schoener found that more than 200 species of marine organisms colonized tiny artificial islands (plastic kitchen scrubbers) within 30 days after she placed them in a Bahamian lagoon. Her research also confirmed that immigration rate increases with island size. Daniel Simberloff and Edward O. Wilson exterminated insects on tiny islands in the Florida Keys and

monitored subsequent immigration and extinction (see "Experimenting with Islands," p. 744). Their research confirmed the equilibrium theory's predictions that an island's size and distance from the mainland influence how many species will occupy it.

The equilibrial view of species richness can also apply to mainland communities that exist as islands in a metaphorical sea of dissimilar habitat. Lakes are "islands" in a "sea" of dry land, and mountaintops are habitat "islands" in a "sea" of low terrain. Species richness in these communities is partly governed by the immigration of new species from distant sources and the extinction of species already present. As human activities disrupt environments across the globe, undisturbed sites function as islandlike refuges for threatened and endangered species. Conservation biologists apply the general lessons of MacArthur and Wilson's theory to the design of nature preserves (see Chapter 32).

The study of community ecology promises to keep biologists busy for some time to come.

STUDY BREAK

1. How does species richness change with increasing latitude?
2. In the island biogeography model proposed by MacArthur and Wilson, what processes govern the number of species on an island? What happens to the number of species once equilibrium is reached?
3. What effect do island size and distance from the mainland have on immigration and extinction of colonizing species?

Bridget Stutchbury studies the behaviour and ecology of songbirds, working at sites in eastern North America (United States and Canada), as well as sites in the Neotropics. One aspect of her research is documenting the reproductive behaviour of birds. Although songbirds were thought to be monogamous over at least a breeding season, using genetic techniques, Stutchbury and others are discovering that both males and females often mate with a bird that is not their mate. This behaviour is called extrapair copulation if it is just mating or extrapair fertilization when young result from the matings.

Using radio tracking to follow individual birds combined with DNA fingerprinting, Stutchbury and her colleagues were able to look at the movement patterns of Acadian Flycatchers (*Empidonax virescens*) and determine how far males and females travelled to meet their extrapair partners. Males travelled 50 to 1500 m from their nests to meet partners.

Work with other species, such as Hooded Warblers (*Wilsonia citrina*), demonstrated that when these birds lived in small forest fragments, their mating behaviours were disturbed compared with the behaviours of those nesting in larger tracts of forest.

Overall, her research has demonstrated that whereas some songbirds in the eastern United States and eastern Canada depend on corridors connecting habitat fragments, other species cross open habitats to use different patches of forest.

In 2007, her book *Silence of the Songbirds* reported declines in numbers of migrating songbirds and raised concerns about their future.

a. Distance effect

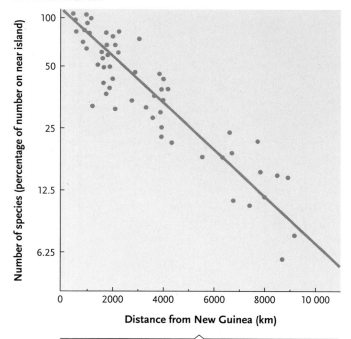

The number of lowland bird species on islands of the South Pacific declines with the islands' distance from the species source, the large island of New Guinea. Data in this graph were corrected for differences in the sizes of the islands. The number of bird species on each island is expressed as a percentage of the number of bird species on an island of equivalent size close to New Guinea.

b. Area effect

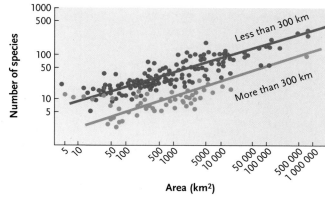

The number of bird species on tropical and subtropical islands throughout the world increases dramatically with island area. The data for islands near to a mainland source and islands far from a mainland source are presented separately to minimize the effect of distance. Notice that the distance effect reduces the number of bird species on islands that are more than 300 km from a mainland source.

Figure 30.32
Factors that influence bird species richness on islands. **(a)** Evidence that fewer bird species colonize islands that are distant from the mainland source. **(b)** Evidence that more bird species colonize large islands than small ones.

Experimenting with Islands

Shortly after Robert MacArthur and Edward O. Wilson published the equilibrium theory of island biogeography in the 1960s, Wilson and Daniel Simberloff, one of Wilson's graduate students at Harvard University, undertook an ambitious experiment in community ecology. Simberloff reasoned that the best way to test the theory's predictions was to monitor immigration and extinction on barren islands.

Simberloff and Wilson devised a system for removing all the animals from individual red mangrove trees in the Florida Keys. The trees, with canopies that spread from 11 to 18 m in diameter, grow in shallow water and are isolated from their neighbours. Thus, each tree is an island that harbours an arthropod community. The species pool on the Florida mainland includes about 1000 species of arthropods, but each mangrove island contains no more than 40 species at one time.

After cataloguing the species on each island, Simberloff and Wilson hired an extermination company to erect large tents over each mangrove island and fumigate them to eliminate all arthropods on them **(Figure 1)**. The exterminators used methylbromide, a pesticide that does not harm trees or leave any residue. The tents were then removed.

Simberloff then monitored both the immigration of arthropods to the islands and the extinction of species that became established on them. He surveyed four islands regularly for two years and at intervals thereafter.

The results of this experiment confirm several predictions of MacArthur and Wilson's theory **(Figure 2)**. Arthropods rapidly recolonized the islands, and within eight or nine months, the number of species living on each island had reached an equilibrium that was close to the original species number. The island nearest the mainland had more species than the most distant island. However, immigration and extinction were rapid, and Simberloff and Wilson

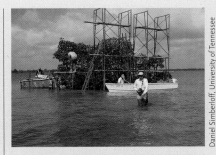

Daniel Simberloff, University of Tennessee

Figure 1
After cataloguing the arthropods, Simberloff and Wilson hired an extermination company to eliminate all living arthropods.

suspected that some species went extinct even before they had noted their presence. The researchers also discovered that three years after the experimental treatments, the species composition of the islands was still changing constantly and did not remotely resemble the species composition on the islands before they were defaunated.

Simberloff and Wilson's research was a landmark study in ecology because it tested the predictions of an important theory using a field experiment. Although such efforts are now almost routine in ecological studies, this project was one of the first to demonstrate that large-scale experimental manipulations of natural systems are feasible and that they often produce clear results.

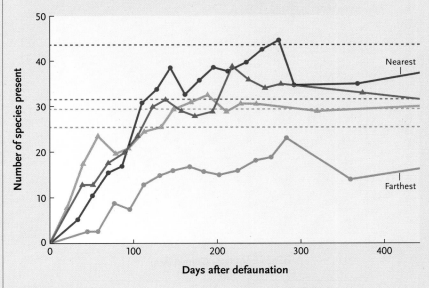

Figure 2
On three of four islands, species richness slowly returned to the predefaunation level (indicated by colour-coded dotted lines). The most distant island had not reached its predefaunation species richness after two years.

Review

 To access course materials such as Aplia and other companion resources, please visit www.NELSONbrain.com.

30.1 Interspecific Interactions

- Coevolution involves genetically based, reciprocal adaptations in two or more interacting species. Coevolution is not restricted to two species but often involves complex interactions among several species in a community.

30.2 Getting Food

- Predators eat animal prey, whereas herbivores eat plants. Predators and herbivores are animals with characteristics allowing them to feed efficiently. Predation and herbivory are the most conspicuous relationships in ecological communities.
- A koala is a specialist because it eats the leaves of only a few of the available species of *Eucalyptus*. Specialists tend to eat only a few types of food, whereas generalists take a broader diet.
- Optimal foraging theory predicts that an animal's diet is a compromise between the costs and the benefits associated with different types of food. Costs include the time and energy it takes to pursue, capture, and consume a particular kind of food. Benefits are the energy that the food provides.

30.3 Defence

- Animals use seven types of defence: size, eternal vigilance, avoiding detection, counterattack, chemical defence, warnings, and mimicry. Below are examples of each.
- The large size of elephants means that they have few natural predators.
- Meerkats are constantly on the lookout for potential predators.
- Caterpillars that look like bird droppings are not recognized as edible.
- A bee or a scorpion has a sting, and porcupines and other mammals have spines.
- Skunks spray a noxious chemical at potential predators.
- Skunks have black and white coloration and monarch butterflies have orange and black coloration.
- The harmless drone fly mimics the coloration and behaviour of the stinging bee or wasp.
- Animals using chemical defences either synthesize the chemicals themselves or sequester them from other sources. This can include plants that the organism eats.
- Aposematic displays teach would-be predators to avoid the signaller.
- Batesian mimicry occurs when an edible or harmless species mimics an inedible or a poisonous one. In Müllerian mimicry, two or more unpalatable or poisonous species have a similar appearance.

30.4 Competition

- Intraspecific competition can occur between two different species. Two types of competition are interference and exploitation. The competitive exclusion principle states that two or more species cannot coexist indefinitely if both rely on the same limiting resources and exploit them in the same way. One species will be able to harvest the available resources better and eventually outcompete the other species.
- A population's ecological niche is defined as the resources it uses and the environmental conditions it requires over its lifetime. A fundamental niche, larger than a realized niche, includes all conditions and resources a population can tolerate. A realized niche is the range of conditions and resources that a population actually encounters in nature.
- Resource partitioning occurs when sympatric species use different resources or the same resources in different ways. Plants may position their root systems at different levels, avoiding competition for water and nutrients.
- Character displacement results in sympatric species that differ in morphology and use different resources even though they would not do so in allopatric situations. An example of character displacement is the honey-eaters of Australia.

30.5 The Nature of Ecological Communities

Type of Interaction	Effect on Species Involved	Example
Commensalism	One species benefits; the other is unaffected (+/0)	Egrets and the large grazers that flush insects out of grasses during feeding
Mutualism	Both species benefit (+/+)	Bull's horn acacia tree and a species of small ants
Parasitism	One species benefits (parasite); the other is harmed (host) (+/−)	Ectoparasites such as mosquitoes and leeches and their mammalian hosts

- Two hypotheses about ecological communities have been developed by ecologists. The interactive hypothesis predicts that mature communities are at equilibrium and, if disturbed, will return to the predisturbed state. The individualistic hypothesis predicts that communities do not achieve equilibrium but rather are in a steady state of flux in response to disturbance and environmental change.
- Ecotones are generally species rich because they contain species from both communities, as well as species that occur only in transition zones.

30.6 Community Characteristics

- Indices allow population biologists to objectively compare the diversity of communities. Shannon's indices provide a measure of diversity (H') and evenness (E_H).
- Alpha (α) diversity is the number of species living in a single community. Beta (β) diversity is the number of species living in a collection of communities.

- Herbivores are primary consumers and form the second trophic level. Omnivores can be primary, secondary, and tertiary consumers (second, third, and fourth trophic levels, respectively) in a single meal.
- Generally, communities that support complex food webs are more stable. The disappearance of one or even two species does not have a major impact on the food web and thus community structure.

30.7 Effects of Population Interactions on Community Structure

- Species distribution and resource use in *K*-selected species are profoundly affected by competition. However, competition seems to have little effect on the community structure of *r*-selected species.
- Predation and herbivory can increase and/or decrease species richness, depending on the circumstances. Species richness can increase if a predator eliminates a strong competitor, allowing other organisms to exploit the available resources, for example, predatory sea stars reducing populations of mussels. Species richness can decrease when a predator eats less abundant species, further reducing their numbers.
- A keystone species has a much greater effect on the community than its numbers might suggest. Only a few individuals can have a profound impact on community structure.

30.8 Effects of Disturbance on Community Characteristics

- A community may never attain equilibrium because of disturbances such as cyclones, mortality caused by internal processes, and the recruitment of new colonies.
- The intermediate disturbance hypothesis states that species richness is greatest in communities experiencing fairly frequent disturbances of moderate intensity. Data gathered about a river system in New Zealand revealed that areas with moderate disturbance (e.g., moved rocks, soil, and sand in the streambed) had the highest species diversity.
- Generally, communities with a higher species richness recover from disturbance much more quickly than those with a low species richness.

30.9 Succession

- Primary succession begins when organisms first colonize habitats without soil, whereas secondary succession occurs after existing vegetation is destroyed or disrupted by an environmental disturbance.
- A climax community is a late successional stage that can be found in both primary and secondary succession. Climax communities are dominated by a few species that replace themselves and persist until a disturbance eliminates them. Species composition of a climax community is determined by local climate and soil conditions, surrounding vegetation, and chance events.
- The facilitation hypothesis holds that species modify the environment in a way that makes it less suitable for themselves but more suitable for those species that follow them in succession. The inhibition hypothesis contends that species currently occupying a successional stage prevent new species from occupying the same community. The tolerance hypothesis holds that early-stage species neither facilitate nor inhibit the growth of new species. Instead, succession proceeds because new species are able to outcompete and replace early-stage species.

30.10 Variation in Species Richness among Communities

- Species richness generally decreases with increasing latitude.
- The numbers of species on an island is governed by immigration of new species and extinction of species already there. Once equilibrium between immigration and extinction is reached, the number of species on an island remains relatively constant. As one species goes extinct, it is replaced by a newly arrived immigrant species.
- Large islands have higher immigration rates and lower extinction rates than small islands. Islands near the mainland have higher immigration rates than distant islands. Distance does not affect extinction rates. As a result, at equilibrium, near islands have more species than far islands.

Questions

Self-Test Questions

1. According to optimal foraging theory, what do predators do?
   a. always eat the largest prey possible
   b. always eat the prey that are easiest to catch
   c. choose prey based on the costs of consuming it compared to the energy it provides
   d. eat plants when animal prey are scarce
   e. have coevolved mechanisms to overcome prey defences

2. What term refers to the use of the same limiting resource by two species?
   a. brood parasitism
   b. interference competition
   c. exploitative competition
   d. mutualism

3. What is the range of resources that a population of one species can possibly use called?
   a. its fundamental niche
   b. its realized niche
   c. resource partitioning
   d. its relative abundance

4. Differences in molar (tooth) structure of sympatric mammals may reflect which of the following?
   a. predation
   b. character displacement
   c. interference competition
   d. cryptic coloration

5. Bacteria that live in the human intestine assist human digestion and eat nutrients that the human consumes. Which term best describes this relationship?
   a. commensalism
   b. mutualism
   c. endoparasitism
   d. ectoparasitism

6. In the table below, the letters refer to four communities, and the numbers indicate how many individuals were recorded for each of five species. Which community has the highest species diversity?

	Species 1	Species 2	Species 3	Species 4	Species 5
a.	80	10	10	0	0
b.	25	25	25	25	0
c.		4	6	8	80
d.	20	20	20	20	20

7. Which sentence best describes a keystone species?
   a. It is usually a primary producer.
   b. It has a critically important role in determining the species composition of its community.
   c. It is always a predator.
   d. It usually exhibits aposematic coloration.

8. Species richness can be highest in communities with this type of disturbances.
   a. very frequent and severe
   b. very frequent and of moderate intensity
   c. very rare and severe
   d. of intermediate frequency and moderate intensity

9. Which term refers to a community's change in species composition from bare and lifeless rock to climax vegetation?
   a. competition
   b. secondary succession
   c. primary succession
   d. facilitation

10. What does the equilibrium theory of island biogeography predict about the number of species found on an island?
    a. It increases steadily until it equals the number in the mainland species pool.
    b. It is greater on large islands than on small ones.
    c. It is smaller on islands near the mainland than on distant islands.
    d. It is greater for islands near the equator than for islands near the poles.

Questions for Discussion

1. Many landscapes dominated by agricultural activities also have patches of forest of various sizes. What is the minimum amount of habitat required by different species? Focus on 10 species —5 animals and 5 plants. For each species, can you estimate the minimum viable population?

2. Using the terms and concepts introduced in this chapter, describe the interactions that humans have with 10 other species. Try to choose at least 8 species that we do not eat.

3. After reading about the two potential biases in the scientific literature on competition, describe how future studies of competition might avoid such biases.

4. What are the primary producers in a community of parasites?

5. What influence does agriculture have on population interactions and community development?

31

Among the fastest growing ecosystems in the world—shopping malls and residential areas sprawl in the north part of London, Ontario.

Ecosystems

WHY IT MATTERS

As shown in the chapter-opening photograph, urban ecosystems are the most rapidly growing habitat on the planet, replacing existing habitats at an astonishing rate, partly because of growth in populations and in economies. The system portrayed in the photograph is low-density housing with services (water and electricity), but in many parts of the world, housing expansions are high density, with few, if any, services. Apart from humans and our domesticated plant and animal species, which components of the original flora and fauna persist? Walk around your neighbourhood and check it out.

How does the urban ecosystem differ from what was there before? In what ways does it differ? Think of runoff from rain and snow, of the heat-absorbing and reflecting properties of buildings, concrete, and asphalt. What are the effects of gardeners and landscapers, however well meaning? The urban ecosystem offers people in general, and biologists in particular, many opportunities for research and study. We must determine what changes we can effect in the construction and design of neighbourhoods to maximize their compatibility with native organisms. How can we make urban neighbourhoods more useful to migrating songbirds?

Archaeological evidence indicates that urban sprawl occurred around 6000 years ago around the site of Tell Brak in what is now northern Syria. Then the "city" that stood at Tell Brak occupied about 55 ha when other contemporary settlements rarely exceeded 3 ha and the largest of its neighbours was just 15 ha. There is evidence of spatial separation between subcommunities at Tell Brak, where neighbourhoods were divided by walls and limited points of access.

Urban sprawl may not be new, but the current scale makes it a frontier for action to achieve conservation of biodiversity. The purpose of this chapter is to explore some aspects of ecosystems and introduce them as objects of biological study.

Ecosystems are often studied by following the movement of energy from one level to another. Photosynthetic organisms form the energetic basis for ecosystems, providing sources of food for other organisms (usually animals). Levels of biomass at different trophic levels (primary producers; primary, secondary, and tertiary consumers) generally reflect the movement of energy. The movement of markers such as DDT (see "Molecule behind Biology," Box 31.2, p. 758) through ecosystems provided an all-too-sobering view of the connectedness among the ecosystems of the globe.

31.1 Energy Flow and Ecosystem Energetics

Ecosystems receive a steady input of energy from an external source, usually the Sun. Energy flows through an ecosystem, but, as dictated by the laws of thermodynamics, much of that energy is lost without being used by organisms. In contrast, materials cycle between living and nonliving reservoirs, both locally and on a global scale. The flow of energy through and

the cycling of materials around an ecosystem make resident organisms highly dependent on one another and on their physical surroundings.

Food webs define the pathways by which energy and nutrients move through an ecosystem's biotic components. In most ecosystems, nutrients and energy move simultaneously through a grazing food web and a detrital food web **(Figure 31.1)**. The grazing food web includes the producer, herbivore, and secondary consumer trophic levels. The detrital food web includes detritivores and decomposers. Because detritivores and decomposers subsist on the remains and waste products of organisms at every trophic level, the two food webs are closely interconnected. Detritivores also contribute to the grazing food web when carnivores eat them.

All organisms in a particular trophic level are the same number of energy transfers from the ecosystem's ultimate energy source. Photosynthetic plants are one energy transfer removed from sunlight, herbivores

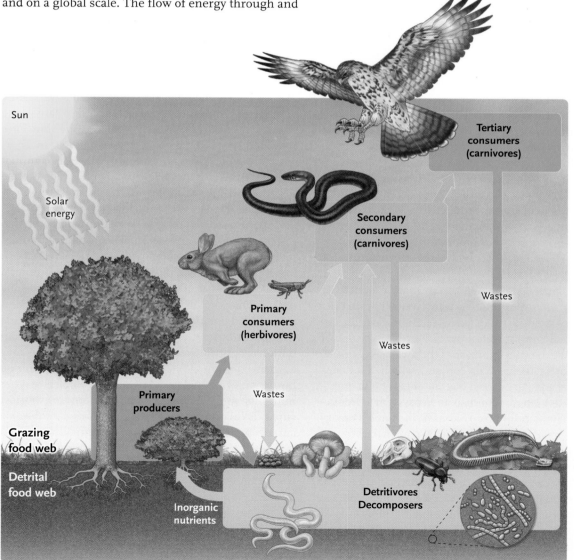

Figure 31.1
Grazing and detrital food webs. Energy and nutrients move through two parallel food webs in most ecosystems. The grazing food web includes producers, herbivores, and carnivores. The detrital food web includes detritivores and decomposers. Each box in this diagram represents many species, and each arrow represents many arrows.

(primary consumers) are two, secondary consumers are three, and tertiary consumers are four.

31.1a Primary Productivity Involves Fixing Carbon

Virtually all life on Earth depends on the input of solar energy. Every minute of every day, Earth's atmosphere intercepts roughly 80 kJ (kilojoules) of energy per square metre (see Chapter 1). About half of that energy is absorbed, scattered, or reflected by gases, dust, water vapour, and clouds before it reaches the planet's surface (see *The Purple Pages*). Most energy reaching the surface falls on bodies of water or bare ground, where it is absorbed as heat or reflected back into the atmosphere. Reflected energy warms the atmosphere. Only a small percentage contacts primary producers, and most of that energy evaporates water, driving transpiration in plants (see Chapter 7).

Ultimately, photosynthesis converts less than 1% of the solar energy arriving at Earth's surface into chemical energy. But primary producers still capture enough energy to produce an average of several kilograms of dry plant material per square metre per year. On a global scale, they produce more than 150 billion tonnes of new biological material annually. Some of the solar energy that producers convert into chemical energy is transferred to consumers at higher trophic levels.

The rate at which producers convert solar energy into chemical energy is an ecosystem's **gross primary productivity**. But, like other organisms, producers use energy for their own maintenance functions. After deducting energy used for these functions (see Chapter 6), whatever chemical energy remains is the ecosystem's **net primary productivity**. In most ecosystems, net primary productivity is 50 to 90% of gross primary productivity. In other words, producers use between 10 and 50% of the energy they capture for their own respiration.

Ecologists usually measure primary productivity in units of energy captured ($kJ \cdot m^{-2} \cdot year^{-1}$) or in units of biomass created ($kg \cdot m^{-2} \cdot year^{-1}$). **Biomass** is the dry mass of biological material per unit area or volume of habitat. (We measure biomass as the *dry* mass of organisms because their water content, which fluctuates with water uptake or loss, has no energetic or nutritional value.) Do not confuse an ecosystem's productivity with its **standing crop biomass**, the total dry mass of plants present at a given time. Net primary productivity is the *rate* at which the standing crop produces *new* biomass (see Chapter 7).

Energy captured by plants is stored in biological molecules, mostly carbohydrates, lipids, and proteins. Ecologists can convert units of biomass into units of energy or vice versa as long as they know how much carbohydrate, protein, and lipid a sample of biological material contains. For reference, 1 g of carbohydrate and

1 g of protein each contains about 17.5 kJ of energy. Thus, net primary productivity indexes the rate at which producers accumulate energy as well as the rate at which new biomass is added to an ecosystem. Ecologists measure changes in biomass to estimate productivity because it is far easier to measure biomass than energy content. New biomass takes several forms, including

- growth of existing producers,
- creation of new producers by reproduction, and
- storage of energy as carbohydrates.

Because herbivores eat all three forms of new biomass, net primary productivity also measures how much new energy is available for primary consumers.

The potential rate of photosynthesis in any ecosystem is proportional to the intensity and duration of sunlight, which varies geographically and seasonally (see Chapter 6, Chapter 7, and *The Purple Pages*). Sunlight is most intense and day length is least variable near the equator. In contrast, the intensity of sunlight is weakest and day length is most variable near the poles. This means that producers at the equator can photosynthesize for nearly 12 hours a day, every day of the year, whereas near the poles, photosynthesis is virtually impossible during the long, dark winter. In summer, however, photosynthesis occurs virtually around the clock.

Sunlight is not the only factor influencing the rate of primary productivity. Temperature and availability of water and nutrients also affect this rate. Many of the world's **deserts** receive plenty of sunshine but have low rates of productivity because water is in short supply and the soil is poor in nutrients. Mean annual primary productivity varies greatly on a global scale **(Figure 31.2)**, reflecting variations in these environmental factors (see *The Purple Pages*).

On a finer geographic scale, within a particular terrestrial ecosystem, mean annual net productivity often increases with the availability of water **(Figure 31.3)**. In systems with sufficient water, a shortage of mineral nutrients may be limiting. All plants need specific ratios of macronutrients and micronutrients for maintenance and photosynthesis (see Chapter 7). But plants withdraw nutrients from soil, and if nutrient concentration drops below a critical level, photosynthesis may decrease or stop altogether. In every ecosystem, one nutrient inevitably runs out before the supplies of other nutrients are exhausted. The element in shortest supply is called a **limiting nutrient** because its absence curtails productivity. Productivity in agricultural fields is subject to the same constraints as productivity in natural ecosystems. Farmers increase productivity by irrigating (adding water to) and fertilizing (adding nutrients to) their crops.

In freshwater and marine ecosystems, where water is always readily available, the depth of water and combined availability of sunlight and nutrients govern the rate of primary productivity. Productivity

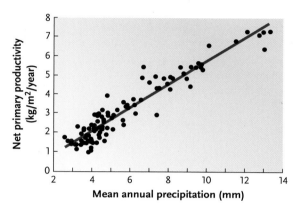

Figure 31.2
Global variation in primary productivity. Satellite data from 2002 provide a visual portrait of net primary productivity across Earth's surface. High-productivity regions on land are dark green; low-productivity regions are yellow. For aquatic environments, the highest productivity is red, down through orange, yellow, green, blue, and purple (lowest).

Figure 31.3
Water and net primary productivity. Mean annual precipitation at 100 sites in the Great Plains of North America. These data include only above-ground productivity.

(Figure 31.3: scatter plot of Net primary productivity (kg/m²/year) on y-axis (0–8) versus Mean annual precipitation (mm) on x-axis (2–14), showing a positive linear relationship.)

Table 31.1	Standing Crop Biomass and Net Primary Productivity of Different Ecosystems	
Ecosystem	Mean Standing Crop Biomass (kg/m²)	Mean Net Primary Productivity (kg/m²/y¹)
Terrestrial Ecosystems		
Tropical rain forest	450	22.0
Tropical deciduous forest	350	16.0
Temperate rain forest	350	13.0
Temperate deciduous forest	300	12.0
Savannah	40	9.0
Boreal forest (**taiga**)	200	8.0
Woodland and shrubland	60	7.0
Agricultural land	10	6.5
Temperate grassland	16	6.0
Tundra and alpine tundra	6.0	1.4
Desert and thornwoods	7.0	0.9
Extreme desert, rock, sand, ice	0.2	0.03
Freshwater Ecosystems		
Swamp and marsh	150	20
Lake and stream	0.2	2.5
Marine Ecosystems		
Open ocean	0.03	1.3
Upwelling zones	0.2	5.0
Continental shelf	0.1	3.6
Kelp beds and reefs	20	25
Estuaries	10	15
World Total	**36**	**3.3**

Source: Based on Whittaker, R.H. 1975. Communities and Ecosystems. 2nd ed. Macmillan.

is high in near-shore ecosystems, where sunlight penetrates shallow, nutrient-rich waters. Kelp beds and coral reefs along temperate and tropical marine coastlines, respectively, are among the most productive ecosystems on Earth (**Table 31.1;** see also Figure 31.2). In contrast, productivity is low in the open waters of a large lake or ocean. There sunlight penetrates only the upper layers, and nutrients sink to the bottom; thus, the two requirements for photosynthesis—sunlight and nutrients—are available in different places.

Although ecosystems vary in their rates of primary productivity, these differences are not always proportional to variations in their standing crop biomass (see Table 31.1). For example, biomass amounts in temperate deciduous forests and **temperate grasslands** differ by a factor of 20, but the difference in their rates of net primary productivity is much smaller. Most biomass in trees is present in nonphotosynthetic tissues such as wood, so their ratio of productivity to biomass is low ($12\,\text{kg} \cdot \text{m}^{-2}/300\,\text{kg} \cdot \text{m}^{-2} = 0.04$). By contrast, grasslands do not accumulate much biomass because annual mortality, herbivores, and fires remove plant material

as it is produced, so their productivity to biomass ratio is much higher ($6.0\,\text{kg} \cdot \text{m}^{-2}/16\,\text{kg} \cdot \text{m}^{-2} = 0.375$).

Some ecosystems contribute more than others to overall net primary productivity (**Figure 31.4, p. 752**).

Ecosystems covering large areas make substantial total contributions, even if their productivity per unit area is low. Conversely, geographically restricted ecosystems make large contributions if their productivity is high. Open ocean and tropical rain forests contribute about equally to total global productivity, but for different reasons. Open oceans have low productivity, but they cover nearly two-thirds of Earth's surface. Tropical rain forests are highly productive but cover only a relatively small area.

Net primary productivity ultimately supports all consumers in grazing and detrital food webs. Consumers in the grazing food web eat some biomass at every trophic level except the highest. Uneaten biomass eventually dies and passes into detrital food webs. Moreover, consumers assimilate only a portion of the material they ingest, and unassimilated material passed as feces also supports detritivores and decomposers.

31.1b Secondary Productivity Involves Animals Eating Plants, Thus Moving Up the Trophic Scale

As energy is transferred from producers to consumers, some is stored in new consumer biomass, called **secondary productivity**. Nevertheless, two factors cause energy to be lost from the ecosystem every time it flows

from one trophic level to another. First, animals use much of the energy they assimilate for maintenance and locomotion rather than for production of new biomass. Second, as dictated by the second law of thermodynamics, no biochemical reaction is 100% efficient, so some of the chemical energy liberated by cellular respiration is converted to heat, which most organisms do not use.

31.1c Ecological Efficiency Is Measured by Use of Energy

Ecological efficiency is the ratio of net productivity at one trophic level to net productivity at the trophic level below. If plants in an ecosystem have a net primary productivity of $1.0 \ kg \cdot m^{-2} \cdot year^{-1}$ of new tissue, and the herbivores that eat these plants produce 0.1 kg of new tissue per square metre per year, the ecological efficiency of the herbivores is 10%. The efficiencies of three processes (harvesting food, assimilating ingested energy, and producing new biomass) determine the ecological efficiencies of consumers.

Harvesting efficiency is the ratio of the energy content of food consumed to the energy content of food available. Predators harvest food efficiently when prey are abundant and easy to capture (see Chapter 30).

Assimilation efficiency is the ratio of the energy absorbed from consumed food to the total energy content

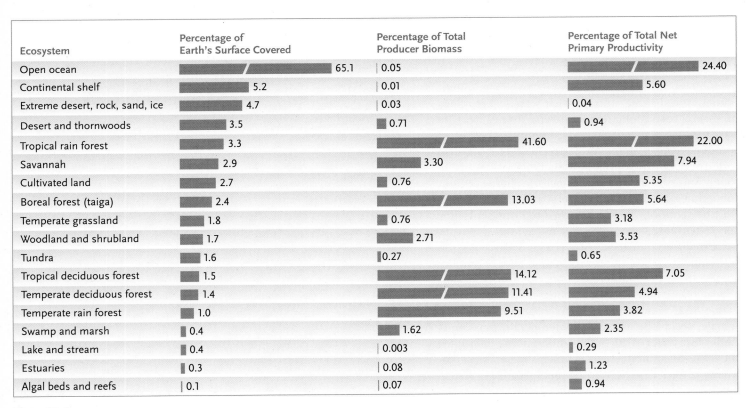

Figure 31.4
Biomass and net primary productivity. An ecosystem's percentage coverage of Earth's surface is not proportional to its contribution to total biomass of producers or its contribution to the total net primary productivity.

of the food. Because animal prey is relatively easy to digest, carnivores absorb between 60 and 90% of the energy in their food. Assimilation efficiency is lower for prey with indigestible parts such as bones or exoskeletons. Herbivores assimilate only 15 to 80% of the energy they consume because cellulose is not very digestible. Herbivores lacking cellulose-digesting systems are on the low end of the scale, whereas those that can digest cellulose are at the higher end.

Production efficiency is the ratio of the energy content of new tissue produced to the energy assimilated from food. Production efficiency varies with maintenance costs. Endothermic animals often use less than 10% of their assimilated energy for growth and reproduction because they use energy to generate body heat (see Chapter 50). Ectothermic animals channel more than 50% of their assimilated energy into new biomass.

The overall ecological efficiency of most organisms is 5 to 20%. As a rule of thumb, only about 10% of energy accumulated at one trophic level is converted into biomass at the next higher trophic level, as illustrated by energy transfers at Silver Springs, Florida **(Figure 31.5)**. Silver Springs is an ecosystem that has been studied for many years. Producers in the Silver Springs ecosystem convert 1.2% of the solar energy they intercept into chemical energy (represented by 86 986 kJ·m^{-2}·year^{-1} of gross primary productivity). However, plants use about two-thirds of this energy for respiration, leaving a net primary productivity, one-third of which is to be included in new plant biomass. All consumers in the grazing food web (on the right in Figure 31.5) ultimately depend on this energy source, which diminishes with each transfer between trophic levels. Energy is lost to respiration and export at each trophic level. In addition, organic

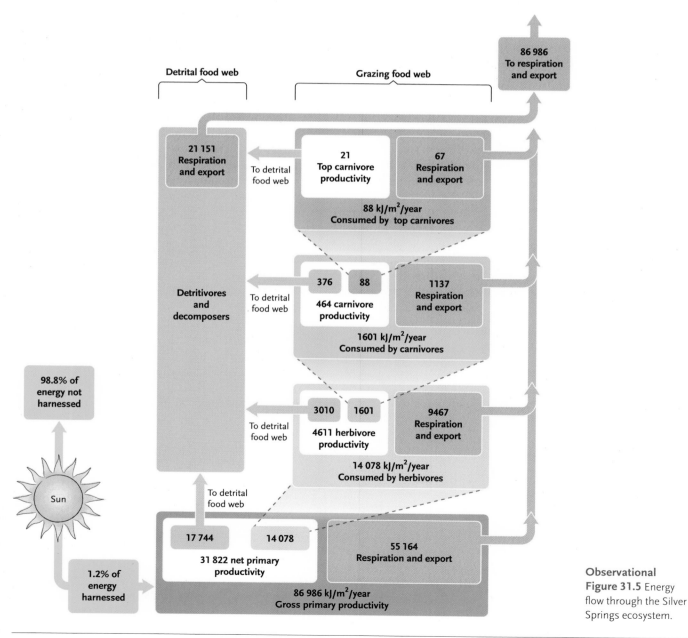

Observational Figure 31.5 Energy flow through the Silver Springs ecosystem.

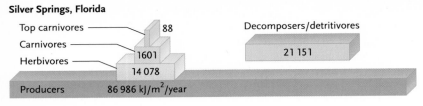

Silver Springs, Florida

Top carnivores — 88

Carnivores —

Herbivores — 1601

14 078

Producers — 86 986 kJ/m²/year

Decomposers/detritivores

21 151

Figure 31.6

Pyramids of energy. The pyramid of energy for Silver Springs, Florida, shows that the amount of energy (kJ·m⁻²·year⁻¹) passing through each trophic level decreases as it moves up the food web.

wastes and uneaten biomass represent substantial energy that flows into the detrital food web (on the left in Figure 31.5). To determine the ecological efficiency of any trophic level, we divide its productivity by the productivity of the level below it. The ecological efficiency of midlevel carnivores at Silver Springs is 10.06%, 464 kJ·m⁻²·year⁻¹/4611 kJ·m⁻²·year⁻¹.

31.1d Three Types of Pyramids Describe Ecosystems: Energy, Biomass, and Numbers

As energy works its way up a food web, energy losses are multiplied in successive energy transfers, greatly reducing the energy available to support the highest trophic levels (see Figure 31.5). Consider a hypothetical example in which ecological efficiency is 10% for all consumers. Assume that the plants in a small field annually produce new tissues containing 100 kJ of energy. Because only 10% of that energy is transferred to new herbivore biomass, the 100 kJ in plants produces 10 kJ of new herbivorous insects, 1 kJ of new songbirds that eat insects, and only 0.1 kJ of new falcons that eat songbirds. About 0.1% of the energy from primary productivity remains after three trophic levels of transfer. If the energy available to each trophic level is depicted graphically, the result is a **pyramid of energy**, with primary producers on the bottom and higher-level consumers on the top **(Figure 31.6).**

The low ecological efficiencies that characterize most energy transfers illustrate one advantage of eating "lower on the food chain." This reality is reflected by major adaptive radiations of lineages of

animals whose ancestors were secondary consumers when they switched to being primary consumers. Good examples of such radiations occur, for example, among insects, fish, dinosaurs, and mammals.

Even though humans digest and assimilate meat more efficiently than vegetables, we could feed more people if we all ate more primary producers directly instead of first passing them through another trophic level, such as cattle or chickens, to produce meat. Production of animal protein is costly because much of the energy fed to livestock is used for their own maintenance rather than production of new biomass. But despite the economic and health-related logic of a more vegetarian diet, changing our eating habits alone will not eliminate food shortages or the frequency of malnutrition. Many regions of Africa, Australia, North America, and South America support vegetation that is suitable only for grazing by large herbivores. These areas could not produce significant quantities of edible grains and vegetables without significant additions of water and fertilizer (see Chapter 33).

Inefficiency of energy transfer from one trophic level to the next has profound effects on ecosystem structure. Ecologists illustrate these effects in diagrams called **ecological pyramids.** Trophic levels are drawn as stacked blocks, with the size of each block proportional to the energy, biomass, or numbers of organisms present. Pyramids of energy typically have wide bases and narrow tops (see Figure 31.6) because each trophic level contains only about 10% as much energy as the trophic level below it.

Progressive reduction in productivity at higher trophic levels usually establishes a **pyramid of biomass (Figure 31.7).** The biomass at each trophic level is proportional to the amount of chemical energy temporarily stored there. Thus, in terrestrial ecosystems, the total mass of producers is generally greater than the total mass of herbivores, which is, in turn, greater than the total mass of predators (see Figure 31.7a). Populations of top predators, from killer whales to lions and crocodiles, contain too little biomass and energy to support another trophic level; thus, they have no nonhuman predators.

Freshwater and marine ecosystems sometimes exhibit inverted pyramids of biomass (see Figure 1.7b). In the open waters of a lake or ocean, primary consumers

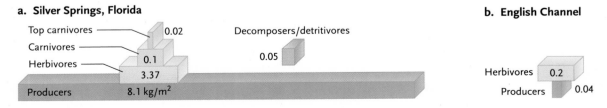

a. Silver Springs, Florida

Top carnivores — 0.02

Carnivores —

Herbivores — 0.1

3.37

Producers — 8.1 kg/m²

Decomposers/detritivores

0.05

b. English Channel

Herbivores 0.2

Producers 0.04

Figure 31.7

Pyramids of biomass. **(a)** The pyramid of standing crop biomass for Silver Springs is bottom heavy, as it is for most ecosystems. **(b)** Some marine ecosystems, such as that in the English Channel, have an inverted pyramid of biomass because producers are quickly eaten by primary consumers. Only the producer and herbivore trophic levels are illustrated here. The data for both pyramids are given in kilograms per square metre of dry biomass.

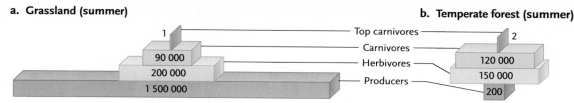

a. Grassland (summer)

Top carnivores — 1

Carnivores — 90 000

Herbivores — 200 000

Producers — 1 500 000

b. Temperate forest (summer)

Top carnivores — 2

Carnivores — 120 000

Herbivores — 150 000

Producers — 200

Figure 31.8

Pyramids of numbers. **(a)** The pyramid of numbers (numbers of individuals per 1000 m²) for temperate grasslands is bottom heavy because individual producers are small and very numerous. **(b)** The pyramid of numbers for forests may have a narrow base because herbivorous insects usually outnumber the producers, many of which are large trees. Data for both pyramids were collected in summer. Detritivores and decomposers (soil animals and microorganisms) are not included because they are difficult to count.

(zooplankton) eat primary producers (phytoplankton) almost as soon as they are produced. As a result, the standing crop of primary consumers at any moment in time is actually larger than the standing crop of primary producers. Food webs in these ecosystems are stable because producers have exceptionally high **turnover rates**. In other words, producers divide and their populations grow so quickly that feeding by zooplankton does not endanger their populations or reduce the producers' productivity. However, on an annual basis, the *cumulative total* biomass of primary producers far outweighs that of primary consumers.

The reduction of energy and biomass affects population sizes of organisms at the top of a food web. Top predators can be relatively large animals, so the limited biomass present in the highest trophic levels is concentrated in relatively few animals **(Figure 31.8)**. The extremely narrow top of this **pyramid of numbers** has grave implications for conservation biology (see Chapter 32). Top predators tend to be large animals with small population sizes. And because each individual must patrol a large area to find sufficient food, members of a population are often widely dispersed within their habitats. As a result, they are subject to genetic drift (see Chapter 18) and are highly sensitive to hunting, habitat destruction, and random events that can lead to extinction. Top predators may also suffer from the accumulation of poisonous materials that move through food webs (see the next section). Even predators that feed below the top trophic level often suffer the ill effects of human activities. Consumers sometimes regulate ecosystem processes.

Numerous abiotic factors, such as the intensity and duration of sunlight, rainfall, temperature, and the availability of nutrients, have significant effects on primary productivity. Primary productivity, in turn, profoundly affects the populations of herbivores and predators that feed on them. But what effect does feeding by these consumers have on primary productivity?

Consumers sometimes influence rates of primary productivity, especially in ecosystems with low species diversity and relatively few trophic levels. Food webs in lake ecosystems depend primarily on the productivity of phytoplankton **(Figure 31.9)**. Phytoplankton are, in turn, eaten by herbivorous zooplankton, themselves

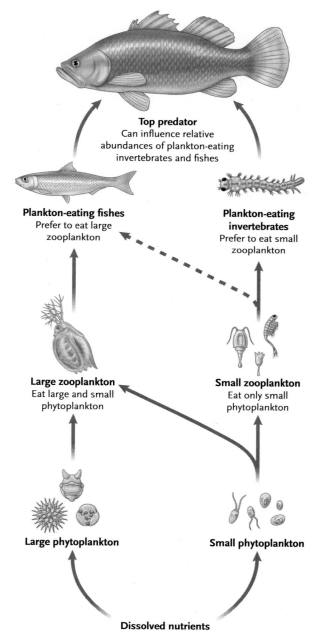

Top predator
Can influence relative abundances of plankton-eating invertebrates and fishes

Plankton-eating fishes
Prefer to eat large zooplankton

Plankton-eating invertebrates
Prefer to eat small zooplankton

Large zooplankton
Eat large and small phytoplankton

Small zooplankton
Eat only small phytoplankton

Large phytoplankton

Small phytoplankton

Dissolved nutrients

Figure 31.9

Consumer regulation of primary productivity. A simplified food web illustrates that lake ecosystems have relatively few trophic levels. The effects of feeding by top carnivores can cascade downward, exerting an indirect effect on the phytoplankton and thus on primary productivity.

Fishing Fleets at Loggerheads with Sea Turtles

Populations of loggerhead sea turtles (*Caretta caretta*) that nest on Western Pacific beaches in Australia and Japan have been in decline. Like other sea turtles, *C. caretta* hatch from eggs that females bury on sandy beaches. Immediately after hatching, the young turtles rush to the surf and the open ocean. Turtles mature at sea and return to their hatching beaches to lay eggs. Using mitochondrial DNA (mtDNA), Bruce Bowen and colleagues explored the situation sea turtles face.

The researchers took mtDNA samples from nesting populations in Australia and Japan, from populations of turtles feeding in Baja California, and from turtles drowned in fishing nets in the north Pacific. One 350-base-pair (bp) segment of mtDNA included sequence variations that are characteristic of different loggerhead populations. After samples were amplified by the polymerase chain reaction, sequencing revealed three major variants of mtDNA, which the researchers designated as sequences A, B, and C. The sequences were distributed among loggerhead turtles, as shown in **Table 1.**

The mtDNA of most *C. caretta* found in Baja California and in fishing nets in the north Pacific matched that of turtles from the Japanese nesting areas. These data support the idea that loggerhead turtles hatched in Japan make the 10 000-km-long migration across the North Pacific to Baja California. The data also indicate that a few turtles that hatched in Australia may follow the same migratory route.

This migration could be aided by the North Pacific Current, which moves from west to east, whereas the return trip from Baja to Japan could be made via the North Equatorial Current, which runs from east to west just north of the equator. Loggerhead turtles have been found in these currents, and further tests will reveal whether they have the mtDNA sequence characteristic of the individuals nesting in Japan and feeding in Baja California.

The nesting population of *C. caretta* in Japan is 2000 to 3000 females. It is uncertain if this population can survive the loss of thousands of offspring to fishing in the North Pacific. The number of female loggerhead turtles nesting in Australia has declined by 50 to 80% in the last decade, so the loss of only a few individuals in fishing nets could also have a drastic impact on this population. To save the loggerhead turtles, wildlife managers and international agencies must establish and enforce limits on the number of migrating individuals trapped and killed in the ocean fisheries.

Like other sea turtles, the loggerheads are severely impacted by fishing. As many as 4000 loggerheads drown in nets every year, and others are caught in longline fisheries. Adoption of a new fish hook **(Figure 1)** could reduce the turtle catch in longline fishing operations.

Table 1	Sources of Turtles by Nesting Grounds		
	Number of Turtles		
Location	Sequence A	Sequence B	Sequence C
Australian nesting areas	26	0	0
Japanese nesting areas	0	23	3
Baja California feeding grounds	2	19	5
North Pacific	1	28	5

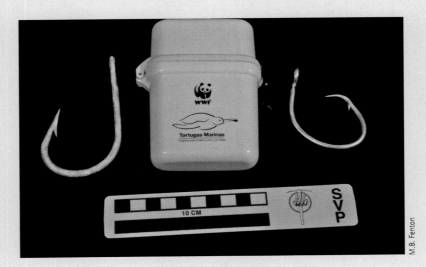

Figure 1
Sea turtles and longlining. Conventional longline hooks (left) readily catch sea turtles, whereas the hook on the right does not. The World Wildlife Fund is promoting the use of hooks (right) that are friendly to sea turtles in an effort to reduce their losses to longline fishing.

M.B. Fenton

consumed by predatory invertebrates and fishes. The top nonhuman carnivore in these food webs is usually a predatory fish.

Herbivorous zooplankton play a central role in regulation of lake ecosystems. Small zooplankton species consume only small phytoplankton. Thus, when *small* zooplankton are especially abundant, large phytoplankton escape predation and survive, and the lake's primary productivity is high. By contrast, large zooplankton are voracious, eating both small and large phytoplankton. When large zooplankton are especially abundant, they reduce the overall biomass of phytoplankton, lowering the ecosystem's primary productivity.

In this **trophic cascade**, predator–prey effects reverberate through population interactions at two or more trophic levels in an ecosystem. Feeding by plankton-eating invertebrates and fishes has a *direct* impact on herbivorous zooplankton populations and an *indirect* impact on phytoplankton populations (the ecosystem's primary producers). Invertebrate predators prefer small zooplankton. And when the invertebrates that eat small zooplankton are the dominant predators in the ecosystem, large zooplankton become more abundant; they consume many phytoplankton, causing a decrease in productivity. But plankton-eating fishes prefer to eat large zooplankton (see Figure 31.9, p. 755), so when they are abundant, small zooplankton become the dominant herbivores, leading large phytoplankton to become more numerous, which raises the lake's productivity.

Large predatory fishes may add an additional level of control to the system because they feed on and regulate the population sizes of plankton-eating invertebrates and fishes. Thus, the effects of feeding by the top predator can cascade downward through the food web, affecting the densities of plankton-eating invertebrates and fishes, herbivorous zooplankton, and phytoplankton. Research in Norway with brown trout (*Salmo trutta*), a top predator, and Arctic char (*Salvelinus alpinus*), the prey, demonstrated how culling prey can promote the recovery of top predators. In this case, Lake Takvatn was the scene of a large-scale experiment. Older, stunted prey species (*S. alpinus*) were removed. These fish had eaten small prey, so an increase in the availability of prey and recovery of the predator resulted. In this case, *S. trutta* was the top predator, and *S. alpinus*, an introduced species, was culled to rejuvenate the system. Another process of bioremediation, the addition of piscivorous fish to a lake, has also been successful in restoring ecosystem balance in other parts of the world.

31.1e Biological Magnification Is the Movement of Contaminants Up the Food Chain

DDT (a formerly popular insecticide; see "Molecule behind Biology," Box 31.1, p. 758) provided a clear demonstration of the interconnectedness of organisms.

Consumers accumulate DDT from all the organisms they eat in their lifetimes. Primary consumers, such as herbivorous insects, may ingest relatively small amounts of DDT, but a songbird that eats many of these insects accumulates all the collected DDT consumed by its prey. A predator such as a raptor, perhaps a Sharp-shinned Hawk (*Accipiter striatus*), that eats songbirds accumulates even more. Whether the food chain (web) is aquatic or terrestrial, the net effect on higher-level consumers is the same **(Figure 31.10)**.

Natural systems have provided many examples of **biological magnification.** In cities where DDT was used in an effort to control the spread of Dutch elm disease, songbirds died from DDT poisoning after eating insects that had been sprayed (whether or not they were involved in spreading the disease). In forests, DDT was used in an effort to control spruce budworm moths (*Choristoneura occidentalis*), and salmon died because runoff carried DDT into their streams and rivers, where their herbivorous prey consumed it.

Despite the ban on the use of DDT in the United States in 1973, in 1990, the California State Department of Health recommended closing a fishery off the coast of California because of DDT accumulating there. DDT discharged in industrial waste 20 years earlier was still moving through the ecosystem. The half-life of DDT in an organism's body fat is eight years.

Other contaminants emulate DDT. Mercury contamination is common in many parts of the world, often as a by-product of the pulp and paper industry. Minamata, the disease humans get from mercury

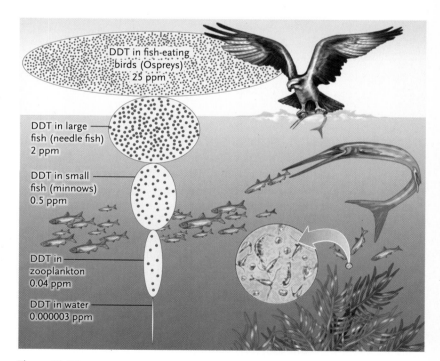

Figure 31.10

Biological magnification. In this marine food web in northeastern North America, DDT concentration (measured in parts per million, ppm) was magnified nearly 10 million times between zooplankton and the Osprey (*Pandion haliaetus*).

DDT: Dichloro-Diphenyl-Trichloroethane

Figure 1
A molecule of DDT.

Originally formulated in 1873, DDT's potential as an insecticide was only recognized in 1939 by Paul Muller of Geigy Pharmaceutical in Switzerland. DDT **(Figure 1),** the first of the chlorinated insecticides, was used extensively in some theatres of World War II, notably in Burma (now Myanmar) in 1944, when the Japanese forces were on the brink of moving into India. There Allied forces suffered from "three m's": mud, morale, and malaria. Meanwhile, in 1943 in southern Italy, DDT was instrumental in controlling populations of lice that plagued Canadian troops there. Widespread application of DDT in Burma reduced the incidence of malaria by killing mosquitoes, the vectors for the disease (see Chapter 27). After World War II, the use of DDT spread rapidly, and the World Health Organization (WHO) credited this molecule with saving 25 million human lives (mainly through control of mosquitoes that carry malaria).

At first, DDT appeared to be an ideal insecticide. In addition to being inexpensive to produce, it had low toxicity to mammals (300 to 500 mg·kg^{-1} is the LD$_{50}$, the amount required to kill half of the target population). But many insects subsequently developed immunity to DDT, reducing its effectiveness.

DDT is chemically stable and soluble in fat, so instead of being metabolized by mammals, it is stored in their fat. The biological half-life of DDT is approximately eight years (it takes about eight years for a mammal to metabolize half the amount of DDT it has assimilated). DDT is released when fat is metabolized, so when mammals metabolize fat (for example, when humans go on a diet), they are exposed to higher concentrations of DDT in their blood. DDT also had dramatic effects on some birds, notably those higher up the food chain. Eggshell thinning was a consequence of exposure to DDT. Populations of birds such as Peregrine Falcons (*Falco peregrinus*) plummeted.

Since 1985, the use of DDT has been totally banned in Canada, and it is now banned in many other countries. But DDT is still produced in countries such as the United States and still used in countries where malaria is a prominent problem because the ecological costs of DDT are considered secondary to the importance of controlling the mosquitoes. WHO estimates that every 30 seconds, a child dies of malaria. Approximately 40% of the world's population of humans is at risk of contracting malaria where they live, mainly in Africa. Malaria also remains a problem in tropical and subtropical Asia and Central and South America. People in southern Europe and the Middle East may also be at risk.

By the early 1970s, cetaceans in the waters around Antarctica had DDT in their body fat even though DDT had never been used there. The movement of DDT up the food chain and through food webs demonstrated the interconnections in biological systems. The movement of DDT also provides a graphic demonstration of the transfer of materials from one trophic level to another.

Removing DDT from the arsenal of products used to control insects has had other impacts. For example, there has been an upsurge recently in the number of houses and apartments infested by bedbugs (*Cimex lectularius*). DDT had been very effective in the control of bedbugs, but in its absence, populations of these insects have rebounded, renewing old challenges that our grandparents had experienced.

poisoning, is usually linked to the consumption of fish taken from contaminated watersheds. Eating fish contaminated with mercury can result in mercury concentrations in people's hair (0.9 to 94 mg·kg^{-1}) and in otters (*Lontra canadensis;* 0.49 to 54.37 mg·kg^{-1}). In southern Ontario, the hair of bats that eat insects that emerge from mercury-contaminated sediments contains concentrations up to 13 mg·kg^{-1}. Fish are obviously not essential to this chain of biomagnification.

Evidence for the impact of pesticides often comes from sources we might not have expected. An excellent example is the accumulation of Chimney Swift (*Chaetura pelagica*) droppings at the bottom of a chimney in Kingston, Ontario. As their name implies, Chimney Swifts often nest in chimneys, and analysis of samples from the accumulated droppings revealed how changes in levels of insecticides coincided with changes in the birds' diets and, presumably, in populations of their prey. This research may help us to understand the reasons for declines in populations of aerial insectivorous birds (such as Chimney Swifts).

STUDY BREAK

1. What is net primary productivity? How does it differ from standing crop biomass? Are the pyramids useful?
2. Many deserts have low levels of productivity despite receiving a lot of sunlight. Why?
3. What are assimilation efficiency and production efficiency?

31.2 Nutrient Cycling in Ecosystems

The availability of nutrients is as important to ecosystem function as the input of energy. Photosynthesis requires carbon, hydrogen, and oxygen, which producers acquire from water and air. Primary producers also need nitrogen, phosphorus, and other minerals (see Chapter 48). A deficiency in any of these minerals can reduce primary productivity.

Earth is essentially a closed system with respect to matter, even though cosmic dust enters the atmosphere. Thus, unlike energy, for which there is a constant cosmic input, virtually all of the nutrients that will ever be available for biological systems are already present. Nutrient ions and molecules constantly circulate between the abiotic environment and living organisms in **biogeochemical cycles**. And, unlike energy, which flows through ecosystems and is gradually lost as heat, matter is conserved in biogeochemical cycles. Although there may be local shortages of specific nutrients, Earth's overall supplies of these chemical elements are never depleted or increased.

Nutrients take various forms as they pass through biogeochemical cycles. Materials such as carbon, nitrogen, and oxygen form gases that move through global *atmospheric cycles*. Geologic processes move other materials, such as phosphorus, through local *sedimentary cycles*, carrying them between dry land and the sea floor. Rocks, soil, water, and air are the reservoirs where mineral nutrients accumulate, sometimes for many years.

Ecologists use a **generalized compartment model** to describe nutrient cycling **(Figure 31.11)**. Two criteria divide ecosystems into four compartments in which nutrients accumulate. First, nutrient molecules and ions are either *available* or *unavailable*, depending on whether they can be assimilated by organisms. Second, nutrients are present in either *organic* material, living or dead tissues of organisms, or *inorganic* material, such as rocks and soil. Minerals in dead leaves on the forest floor are in the available-organic compartment because they are in the remains of organisms that can be eaten by detritivores. Calcium ions in limestone rocks are in the unavailable-inorganic compartment because they are in a nonbiological form that producers cannot assimilate.

Nutrients move rapidly within and between the available compartments. Living organisms are in the available-organic compartment, and whenever heterotrophs consume food, they recycle nutrients within that reservoir (indicated by the circular arrow in the upper left of Figure 31.11). Producers acquire nutrients from the air, soil, and water of the available-inorganic compartment. Consumers acquire nutrients from the available-inorganic compartment when

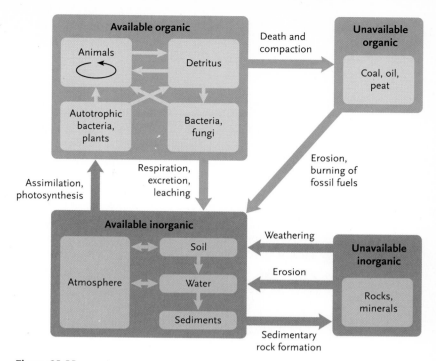

Figure 31.11

A generalized compartment model of nutrient cycling. Nutrients cycle through four major compartments within ecosystems. Processes that move nutrients from one compartment to another are indicated on the arrows. The circular arrow under "Animals" represents animal predation on other animals.

they drink water or absorb mineral ions through their integument. Several processes routinely transfer nutrients from organisms to the available-inorganic compartment. Respiration releases carbon dioxide, moving both carbon and oxygen from the available-organic compartment to the available-inorganic compartment.

By contrast, the exchange of materials into and out of the unavailable compartments is generally slow. Sedimentation, a long-term geologic process, converts ions and particles of the available-inorganic compartment into rocks of the unavailable-inorganic compartment. Materials are gradually returned to the available-inorganic compartment when rocks are uplifted and eroded or weathered. Similarly, over millions of years, the remains of organisms in the available-organic compartment were converted into the coal, oil, and peat of the unavailable-organic compartment.

Except for the input of solar energy, we have described energy flow and nutrient cycling as though ecosystems were closed systems. In reality, most ecosystems exchange energy and nutrients with neighbouring ecosystems. Rainfall carries nutrients into a forest ecosystem, and runoff carries nutrients from a forest into a lake or river. Ecologists have mapped biogeochemical cycles of important elements, often by using radioactively labelled molecules that they can follow in the environment.

31.2a Water Is the Staff of Life

Water is the universal intracellular solvent for biochemical reactions, but only a fraction of 1% of Earth's total water is present in biological systems at any time.

The cycling of water, the **hydrogeologic cycle**, is global, with water molecules moving from oceans into the atmosphere, to land, through freshwater ecosystems, and back to the oceans **(Figure 31.12)**. Solar energy causes water to evaporate from oceans, lakes, rivers, soil, and living organisms, entering the atmosphere as a vapour and remaining aloft as a gas, as droplets in clouds, or as ice crystals. Water falls as precipitation, mostly in the form of rain and snow. When precipitation falls on land, water flows across the surface or percolates to great depths in soil, eventually reentering the ocean reservoir through the flow of streams and rivers.

The hydrogeologic cycle maintains its global balance because the total amount of water entering the atmosphere is equal to the amount that falls as precipitation. Most water that enters the atmosphere evaporates from the oceans, which are the largest reservoir of water on the planet. A much smaller fraction evaporates from terrestrial ecosystems, and most of that is through transpiration by green plants.

Constant recirculation provides fresh water to terrestrial organisms and maintains freshwater ecosystems such as lakes and rivers. Water also serves as a transport medium that moves nutrients within and between ecosystems, as demonstrated in a series of classic experiments in the Hubbard Brook Experimental Forest (see "Studies of the Hubbard Brook Watershed").

31.2b Carbon Is the Backbone of Life

Carbon atoms provide the backbone of most biological molecules, and carbon compounds store the energy captured by photosynthesis (see Chapter 7). Carbon enters food webs when producers convert atmospheric carbon dioxide (CO_2) into carbohydrates. Heterotrophs acquire carbon by eating other organisms or detritus. Although carbon moves somewhat independently in sea and on land, a common atmospheric pool of CO_2 creates a global **carbon cycle** **(Figure 31.13, p. 762)**.

The largest reservoir of carbon is sedimentary rock, such as limestone. Rocks are in the unavailable-inorganic compartment, and they exchange carbon with living organisms at an exceedingly slow pace. Most *available* carbon is present as dissolved bicarbonate

a. The water cycle

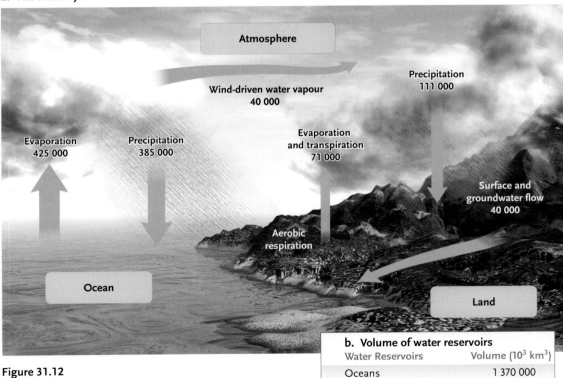

Figure 31.12

The hydrogeologic cycle. Water cycles through marine, atmospheric, and terrestrial reservoirs. **(a)** Data next to the arrows list the amount of water (in km³·year⁻¹) moved among the reservoirs by various processes. **(b)** The oceans are by far the largest of the six major reservoirs of water on Earth.

b. Volume of water reservoirs

Water Reservoirs	Volume (10³ km³)
Oceans	1 370 000
Polar ice, glaciers	29 000
Groundwater	4 000
Lakes, rivers	230
Soil moisture	67
Atmosphere (water vapour)	14

Studies of the Hubbard Brook Watershed Box

Water flows downhill, so local topography affects the movement of dissolved nutrients in terrestrial ecosystems. A **watershed** is an area of land from which precipitation drains into a single stream or river. Each watershed represents a part of an ecosystem from which nutrients exit through a single outlet. When several streams join to form a river, the watershed drained by the river encompasses the smaller watersheds drained by the streams. The Mackenzie River watershed covers roughly 20% of Canada and includes the watersheds of the Peace and Athabasca rivers, as well as many other watersheds drained by smaller streams and rivers.

Watersheds are ideal for large-scale field experiments about nutrient flow in ecosystems because they are relatively self-contained units. Herbert Bormann and Gene Likens conducted a classic experiment on nutrients in watersheds in the 1960s. Bormann and Likens manipulated small watersheds of temperate deciduous forest in the Hubbard Brook Experimental Forest in the White Mountain National Forest of New Hampshire. They measured precipitation and nutrient input into the watersheds, the uptake of nutrients by vegetation, and the amount of nutrients leaving the watershed via streamflow. They monitored nutrients exported in streamflow by collecting water samples from V-shaped concrete weirs built into bedrock below the streams that drained the watersheds **(Figure 1).** Impermeable bedrock underlies the soil, preventing water from leaving the system by deep seepage.

Bormann and Likens collected several years of baseline data on six undisturbed watersheds. Then, in 1965 and 1966, they felled all of the trees in one small watershed and used herbicides to prevent regrowth. After

Gene E. Likens, from Gene E. Likens et al., *Ecology Monograph*, 40(1): 23–47, 1970

Figure 1
A weir used to measure the volume and nutrient content of water leaving a watershed by streamflow.

these manipulations, they monitored the output of nutrients in streams that drained experimental and control watersheds. They attributed differences in nutrient export between undisturbed watersheds (controls) and the clear-cut watershed (experimental treatment) to the effects of deforestation.

Bormann and Likens determined that vegetation absorbed substantial water and conserved nutrients in undisturbed watersheds. Plants used about 40% of the precipitation for transpiration. The rest contributed to

runoff and groundwater. Control watersheds lost only about 8 to 10 kg of calcium per hectare each year, an amount replaced by erosion of bedrock and input from rain. Moreover, control watersheds actually accumulated about 2 kg of nitrogen per hectare per year and slightly smaller amounts of potassium.

The experimentally deforested watershed experienced a 40% annual increase in runoff, including a 300% increase during a four-month period in summer. Some mineral losses were similarly large. The net loss of calcium was 10 times as high **(Figure 2)** as in the control watersheds and of potassium was 21 times as high. Phosphorus losses did not increase because this mineral was apparently retained by the soil. The loss of nitrogen, however, was very large—120 kg $\cdot$ ha^{-1} $\cdot$ year^{-1}. The washing out of nitrogen meant that the stream draining the experimental watershed became choked with algae and cyanobacteria. The Hubbard Brook experiment demonstrated that deforestation increases flooding and decreases the fertility of ecosystems.

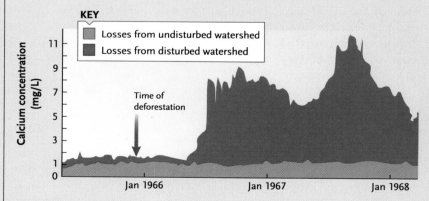

Figure 2
Calcium losses from the deforested watershed were much greater than those from controls. The arrow indicates the time of deforestation in early winter. Mineral losses did not increase until after the ground thawed the following spring. Increased runoff also caused large water losses from the watershed.

a. Amount of carbon in major reservoirs

Carbon Reservoirs	Mass (10^{12} g)
Sediments and rocks	770 000 000
Ocean (dissolved forms)	397 000
Soil	15 000
Atmosphere	7 500
Biomass on land	7 150

b. Annual global carbon movement between reservoirs

Direction of Movement	Mass (10^{12} kg)
From atmosphere to plants (carbon fixation)	1200
From atmosphere to ocean	1070
To atmosphere from ocean	1050
To atmosphere from plants	600
To atmosphere from soil	600
To atmosphere from burning fossil fuel	50
To atmosphere from burning plants	20
To ocean from runoff	4
Burial in ocean sediments	1

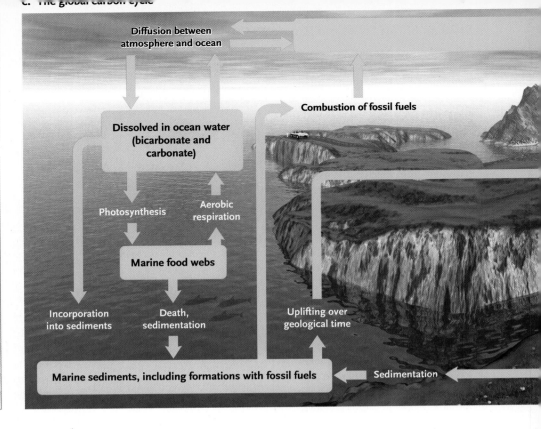

c. The global carbon cycle

Figure 31.13

The carbon cycle. Marine and terrestrial components of the global carbon cycle are linked through an atmospheric reservoir of carbon dioxide. **(a)** By far the largest amount of Earth's carbon is found in sediments and rocks. **(b)** Earth's atmosphere mediates most of the movement of carbon. **(c)** In this illustration of the carbon cycle, boxes identify major reservoirs and labels on arrows identify the processes that cause carbon to move between reservoirs.

ions (HCO_3^-) in the ocean. Soil, atmosphere, and plant biomass are significant, but much smaller, reservoirs of available carbon. Atmospheric carbon is mostly in the form of molecular CO_2, a product of aerobic respiration. Volcanic eruptions also release small quantities of CO_2 into the atmosphere.

Carbon atoms sometimes leave organic compartments for long periods of time. Some organisms in marine food webs build shells and other hard parts by incorporating dissolved carbon into calcium carbonate ($CaCO_3$) and other insoluble salts. When shelled organisms die, they sink to the bottom and are buried in sediments. Other animals, notably vertebrates, store calcium in bone. Insoluble carbon that accumulates as rock in deep sediments may remain buried for millions of years before tectonic uplifting brings it to the surface, where erosion and weathering dissolve sedimentary rocks and return carbon to an available form.

Carbon atoms are also transferred to the unavailable-organic compartment when soft-bodied organisms die and are buried in habitats where low oxygen concentration prevents decomposition. In the past, under suitable geologic conditions, these carbon-rich tissues were slowly converted to gas, petroleum, or coal, which we now use as fossil fuels. Human activities, especially burning fossil fuels, are transferring carbon into the atmosphere at an unnaturally high rate. The resulting change in the worldwide distribution of carbon is having profound consequences for Earth's atmosphere and climate, including a general warming of the climate and a rise in sea level (see "Disruption of the Carbon Cycle," p. 764).

31.2c Nitrogen Is a Limiting Element

All organisms require nitrogen to construct nucleic acids, proteins, and other biological molecules (see Chapter 48). Earth's atmosphere had a high nitrogen concentration long before life began. Today, a global **nitrogen cycle** moves this element between the huge atmospheric pool of gaseous molecular nitrogen (N_2) and several much smaller pools of nitrogen-containing compounds in soils, marine and freshwater ecosystems, and living organisms **(Figure 31.14)**.

Molecular nitrogen is abundant in the atmosphere, but triple covalent bonds bind its two atoms so tightly that most organisms cannot use it. Only certain microorganisms, volcanic action, and lightning can convert N_2 into ammonium (NH_4^+) and nitrate (NO_3^-) ions. This conversion is called **nitrogen fixation** (see Chapter 37). Once nitrogen is fixed,

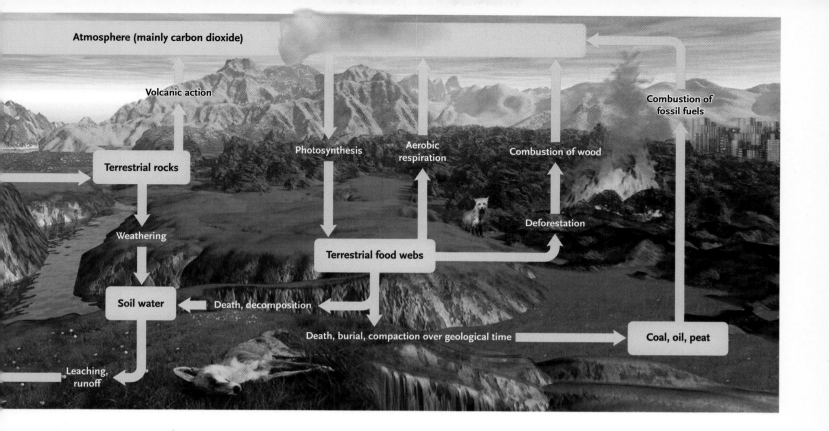

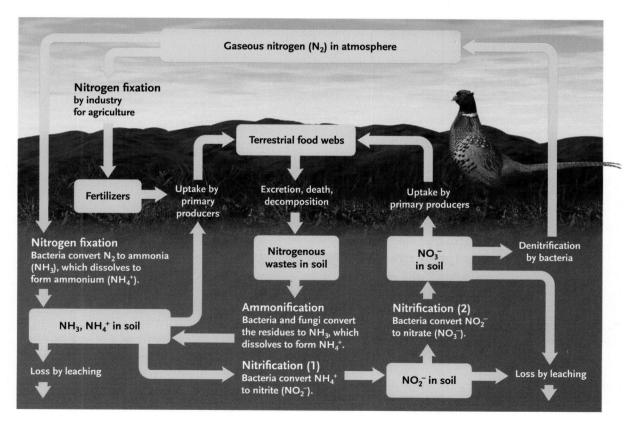

Figure 31.14

The nitrogen cycle in a terrestrial ecosystem. Nitrogen-fixing bacteria make molecular nitrogen available in terrestrial ecosystems. Other bacteria recycle nitrogen within the available-organic compartment through ammonification and two types of nitrification, converting organic wastes into ammonium ions and nitrates. Denitrification converts nitrate to molecular nitrogen, which returns to the atmosphere. Runoff carries nitrogen from terrestrial ecosystems into aquatic ecosystems, where it is recycled in freshwater and marine food webs.

Disruption of the Carbon Cycle

The concentrations of gases in the lower atmosphere have a profound effect on global temperature, in turn affecting global climate. Molecules of CO_2, water vapour, ozone, methane, nitrous oxide, and other compounds collectively act like a pane of glass in a greenhouse (hence the term *greenhouse gases*). They allow short wavelengths of visible light to reach Earth's surface while impeding the escape of longer, infrared wavelengths into space, trapping much of their energy as heat **(Figure 1).** Greenhouse gases foster the accumulation of heat in the lower atmosphere, a warming action known as the **greenhouse effect.** This natural process prevents Earth from being a cold and lifeless planet.

Data from air bubbles trapped in glacial ice indicate that atmospheric CO_2 concentrations have fluctuated widely over Earth's history **(Figure 2).** Since the late 1950s, scientists have measured atmospheric concentrations of CO_2 and other greenhouse gases at remote sampling sites such as the top of Mauna Loa in the Hawaiian Islands. These sites are free of local contamination and reflect average global conditions. Concentrations of greenhouse gases have increased steadily for as long as they have been monitored **(Figure 3).**

The graph for atmospheric CO_2 concentration has a regular zigzag pattern that follows the annual cycle of plant growth (see Figure 3). The concentration of CO_2 decreases during the summer because photosynthesis withdraws so much from the atmospheric available-inorganic pool. The concentration of CO_2 is higher during the winter when photosynthesis slows while aerobic respiration continues, returning carbon to the atmospheric available-inorganic pool. Whereas the zigs and zags in the data for CO_2 represent seasonal highs and lows, the midpoint of the annual peaks and troughs has increased steadily for 40 years. These data are evidence of a rapid buildup of atmospheric CO_2, representing a shift in the distribution of carbon in the major reservoirs on Earth. The best estimates suggest that CO_2 concentration has increased by 35% in the last 150 years and by more than 10% in the last 30 years.

The increase in the atmospheric concentration of CO_2 appears to result from combustion, whether we burn fossil fuels or wood. Today, humans burn more wood and fossil fuels than ever before. Vast tracts of tropical forests are being cleared and burned (see Chapter 32). To make matters worse, deforestation reduces the world's biomass of plants that assimilate CO_2 and help maintain the carbon cycle as it existed before human activities disrupted it.

The increase in the concentration of atmospheric CO_2 is alarming because plants with C_3 metabolism respond to increased CO_2 concentrations with increased growth rates. This is not true of C_4 plants (see Chapter 7). Thus, rising atmospheric levels of CO_2 will probably alter the relative abundances of many plant species, changing the composition and dynamics of their communities.

Simulation models suggest that increasing concentrations of any greenhouse gas may intensify the greenhouse effect, contributing to a trend of global warming. Should we be alarmed about the prospect of a warmer planet? Some models predict that the mean temperature of the lower atmosphere will rise by 4°C, enough to increase ocean surface temperatures. In some areas, such as

Figure 1
The greenhouse effect.

Sunlight penetrates the atmosphere and warms Earth's surface.

Earth's surface radiates heat (infrared wavelengths) to the atmosphere. Some heat escapes into space. Greenhouse gases and water vapour absorb some infrared energy and reradiate the rest of it back toward Earth.

When atmospheric concentrations of greenhouse gases increase, the atmosphere near Earth's surface traps more heat. The warming causes a positive feedback cycle in which rising ocean temperatures cause increased evaporation of water, which further enhances the greenhouse effect.

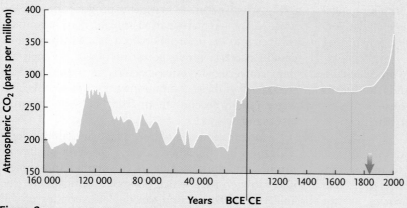

Figure 2

Carbon dioxide levels over time. The amount of atmospheric CO_2 has risen dramatically since about 1850 (arrow).

the Canadian Arctic and the Antarctic, warming has occurred much more rapidly than predicted or expected. Water expands when heated, and global sea levels could rise as much as 0.6 m just from this expansion. In addition, atmospheric temperature is rising fastest near the poles. Thus, global warming may also foster melting of glaciers and the Antarctic ice sheet, which might raise sea levels as much as 50 to 100 m, inundating low coastal regions. Waterfronts in

Vancouver, Los Angeles, Hong Kong, Durban, Rio de Janeiro, Sydney, New York, and London would be submerged. So would agricultural lands in India, China, and Bangladesh, where much of the world's rice is grown. Moreover, global warming could disturb regional patterns of precipitation and temperature. Areas that now produce much of the world's grains would become arid scrub or deserts, and the now-forested areas to their north would become dry grasslands.

Many scientists believe that atmospheric levels of greenhouse gases will continue to increase at least until the middle of the twenty-first century and that global temperature may rise by several degrees. At the Earth Summit in 1992, leaders of the industrialized countries agreed to try to stabilize CO_2 emissions by the end of the twentieth century. We have already missed that target, and some countries, including the United States (then the largest producer of greenhouse gases), have now forsaken that goal as too costly. Stabilizing emissions at current levels will not reverse the damage already done, nor will it stop the trend toward global warming. We should begin preparing for the consequences of global warming now. We might increase reforestation efforts because a large tract of forest can withdraw significant amounts of CO_2 from the atmosphere. We might also step up genetic engineering studies to develop heat-resistant and drought-resistant crop plants, which may provide crucial food reserves in regions of climate change.

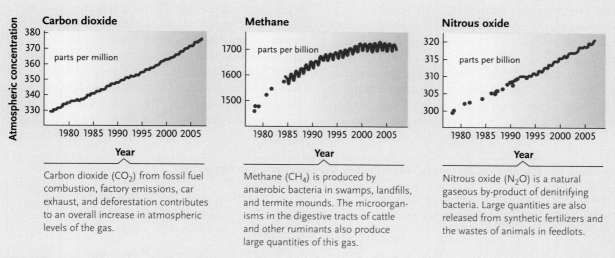

Carbon dioxide (CO_2) from fossil fuel combustion, factory emissions, car exhaust, and deforestation contributes to an overall increase in atmospheric levels of the gas.

Methane (CH_4) is produced by anaerobic bacteria in swamps, landfills, and termite mounds. The microorganisms in the digestive tracts of cattle and other ruminants also produce large quantities of this gas.

Nitrous oxide (N_2O) is a natural gaseous by-product of denitrifying bacteria. Large quantities are also released from synthetic fertilizers and the wastes of animals in feedlots.

Figure 3

Increases in atmospheric concentrations of three greenhouse gases, mid-1970s through 2004. The data were collected at a remote monitoring station in Australia (Cape Grim, Tasmania) and compiled by scientists at the Commonwealth Scientific and Industrial Research Organization, an agency of the Australian government.

primary producers can incorporate it into biological molecules such as proteins and nucleic acids. Secondary consumers obtain nitrogen by consuming these molecules.

Several biochemical processes produce different nitrogen-containing compounds and thus move nitrogen through ecosystems. These processes are nitrogen fixation, ammonification, nitrification, and denitrification (Table 31.2).

In nitrogen fixation, several kinds of microorganisms convert molecular nitrogen (N_2) to ammonium ions (NH_4^+). Certain bacteria, which collect molecular nitrogen from the air between soil particles, are the major nitrogen fixers in terrestrial ecosystems (see Table 31.2). The cyanobacteria partners in some lichens (see Chapter 25) also fix molecular nitrogen. Other cyanobacteria are important nitrogen fixers in aquatic ecosystems, whereas the water fern (genus *Azolla*) plays that role in rice paddies. Collectively, these organisms fix an astounding 200 million tonnes of nitrogen each year. Plants and other primary producers assimilate and use this nitrogen in the biosynthesis of amino acids, proteins, and nucleic acids, which then circulate through food webs.

Some plants, including legumes (such as beans and clover), alders (*Alnus* species), and some members of the rose family (Rosaceae), are mutualists with nitrogen-fixing bacteria. These plants acquire nitrogen from soils much more readily than plants that lack such mutualists. Although these plants have the competitive edge in nitrogen-poor soil, nonmutualistic species often displace them in nitrogen-rich soil. In an interesting twist on the usual predator–prey relationships, several species of flowering plants living in nitrogen-poor soils capture and digest insects (see "Pitcher Plant Ecosystems," pp. 767–768).

In addition to nitrogen fixation, other biochemical processes make large quantities of nitrogen available to producers. **Ammonification** of detritus by bacteria and fungi converts organic nitrogen into ammonia (NH_3), which dissolves in water to produce ammonium ions (NH_4^+) that plants can assimilate. Some ammonia escapes into the atmosphere as a gas. **Nitrification** by certain bacteria produces nitrites (NO_2^-), which are then converted by other bacteria to usable nitrates (NO_3^-). All of these compounds are water soluble, and water rapidly leaches them from soil into streams, lakes, and oceans.

Under conditions of low oxygen availability, **denitrification** by still other bacteria converts nitrites or nitrates into nitrous oxide (N_2O) and then into molecular nitrogen (N_2), which enters the atmosphere (see Table 31.2). This action can deplete supplies of soil nitrogen in waterlogged or otherwise poorly aerated environments, such as bogs and swamps.

In 1909, Fritz Haber developed a process for fixing nitrogen, and with the help of Carl Bosch, the process was commercialized for fertilizer production. The Haber–Bosch process has altered Earth's nitrogen cycles and is said to be responsible for the existence of 40% of the people on Earth. Before the implementation of the Haber–Bosch process, the amount of nitrogen available for life was limited by the rates at which N_2 was fixed by bacteria or generated by lightning strikes. Today, spreading fertilizers rich in nitrogen is the basis for most of agriculture's productivity. This practice has quadrupled some yields over the past 50 years (see Chapter 33). Of all nutrients required for primary production, nitrogen is often the least abundant. Agriculture routinely depletes soil nitrogen, which is removed from fields through the harvesting of plants that have accumulated nitrogen in their tissues. Soil erosion and leaching remove more. Traditionally, farmers rotated their crops, alternately planting legumes and other crops in the same fields. In combination with other soil conservation practices, crop rotation stabilized soils and kept them productive, sometimes for hundreds of years. Some of the most arable land in New York State was farmed by members of the Mohawk Iroquois First Nations. The evidence of this comes from the locations of palisaded villages. The people moved their villages

Table 31.2	Biochemical Processes That Influence Nitrogen Cycling in Ecosystems		
Process	Organisms Responsible	Products	Outcome
Nitrogen fixation	Bacteria: *Rhizobium, Azotobacter, Frankia* Cyanobacteria: *Anabaena, Nostoc*	Ammonia (NH_3), ammonium ions (NH_4^+)	Assimilated by primary producers
Ammonification of organic detritus	Soil bacteria and fungi	Ammonia (NH_3), ammonium ions (NH_4^+)	Assimilated by primary producers
Nitrification			
(1) Oxidation of NH_3	Bacteria: *Nitrosomonas, Nitrococcus*	Nitrite (NO_2^-)	Used by nitrifying bacteria
(2) Oxidation of NO_2^-	Bacteria: *Nitrobacter*	Nitrate (NO_3^-)	Assimilated by primary producers
Denitrification of NO_3^-	Soil bacteria	Nitrous oxide (N_2O), molecular nitrogen (N_2)	Released to atmosphere

Pitcher Plant Ecosystems

Pitcher plants have modified leaves (pitchers) that act as pitfall traps for drowning and digesting insect prey. Pitchers have developed in at least five different evolutionary lines of vascular plants (see Chapter 35). Throughout much of North America, pitcher plants (the provincial flower of Newfoundland and Labrador; **Figure 1**) are common in bogs. *Sarracenia purpurea*, like other carnivorous plants, obtain much of their nitrogen from the insects they capture.

The captured arthropod prey, mainly ants and flies, is the base of a food web inside the pitchers. These are shredded and partly consumed by larvae of midges (*Metriocnemus knabi*) and sarcophagid flies (*Fletcherimyia fletcheri*; **Figure 2**). A subweb of bacteria and protozoa processes shredded prey, which are eaten by filter-feeding rotifers (*Habrotrocha rosa*; **Figure 3, p. 768**) and mites (*Sarraceniopus gibsonii*). Mosquito larvae (*Wyeomyia smithii*) eat the bacteria, protozoa, and rotifers, whereas the larger sarcophagid fly larvae eat the rotifers and smaller mosquito larvae. Populations of bacteria, protozoa, and rotifers grow much more rapidly than populations of mosquito or midge larvae, making the system sustainable.

Pitchers are essential to the life cycles of two species of insects whose larvae live in them. A mosquito and a midge coexist in the same pitchers, and their populations are limited by

Figure 1
Sarracenia purpurea, *a pitcher plant. The flower on a long stalk extends above the pitchers. One pitcher is shown in the photo on the right.*

the availability of insect carcasses. In any pitcher, growth in populations of the midge larvae is not affected by increases in the numbers of mosquito larvae. But populations of mosquito larvae increase as populations of midge larvae increase (see Figure 2).

The situation is an example of processing-chain commensalism because the action of one species creates opportunities for another. In this case, midge larvae feed on the

hard parts of insect carcasses and break them up in the process. Mosquito larvae are filter feeders, consuming particles derived from the decaying matter. The feeding of the midges generates additional food for the mosquito larvae. Although the populations of midge and mosquito larvae can be large in any pitcher, only a single sarcophagid fly larva occurs in any pitcher. *F. fletcheri* is a *K*-strategist (see Chapter 29) and gives birth to

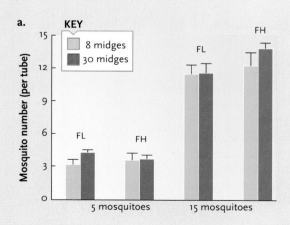

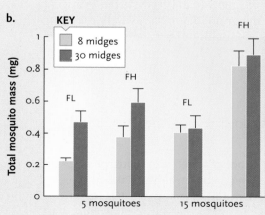

Figure 2
Midge and mosquito larvae in pitchers. (a) The density and (b) total dry mass of mosquito larvae are the same whether the population of midges is low (8 midges) or high (30 midges). FH = high food availability; FL = low food availability. Error bars show standard errors of the mean.

(Continued)

Pitcher Plant Ecosystems (*Continued*)

Figure 3
A bdelloid rotifer, Habrotrocha rosa, *from a* Sarracenia purpurea *pitcher.*

a.

Courtesy of the Biodiversity Institute of Ontario

b.

Courtesy of Daniel Handfield

Figure 4
Moths whose caterpillars eat Sarracenia purpurea. *The caterpillars of* **(a)** Exyra fax *and* **(b)** Papaipema appassionata *feed on pitcher plants, either (a) the lining of pitchers or (b) the rhizomes.*

larvae. If you place more than one *F. fletcheri* larva in a pitcher, a fight ensues. The larger larva either wins or leaves the pitcher to pupate in the sphagnum around it.

These insects do not appear to compete with their hosts, the pitcher plants. The abundance of rotifers living in the pitchers of *S. purpurea* is negatively associated with the presence of midge and mosquito larvae (which eat the rotifers). Rotifers are detritivores, and their excretory products (NO_3^--N, NH_4OH, P) account for a major portion of the N acquired by the plants from their insect prey.

Two species of moths also exploit *S. purpurea* **(Figure 4)**. *Exyra fax* and *Papaipema appassionata* do not live in the pitchers. *Exyra fax* caterpillars eat the interior surface of the pitcher chambers, whereas *P. appassionata* caterpillars consume the rhizomes. Although predation by *E. fax* caterpillars does not kill the plants, predation by *P. appassionata* does. To what trophic level does one assign moths whose caterpillars are herbivores feeding on primary producers that eat insects?

and farming operations every 10 to 20 years, changing fields repeatedly over hundreds of years.

The production of synthetic fertilizers is expensive, using fossil fuels as both raw material and an energy source. Fertilizer becomes increasingly costly as supplies of fossil fuels dwindle. Furthermore, rain and runoff leach excess fertilizer from agricultural fields and carry it into aquatic ecosystems. Nitrogen has become a major pollutant of freshwater ecosystems, artificially enriching the waters and allowing producers to expand their populations. Human activities have disrupted the global nitrogen cycle **(Figure 31.15)**.

31.2d Phosphorus Is Another Essential Element

Phosphorus compounds lack a gaseous phase, and this element moves between terrestrial and marine ecosystems in a sedimentary cycle **(Figure 31.16)**. Earth's crust is the main reservoir of phosphorus, as it is for other minerals, such as calcium and potassium, that also undergo sedimentary cycles.

Phosphorus is present in terrestrial rocks in the form of phosphates (PO_4^{3-}). In the **phosphorus cycle**, weathering and erosion add phosphate ions to soil and carry them into streams and rivers, which eventually transport them to the ocean. Once there, some phosphorus enters marine food webs, but most of it precipitates out of solution and accumulates for millions of years as insoluble deposits, mainly on continental shelves. When parts of the sea floor are uplifted and exposed, weathering releases the phosphates.

Plants absorb and assimilate dissolved phosphates directly, and phosphorus moves easily to higher trophic levels. All heterotrophs excrete some phosphorus as a waste product in urine and feces; the phosphorus becomes available after decomposition. Primary producers readily absorb the phosphate ions, so phosphorus cycles rapidly *within* terrestrial communities.

Supplies of available phosphate are generally limited, however, and plants acquire it so efficiently that they reduce soil phosphate concentration to extremely low levels. Thus, like nitrogen, phosphorus is a common ingredient in agricultural fertilizers, and excess phosphates are pollutants of freshwater

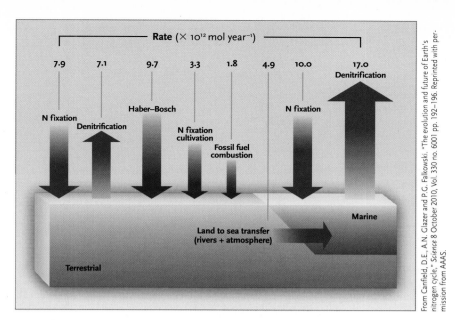

Figure 31.15
Modern global nitrogen flux depends upon the efficiency of transfer of N between reservoirs. Thickness of arrows indicates relative size of flux. Anthropogenic inputs are shown as dark brown arrows.

From Canfield, D.E., A.N. Glazer and P.G. Falkowski: "The evolution and future of Earth's nitrogen cycle," *Science* 8 October 2010, Vol. 330 no. 6001 pp. 192–196. Reprinted with permission from AAAS.

ecosystems. A particularly good example is Lake Erie, a Great Lake that was heavily affected by accumulations of phosphorus. The example here is more convincing because the problem has largely been resolved over the years.

For many years, phosphate for fertilizers was obtained from guano (the droppings of seabirds that consume phosphorus-rich food), which was mined on small islands that hosted seabird colonies, for example, in Polynesia and Micronesia. We now obtain most phosphate for fertilizer from phosphate rock mined in

places such as Saskatchewan that have abundant marine deposits.

STUDY BREAK

1. How is balance maintained in the hydrogeologic cycle?
2. How do consumers obtain carbon?
3. What is the role of cyanobacteria in the nitrogen cycle? Why is their role important?

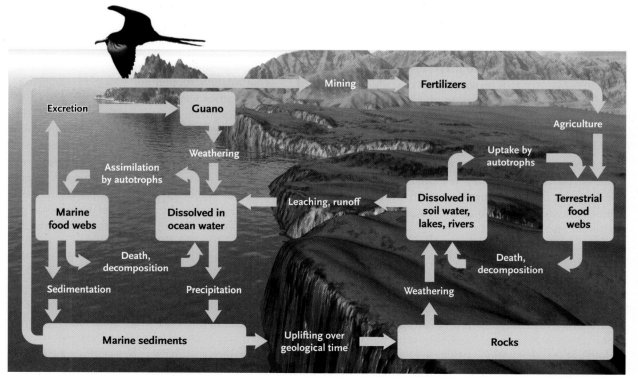

Figure 31.16
The phosphorus cycle. Phosphorus becomes available to biological systems when wind and rainfall dissolve phosphates in rocks and carry them into adjacent soil and freshwater ecosystems. Runoff carries dissolved phosphorus into marine ecosystems, where it precipitates out of solution and is incorporated into marine sediments.

Lenore Fahrig, *Carleton University*

The fragmentation of habitats is a ubiquitous effect of human activity on landscapes. In many parts of the world, land areas that used to be continuous forest are now large expanses of agricultural or urban landscapes dotted with small fragments of forest **(Figure 1)**. Lenore Fahrig examines the impact of landscape structure on the abundance, distribution, and persistence of organisms.

In her research, Fahrig uses a variety of organisms, from beetles to plants and birds. She considers habitats and the impacts of roads and fence lines. She and her students try to identify the habitat features associated with the persistence of species after fragmentation and the role of connectivity between fragments in the persistence of populations in the fragments.

Using a combination of theoretical work and fieldwork, she has assessed the responses of species in different trophic roles to the fragmentation of habitat. Her work demonstrates that not all species respond in the same way and that some benefit from fragmentation.

The connections between theoretical work and reality emerge clearly from her research, and the implications for conservation of biodiversity (see Chapter 32) are clear.

Figure 1

An aerial view of farmland in southwestern Ontario illustrates isolated patches of forest (woodlots) and bands of woodland (riparian) along the edges of a creek. The woodlots are varied in their size and shape and in the degree of their isolation or connection to other woodlots.

31.3 Ecosystem Modelling

Ecologists use modelling to make predictions about how an ecosystem will respond to specific changes in physical factors, energy flow, or nutrient availability. Analyses of energy flow and nutrient cycling allow us to create a *conceptual model* of how ecosystems function **(Figure 31.17)**. Energy that enters ecosystems is gradually dissipated as it flows through a food web. By contrast, nutrients are conserved and recycled among the system's living and nonliving components. This general model does not include processes that carry nutrients and energy out of one ecosystem and into another.

More important, the model ignores the nuts-and-bolts details of exactly how specific ecosystems function. Although it is a useful tool, a conceptual model does not really help us predict what would happen, say, if we harvested 10 million tonnes of introduced salmon from Lake Erie every year. We could simply harvest the fishes and see what happens. But ecologists prefer less intrusive approaches to studying the potential effects of disturbances.

One approach to predicting "what would happen if..." is **simulation modelling**. Using this approach, researchers gather detailed information about a specific ecosystem. They then derive a series of mathematical equations that define its most important relationships. One set of equations might describe how nutrient availability limits productivity at various trophic levels. Another might relate the population growth of zooplankton to the productivity of phytoplankton. Other equations would relate the population dynamics of primary carnivores to the availability of their food, and still others would describe how the densities of primary carnivores influence reproduction in populations at both lower and higher trophic levels. Thus, a complete simulation model is a set of interlocking equations that collectively predict how changes in one feature of an ecosystem might influence other features.

Creating a simulation model is a challenge because the relationships within every ecosystem are complex.

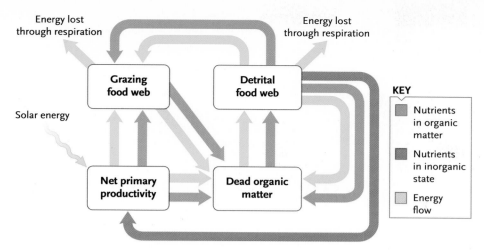

Energy lost through respiration

Energy lost through respiration

Solar energy

Grazing food web

Detrital food web

Net primary productivity

Dead organic matter

KEY

Nutrients in organic matter

Nutrients in inorganic state

Energy flow

Figure 31.17
A conceptual ecosystem model. A simple conceptual model of an ecosystem illustrates how energy flows through the system and is lost from both detrital and grazing food webs. Nutrients are recycled and conserved.

First, you must identify the important species, estimate their population sizes, and measure the average energy and nutrient content of each. Next, you must describe the food webs in which they participate, measure the quantity of food each species consumes, and estimate the productivity of each population. And, for the sake of completeness, you must determine the ecosystem's energy and nutrient gains and losses caused by erosion, weathering, precipitation, and runoff. You must repeat these measurements seasonally to identify annual variation in these factors. Finally, you might repeat the measurements over several years to determine the effects of year-to-year variation in climate and chance events.

After collecting these data, you must write equations that quantify the relationships in the ecosystem, including information about how temperature and other abiotic factors influence the ecology of each species. Having completed that job, you would begin to predict, for example, possibly in great detail, the effects of adding 1000 new housing units to an area of native prairie or boreal forest. Of course, you must refine the model whenever new data become available.

Some ecologists devote their professional lives to studying ecosystem processes and creating simulation models. The long-term initiative at the Hubbard Brook Forest provides a good example (see "Studies of the Hubbard Brook Watershed"). As we attempt to understand larger and more complex ecosystems (and as we create larger and more complex environmental problems), modelling becomes an increasingly important tool. If a model is based on well-defined ecological relationships and good empirical data, it can allow us to make accurate predictions about ecosystem changes without the need for costly and environmentally damaging experiments. But, like all ideas in science, a model is only as good as its assumptions, and models must constantly be adjusted to incorporate new ideas and recently discovered facts.

STUDY BREAK

1. Briefly describe the process of simulation modelling.
2. Why is simulation modelling necessary?

31.4 Scale, Ecosystems, Species

As we have seen, the complex interactions between and among species combine with abiotic and biotic factors to produce even more complex situations. Several questions emerge from this situation: What determines which species occur in an ecosystem? What controls the size of the populations of species in an ecosystem? How do species in an ecosystem interact? What effect does scale have on the situation?

Ecosystems span scales from millimetres to kilometres. Consider the microorganisms in a biofilm of water compared to the species, some of them microorganisms, in the water contained in the pitcher of a pitcher plant. Furthermore, the community of organisms may vary among pitchers on one plant. Like the pitcher-based community, terrestrial organisms living on an island may be relatively isolated. Consider the differences among islands, such as the British Isles, the Hawaiian Islands, or the Galapagos Islands, with respect to the combination of size (area), degree of isolation (distance from mainland), and range of habitats.

Variations in the scale of interactions **(Figure 31.18, p. 772)** help to put the nature of ecosystems in context.

Compare Figure 31.18 to Figure 3 in "Dispersal," Box 30.7, Chapter 30, about pollen and seed dispersal. More obvious in the latter is the influence of mobility on patterns of dispersal and connections. Large animals disperse seeds farther than small ones. Animals that can fly have greater potential as dispersers of seeds and pollen than those that walk or run. Data on

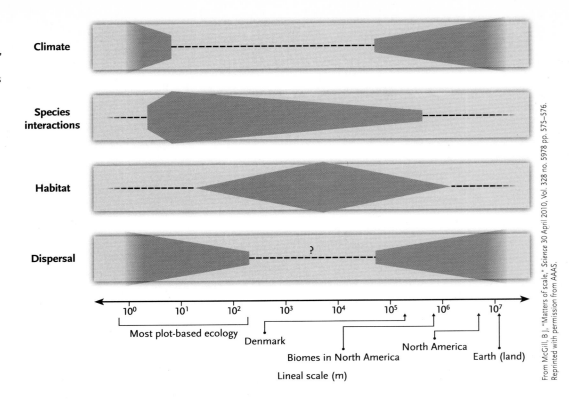

Figure 31.18
Climate, species interactions, habitat, and dispersal are four main factors controlling the distribution of species (vertical axis). Note the variation in scale (horizontal axis) across which these factors can act.

Climate

Species interactions

Habitat

Dispersal

Most plot-based ecology

Denmark

Biomes in North America

North America

Earth (land)

Lineal scale (m)

From McGill, B.J., "Matters of scale," *Science* 30 April 2010, Vol. 328 no. 5978 pp. 575–576. Reprinted with permission from AAAS.

the distribution and habitat associations of terrestrial birds in Denmark reveal how species in the same genus and those filling similar niches have more influence on the patterns of distribution of one another than less similar species.

Studies of salmon along the northwest coast of North America (Great Bear Rainforest in British Columbia and sites around Bristol Bay in Alaska) reveal how species in the genus *Oncorhynchus* can influence the plant communities bordering the streams in which they spawn. Nutrients from salmon enter these communities when salmon die after spawning, or when they are taken and eaten by predators such as bears. Healthy populations of salmon affect the nutrient loading in terrestrial plants along the rivers and streams. The density of salmon and the characteristics of the watershed (steep versus shallow banks) influence the situation. Nutrient input from salmon leads to an increase in plants such as salmonberry, associated with nutrient-rich soils. Lower input from salmon is associated with plants associated with nutrient-poor soils (e.g., blueberries). Increases in nutrient-rich soils coincide with decreases in plant diversity **(Figure 31.19)**.

The above data from salmon at sites along 50 watersheds in British Columbia demonstrate the local impact that species in one genus can have. Work from sites in Alaska shows how the inherent diversity

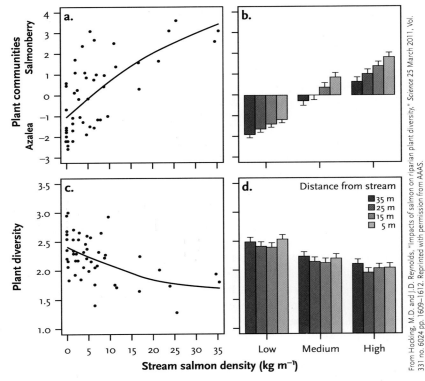

From Hocking, M. D. and J.D. Reynolds, "Impacts of salmon on riparian plant diversity," *Science* 25 March 2011, Vol. 331 no. 6024 pp. 1609–1612. Reprinted with permission from AAAS.

Figure 31.19
The influence of stream level density of spawning salmon (horizontal axis) on the structure of the community of understorey plants **(a, b)**, and on the diversity of plants **(c, d)** (vertical axis). Note the impact of distance from the stream.

of populations of sockeye salmon (*Oncorhynchus nerka*) is vital to the survival of the species and, by extension, the ecosystem. The data from sockeye salmon show how damping variance in the population provides stability. One example of variance is the timing of returns of salmon to the streams in which they hatched. The diversity is part of the *portfolio effect,* named because it is analogous to the impact of asset diversity on the stability of financial portfolios.

The diversity inherent in several hundred discrete watershed-based populations of sockeye is less than half the diversity that would occur if the sockeye were a single homogeneous population. The diversity also makes the sockeye more resilient to pressures of fishing. Studies of food webs in the watersheds provide further evidence of diversity and extend the portfolio effect.

In short, work with salmon advances our knowledge of the fundamental nature of ecosystems and helps us to appreciate the importance of maintaining biodiversity, setting the stage for Chapter 32.

STUDY BREAK

1. Why do ecosystems range so greatly in scale?
2. How do salmon have such a great influence on ecosystems around the streams in which they spawn?

31.5 Biomes and Ecozones

Biomes and ecozones are clusters of ecosystems that can occur anywhere in the world where the environmental conditions are appropriate. Biomes are **biota** (groups of organisms) characteristic of specific climatic and topographic situations (e.g., rain forest versus montane rain forest). Ecozones, or extensive ecological regions, are generally reflected by their dominant biotas or groupings of species. Like biomes, the occurrence of ecozones generally reflects climatic and topographic conditions. Both biomes and ecozones are categories of habitats that encompass a variety of ecosystems. **Table 31.3, p. 774** and **Figure 31.20, p. 776** illustrate the ecozones of Canada.

The view from the ground provides a better impression of ecozones. For example, continental shelf biome waters **(Figure 31.21a, p. 776)** are oceanic waters above the continental shelf within the Pacific Maritime ecozone. The coastal rain forest in Figure 31.21a occurs in the Pacific Maritime Biome and the Pacific Maritime ecozone. In Figure 31.21b, Lake Erie is a lentic biome, and the adjacent terrestrial biome is in the Mixedwood Biome, which here consists of a patchwork of habitat including forest (woodlots) and an open agroecosystem.

For instance, although we often think of rain forests as tropical, they also occur in temperate areas or anywhere in the world where there are forests and high levels of rainfall. For any biome, similarity of structure (e.g., plains or mountains) and dominant life forms (e.g., trees or grasses) are the key features, whether the focus is rain forest or desert. The range of terrestrial biomes, in addition to freshwater and marine biomes, in Canada illustrates their diversity. Two other Canadian biomes are shown in **Figure 31.22a and b, p. 777.**

Biomes reflect the climatic and biotic qualities that currently prevail, as well as some aspects of the history of the areas where they occur. Biomes in most of Canada resulted from changes that occurred over the last 12 thousand to 8 thousand years with the disappearance of continental ice sheets. The fossil record from sites in southern Canada and the adjacent United States reveals that at the peak of glaciation, organisms that currently occur in Arctic biomes occurred much farther south.

After settlement by Europeans, extensive clearing of forests (mixedwood plains) in what is now southwestern Ontario meant that over about 100 years biomes changed substantially to more prairielike conditions. More recently, changes in agricultural practices (e.g., time of ploughing) and land use (taking land out of agricultural production) mean that many areas cleared by 1900 CE are now reforested. Before the arrival of European settlers, some First Nations peoples also cleared forests to grow crops, but apparently not as extensively as the new arrivals. Changes in habitat are also reflected by changes in the distribution of species. For example, birds of open grasslands, such as Eastern Meadowlarks (*Sturnella magna*) and Bobolinks (*Dolichonyx oryzivorus*), became widespread in southern Ontario. Populations of some grassland bird species are now in decline in southern Ontario, reflecting changes in land use and habitat.

Turbines associated with commercial wind energy projects are prominent features of three of the terrestrial biomes shown in Figures 31.21 and 31.22. Wind farms are another example of technical modifications of existing habitats. The size and density of turbines may be less disruptive of natural systems than agricultural operations. However, both birds and bats are killed by turbines, and there are allegations that wind farms have a negative impact on human health. The longer-term impact of turbines on biomes and ecozones remains to be determined.

31.5a Cities Are Urban Techno Ecosystems

Cities are the fastest-growing ecosystems in the world **(Figure 31.23, p. 777).** In many countries, more people now live in cities (urban) than in the countryside (rural). By the year 2 CE, Rome was the first western city to reach a population of one million people, reflecting an extensive infrastructure for delivering food and services and removing wastes. In size, human populations in cities range from many with tens of

Table 31.3 — Characteristics of Some of the Terrestrial Ecozones of Canada

Each of the terrestrial ecozones is distinct in terms of its geomorphology, climate, and prevailing ecological communities. MAT is the mean annual temperature; MST is the mean summer temperature.

Ecozone	Region	Geomorphology	Climate	Soil	Prevailing Vegetation
Northern Arctic	Northern Quebec; mainland and eastern islands of Nunavut	Rocky lowlands; much exposed bedrock and glacial debris	High-Arctic; long cold winters, short cool summers; MAT −17 to −11°C; MST −2 to 4°C; precipitation 10–20 cm/yr	Regosol; permafrost throughout; active layer 30–50 cm; generally moist soil	High-Arctic tundra (Arctic desert or semidesert); dominant cover is lichens, mosses, and low-growing vascular plants
Arctic Cordillera	Northern Labrador, eastern Arctic islands of Nunavut	Mountains and rocky uplands, glacial ice fields; much exposed bedrock and glacial debris	High-Arctic; long cold winters, short cool summers; MAT −20 to −6°C; MST −2 to 6°C; precipitation 10–60 cm/yr	Regosol; permafrost throughout; active layer 20–40 cm; generally moist soil	75% of terrain is rocks or ice; elsewhere high-Arctic tundra of lichens, mosses, and low-growing vascular plants
Southern Arctic	Northern Quebec and across mainland Nunavut and northern Northwest Territories	Extensive rolling terrain and lowlands; much exposed bedrock and glacial debris	Low-Arctic; long cold winters, short cool–warm summers; MAT −11 to −7°C; MST 4 to 6°C; precipitation 20–40 cm/yr	Regosol; permafrost throughout (except under large lakes and rivers); active layer 50–70 cm; generally moist soil	Low-Arctic tundra, with more continuous cover than in high Arctic; low-shrub heath and graminoid meadows
Taiga Plains	Western Northwest Territories, northern British Columbia, northwestern Alberta; taiga watershed of Mackenzie River	Rolling plains and uplands; postglacial sediment and debris abundant; lakes and wetlands common	Subarctic; long cold winters, short warm summers; MAT −10 to −1°C; MST 7 to 14°C; precipitation 20–50 cm/yr	Podsol; permafrost discontinuous; active layer > 80 cm; generally moist soil	Open boreal forest (taiga) of relatively short, well-spaced, slow-growing trees, mostly spruce and pine with aspen, poplar, and birch; periodic wildfires
Taiga Shield	Central Quebec, Labrador, southeastern Northwest Territories, northern Saskatchewan and Manitoba	Rolling terrain on quartzitic shield bedrock; much exposed bedrock and glacial debris; lakes and wetlands common	Boreal continental; long cold winters, short warm summers; MAT −8 to 0°C; MST 6 to 11°C; precipitation 20–50 cm/yr	Thin podsol, permafrost discontinuous; active layer > 80 cm; generally moist soil	Open boreal forest (taiga) of relatively short, well-spaced, slow-growing trees, mostly spruce and pine with aspen, poplar, and birch; also open tundralike areas of low shrubs; periodic wildfires
Boreal Shield	Newfoundland and southern Labrador, southern Quebec, northern Ontario, central Manitoba, northern Saskatchewan	Rolling terrain on quartzitic shield bedrock; much exposed bedrock and glacial debris; lakes and wetlands common	Boreal continental; long cold winters, short warm summers; MAT −4°C in continental areas to 5.5°C in maritime Newfoundland; MST 11 to 15°C; precipitation 10–50 cm/yr in continental and 90–160 cm/yr in maritime	Podsol; generally moist soil	Closed boreal forest, mostly of spruce, pine, fir with aspen, poplar, and birch; periodic wildfires
Atlantic Maritime	New Brunswick, Nova Scotia, Prince Edward Island, adjacent Gaspé of Quebec	Rolling terrain on various bedrock, from quartzitic to sedimentary; abundant glacial debris; lakes and wetlands common	Temperate coastal to continental; cold winters, long warm summers; MAT 4 to 7°C; MST 13 to 16°C; precipitation 90–150 cm/yr	Complex soils, from podsol to brunisol; generally moist soil	Mixed-species forests of temperate trees, ranging from angiosperm dominated to coniferous dominated
Mixedwood Plains	Southern Quebec and Ontario within Great Lakes–St. Lawrence valley	Gently rolling terrain over sedimentary, often limestone bedrock	Temperate continental; cold winters, long hot summers; MAT 5 to 8°C; MST 16 to 18°C; precipitation 70–100 cm/yr	Deep, base-rich (high-calcium) brunisol, especially on postglacial lakebed parent materials; generally moist soil	Mixedwood forest, mostly angiosperm dominated; most of the natural cover is converted to agriculture and urban uses

Ecozone	Region	Geomorphology	Climate	Soil	Prevailing Vegetation
Boreal Plains	Central Manitoba and Saskatchewan to northern Alberta and northeastern British Columbia	Rolling terrain over moraine and flatter areas of postglacial lake sediment; abundant wetlands	Boreal continental; long cold winters, short hot summers; MAT −2 to 2°C; MST 13 to 16°C; precipitation 30–63 cm/yr	Deep podsol to brunisol; generally moist soil	Mostly conifer-dominated forest, with angiosperm dominated in the south
Prairies	Southern and central Manitoba, Saskatchewan, Alberta	Rolling terrain over moraine and flatter areas of postglacial lake sediment; abundant ponds and wetlands	Temperate continental; cold winters, long hot summers; MAT 2 to 4°C; MST 14 to 16°C; precipitation 25–70 cm/yr	Chernozem; soil dry to moist	Prairie dominated by grasses and forbs, ranging from tallgrass to mixedgrass to shortgrass types; most of the natural cover is converted to agriculture and urban uses
Taiga Cordillera	Western Northwest Territories and northern Yukon Territory	Steep to rolling terrain of northern Rocky Mountains and foothills; streams and rivers, fewer lakes	Subarctic coastal to continental; long cold winters, short warm summers; MAT −10 to −5°C; MST 7 to 10°C; precipitation 30–70 cm/yr	Podsol; generally moist soil	Because of altitudinal range, vegetation ranges from alpine tundra to subarctic boreal forest of spruce and birch
Boreal Cordillera	Northern British Columbia and southern Yukon Territory	Rugged mountainous terrain with foothills and deep wide valleys; streams and rivers, fewer lakes	Boreal continental; long cold winters, short warm summers; MAT 1 to 6°C; MST 10 to 12°C; precipitation < 30 cm/yr in rain shadow to > 150 cm/yr of orographic precipitation (i.e., increased by mountainous terrain that forces moist air masses to rise in altitude, cool, and precipitate their water content)	Podsol; permafrost discontinuous in north, with active layer > 80 cm; generally moist soil	Because of altitudinal range, vegetation ranges from alpine tundra to montane forest of spruce and aspen; open forest and grasslands in southern areas
Pacific Maritime	Coastal British Columbia	Rugged mountainous terrain with foothills and narrow coastal lowlands; streams and rivers, fewer lakes	Temperate coastal; short cool winters, long warm summers; MAT 5 to 9°C; MST 10 to 16°C; precipitation 60 cm/yr in dry Gulf islands to 400 cm/yr if orographic precipitation; generally 150–300 cm/yr	Podsol; moist soil	Because of altitudinal range and variable rainfall, vegetation ranges from alpine tundra to open dry forest to old-growth mixed-species conifer rain forest
Montane Cordillera	Southwestern Alberta and southern British Columbia	Rugged mountainous terrain with foothills; streams and rivers, fewer lakes	Temperate continental; short cold winters, long warm summers; MAT 1 to 8°C; MST 11 to 17°C; precipitation 30 cm/yr in rain shadow to 120 cm/yr if orographic precipitation	Podsol; drier soil	Because of altitudinal range and variable rainfall, vegetation ranges from alpine tundra to open grassland–forest to closed mesic forest
Hudson Plains	Northwestern Quebec, northern Ontario, northeastern Manitoba	Lowlands of postglacial James and Hudson Bays; surface waters abundant	Boreal coastal to continental; long cold winters, short cool summers; MAT −4 to −2°C; MST 11 to 12°C; precipitation 40–80 cm/yr	Regosol and podsol; permafrost discontinuous, with active layer > 80 cm; generally moist soil	Coastal areas have salt marsh, then tundra farther inland, and then open boreal coniferous forest

Source: From FREEDMAN. Ecology, 2E. © 2015 Nelson Education Ltd. Reproduced by permission. www.cengage.com/permissions

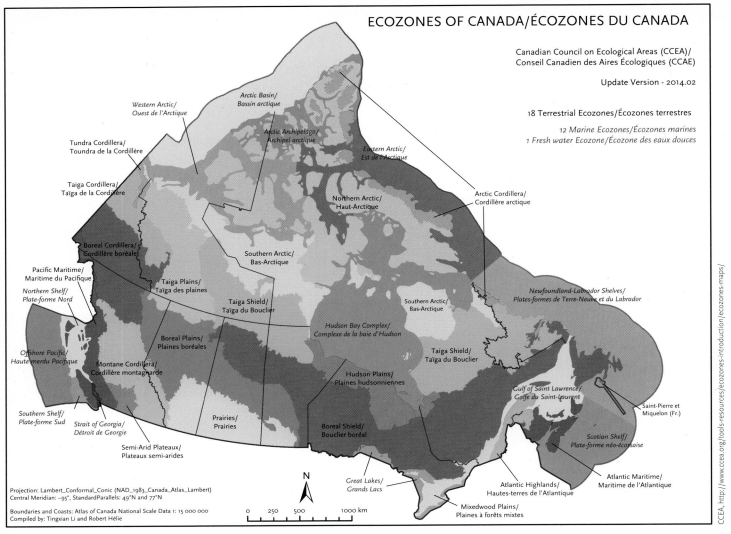

Figure 31.20
Distribution of the terrestrial ecozones and marine and freshwater ecozones of Canada.

Figure 31.21
Four Canadian examples of distinct ecozones are shown in these two pictures. In **(a)**, the Pacific ecozone, inhabited by humpback whales (*Megaeptera novae-angliae*), along with the Pacific maritime biome in the background. The setting is Haida Gwaii off the west coast of British Columbia. In **(b)**, along the north shore of Lake Erie, what had been mixedwood plains is now a mosaic of open agroecosystems and woodlots.

Figure 31.22

(a) A boreal shield ecozone east of Thunder Bay in Northern Ontario. **(b)** A view of prairie near Head-Smashed-In Buffalo Jump in southwestern Alberta. Turbines of commercial wind energy projects are present in both (a) and (b).

thousands of citizens, to Tokyo with over 35 million. Like other biomes, cities are dynamic and changing.

In a 2013 model, Luís Bettencourt identified four simple assumptions that could be used to advantage in understanding the dynamics of cities. First, cities have the capacity for mixing populations of citizens that live there. This means that citizens can afford to fully explore the city and use it to advantage. Second, infrastructure develops gradually and incrementally, accommodating the expanding population. The mainstay of this assumption is a city's network of roads. Third, G (the product of gross domestic product and road volume per capita) is mainly independent of N (population size). G reflects an increasing demand by cities on the mental and physical efforts of their citizens. This also involves communication networks. Fourth, socioeconomic outputs are proportional to local social interactions. In other words, cities are concentrations of social interactions, not just of people. Unlike biological systems that appear to minimize dissipation of energy, cities are systems in which energy dissipation is maximized. This reflects ongoing processes such as transportation, as well as heating and cooling.

Cities share many features with more biological ecosystems, and thinking of them in this way may teach us more about these different kinds of systems and how they operate.

Figure 31.23

A view of an urban techno ecosystem, specifically the campus of Western University as seen in 2007. Note that expanses of woodland may be more extensive than those in agroecosystems (Figure 31.21b).

Review

To access course materials such as Aplia and other companion resources, please visit www.NELSONbrain.com.

31.1 Energy Flow and Ecosystem Energetics

- Net primary productivity is the chemical energy remaining in a system after energy has been used by producers to complete life processes and cellular respiration. Net primary productivity differs from standing crop biomass in that net primary productivity is a measure of energy, whereas standing crop biomass is a measure of dry mass.
- Other factors affect primary productivity, such as water and access to nutrients.
- Assimilation efficiency refers to energy absorbed from eating compared with the total energy in the food. Production efficiency is the energy content of new tissue material compared with the energy absorbed from food intake.
- Some energy is lost during transfer by consumption. The process becomes less efficient as the number of transfers increases, meaning that less energy is transferred to the final consumer.
- Biological magnification occurs when material (e.g., DDT) present in small amounts in a producer or low-level organism is consumed by another organism, transferring the material to the predator. DDT accumulates with each successive transfer. Top predators exhibit the highest concentrations of contaminants such as DDT.

31.2 Nutrient Cycling in Ecosystems

- The amount of water that leaves Earth and enters the atmosphere through evaporation is equal to the amount of water reaching Earth by precipitation.
- Producers transform atmospheric carbon (CO_2) into carbohydrates. Consumers then eat the producers and take in the carbohydrates.
- Cyanobacteria can fix nitrogen, which is crucial because although atmospheric nitrogen levels are high, this nitrogen is not accessible to plants or animals. Atmospheric nitrogen must be converted or fixed into a usable form such as ammonium or nitrate.

- Phosphate ions are carried to bodies of water through weathering and erosion, where most of the phosphate precipitates. Eventually, weathering releases phosphates, which are then directly absorbed by plants.

31.3 Ecosystem Modelling

- Modelling involves collecting data about an ecosystem and deriving mathematical equations about the relationships in the ecosystem. Data collected over different seasons and different years can be used to simulate the effects of a disruption on various levels of the ecosystem in question.
- Simulation modelling helps us understand and predict the impact of influences on certain ecosystems without actually conducting an experiment. Altering anything in an ecosystem without knowledge of its possible effects can be devastating on many or all levels.

31.4 Scale, Ecosystems, Species

- Biotic and abiotic factors interact in ecosystems, and these can occur across scales from millimetres to kilometres.
- At one end, the microorganisms in a biofilm of water provide an example of ecological interactions, as do the organisms living in the water in a pitcher plant.
- Islands provide another example of a range of scales and interactions, from the Galápagos Islands as a group to individual islands.
- Plants disperse using different mechanisms, from wind to the activities of animals.
- In the west coast rain forest of North America, salmon (*Oncorhynchus*) influence plant communities along the rivers and streams in which they spawn.

31.5 Biomes and Ecozones

- *Biota* refers to species, so the biota of an area is the species that occur there.
- Biomes are defined by their biota, perhaps as a list of species.
- *Ecozones* are extensive ecological regions (typically consisting of more than one biome).
- *Biome* and *ecozone* are terms for identifying categories of habitats, often connecting the species to the prevailing climatic conditions.

Questions

Self-Test Questions

1. Which of the following events moves energy and material from a detrital food web into a grazing food web?
   a. an earthworm eating dead leaves on the forest floor
   b. a robin catching and eating an earthworm
   c. a crow eating a dead robin
   d. a bacterium decomposing the feces of an earthworm

2. What is the definition of the total dry mass of plant material in a forest?
   a. a measure of the forest's gross primary productivity
   b. a measure of the forest's net primary productivity
   c. a measure of the forest's standing crop biomass
   d. a measure of the forest's ecological efficiency

3. Which ecosystem has the highest rate of net primary productivity?
   a. open ocean
   b. temperate deciduous forest
   c. tropical rainforest
   d. agricultural land

4. Endothermic animals exhibit a lower ecological efficiency than ectothermic animals for which reason?
   a. Endotherms are less successful hunters than ectotherms.
   b. Endotherms eat more plant material than ectotherms.
   c. Endotherms are larger than ectotherms.
   d. Endotherms use more of their energy to maintain body temperature than ectotherms.

5. What determines the amount of energy available at the highest trophic level in an ecosystem?
   a. only the gross primary productivity of the ecosystem
   b. only the net primary productivity of the ecosystem
   c. the net primary productivity and the ecological efficiencies of herbivores
   d. the net primary productivity and the ecological efficiencies at all lower trophic levels

6. Which pyramid is inverted in some freshwater and marine ecosystems exhibit?
   a. biomass
   b. energy
   c. numbers
   d. ecological efficiency

7. Which process moves nutrients from the available-organic compartment to the available-inorganic compartment?
   a. respiration
   b. assimilation
   c. sedimentation
   d. photosynthesis

8. Which of the following materials has a sedimentary cycle?
   a. oxygen
   b. nitrogen
   c. phosphorus
   d. carbon

9. Which statement is supported by the results of studies at the Hubbard Brook Experimental Forest?
   a. Most energy captured by primary producers is lost before reaching the highest trophic level in an ecosystem.
   b. Deforested watersheds experience a more significant decrease in runoff than undisturbed watersheds.
   c. Deforested watersheds lose more calcium and nitrogen in runoff than undisturbed watersheds.
   d. Nutrients generally move through biogeochemical cycles very quickly.

10. Biological magnification describes which phenomenon?
    a. Certain materials become increasingly concentrated in the tissues of animals at higher trophic levels.
    b. Certain materials become most concentrated in the tissues of animals at the lowest trophic levels.
    c. Certain materials accumulate only in the tissues of tertiary consumers.
    d. Certain materials accumulate only in the tissues of detritivores.

Questions for Discussion

1. Identify 12 ecosystem changes associated with hydroelectric power projects. Consider upstream and downstream changes as well as those associated with transmission of generated power. How does preparing your answers draw on information presented in this chapter?

2. A lake near your home becomes overgrown with algae and pondweeds a few months after a new housing development is built nearby. What kind of data would you collect to determine whether the housing development might be responsible for the changes in the lake?

3. Some politicians question whether recent increases in atmospheric temperature result from our release of greenhouse gases into the atmosphere. They argue that atmospheric temperature has fluctuated widely over Earth's history, and the changing temperature is just part of a historical trend. What information would allow you to refute or confirm their hypothesis? From another perspective, describe the pros and cons of reducing greenhouse gases as soon as possible versus taking a "wait-and-see" approach to this question.

4. What are the ecological consequences of converting crops from food to biofuels?

5. Look at the birds in your neighbourhood. How many have been introduced? How does their behaviour differ from those of native species of birds living in the same immediate areas?

32

A leopard photographed in the wild in South Africa.

Conservation of Biodiversity

WHY IT MATTERS

Achieving preservation of Earth's biodiversity is one of the most pressing challenges facing our species today. Numbers of species are a simple indicator of **biodiversity**, perhaps the most apparent and easy to grasp. But as we have seen, many species of organisms remain undescribed and unnamed. Without names and descriptions, how can we recognize or count them? As we shall see shortly, being unnamed means being unprotected. The Barcode of Life project (see "Barcode of Life," Chapter 30) is one promising effort to better catalogue biodiversity by identifying and allowing us to name its components.

Research on ecosystems shows repeatedly how the numbers of species are associated with stability and productivity, and, as we have seen (Chapter 30), antagonistic and mutualistic systems differ in community composition. The natural order (association between productivity and biodiversity), however, does not coincide with the productivity that our own species must achieve to feed our ever-expanding populations. Creating and maintaining agricultural monocultures is a way for us to maximize food production and efficiency of harvest. In many areas, this approach leads to the disappearance of family-operated farms. Is this progress justified by efforts to increase efficiency and yield? Humans also use genetically modified organisms to increase productivity and marketability, as well as other features, such as shelf life and portability. All too often, increased agricultural productivity is achieved by the use of more fertilizer, water, and energy. Does agriculture have to be the enemy of biodiversity?

Some people have connected humans' attitude to Earth and its riches with religious teachings. In 1967, Lynn White Jr., a professor of medieval history, explored the historical roots of our ecological crisis. He focused on the Christian view of creation, the importance of science, and the separation of humans from their environmental roots. A dualism between humans and nature had emerged in some Christian societies more than in others. Inherent in these societies was the prevailing idea that it is God's will that humans exploit nature for their own ends. White

Figure 32.1
St. Francis of Assisi.

nominated St. Francis of Assisi **(Figure 32.1)** as the patron saint of ecologists. He said that appreciating the virtue of humility was key to understanding the teachings of St. Francis. His point was that as soon as an animal or a plant (or a meadow, lake, or grove of trees) has its own place in nature (in God's eyes), then it can become as important as we believe we are.

The onus is on us as citizens of the planet to conserve biodiversity, whether the focus is species or habitats. One of the main problems we must overcome is the attitude of many humans, as reviewed by White. Today a common reflection of this attitude is that being able to do something (afford to, have the means to) is justification enough for doing it—whether the project involves making space for a shopping mall by draining a wetland or cutting down the trees in a woodlot.

If we as a species can recognize the importance of biodiversity and accept that the world is not ours to do with as we please, what is the best route to protecting and conserving biodiversity? Should we focus on species? On genetic diversity? On ecosystems? How should we blend these approaches to achieve the best support for the endeavour? How can we engage people in this important activity and perhaps move them away from a human-centric view of the world?

As we shall see, at almost every turn there are examples of human activities driving other species to extinction. The motivations for human actions range from

little more than greed to the daily effort to survive. The purpose of this chapter is to introduce you to a range of situations and examples associated with the reduction of biodiversity by extinction and the threat of extinction. We also consider steps that can be taken to protect biodiversity, including some successes and some failures.

Although extinction has been a recurring phenomenon in the history of life on Earth, the current extinction of species as a result of human activities is of grave concern to biologists and others concerned about our future. Changes wrought by humans include introduction of alien species and overharvesting of natural resources. These changes often have unexpected impacts on other species. Adoption of standardized criteria for identifying species at risk is an important part of conservation. Efforts to conserve biodiversity are focused at the species and/or habitat levels. In the final analysis, many of them depend upon taking steps to control the population of our own species and the related consumption of resources and habitat. We also need to change our attitudes and recognize that we are but one of millions of species on the planet and that, ultimately, our survival will depend upon theirs.

32.1 Extinction

Extinction is part of the process of evolution. Given that life has been on Earth for about 3 billion years, today there are more extinct than living species. Occasionally, the fossil record demonstrates a continuum in time from one species to another, sometimes blurring the boundaries between taxa (see Chapter 19). In this case, one could argue that the original species in a series lives on in its descendants. For example, the discovery that the genome of *Homo sapiens sapiens* contains some genes from *Homo sapiens neanderthalensis* leaves open the question about the distinctness of the two taxa. Although data for fossil species usually do not permit us to assess the levels of gene flow between populations, the Neanderthals provide an interesting exception. The difficulties inherent in applying the species concept to fossil material are familiar to paleontologists but less so to biologists.

Species and lineages have been going extinct since life first appeared. We should expect species to disappear at some low rate, the **background extinction rate;** as environments change, poorly adapted organisms do not survive or reproduce. In all likelihood, more than 99.9% of the species that have ever lived are now extinct. David Raup has suggested that, on average, as many as 10% of species go extinct every million years and more than 50% go extinct every 100 million years. Thus, the history of life has been characterized by an ongoing turnover of species.

The fossil record indicates that extinction rates rose well above the background rate at least five times in Earth's history. These events are referred to as mass extinctions. One extinction occurred at the end of the Ordovician and the beginning of the Devonian, the next at the end of the Devonian, then the end of the Permian, the end of the Triassic, and the end of the Cretaceous. The Permian extinction was the most severe, and more than 85% of the species alive at that time disappeared forever. This extinction was the end of the trilobites, many amphibians, and the trees of the coal swamp forests. During the last mass extinction, at the end of the Cretaceous, half of the species on Earth, including most dinosaurs, disappeared. A sixth mass extinction, potentially the largest of all, is occurring now as a result of human degradation of the environment.

Different factors were responsible for the five **mass extinctions.** Some were probably caused by tectonic activity and associated changes in climate. For example, the Ordovician extinction occurred after Gondwana moved toward the South Pole, triggering a glaciation that cooled the world's climate and lowered sea levels. The Permian extinction coincided with a major glaciation and a decline in sea level induced by the formation of Pangaea (see Chapter 20).

Many researchers believe that an asteroid impact caused the Cretaceous mass extinction. The resulting dust cloud may have blocked the sunlight necessary for photosynthesis, setting up a chain reaction of extinctions that began with microscopic marine organisms. Geologic evidence supports this hypothesis. Rocks dating to the end of the Cretaceous period (65 mya) contain a highly concentrated layer of iridium, a metal that is rare on Earth but common in asteroids. The impact from an iridium-laden asteroid only 10 km in diameter could have caused an explosion—equivalent to a billion tonnes of TNT—that scattered iridium dust around the world. Geologists have identified the submarine Chicxulub crater, 180 km in diameter, off Mexico's Yucatán peninsula as the likely site of the impact.

Although scientists agree that an asteroid struck Earth at that time, many question its precise relationship to the mass extinction. Dinosaurs had begun their decline at least 8 million years earlier, but many persisted for at least 40 thousand years after the impact. Moreover, other groups of organisms did not suddenly disappear, as one would expect after a global calamity. The Cretaceous extinction took place over tens of thousands of years. Furthermore, some organisms survived periods of extinction, such as ginkgo trees (*Ginkgo biloba*), horseshoe crabs (*Limulus polyphemus*), and coelocanths (*Latimeria chalumnae*).

Even today we cannot blame the extinctions of most species on the activities of humans. But our increasing technological capability and prowess coincide with a burgeoning population of people. This situation is exacerbated by the philosophical view that humans are disconnected from nature. Thus, we are becoming better and better at destroying the biota of the planet. Taking action requires identifying root causes and then trying to make changes that will alleviate the problems.

First, we consider extinctions not linked to humans, and then we review examples of situations in which our actions have either directly or indirectly led to the extinction of species. The fact that extinction is integral to the process of evolution is hardly justification or rationalization for our driving so many species there. Put another way, invoking "survival of the fittest" may not be adequate justification for eradicating other species.

32.1a Dinosaurs: A Most Notable Extinction

Why do species go extinct? There could be as many theories as there are extinct species! The disappearance of the dinosaurs is one of the best-known extinction events in Earth's history. At the end of the Cretaceous about 65.5 mya, the dinosaurs disappeared. The ancestors of dinosaurs had appeared in the Triassic, and the group underwent extensive adaptive radiation reflected in body size, lifestyle, and distribution. Although people think of large and spectacular carnivorous dinosaurs such as *Tyrannosaurus rex* or the huge herbivore *Apatosaurus* (previously known as *Brontosaurus*), in reality, many species of dinosaurs were small and delicate.

Evidence from deposits in Alberta suggests that the carnivorous dinosaur *Albertosaurus* **(Figure 32.2)** showed age-specific mortality and high juvenile survival **(Figure 32.3a).** Indeed, the survivorship curves (see Chapter 29) for *Albertosaurus* resemble those for

Figure 32.2
Mounted skeleton of *Albertosaurus* on display in the Royal Tyrrell Museum, Drumheller, Alberta. This late Cretaceous carnivore is abundant in fossil beds in Alberta and elsewhere.

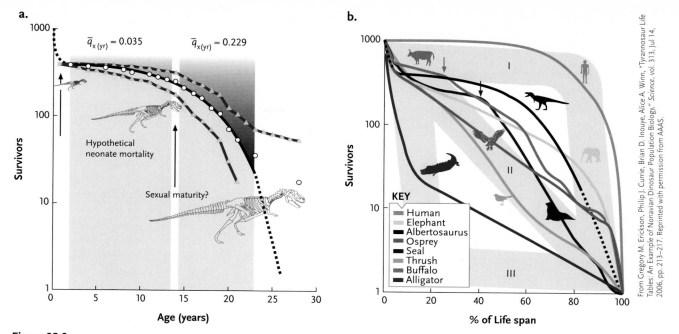

a.

$\bar{q}_{x(yr)} = 0.035$

$\bar{q}_{x(yr)} = 0.229$

Survivors

Hypothetical neonate mortality

Sexual maturity?

Age (years)

b.

Survivors

KEY
- Human
- Elephant
- Albertosaurus
- Osprey
- Seal
- Thrush
- Buffalo
- Alligator

I

II

III

% of Life span

From Gregory M. Erickson, Philip J. Currie, Brian D. Inouye, Alice A. Winn, "Tyrannosaur Life Tables: An Example of Nonavian Dinosaur Population Biology," *Science*, vol. 313, Jul 14, 2006, pp. 213–217. Reprinted with permission from AAAS.

Figure 32.3

(a) Survivorship curve for a hypothetical cohort of 1000 Albertosaurus with presumed neonatal mortality of 60%. **(b)** Survivorship of *Albertosaurus* compared with that of other animals, including humans from developed countries, short-lived birds, mammals, and lizards, as well as crocodilians and some captive mammals.

humans (Figure 32.3b). The data do not provide any indication of a flaw that predisposed dinosaurs to extinction. This is sobering, given the similarity between some aspects of dinosaur population biology and those of our own. The prevailing view today is that the disappearance of the dinosaurs is linked to the impact of an asteroid. Many other theories have been proposed to explain extinction, but the fact that birds and mammals and many other groups of organisms showed widespread extinctions at the end of the Cretaceous implies a pervasive catastrophic event.

32.1b Multituberculates Are a Mammalian Example of Extinction

There is more to extinction than dinosaurs. Competition (see Chapter 30) has been proposed as a mechanism that can lead to extinction. Among mammals, the adaptive radiation of rodents (order Rodentia; **Figure 32.4a**) in the early Oligocene coincides with the disappearance of multituberculates (order Multituberculata; Figure 32.4b). As a group, multituberculates were prominent and persisted for 100 million years (compared with 150 million years for dinosaurs), making them the most successful mammals to date.

Multituberculates ranged from small (~20 g) to medium (5 to 10 kg) in size and exhibited both terrestrial and arboreal lifestyles. We can only speculate what happened to them, and why they became extinct. The widespread success of rodents almost worldwide could lend credibility to competition as the reason for

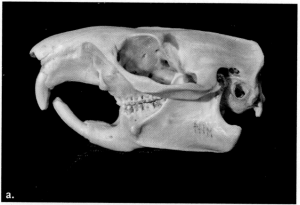

a.

b.

M.B. Fenton

DigiMorph.org

the multituberculates' demise. However, the fossil record does not tell us what rodents and multituberculates competed for: food? nest sites?

STUDY BREAK

1. What is extinction?
2. What were multituberculates?

Figure 32.4

(a) The skull of a groundhog (*Marmota monax*), a North American rodent, compared with **(b)** the skull of a multituberculate (*Kryptobaatar dashzevegi*) from the Mongolian late Cretaceous.

Sex Determination and Global Warming

Failure to reproduce puts the survival of a species on the edge, so anything that interferes with reproduction can be threatening. Genetic recombination is a fundamental benefit of sexual reproduction, enabling increases in genetic diversity and elimination of deleterious mutants. Effective sexual reproduction means having male and female systems, sometimes in one individual (hermaphrodites) and perhaps more often in different individuals. Males and females differ in many fundamental ways—genetically, hormonally, physiologically, and anatomically.

In humans and many other animals, gender is determined by genotype, with males having an X and a Y chromosome and females having two X chromosomes. The reverse is true in many other animals, for example birds. But in many reptiles gender is determined environmentally. Eggs incubated at some temperatures develop into males; when incubated at other temperatures, they produce females.

In 2008, D. A. Warner and R. Shine reported the results of experiments done with jacky dragons (*Amphibolurus muricatus*), an Australian lizard in which gender is determined by temperature. Eggs incubated at 23 to 26°C or 30 to 33°C produce females; those incubated from 27 to 29°C produce males. Warner and Shine tested the hypothesis that temperature-dependent sex determination ensured production of females when they had an advantage and males when the advantage was to them. Using a combination of temperature and hormonal manipulations, Warner and Shine could produce males or females at any temperature. They analyzed paternity to assess the reproductive output of these males and observed eggs laid and hatched to document these females' reproductive output.

In female jacky dragons, larger body sizes occur at higher temperatures, and larger females have higher fecundity than smaller ones. Higher temperatures also correlate with larger body size in males. However, males hatched from eggs incubated between 27 and 29°C sired more offspring than those hatched from eggs incubated at lower or higher temperatures.

Change in climate, such as global warming, could put species with temperature-dependent sex determination at risk by effectively eliminating males or females from the population. Eggs incubated at the wrong temperatures would fail to hatch. The importance of variation in temperature during development in ectothermic organisms could explain the prevalence of genotypic-dependent sex determination in euthermic (homeothermic) viviparous animals. Viviparous or ovoviviparous ectotherms (fish, amphibians, reptiles, other animals) could also rely on temperature-dependent gender determination, provided that their developing young experience an appropriate range of temperatures.

Recent work from Mexico reveals that since 1975, 12% of local populations of lizards have disappeared, likewise 4% of worldwide local populations. Like other ectotherms, lizards have a narrow thermal range in which they thrive, and climate change has altered the thermal niches available to them. This and temperature-dependent gender determination put lizards in double jeopardy.

32.2 The Impact of Humans

When it comes to extinctions, we know most about those resulting from our activities, usually because these records are relatively recent and accessible. If you recently visited Mauritius, you might have noticed that the few remaining Mauritian calvaria trees are slowly dying of old age. Their passing will mark the extinction of this species, which has occurred even though the trees continued to bloom and produce seeds. The key to the pending extinction of *Sideroxylon majus* is the earlier extinction of Dodos. To germinate, seeds of Mauritian calvaria trees had to pass through the Dodo's digestive tract. The Dodo **(Figure 32.5)** was a medium-sized flightless bird that lived on the island of Mauritius. When European sailors first visited the island, they used Dodos as a source of fresh meat. Then, as the island was settled, the birds were exposed to introduced predators (cats, dogs, rats) and an expanding human population. Dodos vanished by 1690.

Species confined to islands often have small populations and are unaccustomed to terrestrial predators, making them

Figure 32.5
(a) A reconstruction of a *Raphus cucullatus*, the Dodo, an extinct flightless bird from **(b)** Mauritius.

Raphus cucullatus by Roelandt Savery, 1626

Reprinted by permission from Macmillan Publishers Ltd: NATURE, Vol. 443: pp. 138–140, "Ornithology: Digging for dodo" by Henry Nicholls, copyright (2006).

Amanda Vincent, *University of British Columbia*

Sea horses are a central focus for Amanda Vincent's research. She holds a Canada Research Chair in marine conservation and is the director of Project Seahorse. Sea horses are notable for the details of their biology (see Chapter 28) and because they are big business.

Vincent has studied the behaviour of sea horses. During mating, the male's sperm fertilizes the female's eggs, but the female then transfers the fertilized eggs to the male's brood pouch; thus, the male gets "pregnant." Males and females are in regular contact during the period of preg-

nancy. These contact behaviours may be key to the monogamy that appears to be typical of male–female relationships in sea horses.

The world trade in sea horses involves an estimated 20 million of them each year. In Asia, millions of dried sea horses are traded each year, mainly for use in traditional Chinese medicines. Remedies that include sea horses are said to be useful in treating symptoms from asthma and skin problems to incontinence and disorders of the thyroid. Dried sea horses may also be ingredients in aphrodisiacs.

Vincent has been very active in efforts to conserve sea horses, including working with local fishing communities to conserve them and maintain a sustainable harvest. This means establishing local protected areas and growing some sea horses in controlled conditions so that they, rather than wild stock, are harvested.

Vincent is an example of a biologist who combines an academic interest with its practical applications. Her work in conservation connects the realities of harvesting animals with the demands for their conservation.

vulnerable to extinction. The fossil and subfossil records show that many species of birds disappeared from islands in the South Pacific as Polynesians arrived there from the west. This occurred from Tonga to Easter Island and beyond **(Figure 32.6).** The Galápagos, only discovered by people in 1535, was sheltered from the wave of human-induced extinctions. On Easter Island, **endemic species** of sea birds and other species disappeared soon after people settled there. These examples demonstrate that humans do not have to be industrial or "high tech" to effect extinctions.

Meanwhile, in the North Atlantic, people hunted *Pinguinus impennis,* the Great Auk, to extinction. However, land birds with large distributions and huge populations have also disappeared, such as *Ectopistes migratorius,* the Passenger Pigeon, in eastern North America. Large-scale harvesting of these birds, combined with their low reproductive rate (clutch size: one egg), made the birds vulnerable in spite of their enormous populations. Animals that produce one young per year and suffer "normal" mortality must live at least 10 years to replace themselves in the population (see Chapter 29).

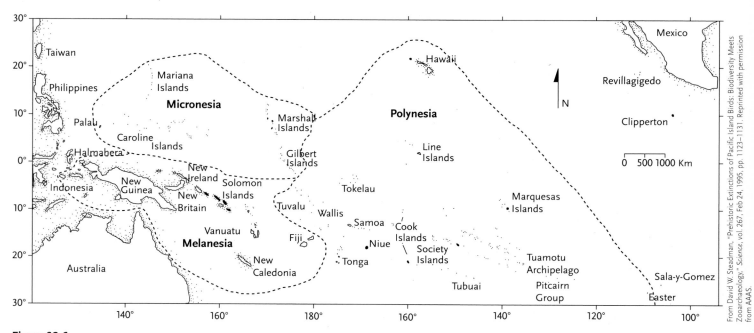

Figure 32.6

Islands in the South Pacific where the arrival of Polynesians coincided with the extinction of many island species of birds.

From David W. Steadman, "Prehistoric Extinctions of Pacific Island Birds: Biodiversity Meets Zooarchaeology," *Science,* vol. 267, Feb 24, 1995, pp. 1123–1131. Reprinted with permission from AAAS.

STUDY BREAK

CONCEPT FIX It is easy to believe that we must focus conservation efforts on large, charismatic animals and plants because these can be the poster images of conservation. We now realize that it is often more important to protect ecosystems, recognizing that the many components of ecosystems play a vital role in maintaining biodiversity, including the large and charismatic members of ecological communities. ⬡

32.3 Introduced and Invasive Species

Humans cause extinction through hunting and by the introduction of other species. House cats, *Felis domesticus,* are among the worst introductions people have made. Anecdotal records suggest that in 1894, one house cat (named Tibbles) exterminated an entire population of flightless wrens **(Figure 32.7)** on Stephen's Island, a 2.6 km² island off the north shore of New Zealand. Fossils indicate that the wrens had occurred widely in New Zealand. This record stands for one individual, Tibbles, taking out the remaining approximately 10 pairs and exterminating the species.

Figure 32.7
Stephen's Island Wren, *Xenicus lyalli.* This species was exterminated by one cat.

Figure 32.8
(a) Zebra mussels, *Dreissena polymorpha*, were introduced to the Great Lakes in North America, where they have spread rapidly. **(b)** They are directly responsible for the declines in eastern pond mussels (*Lampsilis radiata*), a local mussel species.

It should be obvious that moving species from one part of the world to another, whether done willfully or by accident, can have calamitous impacts. The invaders, once arrived and established, may outcompete resident species, laying waste to species and ecosystems. The list of introduced organisms is very long and includes many domesticated or commensal species of animals and plants. The arrival of zebra mussels **(Figure 32.8a)** in the Great Lakes is the main reason for the decline of the now endangered eastern pond mussels (Figure 32.8b). The immigrant mussels outcompeted and overgrew the native ones, reducing their range and populations to levels that resulted in eastern pond mussels being recommended for listing as endangered in Canada in 2007.

Meanwhile, in parts of the British Isles, flatworms (*Arthurdendyus triangulatus;* **Figure 32.9, p. 788;** see also Chapters 27 and 48) introduced from New Zealand are deadly predators of earthworms. Since their arrival in garden pots, the flatworms have thrived and spread rapidly, coinciding with the demise of earthworms. Ironically, although we often think of gardeners as individuals in touch with nature, their propensity to introduce exotic species may not be compatible with conservation. Earthworms themselves have been introduced widely to places around the world.

Some organisms move about in ballast water. Since about 1880, ships have regularly used water for ballast. In the early 1990s, a survey of ballast water in 159 cargo ships in Coos Bay, Oregon, revealed 367 taxa

The 2,4-D Molecule and Resiliency

Resiliency is one of the most impressive features of life at the species and/or ecosystem levels. In one respect, this feature complicates the challenges of conserving biodiversity because introduced species can be so invasive, reflecting their adaptability. Humans first identified 2,4-D (2,4-dichlorophenoxyacetic acid; **Figure 1**) in 1942, and from 1944, it was marketed as a herbicide more effective against broad-leaved plants than against grasses. Technically, 2,4-D is a hormone absorbed by the plant and translocated to the growing points of roots and shoots. 2,4-D kills weeds by inhibiting growth. The global market for 2,4-D is probably more than U.S.$300 million, and it is mainly used to control broad-leaved weeds in cereal crops. According to the World Health Organization (WHO), 2,4-D is a "moderately hazardous pesticide" known to affect a variety of animals (e.g., dogs but not rats). Curiously, it turns out that other animals may use 2,4-D for their own ends.

In 1971, Thomas Eisner and colleagues reported that a grasshopper (*Romalea microptera;* **Figure 2**) produced a froth of chemicals **(Figure 3)** for protection against ants. One of the main ingredients in the froth was 2,5-dichlorophenol, apparently derived from 2,4-D. This is an astonishing demonstration of adaptability that can underlie resiliency.

Resiliency and the recuperative powers of ecosystems are demonstrated by stories of "lost cities," for example, structures built by Maya in Central America, being found in a jungle. Archaeological evidence reveals that in some habitats, these buildings and pyramids were overgrown by the rain forest in about 100 years. The Great Zimbabwe Ruins in southern Africa were overgrown by savannah woodland in a period of 100 to 200 years and only latterly "discovered" by European explorers.

2, 4-D

(2,4-dichlorophenoxy)acetic acid

Figure 1
2,4-dichlorophenoxyacetic acid, 2,4-D.

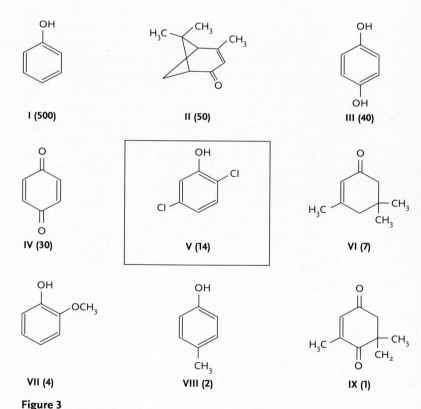

© KHALED KASSEM/Alamy

Figure 2
Romalea microptera, *a grasshopper that uses an ant repellent with a 2,4-D derivative.*

I (500)   II (50)   III (40)

IV (30)   V (14)   VI (7)

VII (4)   VIII (2)   IX (1)

Figure 3
Active ingredients in the defensive froth of the grasshopper, Romalea microptera. *2,5-dichlorophenol (boxed) is apparently derived from 2,4-D.*

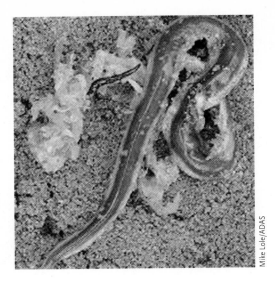

Figure 32.9
This earthworm-eating planarian (*Arthurdendyus triangulatus*) was introduced to the British Isles from New Zealand. It has had a devastating effect on local populations of earthworms.

Mike Lole/ADAS

representing 16 animal and 3 protist phyla, as well as 3 plant divisions. The samples included all major and most minor phyla. Organisms in the ballast water included carnivores, herbivores, omnivores, deposit feeders, scavengers, suspension feeders, primary producers, and parasites. Ballast water is taken on in one port and discharged in another, providing many species with almost open access to waters around the world.

Meanwhile, introduced diseases (and the organisms that cause them) have decimated, if not obliterated, resident species. When Europeans arrived in the New World, *Castanea dentata,* the American chestnut tree, was widespread in forests from southern Ontario to Alabama. This large tree of the forest canopy grew to heights of 30 m. Often most abundant on prime agricultural soils, the species' distribution and density were reduced as settlers from Europe cleared more and more land for agriculture. *Endothia parasitica,* the chestnut blight, was introduced perhaps around 1904 from Asian nursery stock. This introduced blight killed the American chestnut trees by the 1930s. By 2000, only scattered American chestnut trees remained, most of them stump sprouts.

Why are invading species so successful? Does the spread of Starlings (*Sturnus vulgaris*) or dandelions (*Taraxacum officinale*) after introduction to new continents suggest that they moved into vacant niches? Does it mean that they are better competitors? In the case of Starlings, 13 birds were introduced to Central Park in New York City in 1890, and they have spread far and wide. Once they are established, invading or introduced species can pose huge conservation problems because of their effects on ecosystems and diversity.

Although many invaders arrive, only a few are widely successful and become large-scale problems in their new settings. Invading plants are most often successful in nutrient-rich habitats, where they can achieve high growth rates, early reproduction, and maximal production of offspring. What happens in resource-poor settings? In the past, conventional wisdom has suggested that low-resource settings could be reservoirs for native species that could outcompete invaders.

However, an experimental examination of the responses of native and introduced species to challenging conditions revealed that invasive plant species almost always fared better (**Figure 32.10**). Resource use efficiency (RUE), calculated by measuring carbon assimilation per unit of resource, provides an indicator of success. Many invasive species, such as ferns, C_3 and C_4 grasses, herbs, shrubs, and trees, were more successful in low-resource systems than native species were.

This research was conducted in Hawaii, an excellent place for studying invasive species because so many are there. Among the invaders were *Bromus tectorum* (cheatgrass), *Heracleum mantegazzianum* (cartwheel flower or giant hogweed), and *Pinus radiata* (Monterey pine). Humans have introduced these plants for gardening (cheatgrass and cartwheel flower) or commercial timber production (Monterey pine). The data demonstrate that attempting to restore ecosystems and exclude invading species by reducing resource availability does not succeed because of the efficiency with which some species use resources.

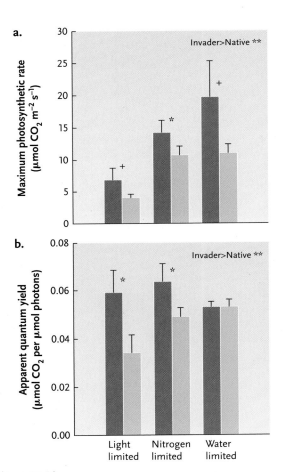

Figure 32.10
(a) Photosynthetic rates (RUE) and **(b)** light-use efficiency of invasive plant species (blue bars) make them more competitive than native ones (yellow bars). The plants were from three different habitats in Hawaii.

In the graphs, + denotes P < 0.01, while * denotes P < 0.05.

** Indicates that in both instances, invaders are significantly more efficient than native species

1. Give two examples of invasive species, one animal and one plant. Describe the history of the invasion.
2. Are invasive species more apt to be successful than native species? Why or why not?
3. What is ballast water? Why does it figure in transporting species from one place to another?

a.

32.4 How We Got/Get There

Lamentably, we know that humans can exterminate species that are populous and widespread as well as ones that have small populations and occur in a small area.

32.4a Poaching Caused the Demise of the Black Rhinoceros

It is estimated that 60 thousand black rhinos (*Diceros bicornis*) lived in the wild in Africa in 1960 **(Figure 32.11a)**. This large (1.5 m at the shoulder, 1400 kg) browsing mammal was widespread in sub-Saharan Africa (Figure 32.11b). Adult males and females have two distinctive "horns" **(Figure 32.12a;** see also Figure 32.11a), actually formed from hair. Rhinos use the horns to protect themselves and their young from predators and other rhinos. By 1981, the populations in the wild had been reduced to between 10 thousand and 15 thousand, and again reduced to about 3500 by 1987. Today only a few individuals survive in some protected areas in Africa. In less than 30 years, the species was almost exterminated in the wild.

In 1960, black rhinos were one of the "big five" on the list of big game for which hunters made safaris to Africa to shoot as trophies. Others on the list were the African lion, African elephant, Cape buffalo (*Syncerus caffer*), and leopard. Safari hunters then paid large sums of money to go to Africa and obtain licences to kill trophy specimens of each of the big five. But this hunting pressure, which has since stopped, did not lead to the extermination of black rhinos.

b.

Figure 32.11
(a) Black rhinos (*Diceros bicornis*) were widespread and common in Africa in 1960 (orange area on the map in (b)). **(b)** Today their range (dark spots in orange areas) is much reduced, reflecting diminished populations. Note the oxpecker (*Buphagus africanus*) sitting on the rhino.

People have long used the horns of all species of rhino in different ways. In China, bowls made from rhino horn (Figure 32.12b) were believed to have magical properties in that they could remove or neutralize poisons. Travelling nobles were served wine in their own rhino horn bowls to minimize the chances of their being poisoned. In India and some other areas from India to Korea, powdered rhino horn was used as a fever suppressant. Contrary to popular belief, rhino

a.

b.

c.

Figure 32.12
(a) A horn from a black rhino in Zimbabwe is shown with **(b)** a rhino horn bowl from China and **(c)** a jambiya with a rhino horn handle.

Figure 32.13

A Kalashnikov assault rifle (an AK), a weapon widely used in the poaching of animals in many parts of the world.

horn does not appear to have been used as an aphrodisiac, an early version of Viagra.

A tradition among some people of the Arabian Peninsula is the carrying of a jambiya or ceremonial dagger. Jambiyas with rhino horn handles (Figure 32.12c) were highly prized. In 1973, when the price of oil jumped from U.S.$4 to U.S.$12 a barrel, the ensuing "energy crisis" meant a larger market for jambiyas because more people could afford them. Increased cash flow and easy access to military weapons such as Kalashnikov assault rifles **(Figure 32.13)** provided an incentive and a means to kill rhinos. The epidemic of poaching started in northern Kenya and spread southward throughout the continent. Thus, poaching for their horns led to the catastrophic reduction in the populations of black rhinos. The large population of rhinos that had long survived in the presence of predators, including *Homo sapiens,* was not protected from extermination. In 1984, going for a walk at night around the headquarters of Mana Pools National Park in Zimbabwe almost always meant meeting a black rhino. By 1987, the rhinos were very scarce, and by 1990 they did not exist in the area.

The demise of black rhinos can only be attributed to human greed.

32.4b White-Nose Syndrome and Bats: A Different Challenge

In March 2006, at sites near Albany, New York, biologists counting bats hibernating in caves and abandoned mines were shocked to find thousands of dead bats where they had expected thousands of live ones. The bats had died from White-Nose Syndrome (WNS), which is caused by a cold-loving fungus (*Pseudogymnoascus destructans* (formerly *Geomyces destructans*)) that interrupted their rhythm of hibernation. Some infected bats were easy to recognize by the white funguslike

Figure 32.14

Three Little Brown Myotis, one (middle) showing characteristic signs of infection by the fungus *Pseudogymnoascus destructans* (formerly *Geomyces destructans*) that causes white-nose syndrome (WNS) in bats that hibernate underground in the United States and Canada.

structures around their nostrils **(Figure 32.14)**. To survive hibernation, bats minimize the number of times they arouse from torpor because of the metabolic cost of waking up (raising the body temperature from 2–5°C to over 35°C). At most Canadian hibernation sites bats normally go about 90 days between arousals because each arousal costs them energy that they could use in 60 days of hibernation. WNS causes them to arouse much more often and exhaust their stores of body fat in January or February, well before spring and the re-emergence of insect prey.

The initial focal area for WNS in North America was specific sites around Albany. By March 2010, WNS had spread to underground hibernation sites in Ontario and Quebec, as well as to many other sites in the United States. In the intervening years WNS continued to spread. In June 2014, in Canada, WNS had not been found west of Wawa in Ontario and had not been reported from sites in Newfoundland and Labrador. But WNS has continued to spread west and south in the United States. The spread of WNS surely reflects the movements and behaviour of the bats.

At known hibernation sites, populations of Little Brown Bats (*Myotis lucifugus*) declined by over 95%. Professor Craig Willis from the University of Winnipeg and his colleagues demonstrated that the strain of the fungus causing WNS in North American bats originated in Europe. It was presumably inadvertently transferred to the sites near Albany by cave explorers or bat biologists. The European strain did not cause WNS in bats there, just as the North American strain did not cause WNS in bat species from North America.

Can the populations of Little Brown Bats recover? Probably not, because like most bats, Little Browns live in the slow life history lane (see Chapter 29). They reproduce slowly (a single young per year) and like most species of bats of the temperate regions, up to 60% of young do not survive their first year. The combination of low reproductive output and low survival of first year translates into low potential for increase of their populations.

There are 19 species of bats in Canada, not all of which are exposed to WNS because not all of them hibernate in underground sites (usually caves and abandoned mines). So, perhaps WNS does not mean the end of our bats, but likely the loss of more than half of the species. WNS is a stark example of how a widespread and abundant species can become endangered.

32.4c The Bay Scallop Is Affected by Overfishing of Sharks

Populations of organisms we harvest for food often show marked declines. The annual harvest of bivalve molluscs has been a local fishery in Chesapeake Bay in the United States and elsewhere along the eastern seaboard for hundreds of years. In 1999, populations of

Figure 32.15

A handful of bay scallops (*Argopecten irradians*).

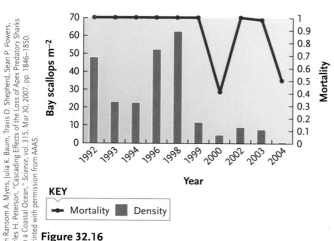

Figure 32.17

A cownose ray (*Rhinoptera bonasus*).

bay scallops (*Argopecten irradians;* **Figures 32.15** and **32.16**), a main target of the fishery, were very low. The immediate reason for the low populations was the impact of predation by skates and rays that feed heavily on bivalve molluscs. Skates and rays are tertiary consumers and in turn are eaten by larger elasmobranchs, specifically various species of sharks.

Among tertiary consumers, the cownose ray **(Figure 32.17)** showed a marked increase in population. Evidence from surveys on the U.S. Atlantic coast estimates an order-of-magnitude increase in populations of cownose rays, and the total population of 14 species of rays and skates exceeds 40 million. So the decline in scallop (and other bivalve) populations can be explained by the increase in predation by tertiary consumers, especially skates and rays.

The picture becomes clearer when the population data for the local great sharks are added to the mix **(Figure 32.18, p. 792).** Prolonged and intensive fishing of 12 species of sharks accounts for a 35-year decline in their populations (see Figure 32.18, top row). The sharks have been taken primarily for their fins and meat. In some parts of the world, shark fins sell for around U.S.$700 per kilogram and are used to make shark fin soup.

The data demonstrate how a century-old scallop fishery was effectively destroyed because of predation by tertiary consumers, whose populations, in turn, had been enhanced (see Figure 32.18, middle row) by the removal of top predators, the great sharks. The data illustrate a cascading ecological effect and demonstrate the potential long-term harm that our species can do to ecosystems and the species inhabiting them. The demise of bay scallops and other bivalves can be attributed to the impact of large-scale harvesting of marine resources. The late Ransome Myers and his colleagues documented this cascade of effects.

The examples above are merely samples from a long list of species. Evidence of declines of populations of native species can be found almost everywhere. Whether the root cause is overharvesting, introduced species, or destruction of habitat, species from whales to songbirds are threatened by human activity. What can we do about it?

STUDY BREAK

1. What is the significance of rhino horn in the conservation of these animals?
2. What connects the decline in the scallop fishery to sharks?
3. What is White-Nose Syndrome?

32.5 Protecting Species

The widespread recognition of trademarks such as the World Wildlife Fund (WWF) panda demonstrates how associating a cause with an icon can be very successful. It is not surprising that many conservation efforts began with a focus on one species—such as giant pandas (*Ailuropoda melanoleuca*), polar bears (*Ursus maritimus*), or redwood trees (*Sequoia sempervirens*).

From Ransom A. Myers, Julia K. Baum, Travis D. Shepherd, Sean P. Powers, Charles H. Peterson, "Cascading Effects of the Loss of Apex Predatory Sharks from a Coastal Ocean," *Science,* vol. 315, Mar 30, 2007, pp. 1846–1850. Reprinted with permission from AAAS.

KEY

◆ Mortality ▬ Density

Figure 32.16

Numbers of bay scallops off the east coast of the United States.

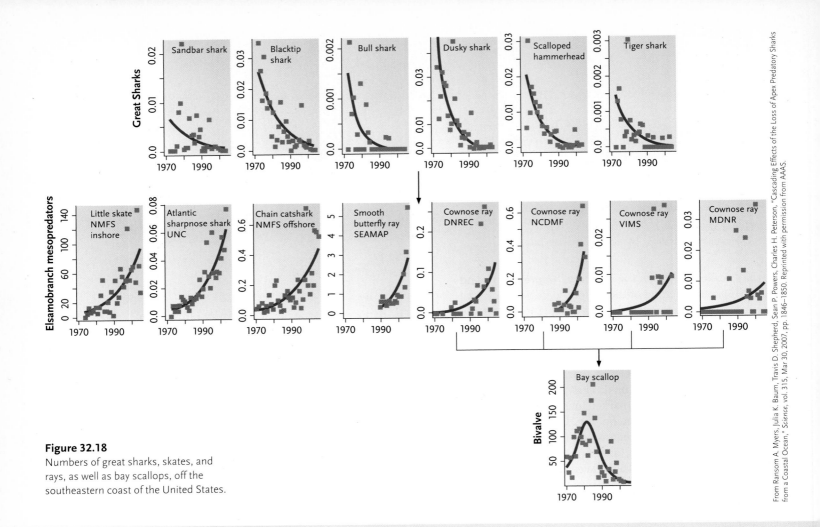

From Ransom A. Myers, Julia K. Baum, Travis D. Shepherd, Sean P. Powers, Charles H. Peterson, "Cascading Effects of the Loss of Apex Predatory Sharks from a Coastal Ocean," *Science*, vol. 315, Mar 30, 2007, pp. 1846–1850. Reprinted with permission from AAAS.

Figure 32.18
Numbers of great sharks, skates, and rays, as well as bay scallops, off the southeastern coast of the United States.

Hunting: Threat or Salvation?

We saw earlier (see Chapter 19) how the Linnaean system of nomenclature is used to name species. Once a species has a name, however acquired, it may benefit from protection under CITES, the Convention on International Trade in Endangered Species of Wild Fauna and Flora. But will data-based decisions about what counts as endangered be consistent and predictable? The answer is "yes" and "no." The example of black rhinos showed one situation in which protection under CITES did not work. There are others.

Also in Africa, the leopard (see the chapter-opening photograph) was accorded protection under CITES. The passing of the Endangered Species Act (ESA) in the United States (1972) precipitated an interesting situation: it

obliged Americans to "obey" the listing of leopards on CITES Appendix 1, which banned the importation of leopard skins, including those shot on safari hunts. The rationale for the listing was the belief that leopards were endangered and their survival was threatened by hunting.

There were quick, negative responses to the ban on importing leopard skins into the United States from two different groups. First were the hunting and related associations and lobbies whose members were anxious to be able to bring home trophies. Second, leaders and governments in many African countries that benefited from the hunts objected to the ban because safaris were (and still are) an important source of foreign

exchange. In many of these countries, "safari hunting areas" were set aside to accommodate visitors, and these large tracts of land also protected populations of nongame species and appropriate habitat.

What do the data show? Leopards are 40 to 80 kg, solitary cats that hunt by stealth. They are widespread in Africa but have been little studied. The estimate is that there are more than 700 000 leopards in the wild in Africa, with resident populations in all but very small countries with high human population densities. In 2000, Zimbabwe alone had a population of more than 16 000 leopards in the wild. The 1969 safari harvest of 6100 leopards throughout Africa and the export of their skins were not a threat to the population in

Zimbabwe, let alone to leopards in the whole continent.

Ecologists studied the population of leopards in the Matetsi Safari Area in Zimbabwe. Before 1974, the 4300 km² area was a cattle ranch whose operators made strong efforts to eradicate leopards to protect their livestock. After conversion to a hunting area, people on the first safaris rarely succeeded in shooting leopards. By 1984, the leopard population in the Matetsi Safari Area was 800 to 1000, and in 1988, the annual safari quota there was 3.6% (12 to 28 leopards). When leopards shot in the mid-1980s were compared with those taken in the 1970s, no change in leopard size was found. But by 1986, the average age of leopards taken as trophies was 5.4 years, compared with 3.2 years from the earlier period. These data show that leopards can persist even when subjected to heavy hunting pressure. On average, leopards live longer in a safari hunting regime than when they are being hunted in the context of predator control operations. Other evidence suggests that populations of leopards persist even in urban areas—trapping evidence suggests that resident leopards live in Nairobi, the capital of Kenya.

Leopards are an interesting example of human responses to conservation. Hunting or some other form of harvesting is not necessarily a threat to the survival of some species. Indeed, some harvesting may be critical to the livelihood of some people and can advance efforts to protect some species. But decisions about harvesting made in one part of the world can influence what happens elsewhere.

Today there are quotas for the numbers of leopards that can be harvested in different countries in Africa. Safari hunters must obtain licences to take trophies, and skins exported must be accompanied by paperwork showing that the harvest was legal. The documentation allows a citizen, for example, of Canada or of a European Union country, to import a leopard skin. This was not possible in the United States in the 1970s, but it is now. In Africa, local farmers are permitted to kill "problem" animals that threaten their livestock or themselves and their families and may be supported in this by government officials.

Key elements in the success of harvesting include having data about the population of organisms, the rates of reproduction, and the rates of harvest. Enforcement of quotas is essential if this approach is to succeed. Legal harvest quotas do not require people who object to hunting to be hunters. Trophy hunting is not the exclusive preserve of countries in Africa. On April 3, 2007, *The Globe and Mail* reported that the economy of Nunavut received C$2.9 million from polar bear **(Figure 1)** hunting. Hunters can pay U.S.$20 000 for a polar bear hunt.

In 1992, saola **(Figure 2)** made the news as one of the first "new" species of large mammals to be discovered in recent times. These goatlike animals live in a restricted area of Vietnam, where they have been and are hunted by local people. Saolas are rare, and little is known about them. There are no quotas for the local hunters, and it is not practical to enforce a ban on their harvest. In reality, we probably lack critical information about the biology of many species of wildlife today. However, once they have names, they have a chance of being protected.

Figure 1
Polar bear, Ursus maritimus.

a.

b.

From Richard Stone, "The Saola's Last Stand," *Science,* vol. 314, Dec 1, 2006, pp. 1380–1383. Reprinted with permission from AAAS.

Figure 2
(a) Saola (Pseudoryx nghetinhensis) *and (b) its distribution.*

The lure of conservation movements that focus on charismatic species is very strong. But charismatic organisms may not need protection, whereas some species that are unattractive, dangerous, or mundane are in desperate need of our assistance. Unfortunately, mundane, ugly, and dangerous (to us) species are unlikely to serve as a call to arms (or to attract financial support). Worldwide, the WWF panda is one of the most recognized logos, whether or not pandas are in the neighbourhood.

A critical first step toward conservation is the development and adoption of objective, data-based criteria for assessing the risk posed to different species. This process has been developed on several fronts around the world. The criteria and assessment procedures perfected by the International Union for Conservation of Nature (IUCN) are used widely. There are many records of success, but there also are many examples of species with which and situations in which we have failed. Making arguments based on data does not guarantee success. Using a data-based approach, some species emerge as being in need of protection, but others do not. Being rare or unusual, by itself, will not warrant protection. The species concept and the Linnaean system of nomenclature (see Chapter 19) are fundamental to conservation.

In Canada, recommendations about the conservation status of species involve the Committee on the Status of Endangered Wildlife in Canada (COSEWIC). The definition of wildlife includes plants and animals. Like IUCN, COSEWIC recognizes six categories for assessing species at risk:

- *Extinct* wildlife species no longer exist.
- *Extirpated* species no longer exist in one location in the wild but occur elsewhere.
- *Endangered* species face imminent extirpation or extinction.
- *Threatened* species are likely to become endangered if limiting factors are not reversed.
- *Special concern* species may become threatened or endangered because of a combination of biological characteristics and identified threats.
- *Data deficient* is a category used when available information is insufficient either to resolve a wildlife species' eligibility of assessment or to permit an assessment of its risk of extinction.

A seventh category—*not at risk*—is used to identify species not at risk of extinction under current circumstances.

COSEWIC members vote on the appropriate conservation category for each species whose status they review. The members consider the area of occupancy, which is an indication of the range of a species and the availability of suitable habitat. They take into consideration population information, including trends in the numbers of organisms, correcting for species that show extreme fluctuations in numbers from year to year. They consider the demographics of the species and the variability in the habitat where the species occurs. Generation time is also considered, along with specific habitat features that may be essential for the species' survival. Data on population size, particularly the numbers of reproducing adults, are important, as well as risks to the species' survival.

In a biological context, the criteria used by COSEWIC (and similar agencies elsewhere) are familiar to population biologists (see Chapter 29). The data describe the numbers of individuals in the population, fecundity, mortality, and the intrinsic rate of increase. Carrying capacity is also important, as is the area (range) over which the species occurs. These criteria are designed to promote data-based decisions about the conservation status of species.

STUDY BREAK

1. What is IUCN? What role does it play in conservation?
2. What criteria would identify a species as endangered? Give an example.

32.6 Protecting What?

Before data are used to address questions of species-at-risk status, conservation biologists must decide about eligibility. The conservation jargon for this is "designatable unit." Are the organisms "real" species? Are they subspecies? Are they distinct populations? Are they really Canadian? Do they regularly occur in Canada or perhaps turn up here by accident? If the species does not breed here, is the habitat they use in Canada essential to their survival? Most species of wildlife in Canada occur close to the border with the United States, and many species widespread in the United States just make it into Canada. In some cases, a distinct population is treated as a designatable unit. Distinct populations may be recognized by their geographic distribution and/or their genetic structure.

Questions about what units are designatable hearken back to the definition of species (see Chapter 19). Off the west coast of Canada, striking differences in behaviour can be used to distinguish between two "kinds" of killer whales. The *resident* killer whales eat mainly fish and often echolocate. The *transient* killer whales eat mainly marine mammals and rarely produce echolocation signals. Furthermore, repeated sightings of recognizable individual whales indicate that different groups of these animals live in different areas along the coast (Figure 32.19).

a.

b.

c.

M.B. Fenton

M.B. Fenton

M.B. Fenton

Figure 32.19
Three views of a killer whale (*Orcinus orca*). **(a)** A captive animal in Vancouver, **(b)** a wild orca swimming off the Queen Charlotte Islands, and **(c)** a Haida representation.

In reviewing the conservation status of killer whales, COSEWIC recognized different designatable units based on behaviour and geography **(Figure 32.20, p. 796)**. The different units faced different threats to their survival.

Questions about what to protect often reflect different realities of biology. Migrating birds may be blown off course and end up in southern Ontario instead of their usual habitat much farther south. Marine birds or mammals may feed in Canadian

An Endangered Species

Banff Springs snails, *Physella johnsoni* **(Figure 1)**, live and eat algae in five hot springs on Sulphur Mountain in Banff National Park, Alberta. Not very long ago, Banff Springs snails were found in nine springs. In 1996, the total population of snails was about 5000. Water temperatures in the springs occupied by the snails range from 26 to 48°C, but temperatures less than 44°C seem best for them. Their very limited occurrence makes them vulnerable to extinction (COSEWIC, 2000).

Humans appear to be the main threat to the survival of Banff Springs snails. By discarding unsightly (to humans) accumulations of algae from pools, people have killed some snails that were in the algal mats. Changes in the patterns of water circulation may subject some snails to high temperatures that could be lethal. Well-wishers

that throw copper coins into the pools may have harmed snails because of contamination arising from the interaction of copper with sulphurous water in the springs.

Other impacts of people are not clear, but one threat is entertaining to contemplate—people "skinny-dipping" in the pools are thought to threaten the snails. Some skinny-dippers have been caught and charged. Bathers in the pools, clad or unclad, may have crushed snails while getting into or out of the water. Bathers doused in sunscreen or insect repellents may have

introduced chemicals into the snails' habitat and further reduced their populations.

Banff Springs snails are neither charismatic nor prominent, but the data-based approach to decision making has provided the basis for identifying them as endangered.

Degner, M. & L. used with permission of Parks Canada

Figure 1
Banff Springs snail, Physella johnsoni.

CHAPTER 32 CONSERVATION OF BIODIVERSITY

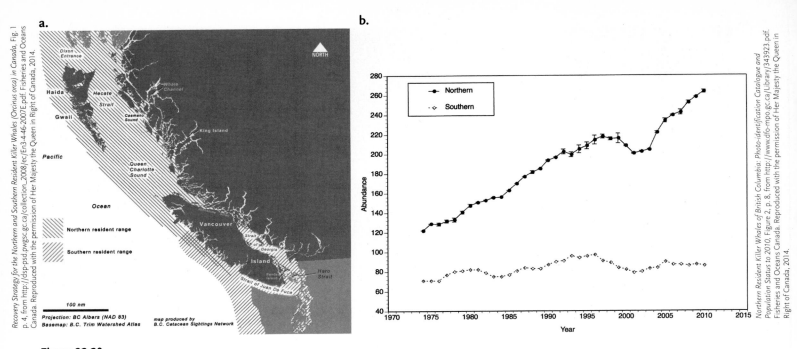

Recovery Strategy for the Northern and Southern Resident Killer Whales (Orcinus orca) in Canada, Fig. 1 p. 4, from http://dsp-psd.pwgsc.gc.ca/collection_2008/ec/En3-4-46-2007E.pdf. Fisheries and Oceans Canada. Reproduced with the permission of Her Majesty the Queen in Right of Canada, 2014.

Northern Resident Killer Whales of British Columbia: Photo-identification Catalogue and Population Status to 2010, Figure 2, p. 8, from http://www.dfo-mpo.gc.ca/Library/343923.pdf. Fisheries and Oceans Canada. Reproduced with the permission of Her Majesty the Queen in Right of Canada, 2014.

Figure 32.20

(a) The population distribution (designatable units) of killer whales off the coast of British Columbia and **(b)** estimates of population sizes of northern and southern resident killer whales (British Columbia).

waters but breed elsewhere. Many organisms commonly hitchhike, using ocean vessels, aircraft, or automobiles as vehicles of dispersal. But some hitchhikers, for example, some snails, travel with birds, making the association and the dispersal more "natural."

People can be quick to try to protect species they consider to be important or distinctive. In 2003, the Ontario Ministry of Natural Resources reported four to six white-coloured moose (*Alces alces*) among the approximately 1900 moose in two wildlife management areas near Foleyet in northeastern Ontario. Should white-coloured moose be protected? There was local support for protecting the moose, animals that have cultural and spiritual significance for First Nations communities. White moose have been reported from other places in northern Ontario, Newfoundland and Labrador, and elsewhere. Although the population of white moose is small and widespread, there is no evidence that they are a designatable unit. In Canada, they have not been accorded special protection.

There is protection at the international level. CITES, the Convention on International Trade in Endangered Species of Wild Fauna and Flora, plays a pivotal role in protecting species. International trade in wildlife is a leading threat to conserving biodiversity because in addition to directly affecting local populations of threatened species, it can also spread infectious diseases and promote the spread of invasive species. Membership in CITES includes 180 countries, and CITES tries to regulate trade in almost

36 000 species. Basic, accurate, and reliable biological data about species are essential for informing decisions about which species should be protected. Yet, decisions about what species are protected by CITES are political and not necessarily uniformly acclaimed. Between 2014 and 2016, the annual budget of the secretariat of CITES averages U.S.$6.2 million, coming from donations. Budget restrictions influence the effectiveness of CITES at the secretariat level by affecting the capacity for detailed collection and analysis of basic data.

At local levels, however, unmonitored trade in wildlife continues to occur openly, often in clear violation of CITES. Orchids are a clear example. In a period when CITES records showed 20 cases of orchids listed by CITES coming into Thailand from four neighbouring countries (Lao People's Democratic Republic, Myanmar, Cambodia, and Vietnam), 168 cases were observed in local markets. At one market in the Mekong Delta, one trader can sell more orchids in one day than reported by CITES over a period of nine years!

STUDY BREAK

1. Give an example of a designatable unit that is a species.
2. Give an example of a designatable unit that is a population.
3. Is designatable unit synonymous with species?

Who Gets Protection?

Being recognized as rare and considered to be endangered does not necessarily translate into protection.

Because the Endangered Species Act (ESA) in the United States does not protect hybrids, this can affect the conservation of, for example, the Florida panther **(Figure 1),** a subspecies of cougar. Cougars, also known as panthers, used to occur widely in North, South, and Central America. Although still widespread in some areas, the current range of cougars in most of the United States and Canada is much less than it was when Columbus arrived in the New World in 1492. Florida panthers, a small population recognized as a subspecies, occur mainly in the Florida keys. Florida panthers were protected under the ESA.

Using techniques of molecular genetics, biologists determined that Florida panthers carried the genes of cougars from South America. This situation probably arose when panthers originally caught in South America were brought to the United States as zoo animals or for display in circuses or animal shows. Some of these animals escaped and interbred with local Florida panthers. Florida panthers with genes from South American cougars are technically hybrids and are therefore not protected by the ESA.

There are many other examples of situations in which genetic tools allow clearer delineation of boundaries between populations (designatable units) and species. In some cases, however, removal of protection from other "species" because of their genetic status can lead to their extinction. *Ammodramus maritimus nigrescens,* or the Dusky Seaside Sparrow, was previously considered to be a distinct form living in Florida. When genetic evidence showed that these darker animals were not genetically distinct, they lost their protected status and have virtually disappeared.

Other species, such as roundnosed grenadier, have suffered calamitous declines in population. These codlike fish **(Figure 2)** were taken in large numbers after cod populations had declined **(Figure 3).** The species was on the verge of extinction even before much was known about it. We do know that round-nosed grenadiers are late to mature, and their populations are slow to recover.

Although round-nosed grenadiers and at least four other species meet the IUCN criteria for listing as endangered, these fish have not been

Figure 2
Coryphaenoides rupestris, *a round-nosed grenadier.*

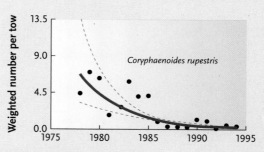

Figure 3
Captures of round-nosed grenadiers.

and are not protected. Fisheries and Oceans Canada has not supported a move to protect round-nosed grenadiers. Changing fishing practices to avoid catching the few remaining round-nosed grenadiers is not economically feasible when other species are still being caught in sufficient numbers to justify a continued fishery. The situation differs only from that facing the Little Brown Bat in that the round-nosed grenadiers have been the targets of an active fishery.

We have seen that the hunt for polar bears can bring significant income to the economy of Nunavut, and the same is true of other jurisdictions within the bear's range.

(Continued)

Figure 1
A Florida panther (Felis concolor coryi).

Who Gets Protection? (*Continued*)

There are distinct populations of polar bears **(Table 1)** within Canada's jurisdiction. The occurrence of bears in political jurisdictions including Canada, the United States, Russia, Iceland, Denmark (Greenland), Norway, Finland, and Sweden makes it more difficult to protect them. The apparent vulnerability of the bears to climate change and their value as trophies may combine to hasten their demise.

Table 1   **Status of Canadian Polar Bear Populations (January 1997)**

Population	% Females in Harvest	Number	Sustainable Annual Kill	Mean Annual Kill	Environ. Concern	Status	Quality of Estimate	Degree of Bias	Age of Estimate	Harvest/ Capture Data
Western Hudson Bay	31	1200	54	44	None	S[a]	Good	None	Current	Good (>15 yr)
Southern Hudson Bay	35	1000	43	45	None	S[a]	Fair	Moderate	Old	Fair (5–10 yr)
Foxe Basin	38	2300	91	118	None	S[a]	Good	None	Current	Good (>15 yr)
Lancaster Sound	25	1700	77	81	None	S[a]	Fair	None	Current	Good (>15 yr)
Baffin Bay	35	2200	94	122	None	D?[b]	Fair	None	Current	Fair (>15 yr)
Norwegian Bay	30	100	4	4	None	S[a]	Fair	None	Current	Good (>15 yr)
Kane Basin	37	200	8	6	None	S	Fair	None	Current	Fair (>15 yr)
Queen Elizabeth	–	(200?)	9?	0	Possible	S?[b]	None	–	–	–
Davis Strait	36	1400	58	57	None	S?[b]	Fair	Moderate	Outdated	Good (>15 yr)
Gulf of Boothia	42	900	32	37	None	S[a]	Poor	Moderate	Outdated	Good (>15 yr)
M'Clintock Channel	33	700	32	25	None	S[a]	Poor	Moderate	Outdated	Good (>15 yr)
Viscount Melville Sound	0	230	4	0	None	I	Good	None	Current	Good (>15 yr)
Northern Beaufort Sea	43	1200	42	29	None	S	Good	None	Recent	Good (>15 yr)
Southern Beaufort Sea	36	1800	75	56	None	S	Good	Moderate	Recent	Good (>15 yr)

[1] D = decreasing; I = increasing; S = stationary; ? = indicated trend uncertain.
[a] Population is managed with a flexible quota system in which overharvesting in a given year results in a fully compensatory reduction to the following year's quota.
[b] See text at link below, "Population Size and Trend," for discussion.

Source: *COSEWIC Assessment and Update Status Report on the Polar Bear Ursus maritimus in Canada*, 2002. http://www.sararegistry.gc.ca/document/default_e.cfm?documentID=248. © Her Majesty The Queen in Right of Canada, Environment Canada, 2014. Reproduced with the permission of the Minister of Public Works and Government Services Canada.

32.7 The Downside of Being Rare

Whether the commodity is coins, stamps, antiques, or endangered species, as soon as something is rare enough, there is a market for it. This "get them while they last" attitude is exemplified by trade in *Leucopsar rothschildi*, Bali Starlings **(Figure 32.21)**. This bird, another island species, faces immediate extinction, but it is in high demand as an exotic pet. In 1982, when there were fewer than 150 individuals in the wild, 35 were for sale as pets, 19 in Singapore, and 16 in Bali.

Rare species may also be in demand for use of their body parts in traditional medicine. One stark example is the swim bladders of *Bahaba taipingensis,* the Chinese bahaba. At a time when fewer than 6 individuals are caught each year, more than 100 boats are trying to catch them. The swim bladders are used in traditional medicine. They are worth at least seven times their weight in gold. Shark fins are even more valuable. These are extreme examples similar to the earlier story about rhino horns and jambiyas.

Before criticizing and condemning the users or consumers of jambiyas or Chinese bahaba swim bladders, think about the overall impact of our lifestyle on other species of animals and plants. Of particular note is an insatiable demand for energy. Are sport utility vehicles necessary? Personal watercraft? Snowmobiles? All-terrain vehicles? The list goes on. Is a Canadian as justified in buying a large SUV as a North Yemenese a jambiya with a rhino horn handle? Once again, might (the ability or capacity to do something) may not be right.

Figure 32.21
A Bali Starling (*Leucopsar rothschildi*).

STUDY BREAK

1. What is the value of the International Union for Conservation of Nature (IUCN)?
2. Why do species description and formal naming affect the Convention on International Trade in Endangered Species of Wild Fauna and Flora (CITES)?
3. What is the difference between an extinct species and an extirpated species? Give an example of each.

32.8 Protecting Habitat

It is obvious from many of the examples above that protecting species has not been entirely successful as a conservation strategy. As a species, we are much better at killing than we are at conserving. Whether this is direct or indirect, the end result can be the same. It is also clear that destruction of habitat is an effective way to remove a species. For example, populations of mosquitoes can be limited by denying them places to lay their eggs. This is a common theme in public education programs designed to reduce the incidence of West Nile virus (or other mosquito-borne diseases).

Is protection of habitat an effective strategy? The answer can be "yes," particularly for species that are not motile. Many species of plants have specific habitat requirements. From trees to shrubs, forbs, ferns, and mosses, we know that we can protect species by protecting habitat. Furthermore, protecting large tracts of habitat can also protect large, mobile species. Rainforests, whether tropical or temperate, are examples of habitats that can be flagships for protection and conservation. They are also considered by many to be storehouses of wealth associated with biodiversity, from building materials to compounds of pharmacological value.

The case of the black rhino demonstrated how a species targeted for harvesting can be driven to the brink of extinction even when it is protected (or lives in national parks or game reserves). *Panax quinquefolius,* American ginseng, is another target species, now endangered in Canada because of harvesting. The species used to grow wild from southwestern Quebec and southern Ontario south to Louisiana and Georgia. This 20- to 70-cm-tall perennial is long lived in rich, moist, mature, sugar maple–dominated woods. Although the species has been listed on Appendix II of COSEWIC since 1973, populations have continued to decline. In 2000, there were 22 viable populations in Ontario and Quebec, but none were secure. Black rhinos and ginseng were common about 50 years ago, but by 2008, both demonstrated the risks of being rare and expensive. They are also examples of the need for immediate on-the-ground enforcement of regulations and laws protecting species and habitats.

Protecting habitats can be most challenging in areas with larger human populations. All of the viable populations of American ginseng in Ontario and Quebec were close to roads, making the plants vulnerable to anyone who knew about them and wished to take advantage of the economic opportunity they presented. *Sorex bendirii*, the Pacific water shrew, is another example of a species whose future in Canada is threatened by expanding human populations and the associated value of real estate **(Figure 32.22)**.

In British Columbia, expanding human population and the wine industry in the southern Okanagan

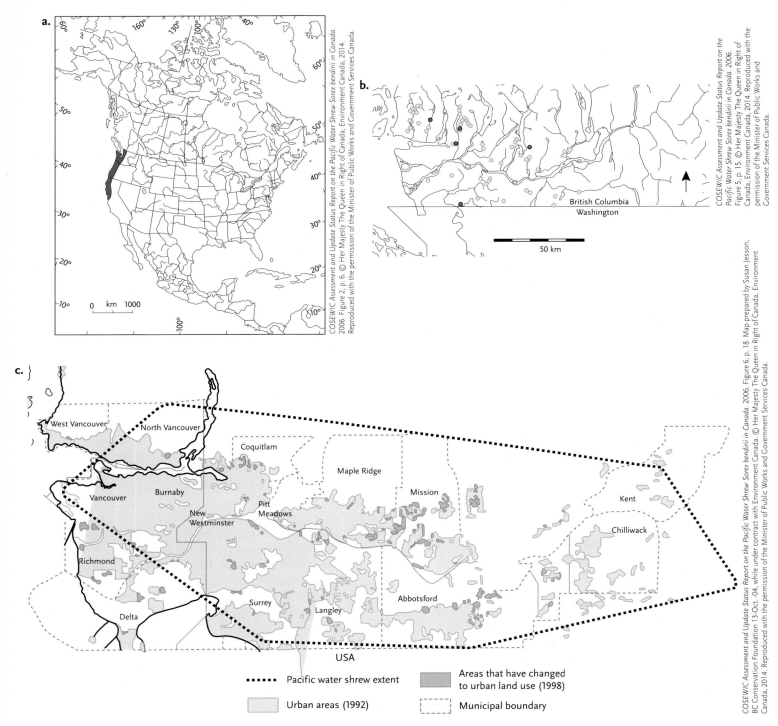

Figure 32.22
(a) The distribution of *Sorex bendiri*, the Pacific water shrew. **(b)** Lower Fraser Valley locations where it was found (solid circles) or not found (open circles).
(c) For comparison, the same area is shown with changes in the availability of urban lands in 1992 and 1998. Data on the map are from 2004.
Baseline Thematic Mapping Present Land Use Mapping at 1:220 000.

Figure 32.23

Antelope bush, *Purshia tridentata*, showing **(a)** the bush and **(b)** a cross-section of the stem. These woody shrubs have long life spans, and the ecosystem they typify is home to a variety of species.

Valley have combined to dramatically reduce a local ecosystem dominated by antelope bush **(Figure 32.23)**. The antelope bush system, one of the most endangered ecosystems in Canada, is home to a number of species of plants and animals whose future is now threatened by the demise of the habitat they require. The boom in real estate for people looking for retirement properties, more than just the density of human populations, is a key factor in this situation. Meanwhile, in southern Ontario, the demand for real estate to accommodate the expanding housing and business market is reducing both the available natural habitats and farmland.

STUDY BREAK

1. Give examples of how protecting a habitat can work.
2. How do market forces influence the abundance of species?

32.9 Effecting Conservation

Today we face many challenges when trying to protect biodiversity. Too many of the immediate threats are the direct or indirect consequences of human activities. We must protect species by acting at levels ranging from species to populations and habitats.

32.9a Human Population Is a Root Problem in Conservation

One fundamental root cause of declining biodiversity is the human population and the energy and habitat consumed in trying to feed, house, and protect our flourishing species. Visit the website http://www.ined.fr/en/everything_about_population/world_population_me/ and use it to determine the estimated human population in the year you were born and then for the years in which your parents and grandparents were born. Even when many people are killed, the momentum of our population increase does not slow down. The December 2004 tsunami killed approximately 250 000 people, at a time when the world population was estimated at 6 billion. By comparison, the 1883 explosion of Krakatoa (and resulting tsunamis) is thought to have killed 35 000 people when the global human population was about 1.5 billion. If these estimates are correct, $4.1 \times 10^{-3}\%$ of the human population at the time was killed by the 2004 tsunami and $2.3 \times 10^{-3}\%$ by the explosion of Krakatoa. Neither calamity caused the human population growth curve (see Chapter 29) to waver.

If human population growth continues at the same rate as it is growing now, it will double in 40 years. However, studies show that our population is not growing as quickly as it did during much of the twentieth century. The United Nations Development Program (UNDP) has released data on human fertility (the total number of births per woman) for 162 countries **(Table 32.1, p. 802)**. Compared with 1970–75, 152 countries had lower human fertility in 2000–05, 3 countries showed increases in fertility, and 7 showed no change.

Concerned about the global population and its effect on Earth, world leaders adopted the United Nations Millennium Development Goals in 2000, committing their nations to achieving the following goals by 2015:

- ending poverty and hunger,
- universal education,

Table 32.1	Variations in Fertility Rate (Total Births per Woman): A Sample of UNDP Data for 162 Countries		
Country	Human Development Index (HDI) Rank	1970–75	2000–05
Norway	2	2.2	1.8
Canada	4	2.0	1.5
United States	12	2.0	2.0
Portugal	29	2.7	1.5
Brazil	70	4.7	2.3
China	81	4.9	1.7
Indonesia	107	5.2	2.4
India	128	5.4	3.1

- gender equality,
- child health,
- maternal health,
- combatting HIV/AIDS,
- environmental sustainability, and
- global partnerships.

These goals can be achieved only if reproduction is controlled (see Chapter 29). Go to the United Nations Millennium Goals website at http://www.un.org/millenniumgoals/bkgd.shtml to see how we are faring. In 1994, the United Nations held the International Conference on Population and Development (ICPD), which set a target for global investment in family planning. By 2004, the amount spent had fallen to 13% of this target. Consequently, family planning information and devices (usually for fertility control) are not readily available in many of the lowest-income countries. In 1950, Sri Lanka and Afghanistan had the same population. Sri Lanka began strong efforts to make family planning available in culturally acceptable ways. This did not happen in Afghanistan. By 2050, Afghanistan will have four times as many people as Sri Lanka. The solution centres around controlling the fertility of women, but more particularly on giving them the power to control their own fertility in culturally acceptable ways. As seen in Chapter 29, the growth potential of a population is determined by the numbers of females of reproductive age. Why females? Because females are the limiting step in reproduction—the ones who produce the eggs or young.

32.9b Signs of Stress Show Up on Systems and on Species

People's demand for food, water, and energy puts thousands of other species at risk. We do not have to look far to see examples of species and ecosystems under stress (see Chapter 31). For example, we are losing birds. We know this because for years, bird-watchers and ornithologists have counted them and monitored their behaviour and activity. Locally, birds are affected by changes in habitat availability as cities and towns and their suburbs expand into adjoining land. Birds also lose habitat when agricultural operations expand to increase productivity. Birds that make annual migrations from temperate areas of the world to tropical and subtropical ones must survive the changes that accumulate across their entire circuit of habitats, each one essential to their survival.

Avian influenza (also called bird flu) is a looming crisis for humans, one that appears to involve birds as central players. The issue here is another one involving basic biology, namely the outcome when a disease-causing organism jumps from one species (host) to another. Bird flu could have as much to do with our insatiable demand for poultry as food as it does with birds. In 2006, 12 billion chickens were farmed in China. Worldwide, poultry farms housed over 100 billion broiler chickens. Raising organisms at very high densities (see Chapter 29) provides an ideal setting for the spread of disease. Humans have responded to the threat of bird flu by wholesale slaughter of fowl, raising concerns about the roles played by migrating birds, and trying to develop a vaccine that will protect humans from bird flu. All involve basic biology.

Drylands are arid, semiarid, and subhumid areas where precipitation is scarce and more or less unpredictable. In drylands, the combination of high temperatures, low relative humidities, and abundant solar radiation means high potential evapotranspiration. Drylands cover approximately 41% of Earth's land surface and are home to about 38% of the human population. Drylands are not just a problem of deserts but cover large expanses, for example, of Canada's prairie provinces. However, between 10 and 20% of the drylands are subject to some form of severe land degradation, directly affecting the lives of at least 250 million people. Climate change, combined with increasing pressure on water resources for these people, their crops, and their animals, compounds the problems that confront them. Competition for limited resources, such as water, can generate local and international strife.

We have seen that complexity is an important and pervasive feature of ecosystems. Biodiversity is intimately associated with complexity, and disruption of this complexity often translates into reduced biodiversity and decay of ecosystems. Ironically, many social and economic systems that humans have developed are also subject to disruption by stress. This places the onus on our species to develop sustainable operations, whether in the area of agriculture, resource use and exploitation, or conservation.

1. How is reproductive effort different between males and females in birds and in mammals?
2. List the United Nations Millennium Development Goals.
3. How are drylands at risk?

32.10 Taking Action

It is easy to believe that nothing can change, that as individuals we have no power. Yet we can also think of things that have changed dramatically in a relatively short time. Two good examples are the abolition of slavery and the emancipation of women, proving humans' capacity for effecting change. On a more local level, the acceptance of the use of tobacco in public has declined remarkably in the last 20 years—in Canada and elsewhere. We also have seen the abolition of capital punishment and much more ready access to abortion in Canada.

But none of these changes is universal. In the daily news we find stories about people living in virtual slavery, of people executed in public, of women with few or no rights in their home countries. To complicate the matter, not everyone agrees that the changes listed above are for the better.

Effecting changes in our approach to conservation means identifying the root causes for the erosion of biodiversity and the things that are impediments to conservation. This means starting by changing our own lifestyles, including the food we eat and our use of energy. We must be wary of simple, and often misleading, solutions and avoid blaming someone else as a way of self-exoneration. We must respect the rights of others; use education and training to become informed; and learn to be objective, to examine and evaluate data or evidence. The outpouring of support for victims of the 2004 tsunami demonstrated that humans have great empathy for their fellows, and we need to extend this concern to the other species with whom we share the planet.

We have seen that action is needed at the species and the habitat level, and there is a propensity to focus more on species. But in the human view, all species are not equal. The 2006 IUCN list of threatened species shows that whereas 20% of the described species of mammals were listed as threatened, only 0.07% of the insect species received this level of attention. Other interesting numbers from this table are 12% of described species of birds listed as threatened, 4% of fish species, 3.5% of dicotyledonous plants, and 0.006% of species of mushrooms. In Canada, the same situation prevails, with mammals and birds dominating the list of threatened species, with other taxa receiving less attention. Do these data about threatened species mean that mammals are more vulnerable than insects? That we care more about mammals than about insects? Or does it mean that there are more "experts" to offer opinions and data about mammals than about insects? Are the possibilities mutually exclusive?

Biology can be at the centre of the movement to achieve conservation of biodiversity while being part of our efforts to achieve sustainable use of the resources we need as a species **(Figure 32.24)**. Conservation begins at home when we modify our lifestyles and become active on any front, from protecting local habitat and species to protecting charismatic species elsewhere. To better appreciate the situation, try to answer the questions posed in **Figure 32.25**. Elephants are an excellent example of how the objectivity that can be inherent in data is vulnerable to emotional responses.

STUDY BREAK

Figure 32.24

Eat yourself out of house and home—like this African elephant (*Loxodonta africana*) trekking across the shore to Lake Kariba.

M.B. Fenton

a.

b.

c.

Figure 32.25
To understand some of the dilemmas facing conservationists, use the Internet to explore the situation of African elephants. **(a)** How many species are there? What are the populations in the wild? What products from elephants do we use **(b)** and **(c)**? Are elephants endangered? How can they be protected? What are the main threats to their survival?

Review

To access course materials such as Aplia and other companion resources, please visit www.NELSONbrain.com.

32.1 Extinction

- A species is said to be extinct when there are no living representatives known on Earth. Conservation organizations usually say that a species is extinct when it has not been seen or recorded for 50 years.

- Mass extinctions occurred at the end of the Ordovician and the beginning of the Devonian, at the end of the Devonian, at the end of the Permian, at the end of the Triassic, and at the end of the Cretaceous. The Permian extinction was the most severe, and more than 85% of the species alive at that time disappeared forever, including the trilobites, many amphibians, and the trees of the coal swamp forests. Dinosaurs did not survive the extinction that occurred at the end of the Cretaceous.

- The extinction at the end of the Cretaceous is believed to have been caused by an asteroid impact. Dust clouds resulting from the impact blocked the sunlight necessary for photosynthesis, setting up a chain reaction of extinctions that began with microscopic marine organisms and finished with dinosaurs (as well as many birds and mammals).

- Measured by time on Earth, multituberculates were the most successful mammals.

32.2 The Impact of Humans

- Species (particularly flightless birds) that are confined to islands often have small populations and are unaccustomed to introduced terrestrial predators (such as cats, dogs, rats), making them vulnerable to extinction when human populations settle and expand.

- The demise of the calvaria trees (*Sideroxylum majus*) on the island of Mauritius will occur even though the trees continue to bloom and produce seeds. The extinction of the tree is linked to the earlier extinction of Dodos, since *S. majus* seeds had to pass through the Dodo's digestive tract to germinate.

32.3 Introduced and Invasive Species

- Stephen's Island wrens were flightless and unaccustomed to predators. The population on the island was small, and it was easy for Tibbles the cat to catch and kill the remaining 20 birds.

- Since about 1880, ships have regularly used ballast water. A survey of ballast water in 159 ships in Coos

Bay, Oregon, revealed 367 species of organisms representing 19 animal phyla and 3 plant divisions. When ships empty their ballast, the organisms in it are introduced to the system where the ship is anchored.

- RUE, resource use efficiency, is measured as carbon assimilation per unit resource. Many invasive plant species in Hawaii are more efficient than native species. This means that conserving native biodiversity in the face of invasive and introduced organisms is a pervasive problem.

32.4 How We Got/Get There

- Horns of rhinos, particularly black rhinos (*Diceros bicornis*), have been used to make handles for ornamental daggers (jambiyas) in some parts of the Arabian peninsula. Increasing oil prices in the early 1970s increased the demand for jambiyas. The main source of rhino horn was from poaching rhinos in Africa.

- Since March 2006, White-Nose Syndrome has killed millions of Little Brown Bats in the northeastern United States and adjacent Canada.

- The demise of the bay scallops occurred because of an increase in the populations of skates and rays that are predators of bay scallops. The increase in skates and rays was attributed to the decline in populations of their predators, sharks. The sharks were extensively fished for their fins. Large-scale harvesting of the scallops for human consumption also compounded the impacts and contributed to their demise.

32.5 Protecting Species and 32.6 Protecting What?

- The International Union for Conservation of Nature (IUCN) has established objective criteria identifying species that are at risk. Extinct means the species no longer exists, extirpated means the species is locally extinct, endangered means the species is facing imminent extirpation or extinction, threatened means the species is likely to become endangered if limiting factors are not reversed, and special concern means the species may become threatened or endangered because of biological characteristics and identified threats. The criteria take into account data on populations, their patterns of distribution, and their population status.

- The Convention on International Trade in Endangered Species of Wild Fauna and Flora (CITES) attempts to prohibit international trade in endangered species. Newly described and as yet undescribed (and therefore unnamed) species are not protected because they have no legal identity.

- Species such as passenger pigeons or Dodos that have been exterminated are extinct. Extirpated species are locally extinct. Black-footed ferrets (see "Black-Footed Ferret, *Mustela nigripes*," Chapter 29) have been extirpated in Canada but still occur in the United States.

- In Canada, recommendations about the conservation status of species involve the Committee on the Status

of Endangered Wildlife in Canada (COSEWIC). COSEWIC members vote on the appropriate conservation category for each species whose status they review and use IUCN criteria to assess the status of species.

32.7 The Downside of Being Rare

- An animal on the list of endangered species is more likely to become a commodity in high demand because it has become rare. The Bali Starling is an example.

32.8 Protecting Habitat

- Protecting habitat can be particularly difficult.
- Networks of protected areas can influence conservation.

32.9 Effecting Conservation

- In 1994, the International Conference on Population and Development (ICPD) outlined a plan for investing in family planning.

- The United Nations Millennium Development Goals of 2000 are ending poverty and hunger, universal education, gender equality, child and maternal health, combatting HIV/AIDS, environmental sustainability, and global partnerships.

- Drylands cover 41% of Earth's land surface, and 10 to 20% of drylands are subject to severe land degradation, affecting, in 2008, the lives of at least 250 million people.

- Climate change and increasing pressure on water supplies negatively affect drylands.

- The case of leopards (*Panthera pardus*) demonstrates how some species persist even in the face of considerable hunting pressure.

- Targeted hunting—selection of trophy or spectacular specimens—can be less threatening to a species' survival than disease or eradication programs (bounties on predators such as wolves). Extensive killing, even of species with large populations, can drive them to the brink of extinction. Black rhinos are a telling example.

- The overgrowth of Mayan cities or ruins in Africa demonstrates the resiliency of ecosystems. A grasshopper's use of 2,4-D to synthesize an ant repellent demonstrates the resiliency of individuals.

- Hybrids are not protected by the U.S. Endangered Species Act, putting species such as Florida panthers at risk because their populations have been genetically contaminated.

32.10 Taking Action

- African elephants epitomize some of the challenges involved in taking action to protect a species and its habitats.

- This species also illustrates the potential importance of hunting and poaching.

Questions

Self-Test Questions

1. Extinction is a natural part of the process of specia-tion. What do some estimates suggest is the per-centage of the species that have ever lived that are now extinct?
   a. more than 20%
   b. more than 30%
   c. more than 50%
   d. more than 80%

2. Some researchers use evidence from a variety of sources to support the suggestion that an asteroid striking Earth in this period largely explains the extinction of the dinosaurs.
   a. Ordovician
   b. Triassic
   c. Cretaceous
   d. Pleistocene

3. If our species first appeared 200 000 years before present, the multituberculates survived this many times as long as we have to date.
   a. 50
   b. 100
   c. 500
   d. 1000

4. Hunting by people is largely responsible for the extinction of which of the following?
   a. multituberculates, Dodos, and passenger pigeons
   b. black-footed ferrets and giant auks
   c. passenger pigeons, giant auks, and Dodos
   d. black rhinos, Bali starlings, and ginseng

5. The ballast water of ships is responsible for the spread of which species?
   a. *Arthurdendyus triangulatus*
   b. *Dreissena polymorpha*
   c. *Rattus norvegicus*
   d. *Lampsilis radiata*

6. In Hawaii, high resource use efficiency (RUE), measured as carbon use, partly explains the success of these invading species.
   a. ferns
   b. C_3 and C_4 grasses
   c. flatworms
   d. Both b and c are correct.

7. Which tertiary consumers have experienced increases in populations, which may explain the demise of scallops off the southeastern coast of the United States?
   a. skates and rays
   b. sharks
   c. killer whales
   d. pelagic seabirds

8. Species such as black-footed ferrets (*Mustela nigripes*) no longer occur in Canada but still live in the United States. Which term describes their status?
   a. extinct
   b. extirpated
   c. highly endangered
   d. not at risk

9. CITES is designed to stop international trade in which of the following species, among others?
   a. passenger pigeons
   b. black rhinos
   c. Canada geese
   d. leopards

10. Differences in government support for family planning explain the differences in the growth of human populations in which two countries?
    a. Great Britain and France
    b. Afghanistan and Sri Lanka
    c. Mexico and Germany
    d. India and South Africa

Questions for Discussion

1. Should gardeners and farmers be exempt from rules concerning the introduction of foreign species? Why or why not?

2. In situations where the behaviour of one endan-gered species threatens the survival of another (or others), how should authorities proceed?

3. What species are "rare" on your campus? What is a good working definition of rare? What steps can you take to protect rare species?

4. How will the cost of food influence efforts to conserve species at risk?

5. How does the biological definition of species influence our efforts to conserve species?

Sunflowers. Originally from the New World, sunflowers (*Helianthus annuus*) are grown as a source of oil. In terms of harvest and area under cultivation, in 1998, sunflowers ranked twelfth in importance among domesticated plants in the world. Domesticated sunflowers often hybridize with local wild species, creating a challenge for those concerned about biodiversity.

Putting Selection to Work

WHY IT MATTERS

In 1960, an estimated 1.8 billion people in the world (60% of the population) did not receive enough food every day to sustain themselves fully over the longer period—they were hungry. This number was reduced to 1.1 billion (17%) in 2000. Even though the world population had grown by 3 billion in the intervening period, about 700 million fewer people were hungry in 2000.

Worldwide in 2000, subsistence farmers accounted for about 66% of the hungry people. The reduction in the numbers of hungry people can be tied to changes in agriculture that have increased yields. Specifically, the combination of new genetic strains, better fertilizers, better irrigation, more effective pest control, and more efficient harvesting and processing means more productivity. One indication of this change is provided by data about corn yields. In Iowa in 1935, corn yields were about 1600 kg·ha^{-1} compared with about 10 700 kg·ha^{-1} in 2000. Changes in crop yield are part of the "green revolution." Agriculture in general and the green revolution in particular have allowed humans to continue to redefine one element of carrying capacity (see Chapter 29): the amount of food available to our populations.

But agricultural improvements are not enough. Climate also influences crop yield. In 2006 in southwestern Ontario (~42° N in Canada), the corn yield was about 10 000 kg·ha^{-1}, whereas in Zimbabwe (~18° S), on commercial farms, it was about 5500 to 6600 kg·ha^{-1}, compared with about 500 to 1000 kg·ha^{-1} on communal lands where farming was low tech. Irrigation also influences yield: in Zimbabwe, irrigated cornfields produce 8500 to 10 000 kg·ha^{-1}, much more than nonirrigated commercial farms.

Although increases in crop yield and a reduced incidence of malnutrition and starvation sound like good news, in 2011, hunger still claimed the lives of about 8500 children a day. Worldwide, one child in three is under-weight and malnourished. Ironically, at the same time in some developed countries, obesity in children reached almost epidemic proportions.

In addition to the *biology* of domestication and increasing crop yields, the *security* of food sources plays a vital role in the survival of tens of thousands of people. In this context, security refers to social and political factors and stability. In 2011, changes in the political structure of several countries in North Africa and the Middle East may have been as attributable to rising costs of food and their impact on average people as to any other single factor. Availability of fresh water is inextricably associated with agriculture and the global food supply.

Changes in diet may have been fundamental to the origin and adaptive radiation of species in the genus *Homo*. The use of fire and tools influenced our ancestors' abilities to obtain food. Diets rich in "brain food," such as many aquatic animals, as well as in starches **(Figure 33.1)** could have heralded important changes in our ancestors.

Our species has turned natural selection to its advantage by domesticating other species, selectively breeding strains with traits that provide us with more food or other desirable commodities. Domestication emerged from cultivation of wild forms, probably in association with a more sedentary lifestyle, and has occurred in different parts of the world, from central Asia to the Far East, from the Middle East to the New World and southeast Asia. We depend on the productivity of many domesticated food plants to sustain our populations, and on domesticated animals for labour, hides, and food. Domestication provides repeated examples of how our species has benefited from selection, and unravelling the history of the process depends heavily on genetic tools.

33.1 Domestication

The purpose of this chapter is to explore how humans have used selection to put biodiversity to work. Biologists and anthropologists believe that our ancestors originally gathered plants and hunted animals in the wild for use as food, building products, or fuel (see "Molecule behind Biology," Box 33.1, p. 811). From gathering, our ancestors progressed to cultivating plants, a process involving the systematic sowing of wild plant seeds. Over time, cultivation improved when people provided more care to their crops and eventually involved repetitive cycles of sowing, collecting, and sowing wild stock **(Figure 33.2). Domestication** is more than just taming. It occurs when people selectively breed individuals of other species (plants and animals) to increase the desirable characteristics in the progeny (e.g., in plants: yield, taste, colour, shelf life). This marked the birth of agriculture. The progression from gathering to cultivation to domestication of plants occurred independently at several locations around the world. The beginning of the Neolithic Period is often defined by the domestication of other species, and this period started at different times in different parts of the world.

But agriculture is not the exclusive domain of humans. Recall that about 50 mya, ants of the tribe Attini were the first to manipulate other species (fungi) to increase food availability (see Chapter 25). Today at least 200 species of ants in this tribe are obligate farmers. These early farmers have lost their own digestive enzymes and rely on fungal enzymes to digest the food for them. The ant farmers propagate their fungal crops asexually, with each colony working with one species. Therefore, any single species of farmer ant may propagate several different species of fungi. These ants have been involved in at least five domestication

Figure 33.1

A diagrammatic presentation of the progression in diet and brain size across 4 million years of hominin history.

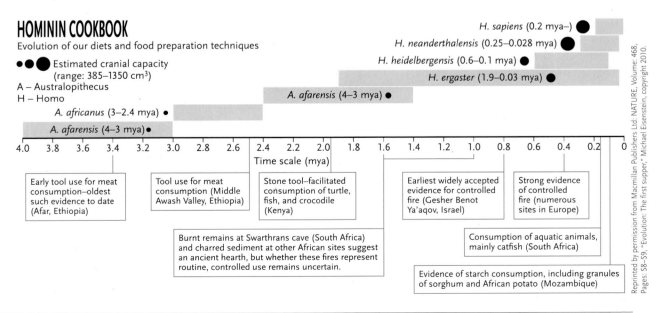

HOMININ COOKBOOK
Evolution of our diets and food preparation techniques

●●● Estimated cranial capacity (range: 385–1350 cm³)
A – Australopithecus
H – Homo

H. sapiens (0.2 mya–) ●
H. neanderthalensis (0.25–0.028 mya) ●
H. heidelbergensis (0.6–0.1 mya) ●
H. ergaster (1.9–0.03 mya) ●
A. afarensis (4–3 mya) ●
A. africanus (3–2.4 mya) ●
A. afarensis (4–3 mya) ●

Time scale (mya)
4.0 3.8 3.6 3.4 3.2 3.0 2.8 2.6 2.4 2.2 2.0 1.8 1.6 1.4 1.2 1.0 0.8 0.6 0.4 0.2 0

Early tool use for meat consumption–oldest such evidence to date (Afar, Ethiopia)

Tool use for meat consumption (Middle Awash Valley, Ethiopia)

Stone tool–facilitated consumption of turtle, fish, and crocodile (Kenya)

Earliest widely accepted evidence for controlled fire (Gesher Benot Ya'aqov, Israel)

Strong evidence of controlled fire (numerous sites in Europe)

Burnt remains at Swarthrans cave (South Africa) and charred sediment at other African sites suggest an ancient hearth, but whether these fires represent routine, controlled use remains uncertain.

Consumption of aquatic animals, mainly catfish (South Africa)

Evidence of starch consumption, including granules of sorghum and African potato (Mozambique)

Reprinted by permission from Macmillan Publishers Ltd: NATURE, Volume: 468, Pages: S8–S9, "Evolution: The first supper," Michael Eisenstein, copyright 2010.

Figure 33.2
The way in which garlic (*Allium sativum*) is grown influences the size and development of the bulbs. From left to right: one domesticated, two cultivated, and two wild garlic bulbs.

M.B. Fenton

events, and there are a number of interesting parallels between these ants and people.

The list of species that humans have domesticated is long. It includes many land plants (~250 species), some yeasts, and terrestrial animals from insects to birds and mammals (~44 species). Biogeographic and genetic evidence shows that domestication of some species by humans occurred in different places and at different times. Domestication was not a one-time (or one-location) event and appears to have arisen independently in 8 to 10 environmentally and biotically diverse areas in the world.

As intriguing as which species were domesticated is the fact that very few available species of animals, plants, and fungi were domesticable. At the same time, humans were exploiting and continue to exploit many species without actually domesticating them.

33.1a When and Where Did Domestication Take Place?

Once they were domesticated, many domesticated plants and animals became widely used, becoming staple foods carried with humans as they moved to occupy many of the land areas on the planet. Data provided by the tools of molecular genetics (see Chapter 17) have made it easier to determine where and when domestication events took place. In the past, archaeologists had to try to recognize the remains of domesticated species and distinguish them from wild species. This was often impossible because individual bones or pieces of plant did not always provide a clear indication

of domestication. In 1973, radiocarbon dates suggested that the first dogs (*Canis familiaris*) were domesticated by 9500 years B.P. (before present), based on remains found in England and elsewhere in Europe. In 2002, mitochondrial DNA (mtDNA) evidence suggested an East Asian origin of domestication of dogs dating from 15 thousand years B.P. But pictures based on genetic evidence can also change. In 2003, morphological and genetic evidence suggested a southeast Asian origin of domesticated pigs (*Sus scrofa*), whereas in 2005, new genetic data indicated multiple origins of domestication of pigs across Eurasia (**Figure 33.3, p. 810**).

Worldwide, domestication of aquatic species has lagged behind that of terrestrial ones. Although there are about 180 species of domesticated freshwater animals, about 250 species of marine animals, and about 19 species of marine plants, all were domesticated in the last 1000 years, and most in the last 100 years (**Figure 33.4, p. 810**).

33.1b How Long Did Domestication Take? Archaeological Evidence

The time it takes to progress from harvesting tended wild crops to cultivating them and then to domesticating them varies with species and situation. When there are clear morphological or chemical differences between cultivated and domesticated stocks, determining the place and time of domestication is possible. Wheat is an example of such a morphological change. Wheat and other cereal crops are grasses that disperse their seed explosively by shattering. The process of

Figure 33.3

Origins of domestication of pigs.
Mitochondrial DNA obtained from pigs indicates 14 clusters of related lineages, each identified by a different colour. The geographic relationships are shown with the phylogeny. Pigs were domesticated in numerous centres.

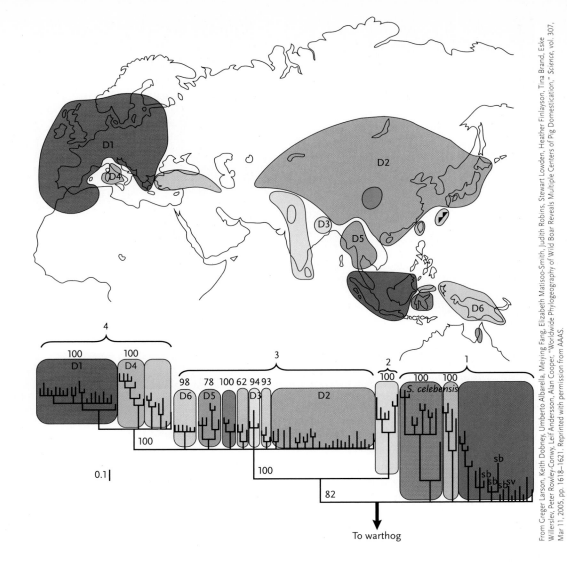

From Greger Larson, Keith Dobney, Umberto Albarella, Meiying Fang, Elizabeth Matisoo-Smith, Judith Robins, Stewart Lowden, Heather Finlayson, Tina Brand, Eske Willerslev, Peter Rowley-Conwy, Leif Andersson, Alan Cooper, "Worldwide Phylogeography of Wild Boar Reveals Multiple Centers of Pig Domestication," *Science*, vol. 307, Mar 11, 2005, pp. 1618–1621. Reprinted with permission from AAAS.

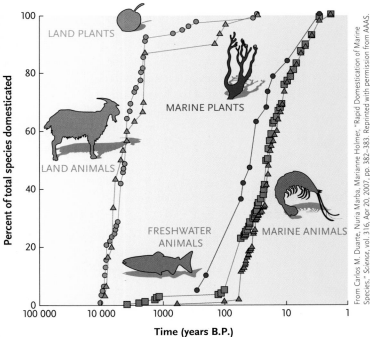

From Carlos M. Duarte, Nuria Marba, Marianne Holmer, "Rapid Domestication of Marine Species," *Science*, vol. 316, Apr 20, 2007, pp. 382–383. Reprinted with permission from AAAS.

Figure 33.4

Most land species were domesticated much earlier than most aquatic ones.

domestication of wheat and cereal crops meant developing stocks that do not shatter (indehiscence) from stocks that shattered (dehiscence) **(Figure 33.5, p. 811)**. Ripe indehiscent grains are easily gathered (harvested) compared with dehiscent ones that naturally scatter. Indehiscence in wheat results from a naturally occurring mutation.

But, in addition to selecting indehiscent stock, the early farms also had to select for plants whose seeds did not go through a period of dormancy. This change would allow repeated sowing of crops when conditions were appropriate. Material recovered from sites in northeastern Syria and Turkey has been radiocarbon dated and shows that wild varieties of wheat were cultivated for at least 1000 years before domestication **(Figure 33.6)**. When sexual reproduction is involved in the breeding process, the time to domestication is partly determined by life cycle, so finding stock that can self-fertilize can accelerate the domestication process.

When organisms reproduce asexually, domestication may occur more rapidly. Common figs (*Ficus*

Salicylic Acid

Figure 1
The molecular structure of salicylic acid.

Salicylic acid
2-OH-C₆C₄CO₂H

The precursor of the main active ingredient in aspirin is salicylic acid, which is obtained from the bark of willow trees (*Salix* species). Over 2500 years B.P., Chinese medical practitioners used an extract of willow bark to relieve pain and fever. The same kinds of extracts were used in medicine as practised in Greece and in Assyria. In Iceland 500 years ago, willow bark extracts were used to treat the symptoms of colds and headaches. Willow extract was widely used among First Nations people in North America, who commonly used it to stanch bleeding. They also used the supple willow twigs in other applications, from snares for catching mammals to nets for catching fish. This is not an example of domestication, but it demonstrates how the spread of traditional knowledge about plants and their products among peoples is pervasive. For more information about salicylic acid, see Chapter 38, p. 932.

a. **b.**

M.B. Fenton

M.B. Fenton

Figure 33.5
Some plants readily shed ripe seeds from the inflorescence; these plants show dehiscence. Indehiscence is the propensity to hold seeds. These two herbarium specimens, **(a)** bottle brush grass (*Elymus hystrix*) and **(b)** riverbank wild rye (*Elymus riparius*), illustrate dehiscence and indehiscence, respectively.

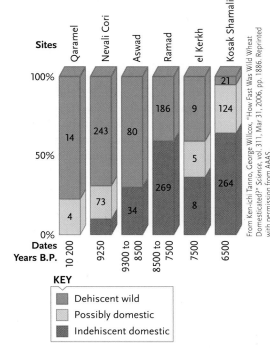

From Ken-ichi Tanno, George Willcox, "How Fast Was Wild Wheat Domesticated?" *Science*, vol. 311, Mar 31, 2006, pp. 1886. Reprinted with permission from AAAS.

Figure 33.6
Timing of domestication of wheat. Data from archaeological digs at six locations in the Middle East demonstrate the transition from wild (dehiscent) to domesticated (indehiscent) wheat from 10 200 to 6500 years B.P.

KEY
- ▨ Dehiscent wild
- ▫ Possibly domestic
- ▩ Indehiscent domestic

carica var. *domestica*) are gynodioecious and provide an example of more rapid domestication. In parthenocarpic female figs, ovaries develop without pollination and fertilization. Parthenocarpic figs can be propagated by cutting branches, sticking them in the ground, and waiting for them to grow into trees. When figs reproduce sexually, symbiotic fig wasps (*Blastophaga psenes;* **Figure 33.7**) serve as pollinators. The absence of access holes for wasps in fossil figs allows biologists to recognize parthenocarpic figs and date early incidences of fig domestication. At one site in the lower Jordan Valley (Middle East), parthenocarpic figs date to between 10 500 and 11 400 years B.P., perhaps preceding the domestication of cereal crops by about 1000 years.

CEFE–UMR 5175. Photo courtesy of Finn Kjellberg.

Figure 33.7
Female flowers on some fig trees (*Ficus carica*) are fertilized by symbiotic wasps, *Blastophaga psenes*.

Figure 33.8

In Ghana (West Africa), fire is still used in slash and burn to clear the underbrush from an area where forest trees have been felled.

M.B. Fenton

33.1c In Which Habitats Did Domestication Occur?

The transition from nomadic hunters and gatherers to people living more localized lives in more permanent dwellings appears to have been a prelude to cultivation and domestication. These changes meant that people would have been available to care for their "crops," whether grown from seeds or from parthenocarpic plants, and whether cultivated or domesticated.

Although controlled burning has been documented at many sites in the last 10 thousand years, there is evidence of it 50 thousand to 55 thousand years ago at sites near Mossel Bay in South Africa, and it still occurs today **(Figure 33.8)**. Archaeological evidence indicates that some humans were increasingly using some plant resources and using local burning to increase productivity. The increased use of plant resources and fire occurred during a period of harsh environmental conditions. These changes in human behaviour coincided with the appearance of more sophisticated tools, the use of marine organisms as food, and the first use of ochre for decoration. These modifications suggest differences in human behaviour that may have assisted the emergence of domestication.

Were changes in habitat associated with domestication? Evidence from pollen shows that from 7500 years B.P. in the lower Yangtze region of China, people used fire to clear alders, which are small woody bushes (*Alnus* species). This element in the process of domestication is called *niche construction* or *ecosystem engineering,* modifying the environment and setting the stage for domestication. At the lower Yangtze Neolithic site, people used fire first to prepare and then to maintain sites in lowland swamps, where they cultivated rice **(Figure 33.9)**. This region in China was a major centre of rice domestication. The evidence suggests that rice cultivation began in coastal

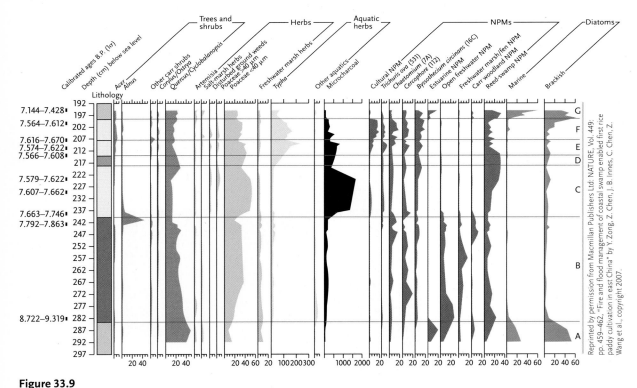

Reprinted by permission from Macmillan Publishers Ltd: NATURE, Vol. 449. pp. 459–462, "Fire and flood management of coastal swamp enabled first rice paddy cultivation in east China" by Y. Zong, Z. Chen, J. B. Innes, C. Chen, Z. Wang et al., copyright 2007.

Figure 33.9

The climatic setting for rice domestication at Kuahugiao in China. Shown here are generalized stratigraphic units and their associated pollen (shown in colours) and microscopic charcoal (black) that indicate conditions of climate and habitat. The data support the use of fire to establish favourable conditions for growing rice. NPMs are nonpollen microfossils that provide paleoecological data. The increases in charcoal, grasses (Poaceae), and reed-swamp microfossils in phases C, E, and F (right edge) indicate the use of fire to establish favourable conditions for growing rice.

wetlands in an ecosystem vulnerable to coastal change. This system was very fertile and productive and used for at least 200 years before the land was inundated by sea water.

Meanwhile, at sites on the coast of Peru, occupied between 3800 and 3500 years B.P., people ate marine organisms as the main animal food, combined with cultivated plants (squashes, *Cucurbita* species; beans, *Phaseolus lunatus* and *Phaseolus vulgaris*; peppers, *Capsicum* species; jicama, *Pachyrhizus tuberosus*) and wild plants (guava, *Psidium guajava*; lacuma, *Lucuma bifera*; and pacay, *Inga feuillei*). Cotton was an important crop used for making fishing tackle and clothing. The findings from these Peruvian sites and many other sites around the world suggest a progression toward domestication, including the range of foods consumed, the development of more sophisticated tools, and the use of materials from plants and animals as tools, as well as in food. Domestication, therefore, involved the spread among peoples of the practice of using plants and animals to advantage.

33.1d Abu Hureyra on the Euphrates Is an Example of a Setting for Domestication

The prehistoric settlement of Abu Hureyra (a recent photograph is shown in **Figure 33.10**) on the south side of the Euphrates River (35° 52 N, 38° 24 E) about 130 km from Aleppo (a modern Syrian city) also illustrates progression toward domestication. The first habitations that we know of in Abu Hureyra date from about 12 thousand years ago. Its population was estimated at 100 to 200 people who lived in semi-subterranean pit dwellings clustered together on a low promontory overlooking the river. By 7000 to 9400 years ago, 4000 to 6000 people lived at the same site, now in multiroomed family dwellings made of mud and brick. This settlement was built over the remains of the earlier one.

People living at Abu Hureyra about 12 thousand years B.P. ate the fruits and seeds of over 100 species of local plants as well as local animals such as gazelles. Many of the plants and animals appear to have come from the adjoining oak-dominated park woodland. It appears to have been a time of plentiful food. The situation changed, however, and by 9400 years ago, the climate was cooler and drier, and the people relied more on cultivated plants and less on wild ones. By this time, there was little evidence of use of plants from the oak-dominated parkland, which by then was at least 14 km from the settlement. These changes were evident in pollen records and in plant and animal remains associated with the dwellings. The climate change likely triggered the start of cultivation of foods that could serve as caloric staples. Despite the changing climate and the focus on fewer food staples, the human population at the site dramatically increased.

33.2 Why Some Organisms Were Domesticated

We can surmise, perhaps accurately, that securing a sustainable food supply provided an initial motivation for cultivation and domestication. It is certainly true that cultivated plants such as beans, squash, corn, rice, and cereal grains all help feed many, many people worldwide. People eat different parts of plants, from flowers and fruits to seeds, leaves, stems, roots, and tubers. Plants may be a source of energy (calories), or their products may be used to enhance flavours, to control and repel pests, or as medicines. Still others, such as the bottle gourd (*Lagenaria siceraria*), are used as containers. Domesticated cereal grains have been derived from variants (local varieties, sometimes known as breeds, cultivars, or landraces) with four important features: (a) nonshattering, (b) large seeds, (c) self-compatibility, and (d) no required dormancy. This suite of characters makes them valuable because they can be readily fertilized, harvested, and planted whenever conditions permit. Domesticated animals provide food, but many are also used as a source of labour. The following are examples of four very different domesticated species and how people use them.

33.2a Cattle

Cattle were among the first of the large herd mammals to be domesticated, at least 9000 years B.P. One theory proposes that the domesticators of cattle were sedentary farmers, not nomadic hunters. Some anthropologists maintain that a religious motivation was behind the domestication of cattle because the curve of their horns resembled the crescent of the Moon and hence the mother-goddess. Imposing horns were particularly prominent in some male *Bos primigenius* (called *urus*), the apparent Pleistocene ancestor of domesticated cattle. Whatever the original impetus, today there are two basic stocks of cattle (Figure 33.10), the humped *Bos indicus* and the humpless *Bos taurus*. Cattle provide us with labour, milk, meat, hides, and blood, and in some societies, they are symbols of wealth. At the root of the domestication of cattle are some biological realities: they are relatively docile animals that live in herds and can be useful in many ways.

Figure 33.10
Today there are two varieties of domesticated cattle: **(a)** the humped *Bos indicus* and **(b)** the humpless *Bos taurus*.

33.2b Honeybees

Domestic honeybees provide us with honey and pollination services. Steps to domestication of honeybees included changes in their behaviour compatible with large population size in hives. Features of honeybees that make them suitable for domestication are their colonial and food-storing behaviours, unlike other species of bees that are solitary and do not store honey. The changes could have involved hygiene, aggression, and foraging. Although everyone recognizes honey as a product of bees, the service provided by bees is often overlooked. In 2000 in the United States, it is estimated that bees contributed about U.S.$14.5 billion through their role as pollinators. Plants such as alfalfa, apples, almonds, onions, broccoli, and sunflowers are exclusively pollinated by insects, usually more than 90% by honeybees. Many beekeepers earn significant income by moving their bees from location to location, thus providing a mobile pollinator service for farmers. Declines in populations of honeybees have serious economic implications throughout the world, but many conservationists are also concerned about the impact of populations of honeybees on native bee species.

33.2c Cotton

At least four species of cotton **(Figure 33.11)** have been domesticated: two diploid species from the Old World (*Gossypium arboreum, Gossypium herbaceum*) and two tetraploids from the New World (*Gossypium hirsutum* and *Gossypium barbadense*). The domestication events appear to have been independent, and one site on the Mexican gulf coast of Tabasco shows evidence of people growing cotton by 4400 years B.P. Cotton seeds were a source of oil, whereas fibre was and is still used in applications ranging from clothing to implements such as ropes and nets.

Figure 33.11
Cotton, *Gossypium herbaceum*, showing flowers and cotton bolls.

33.2d Yeast

Strains of the yeast *Saccharomyces cerevisiae* have been used by people in bread-making beginning at least 6000 years B.P. Evidence of this is in archaeological finds in Egypt, indicating the presence of bakeries and breweries, two yeast-based operations. Analysis of 12 DNA microsatellites obtained from 651 strains of *S. cerevisiae* collected at 56 locations around the world revealed 575 distinct genotypes. Yeasts associated with

PEOPLE BEHIND BIOLOGY 33.2

Richard Keith Downey, Research Scientist, *Agriculture and Agri-Food Canada*

Richard Keith Downey was born in 1927 in Saskatchewan and attended the University of Saskatchewan and Cornell University. He is best known for his work on rapeseed. In 1928, rapeseed had been brought to Shellbrook, Saskatchewan, by Fred and Olga Solovonuk when they immigrated to Canada from their native Poland. They had brought seeds of *Brassica rapa*, although *Brassica napus* from Argentina was the main traditional source of rapeseed oil. During World War II, rapeseed was grown as a source of industrial lubricants because erucic acid and glucosinolates in the oil precluded its use as food for people and livestock.

Downey was the leader of a team of researchers at Agriculture and Agrifoods Canada that developed canola (the name comes from "Canada" and "oil"), an edible, high-value crop that is high in proteins. Canola is well known as a vegetable oil, but the meal derived from the plants is the source of protein, making canola useful as animal feed. The team's work resulted in a crop that went from 2600 to 4.25 million hectares under cultivation in Canada. Canola is now grown around the world, and the high crop is used in cooking oils and feed for livestock.

Downey and his team developed the *half-seed* method, partitioning a seed so that half could be tested for composition using gas–liquid chromatography. The researchers could then select and germinate the half-seeds with the most promising features. Using this approach, Downey and his team developed 18 varieties of canola and several of table mustard.

This work had a huge impact on the economy of Canada in general and the prairies in particular, as well as an influence on food availability throughout the world.

Downey is also known for a program that introduced children in elementary school to the scientific method. The approach involved using canola seeds that had travelled into space aboard the space shuttle *Columbia* in 1996. Children in over 2000 classrooms in Canada received space seeds along with control seeds and germinated them. Downey helped analyze the results of the experiments. Space seeds germinated faster and at higher rates than control seeds and grew more rapidly. In this way, Downey introduced elementary students to plant science and the space environment.

---

bread were intermediate between wild types and those used in making beer and wine, whereas those used in the production of rice wine and sake were more similar to those used for beer. About 28% of the genetic variation in yeast genotypes was associated with geographic location. The basal group of these 12 DNA microsatellites was samples from Lebanon, suggesting a Mesopotamian origin and a spread of yeast types along the Danube River and around the Mediterranean. Different strains of yeast have different capacities for maltose fermentation. Commercial bakers' yeast strains are more effective at maltose fermentation than nonindustrial strains. Domesticated yeast makes important contributions to providing humans with food and drink and supports lucrative industries.

33.2e Rice

Rice (*Oryza sativa*) is one of the world's most important food crops, and its domestication depended on the change from dehiscence (shattering) to indehiscence (nonshattering). Domesticated rice is derived from two wild species: *Oryza rufipogon* and *Oryza indica*. In 2006, Changbao Li and his colleagues reported that three quantitative trait loci (QTL) in F_1 hybrids between these two species were responsible for a reduction of grain shattering (dehiscence) in rice. Specifically, *sh3*, *sh4*, and *sh8* were involved, with *sh4* explaining 64% of the phenotypic variance. In the wild species, *sh4* was

dominant and caused the shattering. The genetic changes in *sh3*, *sh4*, and *sh8* affected normal development of the abscission layer, explaining the change to indehiscence **(Figure 33.12)**. We can now better understand the genetics of changes associated with the domestication of rice.

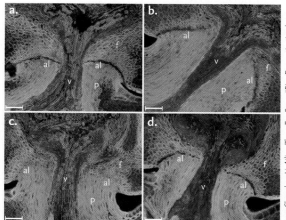

From Changbao Li, Ailing Zhou, Tao Sang, "Rice Domestication by Reducing Shattering," *Science*, vol. 311, Mar 31, 2006, pp. 1936–1939. Reprinted with permission from AAAS.

Figure 33.12

Rice dehiscence and indehiscence. Under a fluorescence microscope, a longitudinal section of the junction between the rice flower and its pedicel shows a complete abscission layer (al in **(a)**) and an incomplete one (al in **(b)**). In these figures, f = flower side; p = pedicel side; and v = vascular bundles. *Oryza nivara* is shown in **(a)**, *Oryza sativa* in **(b)**, *Oryza sativa japonica* in **(c)**, and transformed *O. sativa japonica* in **(d)**.

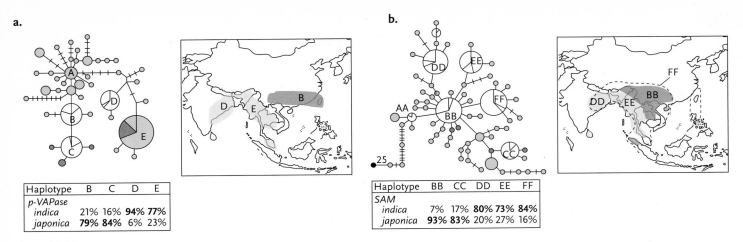

a.

b.

Haplotype	B	C	D	E
p-VAPase				
indica	21%	16%	**94%**	**77%**
japonica	**79%**	**84%**	6%	23%

Haplotype	BB	CC	DD	EE	FF
SAM					
indica	7%	17%	**80%**	**73%**	**84%**
japonica	**93%**	**83%**	20%	27%	16%

Figure 33.13

Geography of rice domestication. Genetic haplotype information allows identification of major domestication regions for rice. The haplotype network of the neutral nuclear p-VAPase region is shown in **(a)**, compared with **(b)**, the haplotype of the functional nuclear SAM region. In this diagram, orange represents *Oryza japonica;* blue, *Oryza indica* (Aus cultivar); and yellow, *O. indica*. Lines joining haplotypes represent mutational steps.

Londo, J. Y-C Chiang, K-H Hung, T-Y Chiang, and B. Schaal. (2006) "Phylogeography of Asian wild rice, Oryza rufipogon reveals multiple independent domestications of cultivated rice, Oryza sativa," PNAS, 103: 9578–9583. Copyright 2006 National Academy of Sciences, U.S.A.

These changes occur in the two major varieties of domesticated rice, *Oryza sativa japonica* and *Oryza sativa indica*. Jason Londo and his colleagues used DNA sequence variation to demonstrate multiple independent domestication events. Although *O. sativa indica* was probably domesticated in eastern India, Myanmar, and Thailand south of the Himalayan mountains, *O. sativa japonica* was domesticated in southern China **(Figure 33.13)**.

33.2f Wheat

Other major domestications appear to stem from single domestication events, including wheat, barley **(Figure 33.14)**, and corn. DNA fingerprinting was used by Manfred Heun and his colleagues to identify the Karacadağ mountains **(Figure 33.15)** in today's Turkey as *the* site of domestication of einkorn wheat. Einkorn wheat is derived from the wild *Triticum monococcum boeoticum*. Einkorn wheat is selfing and nonshattering and has a firm stalk, heavier seeds, and denser seed masses than its progenitor (see Figure 19.16, Chapter 19), all contributing to the ease and efficiency of harvest. The genetic changes between einkorn wheat and the wild stock from which it was derived were relatively minor, and domestication probably occurred in a short period of time.

Archaeological evidence reveals that einkorn wheat spread rapidly throughout the immediate area and beyond. The same single domestication event and rapid spread of the new crop also appear to be true of barley **(Figure 33.16b)**; *Hordeum vulgare* derived from *Hordeum spontaneum*). More recent wheat domestication is discussed in "Marquis Wheat," p. 819.

33.2g Lentils

Lentils (*Lens culinaris* from *Lens orientalis;* Figure 33.16a) were more difficult to domesticate than wheat or rice. Wild lentils are small plants, producing, on

Figure 33.14

A barley field near London, Ontario, illustrates the consistency of a monoculture, one plant dominating an area, in this case for artificial reasons.

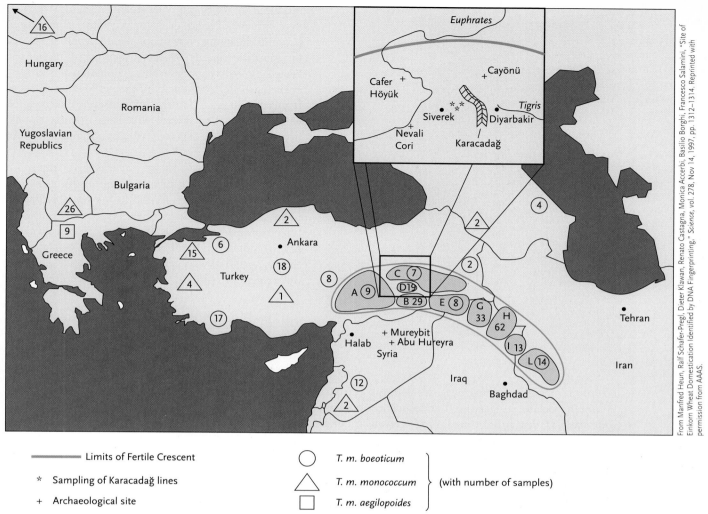

From Manfred Heun, Ralf Schafer-Pregl, Dieter Klawan, Renato Castagna, Monica Accerbi, Basilio Borghi, Francesco Salamini, "Site of Einkorn Wheat Domestication Identified by DNA Fingerprinting." *Science*, vol. 278, Nov 14, 1997, pp. 1312–1314. Reprinted with permission from AAAS.

─────── Limits of Fertile Crescent   ◯   *T. m. boeoticum*

*   Sampling of Karacadağ lines   △   *T. m. monococcum*   } (with number of samples)

\+   Archaeological site   ☐   *T. m. aegilopoides*

A–L: areas of wild *T. m. boeoticum* sampling in the Fertile Crescent

Figure 33.15

Site of wheat domestication. A phylogenetic analysis based on allelic frequency at 288 amplified fragment length polymorphism marker loci revealed that the progenitor of *T. m. boeoticum* came from what is now southeastern Turkey.

M.B. Fenton

Figure 33.16

Two important domesticated crops are **(a)** lentils (*Lens culinaris*) and **(b)** barley (*Hordeum vulgare*). In each case, individual grains are about 6 mm long.

average, 10 seeds per plant. Furthermore, these seeds go through a period of programmed dormancy. The combination of low yield and dormancy means that relatively few seeds germinate. Lentils could have been

cultivated only after a dormancy-free mutant had appeared. Archaeological evidence suggests that the first stages of lentil domestication (loss of dormancy) occurred in what is now southeastern Turkey and northern Syria, suggesting a single initial domestication event. The second phase was selection for strains that produced large numbers of seeds. This change may have occurred some distance (hundreds or even thousands of kilometres) from the original sites of domestication.

33.2h Corn

Corn (*Zea mays*), like wheat and barley, appears to have been domesticated at one location. Unlike wheat or barley, domestication of corn required drastic changes from the ancestral teosinte (*Zea mays parviglumis*; **Figure 33.17, p. 818**). Corn kernels do not dissociate from the cob, presumably reflecting changes associated with domestication (and analogous to indehiscence). Analysis of some of the most ancient inflorescences of

Figure 33.17
Teosinte and domesticated corn. Ears of domesticated corn (*Zea mays*, right) are larger than those of one species of its wild relative teosinte (*Zea diploperennis*, left). Crossing domesticated corn and teosinte produces intermediate forms (centre).

John Doebley

teosinte indicates that by 6250 years B.P. the kernels did not dissociate, suggesting that domestication was under way at that time. Corn appears to have been domesticated in the Central Balsas Valley of south central Mexico or in the Mexican State of Puebla at altitudes of 1000 to 1500 m. The deposits that contained the undissociated teosinte kernels also had the remains of squash (*Cucurbita pepo*). Although squash appears to have been domesticated in the same area, remains from a cave in Puebla and Oxa States in Mexico suggest that the people there were cultivating it by about 4000 years before corn. These data again demonstrate that some people were responsible for multiple domestication events. Corn, beans, and squash, the "three sisters" of early farmers in the New World, together provided different foods. The beans climbed on the corn stalks and provided nitrogen fixation services (see Chapter 31), while the squash leaves shaded out the weeds.

33.2i Grapes

The Eurasian grape (*Vitis vinifera sylvestris*), a dioecious plant, is widespread from the Atlantic coasts of Europe to the western Himalayas, and its fruits were often eaten by Paleolithic hunter–gatherers. Domestication of *V. vinifera* involved the selection of hermaphroditic genotypes that

produced larger, more colourful fruit, along with the development of techniques for vegetative propagation. Grafting, attaching parts of one plant to another, further increased the capacity for vegetative propagation. Domesticated grapes have a higher sugar content than wild ones, ensuring better fermentation, greater yield, and more regular production. Domestication of grapes is associated with the production of wine, which also required storage in containers made of pottery, which appeared only about 10 500 years B.P.

Genetic analyses suggest at least two important origins of grape domestication: one in the near East and the other in the region of the western Mediterranean. Many wine-grape cultivars from Europe can be traced to western Mediterranean stock, as can over 70% of the cultivars on the Iberian Peninsula. It is possible that the original wild stock of *V. vinifera* has vanished due to genetic contamination by various cultivars.

33.2j Plants in the Family Solonaceae

Solanaceous plants include species such as deadly nightshade (*Atropa belladonna*) that produce virulent poisons, as well as food species (tomatoes, *Solanum lycopersicum*; potatoes, *Solanum tuberosum*; and eggplant, *Solanum melongena*) that are staples of many meals worldwide **(Figure 33.18)**.

Potatoes originated in western South America, specifically in areas of what is now Chile. Today, local varieties (also known as breeds, cultivars, or landraces) of potatoes are adapted to the local conditions where they grow. These landraces form the *Solanum brevicaule* complex. Genotypes from multilocus amplified

M.B. Fenton

Figure 33.18
Solonaceae. Domesticated species include **(a)** capsicum peppers, **(b)** eggplant, **(c)** potato, and **(d)** tomato.

Marquis Wheat

Different strains of domesticated crops, such as wheat and corn, have different features that make them better suited to some areas (climates) than to others. But wheat crops provide both flour for bread and straw for thatching and animal bedding. Whereas Red Fife wheat matures in 130 days and produces 3270 kg·ha^{-1}, Hard Red Calcutta wheat matures in 110 days but yields only 1240 kg·ha^{-1}. In an effort to find a wheat variety that would grow well and produce a good yield in the Canadian Prairies, Sir Charles E. Saunders made extensive crosses and developed Marquis wheat by 1906. To do this involved selective breeding: a planned program of hybridization of different wheat varieties, rigid selection of the best available material, preliminary and final evaluations of the results from replicated trials, and extensive testing of the new varieties. In the 1880s, Dr. William Saunders (Sir Charles's father) had introduced and tested many strains of wheat, often from Russia and India.

Crosses that led to the emergence of Marquis wheat were mainly focused on Hard Red Calcutta and Red Fife. The products of the crosses were tested at stations near Agassiz, British Columbia; Indian Head, Saskatchewan; and Brandon, Manitoba. At the latter two locations, these two varieties differed by three weeks in reaching maturity.

Sir Charles is famous for the "chewing test" he used to test the products of the crosses. He observed that chewing allowed him to determine the elasticity of the gummy substance produced (gluten). Other tests included baking bread with flour ground from the different strains to ensure that the product was satisfactory.

In 1906, Marquis wheat emerged as the product of a cross between a Hard Red Calcutta female and a Red Fife male. The kernels were dark red and hard, medium in size, and short. Heads were medium in length and bearded. Marquis ripens a few days before Red Fife and produces flour that is strong and of good colour. Tested at Brandon in 1908, Marquis wheat was the earliest to ripen and yielded 4336 kg·ha^{-1}—the best among the strains compared. Note that in Ottawa, Marquis wheat was not as productive, yielding only 2522 kg·ha^{-1}, reflecting the effect of climate and conditions on yield.

---

fragment length polymorphisms were determined from 261 wild and 98 landrace samples. The resulting data suggest that potatoes (*S. tuberosum*) arose from one domestication event. Today, wild potatoes still occur in Chile, along with eight cultivar groups of potatoes—some diploid and some triploid, others tetraploid or pentaploid (see Chapter 19). They vary noticeably in leaf shape, floral patterns, and tuber colour. Domesticated potatoes have been selected for short stolons, large tubers, and various colours of tubers from white to yellow to black. After being introduced to Europe, potatoes became an important food staple, supplanting traditional crops such as grains, and millions of people were adversely affected when a blight caused widespread failure of the potato crop. The blight was caused by the oomycete protist *Phytophthora infestans*. The "potato famine" in Ireland had huge social repercussions: many people died, and others emigrated to Canada, the United States, and Australia.

Tomatoes (*Solanum lycopersicum*) were domesticated from plants that grew in South and Central America, but there is considerable debate about when they were domesticated. Two modern forms, wild cherry tomatoes and currant tomatoes, were recently domesticated from stock native to eastern Mexico. Eaten raw or cooked, tomatoes come in different sizes, shapes, and colours. Different strains grow well in a variety of situations, meaning that tomatoes can be grown in many different climate zones. When grown in greenhouses, tomatoes are often pollinated by resident bumblebees, although many cultivars are self-pollinating.

Eggplants include three closely related cultivated species: *Solanum melongena,* the brinjal eggplant or aubergine; *Solanum aethiopicum,* the scarlet eggplant; and *Solanum macrocarpum,* the gboma eggplant. All cultivated species are native to the Old World, with *S. macrocarpum* and *S. aethiopicum* having been domesticated in Africa. The origin of the brinjal eggplant is less certain, but it may have originated in Africa and been domesticated in India and southeast China. During the Arab conquests, it spread from there to the Mediterranean and today is cultivated around the world. Brinjal eggplants and tomatoes are autogamous diploids.

Chili peppers, *Capsicum* species, are another member of the Solonaceae that originated in the New World. Known as producers of capsaicin ("Molecule behind Biology," Box 45.2, Chapter 45), chili peppers are often used to spice food. Cultivation of *Capsicum* species was well advanced and widespread in the Americas by 6000 years B.P. Then, as now, they were used as condiments.

People use other members of this family as the source of hallucinogenic compounds. Notable examples are tobacco and species in the genus *Datura*. Jimson weed or locoweed (*Datura stramonium*) contains strong poisons, including belladonna alkaloids, atropine, and scopalamine. It grows in many parts of the world and has often been used as a hallucinogenic drug because one active ingredient interferes with neurotransmitters (see Chapter 44) and can induce violent hallucinations. The name "locoweed" is a useful, important warning.

Stepping outside the Solonaceae to the family Brassicaceae, species in the genus *Brassica* also include many varieties seen on dinner tables worldwide (see "Domesticated Plants in the Genus *Brassica*," p. 823).

33.2k Squash

In the family Cucurbitaceae, at least five species of squash (*Cucurbita;* **Figure 33.19**) were domesticated in the Americas before European settlers arrived, some of them at least 10 thousand years B.P. Squash, beans, and corn were the "three sisters," staple foods farmed by many First Nations peoples in the New World (see Section 33.2h). Genetic data obtained from an intron region of the mitochondrial *nad1* gene suggest that at least six independent domestication events occurred. *Cucurbita argyrosperma* appears to have been domesticated from *Cucurbita sororia,* a wild Mexican gourd that grew in the same general area of Mexico as teosinte. *Cucurbita moschata* was probably domesticated somewhere in lowland South America and *Cucurbita maxima* in the humid lowlands of Bolivia from *Cucurbita andreana.* The *Cucurbita pepo* complex seems to be derived from at least two domestication events, one in eastern North America and one in northeastern Mexico.

Many people are familiar with *C. pepo* as the pumpkin, but the species also includes summer squashes and zucchinis. *Cucurbita maxima* is the Hubbard and other winter squashes, which also include some *C. pepo* and *C. moschata.* The diversity of these cultivars is astonishing. The intraspecific variations provide another example of the difficulty of applying the species concept to the diversity of life (see Chapter 19).

Figure 33.19
Many cultivars of squash, *Cucurbita* species, are New World domesticates, some first domesticated in southwestern Mexico at least 10 thousand years B.P.

33.2l Dogs

Behaviour provides clues to important aspects about the domestication of dogs. Researchers assessed the abilities of dogs, chimps, and wolves to read human signals indicating the location of food. Even young puppies with little human contact were more skillful at these social cognition skills than chimpanzees and wolves. Interspecific communication appears to have been strongly selected for during the domestication of dogs, building on the evolutionary history of social skills associated with cooperative hunting inherited from their wolf ancestors.

Dogs were among the first animals to have been domesticated, presumably to help people with hunting. Genetic, behavioural, and morphological evidence indicates that dogs were derived from wolves. The earliest morphological evidence suggests domestication of dogs by 14 500 years B.P. whereas mtDNA data suggest a date of 15 000 years B.P. Genetic data suggest an East Asian origin for dogs. Other mtDNA data from specimens in Latin American and Alaska indicate that dogs crossed into North America via the Bering Land Bridge, with people producing a group (clade) of dogs unique to the New World. The mtDNA data imply that either European colonists or native aboriginals actively prevented dogs that they brought with them from interbreeding with dogs already present in the New World.

33.2m *Salmo salar*, Atlantic Salmon

Atlantic salmon naturally occurs around the North Atlantic (locations in North America, Greenland, and Europe), but intensive fishing has reduced its natural stocks to the brink of extinction in some areas. However, Atlantic salmon have also been introduced to many sites around the world, from Jordan and Greece to Australia, New Zealand, Chile, Argentina, Brazil, and the Falkland Islands. There are both landlocked natural and introduced freshwater populations of Atlantic salmon. This species has been a traditional target of subsistence, sport, and commercial fishing and more recently the focus of aquaculture operations. The farmed fish have been selectively bred; thus, they are domesticated. Farmed fish are larger and more aggressive than those from wild stock, and they mature later.

Aquaculture operations have proved to be very lucrative, leading to a proliferation of these facilities in many areas. But some aquacultural operations have negative impacts. For example, escaped fish are thought to interbreed with local species (on the west coast of North America), threatening their genetic survival. This threat may be reduced by using sterile triploid Atlantic salmon for aquaculture. Sterile triploids can be mass produced, making it relatively feasible to use them in many areas. But concerns about the productivity and survival of triploid fish compared with diploid individuals have slowed the spread of their use.

Aquaculture can bring other problems. Sea lice, such as *Lepeophtheirus salmonis,* are parasitic copepod crustaceans that can cause serious problems for salmon aquacultural facilities. Recurrent sea lice infestations of aquacultured populations have spread to and decimated some wild salmon populations. Infestations by sea lice originating from cultured salmon have also caused a 99% collapse of some wild populations of pink salmon (*Oncorhynchus gorbuscha*) in coastal British Columbia.

The scale of aquaculture operations involving Atlantic salmon is astounding. In 2006, in Nova Scotia alone, 35 thousand tonnes of Atlantic salmon were produced by aquaculture. Atlantic salmon also dominate farmed stock in British Columbia. The scale of production has wide environmental, social, and economic implications for human nutrition and employment.

33.2n Some Organisms Were Domesticated for More Than One Use

Some animals and plants provide more than one crop. Cotton, as noted above, provides oilseed and fibre. Cattle are sources of meat, milk, blood, hides, and labour. Sheep provide wool, milk, meat, and hides. Sheepskins with the wool attached are used to make clothing.

33.2o Some Organisms Were Cultivated but Not Domesticated

People cultivate many species that have not been domesticated because there is no evidence of selective breeding. Mushrooms are examples because they have been cultivated and used as a source of protein for several thousand years without selective breeding of specific lines (= domestication).

Ostriches (*Struthio camelus*) are ranched (= cultivated) for their meat, hides, feathers, and eggs, but they are not domesticated. Crocodiles (*Crocodylus* species) are also ranched for their hides and meat. Oysters (*Pinctada fucata*) and other species of molluscs have been cultivated for hundreds of years mainly for pearls **(Figure 33.20).**

Figure 33.20
Pearls for sale in the Pearl Market in Beijing, China.

Other animals, such as *Python regius* (ball pythons), are bred and sold to snake fanciers. Breeders may select individuals with specific traits in their breeding programs, technically making the animals domesticated because the definition does not speak to use.

STUDY BREAK

1. Distinguish between cultivation and domestication. Give examples.
2. How has domestication affected K, the carrying capacity?
3. Use two of the examples in this section to compare the timing, location, and path of domestication. Be sure to explain how the domesticated organisms are used.
4. How did the domestication of rice, wheat, barley, and lentils differ from that of squash and potatoes?
5. Use an example to show how domestication of animals depends on their behaviour.

33.3 Yields

In his book *The Upside of Down: Catastrophe, Creativity and the Renewal of Civilization,* Thomas Homer-Dixon calculated the amount of energy it would have taken to build the Coliseum in Rome. He estimated that it would have taken 44 billion kilocalories of energy: 34 billion for oxen and 10 billion for human workers (assuming that both worked 220 days a year for 5 years and that the humans received 12 500 $kJ \cdot day^{-1}$). The Roman workers would have eaten grain, mainly wheat, as well as legumes, vegetables, wine, and a little meat. The oxen would have been fed hay, mainly alfalfa, as well as legumes, millet, clover, tree foliage, and wheat chaff. Records from the time indicate a yield of wheat of about 1160 $kg \cdot ha^{-1}$ and alfalfa of about 2600 $kg \cdot ha^{-1}$. Wheat delivers $1.0 \times 10^7 \ kJ \cdot ha^{-1}$ and alfalfa $1.6 \times 10^7 \ kJ \cdot ha^{-1}$. Growing wheat would have required 58 days of slave labour per hectare per year. If farmers had had to pay labourers, the cost of production would have been higher.

Based on these data, Homer-Dixon calculated that building the Coliseum would have required the wheat grown on 19.8 km² of land and the alfalfa on 35.2 km². At its peak around 1 and 2 CE the population of the city of Rome was 1 million, and that number would have required the food produced on 8800 km² of land, equivalent to the area of Lebanon today. Much of the wheat that fed Rome came from North Africa, as well as Sicily and other parts of Italy.

On average, one adult human needs 8300 $kJ \cdot day^{-1}$ or $3.0 \times 10^6 \ kJ \cdot yr^{-1}$ to maintain a stable body mass. This assumes a much lower level of exercise and

Figure 33.21
The field marked with the * is 20 ha and, if planted in wheat, should yield 13 200 kg in southwestern Ontario. Note that the area is also being used to harvest wind energy.

M.B. Fenton

10 000 kg·ha^{-1} (Table 33.1). But, even without considering the costs of transportation (from farm to market), the data illustrate the impact of the cost of fuel. Spraying, planting, and trucking all require diesel fuel, so any change in the price of this commodity influences the costs of farm operations.

Terrain influences the level of technology that can be used in farming practices. Small terraced plots **(Figure 33.23, p. 824)** must be worked by hand or with

physical exertion than the Roman worker in Homer-Dixon's calculations. If people were to meet their caloric demands from wheat alone, and if wheat delivers about 8700 kJ·kg^{-1}, at 8300 kJ·day^{-1}, each person would need to consume about 350 kg of wheat a year. In 2007 in southwestern Ontario, a wheat yield of 6600 kg·ha^{-1} meant that 1 ha of land would support 18.8 people for a year. A city of 50 thousand, eating only wheat, would need the wheat produced on 2660 ha **(Figures 33.21** and **33.22).**

The data in **Table 33.1,** demonstrate that it takes energy input to generate energy, in this case food. So variations in crop yields noted at the beginning of this chapter have huge repercussions. From a farmer's perspective, the difference in farm income would be substantial if the yield of corn were 1600 kg·ha^{-1} versus

Table 33.1	Balancing Cost and Yield, a Farm in Southwestern Ontario (Costs as of Autumn, 2007)	
Crop/Material/Process	Yield kg·ha^{-1}	C$·ha^{-1}
Corn	10 000	1650
Soy beans	3 500	1300
Wheat	6 600	900
Costs		
Seed		150
Fertilizer		125–225
Spray		50–100
Planting		38
Labour		40–190
Combine		100
Trucking		25
Crop insurance		20
Field rental		325

a.

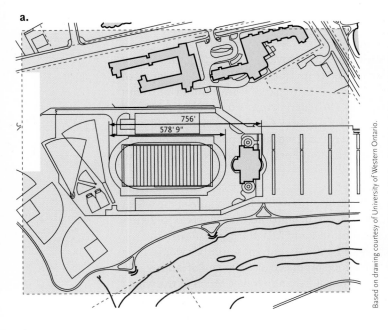

756'
578' 9"

Based on drawing courtesy of University of Western Ontario.

b.

M.B. Fenton

Figure 33.22
Envisioning 20 ha. Two views of the same area: a drawing **(a)** in which the coloured area represents 20 ha and an aerial photograph **(b)** showing the same football stadium on the campus of the Western University, in London, Ontario.

Domesticated Plants in the Genus *Brassica*

Some plants have provided humans with an embarrassment of riches. Imagine plants in one genus (*Brassica*) whose flowers, roots, stems, leaves, and seeds are all important food crops **(Figure 1)**. Foods from these vegetables are high in vitamin C and soluble fibre. They contain a rich mixture of nutrients, including some (diindolylmethane, sulforaphane, and selenium) thought to have anticancer effects.

A sample of an all-*Brassica* meal could include broccoli, cauliflower, Brussels sprouts, kale, and cabbage, all variations of one species, *Brassica oleracea*. You could add rutabaga (*Brassica napus*) along with turnip and rapini (*Brassica rapa*) to the menu and use canola oil (see "People behind Biology," Box 33.2) and mustard (*B. rapa, Brassica carinata, v elongata, Brassica juncea, Brassica nigra, Brassica ruprestris*) to dress the parts of the meal presented as salad.

Nuclear restriction fragment length polymorphisms (RFLPs) obtained from 10 different *Brassica rapa*, 9 cultivated types of *B. oleracea*, and 6 other species in *Brassica* and related genera suggest two basic evolutionary pathways **(Figure 2)** for diploid species: one that gave rise to *Brassica fruticulosa, B. nigra*, and

M.B. Fenton

Figure 1

***Brassicas** in our diets. Shown here are cultivars of brassicas that commonly appear in people's diets, including **(a)** radishes, **(b)** Brussels sprouts, **(c)** cauliflower, **(d)** broccoli, **(e)** turnip, **(f)** rutabaga, **(g)** mustard, **(h)** cabbage, and **(i)** canola oil.*

Sinapis arvensis (*Brassica adpressa* is a close relative), and the other to *B. oleracea* and *B. rapa* **(Figure 3)**. *Raphanus sativus* and *Eruca sativa* appear to be intermediate between the two lineages (see Figure 2). Europe and East Asia appear to have been centres of domestication for *Brassica* species. The related *Arabidopsis* is an important experimental tool used in understanding the genetics and selection of desirable traits in *Brassica*.

Figure 2

The shortest phylogenetic tree showing the relationships between different species of Brassica, *including cultivars and wild species.* A1–A5 B. rapa *cultivars:* 1 = *flowering pak choi;* 2 = *pak choi;* 3 = B. narinosa; 4 = *Chinese cabbage;* 5 = *turnip;* A6–A10 B. rapa *wild:* Bal = B. alboglabra; Bc 1–4 = B. cretica; Bd = B. drepanensis; Bia = B. incana; Bis = B. isularis Bma = B. macrocarpa; Bmo = B. montana; Bol = B. oleracea; Br = B Rupestris; BBv = B. villosa; C2–C23 B. oleracea *cultivars:* 2 = *broccoli;* 3 = *broccoli (packman);* 4 = *cabbage;* 8 = *Portuguese tree kale;* 12 = *Chinese kale;* 15 = *kohlrabi;* 19 = *borecole;* 23 = *cauliflower;* Bf = B. fruticulosa; Bn = B. nigra; Bt = B. tournefortii; De = Diplotaxis erucoides; Es = Eruca sativa; Rs = Raphanus sativus.

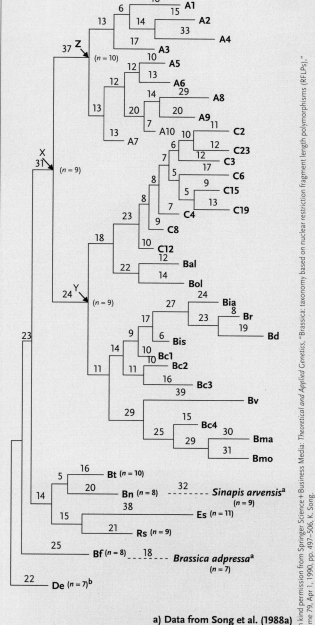

With kind permission from Springer Science+Business Media: *Theoretical and Applied Genetics*, "Brassica: taxonomy based on nuclear restriction fragment length polymorphisms (RFLPs)," volume 79, Apr 1, 1990, pp. 497–506, K. Song.

a) Data from Song et al. (1988a)
b) Outgroup

(Continued)

the help of animals. Large expanses of relatively flat land can be worked effectively and efficiently with machinery **(Figure 33.24, p. 824)**, increasing the energy consumption associated with farming but also the yield. Modern agriculture relies on energy from petroleum products, so increases in the price of oil can undermine food production, especially if crops originally grown for food are used in the production of biofuels.

STUDY BREAK

1. How must crop yield be balanced against the cost of achieving it? Does this apply to a kitchen garden (as opposed to a functioning farm)?
2. What factors were fundamental in allowing the city of Rome to have a population of over 1 million people in 2 CE ?

Domesticated Plants in the Genus *Brassica* (Continued)

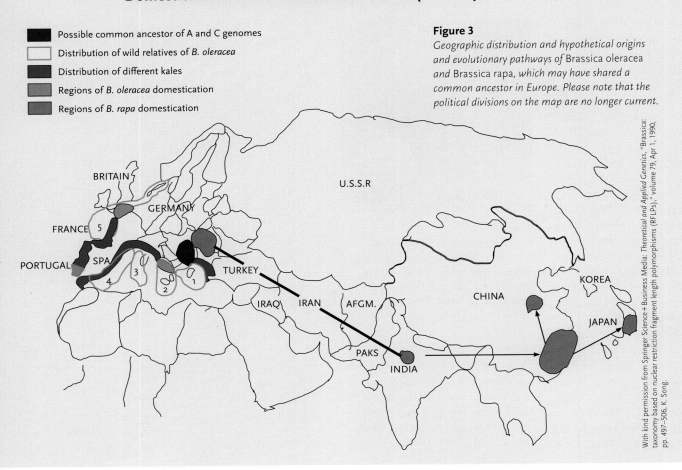

- ⬛ Possible common ancestor of A and C genomes
- ⬜ Distribution of wild relatives of *B. oleracea*
- ⬛ Distribution of different kales
- ⬛ Regions of *B. oleracea* domestication
- ⬛ Regions of *B. rapa* domestication

Figure 3

Geographic distribution and hypothetical origins and evolutionary pathways of Brassica oleracea and Brassica rapa, which may have shared a common ancestor in Europe. Please note that the political divisions on the map are no longer current.

With kind permission from Springer Science+Business Media: *Theoretical and Applied Genetics*, "Brassica: taxonomy based on nuclear restriction fragment length polymorphisms (RFLPs)," volume 79, Apr 1, 1990, pp. 497–506, K. Song.

Figure 33.23
A series of rock walls creates terraced areas for growing crops near Beijing in China. The terraces hold soil, but the setting precludes extensive use of mechanized farming equipment.

Figure 33.24
Large expanses of relatively flat land lend themselves to mechanized farming, allowing more uniform conditions and crops.

33.4 Complications

As anyone who has ever gardened or worked a farm knows well, there is more to growing crops than putting seeds or small plants in the ground and then harvesting the crops.

33.4a Fertilizer, Water, Yield, and Pests Are Factors in the Care of Crops

In the course of operating an experimental farm in the Negev Desert in southern Israel **(Figure 33.25)**, researchers established several basic truths. By providing 20 m³ of manure (sheep and goat) and 600 kg of ammonium sulfate per hectare, they could obtain good yields: 4800 kg·ha⁻¹ of barley (where less-tended crops yielded 400 to 600 kg·ha⁻¹) and 4400 kg·ha⁻¹ (nanasit strain) or 2700 kg·ha⁻¹ (Florence strain) of wheat. They were able to produce 750 kg·ha⁻¹ of carrots and 650 kg·ha⁻¹ of onions. Achieving these yields required cultivation of the soil, irrigation, and dealing with a variety of pests. Their farm became an oasis of green that attracted hares, gazelles, porcupines, desert partridges, and a host of insects, meaning that control of pest species had to be routine.

Irrigation was a key to good crops, and the experiment had been designed to test the prediction that by collecting runoff water and storing it in cisterns, the people there could farm in an area with little and highly seasonal rainfall. In 2006–07, the 60 mm total rainfall in the area occurred between November 20 and April 16. Water collected as runoff and stored in a 1400 m³ cistern could last the farm (people, animals, and crops) over two years.

33.4b Cats Are Sometimes Workers

The need to control rodent pests in areas where grain is stored **(Figure 33.26)** might be one factor explaining the domestication of cats, *Felis silvestris catus*. Carlos A. Driscoll and colleagues examined short tandem repeat (STR) and mtDNA data from 979 cats and wild progenitors to examine relationships among them. The evidence suggested at least five founder populations, including the European wildcat (*Felis silvestris silvestris*), near Eastern wildcat (*Felis silvestris lybica*), central Asian wildcat (*Felis silvestris ornata*), southern African wildcat (*Felis silvestris cafra*), and Chinese desert cat (*Felis silvestris bieti*). Each of these populations represents a distinct subspecies. Cats were thought to have been domesticated in the Near East, and their descendants were transported across the world with assistance from humans **(Figure 33.27 , p. 826)**. Driscoll and his colleagues proposed that the domestication of cats coincided with the development of agriculture in different locations.

33.4c Crops Can Become Contaminated

People living in rural parts of Bosnia, Bulgaria, Croatia, Romania, and Serbia exhibit a high incidence of a devastating renal disease termed *endemic Balkan nephropathy* (EN). People afflicted with EN progress

Figure 33.25

An experimental farm plot near Avdot in the Negev Desert in southern Israel. The green areas are irrigated with water stored in an underground cistern. The experimental farm was established on an ancient farm site. Other fields previously under cultivation are shown. By collecting runoff during and after rainfall, farmers have stored water for their families and crops for hundreds of years.

Figure 33.26

Harvested crops can be stored in different ways. Near Tien in China, farmers hang collections of corn cobs after harvest. This approach to crop storage suggests a dearth of local birds and rodents that might consume the corn.

Figure 33.27

The origins of domestic cats. **(a)** Genetic assessment of 979 cats (*Felis silvestris catus*) based on short tandem repeats (STR) and mtDNA identifies different contributors to cat genotypes. **(b)** The accompanying phenogram of 851 domesticated and wild cats illustrates the relationships between domestic cats and wild genetic contributors.

From Carlos A. Driscoll, Marilyn Menotti-Raymond, Alfred L. Roca, Karsten Hupe, Warren E. Johnson, Eli Geffen, Eric H. Harley, Miguel Delibes, Dominique Pontier, Andrew C. Kitchener, Nobuyuki Yamaguchi, Stephen J. O'Brien, David W. Macdonald, "The Near Eastern Origin of Cat Domestication," *Science*, vol. 317, Jul 27, 2007, pp. 519–523. Reprinted with permission from AAAS.

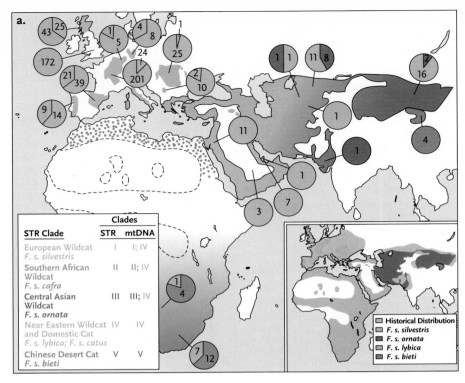

STR Clade	Clades	
	STR	mtDNA
European Wildcat *F. s. silvestris*	I	I; IV
Southern African Wildcat *F. s. cafra*	II	II; IV
Central Asian Wildcat *F. s. ornata*	III	III; IV
Near Eastern Wildcat and Domestic Cat *F. s. lybica; F. s. catus*	IV	IV
Chinese Desert Cat *F. s. bieti*	V	V

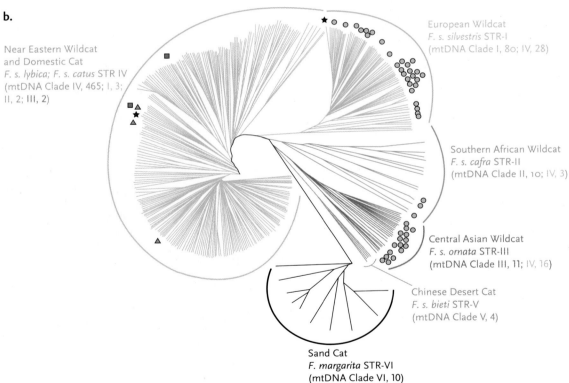

b.

Near Eastern Wildcat and Domestic Cat
F. s. lybica; F. s. catus STR IV
(mtDNA Clade IV, 465; I, 3; II, 2; III, 2)

European Wildcat
F. s. silvestris STR-I
(mtDNA Clade I, 80; IV, 28)

Southern African Wildcat
F. s. cafra STR-II
(mtDNA Clade II, 10; IV, 3)

Central Asian Wildcat
F. s. ornata STR-III
(mtDNA Clade III, 11; IV, 16)

Chinese Desert Cat
F. s. bieti STR-V
(mtDNA Clade V, 4)

Sand Cat
F. margarita STR-VI
(mtDNA Clade VI, 10)

from chronic renal failure to a high incidence of cancer of the upper urinary tract. EN and its associated cancer can be related to chronic dietary poisoning by aristolochic acid. The source of the poisoning is contamination of grain crops with the plant *Aristolochia clematitis*. A clue to this situation came from horses that developed renal failure after being fed hay contaminated with *A. clematitis*. The presence of weeds or other contaminants in crops **(Figure 33.28)** poses an important challenge to farmers, one that goes beyond how the productivity of the crop is affected by weeds' needs for water and nutrients.

Ironically, the medicinal virtues of extracts of *A. clematitis* are extolled on some websites selling homeopathic remedies. A first step in solving the mystery of EN came from case studies of some Belgian women who had developed renal problems after taking extracts of *A. clematitis* as part of a weight loss program.

Figure 33.28
Weeds, such as the milkweeds (*Asclepias* species) growing in this barley field in southwestern Ontario, can pose a problem at harvesting. If the weeds are toxic, they must be extracted from the harvested crop to ensure the safety of animals and people that consume the barley.

Plants that look the same often produce quite different chemicals. Poison hemlock, for example, superficially resembles Queen Anne's lace (*Daucus carota*), a common weed. It can also be confused with fennel (*Foeniculum vulgare*) and parsley (*Petroselinum crispum*), two herbs often used in cooking.

The speed with which changes can occur in biological systems can be astonishing, exemplified by the assimilation of new foods into our own diets whether the product is natural or synthetic. This is not a feature unique to humans. Work on ant farmers revealed that one species (*Cyphomyrmex rimosus*) introduced to Florida quickly acquired a crop cultivated by another species of ant indigenous there (*Cyphomyrmex minutus*).

STUDY BREAK

1. What factors influence crop yield?
2. How does crop yield vary?
3. What are some of the problems associated with crops?

33.5 Chemicals, Good and Bad

Plants are often treasured as much for the chemicals they produce as for their use as food, and until the time of Linnaeus (Chapter 19), plants were mainly classified by their medicinal properties. The chef who adds rosemary (*Rosmarinus officinalis;* **Figure 33.29**) to a dish as it is cooking knows that the flavour will enhance the final product. Other plant products are used as medicines. Phenolic compounds are responsible for the distinctive flavours of coffee, cinnamon, cloves, and nutmeg, which add flavour and aroma to food. Some of these compounds, such as caffeine, can be addictive.

The secondary compounds of plants may add taste to our food but can also be lethal. Consider how the plants use these compounds. Ginsenosides (obtained from ginseng, *Panax quinquefolius*), for example, appear to be used by ginseng plants as fungicides, whereas humans take them to stimulate the immune system. Other plant products are toxic. Conine, produced by poison hemlock (*Conium maculatum;* not to be confused with the hemlock tree), is an active ingredient used by the plant to defend against herbivores (**Figure 33.30**). It was the hemlock used to kill Socrates.

Figure 33.29
Herbs such as rosemary (*Rosmarinus officinalis*) are used to add flavour during cooking.

Figure 33.30
Conine, the poison from poison hemlock.

Also in Florida, the grasshopper *Romalea microptera* uses a mix of chemicals to repel ants (see "Molecule behind Biology," Box 32.3, Chapter 32). Soon after people began to use 2,4-D (2,4-dichlorophenoxyacetic acid) to kill weeds, the grasshopper added 2,5- dichlorophenol to its ant-repellent mixture. This situation is analogous to humans cultivating and domesticating new crop species.

The pharmacological potential of plant products for treating and preventing human ailments has not been lost on three groups of people: those interested in conserving biodiversity, those concerned about alleviating human suffering, and those anxious to make money by selling biopharmaceuticals to others.

STUDY BREAK

1. Why do humans use 2,4-D? What other uses could it have?
2. Why are spices important?
3. What is poison hemlock?

33.6 Molecular Farming

CONCEPT FIX Some people believe that genetically modified organisms pose a serious risk to those who consume them. The reality is that genetically modified food organisms, such as rice, are not dangerous and may be beneficial (e.g., golden rice with enriched Vitamin A). The other reality is that some genetically modified organisms pose a risk to wild stock of the same species. ⬡

Many native peoples in the New World used tobacco (*Nicotiana tabacum*) as a traditional medicine to ease the pain of childbirth, stave off hunger, and treat various ailments **(Figure 33.31)**. Tobacco was dried and smoked in ceremonies; used in poultices; brewed as a tea; and used as an emetic, an expectorant, and a laxative. By 1540, tobacco reached Europe, where it was said to cure illnesses ranging from epilepsy to plague. People have also used nicotine, an alkaloid in tobacco, in the same way as tobacco plants do, as an insecticide. The recreational use of tobacco has given the plant a bad name, even though other *Nicotiana* species are commonly used in gardens for their aromatic white or pink flowers.

Tobacco, however, is undergoing a rebirth as a positive contributor. *Molecular farming* involves the use of plants as producers of specific proteins useful as human medications. Plants can be genetically modified to produce large amounts of these proteins at low cost. Concerns about using plants as molecular farms reflect the possibility of biologically active products entering ecosystems and dispersing through food webs, via pollen or seeds, all of which could negatively affect existing crops. Enter tobacco, a nonfood plant that is harvested before flowering. Tobacco is not cold hardy (in Canada), but it is easy to genetically modify and is highly productive

Figure 33.31
A tobacco field in southwestern Ontario.

M.B. Fenton

(40-day production cycle). Using appropriate DNA technology (see Chapter 15), T-DNA containing the human *IL-10* gene was inserted into the tobacco genome by connecting it to a plant promoter and an *Agrobacterium* terminator. In this way, tobacco plants are modified to produce interleukin-10 (*IL-10*), which can be used to treat irritable bowel disease in humans. Interleukin is a cytokine involved in the regulation of inflammatory diseases, reducing the production of necrosis factor by tumours. The IL-10 was found not to enter the soil in which the tobacco was grown or the aphids or other insects that fed on the tobacco plants. The vast amount of agricultural land that was used to grow tobacco for recreational use can now be used for molecular farming. This change to molecular farming of tobacco could produce many different useful compounds.

STUDY BREAK

1. How are plant products used in addition to their role as food?
2. What is molecular farming?
3. What conservation risks are associated with molecular farming? With agriculture in general?

Feeding People and Keeping Them Healthy

We have seen how humans' ability to harness nature has allowed us to reduce the number of hungry people in the world. Nevertheless, in 2008, converting food crops to biofuels in the interest of being "green" meant less food. This development coincided with climate change, which further reduced the availability of food worldwide. This impact was amplified by increases in the prices of food crops, in turn making food less accessible to many poor people. This brings us back to the price of oil, the energy used in agriculture to produce food, and the cost of food.

We use plants for more than food. In 2008, over 1 million people died from malaria, caused by a blood parasite (see Chapter 27). Some 200 million people worldwide suffer from malaria. Most (over 75%) of the deaths were of children under 5 years of age and living in Africa. For years, alkaloids (quinine, quinidine, cichonidine, and cinchonine) extracted from the bark of four species of tree in the genus *Cinchona* have been used to treat agues and periodic fevers. Cinchona alkaloids have been particularly effective against malaria, but some strains of malaria are resistant to quinines and other treatments, suggesting that they are not the solution to malaria.

Enter qinghao, also known as huang hua hao, an extract from a ubiquitous shrub, *Artemisia annua*. Extracts of qinghao (artemisinin) have long been used in traditional Chinese medicine. In 1971, Chinese scientists discovered its effectiveness in treating malaria and reported it to the world in 1979. In 2008, artemisinin combination treatments became first-line drugs for some forms of malaria (uncomplicated falciparum malaria), but they were not available worldwide. Using improved agriculture techniques—selection of high-yielding hybrids, microbial production, and synthetic peroxidases—could lower prices, increase the availability of artemisinin, and save many lives.

In 2007, about one-third of the annual U.S.$30 billion worldwide investment in agriculture research was aimed at solving the problems of agriculture in developing countries—home to about 80% of the global population. This investment is less than 3% of the amount that countries of the Organization for Economic Co-operation and Development (OECD) spend to subsidize their own agricultural production. To ensure continued reduction of human hunger, OECD countries must invest more to solve the agriculture problems in the developing world. One important first step in this process is recognizing the energy input required to grow enough food to feed the human population (see Section 33.3).

We know that our agricultural prowess can be used to increase production of food and medicines such as artemisinin. Golden rice provides another example. In 1984, the World Health Organization estimated that 250 thousand to 500 thousand children a year had diets lacking in vitamin A. This deficiency damaged their retinas and corneas, so many of them went blind and half of them died.

By the early 2000s, Ingo Potrykus and Peter Byer had used genetic engineering to splice two daffodil genes and a bacterial gene into the rice genome. The genetically modified rice, **golden rice**, was golden in colour and produced precursors to vitamin A in its endosperm. Golden rice offered a way to address the results of vitamin A deficiency. Growing golden rice was field tested in Louisiana in 2004 and 2005.

In 2008, it appeared that even though golden rice offered the ability to prevent the suffering arising from vitamin A deficiency, it was unlikely that any would be planted before 2012. The delay reflects a combination of widespread public suspicion about genetically modified organisms (GMOs) and the high cost of obtaining approval to use GMOs. The costs associated with getting approval mean that only large companies with large budgets are likely to succeed in getting GMO products approved. The controversy persists today.

Moving more people away from "the edge" means investing more in finding solutions to agricultural operations in the developing world to ensure that discoveries such as artemisinin and golden rice are used to advantage. This requires policies and investments that support small-scale operations, probably subsidizing the costs of energy and water necessary to grow enough food. Our ancestors, who domesticated crops such as wheat and corn, lentils and squash, and potatoes and rice, changed the world for us. We should ensure that their legacy lives on.

33.7 The Future

It is tempting to believe that the diversity of life will continue to provide humans with solutions to many of their problems in the world: food for the hungry, poisons to selectively control pests, and biopharmaceuticals to cure diseases. Molecular farming potentially allows new approaches to solving old problems. Although we can continue to domesticate other species, effectively taking evolution in directions that suit us, we must remember that when new forms require higher investments in energy and fertilizer, costs can outweigh benefits.

In 2008, global increases in the price of food were partly due to our using traditional food crops (e.g., corn) to produce ethanol to fuel vehicles. In 2007, the energy return on investment (EROI) for biofuel (ethanol, or food for internal combustion engines) from corn was about 1. This means that every litre of ethanol produced from corn consumes about a litre of petroleum fuel.

Other crops, for example, poplar trees, have much better EROIs. Although biofuels are potentially low-carbon-emitting energy sources, the means of production dramatically influences their "greenness." In many cases, production of biofuels generates a *biofuel carbon debt*. In Brazil, Southeast Asia, and the United States, converting native habitat (rain forest, peatlands, savannah, and grasslands) to produce biofuels generates 17 to 420 times as much CO_2 as the habitats normally produce. When these costs are taken into account, biofuels are not feasible environmentally friendly alternatives.

Calculating EROI and tracking it over time helps put energy use in perspective. In Rome in 1 CE, the EROI for wheat was 12:1, and for alfalfa, it was 27:1. In 2007, the EROI value for gasoline was about 17:1, whereas in the 1930s, it was about 100:1. The values of EROI speak to the sustainability of a process, so Canadians must be concerned that the EROI on the tar sands is less than 4:1 without taking into consideration the water consumed by the process.

The development of resistance to toxins, whether of bacteria to antibiotics or of insect pests to insecticides or of weeds to herbicides, demonstrates that evolution works both ways. Perfecting genetic strains of crops protected by resistance to a pathogen or pest and using them exclusively can make the crops vulnerable to pathogens or pests that are resistant to the defence(s). The potato famine, discussed in Section 33.2j, was exacerbated by the lack of genetic diversity of the potatoes used in Europe: they were vulnerable to blight.

Over evolutionary history, individuals able to exploit other species had an advantage over those that did not, whether within a society or between societies. The same principles apply to our own species. Our advantages of exploiting other species include increased access to food (quantity and quality), labour, materials for constructing things, and chemicals for treating disorders or controlling pests. These advantages were amplified through domestication, which meant increasing control over the other species, leading to several net effects. One effect was achieving larger populations of humans because of better access to food and/or protection from disease. Ironically, living closer to more animals also exposes us to diseases such as avian flu, swine flu, and smallpox. Another factor was probably the increase in available time for the development and perfection of new tools and techniques and the emergence of groups of people in society who did not contribute directly by gathering or processing food. Such people could have contributed to society through their talents as artisans, soldiers, or even politicians.

The range of possibilities seems endless, particularly with the advent of the ability to directly modify genotypes and thus phenotypes. Perhaps we should be glad that the Attine ants have domesticated only fungi.

STUDY BREAK

What is artemisinin? What is golden rice? Why are they important?

Review

To access course materials such as Aplia and other companion resources, please visit www.NELSONbrain.com.

33.1 Domestication

- The green revolution is credited with reducing the number of hungry people in the world by increasing agricultural productivity. Increased productivity reflects the use of improved strains of crops, increased applications of fertilizers, and more extensive irrigation. Higher productivity (crop yields) also reflects more dependence on mechanized (fossil fuel–based) farming operations.
- There are three stages in the exploitation of biodiversity for human benefit. The initial stage involves hunting and gathering, collecting organisms in the wild. The second stage is cultivation or caring for organisms under progressively controlled conditions. The third stage, domestication, involves selective breeding to enhance desired characteristics and features. The domestication process can apply to animals; plants; or other organisms, such as yeast and fungi.
- Domestication of asexually reproducing organisms does not involve selective breeding, which provides more control, allows shorter generation times, or speeds up the process of selection.
- Sexually reproducing organisms that are successfully domesticated are usually capable of self-fertilization and/or readily propagated by grafting.
- Living in the same sites year-round allowed people to tend and protect their crops. Increased time in one place would also have facilitated the process of selective breeding.
- At least 200 species of ants in the tribe Attini are obligate farmers. Other ants also tend seed gardens.
- The first evidence of domestication appears to be figs by about 12 000 years B.P. Cultivation may have been practised for 1000 years before domestication. Many crops had been domesticated by 6000 years B.P.

33.2 Why Some Organisms Were Domesticated

- Yeast appears to have been domesticated in Egypt by 6000 years B.P., when it was used in making bread and beer. Cotton may have been domesticated in at least four sites, two in the Old World and two in the New World. Cotton is used as a source of oil seed and fibre, and some domestication had occurred by 4400 years B.P.

- The change from dehiscence (shattering) to indehiscence (nonshattering) was critical in the domestication of plants, such as grasses, whose seeds were the crop to be harvested.
- Squash and potatoes provide carbohydrates from fruits or tubers; the seeds are not the target of domestication.
- The domestication of dogs appears to have been based, in part, on their social behaviour, including their ability to communicate with people. Some experiments show that dogs are better at reading communication signals from people than other animals, such as chimps.
- Some animals are farmed for meat and hides (and for feathers and eggs), but there is no evidence of selective breeding.

33.3 Yields

- The level of farming intensity, the strains of crops, the application of fertilizers, and irrigation affect crop yields. Terrain can influence the level of mechanization.

- The data on corn and wheat yields clearly demonstrate variation over time and location.

33.4 Complications

- Rich patches of food reflect higher productivity but can attract pests.

33.5 Chemicals, Good and Bad

- Various plant chemicals are used as stimulants, medicines, hallucinogens, pesticides, and flavourings.

33.6 Molecular Farming

- Molecular farming is the use of plants to produce various useful proteins in large amounts at lower cost. Molecular farming may prove to be an inexpensive way to manufacture medicines and other substances.

33.7 The Future

- Energy return on investment (EROI) is an important concept whether the topic is crops or oil. When the cost of seeing a crop from planting to harvest exceeds the return, there is little point in continuing to use the crop. The same should be true for harvesting energy, for example, oil from the tar sands.

Questions

Self-Test Questions

1. According to available evidence, when did humans first domesticate other organisms?
   a. 6000 years ago
   b. 8000 years ago
   c. 12 000 years ago
   d. 20 000 years ago

2. The process of moving from cultivation to domestication involved dehiscence (shattering) to indehiscence (nonshattering) in the following crops.
   a. rice
   b. wheat
   c. squash
   d. Both a and b are correct.

3. Parthenocarpy was important in the domestication of which species?
   a. dogs
   b. figs
   c. lentils
   d. cotton

4. At Abu Hureyra, which of the following did settlers domesticate?
   a. olives
   b. some grains
   c. cattle
   d. hot peppers

5. Which group includes only domesticated organisms?
   a. yeast, mushrooms, pigs, and oysters
   b. honeybees, yeast, rice, and ostriches
   c. cattle, pigs, cats, and dogs
   d. lentils, mushrooms, yeast, and crocodiles

6. Plants in the family Solonaceae include which of the following?

   a. tomatoes, potatoes, and eggplant
   b. potatoes, hot peppers, and deadly nightshade
   c. corn, tobacco, and hot peppers
   d. Both a and b are correct.

7. If a person needs 8300 kJ·day^{-1}, and wheat yields 6638 kg·ha^{-1}, how much land would be needed to produce enough wheat for 1 year for a city of 20 thousand people who eat only wheat?
   a. 155 ha
   b. 1055 ha
   c. 2055 ha
   d. 5055 ha

8. Endemic Balkan nephropathy is an example of a disorder arising when people eat crops that have been subjected to the following.
   a. irrigation with polluted water
   b. contamination with *Aristolochia clematitis*
   c. contamination with *Rosmarinus officinalis*
   d. contamination with *Asclepias exultata*

9. *Nicotiana tabacum* is being used to produce which of the following medications?
   a. nicotine
   b. interleukin
   c. acetylcholine
   d. conine

10. EROI, energy return on investment, suggests that alfalfa in Roman times was this many times as efficient at energy production as the tar sands in Alberta in 2008.
    a. 4
    b. 6
    c. 10
    d. 50

Questions for Discussion

1. When is domestication complete? At what point is a domesticated stock a separate species? Is domestication ever complete?

2. Why have so few aquatic species been domesticated? Why has there been a recent increase in the numbers of domesticated aquatic organisms?

3. How do domesticated populations threaten native species? Should this be a concern for conservation biologists? What can be done to minimize this threat to native species?

4. Why is the price of oil so important in food production?

5. How can we ensure continued supplies of water for food production?

Appendix A
Answers to Self-Test Questions

A-1

Chapter 17

1. c 2. c 3. d 4. b 5. c 6. b 7. a 8. b 9. d 10. d

Chapter 18

1. c 2. b 3. c 4. d 5. b 6. e 7. a 8. b 9. c 10. d

Chapter 19

1. a 2. e 3. e 4. d 5. d 6. e 7. b 8. c 9. a 10. b

Chapter 20

1. a 2. c 3. e 4. d 5. b 6. c 7. a 8. d 9. c 10. b

Chapter 21

1. d 2. c 3. d 4. d 5. c 6. a 7. c 8. d 9. b 10. c

Chapter 22

1. b 2. c 3. c 4. b 5. d 6. a 7. a 8. c 9. b 10. a

Chapter 23

1. c 2. a 3. a 4. c 5. d 6. b 7. c 8. c 9. d 10. d

Chapter 24

1. b 2. a 3. d 4. b 5. d 6. c 7. b 8. c 9. d 10. c

Chapter 25

1. c 2. b 3. d 4. b 5. b 6. c 7. d 8. c 9. a 10. c

Chapter 26

1. c 2. d 3. c 4. c 5. b 6. b 7. d 8. a 9. d 10. d

Chapter 27

1. c 2. d 3. d 4. d 5. b 6. d 7. c 8. b 9. c 10. d

Chapter 28

1. a 2. b 3. c 4. a 5. c 6. c 7. c 8. b 9. c 10. a

Chapter 29

1. a 2. b 3. b 4. c 5. a 6. a 7. a 8. d 9. d 10. a

Chapter 30

1. c 2. b 3. a 4. a 5. b 6. d 7. b 8. d 9. c 10. b

Chapter 31

1. d 2. c 3. a 4. d 5. d 6. a 7. c 8. d 9. c 10. a

Chapter 32

1. d 2. c 3. d 4. c 5. b 6. b 7. a 8. b 9. b 10. b

Chapter 33

1. c 2. d 3. b 4. b 5. c 6. d 7. b 8. b 9. b 10. b

NEL

APPENDIX A   ANSWERS TO SELF-TEST QUESTIONS | A-1

Glossary

abdomen In insects, the region behind the thorax. p. 622

absorptive nutrition Mode of nutrition in which an organism secretes digestive enzymes into its environment and then absorbs the small molecules thus produced. p. 541

acetabulum Socket of hip joint, receives head of femur. p. 657

acoelomate A body plan of bilaterally symmetrical animals that lack a body cavity (coelom) between the gut and the body wall. p. 596

acorn worms Sedentary marine animals living in U-shaped tubes or burrows in coastal sand or mud. p. 639

active parental care Parents' investment of time and energy in caring for offspring after they are born or hatched. p. 687

adaptation, evolutionary Characteristic or suite of characteristics that helps an organism survive longer or reproduce more under a particular set of environmental conditions; the accumulation of adaptive traits over time. p. 423

adaptive radiation (diversification) A cluster of closely related species that are each adaptively specialized to a specific habitat or food source. p. 451

adaptive trait A genetically based characteristic, preserved by natural selection, that increases an organism's likelihood of survival or its reproductive output. pp. 398, 423

adaptive zone A part of a habitat that may be occupied by a group of species exploiting the same resources in a similar manner. p. 661

adductor muscle A muscle that pulls inward toward the median line of the body; in bivalve molluscs, it pulls the shell closed. p. 615

aerobe An organism that requires oxygen for cellular respiration. p. 487

African emergence hypothesis A hypothesis proposing that modern humans first evolved in Africa and then dispersed to other continents. p. 477

agar A gelatinous product extracted from certain red algae or seaweed used as a culture medium in the laboratory and as a gelling or stabilizing agent in foods. p. 532

age structure A statistical description or graph of the relative numbers of individuals in each age class in a population. p. 683

age-specific fecundity The average number of offspring produced by surviving females of a particular age. p. 685

age-specific mortality The proportion of individuals alive at the start of an age interval that died during that age interval. p. 685

age-specific survivorship The proportion of individuals alive at the start of an age interval that survived until the start of the next age interval. p. 685

albumin The most abundant protein in blood plasma, important for osmotic balance and pH buffering; also, the portion of an egg that serves as the main source of nutrients and water for the embryo. p. 654

algin Alginic acid, found in the cell walls of brown algae. p. 526

allele frequency The abundance of one allele relative to others at the same gene locus in individuals of a population. p. 414

allopatric speciation The evolution of reproductive isolating mechanisms between two populations that are geographically separated. p. 435

allopolyploidy The genetic condition of having two or more complete sets of chromosomes from different parent species. p. 441

alpine tundra A biome that occurs on high mountaintops throughout the world, in which dominant plants form cushions and mats. p. 572

alternation of generations The regular alternation of mode of reproduction in the life cycle of an organism, such as the alternation between diploid (sporophyte) and haploid (gametophyte) phases in plants. p. 563

altricial Helpless at birth. p. 668

ammocoetes Larval lamprey eel. p. 645

ammonification A metabolic process in which bacteria and fungi convert organic nitrogen compounds into ammonia and ammonium ions; part of the nitrogen cycle. p. 766

amnion In an amniote egg, an extraembryonic membrane that encloses the embryo, forming the amniotic cavity and secreting amniotic fluid, which provides an aquatic environment in which the embryo develops. p. 654

amniote (amniotic) egg A shelled egg that can survive and develop on land. p. 654

amoeboid Similar to an amoeba, particularly in type of movement. p. 601

anaerobic respiration The process by which molecules are oxidized to produce ATP via an electron transport chain and ATP synthase, but unlike aerobic respiration, oxygen is not the final electron acceptor. p. 487

anapsid (lineage Anapsida) A member of the group of amniote vertebrates with no temporal arches and no spaces on the sides of the skull (includes turtles). p. 655

Anapsida An extinct group of fossil fishes. p. 655

ancestral character A trait that was present in a distant common ancestor. p. 458

angiosperm A flowering plant. Its egg-containing ovules mature into seeds within protected chambers called ovaries. p. 586

Animalia The taxonomic kingdom that includes all living and extinct animals. p. 594

annulus In ferns, a ring of thick-walled cells that nearly encircles the sporangium and functions in spore release. p. 579

antennal glands Excretory structures at the base of the antennae in some crustaceans. p. 624

anterior Indicating the head end of an animal. p. 596

antheridium (plural, antheridia) In plants, a structure in which sperm are produced. p. 572

Anthocerophyta The phylum comprising hornworts. p. 575

Anthophyta The phylum comprising flowering plants. p. 587

antibiotic A natural or synthetic substance that kills or inhibits the growth of bacteria and other microorganisms. p. 489

apical growth Growth from the tip of a cell or tissue. p. 542

apical meristem A region of unspecialized dividing cells at the shoot tips and root tips of a plant. p. 568

apicomplexan A group of parasitic organisms with specific structures in their apical complex to penetrate and enter the cells they parasitize. p. 522

aposematic Refers to bright, contrasting patterns that advertise the unpalatability of poisonous or repellent species. p. 715

applied ecology Application of ecological theory and principles to management of natural resources. p. 679

aquatic succession A process in which debris from rivers and runoff accumulates in a body of fresh water, causing it to fill in at the margins. p. 737

arbuscular mycorrhizas Symbiotic association between a glomeromycete fungus and the roots of a wide range of plants, including nonvascular, nonseed, and seed plants. p. 557

arbuscule Highly branched hypha produced inside root cells by arbuscular mycorrhizal fungi; nutrient exchange site between plant and fungus. p. 548

Archaea One of two domains of prokaryotes; archaeans have some unique molecular and biochemical traits, but they also share some traits with Bacteria and other traits with Eukarya. p. 482

archaeocytes A major group of the domain Archaea, members of which are found in different extreme environments. They include methanogens, extreme halophiles, and some extreme thermophiles. *See* Euryarchaeota. p. 601

archegonium (plural, archegonia) The flask-shaped structure in which bryophyte eggs form. p. 572

archenteron The central endoderm-lined cavity of an embryo at the gastrula stage, which forms the primitive gut. p. 597

Archosauromorpha A diverse group of diapsids that comprises crocodilians, pterosaurs, and dinosaurs (including birds). p. 655

arctic tundra A treeless biome that stretches from the boreal forests to the polar ice cap in Europe, Asia, and North America. p. 572

artificial selection Selective breeding of animals or plants to ensure that certain desirable traits appear at higher frequency in successive generations. p. 398

ascocarp A reproductive body that bears or contains asci. p. 548

ascospore Spore formed by meiosis in the ascus, a saclike cell produced by ascomycete fungi. p. 549

ascus (plural, asci) A saclike cell in ascomycetes (sac fungi) in which meiosis gives rise to haploid sexual spores (meiospores). p. 548

assimilation efficiency The ratio of the energy absorbed from consumed food to the total energy content of the food. p. 752

assumption of parsimony Assumption that the simplest explanation should be the most accurate. p. 459

asymmetrical Characterized by a lack of proportion in the spatial arrangement or placement of parts. p. 596

atrial siphon A tube through which invertebrate chordates expel digestive and metabolic wastes. p. 641

atriopore The hole in the body wall of a cephalochordate through which water is expelled from the body. p. 641

atrium (plural, atria) A body cavity or chamber surrounding the perforated pharynx of invertebrate chordates; also one of the chambers that receive blood returning to the heart. p. 641

autopolyploidy The genetic condition of having more than two sets of chromosomes from the same parent species. p. 441

axial skeleton The bones constituting the head and trunk of a vertebrate: the cranium, vertebral column, ribs, and sternum (breastbone). p. 641

axopods Slender, raylike strands of cytoplasm supported internally by long bundles of microtubules. p. 527

backbone (spine) Vertebral column of vertebrates. p. 641

background extinction rate The average rate of extinction of taxa through time. p. 781

Bacteria One of the two domains of prokaryotes; collectively, bacteria are the most metabolically diverse organisms. p. 482

balanced polymorphism The maintenance of two or more phenotypes in fairly stable proportions over many generations. p. 421

basidiocarp A fruiting body of a basidiomycete; mushrooms are examples. p. 553

basidiospore A haploid sexual spore produced by basidiomycete fungi. p. 553

basidium (plural, **basidia**) A small, club-shaped structure in which sexual spores of basidiomycetes arise. p. 550

Batesian mimicry The form of defence in which a palatable or harmless species resembles an unpalatable or poisonous one. p. 716

behavioural isolation A prezygotic reproductive isolating mechanism in which two species do not mate because of differences in courtship behaviour; also known as ethological isolation. p. 431

bilateral symmetry The body plan of animals in which the body can be divided into mirror image right and left halves by a plane passing through the midline of the body. p. 596

binomial nomenclature The naming of species with a two-part scientific name, the first indicating the genus and the second indicating the species. p. 427

binomial Relating to or consisting of two names or terms. p. 427

biodiversity The richness of living systems as reflected in genetic variability within and among species, the number of species living on Earth, and the variety of communities and ecosystems. p. 780

biofilm A microbial community consisting of a complex aggregation of microorganisms attached to a surface. p. 490

biogeochemical cycle Any of several global processes in which a nutrient circulates between the abiotic environment and living organisms. p. 488

biogeography The study of the geographic distributions of plants and animals. p. 392

biological evolution The process by which some individuals in a population experience changes in their DNA and pass those modified instructions to their offspring. p. 392

biological lineage An evolutionary sequence of ancestral organisms and their descendants. p. 401

biological magnification The increasing concentration of nondegradable poisons in the tissues of animals at higher trophic levels. p. 757

biological species concept The definition of species based on the ability of populations to interbreed and produce fertile offspring. p. 429

bioluminescent Refers to an organism that glows or releases a flash of light, particularly when disturbed. p. 521

biomass The dry weight of biological material per unit area or volume of habitat. pp. 482, 750

bioremediation Applications of chemical and biological knowledge to decontaminate polluted environments. p. 481

biota The total collection of organisms in a geographic region. p. 773

bipedalism The habit in animals of walking upright on two legs. p. 469

blastopore The opening at one end of the archenteron in the gastrula that gives rise to the mouth in protostomes and the anus in deuterostomes. p. 597

book lungs Pocketlike respiratory organs found in some arachnids consisting of several parallel membrane folds arranged like the pages of a book. p. 622

boreal forest A biome that is a circumpolar expanse of evergreen coniferous trees in Europe, Asia, and North America. p. 557

Bryophyta The phylum of nonvascular plants, including mosses and their relatives. p. 574

bryophyte A general term for plants (such as mosses) that lack internal transport vessels. p. 572

budding A mode of asexual reproduction in which a new individual grows and develops while attached to the parent. p. 542

capsid *See* coat. p. 501

carapace A protective outer covering that extends backward behind the head on the dorsal side of an animal, such as the shell of a turtle or lobster. p. 624

carbon cycle The global circulation of carbon atoms, especially via the processes of photosynthesis and respiration. p. 760

carnivore An animal that primarily eats other animals. p. 670

carrageenan A chemical extracted from the red alga *Eucheuma* that is used to thicken and stabilize paints, dairy products such as pudding and ice cream, and many other creams and emulsions. p. 532

carrying capacity The maximum size of a population that an environment can support indefinitely. p. 691

catastrophism The theory that Earth has been affected by sudden, violent events that were sometimes worldwide in scope. p. 393

cellular slime mould Any of a variety of primitive organisms of the phylum Acrasiomycota, especially of the genus *Dictyostelium*; the life cycle is characterized by a slimelike amoeboid stage and a multicellular reproductive stage. p. 529

cephalization The development of an anterior head where sensory organs and nervous system tissue are concentrated. p. 596

cephalothorax The anterior section of an arachnid, consisting of a fused head and thorax. p. 622

character displacement The phenomenon in which allopatric populations are morphologically similar and use similar resources, but sympatric populations are morphologically different and use different resources; may also apply to characters influencing mate choice. p. 721

charophyte A member of the group of green algae most similar to the algal ancestors of land plants. p. 533

chelicerae The first pair of fanglike appendages near the mouth of an arachnid, used for biting prey and often modified for grasping and piercing. p. 622

chemical signal Any secretion from one cell type that can alter the behaviour of a different cell that bears a receptor for it; a means of cell communication. p. 529

chemoautotroph An organism that obtains energy by oxidizing inorganic substances such as hydrogen, iron, sulfur, ammonia, nitrites, and nitrates and uses carbon dioxide as a carbon source. p. 487

chemoheterotroph An organism that oxidizes organic molecules as an energy source and obtains carbon in organic form. p. 487

chemotroph An organism that obtains energy by oxidizing inorganic or organic substances. p. 487

choanocyte One of the inner layer of flagellated cells lining the body cavity of a sponge. p. 601

choanoflagellata A group of minute, single-celled protists found in water; the flask-shaped body has a collar of closely packed microvilli that surrounds the single flagellum by which it moves and takes in food. p. 531

chromosome alterations Changes in the structure of chromosomes involving insertion, deletion, inversion, or translocation of significant amounts of DNA sequence. p. 439

circulatory vessel An element of the circulatory system through which fluid flows and carries nutrients and oxygen to tissues and removes wastes. p. 611

clade A monophyletic group of organisms that share homologous features derived from a common ancestor. p. 461

cladistics An approach to systematics that uses shared derived characters to infer the phylogenetic relationships and evolutionary history of groups of organisms. p. 461

cladogenesis The evolution of two or more descendent species from a common ancestor. p. 457

cladogram A branching diagram in which the endpoints of the branches represent different species of organisms, used to illustrate phylogenetic relationships. p. 461

claspers A pair of organs on the pelvic fins of male crustaceans and sharks, which help transfer sperm into the reproductive tract of the female. p. 648

class A Linnaean taxonomic category that ranks below a phylum and above an order. p. 428

classification An arrangement of organisms into hierarchical groups that reflect their relatedness. p. 428

cleavage Mitotic cell divisions of the zygote that produce a blastula from a fertilized ovum. p. 596

climax community A relatively stable, late successional stage in which the dominant vegetation replaces itself and persists until an environmental disturbance eliminates it, allowing other species to invade. p. 735

cline A pattern of smooth variation in a characteristic along a geographic gradient. p. 434

closed circulatory system A circulatory system in which the fluid, blood, is confined in blood vessels and is distinct from the interstitial fluid. p. 616

clumped dispersion A pattern of distribution in which individuals in a population are grouped together. p. 681

cnidocyte A prey-capturing and defensive cell in the epidermis of cnidarians. p. 603

coat The protective layer of protein that surrounds the nucleic acid core of a virus in free form. Also called a capsid. p. 501

coccoid Spherical prokaryotic cell. p. 482

coelom A fluid-filled body cavity in bilaterally symmetrical animals that is completely lined with derivatives of mesoderm. p. 596

coelomate A body plan of bilaterally symmetrical animals that have a coelom. p. 596

coevolution The evolution of genetically based, reciprocal adaptations in two or more species that interact closely in the same ecological setting. pp. 588, 711

cohort A group of individuals of similar age. p. 685

colony Multiple individual organisms of the same species living in a group. p. 516

commensalism A symbiotic interaction in which one species benefits and the other is unaffected. p. 722

community ecology The ecological discipline that examines groups of populations occurring together in one area. p. 680

comparative morphology Analysis of the structure of living and extinct organisms. p. 393

competitive exclusion principle The ecological principle stating that populations of two or more species cannot coexist indefinitely if they rely on the same limiting resources and exploit them in the same way. p. 717

complete digestive system A digestive system with a mouth at one end, through which food enters, and an anus at the other end, through which undigested waste is voided. p. 610

complete metamorphosis The form of metamorphosis in which an insect passes through four separate stages of growth: egg, larva, pupa, and adult. p. 629

compound eye The eye of most insects and some crustaceans, composed of many-faceted, light-sensitive units called ommatidia fitted closely together, each with its own refractive system and each forming a portion of an image. p. 622

cone In the vertebrate eye, a photoreceptor in the retina that is specialized for detection of different wavelengths (colours). In cone-bearing plants, a cluster of sporophylls. p. 577

conidium (plural, conidia) An asexually produced fungal spore. p. 549

Coniferophyta The major phylum of cone-bearing gymnosperms, most of which are substantial trees; includes pines, firs, and other conifers. p. 583

conodont An abundant, bonelike fossil dating from the early Paleozoic era through the early Mesozoic era, now described as a feeding structure of some of the earliest vertebrates. p. 645

conservation biology An interdisciplinary science that focuses on the maintenance and preservation of biodiversity. p. 419

consumer An organism that consumes other organisms in a community or ecosystem. p. 728

contractile vacuole A specialized cytoplasmic organelle that pumps fluid in a cyclical manner from within the cell to the outside by alternately filling and then contracting to release its contents at various points on the surface of the cell. p. 517

convergent evolution The evolution of similar adaptations in distantly related organisms that occupy similar environments. p. 403

coral reef A structure made from the hard skeletons of coral animals or polyps; found largely in tropical and subtropical marine environments. p. 604

corona The ciliated crownlike organ at the anterior end of rotifers used for feeding or locomotion. p. 610

courtship display A behaviour performed by males to attract potential mates or to reinforce the bond between a male and a female. pp. 397, 431

cranial nerve A nerve that connects the brain directly to the head, neck, and body trunk. p. 642

cranium The part of the skull that encloses the brain. p. 641

Crenarchaeota A major group of the domain Archaea, separated from the other archaeans based mainly on rRNA sequences. p. 497

crop Of birds, an enlargement of the digestive tube where the digestive contents are stored and mixed with lubricating mucus. p. 823

cuticle The outer layer of plants and some animals, which helps prevent desiccation by slowing water loss. p. 564

Cycadophyta A phylum of palmlike gymnosperms known as cycads; the pollen-bearing and seed-bearing cones (strobili) occur on separate plants. p. 583

demographic transition model A graphic depiction of the historical relationship between a country's economic development and its birth and death rates. p. 701

demography The statistical study of the processes that change a population's size and density through time. p. 684

denitrification A metabolic process in which certain bacteria convert nitrites or nitrates into nitrous oxide and then into molecular nitrogen, which enters the atmosphere. p. 766

density dependent Description of environmental factors for which the strength of their effect on a population varies with the population's density. p. 694

density independent Description of environmental factors for which the strength of their effect on a population does not vary with the population's density. p. 696

derived character A new version of a trait found in the most recent common ancestor of a group. p. 458

descent with modification Biological evolution. p. 400

desert A sparsely vegetated biome that forms where precipitation averages less than 25 cm per year. p. 750

determinate cleavage A type of cleavage in protostomes in which each cell's developmental path is determined as the cell is produced. p. 597

detritivore An organism that extracts energy from the organic detritus (refuse) produced at other trophic levels. p. 594

deuterostome A division of the Bilateria in which blastopore forms the anus during development and the mouth appears later (includes Echinodermata and Chordata). p. 596

diapsid (lineage Diapsida) A member of a group within the amniote vertebrates with a skull with two temporal arches. Their living descendants include lizards and snakes, crocodilians, and birds. p. 655

diatom Photosynthetic single-celled organisms with a glassy silica shell; also called bacillariophytes. p. 524

dikaryon The life stage in certain fungi in which a cell contains two genetically distinct haploid nuclei. p. 553

dikaryotic hyphae Hyphae containing two separate nuclei in one cell. p. 549

diphyodont Having two generations of teeth, milk (baby) teeth and adult teeth. p. 667

diploblastic An animal body plan in which adult structures arise from only two cell layers, the ectoderm and the endoderm. p. 595

directional selection A type of selection in which individuals near one end of the phenotypic spectrum have the highest relative fitness. p. 412

dispersal 1. The movement of organisms away from their place of origin, as well as the movement from one breeding site to another; 2. The movement of material that is used by an organism to move to the next stage in their life cycle. p. 601

dispersion The spatial distribution of individuals within a population's geographic range. p. 681

disruptive selection A type of natural selection in which extreme phenotypes have higher relative fitness than intermediate phenotypes. p. 414

disturbance climax (disclimax) community An ecological community in which regular disturbance inhibits successional change. p. 740

domestication Selective breeding of other species to increase desirable characteristics in progeny. p. 808

dorsal Indicating the back side of an animal. p. 596

ecdysis Shedding of the cuticle, exoskeleton, or skin; moulting. p. 600

ecological community An assemblage of species living in the same place. p. 680

ecological efficiency The ratio of net productivity at one trophic level to net productivity at the trophic level below it. p. 752

ecological isolation A prezygotic reproductive isolating mechanism in which species that live in the same geographic region occupy different habitats. p. 431

ecological niche The resources a population uses and the environmental conditions it requires over its lifetime. p. 717

ecological pyramid A diagram illustrating the effects of energy transfer from one trophic level to the next. p. 754

ecological succession A somewhat predictable series of changes in the species composition of a community over time. p. 561

ecology The study of the interactions between organisms and their environments. p. 679

ecosystem ecology An ecological discipline that explores the cycling of nutrients and the flow of energy between the biotic components of an ecological community and the abiotic environment. p. 680

ecosystem A group of biological communities interacting with their shared physical environment. p. 749

ecotone A wide transition zone between adjacent communities. p. 725

ectoderm The outermost of the three primary germ layers of an embryo, which develops into epidermis and nervous tissue. p. 595

ectomycorrhiza A mycorrhiza that grows between and around the young roots of trees and shrubs but does not enter root cells. p. 557

ectoparasite A parasite that lives on the exterior of its host organism. p. 609

Elasmobranchii Cartilaginous fishes, including the skates and rays. p. 647

electroreceptor A specialized sensory receptor that detects electrical fields. p. 648

emigration The movement of individuals out of a population. p. 684

endangered species A species in immediate danger of extinction throughout all or a significant portion of its range. p. 419

endemic species A species that occurs in only one place on Earth. p. 785

endoderm The innermost of the three primary germ layers of an embryo, which develops into the gastrointestinal tract and, in some animals, the respiratory organs. p. 595

endoparasite A parasite that lives in the internal organs of its host organism. p. 723

endosporous Pattern of development in some plants (e.g., seed plants) in which the gametophyte develops inside the spore wall. p. 580

endotoxin A lipopolysaccharide released from the outer membrane of the cell wall when a bacterium dies and lyses. p. 489

energy budget The total amount of energy that an organism can accumulate and use to fuel its activities. p. 686

enterocoelom In deuterostomes, the body cavity pinched off by outpocketings of the archenteron. p. 598

envelope Outer glycoprotein layer surrounding the capsid of some viruses, derived in part from host cell plasma membrane. p. 501

enveloped virus A virus that has a surface membrane derived from its host cell. p. 502

epidermis A complex tissue that covers an organism's body in a single continuous layer or sometimes in multiple layers of tightly packed cells. p. 603

equilibrium theory of island biogeography A hypothesis suggesting that the number of species on an island is governed by a give and take between the immigration of new species to the island and the extinction of species already there. p. 741

esophagus A connecting passage of the digestive tube. p. 638

eudicot A plant belonging to the Eudicotyledones, one of the two major classes of angiosperms; their embryos generally have two seed leaves (cotyledons), and their pollen grains have three grooves. p. 587

Eukarya The domain that includes all eukaryotes, organisms that contain a membrane-bound nucleus within each of their cells; all protists, plants, fungi, and animals. p. 482

Euryarchaeota A major group of the domain Archaea, members of which are found in different extreme environments. They include methanogens, extreme halophiles, and some extreme thermophiles. p. 497

evolution The main unifying concept in biology, explaining how the diversity of life on Earth arose and how species change over time in response to changes in their abiotic and biotic environment. p. 392

evolutionary developmental biology A field of biology that compares the genes controlling the developmental processes of different animals to determine the evolutionary origin of morphological novelties and developmental processes. p. 456

evolutionary divergence A process whereby natural selection or genetic drift causes populations to become more different over time. p. 399

excretion The process that helps maintain the body's water and ion balance while ridding the body of metabolic wastes. p. 587

exoskeleton A hard external covering of an animal's body that blocks the passage of water and provides support and protection. p. 620

exotoxin A toxic protein that leaks from or is secreted from a bacterium and interferes with the biochemical processes of body cells in various ways. p. 489

experimental data Information that describes the result of a careful manipulation of the system under study. p. 415

exploitative competition Form of competition in which two or more individuals or populations use the same limiting resources. p. 717

exponential (model of population growth) Model that describes unlimited population growth. p. 689

extinction The death of the last individual in a species or the last species in a lineage. p. 781

extracellular digestion Digestion that takes place outside body cells, in a pouch or tube enclosed within the body. p. 603

facilitation hypothesis A hypothesis that explains ecological succession, suggesting that species modify the local environment in ways that make it less suitable for themselves but more suitable for colonization by species typical of the next successional stage. p. 737

family planning program A program that educates people about ways to produce an optimal family size on an economically feasible schedule. p. 702

family A Linnaean taxonomic category that ranks below an order and above a genus. p. 428

feather A sturdy, lightweight structure of birds, derived from scales in the skin of their ancestors. p. 662

flame cell The cell that forms the primary filtrate in the excretory system of many bilateria. The urine is propelled through ducts by the synchronous beating of cilia, resembling a flickering flame. p. 608

flower The reproductive structure of angiosperms, consisting of floral parts grouped on a stem; the structure in which seeds develop. p. 586

food web A set of interconnected food chains with multiple links. p. 572

founder effect An evolutionary phenomenon in which a population that was established by just a few colonizing individuals has only a fraction of the genetic diversity seen in the population from which it was derived. p. 419

fruit A mature ovary, often with accessory parts, from a flower. p. 586

fruiting body In some fungi, a stalked, spore-producing structure such as a mushroom. p. 529

fundamental niche The range of conditions and resources that a population can possibly tolerate and use. p. 721

furculum Wishbone in birds. p. 661

gametangium (plural, gametangia) A cell or organ in which gametes are produced. pp. 546, 572

gametic isolation A prezygotic reproductive isolating mechanism caused by incompatibility between the sperm of one species and the eggs of another; may prevent fertilization. p. 432

ganglion A functional concentration of nervous system tissue composed principally of nerve cell bodies, usually lying outside the central nervous system. pp. 608, 640

gastrodermis The derivative of endoderm that lines the gastrovascular cavity of radially symmetrical animals and forms the epithelial lining of the midgut in bilaterally symmetrical anmals. p. 603

gastrovascular cavity A saclike body cavity with a single opening, a mouth, which serves both digestive and circulatory functions. p. 602

gemma (plural, gemmae) Small cell mass that forms in cuplike growths on a thallus. p. 574

gemmules Clusters of cells with a resistant covering that allows them to survive unfavourable conditions. p. 602

gene flow The transfer of genes from one population to another through the movement of individuals or their gametes. p. 418

gene pool The sum of all alleles at all gene loci in all individuals in a population. p. 414

generalized compartment model A model used to describe nutrient cycling in which two criteria—organic versus inorganic nutrients and available versus unavailable nutrients—define four compartments where nutrients accumulate. p. 759

generation time The average time between the birth of an organism and the birth of its offspring. p. 684

genetic drift Random fluctuations in allele frequencies as a result of chance events; usually reduces genetic variation in a population. p. 418

genetic equilibrium The point at which neither the allele frequencies nor the genotype frequencies in a population change in succeeding generations. p. 415

genotype frequency The percentage of individuals in a population possessing a particular genotype. p. 414

genus A Linnaean taxonomic category ranking below a family and above a species. p. 427

geographic range The overall spatial boundaries within which a population lives. p. 680

germ layer The layers (up to three) of cells produced during the early development of the embryo of most animals. p. 595

gestation The period of mammalian development in which the embryo develops in the uterus of the mother. p. 668

gill arch One of the series of curved supporting structures between the slits in the pharynx of a chordate. p. 646

gill slit One of the openings in the pharynx of a chordate through which water passes out of the pharynx. p. 640

gill A respiratory organ formed as an evagination of the body that extends outward into the respiratory medium. p. 611

Ginkgophyta A plant phylum with a single living species, the ginkgo (or maidenhair) tree. p. 583

Gnathostomata The group of vertebrates with movable jaws. p. 644

gradualism The view that Earth and its living systems changed slowly over its history. p. 395

Gram stain procedure A procedure of staining bacteria to distinguish between types of bacteria with different cell wall compositions. p. 484

Gram-negative Describing bacteria that do not retain the stain used in the Gram stain procedure. p. 485

Gram-positive Describing bacteria that appear purple when stained using the Gram stain technique. p. 485

greenhouse effect A phenomenon in which certain gases foster the accumulation of heat in the lower atmosphere, maintaining warm temperatures on Earth. p. 764

gross primary productivity The rate at which producers convert solar energy into chemical energy. p. 750

gymnosperm A seed plant that produces "naked" seeds not enclosed in an ovary. p. 581

habitat fragmentation A process in which remaining areas of intact habitat are reduced to small, isolated patches. p. 653

habitat The specific environment in which a population lives, as characterized by its biotic and abiotic features. p. 680

Hardy–Weinberg principle An evolutionary rule of thumb that specifies the conditions under which a population of diploid organisms achieves genetic equilibrium. p. 415

harvesting efficiency The ratio of the energy content of food consumed to with the energy content of food available. p. 752

head-foot In molluscs, the region of the body that provides the major means of locomotion and contains concentrations of nervous system tissues and sense organs. p. 611

helical virus A virus in which the protein subunits of the coat assemble in a rodlike spiral around the genome. p. 502

hemocoel A cavity in the body of some coelomic invertebrates (arthropods and some molluscs) filled with blood. The hemocoel displaces the coelom, which persists as a small chamber surrounding the gonads or heart. p. 596

hemolymph The circulatory fluid of invertebrates with open circulatory systems, including molluscs and arthropods. p. 611

Hepatophyta The phylum that includes liverworts and their bryophyte relatives. p. 573

herbivore An animal that obtains energy and nutrients primarily by eating plants. p. 712

herbivory The interaction between herbivorous animals and the plants they eat. p. 712

heterodont Having different teeth specialized for different jobs. p. 667

heterosporous Producing two types of spores, "male" microspores and "female" megaspores. p. 570

heterozygote advantage An evolutionary circumstance in which individuals that are heterozygous at a particular locus have higher relative fitness than either homozygote. p. 421

Holocephali The chimeras, another group of cartilaginous fishes. p. 647

homeobox A region of a homeotic gene that corresponds to an amino acid section of the homeodomain. p. 394

hominin A member of a monophyletic group of primates, characterized by an erect bipedal stance, that includes modern humans and their recent ancestors. p. 469

hominoid (Hominoidea) The monophyletic group of primates that includes apes and humans. p. 469

homologous traits Characteristics that are similar in two species because they inherited the genetic basis of the trait from their common ancestor. p. 402

homoplasies Characteristics shared by a set of species, often because they live in similar environments, but not present in their common ancestor; often the product of convergent evolution. p. 458

homosporous Producing only one type of spore. p. 570

host race A population of insects that may be reproductively isolated from other populations of the same species as a consequence of their adaptation to feed on a specific host plant species. p. 438

hybrid breakdown A postzygotic reproductive isolating mechanism in which hybrids are capable of reproducing, but their offspring have either reduced fertility or reduced viability. p. 433

hybrid inviability A postzygotic reproductive isolating mechanism in which a hybrid individual has a low probability of survival to reproductive age. p. 432

hybrid sterility A postzygotic reproductive isolating mechanism in which hybrid offspring cannot form functional gametes. p. 432

hybrid zone A geographic area where the hybrid offspring of two divergent populations or species are common. p. 436

hybridization When two species interbreed and produce fertile offspring. p. 427

hydrogeologic cycle The global cycling of water between the ocean, the atmosphere, land, freshwater ecosystems, and living organisms. p. 760

hydrostatic skeleton A structure consisting of muscles and fluid that, by themselves, provide support for the animal or part of the animal; no rigid support, such as a bone, is involved. p. 596

hyomandibular bones Bones that support the hyoid and throat. p. 646

hypha (plural, hyphae) Any of the threadlike filaments that form the mycelium of a fungus. pp. 524, 542

immigration Movement of organisms into a population. p. 684

inbreeding A special form of nonrandom mating in which genetically related individuals mate with each other. p. 420

incisors Flattened, chisel-shaped teeth of mammals, located at the front of the mouth, that are used to nip or cut food. p. 464

incomplete metamorphosis In certain insects, a life cycle characterized by the absence of a pupal stage between the immature and adult stages. p. 629

incurrent siphon A muscular tube that brings water containing oxygen and food into the body of an invertebrate. p. 615

indeterminate cleavage A type of cleavage, observed in many deuterostomes, in which the developmental fates of the first few cells produced by mitosis are not determined as soon as cells are produced. p. 597

inhibition hypothesis A hypothesis suggesting that new species are prevented from occupying a community by whatever species are already present. p. 737

instar The stage between successive moults in insects and other arthropods. p. 628

interference competition Form of competition in which individuals fight over resources or otherwise harm each other directly. p. 717

intermediate disturbance hypothesis Hypothesis proposing that species richness is greatest in communities that experience fairly frequent disturbances of moderate intensity. p. 735

interspecific competition The competition for resources between species. p. 717

intertidal zone The shoreline that is alternately submerged and exposed by tides. p. 563

intracellular digestion The process in which cells take in food particles by endocytosis. p. 603

intraspecific competition The dependence of two or more individuals in a population on the same limiting resource. p. 693

intrinsic rate of increase The maximum possible per capita population growth rate in a population living under ideal conditions. p. 691

karyogamy In plants, the fusion of two sexually compatible haploid nuclei after cell fusion (plasmogamy). p. 543

keeled sternum The ventrally extended breastbone of a bird to which the flight muscles attach. p. 661

keystone species A species that has a greater effect on community structure than its numbers might suggest. p. 732

kingdom Animalia The taxonomic kingdom that includes all living and extinct animals. p. 594

kingdom Fungi The taxonomic kingdom that includes all living or extinct fungi. p. 543

kingdom Plantae The taxonomic kingdom encompassing all living or extinct plants. p. 564

kingdom A Linnaean taxonomic category that ranks below a domain and above a phylum. p. 428

Korarchaeota A group of Archaea recognized solely on the basis of rRNA coding sequences in DNA taken from environmental samples. p. 497

K-selected species Long-lived, slow-reproducing species that thrive in more stable environments. p. 697

larva (larval form) A sexually immature stage in the life cycle of many animals that is morphologically distinct from the adult. p. 595

latent phase The time during which a virus remains in the cell in an inactive form. p. 506

lateral line system The complex of mechanoreceptors along the sides of some fishes and aquatic amphibians that detect vibrations in the water. p. 648

left aortic arch In mammals, leads blood way from the heart to the aorta. p. 667

Lepidosauromorpha A monophyletic lineage of diapsids that includes both marine and terrestrial animals, represented today by sphenodontids, lizards, and snakes. p. 655

lichen A single vegetative body that is the result of an association between a fungus and a photosynthetic partner, often an alga. p. 555

life cycle The sequential stages through which individuals develop, grow, maintain themselves, and reproduce. p. 517

life history The lifetime pattern of growth, maturation, and reproduction that is characteristic of a population or species. p. 686

life table A chart that summarizes the demographic characteristics of a population. p. 685

limiting nutrient An element in short supply within an ecosystem, the shortage of which limits productivity. p. 750

lipopolysaccharide (LPS) A large molecule that consists of a lipid and a carbohydrate joined by a covalent bond. p. 485

logistic (model of population growth) Model of population growth that assumes that a population's per capita growth rate decreases as the population gets larger. p. 689

lophophore The circular or U-shaped fold with one or two rows of hollow, ciliated tentacles that surrounds the mouth of brachiopods, bryozoans, and phoronids and is used to gather food. p. 607

lumbar vertebrae In mammals, the vertebrae from the thoracic (bearing ribs) to the sacral (junction with pelvis). p. 670

Lycophyta The plant phylum that includes club mosses and their close relatives. p. 577

lysed Refers to a cell that has ruptured or undergone lysis. p. 504

macroevolution Large-scale evolutionary patterns in the history of life, producing major changes in species and higher taxonomic groups. p. 401

macronucleus In ciliophorans, a single large nucleus that develops from a micronucleus but loses all genes except those required for basic "housekeeping" functions of the cell and for ribosomal RNAs. p. 521

Malpighian tubule The main organ of excretion and osmoregulation in insects, helping them maintain water and electrolyte balance. p. 626

mandible In arthropods, one of the paired head appendages posterior to the mouth used for feeding. In vertebrates, the lower jaw. p. 624

mantle cavity The protective chamber produced by the mantle in many molluscs. p. 611

mantle One or two folds of the body wall that lines the shell and secretes the substance that forms the shell in molluscs. p. 611

mastax The toothed grinding organ at the anterior of the digestive tract in rotifers. p. 610

mating type A genetically defined strain of an organism (such as a fungus) that can only mate with an organism of the opposite mating type; mating types are often designated + and −. p. 540

maxilla (plural, **maxillae**) One of the paired head appendages posterior to the mouth used for feeding in arthropods. pp. 624, 646

mechanical isolation A prezygotic reproductive isolating mechanism caused by differences in the structure of reproductive organs or other body parts. p. 432

medusa (plural, **medusae**) The tentacled, usually bell-shaped, free-swimming sexual stage in the life cycle of a coelenterate. p. 603

megaspore A plant spore that develops into a female gametophyte; usually larger than a microspore. p. 570

mesenteries Sheets of loose connective tissue, covered on both surfaces with epithelial cells, which suspend the abdominal organs in the coelom and provide lubricated, smooth surfaces that prevent chafing or abrasion between adjacent structures as the body moves. p. 596

mesoderm The middle layer of the three primary germ layers of an animal embryo, from which the muscular, skeletal, vascular, and connective tissues develop. p. 595

mesoglea A layer of gel-like connective tissue separating the gastrodermis and epidermis in radially symmetrical animals. It contains widely dispersed amoeboid cells. p. 603

mesohyl The gelatinous middle layer of cells lining the body cavity of a sponge. p. 601

metamorphosis A reorganization of the form of certain animals during postembryonic development. p. 602

metanephridium (plural, **metanephridia**) The excretory tubule of most annelids and molluscs. p. 617

microclimate The abiotic conditions immediately surrounding an organism. p. 737

microevolution Small-scale genetic changes within populations, often in response to shifting environmental circumstances or chance events. pp. 401, 410

micronucleus In ciliophorans, one or more diploid nuclei that contain a complete complement of genes, functioning primarily in cellular reproduction. p. 521

microspore A plant spore from which a male gametophyte develops; usually smaller than a megaspore. pp. 570, 583

microvilli Fingerlike projections forming a brush border in epithelial cells that cover the villi. p. 601

mimic The species in Batesian mimicry that resembles the model. p. 716

mimicry A form of defence in which one species evolves an appearance resembling that of another. p. 716

model The species in Batesian mimicry that is resembled by the mimic. p. 716

modern synthesis A unified theory of evolution developed in the middle of the twentieth century. p. 401

molars Posteriormost teeth of mammals, with a broad chewing surface for grinding food. p. 464

molecular clock A technique for dating the time of divergence of two species or lineages, based on the number of molecular sequence differences between them. p. 471

monocot A plant belonging to the Monocotyledones, one of the two major classes of angiosperms; monocot embryos have a single seed leaf (cotyledon) and pollen grains with a single groove. p. 587

monoecious Having both "male" flowers (which possess only stamens) and "female" flowers (which possess only carpels). p. 601

monophyletic taxon A group of organisms that includes a single ancestral species and all of its descendants. p. 459

monotreme A lineage of mammals that lay eggs instead of bearing live young. p. 668

morphological species concept The concept that all individuals of a species share measurable traits that distinguish them from individuals of other species. p. 428

mosaic evolution The tendency of characteristics to undergo different rates of evolutionary change within the same lineage. p. 458

motile Capable of self-propelled movement. p. 595

Müllerian mimicry A form of defence in which two or more unpalatable species share a similar appearance. p. 716

multiregional hypothesis A hypothesis proposing that after archaic humans migrated from Africa to many regions on Earth, their different populations evolved into modern humans simultaneously. p. 477

mutualism A symbiotic interaction between species in which both partners benefit. pp. 541, 722

mycelium A network of branching hyphae that constitutes the body of a multicellular fungus. pp. 524, 542

mycobiont The fungal component of a lichen. p. 555

mycorrhiza A mutualistic symbiosis in which fungal hyphae associate intimately with plant roots. p. 548

natural selection The evolutionary process by which alleles that increase the likelihood of survival and the reproductive output of the individuals that carry them become more common in subsequent generations. p. 398

natural theology A belief that knowledge of God may be acquired through the study of natural phenomena. p. 392

nematocyst A coiled thread, encapsulated in a cnidocyte, that cnidarians fire at prey or predators, sometimes releasing a toxin through its tip. p. 603

nerve net A simple nervous system that coordinates responses to stimuli but has no central control organ or brain. p. 603

net primary productivity The chemical energy remaining in an ecosystem after a producer's cellular respiration is deducted. p. 750

neutral mutation hypothesis An evolutionary hypothesis that some variation at gene loci coding for enzymes and other soluble proteins is neither favoured nor eliminated by natural selection. p. 425

nitrification A metabolic process in which certain soil bacteria convert ammonia or ammonium ions into nitrites that are then converted by other bacteria to nitrates, a form usable by plants. pp. 488, 766

nitrogen cycle A biogeochemical cycle that moves nitrogen between the huge atmospheric pool of gaseous molecular nitrogen and several much smaller pools of nitrogen-containing compounds in soils, marine and freshwater ecosystems, and living organisms. p. 762

nitrogen fixation A metabolic process in which certain bacteria and cyanobacteria convert molecular nitrogen into ammonia and ammonium ions, forms usable by plants. pp. 488, 762

nonvascular plant *See* bryophyte. p. 567

notochord A flexible rodlike structure constructed of fluid-filled cells surrounded by tough connective tissue, which supports a chordate embryo from head to tail. p. 640

null model A conceptual model that predicts what one would see if a particular factor had no effect. p. 415

obligate aerobe A microorganism that must use oxygen for cellular respiration and requires oxygen in its surroundings to support growth. p. 487

obligate anaerobe A microorganism that cannot use oxygen and can grow only in the absence of oxygen. p. 487

oocyte A developing gamete that becomes an ootid at the end of meiosis. p. 601

open circulatory system An arrangement of internal transport in some invertebrates in which the vascular fluid, hemolymph, is released into sinuses, bathing organs directly, and is not always retained within vessels. p. 611

operculum A lid or flap of the bone serving as the gill cover in some fishes. pp. 615, 649

optimal foraging theory A set of mathematical models that predict the diet choices of animals as they encounter a range of potential food items. p. 712

oral hood Soft fleshy structure at the anterior end of a cephalochordate that frames the opening of the mouth. p. 641

order A Linnaean taxonomic category of organisms that ranks above a family and below a class. p. 428

organismal ecology An ecological discipline in which researchers study the genetic, biochemical, physiological, morphological, and behavioural adaptations of organisms to their abiotic environments. p. 680

osculum (plural, **oscula**) An opening in a sponge through which water is expelled. p. 601

ostracoderm One of an assortment of extinct, jawless fishes that were covered with bony armour. p. 645

outer membrane In Gram-negative bacteria, an additional boundary membrane that covers the peptidoglycan layer of the cell wall. p. 485

outgroup comparison A technique used to identify ancestral and derived characters by comparing the group under study with more distantly related species that are not otherwise included in the analysis. p. 458

ovule In plants, the structure in a carpel in which a female gametophyte develops and fertilization takes place. p. 582

parapatric speciation Speciation between populations with adjacent geographic distributions. p. 436

paraphyletic taxon A group of organisms that includes an ancestral species and some, but not all, of its descendants. p. 459

parapodium (plural, parapodia) A fleshy lateral extension of the body wall of aquatic annelids, used for locomotion and gas exchange. p. 618

parasite An organism that feeds on the tissues of or otherwise exploits its host. p. 516

parasitism A symbiotic interaction in which one species, the parasite, uses another, the host, in a way that is harmful to the host. pp. 541, 723

parthenogenesis A mode of asexual reproduction in which animals produce offspring by the growth and development of an egg without fertilization. p. 610

passive parental care The amount of energy invested in offspring—in the form of the energy stored in eggs or seeds or energy transferred to developing young through a placenta—before they are born. p. 687

pectoral girdle A bony or cartilaginous structure in vertebrates that supports and is attached to the forelimbs. p. 641

pedicellariae Small pincers at the base of short spines in starfishes and sea urchins. p. 638

pedipalps The second pair of appendages in the head of chelicerates. p. 622

pellicle A layer of supportive protein fibres located inside the cell, just under the plasma membrane, providing strength and flexibility instead of a cell wall. p. 517

pelvic girdle A bony or cartilaginous structure in vertebrates that supports and is attached to the hindlimbs. p. 641

peptidoglycan A polymeric substance formed from a polysaccharide backbone tied together by short polypeptides, which is the primary structural molecule of bacterial cell walls. p. 484

per capita growth rate The difference between the per capita birth rate and the per capita death rate of a population. p. 690

peritoneum The thin tissue derived from mesoderm that lines the abdominal wall and covers most of the organs in the abdomen. p. 596

pharynx The throat. In some invertebrates, a protrusible tube used to bring food into the mouth for passage to the gastrovascular cavity; in mammals, the common pathway for air entering the larynx and food entering the esophagus. p. 609

phenotypic variation Differences in appearance or function between individual organisms. p. 410

pheromone A distinctive volatile chemical released in minute amounts to influence the behaviour of members of the same species. p. 405

phloem The food-conducting tissue of a vascular plant. p. 567

phosphorus cycle A biogeochemcial cycle in which weathering and erosion carry phosphate ions from rocks to soil and into streams and rivers, which eventually transport them to the ocean, where they are slowly incorporated into rocks. p. 768

photobiont The photosynthetic component of a lichen. p. 555

photoheterotroph An organism that uses light as the ultimate energy source but obtains carbon in organic form rather than as carbon dioxide. p. 487

phototroph An organism that obtains energy from light. p. 487

PhyloCode A formal set of rules governing phylogenetic nomenclature. p. 461

phylogenetic species concept A concept that seeks to delineate species as the smallest aggregate population that can be united by shared derived characters. p. 429

phylogenetic tree A branching diagram depicting the evolutionary relationships of groups of organisms. p. 455

phylogeny The evolutionary history of a group of organisms. p. 455

phylum (plural, phyla) A major Linnaean division of a kingdom, ranking above a class. p. 428

phytoplankton Microscopic, free-flowing aquatic plants and protists. p. 516

pinacoderm In sponges, an unstratified outer layer of cells. p. 601

placenta A specialized temporary organ that connects the embryo and fetus with the uterus in mammals, mediating the delivery of oxygen and nutrients. Analagous structures occur in other animals. p. 668

plasmodial slime mould A slime mould of the class Myxomycetes. p. 529

plasmodium The composite mass of plasmodial slime moulds consisting of individual nuclei suspended in a common cytoplasm surrounded by a single plasma membrane. p. 529

plasmogamy The sexual stage of fungi during which the cytoplasms of two genetically different partners fuse. p. 543

plastron The ventral part of the shell of a turtle. p. 659

poikilohydric Having little control over internal water content. p. 565

pollen grain The male gametophyte of a seed plant. p. 582

pollen sac The microsporangium of a seed plant, in which pollen develops. p. 589

pollen tube A tube that grows from a germinating pollen grain through the tissues of a carpel and carries the sperm cells to the ovary. p. 582

pollination The transfer of pollen to a flower's reproductive parts by air currents or on the bodies of animal pollinators. p. 582

polyhedral virus A virus in which the coat proteins form triangular units that fit together like the parts of a geodesic sphere. p. 502

polymorphic development The production during development of one or more morphologically distinct forms. p. 595

polymorphism The existence of discrete variants of a character among individuals in a population. p. 411

polyp The tentacled, usually sessile stage in the life cycle of a coelenterate. p. 603

polyphyletic taxon A group of organisms that belong to different evolutionary lineages and do not share a recent common ancestor. p. 459

polyploidy The condition of having one or more extra copies of the entire haploid complement of chromosomes. p. 439

population bottleneck An evolutionary event that occurs when a stressful factor reduces population size greatly and eliminates some alleles from a population. p. 418

population density The number of individuals per unit area or per unit volume of habitat. p. 680

population ecology The ecological discipline that focuses on how a population's size and other characteristics change in space and time. p. 680

population genetics The branch of science that studies the prevalence and variation in genes among populations of individuals. p. 401

population size The number of individuals in a population at a specified time. p. 680

population All individuals of a single species that live together in the same place and time. p. 410

posterior Indicating the tail end of an animal. p. 596

postzygotic isolating mechanism A reproductive isolating mechanism that acts after zygote formation. p. 431

precocial Born with fur and quickly mobile. p. 668

predation The interaction between predatory animals and the animal prey they consume. p. 712

premolars Teeth located in pairs on each side of the upper and lower jaws of mammals, positioned behind the canines and in front of the molars. p. 668

prezygotic isolating mechanism A reproductive isolating mechanism that acts prior to the production of a zygote, or fertilized egg. p. 431

primary endosymbiosis In the model for the origin of plastids in eukaryotes, the first event in which a eukaryotic cell engulfed a photosynthetic cyanobacterium. p. 535

primary succession Predictable change in species composition of an ecological community that develops on bare ground. p. 735

principle of monophyly A guiding principle of systematic biology that defines monophyletic taxa, each of which contains a single ancestral species and all of its descendants. p. 459

prion An infectious agent that contains only protein and does not include a nucleic acid molecule. p. 509

production efficiency The ratio of the energy content of new tissue produced to the energy assimilated from food. p. 753

proglottid One of the segmentlike repeating units that constitute the body of a tapeworm. p. 609

protist Organism currently classified in the kingdom Protista. p. 513

protonema The structure that arises when a liverwort or moss spore germinates and eventually gives rise to a mature gametophyte. p. 574

protostome A division of the Bilateria in which the blastopore forms the mouth during development of the embryo and the anus appears later. p. 596

pseudocoelom A fluid- or organ-filled body cavity between the gut (a derivative of endoderm) and the muscles of the body wall (a derivative of mesoderm). p. 596

pseudocoelomate A body plan of bilaterally symmetrical animals with a body cavity that lacks a complete lining derived from mesoderm. p. 596

pseudopod (plural, **pseudopodia)** A temporary cytoplasmic extension of a cell. p. 517

psychrophile An archaean or bacterium that grows optimally at temperatures in the range of −10 to −20°C. p. 497

Pterophyta The plant phylum of ferns and their close relatives. p. 577

pupa The nonfeeding stage between the larva and adult in the complete metamorphosis of some insects, during which the larval tissues are completely reorganized within a protective cocoon or hardened case. p. 629

pyramid of biomass A diagram that illustrates differences in standing crop biomass in a series of trophic levels. p. 754

pyramid of energy A diagram that illustrates the amount of energy that flows through a series of trophic levels. p. 754

pyramid of numbers A diagram that illustrates the number of individual organisms present in a series of trophic levels. p. 755

qualitative variation Variation that exists in two or more discrete states, with intermediate forms often being absent. p. 411

quantitative variation Variation that is measured on a continuum (such as height in human beings) rather than in discrete units or categories. p. 410

quorum sensing The use of signalling molecules by prokaryotes to communicate and to coordinate their behaviour. p. 491

radial cleavage A cleavage pattern in deuterostomes in which newly formed cells lie directly above and below other cells of the embryo. p. 596

radial symmetry A body plan of organisms in which structures are arranged regularly around a central axis, like spokes radiating out from the centre of a wheel. p. 602

radula The tooth-lined "tongue" of molluscs that scrapes food into small particles or drills through the shells of prey. p. 611

random dispersion A pattern of distribution in which the individuals in a population are distributed unpredictably in their habitat. p. 682

realized niche The range of conditions and resources that a population actually uses in nature. p. 721

recognition protein Protein in the plasma membrane that identifies a cell as part of the same individual or as foreign. p. 504

red tide A growth in dinoflagellate populations that causes red, orange, or brown discoloration of coastal ocean waters. p. 521

reinforcement 1. The enhancement of reproductive isolation that had begun to develop while populations were geographically separated; 2. Encouraging or establishing a pattern of behaviour using a positive or negative stimulus. p. 436

relative abundance The relative commonness of populations within a community. p. 414

relative fitness The number of surviving offspring that an individual produces compared with the number left by others in the population. p. 420

reproductive isolating mechanism A biological characteristic that prevents the gene pools of two species from mixing. p. 430

resource partitioning The use of different resources or the use of resources in different ways by species living in the same place. p. 721

rhizoid A modified hypha that anchors a fungus to its substrate and absorbs moisture. p. 572

rhizome A horizontal, modified stem that can penetrate a substrate and anchor the plant. p. 577

rhynchocoel A coelomic cavity that contains the proboscis of nemerteans. p. 611

ring species A species with a geographic distribution that forms a ring around uninhabitable terrain. p. 433

rod In the vertebrate eye, a type of photoreceptor in the retina that is specialized for detection of light at low intensities. p. 482

root system An underground (or submerged) network of roots with a large surface area that favours the rapid uptake of soil water and dissolved mineral ions. p. 568

root An anchoring structure in land plants that also absorbs water and nutrients and (in some plant species) stores food. p. 568

r-selected species A short-lived species adapted to function well in a rapidly changing environment. p. 697

saprotroph An organism nourished by dead or decaying organic matter. p. 540

savannah A biome comprising grasslands with few trees, which grows in areas adjacent to tropical deciduous forests. p. 727

schizocoelom In protostomes, the body cavity that develops as inner and outer layers of mesoderm separate. p. 598

sclerotium Tough mass of hyphae, often serving as a survival or overwintering structure. p. 550

scolex The anterior (head) of a tapeworm, adapted for fastening the worm to the intestinal epithelium of its host. p. 609

secondary endosymbiosis In the model for the origin of plastids in eukaryotes, the second event, in which a nonphotosynthetic eukaryote engulfed a photosynthetic eukaryote. p. 535

secondary metabolite Organic compound not required for the growth or survival of an organism; tends to be biologically active. p. 543

secondary productivity Energy stored in new consumer biomass as energy is transferred from producers to consumers. p. 752

secondary succession Predictable changes in species composition in an ecological community that develops after existing vegetation is destroyed or disrupted by an environmental disturbance. p. 737

seed The structure that forms when an ovule matures after a pollen grain reaches it and a sperm fertilizes the egg. p. 582

segmentation The production of body parts and some organ systems in repeating units. p. 599

self-fertilization (self-pollination) Fertilization in which sperm nuclei in pollen produced by anthers fertilize egg cells housed in the carpel of the same flower. Self-fertilization can also occur in hermaphroditic animals. p. 420

septum (plural, septa) A thin partition or cross wall that separates body segments. pp. 542, 617

sessile Unable to move from one place to another. p. 595

seta (plural, **setae**) A chitin-reinforced bristle that protrudes outward from the body wall in some annelid worms. p. 617

sex ratio The relative proportions of males and females in a population. p. 684

sexual selection A form of natural selection established by male competition for access to females and by the females' choice of mates. p. 405

shoot system The stems and leaves of a plant. p. 568

simulation modelling An analytical method in which researchers gather detailed information about a system and then create a series of mathematical equations that predict how the components of the system interact and respond to change. p. 770

sinus (plural, **sinuses**) A body space that surrounds an organ. p. 611

social behaviour The interactions that animals have with other members of their species. p. 469

soredium (plural, **soredia**) A specialized cell cluster produced by lichens, consisting of a mass of algal cells surrounded by fungal hyphae; soredia function like reproductive spores and can give rise to a new lichen. p. 555

sorus (plural, **sori**) A cluster of sporangia on the underside of a fern frond; reproductive spores arise by meiosis inside each sporangium. p. 578

source population In metapopulation analyses, a population that is either stable or increasing in size. p. 738

speciation The process of species formation. p. 427

species cluster A group of closely related species recently descended from a common ancestor. p. 436

species composition The particular combination of species that occupy a site. p. 730

species diversity A community characteristic defined by species richness and the relative abundance of species. p. 727

species richness The number of species that live within an ecological community. p. 726

species A group of populations in which the individuals are so closely related in structure, biochemistry, and behaviour that they can successfully interbreed. p. 427

specific epithet The species name in a binomial. p. 427

spermatocyte A developing gamete that becomes a spermatid at the end of meiosis. p. 813

spinneret A modified abdominal appendage from which spiders secrete silk threads. p. 623

spiral cleavage The cleavage pattern in many protostomes in which newly produced cells lie in the space between the two cells immediately below them. p. 596

spiral valve A corkscrew-shaped fold of mucous membrane in the digestive system of elasmobranchs, which slows the passage of material and increases the surface area available for digestion and absorption. p. 648

spongocoel The central cavity in a sponge. p. 601

sporangium (plural, **sporangia**) A single-celled or multicellular structure in fungi and plants in which spores are produced. pp. 545, 570

sporophyll A specialized leaf that bears sporangia (spore-producing structures). p. 577

squalene A liver oil found in sharks that is lighter than water, which increases their buoyancy. p. 648

stability The ability of a community to maintain its species composition and relative abundances when environmental disturbances eliminate some species from the community. p. 725

stabilizing selection A type of natural selection in which individuals expressing intermediate phenotypes have the highest relative fitness. p. 413

standing crop biomass The total dry weight of plants present in an ecosystem at a given time. p. 750

stapes The smallest of three sound-conducting bones in the middle ear of tetrapod vertebrates. p. 652

stoma (plural, **stomata**) The opening between a pair of guard cells in the epidermis of a plant leaf or stem, through which gases and water vapour pass. p. 565

strobilus *See* cone (of a plant). p. 577

subspecies A taxonomic subdivision of a species. p. 433

succession The change from one community type to another. p. 735

survivorship curve Graphic display of the rate of survival of individuals over a species' life span. p. 685

suspension (filter) feeder An animal that ingests small food items suspended in water. p. 601

swim bladder A gas-filled internal organ that helps fish maintain buoyancy. p. 649

symbiont An organism living in symbiosis with another organism; the symbionts are not usually closely related. p. 540

symmetry (adj., symmetrical) Exact correspondence of form and constituent configuration on opposite sides of a dividing line or plane. p. 595

sympatric speciation Speciation that occurs without the geographic isolation of populations. p. 438

sympatric Occupying the same spaces at the same time. p. 680

synapsid One of a group of amniotes with one temporal arch on each side of the head, which includes living mammals. p. 655

taiga *See* boreal forest. p. 751

taxon (plural, **taxa**) A name designating a group of organisms included within a category in the Linnaean taxonomic hierarchy. p. 428

taxonomic hierarchy A system of classification based on arranging organisms into ever more inclusive categories. p. 428

taxonomy The science of the classification of organisms into an ordered system that indicates natural relationships. p. 427

temperate deciduous forest A forested biome found at low to middle altitudes at temperate latitudes, with warm summers, cold winters, and annual precipitation between 75 and 250 cm. p. 726

temperate grassland A nonforested biome that stretches across the interiors of most continents, where winters are cold and snowy and summers are warm and fairly dry. p. 751

temperate rain forest A coniferous forest biome supported by heavy rain and fog, which grows where winters are mild and wet and the summers are cool. p. 525

temporal isolation A prezygotic reproductive isolating mechanism in which species live in the same habitat but breed at different times of day or different times of year. p. 431

tertiary consumer A carnivore that feeds on other carnivores, a member of the fourth trophic level. p. 729

Tetrapoda A monophyletic lineage of vertebrates that includes animals with four feet, legs, or leglike appendages. pp. 461, 644

T-even bacteriophage Virulent bacteriophages, T2, T4, and T6, that have been valuable for genetic studies of bacteriophage structure and function. p. 504

thallus (plural, **thalli**) A plant body not differentiated into stems, roots, or leaves. pp. 555, 573

thorax The central part of an animal's body, between the head and the abdomen. p. 624

time lag The delayed response of organisms to changes in environmental conditions. p. 693

tolerance hypothesis Hypothesis asserting that ecological succession proceeds because competitively superior species replace competitively inferior ones. p. 739

torsion The realignment of body parts in gastropod molluscs that is independent of shell coiling. p. 615

totipotency The ability to develop into any type of cells. p. 666

totipotent Having the capacity to produce cells that can develop into or generate a new organism or body part. p. 601

tracheal system A branching network of tubes that carries air from small openings in the exoskeleton of an insect to tissues throughout its body. p. 626

traditional evolutionary systematics An approach to systematics that uses phenotypic similarities and differences to infer evolutionary relationships, grouping species that share both ancestral and derived characters. p. 460

trichocyst A dartlike protein thread that can be discharged from a surface organelle for defence or to capture prey. p. 521

triploblastic An animal body plan in which adult structures arise from three primary germ layers: endoderm, mesoderm, and ectoderm. p. 595

trochophore The small, free-swimming, ciliated aquatic larva of various invertebrates, including certain molluscs and annelids. p. 614

trophic cascade The effects of predator–prey interactions that reverberate through other population interactions at two or more trophic levels in an ecosystem. p. 757

trophic level A position in a food chain or web that defines the feeding habits of organisms. p. 728

trophozoite Motile, feeding stage of *Giardia* and other single-celled protists. p. 513

tropical deciduous forest A tropical forest biome that occurs where winter drought reduces photosynthesis and most trees drop their leaves seasonally. p. 751

tropical forest Any forest that grows between the Tropics of Capricorn and Cancer, a region characterized by high temperature and rainfall and thin, nutrient-poor topsoil. p. 582

tropical rain forest A dense tropical forest biome that grows where some rain falls every month, mean annual rainfall exceeds 250 cm, mean annual temperature is at least 25°C, and humidity is above 80%. p. 557

turnover rate The rate at which one generation of producers in an ecosystem is replaced by the next. p. 755

tympanum A thin membrane in the auditory canal that vibrates back and forth when struck by sound waves. p. 652

undulating membrane In parabasalid protists, a fin-like structure formed by a flagellum buried in a fold of the cytoplasm that facilitates movement through thick and viscous fluids. An expansion of the plasma membrane in some flagellates that is usually associated with a flagellum. p. 520

uniform dispersion A pattern of distribution in which the individuals in a population are evenly spaced in their habitat. p. 682

uniformitarianism The concept that the geologic processes that sculpted Earth's surface over long periods of time—such as volcanic eruptions, earthquakes, erosion, and the formation and movement of glaciers—are exactly the same as the processes observed today. p. 395

unreduced gamete A gamete that contains the same number of chromosomes as a somatic cell. p. 441

variable An environmental factor that may differ among places or an organismal characteristic that may differ among individuals. p. 412

vascular plant A plant with xylem, phloem, and usually well-developed roots, stems, and leaves. p. 567

vascular tissue In plants, tissue that transports water and nutrients or the products of photosynthesis through the plant body. p. 567

veliger A second larva that occurs after the trochophore in some molluscs. p. 599

ventral Indicating the lower or "belly" side of an animal. p. 596

vestigial structure An anatomical feature of living organisms that no longer retains its function. p. 393

virion A complete virus particle. p. 503

viroid A plant pathogen that consists of strands or circles of RNA, smaller than any viral DNA or RNA molecule, that have no protein coat. p. 509

visceral mass In molluscs, the region of the body containing the internal organs. p. 611

viviparous Referring to animals that retain the embryo within the mother's body and nourish it during at least early embryo development. p. 668

watershed An area of land from which precipitation drains into a single stream or river. p. 761

wetland A highly productive ecotone often at the border between a freshwater biome and a terrestrial biome. p. 586

yolk The portion of an egg that serves as the main energy source for the embryo. p. 654

zero population growth A circumstance in which the birth rate of a population equals the death rate. p. 691

zooplankton Small, usually microscopic, animals that float in aquatic habitats. p. 516

zygospore A multinucleate, thick-walled sexual spore in some fungi that is formed from the union of two gametes. p. 545

Index

The letter i *designates illustration;* t *designates table;*
b *designates box;* **bold** *designates defined or introduced term.*

Birds
    dinosaur lineages, 401, 402*i*
    evolution of, 461–464, 463*i*, 466
    finch bill shape and food habits, 397*i*, 398–399,
        399*b*, 399*i*
    finches, Darwin's, 396
    flightless, 393*i*
    species recognition among, 434
    See also Aves, class
Birth, **475–476**
Bivalvia, class, **615**, 616*f*
Black-footed ferret, 683*f*
Blastophaga psenes, 811, 811*i*
Blastopore, **597**
Blood vessels
Blowfly, 391, 391*i*
Blue jay, 418, 418*i*
Blue-footed booby, 397*i*
Blue-headed wrasse, 435*i*
Blue-stain fungi, 548, 549*i*
Body cavity, **596**, 597*f*
Body language, 477*f*
Bony fish, 648–651
Book lungs, **622**
Boreal forest, **557**
Bos indicus, 813, 813*i*
Bos primigenius, 813
Bos taurus, 678, 813, 813*i*
Bottom-up theory, **665**
Bovine spongiform encephalopathy (BSE), 510*i*
Brachiopoda, phylum, **607**, 607*f*
Bradshaw, H. D., 440*b*
Brain
    capacity related to diet, 808, 808*i*
"Brainscan Atlas," **468**
Branching evolution, **455**, 456*i*
Branchiostoma, 641
Brassica, 815*b*, 823*b*, 823*i*
Bread mould, 546*i*, 547
Breastbone, 664*f*
Breeding frequency, **687**
Brittle star, 638
Bryophytes, 568*t*, 571*i*, **572–575**, 572*i*, 573*i*, 576*t*,
    596*t*
Bucket traps, 390, 390*i*
Budding, **542**, 548
Bullock's oriole, 438*i*
Burgess Shale, **445**, 593–594, 594*f*
Butterfly, 403, 405*i*, 682*f*
Butterwort, 390*i*
Byer, Peter, 829*b*

C
Cactus finch, 399*b*, 399*i*
Caecilian, **652–654**
Caenorhabditis elegans, 620
Cambrian period, 448*t*
Canadian bunchberry, 427
Canadian tiger swallowtail butterfly, 427
Canis familiaris, domestication of, 809, 820
Capsicum, domestication of, 819

Capsid, **501**
Capsule, **485**
Capture-recapture, 682*f*
Carapace, **624**
Carapus bermudensis, 639
Carbon cycle, **760**
Carboniferous period, 448*t*
    called the Coal Age, 567*b*, 578*i*
    as dominated by *Lycophyta*, 577
    as dominated by nonvascular plants, 581
    as dominated by seedless vascular plants, 567*b*, 575
Cardenolides, 403, 405*i*
Cardiac glycosides, 403, 405*i*
Carnivore, **670**
Carpel, **586**, 589*i*
Carrageenan, **532**
Carrying capacity, **807**
Carson, Hampton, 437*b*
Catastrophism, **393**, 395
Cats, 825, 826*i*
Cattle, domestication of, 813–814, 814*i*, 821
Cellular DNA/RNA, viruses evolving from, **509**, 511
Cellular slime mould, **529**
Cellulose, **539**, 551, 563, 564
Cenozoic era, 447*t*, 587
Cephalization, **596**
Cephalochordata, subphylum, **641**
Cephalopoda, class, **615–617**, 616*f*
Cephalothorax, **622**
Certhidea olivacea (warbler finch), 399
Cestoda, class, **609**, 610*f*
Character displacement, **721**
Charophytes, **564**, 564*i*
Chase, Ron, 611
Cheetah, African, 419
Chelicerae, **622**
Chelicerata, subphylum, **622–623**, 623*f*
Chemicals
    from plants, 827–828, 827*i*, 831
    from plants and animals, 827*i*, 831
Chemoautotroph, **487**
Chemoheterotroph, **487**
Chemotroph, **487**
Chili peppers, domestication of, 819
Chitin, **542**, 543
Chlorocebus pygerythrus, 476
Choanocytes, **601**
Choanoflagellates, 595
Chondrichthyes, class, **647–648**
Chondricthye, 647*f*
Chordata, phylum, **640–642**, 640*f*
Chromosomal alterations, 442
Chromosome banding, 472*b*
Chytridiomycota, **544**, 544*i*, 544*t*, 545, 545*i*
Chytrids, 544*t*, 545, 548
Cicada, 621*f*
Circulatory system, in molluscs
    closed, 616
    open, **611**
Circulatory vessel, **611**
Clade, **461**

Insects
  body plan, 627*f*
  cardenolides, toxicity of, 403, 405*i*
  development, 628*f*
  diversity, 626*f*
  juvenile hormone, 629*f*
  mouthparts, 627*f*
  physiology, 628*f*
  sodium-potassium pump, 403
Instar, **628**
Interference competition, **717**
Interleuken, **828**
Intermediate disturbance hypothesis, **735**
Interspecific competition, **717**
Interspecific hybrids, 433*i*
Intertidal zones, **563**
Intracellular digestion, **603**
Intraspecific competition, **693**
Intrinsic rate of increase, **691**
Irrigation, 807, 825, 825*i*
Irritable bowel disease, **828**
Isthmus of Panama, 435, 435*i*

J
Jawed fish, **646–651**, 646*f*
Jimson weed, 819
Junipers, 557*i*
Jurassic period, 448*t*
Juvenile hormone (JH), 629*f*

K
Karyogamy, **543**
Keeled sternum, **661**
Keystone species, **732**
Kingdom Animalia, **594**
Kingdom Fungi, **543**
Kingdom Plantae, **564**
Kitlg gene, 422*b*
Korarchaeota, **497**
K-selected species, **697**

L
Labium, **627**
Lactose intolerance, **477**
Lamarck, Jean Baptiste de, 395–396
Lambda, **505–506**, 505*i*
Lamprey, **644–646**
Lancelet, 641*f*
Land, movement of organisms from, 453
Land plants
  defining characteristics of, 563
  phylogenetic relationships between major groups
     of, 571*i*
Language, **476**
Large ground-finch, 397*i*, 399
Larus, **678**
Larva (larval form), **595**
Latent phase, **506**
Leaf endophytes, 559*i*
Leaf-cutter ants, 540, 540*i*, 541*i*
Leakey, Mary, 473*f*

Leaves, **569**, 569*i*
Left aortic arch, **667**
Leg, **473–474**
Lens culinaris, **816–817**, 816–817*i*
Lens orientalis, **816–817**, 816–817*i*
Lentils, domestication of, 816–817, 817*i*
"Leopard" alarm, **476**
Lepeophtheirus salmonis, 821
Lepidodendron (lycophyte tree), 578*i*
Lepidosaura, infraclass, **660–661**
Lepidosauromorpha, **655**
Lethal mutations, **417**
Lichens, **548**, 555–556, 558*i*
Life cycle, **517**
Life history, **686**
  evolution, 686–688
  guppies, 688–689*f*
Life table, **684–685**, 685*t*
Lightfoot, John, 395
Lignin, **551**, 552*i*, 566–568, 567*b*
Limiting nutrient, **750**
Limpet, common, 615*f*
Linnaeus, Carolus, 392, 427–428
Linné, Carl von. *See* Linnaeus, Carolus
Lipopolysaccharide (LPS), **485**
Liver fluke, 612*f*
Lizards, 660–661
Locoweed, 819
Logistic model of population growth,
     691–693
Londo, Jason, 816
Lophophorate, **607**
Lophophore, **607**
Lophotrochozoan protostome, **606–619**
"Love dart," **611**
LSD, **551**
"Lucy," 469*b*, 474*f*
Lumbar vertebrae, **670**
Lungs
  book, **622**
Lycophyta, 571*i*, **576**, 576*t*, 577
Lycophytes (club mosses), 569*i*, 570*i*, **577**, 578*i*
Lycopodium, 570*i*, **577**, 578*i*
Lyell, Charles, 399
     *Principles of Geology: An Attempt to Explain the Former
       Changes of the Earth's Surface by Reference to Causes
       Now in Operation*, 395, 396
Lysed, **504**
Lysergic acid, **550**, 550*b*
Lysogenic cycle, **504**
Lysozyme, **504**
Lytic cycle, **505**

M
Macroevolution, **401**, 426–427, 426–444
Macronucleus, **521**
Malaria (*Plasmodium falciparum*), **421**, 421*i*, 829*b*
Mallard duck (*Anas platyrhynchos*), 440*i*
Malloch, David, 567*b*
Malpighian tubules, **626**

breeding frequency, 687
courtship displays, 431
geographical speciation, 433–439
inbreeding, 420
isolating mechanisms, 430–433
nonrandom mating, 417t, 420
polyploidy speciation, 441i
proportion, 684
sexual selection, 405–406
See also Speciation
Reproductive isolating mechanism, **430–433**
Resource partitioning, **721**
Retrovirus, 503t
Reverse transcriptase, 507–508i
Rhabdovirus, 503t
Rhamphorynchus meunsteri, 657f
Rhea darwinii, **428**
Rhinovirus, 503t
Rhizoids, **572**, 573, 574i, 577i, 578
Rhizomes, **577**, 577i, 579i
Rhizopus nigricans, 547i
Rhynchocoel, **611**
Rhynia, 577
Rhynia gwynne-vaughnii, 577i
Ribbon worm, 611, 611f
Rice
    domestication of, 812–813, 812i, 815–816
    genetic modification of, 829b
Ring canal, **638**
Ring species, **433–434**, 434i
Ringworm, 549
Rising, James, 438i
Romalea microptera, **828**
Rosa acicularis (wild rose), 587i
Rosmarinus officinalis, **827**, 827i
Rotifera, phylum, **609–611**, 610f
Roundworm, 619–621, 620f
R-selected species, **697**

S
Salamander, 434, 434i, 652–654
Salem witch trials, possible connection of to lysergic
    acid, 550, 551
Salicylic acid (SA), 811b
Salix, 811b
Salmo salar, domestication of, 820–821
Sand dollar, 638–639
Saprotrophs, **540**, 545, 548, 554
Sarcopterygii, class, **649–651**, 651f
Saunders, Charles E., 819b
Saurischian dinosaur, 657f, 658f
Savannah, **727**
Scapula, **475**, 475f
Scarlet monkey-flower, 432, 432i, 440b
Scherer, Stephen, 468
Schizocoelom, **598**
Schluter, Dolf, 422b
Sclerotiz/sclerotium, **550**, 551, 551i
Scolex, **609**
Scyphozoa, class, **604**, 605f
Sea cucumber, 639

Sea lily, 639
Sea squirt, 640–641, 641f
Sea star, 638f
Sea urchins, 638–639
Seasonal change, **392**
Secondary contact, **436**
Secondary endosymbiosis, **535**
Secondary metabolites, **542**
Secondary productivity, **752**
Secondary succession, **737**
Security of food sources, **808**
Sedimentation, 447i
Seed coat, 582i, **583**, 584i
Seed ferns, **577**, 587
Seedless vascular plants, **575–580**, 576t
Segmentation, **599**, 617f
Selaginella densa, **577**, 578i
Selaginella strobilus, 570i
Selection, **403–406**
Selective breeding, **808–811**, 809i, 810i, 811i, 812i
Selective pressures, **391**, 397–398, 401
Self-fertilization (self-pollination), **420**
Sense codon, **392**
Septum (plural, septa), **542**, 542i, 617
Sessile (stationary), **563**, 588
Sessile animal, **595**
Seta (plural, setae), **617**
Sex ratio, **684**
Sexual dimorphism, **405**, 406i
Sexual reproduction
    speed of domestication and, 810
Sexual selection, **405–406**
Sheep, domestication of, 821
Shoot systems, **568**
Shoulder, **475**
Shoulder blade, **475**, 475f
Siamese fighting fish, 405
Sickle cell *(HbS)*, **421**, 421i
Silurian period, 448t, **575**
Simulation modeling, **770**
Sinus (plural, sinuses), **611**
Siphon, **615**
Skeletons
    Australopithecus afarensis, 474f
    human, 473f
    importance of, 452
Slash and burn farming, **812**, 812i
Snails, 410i, 421, 423, 439
Snakes, 402–403, 402i, 660–661
Snapdragon, 414, 414t, 415b–416b
Snow geese, 411i, 420
Social network, **476**
Sodium-potassium pump (Na-K ATPase), 403
Solanum aethiopicum, domestication of, 819
Solanum lycopersicum, domestication of, 818, 818i, 819
Solanum macrocarpum, domestication of, 819
Solanum melongena, domestication of, 818, 818i, 819
Solanum tuberosum, domestication of, 818–819, 818i
Solonaceae family, domestication of, 818–820, 818i
Song thrush, 423
Sordino, Paolo, 394b

Soredia (plural, soredium), **555**, 556*i*
Sorus (plural, sori), **578**, 579*i*
Sossin, Wayne, 615
Source population, **738**
South American rhea, 393*i*
Sparrow, 434*i*
Spear, **475**
Speciation, **427**
  allopatric, 435–436, 435*i*
    genetic mechanisms, 439–443
    geographical variation, 433–435
    parapatric, **436**, 436–438
    sympatric, **438**, 438–439
Species, **427**–429
    concepts, 428–429
    nomenclature, 427–428
Species clusters, **436**, 436*i*
Species composition, **730**
Species concept, **479**
Species diversity, **727**
Species name, **427**
Species richness, **726**
Specific epithet, **427**
Spermatocyte, **813**
Spermatophyta, 571*i*
Sphenodon, **655**, 660
Sphenodon punctatus, 660
Sphenodontid, **660**–661
Spine, **641**
Spinneret, **623**
Spiral cleavage, **596**
Spiral valve, **648**
Sponges. *See* Porifera, phylum
"Spongiform," **509**–510, 510*i*
Spongocoel, **601**
Sporangium (plural, sporangia), **545**, 570, 574*i*, 577, 577*i*, 578, 580*i*, 584*i*
Spores, **543**, 547*i*, 550, 551*i*, 563
Sporophylls, **577**, 583, 586
Sporophytes, 563, 569*i*, 570, 572, 573*i*, 575, 579*i*, 582
Squalene, **648**
Squamate, **660**–661
Squash, domestication of, 818, 820, 820*i*
Stability, **725**
Stabilizing selection, **413**–414, 413*i*
Stalked-eyed fly, 405, 406*i*
Standing crop biomass, **750**
Stapes, **652**
Starfish, 638
Sternum, **661**
Sticky traps, **390**, 390*i*
Stoma (plural, stomata), **565**, 565*i*, 572, 583
Stone axe, **475**
Stone cutting tool, **475**
Strobili/strobilus, 570*i*, **577**, 580*i*
  See also Cones
Subsistence farmers, **807**
Subspecies, **433**, 433*i*
Succession, **735**
Sulfur shelf fungus, 540*i*
Sundew, 390*i*

Survivorship curve, **685**–686, 686*f*
Sus scrofa, **809**, 810*i*
Suspension, **601**
Suspension feeders, **601**
Swallow, John G., 412
Swim bladder, **649**
Swordtail, 439
Symbionts, **540**, 555–559
Symbiosis, **540**
Symbiotic associations, **540**, 544, 548, 555–556, 565
Symbolism, **476**
Symmetry, **595**
Symmetry, body, **595**–596, 596*f*, 602–606
Sympatric speciation, **438**–439
Sympatricism, **680**
Synapsida, **655**
Syntax, **476**
Systematic characters, evaluating, **456**–458, 457*i*, 458*i*, 459*i*
Systematics and the Origin of Species (Mayr), 426

T
Taiga. *See* Boreal forest
Tall goldenrod plant, 413
Tapeworm. *See* Cestoda, class
Taxol (Taxol), **558**–559
Taxon (plural, taxa), **428**
Taxonomic hierarchy, **428**, 428*i*
Taxonomy, **427**, 430*b*
Teeth
  elasmobranch, 648*f*
  mammal, 668*f*, 671–672, 671*f*, 672*f*, 673*f*
  reptile, 659*f*
  teleost, 651*f*
Teleost
  diversity, 650*f*
  teeth, 651*f*
Temperate deciduous forest, **726**
Temperate grassland, **751**
Temperate rain forest, **525**
Temporal isolation, **431**, 431*t*
Termites, 613*f*
Tertiary consumer, **729**
Testudinata, subclass, **659**–660, 659*f*
Tetrapoda, **644**
Tetrapods, 394*b*, 651–654, 652*f*
T-even bacteriophage, **504**, 504*i*
Thallus (plural, thalli), **555**, 556*i*, 573, 573*i*
Thorax, **624**
Thorpe, S.K.S., 474
3-D structure, of fossils, 450*b*
Threespine stickleback, 422*b*, 422*i*
Throwing, **475**
Thrush, 549
Time lag, **693**
Tissues, **595**
  animal, 595
Toad, 652–654
Tobacco
  cultivation of, 828